RECORDS
OF THE AMERICAN-AUSTRALIAN SCIENTIFIC EXPEDITION TO
ARNHEM LAND
4

ARNHEM LAND EXPEDITION

1. *Art, Myth and Symbolism*
   By Charles P. Mountford
2. *Anthropology and Nutrition*
   Edited by C. P. Mountford
3. *Botany and Plant Ecology*
   Edited by R. L. Specht
   and C. P. Mountford
4. *Zoology*
   Edited by R. L. Specht

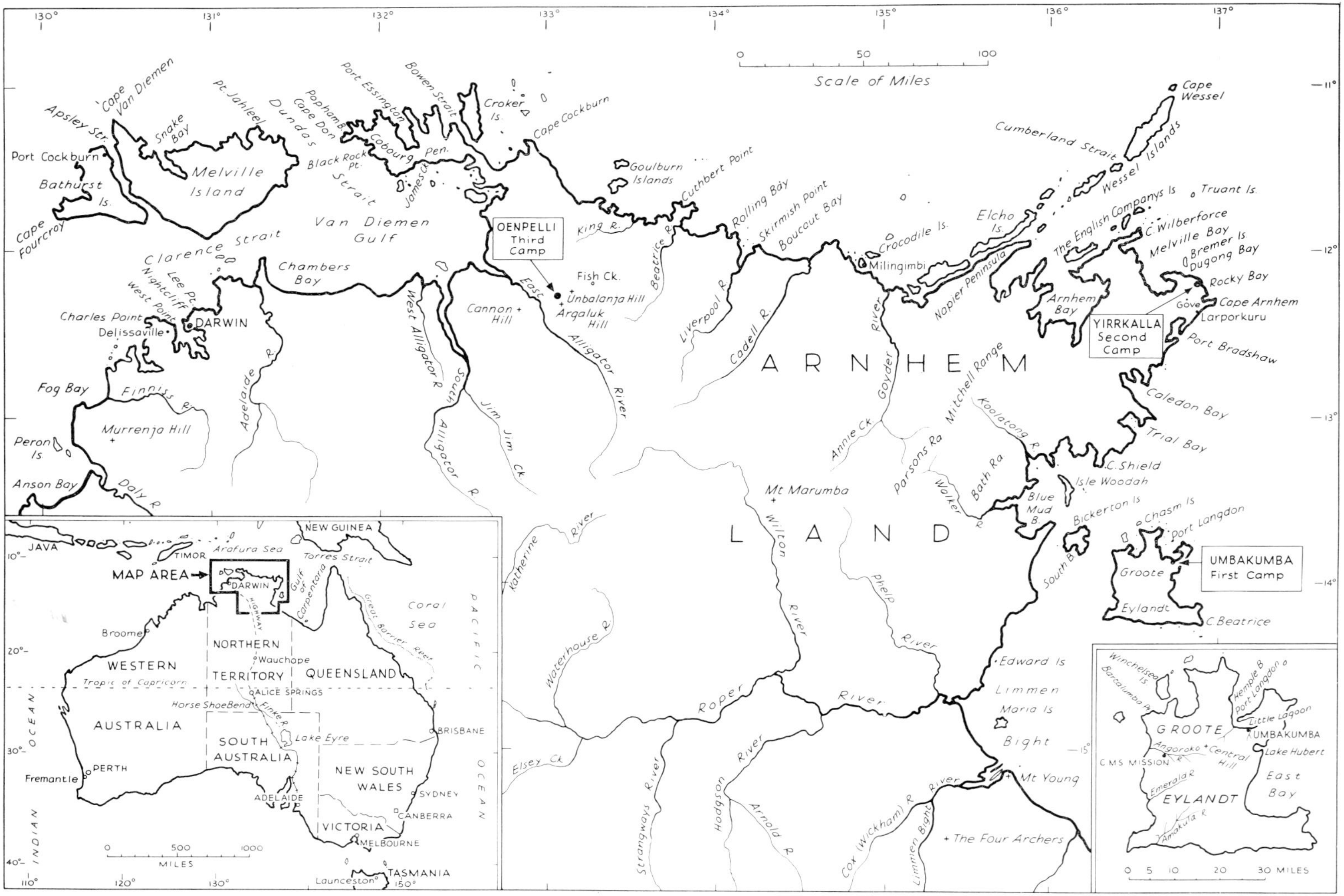

Arnhem Land, showing expedition camps and collecting localities

RECORDS
OF THE AMERICAN-AUSTRALIAN SCIENTIFIC EXPEDITION TO
ARNHEM LAND

GENERAL EDITOR: C. P. MOUNTFORD

4
Zoology

edited by

R. L. SPECHT

MELBOURNE UNIVERSITY PRESS

First published October 1964
Melbourne University Press, Parkville N.2, Victoria
London and New York: Cambridge University Press

Printed and bound in Australia by
Brown, Prior, Anderson Pty. Ltd.,
430 Little Bourke Street, Melbourne

Registered in Australia for transmission by post as a book

EPILOGUE

THIS VOLUME completes the publication of the scientific work of the 1948 Arnhem Land expedition, a joint effort of the National Geographic Society of America, the Smithsonian Institution of Washington and the Commonwealth Government of Australia. It originated from a lecture on aborigines given by the writer twenty years ago before the Washington members of the National Geographic Society of America.

The formation of this expedition and its continued work in the field was due to the enthusiasm and vision of three men: the Honorable A. A. Calwell, then Australian Minister for Information; Dr Gilbert Grosvenor, President of the National Geographic Society of America and Dr Alexander Wetmore, Secretary of the Smithsonian Institution of Washington.

I wish to record my personal thanks to these gentlemen whose loyalty and strong support enabled me to organize and keep the expedition in the field. My thanks are due to Mr Calwell for making available a substantial contribution towards the cost of publishing its scientific results.

After the many negotiations and set-backs that seem inevitable to any major undertaking of this nature, an expedition party of sixteen people left Adelaide on 18 March 1948, returning to base exactly eight months later. Of this time seven months were spent investigating the natural history and ethnology of Arnhem Land, then one of the least known parts of Australia.

The number of specimens collected by the indefatigable members of the expedition during those seven months was impressive; some 30,000 fish, 13,500 plants, 850 birds, 460 animals, several thousand wooden and stone aboriginal artifacts, a large number of string figures, over 300 bark paintings with their associated myths and many hundreds of photographs and sketches of the cave paintings of both Groote Eylandt and western Arnhem Land.

In addition to the physical collection, each specialist had compiled extensive field notes which were later used as a basis for scientific papers.

During the sixteen years since the expedition returned twenty-nine specialists have classified, illustrated and prepared for publication this huge collection of natural history and ethnological objects.

Impressive as the field collections undoubtedly were, the volumes of the *Records of the American-Australian Scientific Expediton to Arnhem Land* are even more outstanding. These four books contain 2,000 pages of descriptive text, 316 plates including 9 in colour, and 442 text figures. It is indeed gratifying to me that, after all the hardships in the field and the countless hours spent in the laboratories, this information is now published and available to the scientific world.

Volume 1 by Charles P. Mountford, a book of 513 pages, 159 plates with 2 in colour, and 68 text figures, deals with the art, myth and symbolism of the aborigines of Arnhem Land.

Volume 2, the combined efforts of eleven specialists in aboriginal nutrition, ethnology and archaeology, consists of 515 pages of text, 47 plates with one in colour, 281 text figures, and 7 graphs.

Volume 3, the work of eleven specialists, is based on the collections of Raymond L. Specht, the botanist of the expedition, and contains 522 pages of text, 18 plates with one in colour, and 80 text figures.

Volume 4, dealing with the zoology of Arnhem Land, is a book of 533 pages, 92 plates with 5 in colour, 13 text figures, the work of six specialists in this field.

In the production of these four volumes the major credit and thanks are due to those who so willingly undertook, often in an honorary capacity, to write up the scientific descriptions of their own speciality. Without the efforts of these men and women the whole expense of the expedition and the work of its members would have been fruitless, nor would the information gathered now be available to the world of science.

I wish to acknowledge my debt to Dr Raymond L. Specht, the botanist of the expedition, for accepting the burden of collecting most of the papers of Volume 3 and editing them for publication. After I had assembled the papers for Volume 4, he also willingly assumed responsibility for editing and supervising their production.

Finally, credit is due to the management and staff of the Melbourne University Press who, over many years, have laboured unceasingly to produce in the best possible manner these four volumes of basic research in the ethnology and natural history of Arnhem Land.

CHARLES P. MOUNTFORD
Leader and General Editor

St Peters, South Australia

PREFACE

THIS VOLUME, the fourth and final volume in the series of reports on the scientific results of the 1948 expedition to the Arnhem Land Aboriginal Reserve, deals with the zoology of the area.

Arnhem Land and Cape York Peninsula are the two most northerly portions of the Australian continent and were undoubtedly connected by continuous land bridges to New Guinea during one or more of the Pleistocene glaciations. Migration of both plants and animals was then possible between these two land masses. The resultant distribution patterns provide a potentially rewarding field for biogeographical studies.

Surprisingly, little interest has been shown in the natural history of the Arnhem Land area. As shown in volume 3 of this series, few extensive botanical collections had been made in the area since Robert Brown visited the area with Captain Matthew Flinders in 1803. Flinders' voyage yielded only a little zoological material but the settlement at Port Essington, in the far north-west corner of Arnhem Land, was a popular collecting ground for many zoologists from 1838 to 1849. Since that date little comprehensive zoological collecting has been attempted; in most cases the small collections were merely secondary to the major pursuits of exploration, of geological or anthropological studies or of pure adventure.

The zoological records of the American-Australian Scientific Expedition to Arnhem Land are thus of inestimable value. Three highly competent zoologists, Herbert G. Deignan (ornithologist), Dr David H. Johnson (mammalogist) and Dr Robert R. Miller (ichthyologist), all of the Smithsonian Institution, Washington, D.C., were members of the expedition. They were assisted by various members of the expedition staff, notably Mrs C. P. Mountford who collected shells, Keith Cordon (shells), John E. Bray (insects and reptiles), and the late William E. Harney who had a unique fund of information and ideas. As well, the zoologists' work was helped by enthusiastic support from the following people: Fred Gray and Jerry Blitner of Umbakumba Settlement, Groote Eylandt; Nick Haritos of Darwin; H. L. Hillier of Oenpelli; K. Davidson of Cape Don Lighthouse; Mr and Mrs Harold Shepherdson of Elcho Island Mission and Mrs Golder of Horse Shoe Bend cattle station. The aboriginal inhabitants of Arnhem Land helped the zoologists in many ways and enriched their work by imparting some of their profound knowledge of animal life.

The expedition spent almost eight months in the field during which time extensive zoological collections were made at Darwin, Groote Eylandt, Yirrkalla and Oenpelli, the main base camps of the expedition. As well, numerous small collections were made at various points around the northern and north-eastern coasts of Arnhem Land. This wealth of material provided, for the first time, a reasonably comprehensive survey of the fauna of coastal and sub-coastal Arnhem

TABLE 1

DETAILS OF ZOOLOGICAL COLLECTIONS MADE ON AMERICAN-AUSTRALIAN EXPEDITION TO ARNHEM LAND, 1948

Animal groups	Collectors	Identifications	Number of specimens	Number of families	Number of species or sub-species	Number of new species	Number of new sub-species	Location of specimens
Insects	J. E. Bray	—	?	?	?	?	?	Aust. Mus.
Molluscs	Mrs C. P. Mountford K. Cordon	B. C. Cotton	More than 175	35	105	2	–	S. Aust. Mus.
Fish	Dr R. R. Miller	Dr W. R. Taylor	14,300	65*	246*	5	–	U.S. Nat. Mus. 5 holotypes in Aust. Mus.
Amphibians	Dr R. R. Miller J. E. Bray	F. J. Mitchell	49	2	10	–	–	U.S. Nat. Mus. S. Aust. Mus. Aust. Mus.
Reptiles	Dr R. R. Miller J. E. Bray	F. J. Mitchell	859	10	53	2	1	U.S. Nat. Mus. S. Aust. Mus. Aust. Mus.
Birds	H. G. Deignan	H. G. Deignan	828 (777 skins, 51 skeletons and alcoholics)	60	191 plus sight records of 31	–	4	U.S. Nat. Mus. 4 holotypes in Aust. Mus. 1 type in Princeton Mus. Zool.
Mammals	Dr D. H. Johnson	Dr D. H. Johnson	501	17	57	3	2	U.S. Nat. Mus. 5 holotypes and set of duplicates in Aust. Mus.

* Families Gobiidae, Eleotridae, Labridae, Clinidae, Blenniidae and a few other small groups are still to be studied and are not included here.

Land. The details of the collections are summarized in Table 1. Almost 17,000 specimens of animal life were collected, representative of at least 194 families of animals and over 693 species or sub-species. So far 12 new species and 7 new sub-species have been recognized in the collections. The collection of insects and of at least five families of fishes has still to be examined.

Descriptions of new interesting material found in the collections have been published in various scientific journals:

Deignan, H. G., *Emu* 50 (1): 17–21, 21–23 (1950)
Deignan, H. G., *Emu* 51 (1): 71–73 (1951)
Johnson, D. H., *Proc. Biol. Soc. Wash.* 67: 77–80 (1954)
Johnson, D. H., *Proc. Biol. Soc. Wash.* 72: 183–8 (1959)
Lachner, E. A., and Taylor, W. R., *Proc. Biol. Soc. Wash.* 73: 29–34 (1960)
Mitchell, F. J., *Rec. S. Aust. Mus.* 11: 373–408 (1955)

A popular account of kangaroos, based on the work of the expedition, was published by Johnson in the *National Geographic Magazine* 108: 487–500 (October 1955).

Altogether, the expedition collections make an impressive contribution to the systematics and biogeography of the fauna of coastal Arnhem Land. The contributors and their assistants deserve congratulations on the success of the expedition and on this volume. Future expeditions should be able to concentrate on the little known centre of the reserve.

Mrs Marion M. Specht assisted with the editing of this volume. Her untiring co-operation is gratefully acknowledged.

R. L. Specht

Botany Department
University of Melbourne
August 1963

CONTENTS

ILLUSTRATIONS

Plates

Figures

1

THE HISTORY OF ZOOLOGICAL COLLECTIONS MADE IN AND ON THE SHORES OF THE ARNHEM LAND ABORIGINAL RESERVE

by Marion M. Specht*

Over the last 150 years zoological collections have been made in Arnhem Land by men of several nationalities, and the fate of those collections cannot always be determined.

Captain Matthew Flinders in the *Investigator* explored the Gulf of Carpentaria in 1803.[1] On the ship was Robert Brown, the botanist, whose botanical collections were extensive. Many animals were collected also; birds, quadrupeds, insects, and shells are mentioned. Flinders in his narrative gives colourful accounts of the animals and birds seen. One of the earliest birds described from Arnhem Land was a parrot, *Psittacus brownii* (= *Platycercus venustus*), collected on this voyage.

Flinders intended to return from England and complete his survey of Australia. However, on his voyage to England he was imprisoned for years at Mauritius, and his death shortly after he reached home meant that it was some years before the coast of Arnhem Land was again visited. In 1818 the British Admiralty gave instructions to Captain Phillip Parker King[2] to continue Flinders' work. In respect of Arnhem Land he was to explore from Arnhem Bay westward and southward as far as North-West Cape, including Van Diemen's Bay and the Rosemary Isles and was 'to endeavour to obtain any information, assisted by your officers, on any occasion which may offer on the animals, whether birds, beasts, or fishes; insects, reptiles, etc., distinguishing those that are wild from those that are domesticated'. Allan Cunningham, the botanist, was on board. Of the ship's company, the mates John Septimus Roe and Frederick Bedwell, and Mr Hunter, the surgeon, collected birds, insects and other animals. During the course of his work King followed the coastline closely and passed some of it more than once. His narrative mentions the variety of birds—egrets, hawks, wild ducks, pelicans and pigeons, and of course, white cockatoos. His zoological collection ranged over the whole animal kingdom from mammals through birds, reptiles, fish, mollusca, annelids, arthropoda (insects, cirripedes), coelenterates, to a collection of 'sea saw-dust' made at Sim's Island. The specimens were identified by J. E. Gray,

* University of Melbourne.

1. Flinders (1814).
2. King (1827).

Captain King assisted by Mr Vigors, and W. S. MacLeay, who dealt with the Annulosa *(sic)*.

Unfortunately the collection sites of the animals are not included in their identification, except for some collected at King George Sound, and those collected at the Isle of France, though King stated that the geographical distribution of animals is of importance.

There was no more zoological activity in Arnhem Iand for twenty years. Captain King had discovered the fine harbour of Port Essington, and had concluded that it 'must at no very distant period become a place of great trade'. In 1824 a British naval settlement at Port Essington endured for three days, and because of the lack of fresh water was moved to Melville Island. Then in 1827, that settlement was moved to Raffles Bay, but in 1829 the area was completely abandoned.

In October 1838, Captain Gordon Bremer brought a second party to establish a settlement at Victoria, Port Essington. Captain John Macarthur was left in charge of the new settlement, and during the eleven years it was in existence much animal collecting was carried out.

The first visitors to the tiny isolated post at Port Essington were the Frenchmen on board *l'Astrolabe* and *La Zélée*[3] who called there on 7–9 April 1839. The presence of those ships in Raffles Bay was made known to the officers at Port Essington by a Malay trepang fishing fleet. Great consternation evidently ensued, for one purpose of the settlement was to prevent French colonization of Northern Australia, and here were two French ships! Lieutenant Stewart was sent to find *l'Astrolabe* and *La Zélée* and order them to proceed immediately to Port Essington. The ships called in some forty-eight hours later when their physical and hydrographical determinations were complete. The visit was a social success, though the Frenchmen's vivid account of the abundance of termites, flies and mosquitoes—and on their leaping into the Bay to escape these pests being bitten by large numbers of tiny crustaceans—leaves the impression that it was not a comfortable one. The Frenchmen gathered birds and other specimens, including many insects, at Raffles Bay. While at Port Essington they made the acquaintance of Lieutenant Priest who had collected many animal specimens, some of which he gave the Frenchmen. He also gave M. Jacquinot a living flying possum, which became his companion and of which he was evidently quite fond; it eventually lost condition, and was killed and pickled.

The *Beagle*, first under Captain King and later under Captains Wickham and then Stokes[4] made several calls at Port Essington, the first in July 1839, after Dumont d'Urville's visit. John Gould[5] has acknowledged the efforts of several of the ship's company as collectors. Among these were Captain Stokes, Captain Bynoe, and Captain Wickham. Lieutenant Emery also collected, and made drawings of fish. John Lort Stokes became Captain in 1841. A description is given of various animals seen; a large mound-building bird, large vampires in the

3. Dumont d'Urville (1844). 4. Stokes (1846). 5. Gould (1845–63).

bamboos, alligators (*sic*), fish, ducks, among them the whistling tree duck. In the narrative of the voyage the ship's collection of birds, fish, reptiles and insects has been described by John Richardson, J. E. Gray, Adam White and Edward Doubleday.

In July 1840 John Gilbert arrived at Port Essington in H.M.S. *Gilmore*. It was unfortunate that a hurricane during the previous year had decimated both mammals and birds in the region. Gilbert several times journeyed into the interior and obtained more than two hundred specimens of birds (more than ninety species) as well as many mammals, reptiles, fish and insects. Gilbert's specimens were sent to England, where they were described by Gould,[6] Gray, and others. He remained at Port Essington until March 1841; while there he found the jungle fowl and its nest mounds, previously believed to be 'tumuli' of aborigines.

Another naval visitor to Port Essington was H.M.S. *Fly*,[7] commanded by Captain F. P. Blackwood, and which carried on board Joseph Beete Jukes as naturalist and John MacGillivray as zoologist. The zoological collection from the voyage, which was through Torres Strait and other islands of the Eastern Archipelago, numbered between four and five thousand specimens. These were placed at the disposal of the zoological department of the British Museum. Many shells were collected and sent to the British Museum; Mr Jukes sent others to a Mr Cuming. Of the shells, which were described by J. E. Gray, a number are listed from Port Essington; about half of those were in Mr Cuming's collection. Mr. Gray also described starfishes from Arnhem Land, and Professor Richard Owen the skeleton of a dugong collected at Port Essington.

In September 1844 the *Fly* left Lieutanent Ince and Mr MacGillivray at Port Essington and called back for both men in February 1845. Lieutenant Ince, who was a nephew of J. E. Gray of the British Museum, is mentioned by Gould (1863, *loc. cit.*) as the man who brought back to England alive a flying opossum *Belideus ariel* Gould. Of this species, then very rare, Mr Gilbert said 'previous to the hurricane at Port Essington in 1839 was very common'.

The *Fly* made its last trip to Port Essington in June 1845. Conditions at the settlement were not good. Farming was not successful, disease was common, and deaths too frequent. The settlement lived on ship fare. When H.M.S. *Rattlesnake*,[8] commanded by Captain Owen Stanley, called there in 1848 the party was shocked by conditions. MacGillivray was again the naturalist, and so was revisiting the area; T. H. Huxley was assistant surgeon.

In December 1845 the German explorer Ludwig Leichhardt and his party struggled into Port Essington to complete their overland journey from Brisbane.[9] During that unfortunate expedition John Gilbert had been killed by aborigines on the Gulf of Carpentaria. The remainder of the party was starving, their clothing in shreds, and completely exhausted. Leichhardt's and Gilbert's plant collections were burnt by Leichhardt at the Roper River, when the horse which

6. Gould (1841–48).
7. Jukes (1847).
8. MacGillivray (1852).
9. Leichhardt (1847); Chisholm (1941).

carried them was lost. But theirs was the first party to complete the overland journey from the east coast settlements to Port Essington.

In London in 1846 the first reports of the voyage of H.M.S. *Erebus* and *Terror* during the years 1839–43[10] were published. These vessels touched only the southern coast of Australia and did not venture within a thousand miles of Northern Australia. Nevertheless, in the reports of the journey fish, amphibia, reptiles and mammals from the Port Essington district are described. Some at least of these must have been sent back to the British Museum by John Gilbert.

The settlement at Port Essington was abandoned in 1849, and for many years subsequently very little collecting was done in the area. Neither F. Cadell's expedition[11] in 1867 nor that led by David Lindsay in 1883–4[12] yielded zoological specimens.

In 1894 Knut Dahl,[13] a young Norwegian, landed at Port Adelaide, and accompanied by Ingel Olsen Holm as taxidermist, travelled north to the Daly River area, where he lived for two years collecting animals. He returned to his native land with many unknown species, including twenty new vertebrates. He did not, however, collect inside the area covered by the Arnhem Land Aboriginal Reserve, though many of his specimens have a distribution that extends into that area.

As years passed specimens were sent at intervals to museums, e.g. mammals collected by J. T. Tunney[14] during 1902–3; birds, etc., collected by Baldwin Spencer at Oenpelli in 1912.[15] Snakes collected by P. Cahill in 1914 are in the National Museum in Melbourne. William McLennan, collecting birds for H. L. White, cruised around the shores of Arnhem Land and set up a camp at King River in 1915–16. A large collection was made near the area covered by Gilbert nearly eighty years previously.[16] He visited Groote Eylandt in May 1921.[17]

While there he met Norman Tindale who, on behalf of the South Australian Museum, spent fifteen months, 1921–22, on field work on Groote Eylandt and on the Roper River. Nearly twelve months of that time was spent on Groote Eylandt itself. Tindale's mammal, bird[18] and reptile collections are in the South Australian Museum. The butterflies were described by Tindale;[19] the Coleoptera were worked on by A. M. Lea, at that time the Museum's entomologist, in various revisions of groups of beetles. R. J. Tillyard described the Odonata, Neuroptera and Trichoptera of the expedition.[20] Types of all the new species described are in the South Australian Museum collections.

The medical officer of H.M.A.S. *Geranium*, W. E. J. Paradice, and his successor, Surgeon Lieutenant K. E. F. D. Hudson, during 1923–24 collected fish around the Arnhem Land coast from the Sir Edward Pellew Group to Darwin. These have been described by Paradice and Whitley.[21]

10. Richardson and Gray (1846-75).
11. Cadell (1867).
12. Lindsay (1883–4).
13. Dahl 1926).
14. Thomas (1904).
15. Whittell (1954).
16. White (1917); Campbell (1919).
17. White (1922); Campbell (1922).
18. Tindale (1925).
19. Tindale (1922, 1923).
20. Tillyard (1925).
21. Paradice and Whitley (1927).

The next activity in the area was under the auspices of the British Museum, who sent Captain Hubert Wilkins[22] to collect natural history specimens in Australia. In 1925 he reached Groote Eylandt and the Roper River. Species collected from the Arnhem Land Reserve are described by Thomas[23] and Parker.[24] In 1932 H. L. Clark of Harvard collected echinoderms at Port Essington.

After the murder of Japanese seamen from a trepanging fleet by aborigines Donald Thomson went in 1935 to Arnhem Land to investigate the situation there. He made an extensive zoological collection, now at Melbourne University, primarily as a background to the study of the ecology and food-gathering habits of the aborigines. The material has been worked upon, but no systematic publication has been made of the collections, mainly mammals, reptiles and amphibians. Dr Thomson's papers on the aborigines[25] include zoological material.

A few years later Australia was at war with Germany, and subsequently with Japan. Troops were stationed in the reserve, and some men made private collections in their spare time. One of these was L. Gilbert, who published an account of his observations.[26] Charles Barrett made a trip into Arnhem Land early in the war before the entry of Japan and in his story of his travels states that some specimens were sent to museums.[27]

Perhaps this summary should have begun with the aborigines' cave paintings, for in those are preserved identifiable drawings of the local fauna. Matthew Flinders[28] was the first white man to see them in Arnhem Land. On Chasm Island he found drawings in holes or chasms underlying cliffs. He described them as being in 'charcoal and something like red paint' and noted that the drawings were of porpoises, turtles, and kangaroos; and that one was of a kangaroo being chased by thirty-two people. In recent years these cave paintings have received a great amount of attention from anthropologists. But zoologists also have used them as a means of identifying fauna. Tindale[29] has named rays from Groote Eylandt paintings, as also has Whitley.[30]

Some collections made in Arnhem Land, particularly those of special groups, have probably not been mentioned. The task of tracing all the collected material would occupy an enormous amount of time. Perusal of Musgrave's[31] bibliography of Australian entomology published thirty years ago, indicates the wide dispersal of these collections. The very exhaustive volume by Whittell[32] on the Australian birds, also includes much historical detail concerning the area. A volume on the reptiles of Australia,[33] containing a check-list of Australian reptiles, has just appeared. And, like the insects of the American-Australian Scientific Expedition to Arnhem Land in 1948, much material is still awaiting description.

22. Wilkins (1928).
23. Thomas (1926a, b, c).
24. Parker (1926).
25. Thomson (1948, 1949a, b).
26. Gilbert (1949).
27. Barrett (1941).
28. Flinders (1814).
29. Tindale (1926).
30. Whitley (1940).
31. Musgrave (1932).
32. Whittell (1954).
33. Worrell (1963).

REFERENCES

Barrett, C. (1941). *Coast of Adventure: Untamed North Australia.*

Cadell, F. (1867). Northern Territory exploration, 1867. *S. Aust. Parl. Paper,* no. 178.

Campbell, A. J. (1919). Notes on Birds from the Gouldian-Gilbert Type-locality, North Australia. *Emu,* 18: 172–88.

——— (1922). Some Birds of Groote Eylandt, Northern Territory. *Emu,* 21: 188–97.

Chisholm, A. H. (1941). *Strange New World: The adventures of John Gilbert and Ludwig Leichhardt.*

Dahl, K. (1926). *In Savage Australia: An account of a hunting and collecting expedition to Arnhem Land and Dampier Land.*

Dumont d'Urville, J. (1844). *Voyage au Pôle Sud et dans l'Océanie sur les corvettes l'Astrolabe et la Zélée . . . pendant les ans 1837–1840.*

Flinders, M. (1814). *A Voyage to Terra Australis.* 2 vols.

Gilbert, L. A. (1949). Zoological notes on the Northern Territory. *Vict. Nat.,* 65: 96–102, 155–9.

Gould, J. (1841–48). *Birds of Australia.*

——— (1845–63). *Mammals of Australia.*

Jukes, J. B. (1847). *Narrative of the surveying voyage of H.M.S. Fly commanded by F. P. Blackwood in Torres Strait, New Guinea, and other islands of the Eastern archipelago 1842–1846.*

King, P. P. (1827). *Narrative of a survey of the intertropical and western coasts of Australia.* 2 vols.

Leichhardt, L. (1847). *Journal of an overland expedition in Australia from Moreton Bay to Port Essington.*

Lindsay, D. (1883–4). Explorations through Arnheim's Land. *S. Aust. Parl. Paper,* no. 239.

MacGillivray, J. (1852). *Narrative of the voyage of* H.M.S. *Rattlesnake, commanded by the late Captain O. Stanley . . . during the years 1846–50.* 2 vols.

Musgrave, A. (1932). *Bibliography of Australian entomology 1775–1930, with biographical notes on authors and collectors.*

Paradice, W. E. J., and Whitley, G. P. (1927). Northern Territory Fishes. *Mem. Qld. Mus.,* 5: 99–126.

Parker, H. W. (1926). New Reptiles and a new Frog from Queensland. *Ann. Mag. Nat. Hist.,* ser. 9, 17: 665–70.

Richardson, Sir John, and Gray, J. E. (1846–75). *Zoology of the voyage of* H.M.S. *Erebus and Terror under the command of Sir James Clark Ross,* R.N., F.R.S.

Stokes, J. Lort (1846). *Discoveries in Australia; with an account of the coasts and rivers explored and surveyed during the voyage of* H.M.S. *Beagle, in the years 1837–43.*

Tillyard, R. J. (1925). Odonata, Neuroptera and Trichoptera from Groote Eylandt, Gulf of Carpentaria. *Rec. S. Aust. Mus.,* 3: 41–4.

Tindale, N. B. (1922). On a new genus and species of Australian Lycaeninae. *Trans. Roy. Soc. S. Aust.,* 46: 537–8.

——— (1923). On Australian Rhopalocera. *Trans. Roy. Soc., S. Aust.,* 47: 342–54.

——— (1925). Notes on the Birds of Groote Eylandt, Gulf of Carpentaria. Birds observed at Roper River, Northern Territory. *S. Aust. Orn.,* 8: 10–21, 21–4.

——— (1926). Natives of Groote Eylandt and of the West Coast of the Gulf of Carpentaria. Part II. *Rec. S. Aust. Mus.,* 3: 103–34.

Thomas, O. (1904). On a collection of mammals made by Mr J. T. Tunney in Arnhem Land, Northern Territory of South Australia. *Novit. Zool.,* 11: 222-9.

——— (1926a). Two new Rock Wallabies (Petrogale) discovered by Captain G. W. Wilkins in Northern Australia. *Ann. Mag. Nat. Hist.,* ser. 9, 17: 184–7.

——— (1926b). On various Mammals obtained during Capt. Wilkins's expedition in Australia. Ibid., ser. 9, 17: 625–35.

——— (1926c). The local races of *Dasyurus hallucatus.* Ibid., ser. 9, 18: 543–4.

Thomson, D. F. (1948). Arnhem Land: Explorations among an unknown people. Part I. *Geogr. J.,* 112: 146–64.

——— (1949a). Arnhem Land: Explorations among an unknown people. Part II. Ibid., 113: 1–8.

——— (1949b). Arnhem Land: Explorations among an unknown people. Part III. Ibid., 114: 53–67.

White, H. L. (1917). North Australian Birds. *Emu,* 16: 117–58, 205–31.

——— (1922). Two new sub-species of birds. *Emu,* 21: 163.

Whitley, G. P. (1940). *The Fishes of Australia.*

Whittell, H. M. (1954). *The literature of Australian Birds: a history and a bibliography of Australian ornithology.*

Wilkins, G. H. (1928). *Undiscovered Australia, being an account of an expedition to tropical Australia to collect specimens of the rarer native fauna for the British Museum, 1923–1925.*

Worrell, E. (1963). *Reptiles of Australia.*

2

MOLLUSCS OF ARNHEM LAND

by Bernard C. Cotton*

THE NOTES recorded here are based on small collections casually made by members of the Arnhem Land Expedition. To add more interest and value to the account, reference is made to shells from northern Australian localities, such as North West Island (W. J. Kimber) and Broome in Western Australia; Port Keats, Darwin, Melville Island (K. Fry) and Groote Eylandt in the Northern Territory; Murray Island (A. M. Lea), Darnley Island, and Thursday Island in Torres Strait, as well as shells from Manus Island (C. M. Deland) in the Admiralty Island Group.

Yirrkalla and Groote Eylandt are included in the faunal area known as the Dampierian Province, extending from Geraldton to Cape York in which lives a tropical suite of molluscs, mostly of wide distribution, but sometimes showing minor geographical variation in different localities. For this reason species from definite localities are discussed and figured. Locality abbreviations used are as follows: N.W.A., North Western Australia; S.W.A., South Western Australia; S.A., South Australia; Vict., Victoria; N.S.W., New South Wales; Q., Queensland; N.A., North Australia.

Thanks are due to Mrs C. P. Mountford and Mrs L. Cordon who permitted an examination of their shells from Yirrkalla and Groote Eylandt respectively. Mrs Mountford donated a selection of specimens to the South Australian Museum.

The plates were prepared by Mrs B. J. Hunter (née Newman) and Miss Audine Cheek of the Museum Staff. Miss Janice Oliver and Miss Valerie Leane both assisted with the typing.

PELECYPODA

MYTILIDAE

Septifer bilocularis (Linné)

Pl. 3, nos. 5, 6

Mytilus bilocularis Linné, *Syst. Nat.* ed. 10: 705 (1758)

Yirrkalla, and Groote Eylandt, figured, height 34 mm., length 20 mm.

The species, originally described from Mauritius, is recorded from Geraldton around North Australia to Queensland, and in the Indo-Pacific. South Australian Museum specimens are from Yallingup, Moreton Bay, Carnarvon, Geraldton, Shark Bay, Leighton, Abrolhos. This variable species lives under coral in the northern waters, but is found in rock crevices in the south. The New South Wales shell has been described recently as distinct, under the name *Septifer australis* Laseron.

* South Australian Museum, Adelaide.

Musculus cumingianus (Reeve)

Pl. 3, no. 9

Modiola cumingiana Reeve, *Conch. Icon.* 10: sp. 50 (1857)

Darwin and Yirrkalla, figured, height 17 mm., length 28 mm. A few small specimens from 'Port Bradshaw on rocks' (R. L. Specht).

Originally described from Moreton Bay, Queensland, a wide range has been given the species since it is said to occur practically all round Australia, generally attached to sponges and ascidians. In the British Science Guild Handbook, Cotton and Godfrey *The Molluscs of South Australia*, p. 125 (1938), the name *Musculus nana* Dunker (*Lanistina*) described from Port Lincoln, South Australia, is preferred for the South Coast-Shell, and this may be more applicable to the North and Western Australian species than *M. cumingianus.* It has been collected at numerous localities in South Australia and from Esperance, Fremantle, Ellenbrook, Albany and Geraldton in Western Australia.

Stavelia horrida (Dunker)

Pl. 3, nos. 7, 8

Mytilus horridus Dunker, *Proc. Zool. Soc. Lond.*: 359 (1857)

Thursday Island, two odd valves, figured, height 135 mm. exterior view, and height 125 mm. interior view.

It was first recorded from the North Coast of New Holland.

PTERIIDAE

Magnavicula bennetti Iredale

Pl. 2, no. 35

Magnavicula bennetti Iredale, *Great Barrier Reef Exped., Mollusca*, pt. 1: 331, 332 (1939)

Yirrkalla, figured, hinge length 150 mm., height 100 mm., depth 40 mm. The left valve is very convex and right valve less convex.

The shell was described but not figured by Iredale, from Queensland, and seems to be closely allied to the Yirrkalla specimen. It may on the other hand be merely the common *M. macroptera* Lamarck to which *M. bennetti* seems closely allied.

Belchlamys newmanae sp. nov.

Pl. 2, no. 33

Elongate-ovate, valves similar, a little convex, ears very unequal, sculpture of twenty-two regular fine equal radials obsoletely scaled, interstices finely punctate, no interstitial riblets. Length 22 mm., height 25 mm.

Holotype—South Australian Museum D.14409; Groote Eylandt.

South Australian Museum specimens come from Groote Eylandt and Melville Island. The holotype is a half-grown red coloured shell. Some are uniform orange, others livid blue speckled with white. The species differs from the Southern Australian *B. atkinos* in the regularity of the radial ribs and the narrow ear.

Excellichlamys spectabilis (Reeve)

Pl. 2, no. 34

Pecten spectabilis Reeve, *Conch. Icon.* 8: sp. 128 (1853)

Yirrkalla, a typical valve, figured, length 25 mm., height 25 mm.

This species was described from an unknown locality. One complete specimen in the South Australian Museum from the Kenyon collection is also without locality.

Q., N.A.

CARDITIDAE

Cardita crassicostata Lamarck

Pl. 2, no. 28

Cardita crassicostata Lamarck, *Anim. S. Vert.* 6: 24 (1819)

Yirrkalla, single specimen, figured, length 36 mm., height 23 mm., small red coloured.

North Australian shells may grow to 85 mm. in length. The species was described by Lamarck from unknown locality, Reeve later gave 'Putao, Island of Luzon, Philippines, found in coarse sand at the depth of six fathoms (Cuming)'. The rose coloured variety fig. 7c of Reeve is remarkably like the Yirrkalla specimen. The shell is not uncommon being found in South Western, North Western Australia and Queensland. The name *Cardita aviculina* is introduced by Lamarck on page 26 of the above work with the locality 'les mers de la Nouvelle hollande a la baie des Chiens-Marins et à l'île King'. Both species are recorded in *A Preliminary Index of the Mollusca of Western Australia* by Charles Hedley (1916). If the two names cover the species, *C. crassicostata* has priority, but they appear to be distinct. Although cream and pink coloured specimens are found in South Australia, I have not seen from that area one of deep red colour like the Yirrkalla specimen.

Q., N.A.

CARDIIDAE

Regozara flava (Linné)

Pl. 3, nos. 10, 11

Cardium flava Linné, *Syst. Nat.* ed. 10: 680 (1758)

Yirrkalla, a small specimen, height 45 mm., is figured showing the external appearance and a larger one from Groote Eylandt, height 57 mm., figured, shows the hinge features.

N.W.A., Q., N.A.

Fulvia laevigata (Linné)

Pl. 3, no. 12

Cardium laevigatum Linné, *Syst. Nat.* ed. 10: 680 (1758)

Port Keats, one specimen, height 50 mm., length 52 mm., figured.

Discors lyratum (Sowerby)

Pl. 3, nos. 13, 14

Cardium lyratum Sowerby, *Proc. Zool. Soc. Lond.*: 109 (1841)

Port Keats, height 44 mm. and length 40 mm. Both valves are figured.

The species was originally recorded from the Philippines.

Q., N.A., Indo-Pacific.

Pratulum probatum Iredale

Pl. 3, nos. 15, 16

Pratulum probatum Iredale, *Aust. Zool.* 6: 333, pl. 46, fig. 8 (1927) . .

Port Keats, two odd valves. The larger height 63 mm., length 63 mm., interior figured and height 50 mm., and length 50 mm., exterior figured.

This is said to differ slightly from *P. bechei* Reeve from the Philippines.

Q., N.A.

Pratulum setosum (Redfield)

Pl. 3, nos. 17, 18

Cardium setosum Redfield, *Ann. N.Y. Lyceum Nat. Hist.* 4: 168 (1846)

Darwin, one specimen, figured, measures height 50 mm., and length 57 mm.

Australian shells of this species were once placed under *Cardium latum* Born.

Q., N.A.

Hemicardium tumoriforum (Lamarck)

Pl. 3, nos. 19, 20

Cardium tumoriforum Lamarck, *Anim. S. Vert.* 6: 15 (1819)

Darwin, one specimen, figured, height 41 mm., and length 36 mm.

It is closely related to the Indo-Pacific and Queensland *Hemicardium hemicardium* Linné.

N.A.

VENERIDAE

Lioconcha castrensis (Linné)

Pl. 2, no. 27

Venus castrensis Linné, *Syst. Nat.* ed. 10: 687 (1758)

Yirrkalla, figured, length 32 mm., height 39 mm.

The species ranges through North Western and North Australia to Queensland. South Australian Museum specimens come from Monte Bello, Groote Eylandt, and Townsville.

Dosinia scalaris (Menke)

Pl. 2, no. 26

Cytherea scalaris Menke, *Moll. Nov. Holl.*: 42 (1843)

Yirrkalla, length 60 mm., height 56 mm.; Groote Eylandt, figured, length 44 mm., height 41 mm.

The South Australian Museum has specimens from Broome, Port Keats, Cottesloe.

N.W.A., S.W.A., N.A.

Circe gibbium (Lamarck)

Pl. 3, nos. 1, 2

Cytherea gibbia Lamarck, *Anim. S. Vert.* 5: 577 (1818)

Groote Eylandt, figured, length 38 mm., height 33 mm.

Specimens in the South Australian Museum Collection come from Thursday Island, Melville Island, Darwin and New Guinea.

N.A., Indo-Pacific.

SANGUINOLARIIDAE

Gari caerulescens (Reeve)

Pl. 3, no. 4

Psammobia caerulescens Reeve, *Conch. Icon.* 10: sp. 60 (1857)

Yirrkalla, one good specimen of this species, figured, height 30 mm., length 60 mm.

Reeve's figure shows a somewhat less elongate shell, though the other characteristics such as sculpture, correspond with the Yirrkalla specimens. The Japanese species figured by Tadashige Habe (1952), *Genera of the Japanese Shells, Pelecypoda* no. 3: 202, fig. 483, is much like ours in shape and that author quotes *G. truncata* Linné as the name and cites *G. bipartita* Philippi and *G. caerulescens* as synonyms.

N.W.A., N.A.

Psammosolen sulcatus (Dunker)

Pl. 3, no. 3

Macha sulcata Dunker, *Proc. Zool. Soc. Lond.*: 424 (1861)

Yirrkalla, one specimen, figured, height 24 mm., length 48 mm.

The type came from Eastern Australia and Reeve mentions Moreton Bay, Queensland. This is the name usually applied to the North Australian shell though according to Reeve's figures, *Conch. Icon.* 19, pl. 3, the Yirrkalla shell resembles *P. philippinarium* Dunker from the Philippines. This and *P. sulcatus* seem closely allied. *P. australis* Dunker described from unknown locality but again recorded from Moreton Bay by Reeve, is a smaller species found in Southern Queensland and Northern New South Wales. Generic names previously used for species of this kind are *Solecurtus, Solenocurtus* Blainville, *Macha* Oken and *Adasius* Leach.

TELLINIDAE

Tellinella virgata (Linné)

Pl. 2, no. 29

Tellina virgata Linné, *Syst. Nat.* ed. 10: 674 (1758)

Yirrkalla, three specimens, one figured, length 63 mm., height 40 mm.

The species was originally described from the Indian Ocean.

N.W.A., Q., Indo-Pacific.

Phylloda foliacea (Linné)

Pl. 2, no. 31

Tellina foliacea Linné, *Syst. Nat.* ed. 10: 675 (1758)

Yirrkalla, one large specimen, figured, length 10 mm., height 56 mm.

The species was originally described from the Indian Ocean and is found in Queensland, North and North Western Australia.

Angulus pharaonis (Hanley)

Pl. 2, no. 30

Tellina pharaonis Hanley, *Proc. Zool. Soc. Lond.*: 148 (1844)

Yirrkalla and Groote Eylandt, figured, length 65 mm., height 23 mm. The colour varies from the typical pink to a light yellow and there are extremely fine radial striae present.

Described from the Red Sea, this species is also represented in the South Australian Museum Collection from Queensland, Port Keats. Names such as *Tellina perna* Spengler (Singapore), *Tellina vulsella* Chemnitz (Philippines) have been incorrectly used for this species in Australian literature; the first is a deeper shell, but the latter is closely allied.

Q., N.A., Indo-Pacific.

SOLENIDAE

Solen timorensis Reeve

Pl. 2, no. 32

Solen timorensis Reeve, *Conch. Icon.* 19: sp. 7 (1874)

Yirrkalla, figured, length 87 mm., height 17 mm.

This species occurs at the following localities: Melville Island, Port Essington, Port Darwin, North Western Australia. It is smaller and lighter coloured than the large *S. grandis* Dunker of the Philippines and Queensland.

N.W.A., N.A., Indo-Pacific.

HALIOTIDAE

Haliotis asinina Linné

Pl. 2, no. 24

Haliotis asinina Linné, *Syst. Nat.* ed. 10: 780 (1758)

Yirrkalla, figured, 41 mm., lesser diameter.

South Australian Museum specimens come from North West Islet, Murray Island, North Queensland.

N.W.A., Q., N.A., Indo-Pacific.

Sanhaliotis squamata (Reeve)

Pl. 2, no. 23

Haliotis squamata Reeve, *Conch. Icon.* 3: sp. 35 (1846)

Yirrkalla, one specimen, and Groote Eylandt, one specimen, figured, 44 mm., lesser diameter.

N.W.A., Q., N.A., Indo-Pacific.

TROCHIDAE

Trochus maculatus Linné

Pl. 2, no. 17

Trochus maculatus Linné, *Syst. Nat.* ed. 10: 756 (1758)

Groote Eylandt, one specimen, Yirrkalla, figured, height 50 mm.

This Indo-Pacific species is not uncommon at Yirrkalla. It is found generally in North Australia from Queensland to Shark Bay, Western Australia. The South Australian Museum has a specimen from Albany, in the Southern Ocean.

N.W.A., Q., N.A., Indo-Pacific.

Tectus obeliscus (Gmelin)
Pl. 4, no. 14

Trochus obeliscus Gmelin, *Syst. Nat.* ed. 13: 3579 (1791)

Yirrkalla, figured, height 60 mm.

Specimens up to 68 mm. in height are in the collection. The shell is dark green with pale green to yellow base. The South Australian Museum collection has specimens taken in Queensland, North Western Australia and as far south as the Abrolhos Islands.

N.W.A., S.W.A., Q., N.A.

ANGARIIDAE

Angaria tyria (Reeve)
Pl. 2, no. 11

Delphinula tyria Reeve, *Conch. Syst.* 2: 157 (1842)

Yirrkalla, one rather small specimen nodulate on the shoulder but not spinose, figured, height 43 mm.

N.A., Indo-Pacific.

TURBINIDAE

Turbo petholatus Linné
Pl. 2, no. 22

Turbo petholatus Linné, *Syst. Nat.* ed. 10: 762 (1758)

Yirrkalla, one small example of the brownish coloured variety, figured, height 27 mm. Further specimens are of normal size and colour.

N.W.A., Q., N.A., Indo-Pacific.

Senectus squamosus (Gray)
Pl. 2, no. 16

Turbo squamosus Gray, *Voy. Fly* 2: 359 (1847)

Yirrkalla, Groote Eylandt, figured, height 47 mm.

Fairly common in North Australia this species ranges through Queensland and North Western Australia.

N.W.A., Q., N.A., Indo-Pacific.

Distellifer stellare (Gmelin)
Pl. 2, no. 13

Turbo stellaris Gmelin, *Syst. Nat.* ed. 13: 3600 (1791)

Yirrkalla and Groote Eylandt, numerous specimens, figured, height 37 mm.

A common species ranging through Northern Australia and the South Pacific. It is difficult to decide what generic name is the correct one for this species. *Stella* H. & A. Adams is antedated by *Stellaria* Schmidt, but I have been unable to refer to Schmidt's work or to find out what species is the true type species. In the meantime *Distellifer* Iredale has been introduced for a rather elevated species with *D. wallisi* Iredale from Middleton Reef, South Pacific, as type species.

NERITIDAE

Amphinerita polita (Linné)

Pl. 2, no. 6

Nerita polita Linné, *Syst. Nat.* ed. 10: 778 (1758)

Yirrkalla, Groote Eylandt, figured, height 24 mm.

The type came from the Amboina, and the species ranges round North Australia from Shark Bay to New South Wales. The colour pattern is very variable.

N.W.A., S.W.A., N.S.W., Q., N.A., Indo-Pacific.

Nerita lineata Gmelin

Pl. 2, no. 10

Nerita lineata Gmelin, *Syst. Nat.* ed. 13: 3684 (1791)

Yirrkalla, figured, height 32 mm.

The species extends right through the Dampierian and Solanderian Region. South Australian Museum specimens are from Groote Eylandt, Darwin, Melville Island, Gladstone, Broome.

N.W.A., Q., N.A., Indo-Pacific.

Theliostyla reticulata (Karsten)

Pl. 2, no. 9

Nerita reticulata Karsten, *Mus. Lesk.*: 296, pl. 2, fig. 8 (1789)

Yirrkalla, figured, height 20 mm.

South Australian Museum specimens are from Darwin, Melville Islands, Gladstone, Port Douglas, Cooktown.

N.W.A., S.W.A., Q., N.A., Indo-Pacific.

Pictoneritina oualanensis (Lesson)

Pl. 2, no. 8

Nerita oualanensis Lesson, *Zool. Coquille* 2: 379 (1830)

Yirrkalla, figured, height 7 mm.

A small, variable species was originally described from Oualay, Sandwich Islands, and a shell closely resembling it occurs in North Queensland. It is apparently numerous at Yirrkalla and a necklace of this shell from that locality was obtained by Mr C. P. Mountford. It is said to resemble *Neritina mertoniana* Récluz, from the Philippines, but it has less distinct teeth on the columella.

N.W.A., N.S.W., Q., N.A., Indo-Pacific.

TURRITELLIDAE

Turritella terebra (Linné)

Pl. 2, no. 21

Turbo terebra Linné, *Syst. Nat.* ed. 10: 766 (1758)

Groote Eylandt, figured, height 110 mm.

The description and figure of *Turritella cerea* Reeve, *Conch. Icon.* 5: sp. 25, pl. 6, fig. 25 (1849), apply very well to North Australian shells which are smaller and less boldly keeled than *T. terebra* Linné, from the Philippines. However the earlier name is used for the species. In South Western Australia occurs *Turritella occidua* Cotton and Woods, dredged in 12-28 fathoms, a thinner and smaller species than either *T. terebra* or *T. cerea.* Melville Island, Maria Island, Port Darwin, Manus, New Guinea.

N.W.A., N.A., Indo-Pacific.

ARCHITECTONIDAE

Architectonica perspectiva (Linné)

Pl. 2, no. 12

Trochus perspectiva Linné, *Syst. Nat.* ed. 10: 757 (1758)

Yirrkalla and Groote Eylandt, figured, major diameter 50 mm.

Good specimens range up to 50 mm. in diameter and 32 mm. in height.

N.W.A., Q., N.A., Indo-Pacific.

CERITHIIDAE

Clava aspera (Linné)

Pl. 1, no. 34

Murex aspera Linné, *Syst. Nat.* ed. 10: 756 (1758)

Yirrkalla, one specimen, figured, height 45 mm.

Other specimens in the South Australian Museum collection are from North Queensland, North West Islet, and Murray Island.

N.W.A., Q., Indo-Pacific.

Clava pulcher (Sowerby)

Pl. 1, no. 32

Cerithium pulcher Sowerby, *Thes. Conch.* 2: 852, No. 15 (1855)

Yirrkalla, the two specimens are typical, one figured, height 40 mm.

South Australian Museum specimens come from Groote Eylandt, Queensland and North West Islet. A third specimen approaches the closely allied *Pseudovertagus obeliscus* from the East Indies.

N.W.A., Q., N.A., Indo-Pacific.

Clava mountfordae sp. nov.

Pl. 5, no. 1

Shell narrow, rather acute of about fourteen whorls; white sculpture of irregular axials crossed by weak, narrow spiral ribs forming weak nodules on top of the whorls; suture simple but impressed; aperture narrow, oblique, anterior canal oblique and rather produced; columella with two weak oblique plaits; length 62 mm., width 15 mm.

Holotype—South Australian Museum, D.14494, Shark Bay, W.A.

The holotype came from Shark Bay but a specimen from Yirrkalla appears to be the same species. There is little variation in the two specimens which seem to be related to *Cerithium pharos* Hinds but *C. mountfordae* is easily separated by the irregular sculpture and pure white colour. It is named after Mrs C. P. Mountford, wife of the explorer.

Pseudovertagus sinensis (Gmelin)

Pl. 1, no. 35

Murex sinensis Gmelin, *Syst. Nat.* ed. 13: 3542 (1791)

Yirrkalla, two specimens, typical though small, one figured, height 33 mm.

Menke 1843 recorded the species from North Western Australia as *Cerithium obeliscus* Bruguière.

N.W.A., Q., N.A., Indo-Pacific.

Ischnocerithium echinatum (Lamarck)

Pl. 1, no. 33

Cerithium echinatum Lamarck, *Anim. S. Vert.* 7: 69 (1822)
Cerithium columna Sowerby, *Genera of Shells* 2: 42 (1834)

Yirrkalla, one specimen, figured, height 38 mm.

Originally described from the Philippines, this species is recorded from Western and North Australia as well as Queensland. South Australian Museum specimens are from Groote Eylandt, Murray and Piper Island, and also Caloundra.

N.W.A., Q., N.A., Indo-Pacific.

PYRAMIDELLIDAE

Longehaeus sulcatus (Adams)

Pl. 4, no. 15

Obeliscus sulcatus Adams in Sowerby, *Thes. Conch.* 2: 807, pl. 171, fig. 34 (1855)

Groote Eylandt, two specimens, one figured, height 36 mm.

The type was said to come from Tahiti. Two specimens from Groote Eylandt appear to belong to this species rather than *L. maculosa* or *L. punctata*.

N.A., Indo-Pacific.

STROMBIDAE

Euprotomus australis (Gray)

Pl. 1, no. 31

Strombus australis Gray, King's *Intertropical Aust.* 11: 489 (1827)

Groote Eylandt, a selected specimen is figured, height 70 mm.

Specimens were also received from Yirrkalla. The shells are close to typical wrinkled within the aperture.

N.W.A., N.A., Indo-Pacific.

Doxander campbelli (Griffiths & Pidgeon)

Pl. 1, nos. 29, 30

Strombus campbelli Griffiths and Pidgeon in Cuvier, *Anim. Kingdom* 12: 600, pl. 25, fig. 6 (1834)

Groote Eylandt, figured, 48 mm. Yirrkalla, figured, height 55 mm.

This species, *Strombus campbelli*, and the following species are closely allied, the present one differing in having a plicate sutural margin. Five specimens were taken showing variation in adult size and colour pattern ranging from nut brown to pink.

N.A., Indo-Pacific.

Doxander vittatus (Linné)

Pl. 1, no. 28

Strombus vittatus Linné, *Syst. Nat.* ed. 10: 745 (1758)

Groote Eylandt, Yirrkalla, figured, height 43 mm.

Two typical specimens have a smooth sutural margin and rather obsolete axial sculpture on the spire whorls.

N.A., Indo-Pacific.

Canarium urceus (Linné)

Pl. 1, no. 27

Strombus urceus Linné, *Syst. Nat.* ed. 10: 745 (1758)

Yirrkalla, one specimen, and Groote Eylandt, one specimen, figured, height 36 mm.

N.W.A., Q., N.A., Indo-Pacific.

NATICIDAE

Polinices powisiana (Récluz)

Pl. 2, no. 1

Natica powisiana Récluz, *Proc. Zool. Soc. Lond.* 11: 210 (1844)

Yirrkalla, an average-sized specimen, white with deep blackish brown central band and golden coloured line around the umbilical region, figured, height 43 mm.

The type came from the Moluccas. South Australian Museum specimens from Darwin had the name *Natica cumingiana* Smith which is a synonym.

N.W.A., Q., N.A., Indo-Pacific.

Polinices columnaris (Récluz)

Pl. 2, no. 3

Natica columnaris Récluz, *J. Conchyliol.* 1: 394 (1850)

Groote Eylandt and Yirrkalla, figured, height 44 mm.

This large, white, North Australian shell has been recorded from Queensland, Torres Straits, Murray Island.

Q., N.A., Indo-Pacific.

Polinices aulacoglossa Pilsbry & Vanatta

Pl. 2, no. 4

Polinices aulacoglossa Pilsbry & Vanatta, *Proc. Acad. Nat. Sci. Philad.* 55: 558, pl. 29, fig. 1-3 (1908)

Albula didyma Bolten, *Mus. Bolten* (2): 20, No. 245 (1798) and *Natica ampla* Philippi (1848) described from unknown locality are synonyms.

Yirrkalla, two specimens alive, one figured, height 42 mm.

South Australian Museum specimens come from Gladstone, Cairns, Wide Bay, Caloundra, Queensland; Port Essington, Port Darwin; Newcastle, Port Jackson, N.S.W.; Altona Bay, Port Phillip, Victoria.

N.W.A., Vict., N.S.W., Q., N.A., Indo-Pacific.

Polinices pyriformis (Récluz)

Pl. 2, no. 5

Natica pyriformis Récluz, *Proc. Zool. Soc. Lond.*: 211 (1844)

Groote Eylandt and Yirrkalla, figured, height 41 mm.

The type came from Panay. Specimens of this pure white 'Natica' come from Claremont Island, Port Essington, Port Douglas, Cape York, Queensland; Murray Island, North Australia, Broome. The species *Natica mammilla* Linné, with which this has been confused, is one of the names allotted to a West Indian species.

N.W.A., Q., N.A., Indo-Pacific.

Mammilla maura (Lamarck)

Pl. 4, no. 3

Natica maura Lamarck, *Ency. Meth. Vers*: pl. 453, fig. 4 (1816)

Yirrkalla, one specimen, figured, height 32 mm.

South Australian Museum specimens come from Gladstone, Darnley Island, Claremont Island. The species has been known as *Natica nuxcastanea* Martyn 1786 but the name is invalid.

N.A., Indo-Pacific.

Natica lineata (Bolten)

Pl. 2, no. 7

Cochlis lineata Bolten, *Mus. Bolten* (2): 147, No. 1846 (1798)

Yirrkalla and Groote Eylandt, figured, height 29 mm.

This Indo-Pacific species has been collected from New Guinea, Cottesloe, Broome, Port Essington.

N.W.A., S.W.A., Q., N.A., Indo-Pacific.

Natica colliei Récluz

Pl. 2, no. 2

Natica colliei Récluz, *Proc. Zool. Soc. Lond.*: 206 (1844)

Groote Eylandt, figured, height 17 mm.

A probable synonym is *Natica sertata* Menke. South Australian Museum specimens are from Bunbury, Abrolhos, Fremantle, Carnarvon, Darwin, Royston Head, and dredged alive in S.W.A., 22 fms. Taken in South Australia from the beach at Royston Head, and dredged down to 40 fms.

S.W.A., S.A., Q., N.A.

TRIVIIDAE

Trivirostra oryza (Lamarck)

Pl. 4, no. 9

Cypraea oryza Lamarck, *Ann. Mus. Nat. Hist. Paris* 16: 104 (1810)

Groote Eylandt and Yirrkalla, three specimens, figured, height 11 mm.

Specimens are inseparable from a large series taken at Moreton Bay, Queensland.

N.W.A., Q., N.A., Indo-Pacific.

AMPHIPERATIDAE

Volva volva (Linné)

Pl. 5, no. 19

Bulla volva Linné, *Syst. Nat.* ed. 10: 725 (1758)

Yirrkalla, one specimen, figured, height 72 mm.

The type came from Ceylon. This is the first definite record of this species from North Australia.

N.W.A., Q., N.A., Indo-Pacific.

CYPRAEIDAE

Talparia talpa (Linné)

Pl. 1, no. 15

Cypraea talpa Linné, *Syst. Nat.* ed. 10: 720 (1758)

Yirrkalla, Groote Eylandt, figured, height 74 mm.

A series of this species ranges from 50 mm. to 70 mm. in the length of adult shells. Both pale and dark coloured varieties with the characteristic colour pattern occur.

N.W.A., N.S.W., Q., N.A., Indo-Pacific.

Mystaponda vitellus (Linné)

Pl. 1, no. 17

Cypraea vitellus Linné, *Syst. Nat.* ed. 10: 721 (1758)

Yirrkalla, Groote Eylandt, figured, 66 mm.

Two varieties of this species may be distinguished. One rather pale in colour with medium-sized white spots and one dark in colour with large spots, but there is intergradation. The species ranges through the Central Indo-Pacific Province to South Western Australia. While in Western Australia in 1949 I saw an average-sized specimen taken at Leighton just south of Cottesloe.

N.W.A., S.W.A., Q., N.A., Indo-Pacific.

Ponda carneola (Linné)

Pl. 1, no. 20

Cypraea carneola Linné, *Syst. Nat.* ed. 10: 719 (1758)

Yirrkalla, two specimens, dark in colour, measured 57 mm. by 35 mm. and 30 mm., the latter figured.

The shell seems to be a typical *P. carneola* rather than the Eastern Australian subspecies *P. carneola thepalea* Iredale.

N.W.A., Q., N.A., Indo-Pacific.

Arabica arabica westralis Iredale

Arabica westralis Iredale, *Aust. Zool.* 8: 108, pl. 28, fig. 3, 4 (1935)

Yirrkalla, one specimen, not typical but approaches *A. eglantina perconfusa* from Groote Eylandt.

Though less common than *A. eglantina perconfusa* Iredale in the South Australian Museum collection, specimens have been collected from Broome, Rowley Shoal, N.W.A. As mentioned by J. Allan (1956) in *Cowry Shells of World Seas,* this subspecies may be no other than the old *Arabica histrio* Gmelin.

N.W.A., Q., N.A., Indo-Pacific.

Arabica eglantina perconfusa Iredale

Pl. 1, no. 19

Arabica perconfusa Iredale, *Aust. Zool.* 8: 108, pl. 28, fig. 1-2 (1935)

Yirrkalla, one specimen, figured, height 65 mm.

Fairly common at Yirrkalla, the species shows a certain amount of variation. Adult size varies from 40 mm. to 80 mm. in length. The base is sometimes a little flattened but is not heavily flanged as in *A. westralis* nor does the colour pattern become distinctly spotted in that species. A pathological specimen is narrow, thick, with a produced spire. J. Allan (1956) *Cowry Shells of World Seas* points out that this subspecies is a synonym of *Arabica eglantina couturieri* Vayssiere.

N.W.A., Q., N.A., Indo-Pacific.

Arabica scurra (Gmelin)

Cypraea scurra Gmelin, *Syst. Nat.* ed. 13: 3409 (1791)

Yirrkalla, two typical specimens, one measures in height 51 mm.

They have a shape intermediate between this species and *A. perconfusa* Iredale and they are suffused with a pinkish colour. A third specimen in the South Australian Museum from Murray Island is somewhat like them.

N.W.A., N.A., Indo-Pacific.

Bistolida stolida (Linné)

Pl. 1, no. 21

Cypraea stolida Linné, *Syst. Nat.* ed. 10; 724 (1758)

Yirrkalla, single specimen, figured, height 28 mm. It is a typical shell, pale in colouring.

Ceylon is regarded as the type locality of this species. In South Australian Museum collection are specimens from Broome, Roebuck Bay, Western Australia.

N.W.A., N.A.

Erosaria erosa (Linné)

Pl. 1, no. 16

Cypraea erosa Linné, *Syst. Nat.* ed. 10: 723 (1758)

Yirrkalla, Groote Eylandt, one figured, height 30 mm.

Represented by a series of typical, good specimens, this species ranges from adult micromorphs of 20 mm. to large specimens of 40 mm. in height. Juveniles have a grey coloured dorsum faintly banded with brownish grey.

N.W.A., Q., N.A., Indo-Pacific.

Erosaria inocellata diversa (Kenyon)

Pl. 4, nos. 7, 8, 10, 11

Cypraea miliaris diversa Kenyon, *J. Conch.* 10: 184 (1902)

Groote Eylandt, Yirrkalla.

Half a dozen specimens show a range of variation. They are figured here.

a. Typical coloration of pale yellowish brown dorsum and pale blue spots, height 37 mm. (no. 7).
b. Yellowish with pale spots, height 36 mm. (no. 8).

c. Yellowish-green, faintly banded, height 35 mm. (no. 10).

d. Flesh-coloured dorsum, cream sides, no spots, height 37 mm. (no. 11).

Cypraea miliaris diversa Kenyon was described from Shark Bay, Western Australia, holotype D.3903 South Australian Museum, and the type description could well apply to one of the variants taken at Yirrkalla. The holotype and paratype from the Kenyon Collection in the South Australian Museum have the interior coloured pink as mentioned in the original description and noticed in the Arnhem Land specimens. *Cypraea miliaris gabrieli* Gatliff from the 'Northern Territory' can again be matched by the flesh-coloured specimen mentioned above. The specific name *Cypraea miliaris* Gmelin has been proved a synonym of *Cypraea erosa* Linné. *Erosaria metavona* Iredale is the East Coast subspecies related to *E. inocellata diversa* Kenyon.

N.W.A.

Erronea nimisserans Iredale

Pl. 1, no. 22

Erronea nimisserans Iredale, *Aust. Zool.* 8: (2), 130, pl. 9, fig. 1a, 1b (1935)

Yirrkalla, figured, height 33 mm.

A common species in Western Australia, ranges north from New South Wales to Shark Bay, Western Australia.

N.W.A., N.S.W., Q., N.A.

Palangerosa cylindrica sista Iredale

Pl. 1, no. 26

Palangerosa cylindrica sista Iredale, *Aust. Zool.* 9: (3), 321, pl. 29, fig. 35-6 (1939)

Yirrkalla, specimens figured, height 31 mm.

The holotype came from North Western Australia and seems closely allied to the true *P. cylindrica* Born from Amboina. Yirrkalla specimens are more deeply coloured and are somewhat smaller in size than those from Broome, North Western Australia.

N.W.A., N.A.

Palmadusta lutea bizonata Iredale

Pl. 1, no. 24

Palmadusta bizonata Iredale, *Aust. Zool.* 8: (2), 126 (1935)

Yirrkalla, figured, height 13 mm.

This banded subspecies of the *Cypraea lutea* type is well figured by Reeve, *Conch. Icon.* 3: pl. 20, fig. 110c (1846), except that the minute dorsal spots are obsolete in our specimens. Iredale described this subspecies, race or species from Nickol Bay, North West Australia.

N.W.A., N.A.

Evenaria asellus (Linné)

Pl. 1, no. 18

Cypraea asellus Linné, *Syst. Nat.* ed. 10: 722 (1758)

Yirrkalla specimens, figured, are like the typical ones from Amboina and are large and deeply coloured. Groote Eylandt, one specimen, figured, height 12 mm.

N.W.A., N.A.

Evenaria hirundo cameroni Iredale

Pl. 1, no. 25

Evenaria hirundo cameroni Iredale, *Aust. Zool.* 9: (3), 314, pl. 28, fig. 29-31 (1939)

Yirrkalla, two specimens are typical; one figured, height 19 mm.

Evenaria punctata atomaria (Gmelin)

Pl. 4, no. 6

Cypraea atomaria Gmelin, *Syst. Nat.* ed. 13: 3412 (1791)

Yirrkalla, two specimens, one figured, height 12 mm.

Two specimens from Yirrkalla agree with Martini's figure, *Conchylien Cab.* 1: 371, pl. 28, fig. 290, 291 (1769). They are not the related species described as *Evenaria carula* Iredale from Yirrkalla.

N.A., Indo-Pacific.

Solvadusta subviridis (Reeve)

Pl. 1, no. 23

Cypraea subviridis Reeve, *Proc. Zool. Soc. Lond.*: 68 (1835)

Yirrkalla, a number of specimens, figured, height 27 mm.

They are a little paler than the typical shell from North Queensland and could bear the name *Solvadusta subviridis dorsalis* Schilder and Schilder, but there is intergradation.

Q., N.A., Indo-Pacific.

Leporicypraea mappa (Linné)

Pl. 4, nos. 12, 13

Cypraea mappa Linné, *Syst. Nat.* ed. 10: 718 (1758)

Yirrkalla, two specimens: one was the deep rose-coloured variety, the other a fairly elongate pale greenish tinged specimen. The former, no. 12, height 70 mm., and the latter, no. 13, height 80 mm.

The first has been given the name *Cypraea rosea* Gray (a Mauritius subspecies) and the second *Cypraea viridis* Kenyon. But rose-tinted and green-tinted *L. mappa* are found among both Australian, Mauritius and Polynesia specimens.

Q., N.A., Indo-Pacific.

Pustularia cicercula (Linné)

Cypraea cicercula Linné, *Syst. Nat.* ed. 10: 725 (1758)

Groote Eylandt and Yirrkalla, a few specimens. One from the latter locality measures in height 19 mm. The specimens are typical.

N.A., Indo-Pacific.

Cribraria cribraria (Linné)

Cypraea cribraria Linné, *Syst. Nat.* ed. 10: 723 (1758)

Yirrkalla, two specimens, height 23 mm., width 13 mm.

The type came from Ceylon. Two specimens from Yirrkalla approach the subspecies of *C. cribraria fallax* Smith from North Western Australia.

N.A., Indo-Pacific.

CASSIDIDAE

Phalium areola (Linné)

Pl. 1, no. 14

Buccinum areola Linné, *Syst. Nat.* ed. 10: 736 (1758)

Yirrkalla, two specimens, one figured. Groote Eylandt, one specimen, figured.

One brightly coloured specimen figured here from Groote Eylandt has a finely nodulate shoulder, dark reddish-brown aperture and dark pink outer lip.

N.A., Indo-Pacific.

Semicassis diuturna (Iredale)

Pl. 1, no. 13

Semicassis diuturna Iredale, *Rec. Aust. Mus.* 15: 335, pl. 32, fig. 9 (1927)

Yirrkalla, one specimen, figured.

The type specimen came from Northern New South Wales but the species extends northwards and a good specimen was taken at Yirrkalla. Localities mentioned in the South Australian Museum collection are Pialba, Moreton Bay, Burleigh Heads, Caloundra, Port Darwin and New South Wales.

N.S.W., Q., N.A.

CYMATIIDAE

Biplex pulchella (Forbes)

Pl. 2, no. 14

Ranella pulchella Forbes, *Voy. Rattlesnake* 2: pl. 3, fig. 6 (1852)

Groote Eylandt, figured, height 16 mm.

One specimen only has been picked out from the Groote Eylandt smaller shells. There are specimens in the South Australian Museum from Port Keats, N.T., and some said to come from Queensland. This 'Kookaburra Shell' has been sometimes confused with the much larger, very different and rare *Biplex perca* Perry, the 'Perch' or 'Fish Shell' from Japan and Philippines.

N.Q., N.A.

FICIDAE

Ficus eospila (Péron)

Pl. 1, no. 36

Pyrula eospila Péron, *Voy. Terre Austral.* 1: 132 (1807)

Connexion Island and Groote Eylandt, several specimens, figured, height 70 mm.

Yirrkalla, where it is not uncommon, one specimen, measures 65 mm. in length.

The type came from Île Dupuch, W.A., and the species extends from Torres Strait to South Western Australia and along the south coast. I have examined a fragment from Rottnest Island and another broken shell taken in 100 fathoms 90 miles west of Eucla, recorded by Verco 1912 as *Ficus tessellatus* Kobelt. It is more delicate and slender than *F. communis* Bolten, the common Indo-Pacific and North Australian species.

N.W.A., S.W.A., N.A.

MURICIDAE

Euphyllon cornucervi (Bolten)

Pl. 2, no. 19

Purpura cornucervi Bolten, *Mus. Bolten* (2): 142 (1798)

Groote Eylandt, an almost perfect young specimen, figured, measures in length 65 mm.

Murex monodon Sowerby 1879, the type species of *Euphyllon*, is a synonym. It ranges from Queensland through North Australia to North Western Australia.

N.W.A., Q., N.A., Indo-Pacific.

Chicoreus torrefactus (Sowerby)

Pl. 2, no. 20

Murex torrefactus Sowerby, *Conch. Illust.*: fig. 120 (1840)

Yirrkalla, one fairly worn specimen, 85 mm. in length, figured, height 82 mm.

South Australian Museum specimens come from Groote Eylandt, Murray Island, North Australia and Broome, North Western Australia.

N.W.A., Q., N.A.

Rapana thomasiana Crosse

Pl. 2, no. 15

Rapana thomasiana Crosse, *J. Conchyliol.* 9: 176 (1861)

Yirrkalla, common, one specimen, figured, height 52 mm. Groote Eylandt, two shells.

These specimens recall the Japanese variety of the *R. bezoar, R. bulbosa, R. rapidformis complex* which extends from China, Japan, Philippines to North Australia and Queensland.

N.W.A., Q., N.A., Indo-Pacific.

SCALIDAE

Scala scalaris (Linné)

Pl. 2, no. 18

Turbo scalaris Linné, *Syst. Nat.* ed. 10: 764 (1758)

The type came from Amboina and a typical series in the South Australian Museum are from Port Keats. A specimen from Queensland appears to be the same species.

N.A., Indo-Pacific.

PYRENIDAE

Pyrene punctata (Bruguière)

Pl. 4, no. 2

Buccinum punctatum Bruguière, *Ency. Meth. Vers.* 1: 281, pl. 374, fig. 4a, b (1789)

Groote Eylandt, two specimens, figured, height 21 mm. Yirrkalla, common.

Synonyms are *P. discors* Gmelin, *P. rhombiferum* Bolten, *Columbella semipunctata* Lamarck.

N.W.A., Q., N.A.

Euplica varians (Sowerby)

Pl. 4, no. 1

Columbella varians Sowerby, *Proc. Zool. Soc. Lond.*: 118 (1832)

Yirrkalla, one specimen, figured, height 11 mm.

Columbella versicolor Sowerby is a synonym.

N.W.A., Q., N.A., Indo-Pacific.

BUCCINIDAE

Phos senticosus (Linné)

Pl. 4, no. 16

Murex senticosus Linné, *Syst. Nat.* ed. 10: 751 (1758)

Groote Eylandt, two specimens occupied by hermit crabs. Yirrkalla, one large specimen, figured, height 36 mm.

Phos senticosus extends throughout North Australia, Japan, and the South Pacific.

N.W.A., Q., N.A., Indo-Pacific.

Cantharus proteus (Reeve)

Pl. 4, no. 18

Buccinum proteus Reeve, *Conch. Icon.* 3: sp. 51, pl. 7, 51, a, b, c (1846)

Yirrkalla, one specimen, figured, height 30 mm.

Reeve gave Ceylon, Philippine Islands and Australia as the localities of this very variable species. Pilsbry mentions *C. fumosus* Dillwyn, *Buccinum strigosum* Gmelin, *C. undosus* Kiener, not Linné, as covering the same species. He also mentions a variety *C. fumosus rubiginosus* Reeve with the type locality 'Red Sea, whence it extends throughout the Indian Ocean into Polynesia, merging into the typical *C. fumosus*'. As synonyms of this variety he cites *C. subrubiginosus* Smith, Japan, *C. biliratus* Reeve, Galapagos and Viti Islands, *C. nigrocostatus* Reeve, Panama, and *Pisania desmoulinsi* Montrouzier to *C. proteus* Reeve.

N.W.A., Q., N.A., Indo-Pacific.

FASCIOLARIIDAE

Peristernia incarnata (Deshayes)

Pl. 4, no. 17

Turbinella incarnata Deshayes, *Laborde Voy. Arabie*: pl. 65, fig. 20-2 (1830)

Yirrkalla, one specimen, figured, height 35 mm.

South Australian Museum specimens are from Darwin, Fannie Bay and Broome.

N.W.A., Q., N.A., Indo-Pacific.

OLIVIDAE

Oliva episcopalis Lamarck

Oliva episcopalis Lamarck, *Ann. Mus. Nat. Hist. Paris* 16: 313 (1810)

The 'Episcopal Olive' was not represented in the Arnhem Land collection but several specimens from Darnley Island (Matthews) are in the South Australian Museum. Height 50 mm. Synonyms are *O. lugubris* Lamarck and *O. emeliodina* Duclos.

N.W.A., Q., N.A., Indo-Pacific.

Oliva sanguinolenta Lamarck

Oliva sanguinolenta Lamarck, *Ann. Mus. Nat. Hist. Paris* 16: 316 (1810)

The 'Blood Olive' has not been recorded from Australia but there are numerous specimens in the South Australian Museum from Darnley Island (Matthews). Height 40 mm. There are also specimens from Thursday Island and 'Carswell, Queensland' and Torres Straits, Verco Collection.

N.A., Q., Indo-Pacific.

Oliva ornata Marrat

Oliva sericea ornata Marrat, *Ann. Mag. Nat. Hist. Paris* 3rd ser., 20: 214 (1867). Holotype from N.A.

The species was further described and figured by Sowerby, *Thes. Conch.* 4: 13, pl. 334, sp. 59, fig. 102, 103 (1855). South Australian Museum specimens are from Broome (Timms), height 40 mm., and one further example from Mr A. Pearce. The aperture within is pale violet in all our specimens.

N.W.A., N.Q., N.A., Indo-Pacific.

Oliva erythrostoma (Lamarck)

Pl. 1, no. 10

Oliva erythrostoma Lamarck, *Ann. Mus. Nat. Hist. Paris* 16: 309 (1810)

Yirrkalla, figured, height 70 mm.

This species, commonly known as the 'Red Mouth Olive', is found living from North Western Australia, north and east to the Great Barrier Reef. It is plentiful and variable in colour pattern. The shell is typically light flesh colour, reticulated with light grey, spirally ornamented with broken dark grey bands, and has a deep pink aperture. Dark brown specimens of this shell are sometimes labelled *Oliva maura*, but the deep pink aperture distinguishes the present species.

N.W.A., N.S.W., Q., N.A., Indo-Pacific.

Oliva maura Lamarck

Oliva maura Lamarck, *Ann. Mus. Nat. Hist. Paris* 16: 311 (1810)

Yirrkalla, one poor specimen; Darwin, one specimen measures in height 55 mm.

'The Chocolate Olive' inhabits the Indo-Pacific. It is dark chocolate in colour, sometimes almost black, and has an indefinite spiral and axial banding of the same or darker colour. The fasciole and aperture may be white or stained with pale brown.

N.W.A., Q., N.A., Indo-Pacific.

Oliva irisans Lamarck

Oliva irisans Lamarck, *Ann. Mus. Nat. Hist. Paris* 16: 512 (1810)

Yirrkalla, one specimen, height 44 mm.

Recorded from Mauritius, Ceylon, Java, Philippines, Australia and Polynesia. The species is a very variable one. The specimen from Yirrkalla is the pale-coloured variety but the shell may be dark chocolate-coloured like *O. maura*. It may also be brightly and heavily coloured, recalling *O. erythrostoma* when the aperture is violet and yellow within the throat.

Specimens of this type are in the South Australian Museum collection from

Queensland and a series from Guam is in the Hartley Collection. A distinguishing feature is the strong set of teeth extending through the whole length of the columella.

N.W.A., N.A., Indo-Pacific.

Oliva ispidula (Linné)

Pl. 1, no. 11

Voluta ispidula Linné, *Syst. Nat.* ed. 10: 730 (1758)

Yirrkalla, Groote Eylandt, figured, height 47 mm.

A few specimens only were noticed in the collections. The one figured has the distinctive zig-zag ash-coloured markings on a yellowish ground colour, fasciola and columella white, aperture dark brown. The aperture is usually dark bluish-olive in typical specimens. A representative series in the South Australian Museum from Thursday Island, Darnley Island, etc., shows considerable colour variations.

N.W.A., Q., N.A., Indo-Pacific.

Carmione inflata (Lamarck)

Oliva inflata Lamarck, *Ann. Mus. Nat. Hist. Paris* 16: 319 (1810)

Two specimens of the 'Inflated Olive' from Broome (Bardwell) are of the '*O. bicincta*' variety, height 40 mm. Numerous synonyms and varieties are recorded. *O. bicincta* Lamarck, *O. bicingulata* Lamarck, *O. fabagina* Lamarck and *O. undata* Lamarck. Localities range from the Red Sea, Ceylon, Australia to Zanzibar.

N.W.A., N.A., Indo-Pacific.

Galeola carneola (Lamarck)

Oliva carneola Lamarck, *Ann. Mus. Nat. Hist. Paris* 16: 321 (1810)

The 'Flesh Olive' we have from many localities. Kings Sound, N.W.A., Fiji, Philippines and 'North Australia' (Wilkinson), height 20 mm.

N.W.A., N.A., Indo-Pacific.

Neocylindrus tessellata (Lamarck)

Oliva tessellata Lamarck, *Ann. Mus. Nat. Hist. Paris* 16: 320 (1810)

The 'Tessellated Olive' is a distinct flesh-coloured species with lilac spots and intense purple violet aperture and columella. It is widely distributed in the Indo-Pacific. Our specimens from Thursday Island (Matthews), height 35 mm. A synonym is *Cylindrus tigrinus* Meuschen.

N.W.A., N.A., Indo-Pacific.

Anazola acuminata (Lamarck)

Oliva acuminata Lamarck, *Ann. Mus. Nat. Hist. Paris* 16: 325 (1810)

Lamarck, Duclos, Jay, give Java as the type locality of this species. Reeve cites North West Africa and Tryon adds Australia and the Philippines to the range. Two specimens in the South Australian Museum collections are D.9362 'N. Queensland' and D. 642 'Northern Territory', height 44 mm.

N.Q., N.A., Indo-Pacific.

Ancillista cingulata (Sowerby)

Pl. 1, no. 12

Ancillaria cingulata Sowerby, *Spec. Conch.* 1: 6, pl. 13, fig. 36, 37 (1830)

Yirrkalla, several specimens, figured, height 60 mm.

The type locality is Cape York, North Queensland. In the South Australian Museum similar shells are represented from North Keppel Island, Port Keats and Darnley Island.

N.W.A., S.W.A., S.A., Q., N.A.

NASSARIIDAE

Nassarius glans (Linné)

Buccinum glans (Linné), *Syst. Nat.* ed. 10: 737 (1758)

Yirrkalla, two specimens (juvenile), height 25 mm.

The species is widely distributed through the Indo-Pacific.

N.W.A., Q., N.A., Indo-Pacific.

MITRIDAE

Mitraria mitra (Linné)

Pl. 5, no. 13

Voluta mitra Linné, *Syst. Nat.* ed. 10: 732 (1758)

Port Keats, one specimen, figured, length 113 mm.

Widely distributed in the Indo-Pacific, Barrier Reef and northwards to Japan.

N.W.A., Q., N.A., Indo-Pacific.

Mitraria pontificalis (Lamarck)

Pl. 5, no. 17

Mitra pontificalis Lamarck, *Ann. Mus. Nat. Hist. Paris* 17: 198 (1811)

Port Keats, one specimen, figured, length 60 mm., and others from Queensland and Barrier Reef.

Q., N.A., Indo-Pacific.

Mitraria variabilis (Reeve)

Pl. 5, no. 11

Mitra variabilis Reeve, *Conch. Icon.* 2: sp. 95, pl. 13, fig. 95 (1844)

Yirrkalla, two specimens, one figured, height 40 mm.

The type came from Torres Straits.

Q., N.A.

Imbricaria conica (Schumacher)

Pl. 5, no. 18

Mitra conica Schumacher in Martini & Chemnitz, *Syst. Conch.* ed. 2: (16), 97 (1817)

Port Keats, one specimen, figured, length 25 mm.

Further specimens are from Queensland and the Indo-Pacific generally.

N.A., Q., Indo-Pacific.

Vexillum vulpecula (Linné)

Pl. 5, no. 16

Voluta vulpecula Linné, *Syst. Nat.* ed. 10: 732 (1758)

Port Keats, one specimen, figured, length 47 mm., and others from Queensland. This is probably a synonym of *V. taeniata* Lamarck.

Q., N.A., Indo-Pacific.

Pulchrimitra stigmataria (Lamarck)

Pl. 5, no. 15

Mitra stigmataria Lamarck, *Ann. Mus. Nat. Hist. Paris*: 17: 99-100 (1811)

Melville Island, one specimen, length 48 mm.

Generally distributed in the Indo-Pacific.

N.A., Indo-Pacific.

Cylindromitra crenulata (Gmelin)

Pl. 5, no. 14

Mitra crenulata Gmelin, *Syst. Nat.* ed. 13: 3452 (1791)

Port Keats, one specimen, figured, length 35 mm.

Widely distributed in the Indo-Pacific. The Queensland sub-species has been named *C. crenulata toleranda* Iredale.

N.A., Indo-Pacific.

CANCELLARIIDAE

Sydaphera semperiana (Crosse)

Pl. 4, no. 19

Cancellaria semperiana Crosse, *J. Conchyliol.* 11: 65, pl. 2, fig. 7 (1863)

Groote Eylandt, a juvenile specimen, height 33 mm. Port Keats, one adult, figured, height 36 mm.

The holotype came from New Caledonia, but the species is widely distributed in the Indo-Pacific and North Australia.

Q., N.A., New Caledonia, Indo-Pacific.

VOLUTIDAE

Aulicina pulchra (Sowerby)

Voluta pulchra Sowerby, *Tankerville Cat. App.*: 28 (1825)

The South Australian Museum collection contains specimens of this volute from North West Islet (Kimber), Gladstone (Verco), Heron Island (Ivor Powell) and North Australia (Kenyon). The subspecies *A. pulchra wisemani* Brazier we have from 'North Queensland' and 'North Australia' (Kenyon). These specimens are not typical examples of the subspecies but appear to be intermediate in form between *A. pulchra* and *A. pulchra wisemani.* Colour photographs of a different shell from Queensland, Wistaria Reef, Capricorn Group were forwarded to me by Ostergaard and Summers, through Mrs T. Hartley of Melbourne. This appears to have the orange colour band within the lip as in *A. pulchra*, but a colour pattern somewhat like *A. pulchra wisemani*. It has since been named *Aulicina perryi* Ostergaard and Summers, 1957, and the holotype D.14605 presented to the South Australian Museum.

CONIDAE

Gastridium geographus (Linné)

Pl. 1, no. 3

Conus geographus Linné, *Syst. Nat.* ed. 10: 718 (1758)

Yirrkalla, one specimen, 115 mm. in length, figured, and one small juvenile, height 110 mm.

Q., N.A., Indo-Pacific.

Hermes nussatella (Linné)

Pl. 1, no. 2

Conus nussatella Linné, *Syst. Nat.* ed. 10: 716 (1758)

Yirrkalla, common, figured, height 85 mm.

South Australian Museum specimens are from New Guinea, Queensland, Philippines and Fiji.

Q., N.A., Indo-Pacific.

Regiconus aulicus (Linné)

Pl. 1, no. 4

Conus aulicus Linné, *Syst. Nat.* ed. 10: 717 (1758)

Yirrkalla, figured, height 112 mm.

This is a wider more coarsely patterned shell than *R. auratus* Bruguière. South Australian Museum specimens were taken at numerous localities, from Ceylon to New Caledonia.

Q., N.A., Indo-Pacific.

Rhizoconus mustelinus (Bruguière)

Pl. 1, no. 1

Conus mustelinus Bruguière, *Ency. Meth. Vers.* 1: 654 (1792)

Yirrkalla, two specimens, one figured, height 60 mm.

This species apparently grades into *R. capitaneus* Linné, widely distributed in the Indo-Pacific.

Q., N.A., Indo-Pacific.

Rhizoconus vexillum (Gmelin)

Pl. 1, no. 6

Conus vexillum Gmelin, *Syst. Nat.* ed. 13: 3397 (1791)

Yirrkalla, figured, height 70 mm.

South Australian Museum specimens are from New Guinea, New Caledonia, Queensland and Fiji.

Q., N.A., Indo-Pacific.

Rhizoconus capitaneus (Linné)

Pl. 1, no. 7

Conus capitaneus Linné, *Syst. Nat.* ed. 10: 713 (1758)

Yirrkalla, one specimen, figured, height 45 mm.

Specimens in the South Australian Museum prove the wide distribution of this species. West Indies, Fiji, New Guinea, Groote Eylandt, Broome and tropical Australia in general.

N.S.W., Q., N.A., Indo-Pacific.

Chelyconus magus (Linné)

Pl. 1, no. 9

Conus magus Linné, *Syst. Nat.* ed. 10: 716 (1758)

Yirrkalla, figured, height 55 mm.

Described without locality, the species ranges throughout the Indo-Pacific and has been recorded under many different names.

N.W.A., Q., N.A., Indo-Pacific.

Virroconus coronatus (Gmelin)

Pl. 4, no. 21

Conus coronatus Gmelin, *Syst. Nat.* ed. 13: 3389 (1791)

Yirrkalla, figured, height 32 mm.

There are a number of synonyms of this species which is widely dispersed in the Indo-Pacific. We have examples from numerous localities including Geraldton, Murray Island, North West Indies, Port Essington, Fiji and Hawaii.

N.W.A., N.S.W., Q., N.A., Indo-Pacific.

Virgiconus virgo (Linné)

Pl. 1, no. 5

Conus virgo Linné, *Syst. Nat.* ed. 10: 713 (1758)

Yirrkalla, figured, height 60 mm.

This species originally described from Amboina is common and widely dispersed. South Australian Museum specimens come from numerous places(Aden, Mauritius, Ceylon, Melville Island, Manus, Brooker Island, Fiji and Queensland.

Q., N.A., Indo-Pacific.

Leporiconus glans (Bruguière)

Pl. 1, no. 8

Conus glans Bruguière, *Ency. Meth. Vers.* 1: 735 (1792)

Yirrkalla, one large specimen, figured, height 41 mm.

South Australian Museum specimens are from the Philippines, Manus, New Caledonia, Fiji, Exmouth Gulf.

N.W.A., Q., N.A., Indo-Pacific.

Cylinder textile (Linné)

Conus textile Linné, *Syst. Nat.* ed. 10: 717 (1758)

Groote Eylandt, two specimens, height 86 mm.

This species is widely dispersed in the Indo-Pacific.

N.W.A., Q., N.A., Indo-Pacific.

Dauciconus planorbis (Born)

Conus planorbis Born, *Mus. Caes.*: 164, pl. 7, fig. 13-14 (1780)

Cambridge Gulf, one specimen, and Yirrkalla, one specimen.

N.W.A., Q., N.A., Indo-Pacific.

TEREBRIDAE

Hastula strigilata (Linné)

Pl. 5, no. 3

Buccinum strigilata Linné, *Syst. Nat.* ed. 10: 741 (1758)

Melville Island, one specimen, figured, length 41 mm., others from Cossack and Broome.

N.W.A., Q., N.A., Indo-Pacific.

Diplomeriza duplicata (Linné)

Pl. 5, no. 4

Buccinum duplicata Linné, *Syst. Nat.* ed. 10: 742 (1758)

Specimens from Cossack, Broome, Darwin, Melville Island, figured, length 43 mm.

N.W.A., Q., N.A., Indo-Pacific.

Oxymeris maculata (Linné)

Pl. 5, no. 12

Buccinum maculata Linné, *Syst. Nat.* ed. 10: 741 (1758)

Melville Island, one specimen, figured, length 152 mm.

Oxymeris felinum (Dillwyn)

Pl. 5, no. 6

Buccinum felina Dillwyn, *Cat. Shells* 2: 644 (1817)

Port Keats, one specimen, figured, length 45 mm; others from Queensland and Indo-Pacific.

Q., N.A., Indo-Pacific.

Terebra muscaria Lamarck

Pl. 5, no. 5

Terebra muscaria Lamarck, *Anim. S. Vert.* 10: 241 (1822)

Melville Island, one specimen, figured, length 115 mm., and others from Broome and the Barrier Reef.

N.W.A., Q., N.A.

Terebra bernhardi Deshayes

Pl. 5, no. 9

Terebra bernhardi Deshayes, *J. Conchyliol.* 6: 84, pl. 4, fig. 10 (1857)

Port Keats, one specimen, figured, length 56 mm., also from Queensland and the South Pacific.

Q., N.A.

Terebra chlorata Lamarck

Pl. 5, no. 7

Terebra chlorata Lamarck, *Anim. S. Vert.* 7: 288 (1822)

Port Keats, one specimen, figured, length 74 mm. Other specimens from Queensland.

Q., N.A.

Dimidacus cinguliferus (Lamarck)

Pl. 5, no. 8

Terebra cingulifera Lamarck, *Anim. S. Vert.* 7: 289 (1822)

Port Keats, one specimen, figured, length 77 mm.

N.A.

Triphostrephanus praelonga (Deshayes)

Pl. 5, no. 2

Terebra praelonga Deshayes, *Proc. Zool. Soc. Lond.*: 315 (1859)

Port Keats, one specimen, figured, length 92 mm., and other specimens from Gourdon Bay and Queensland. Holotype came from Port Curtis, Queensland.

Q., N.A.

Perirhoe circumcincta (Deshayes)

Pl. 5, no. 10

Terebra circumcincta Deshayes, *J. Conchyliol.* 6: 77, pl. 3, fig. 9 (1857)

Port Keats, one specimen, figured, length 45 mm.

The type came from the Red Sea.

ACTEONIDAE

Solidula solidula (Linné)

Pl. 4, no. 4

Bulla solidula Linné, *Syst. Nat.* ed. 10: 728 (1758)

Groote Eylandt, figured, height 23 mm.

The species was originally described from the Philippines. It is variable and has an extensive Indo-Pacific range. A subfossil, probably of Middle Recent Age, closely resembling this species is found on the raised beaches of South Australia.

N.A., Indo-Pacific.

Solidula suturalis Adams

Pl. 4, no. 5

Solidula suturalis Adams, *Proc. Zool. Soc. Lond.*: 61 (1855)

Yirrkalla, one specimen, figured, height 14 mm.

A series in the South Australian Museum collection come from Cape York.

Q., N.A.

Atys naucum (Linné)

Pl. 2, no. 25

Bulla naucum Linné, *Syst. Nat.* ed. 10: 726 (1758)

Yirrkalla, one perfect adult, figured, height 48 mm.

Originally described from the East Indies, South Australian Museum specimens are from Port Essington, Cape York, Monte Bello Island.

N.W.A., Q., N.A., Indo-Pacific.

XANTHOMELONTIDAE

Xanthomelon jannellii (Le Guillou)

Pl. 4, no. 20

Helix jannellii Le Guillou, *Rev. Zool. Soc. Cuv.*: 137 (1842)

Groote Eylandt, Hemple Bay, specimen figured, height 23 mm.

The species was described from Torres Strait, North Australia. The series taken by R. L. Specht at Groote Eylandt are typical and represent the only land shells received by me. A synonym is *Helix pachystyloides* Cox described from Cape York.

Q., N.A.

(Manuscript received, October 1958)

Plates 1-5

PLATE 1

1. *Rhizoconus mustelinus,* Yirrkalla
2. *Hermes nussatella,* Yirrkalla
3. *Gastridium geographus,* Yirrkalla
4. *Regiconus aulicus,* Yirrkalla
5. *Virgiconus virgo,* Yirrkalla
6. *Rhizoconus vexillum,* Yirrkalla
7. *Rhizoconus capitaneus,* Yirrkalla
8. *Leporiconus glans,* Yirrkalla
9. *Chelyconus magus,* Yirrkalla
10. *Oliva erythrostoma,* Yirrkalla
11. *Oliva ispidula,* Groote Eylandt
12. *Ancillista cingulata,* Yirrkalla
13. *Semicassis diuturna,* Yirrkalla
14. *Phalium areola,* Yirrkalla
15. *Talparia talpa,* Groote Eylandt
16. *Erosaria erosa,* Groote Eylandt
17. *Mystaponda vitellus,* Groote Eylandt
18. *Evenaria asellus,* Groote Eylandt
19. *Arabica eglantina perconfusa,* Yirrkalla
20. *Ponda carneola,* Yirrkalla
21. *Bistolida stolida,* Yirrkalla
22. *Erronea nimisserans,* Yirrkalla
23. *Solvadusta subviridis,* Yirrkalla
24. *Palmadusta lutea bizonata,* Yirrkalla
25. *Evenaria hirundo cameroni,* Yirrkalla
26. *Palangerosa cylindrica sista,* Yirrkalla
27. *Canarium urceus,* Groote Eylandt
28. *Doxander vittatus,* Yirrkalla
29. *Doxander campbelli,* dorsal, Yirrkalla
30. *Doxander campbelli,* ventral, Groote Eylandt
31. *Euprotomus australis,* Groote Eylandt
32. *Clava pulcher,* Yirrkalla
33. *Ischnocerithium echinatum,* Yirrkalla
34. *Clava aspera,* Yirrkalla
35. *Pseudovertagus sinensis,* Yirrkalla
36. *Ficus eospila,* Groote Eylandt

B. J. Hunter

PLATE 1

Plate 2

1. *Polinices powisiana*, Yirrkalla
2. *Natica colliei*, Groote Eylandt
3. *Polinices columnaris*, Yirrkalla
4. *Polinices aulacoglossa*, Yirrkalla
5. *Polinices pyriformis*, Yirrkalla
6. *Amphinerita polita*, Groote Eylandt
7. *Natica lineata*, Groote Eylandt
8. *Pictoneritina oualanensis*, Yirrkalla
9. *Theliostyla reticulata*, Yirrkalla
10. *Nerita lineata*, Yirrkalla
11. *Angaria tyria*, Yirrkalla
12. *Architectonica perspectiva*, Groote Eylandt
13. *Distellifer stellare*, Groote Eylandt
14. *Biplex pulchella*, Groote Eylandt
15. *Rapana thomasiana*, Yirrkalla
16. *Senectus squamosus*, Groote Eylandt
17. *Trochus maculatus*, Yirrkalla
18. *Scala scalaris*, Port Keats
19. *Euphyllon cornucervi*, Groote Eylandt
20. *Chicoreus torrefactus*, Yirrkalla
21. *Turritella terebra*, Groote Eylandt
22. *Turbo petholatus*, Yirrkalla
23. *Sanhaliotis squamata*, Groote Eylandt
24. *Haliotis asinina*, Yirrkalla
25. *Atys naucum*, Yirrkalla
26. *Dosinia scalaris*, Groote Eylandt
27. *Lioconcha castrensis*, Yirrkalla
28. *Cardita crassicostata*, Yirrkalla
29. *Tellinella virgata*, Yirrkalla
30. *Angulus pharaonis*, Groote Eylandt
31. *Phylloda foliacea*, Yirrkalla
32. *Solen timorensis*, Yirrkalla
33. *Belchlamys newmanae* sp. nov., Groote Eylandt
34. *Excellichlamys spectabilis*, Yirrkalla
35. *Magnavicula bennetti*, Yirrkalla

PLATE 2

B. J. Hunter

Plate 3

1. *Circe gibbium,* external, Groote Eylandt
2. *Circe gibbium,* internal, Groote Eylandt
3. *Psammosolen sulcatus,* Yirrkalla
4. *Gari caerulescens,* Yirrkalla
5. *Septifer bilocularis,* external, Groote Eylandt
6. *Septifer bilocularis,* internal, Groote Eylandt
7. *Stavelia horrida,* external, Thursday Island
8. *Stavelia horrida,* internal, Thursday Island
9. *Musculus cumingianus,* Yirrkalla
10. *Regozara flava,* external, Groote Eylandt
11. *Regozara flava,* internal, Groote Eylandt
12. *Fulvia laevigata,* Port Keats
13. *Discors lyratum,* external, Port Keats
14. *Discors lyratum,* internal, Port Keats
15. *Pratulum probatum,* external, Port Keats
16. *Pratulum probatum,* internal, Port Keats
17. *Pratulum setosum,* external, Darwin
18. *Pratulum setosum,* internal, Darwin
19. *Hemicardium tumoriforum,* external, Darwin
20. *Hemicardium tumoriforum,* internal, Darwin

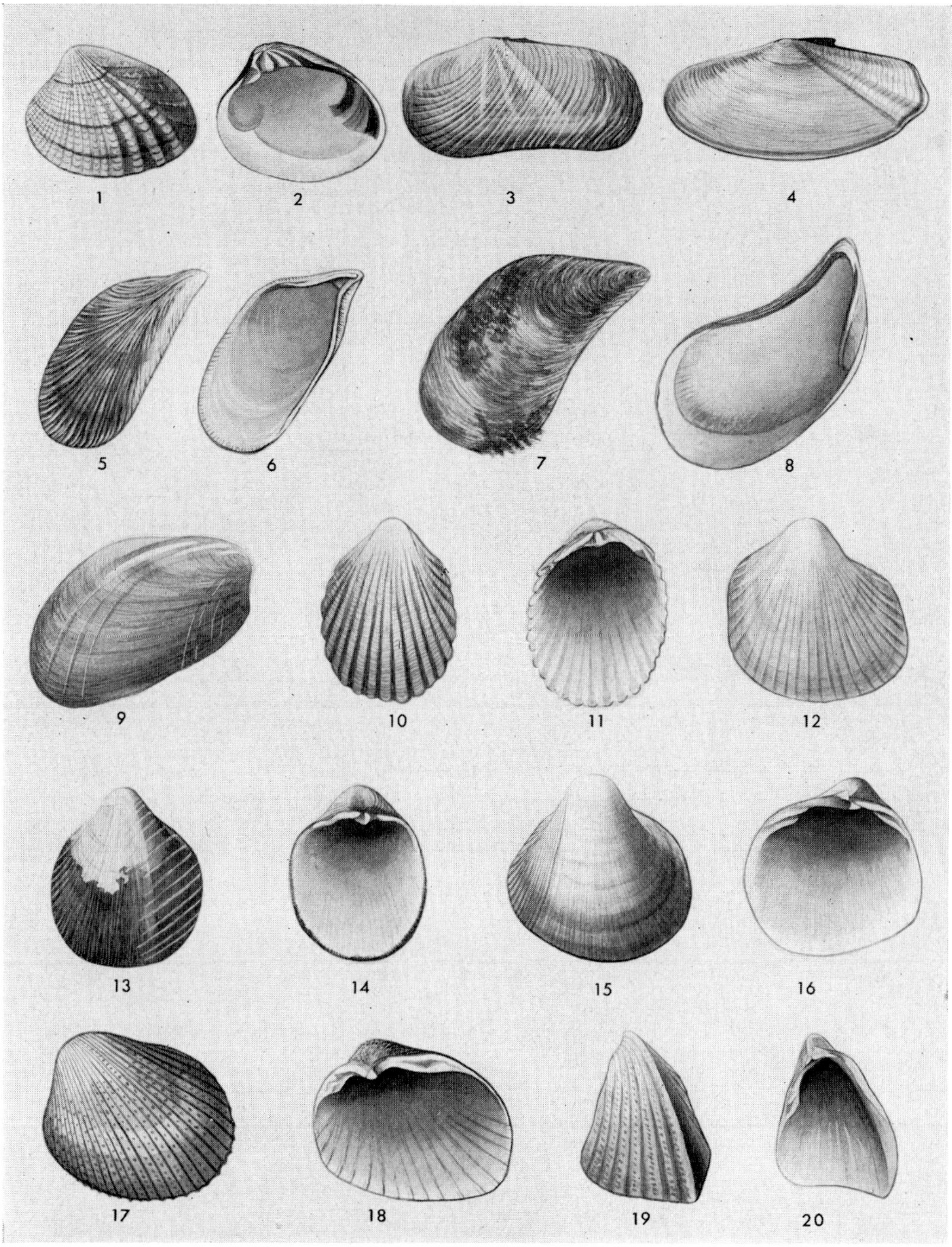

Audine M. Cheek

PLATE 3

PLATE 4

1. *Euplica varians*, Yirrkalla
2. *Pyrene punctata*, Groote Eylandt
3. *Mammilla maura*, Yirrkalla
4. *Solidula solidula*, Groote Eylandt
5. *Solidula suturalis*, Yirrkalla
6. *Evenaria punctata atomaria*, Yirrkalla
7. *Erosaria inocellata diversa*, typical, Yirrkalla
8. *Erosaria inocellata diversa*, pale spotted variant, Yirrkalla
9. *Trivirostra oryza*, Yirrkalla
10. *Erosaria inocellata diversa*, faintly banded variant, Yirrkalla
11. *Erosaria inocellata diversa*, two-coloured variant, Yirrkalla
12. *Leporicypraea mappa*, rose-coloured variant, Yirrkalla
13. *Leporicypraea mappa*, cylindrical green variant, Yirrkalla
14. *Tectus obeliscus*, Yirrkalla
15. *Longehaeus sulcatus*, Groote Eylandt
16. *Phos senticosus*, Yirrkalla
17. *Peristernia incarnata*, Yirrkalla
18. *Cantharus proteus*, Yirrkalla
19. *Sydaphera semperiana*, Port Keats
20. *Xanthomelon jannellii*, Hemple Bay, Groote Eylandt
21. *Virroconus coronatus*, Yirrkalla

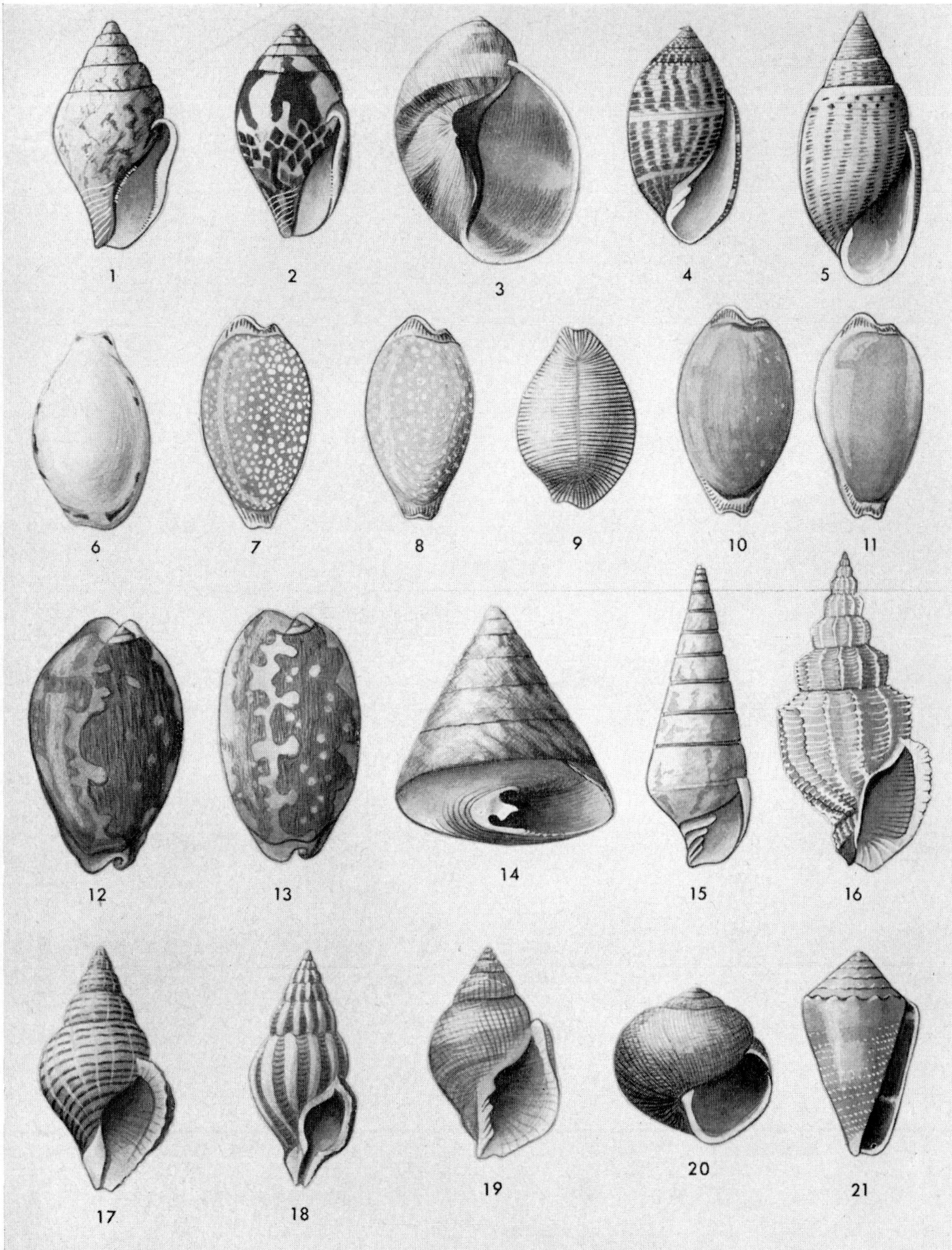

Audine M. Cheek

PLATE 4

PLATE 5

1. *Clava mountfordae* sp. nov., Shark Bay
2. *Triphostrephanus praelonga,* Port Keats
3. *Hastula strigilata,* Melville Island
4. *Diplomeriza duplicata,* Broome
5. *Terebra muscaria,* Melville Island
6. *Oxymeris felinum,* Port Keats
7. *Terebra chlorata,* Port Keats
8. *Dimidacus cinguliferus,* Port Keats
9. *Terebra bernhardi,* Port Keats
10. *Perirhoe circumcincta,* Port Keats
11. *Mitraria variabilis,* Yirrkalla
12. *Oxymeris maculata,* Melville Island
13. *Mitraria mitra,* Port Keats
14. *Cylindromitra crenulata,* Port Keats
15. *Pulchrimitra stigmataria,* Melville Island
16. *Vexillum vulpecula,* Port Keats
17. *Mitraria pontificalis,* Port Keats
18. *Imbricaria conica,* Port Keats
19. *Volva volva,* Yirrkalla

Audine M. Cheek

PLATE 5

3

FISHES OF ARNHEM LAND

by William R. Taylor*

This study is based on the 1948 Arnhem Land Expedition collections made by Dr Robert Rush Miller, now at the University of Michigan. Fishes were preserved from fresh, brackish and marine waters of Arnhem Land and adjacent Northern Territory. The specimens were largely collected in the vicinity of the various camp sites of the expedition (map, frontispiece).

Dr Miller, who was the ichthyologist on the expedition, made extensive field notes which contain data on the native names of fishes, the ecology, numerous colour descriptions and colour sketches. He took colour transparencies of many fishes. Data from his field notes are recorded in this report wherever possible.

Originally it was planned to include all fish families represented in the collections, but lack of time did not permit me to study the Gobiidae, Eleotridae, Labridae, Clinidae, Blenniidae, and a few other small groups.

The descriptive plan adopted for this report is essentially that described by Schultz *et al.* (1953) unless otherwise indicated.

The scientific name of each species is accompanied by the native name if known. A reference to the original description is followed by citations to previously reported Northern Territory fishes. The list of specimens studied includes only the Arnhem Land material.

The descriptive section for each species gives information on the morphology, colour in alcohol, and life colours as noted by Dr Miller. The keys to the species are based chiefly on the Arnhem Land collections but in their construction use was made of the ichthyological literature and other specimens in the United States National Museum as needed.

Most references to names of genera have been deleted from this paper. Unless otherwise noted they may be readily found by consulting Fowler (1931, 1933, 1941), Fowler and Bean (1928, 1929, 1930), Herre (1953), McCulloch (1929-30), and Schultz and Collaborators (1953, 1960). The generic citations of these authors sometimes vary slightly from my findings in pagination, date of publication, or in method of type species designation.

The scientific name of each species is based on a comparison of the Arnhem Land material with the original species descriptions, subsequent studies, and specimens in the United States National Museum. Some identifications are regarded as tentative, since it was impossible to re-examine the respective type specimens in various museums of the world.

* Smithsonian Institution, Washington, D.C.

Much of the difficulty in the identfication of the fishes of the Indo-Pacific region results from a lack of revisions of genera and families.

The marine fish fauna of Arnhem Land is part of the widespread tropical Indo-Pacific faunal region ranging from eastern Africa across the tropical parts of the Indian and Pacific Oceans, at least to Easter Island with some species extending to the western coast of the Americas. Therefore, when monographic studies have been prepared for each family of fishes occurring in this vast area, the scientific names reported here may need to be changed.

The following new species are described in this report: *Melanotaenia solata, Pseudomugil tenellus, Pranesus maculatus, Pomacentrus milleri.* In addition *Archamia melasma* Lachner and Taylor was described in *Proc. Biol. Soc. Wash.* 73: 29-34 (1960).

The faunal relationships of Arnhem Land are beyond the scope of this study. The marine fishes are part of the extensive tropical Indo-Pacific fauna, and show most intimate relationships to those of coastal regions of Australia and New Guinea. The freshwater fauna is related to that of south-western New Guinea and neighbouring tropical Australia. The region has been termed the Leichhardtian Fluvifaunula (Iredale and Whitley, 1938; Whitley, 1947).

Some species characteristic of south-western New Guinea and Arnhem Land are *Melanotaenia rubrostriata, Pseudomugil gertrudae, Zenarchopterus caudovittatus, Cynoglossus heterolepis, Thryssa scratchleyi,* and the catfishes, genera *Porochilus* and *Tandanus.*

Some apparent endemic Arnhem Land species are: *Phyllichthys sclerolepis, Anodontiglanis dahli, Melanotaenia trifasciata, Denariusa bandata* and *Gnathypops darwiniensis.*

ACKNOWLEDGMENTS

I am grateful to Dr Robert Rush Miller for the excellently preserved fish specimens and for his field notes made available for this study. Also, I appreciate the help of the various other members of the Arnhem Land Expedition, including local inhabitants who supplied valuable specimens, especially Jerry Blitner, Fred Gray, Nich Haritos, and Bill Harney.

It is with much pleasure that I express appreciation to my ichthyological colleagues in the United States National Museum, especially to Dr Leonard P. Schultz and Dr Ernest A. Lachner for their valuable suggestions and kind advice when needed.

I gratefully acknowledge the preparation of the numerous drawings reproduced in this publication by the following expert artists: Mrs Patricia Hogue, Mrs Carolyn B. Lutz, and Mrs Dorothea B. Schultz. In the legend for each drawing the name of the artist is given. The drawings were prepared under a grant from the National Geographic Society of Washington, D.C.

LIST OF COLLECTING STATIONS

M48-1. Tide pools on 'ironstone' and red reefs at Nightcliff [lat. 12°23′S., long. 130°50′E.], 7 miles north of Darwin, Northern Territory, 19 to 24 March 1948.

R. R. Miller, Kelvin Hodges and party. Collection made by barbasco root and by hand over a bottom of sand, rocks, and sparse, mostly dead coral, to a depth of 3 feet. Water and air temperatures both 96°F. Water clear to slightly turbid, salty.

M48-2. Rocky coves, with sandy beach, and at edge of reef, at Nightcliff [lat. 12°23'S., long. 130°50'E.], 7 miles north of Darwin, Northern Territory, 19 to 25 March 1948. R. R. Miller, F. M. Setzler, Kelvin Hodges, F. McCarthy, and party. Collection made with 25 foot bag seine and by hook and line over a sand bottom with a few rocks at depths to 12 feet. Water and air temperatures, 20 March 1948, both 90°F. at 3.00 p.m., the water slightly turbid, salty.

M48-3. East Point Reef, about 6 miles north-north-west of Darwin, Northern Territory, 26 March 1948. R. R. Miller and party. Collection made with 10 pounds of barbasco root in pools and crevasses about the coral, to a depth of one and one-half feet. Water clear, salty, its temperature 96°F.

M48-4. Creek, tributary to Woods Inlet, at Delissaville Aboriginal Station, about 12 miles west-south-west of Darwin, Northern Territory, 29 March 1948. R. R. Miller and F. M. Setzler. Collection made with one-fourth inch mesh seine, which 10 feet long, to a depth of 3 feet, over a bottom of hard clay, gravel, sand, and silt. Water clear, but easily roiled, fresh, its temperature 90°F. Creek, ranging from 6 to 20 feet wide, averaging about 15 feet, with a moderate to swift current; matted roots extending from the bank and a few submerged, string-like green plants present.

M48-5. Woods Inlet, at dock, which about 1 mile west of Delissaville Aboriginal Station, Northern Territory, 26 to 30 March 1948. R. R. Miller and Peter Basset-Smith. Collection made with dip net and by hand over a bottom of deep mud and some hard clay. Water brackish; its temperature about 90°F.; current moderate to sluggish; steep banks and mangroves at shore.

M48-6. Billabong on Fred Gray's Camp (Umbakumba), near north-east end of Groote Eylandt, Northern Territory, 8 April 1948. R. R. Miller, F. M. Setzler and K. Hodges. Collection made by 25 foot bag seine in clear, brown, fresh water, to a depth of 3 feet, over a bottom of mud, some sand, and considerable vegetation; water temperatures of 88°F. at the surface and 83°F. at a depth of 5 feet. The billabong arises in a mangrove swamp and flows into Little Lagoon. It occasionally dries up.

M48-7. Mouth of Billabong, at Little Lagoon, Umbakumba, Groote Eylandt, Northern Territory, 8 to 14 April 1948. R. R. Miller, F. M. Setzler, K. Hodges, and party. Collection made with 25 foot bag seine to a depth of 3 feet, over a bottom of sand, silt, debris, and some mud; water brackish, fresh at low tide and salty on surface at high tide.

M48-8. Little Lagoon, at Umbakumba, Groote Eylandt, Northern Territory, 8 to 14 April 1948. R. R. Miller and party. Collection made by 25 foot bag seine to a depth of 3·5 feet over a sand bottom, with shoreline silt, in clear, salty water.

M48-9. Ironstone (laterite) conglomerate reefs and tide pools near point at south side of entrance to Little Lagoon, Groote Eylandt, Northern Territory,

19 April 1948. R. R. Miller, K. Hodges, R. Specht, Peter Basset-Smith and party. Collection made by 9 pounds of barbasco root and dip net, to a depth of 3 feet, over a bottom of rocks, sand, coral, shell fragments and seaweed, in clear salty water with a temperature of 83°F. to 87°F.

M48-9a. Reef, off point at west side of entrance to Little Lagoon, Groote Eylandt, Northern Territory, 25 April 1948. R. R. Miller, Frank Setzler, and party. Collection made at low tide, with 20 pounds of barbasco root and dip net, over a bottom of sand, rocks, coral, beds of seaweed and 'red laterite', which studded with holes. Water clear, salty.

M48-10. Estuary at head of Thompson Bay, south end of Port Langdon, Groote Eylandt, Northern Territory, 1 May 1948. R. R. Miller, Jerry Blitner, and party. Collection made with 25 foot bag seine, to a depth of 4 feet, over a bottom of fine sand, gravel, and shell fragments, in clear water which fresh inland and salty at mouth. Estuary mangrove lined, receiving drainage from a freshwater creek or a billabong.

M48-11. Little Lagoon, north-west end of Groote Eylandt, Northern Territory, various dates, April to July 1948. Collections made by hook and line.

M48-12. Pool along west side of Lake Hubert, about 15 miles south of Umbakumba, Groote Eylandt, Northern Territory, 24 May 1948. R. R. Miller. Collection made by hand. Water clear, easily roiled, salty; no vegetation; bottom of mud, sand, fine silt, and debris.

M48-13 to M48-13f. Various collections were made, by hook and line or spear unless otherwise indicated, in the vicinity of Groote Eylandt, Northern Territory, by R. R. Miller, Jerry Blitner, F. M. Setzler, and others; they were given the number M48-13; all are marine; I have numbered them as follows:

M48-13. South bay of Bickerton Island, 2 June 1948; collection by night light off boat.

M48-13a. South end of Winchelsea Island, off north-west coast of Groote Eylandt, 2 and 6 June 1948; over sandy bottom with loose vegetation and rocks; collection by night light and dip net.

M48-13b. Hemple Bay, west side of Port Langdon, north coast of Groote Eylandt, 23 and 27 May 1948; collected over a sand bottom or from offshore reefs.

M48-13c. Off Bartalumba Bay reef, 7 June 1948; collected over rocks.

M48-13d. Sheltered bay of unnamed, red sandstone island, about 5 miles east of northern end of Chasm Island, 8 June 1948.

M48-13e. Off Chasm Island, 3 and 4 July 1948.

M48-13f. Passage between Winchelsea Island and Groote Eylandt, 9 June 1948; collected by trolling.

M48-14. Emerald River, at old mission site, from 100 yards above road bridge to 200 yards below falls, Groote Eylandt, Northern Territory, 4 June 1948. R. R. Miller, Jerry Blitner, Fred Gray, and natives. Collection made with 35 pounds of derris root over a bottom of weeds, logs, and sandstone rocks to a depth of 6 or 7 feet, in moderate to strong current. Water clear, fresh, rather easily roiled, with a temperature of 79°F. at 11.30 a.m. River ranging from 12 to 45 feet wide, averaging 30 feet; influence of high tide believed to reach to base of falls, which about 3·5 miles above mouth of stream. Approximately three falls at this locality with a total drop of 7 or 8 feet; water above them always fresh; that below brackish; fresh when collection was made.

M48-15. Emerald River, at and near mouth, west side of Groote Eylandt, Northern Territory, 3 to 5 June 1948, R. R. Miller, Jerry Blitner, and natives; water salty, murky green, phosphorescent; bottom sand and mud; collection made by derris root, spearing, and dip net; mouth not well defined, mangrove fringed, approximately 75 yards wide.

M48-16. Billabongs adjacent to Emerald River, at old mission site, Groote Eylandt, Northern Territory, 5 June 1948. R. R. Miller, Jerry Blitner, and natives. Water brownish, fresh, easily roiled, about 6 feet deep; bottom of vegetation, mud, and fallen logs; billabongs up to 50 feet wide, forming a chain of small ponds along north side of river; collection made with 2·5 pounds of derris root.

M48-17. Reef, in Bartalumba Bay, near eastern entrance to Bartalumba Basin, north-west end of Groote Eylandt, Northern Territory, 7 June 1948, R. R. Miller and Jerry Blitner. Collection made with 16 pounds of derris root in water to 4 feet deep, which greenish, fairly clear and salty, over a bottom of coral, sand, rocks, and some vegetation.

M48-18. Rocky pools, tributary in wet season, to north side of South Bay of Bickerton Island, Northern Territory, 8 June 1948, R. R. Miller, Ray Specht, and Jerry Blitner. Collection made with 5 pounds of derris root, in clear, fresh water, to a depth of 3 feet, amongst vegetation, over a sandstone, sand and silt bottom; water temperature 82°F.

M48-19. Creek, tributary to Little Lagoon, about 2 miles east of Umbakumba, Groote Eylandt, Northern Territory, 26 and 27 June 1948. R. R. Miller, Jerry Blitner, Peter Basset-Smith, and natives. Collection made with 40 pounds of barbasco root in brownish, murky water, which brackish and becoming salty at stream mouth; bottom of silt, mud, and sand; creek to 6 feet deep, with slight to moderate current, 25 to 50 feet wide, with dense mangroves along bank.

M48-20. Main jetty at Melville Bay (site of old flying boat base), and old jetty, about 2 miles east-north-east of main jetty, near Cape Arnhem, Northern Territory, various dates, August to September 1948. R. R. Miller, F. M. Setzler, and natives. Collection made by hook and line over a bottom of sand, rocks, and boulders; water murky, salty.

M48-21. Coral reefs at Yirrkalla, Northern Territory, various dates, July and August 1948. R. R. Miller and party. Collections made with derris root in clear, salty water to 4 feet deep over reef containing coral, some sand and ironstone.

M48-22. Bay of Yirrkalla, Northern Territory, various dates, July and August 1948. R. R. Miller and natives. Collection by hook and line and spear in 15 to 20 feet of slightly murky, salty water, over a sand bottom, approximately one-half to three-fourths of a mile off shore.

M48-23. Port Bradshaw, Northern Territory, 22 to 26 July 1948. R. R. Miller and party. Collection made by hook and line and spear, in murky salt water, to a depth of 12 feet over a bottom of sandy silt, with boulders; water temperature 80°F.

M48-24. Beach, at east side of Port Bradshaw, about 2 miles inward from mouth of Port, Northern Territory, 22 to 24 July 1948. R. R. Miller and party. Collections made by 50 foot seines, in clear salty water, over a sand bottom, to a depth of 4 feet.

M48-25. Beach on north side of island in Port Bradshaw; island about 3 miles in from mouth of Port, Northern Territory, 23 July 1948. R. R. Miller and native. Collection made by 25 foot bag seine, to a depth of 3 feet, in clear, salty water, over a bottom of sand, shells, and a few rocks.

M48-26. Sheltered bay on north-west side of island in Port Bradshaw, Northern Territory, 23 July 1948. R. R. Miller and native. Collection made with 8 pounds of derris root, in clear, salty water, to a depth of 3·5 feet, over a flat rocky reef, with offshore sand; water temperature 79°F.

M48-27. Bay and sandy to rocky beach, at first sheltered point along west side, in from entrance to Port Bradshaw, Northern Territory, 24 July 1948. R. R. Miller and natives. Water murky offshore, and clear at shore; salty; no vegetation; tide high; collection made by hook and line and 10 pounds of derris root to a depth of 7 feet; water temperature 79°F.

M48-28. Creek mouth and bay in Port Bradshaw, on west side about 5 miles in from mouth, Northern Territory, 25 July 1948. R. R. Miller, Nick Haritos, Miller and natives. Water murky offshore and clear at shore; salty; no vegetation; bottom of sand in bay and mud to muck at creek mouth; creek mouth densely lined with mangroves; collection made in creek mouth and to 200 yards into bay to a depth of 3 feet with 10 pounds of derris root and spear; water temperature 72° to 79°F.

M48-29. Small pool, cut off in dry weather from small creek at Yirrkalla, Northern Territory, 30 July 1948. R. R. Miller and natives. Water brownish, fresh, very easily roiled; no vegetation; bottom and shore of mud; pool drying up, about 30 feet long x 15 feet wide and a few inches deep; collection made by one-fourth pound of derris root; water temperature 100°F. when air 86°F.

M48-30. Reefs, about 1·5 to 2·5 miles west of Yirrkalla, Northern Territory, August 1948, collected by natives. Water clear, salty, without vegetation; bottom of rocks and sand; collection by three-prong spear and hook and line.

M48-31. Creek about 2 miles west of Yirrkalla, Northern Territory, 8 August 1948. R. R. Miller and natives. Water clear but easily roiled, brackish, cut off by high sand ridge from Gulf of Carpentaria; no vegetation; bottom of coarse sand, rocks and mud; creek 15 to 40 feet wide, averaging 25 and about 3 feet deep; collected with emulsifiable rotenone at a water temperature of 82°F.

M48-32. Rock pools at point about three-fourths of a mile east of Yirrkalla, Northern Territory, 9 August 1948. R. R. Miller and Raymond Specht. Water clear, salty; no vegetation; bottom of granitic ledges, rocks, and sand; pools' dimensions approximately 5 feet x 25 feet and 2 feet deep at low tide, but submerged at high tide by one or more feet of water; collected three pools with one-fourth gallon of 5 per cent emulsifiable rotenone; water temperature 78°F.

M48-33. At mission on south-east side of Milingimbi Island, in the Crocodile Islands, about 275 miles east of Darwin, Northern Territory, 9 to 11 August 1948, collected by natives. Water clear, probably salty; no vegetation except mangroves; bottom of mud and sandy silt; collected by hand in shallow water at the beach and about the mangroves in front of mission.

M48-34. Creek at Yirrkalla Mission, Northern Territory, 21 August 1948. R. R. Miller and natives. Water clear, but easily roiled; brackish at mouth, but fresh upstream, polluted by refuse from native camp; small beds of vegetation on bottom, which of mud, sand, silt, and debris; creek 6 to 12 feet wide, averaging 8 feet, with moderate current and to 4 feet deep; collected with 12 pounds of derris root at a water temperature of 84°F.

M48-35. Large billabong at Oenpelli, Northern Territory, 24 September to 18 October 1948. R. R. Miller, party and natives. This billabong receiving water from Oenpelli Creek and run-off from the adjacent grasslands, marshes and meadows, was described as approximately 4,200 feet long and 600 feet wide, with its long axis north and south; it reportedly had never gone dry. Miller collected from it at the end of the dry season when the water level was falling rapidly; during the wet season it overflows through a series of billabongs to the north-west into the East Alligator River. In extremely wet seasons the surrounding country is submerged except at Oenpelli Mission, and tidal influence has been noted. The billabong is a major source of fish food for the local natives. The water was described as fresh, colourless and turbid, with a bottom visibility of about 10 inches; an algae bloom was noted and *Myriophyllum* and 'water lilies' were stated to be present. The bottom, which ranged to a depth of 5 feet, consisted of firm mud with local areas of sand. Miller collected in the billabong for over two weeks, with a gill net 125 feet long, with a total of 25 pounds of rotenone on two occasions, and along the shore with a 25 foot seine.

M48-36. Oenpelli Creek, above waterfall on sandstone escarpment 5 miles south-south-east of Oenpelli, Northern Territory, 30 September 1948, R. R. Miller. Oenpelli Creek is a tributary to the billabong at Oenpelli. Miller's collection came from above the area where the creek makes two small drops before a final fall of 200 feet. Water clear to slightly turbid due to dense growth of green algae; bottom of sand, rocks, and boulders; creek not flowing; collected pool about 150 feet long by 40 feet wide and 4 feet deep with one-half gallon emulsifiable rotenone; water temperature 86°F.; sandstone bedrock on one bank; other shore wooded.

M48-37. Fish Creek, about 12 miles east of Oenpelli, Northern Territory, 3 October 1948, R. R. Miller and party. This stream apparently a tributary of Oenpelli Creek and flowing when collected, indicating it to be a much larger stream; it cascades off the same escarpment as does Oenpelli Creek. Water slightly brownish with some green algae; bottom of rocks, boulders, sand, and debris. Two pools about one-half mile apart collected with one gallon of emulsified rotenone; upper unshaded pool 100 feet x 35 feet wide and 2·5 feet deep with water temperature 91°F.; lower heavily shaded pool 100 feet long x 15 feet wide and 4 feet deep with water temperature 82°F.

M48-38. Oenpelli Creek, below waterfall, about 4 miles south-south-east of Oenpelli, Northern Territory, 4 October 1948, R. R. Miller. This station, a large pool, below the escarpment mentioned under M48-36, at base of large (then dry) waterfall; pool approximately 300 feet long and 120 feet wide; depth unknown, but collected with one-fourth gallon of emulsifiable rotenone in part of pool to a depth of 4 feet; water green; suspended algae present limiting visibility to 10 inches; bottom of white sand, boulders and rocks; shore wooded; water temperature 96°F. at surface and 86°F. at depth of 3 feet; a source of native food and water.

M48-39. Red Lily Lagoon, about 7 miles west-south-west of Oenpelli or about 2 miles south of East Alligator River, Northern Territory, 10 October 1948, R. R. Miller, Bill Harney, Len Hillier, and Kelvin Hodges. This water body, about three-fourths of a mile long and 600 feet wide, is a tributary of the East Alligator River in wet season. Water clear, easily roiled; bottom visible at 10 feet; maximum depth unknown; bottom of mud, much vegetation and some sand; collection made to a depth of 6 feet at south end of lagoon with 25 pounds of derris root; water temperature 92°F.

M48-40. East Alligator River, at Cahills Landing, about 6 miles west of Oenpelli, Northern Territory, 13 October 1948, R. R. Miller and W. E. Harney. Water brackish with silt in suspension, malt-coloured; no vegetation; bottom and shore of deep, silty mud; low tide and slight current; river 80 to 100 feet wide; few trees along bank; water temperature 86°F.; collection made along about one mile of river, to a depth of 3·5 feet, with 25 pounds of derris root and one gallon of emulsifiable rotenone; a few fish shot with dust pistol; others collected by hook and line.

Freshwater billabong, about 3 miles above Roper River Mission, on north side of Roper River, Northern Territory, 17 August 1948; collection by Jerry Blitner.

ORECTOLOBIDAE

Key to the Arnhem Land Species

1 Caudal long, more than one-third total length . . . . . *Stegostoma tigrinum*
  Caudal less than one-third total length 2
2 Side of head studded with simple dermal lobes; back smooth; nasal cirrus simple . . . . . *Orectolobus wardi*
  Side of head nearly smooth, without lobes 3
3 Mouth nearer front edge of eye than tip of snout; no large black ocellus above pectoral fin; rusty brownish; young with broad dark bars . . . . . *Chiloscyllium punctatum*
  Mouth nearer tip of snout than eye; a large black spot on body above pectoral fin 4
4 A single very large black spot or ocellus above pectoral fin; dark spots on body large and widely scattered . . . . . *Hemiscyllium ocellatum*
  The large black spot or ocellus above pectoral fin closely followed by two

large black spots which may fuse to form a broad black crescent; spots on body small, brownish, close-set . . . . *Hemiscyllium trispeculare*

Orectolobus wardi Whitley

Orectolobus wardi Whitley, *Rec. Aust. Mus.* 20: 264 (1939); type locality, Cape Keith, Melville Island, Northern Territory; Whitley, *The Fishes of Australia*, pt. 1: 82, fig. 72 (1940).

Stegostoma tigrinum (Forster)

Squalis tigrinus Forster, *Zoologia Indica selecta* . . : 24, pl. 34, fig. 1 (1781); type locality, Indian seas.

Recorded as: *Stegostoma tigrinum* Paradice and Whitley (1927): 96, Clarence Strait, Northern Territory.

Chiloscyllium punctatum Müller and Henle

Chiloscyllium punctatum Müller and Henle, *Systematische Beschreibung der Plagiostomen*: 18, pl. 3 (1841); type locality, Java.

Recorded as: *Chiloscyllium punctatum* Klunzinger (1879): 427, Port Darwin; Ogilby (1889): 181; Ogilby and McCulloch (1909): 287, pl. 43, fig. 2.

Hemiscyllium ocellatum (Bonnaterre)

Squalus ocellatus Bonnaterre, *Tableau encyclopédique et méthodique . . . ichthyologie*: 8 (1788); type locality, 'La mer du sud'.

Recorded as: *Chiloscyllium ocellatum* Klunzinger (1879): 427; Macleay (1881): 363; Kent (1889): 240, Port Darwin.

Hemiscyllium trispeculare Richardson

Hemiscyllium trispeculare Richardson, *Icones Piscium or Plates of Rare Fishes*: 5, pl. 1, fig. 2 (1843); type locality, Turtle Island, north-west coast of Australia; Richardson, *Ichthyology of the Voyage of H.M.S. Erebus and Terror*: 43, pl. 28, figs. 3 to 7 (1845).

Recorded as: *Hemiscyllium trispeculare* Whitley (1932b): 321, pl. 36, figs. 2a to 2e and (1940a): 74, fig. 58, Port Darwin; *Chiloscyllium trispeculare* Ogilby and McCulloch (1909): 293, Port Darwin; Paradice and Whitley (1927): 96, Knight Reef, Clarence Strait, Northern Territory.

M48-3, East Point Reef, USNM 174069, 2 specimens, male 568 mm. and female 556 mm. total length.

M48-21, Reef, Yirrkalla, 11 August 1948, USNM 174068, 2 males, smallest immature, 505 and 635 mm. total length.

SCYLIORHINIDAE

Key to the Arnhem Land Species

1 Anterior-medial margin of nostril expanded as a flap, overlapping front edge of mouth; no definite posterior nasal flap; brownish above; plain whitish below; head, body and fins covered with large brown spots; spots on back outlining faint saddles; those on side forming irregular lines with three lines on head extending backward from eye and cheek . . . . *Atelomycterus macleayi*

Anterior nasal flap not reaching mouth; a weak posterior nasal flap; brownish above, lighter below; head, body and fins with large, widely spaced black spots; those on head in about one row across cheek . . *Halaelurus labiosus*

Halaelurus labiosus (Waite)

Catulus labiosus Waite, *Rec. Aust. Mus.* 6: 57, fig. 23 (1905); type locality, Fremantle, Australia.

Aulohalaelurus labiosus Whitley, *The Fishes of Australia*, pt. 1: 89, figs. 78 and 80 (1940).

Records questionably referable: *Scyllium maculatum* Macleay (1878): 366 and (1881): 361; Klunzinger (1879): 427, Port Darwin; *Scylliorhinus maculatus* Ogilby (1889): 179, Port Darwin.

Remarks. The recorded Northern Territory specimens may prove to be referable to *Atelomycterus macleayi* Whitley.

Atelomycterus macleayi Whitley

Atelomycterus macleayi Whitley, *Aust. Zool.* 9: 230, fig. 3 (1939); Melville Island, Northern Territory; type locality, Port Darwin.

Recorded as: *Scyliorhinus marmoratus* McCulloch (1910): 688, Port Darwin; *Atelomycterus marmoratus* Whitley (1932b): 322, pl. 38, fig. 1, Port Darwin; *Atelomycterus macleayi* Whitley (1940a): 92, figs. 78, 86 and 87.

M48-2, Nightcliff, 19 March 1948, USNM 174071, 1 female, 511 mm. total length.

M48-3, East Point Reef, USNM 174070, 1 male, 471 mm. total length.

TRIAKIDAE

Triaenodon apicalis Whitley

Triaenodon apicalis Whitley, *Aust. Zool.* 9: 236, fig. 9 (1939); Pellew Islands; type locality, Lindeman Island, Queensland.

Recorded as: *Triaenodon apicalis* Whitley (1940a): 112, figs. 88 and 114 and (1954): 27, Evans Shoal, Arafura Sea.

CARCHARHINIDAE

Key to the Arnhem Land Species

1 Small spiracles present; a low, longitudinal ridge on each side of caudal peduncle; outer labial groove elongate, about as long as preoral length of snout . . . . *Galeocerdo cuvieri*

No spiracles and no longitudinal ridge on caudal peduncle; labial grooves short or obsolete 2

2 Base and cusps of teeth in both jaws smooth or nearly so 3

Teeth in upper jaw serrate; those in lower jaw serrated or not 5

3 Second dorsal fin large, its base about three-fourths that of first dorsal; snout short and broad; teeth slender and erect . . . . . . *Negaprion queenslandicus*

Second dorsal fin base length less than half that of first dorsal; snout more acute; teeth in sides of jaws oblique 4

4 Outer labial fold elongate, longer than inner fold, about equal to eye length; eye larger than height last gill opening . . . . *Scoliodon longmani*

Outer labial fold short, slightly shorter than inner fold, and projected into eye length about 2·5 times; eye smaller than last gill opening *Scoliodon taylori*

5 Teeth in lower jaw smooth or nearly so, narrow; snout long and pointed . . . . *Carcharhinus nasutus*

Teeth in lower jaw distinctly serrate; snout short and moderately to broadly rounded 6

6 Snout somewhat pointed; anal fin beginning beneath second dorsal fin; teeth notched at base in both jaws, those of lower jaw more slender than in upper; a black tip to second dorsal fin and lower caudal lobe . . . . *Carcharhinus tilstoni*

Snout broadly rounded; second dorsal fin slightly in advance of and equal to anal fin 7

7 Fins pale; teeth broadly triangular, not indented at base *Carcharhinus mckaili*

Fins, especially first dorsal, tipped with black; teeth acutely triangular, shallowly notched at base . . . . *Carcharhinus melanopterus*

Galeocerdo cuvieri (Peron and LeSueur)

Squalus cuvier Peron and LeSueur, *J. Acad. Nat. Sci. Philad.* 2: 351 (1822); type locality, north-west coast of Australia.

Recorded as: *Galeocerdo arcticus* Paradice and Whitley (1927): 78, Pellew Islands.

Negaprion queenslandicus* (Whitley)

Aprionodon acutidens queenslandicus Whitley, *Aust. Zool.* 9: 233, fig. 6 (1939); type locality, Townsville, Queensland.

Recorded as: *Aprionodon acutidens* Paradice and Whitley (1927): 78, Pellew Islands.

Scoliodon† longmani Ogilby

Scoliodon longmani Ogilby, *Mem. Qld. Mus.* 1: 30 (1912); type locality, Moreton Bay, Queensland; Whitley, *The Fishes of Australia*, Pt. 1: 109, figs. 88 and 110 (1940).

M48-11, Little Lagoon, 13 April 1948, USNM 174076, 1 specimen, 500 mm. total length.

M48-22, Bay of Yirrkalla, 20 July 1948, USNM 174077, 1 specimen, 790 mm. total length.

Remarks. Outer labial fold longer than inner labial fold, about equal to eye diameter; eye large, 0·7 to 0·9 in last gill opening; 11 or 12 pores in the series just lateral to and extending backward from rear corner of mouth.

Whitley (1943b) regards *Scoliodon longmani* Ogilby and *Physodon taylori* Ogilby as conspecific. But, judging from his drawings of the types (Whitley, 1940a) and my material, they are not the same species.

Scoliodon taylori (Ogilby)

Physodon taylori Ogilby, *Mem. Qld. Mus.* 3: 117 (1915); type locality, Townsville, Queensland.

Recorded as: *Protozygaena taylori* Whitley (1940a): 107 and 110, figs. 33, 88 and 112, Pellew Islands; *Hypoprion macloti* Paradice and Whitley (1927): 78 and 97, Pellew Islands; *Scoliodon palasorrah* Fowler (1941): 139, Pellew Islands.

M48-11, Little Lagoon, 10 April 1948, USNM 174075, 1 specimen, 565 mm. total length.

Remarks. Outer labial fold short, directed outward, only slightly shorter than inner labial fold; outer fold about 2·3 to 2·5 in eye length; eye small, about 1·1 in last gill opening; 7 or 8 pores in the series just lateral to rear corner of mouth.

* *Negaprion* Whitley (1940a): 111; type species, *Aprionodon acutidens queenslandicus* Whitley, by original designation.

† *Protozygaena* Whitley (1940a): 110; type species, *Physodon taylori* Ogilby, by original designation; here regarded as a synonym of *Scoliodon*.

*Carcharhinus** *nasutus* (Whitley)

Uranga nasuta Whitley, *Proc. Linn. Soc. N.S.W.* 48: 115, fig. 1 (1943); type locality, Urangan, Queensland.

M48-30, Reef, Yirrkalla, 21 August 1948, USNM 174074, 1 female, approximately 1340 mm. total length.

Remarks. This specimen resembles the figure and description given by Whitley of *U. nasuta* in its long, pointed snout; in its rather large second dorsal fin—which only slightly behind origin of anal fin; its slender, serrated upper teeth; smooth, slender lower teeth; and in its small eye and large gill openings. The tips of the fins are dusky to blackish.

Carcharhinus mckaili (Whitley)

Galeolamna greyi mckaili Whitley, *Aust. Zool.* 11: 2 (1945); type locality, Swan River, Western Australia; based on the specimen described by Whitley, *Fishes of Australia*, Pt. 1: 273, fig. 303 (1940).

M48-40, East Alligator River, USNM 174073, 1 head.

Remarks. Miller indicated in his notes that a single species of shark was fairly abundant in the East Alligator River, but unfortunately he did not have facilities for preserving more than the single head. His colour transparencies indicate this to be a relatively plain shark, without obvious fin markings. Judging from the exposures taken, the second dorsal is about the same size as and slightly in advance of the anal fin. The teeth and dermal denticles resemble the illustrations given by Whitley (loc. cit.), but the body depth is not as great before the dorsal fin. The snout is broadly rounded, with the anterior margin of the eyes in advance of the mouth. The distance from the snout tip to a line connecting the anterior edges of the nostrils is projected into the distance between the nares about 3 times.

Carcharhinus melanopterus (Quoy and Gaimard)

Carcharias melanopterus Quoy and Gaimard, *Voyage autour du monde... sur l'Uranie et la Physicienne... Zoologie*: 194, pl. 43, figs. 1 and 2 (1824); type localities, Vaigiou and Mariannes.

Recorded as: *Carcharias hemiodon* Macleay (1878): 366, Port Darwin; *Carcharias melanopterus* Macleay (1881): 353, Port Darwin; Ogilby (1888b): 1768; *Mapolamia spallanzani* (not *Squalus spallanzani* LeSueur) Whitley (1940a): 94, figs. 88 and 89, Northern Territory.

M48-30, Reef, Yirrkalla, 25 August 1948, USNM 174072, 1 specimen, 630 mm. total length.

Carcharhinus tilstoni (Whitley)

Galeolamna pleurotaenia tilstoni Whitley, *W.Aust. Nat.* 2: 100, figs. 1 and 2 (1950); Peron Island; Port Essington; Bathurst Island; Lynedoch Bank; Flinders Reef; Wessel Island; Evans Shoal, all Northern Territory; type locality, Van Cloon Reef, Joseph Bonaparte Gulf, Western Australia.

RHINOBATIDAE

Rhinobatos armatus Gray

Rhinobatus armatus Gray, in Hardwicke, *Illustrations Indian Zoology*, 2: pts. 13 and 14, pl. 99 (1832-4); type locality, India; reference not seen.

* See Garrick (1962): 89-96 for discussion of the status of the generic name *Carcharhinus*.

Rhinobatos batillum Whitley, *Aust. Zool.* 9: 245, fig. 13 (1939); type locality, Sharks Bay, Western Australia.

Recorded as: *Rhinobatos armatus* Norman (1926b): 952, fig. 6, Groote Eylandt, Northern Territory; Paradice and Whitley (1927): 79, Pellew Islands; *Rhinobatus granulatus* Macleay (1878): 366, Port Darwin.

M48-2, Nightcliff, USNM 174079, 1 specimen, 360 mm. total length.

M48-10, Estuary, Thompson Bay, Port Langdon, USNM 174080, 1 specimen, slightly more than 490 mm. total length; tip of tail broken.

M48-11, Little Lagoon, point south of entrance, 29 June 1948, USNM 114628, 1 small dried specimen.

M48-20, Old jetty, Melville Bay, 18 August 1948, USNM 174078, 1 specimen, 1170 mm. total length.

Native names: *Ilyang* (Eelyang), small individuals or *Markapramoorah*, large individuals, Groote Eylandt.

Colour in life. From a colour transparency of the largest individual. Body and head slate grey above, grading to reddish orange below and on pectoral and pelvic fins; margin of pectoral fins narrowly edged with black; rostrum greyish, suffused with pink or orange (with a black spot at anterior tip of lower surface showing on the two largest preserved specimens); both dorsal fins yellowish grey; lower parts of eye and orbit light yellow.

Remarks. In the three specimens, the distance between the spiracles is stepped into the preorbital length 3·1 to 3·5 times; the mouth width is contained 2·6 to 2·9 times in the preoral length; the nostril width is stepped 1·2 times in the mouth width; and the nostril separation is less than half the width of a nostril. Thus I unite *Rhinobatos batillum* Whitley with *Rhinobatus armatus* Gray, since my measurements are intermediate between those given by Whitley and by Norman (loc. cit.).

DASYATIDAE

Key to the Arnhem Land Species

1 Tail with alternating light and dark rings, long and slender, without folds behind spine, two times to much more than two times length of disk; disk slightly broader than long, subquadrangular . . . . . . *Himantura uarnak*

Tail short and thick, less than twice length of disk; caudal fin fold present . . 2

2 Fold along lower side of tail, extending to tip; no dorsal fold; disk suboval, longer than broad; two caudal spines; large spots, blue in life, scattered over disk; a dark, blue in life, stripe along each side of tail . . *Taeniura lymma*

Fold along lower side of tail about one-third length of tail, median, not extending to tail tip; dorsal fold short, above lower fold and behind the single spine; disk broader than long, nearly quadrangular; a few spots (blue in life?) scattered over disk; no dark stripe along tail, which ringed near tip with black alternating with white . . . . *Dasyatis kuhli*

Himantura uarnak (Forskål)

Raja uarnak Forskål, *Descriptiones Animalium*: viii and 18 (1775); type locality, Arabia.

Himantura toshi Whitley, *Aust. Zool.* 9: 258 (1939); type locality, Clarence River estuary, New South Wales.

Recorded as: *Trygon uarnak* Macleay (1878): 366 and (1881): 377, Port Darwin; Ogilby (1888a): 19, Port Essington; Kent (1889): 240; *Himantura arnak* Paradice and Whitley (1927): 78, Pellew Islands; *Himantura toshi* Whitley (1940a): 211, figs. 240-1, Arafura Sea, Pellew Islands and Bedwell Point, Northern Territory; and questionably as: *Dasyatis uarnak* Fowler (1941): 408, Port Essington, but locality may be in error.

M48-23, Port Bradshaw, USNM 174082, 1 male, 1233 mm. total length, the disk 357 mm. long.

M48-24, Beach, Port Bradshaw, USNM 174081, 1 female, 728 mm. total length, the disk 188 mm. long.

Remarks. The specimen recorded by Fowler from Port Essington, USNM 39985, formerly Australian Mus. No. 4792, strongly resembles the figure given by Whitley (1940a) of *Himantura granulata* in shape of disk and disposition of large granules. The tail, perhaps due to long preservation, shows no rings and is devoid of spines and membranes. Of a total length of 955 mm., the disk is 245 mm. long and the tail 710 mm. There is thus a question of identification of the specimen and a possible error in its recorded locality, since I have not found it listed in the Australian literature.

The smaller specimen obtained by Miller has a relatively smooth skin with no sign of granulations, but the median disk ridge is apparent. In the larger specimen, this area is covered by a series of enlarged granules or tubercles. An area of armour or flattened tubercles, the pavement, extends from before the eyes backward into a rectangular area, bounded by the eyes, well out on to the tail, the tubercles becoming spine-like behind the 'sting'. Since the disk shape is like that figured by Whitley (1940a) for *toshi* (equals *uarnak*) rather than that of the Port Essington specimen mentioned above or his figure of *granulata* (Whitley, 1940a), I identify it as *Himantura uarnak.*

Dasyatis kuhli (Müller and Henle)

Trygon kuhlii Müller and Henle, *Systematische Beschreibung der Plagiostomen*: 164, pl. 51 (1841); type localities, Vanicoro, India and New Guinea.

Recorded as: *Trygon kuhlii* Ogilby (1888a): 19, Port Essington; *Neotrygon kuhlii* Whitley (1940a): 208, fig. 235, Arafura Sea; and questionably as: *Trygon pastinaca* Macleay (1878): 366 and (1881): 377, Port Darwin.

M48-23, Port Bradshaw, 16 May 1948, USNM 174084, 1 immature male, 297 mm. total length.

Taeniura lymma (Forskål)

Raja lymma Forskål, *Descriptiones Animalium*: viii and 17 (1775); type locality, Lohaja, Red Sea.

Recorded as: *Taeniura lymma* McCulloch (1920): 41, pl. 10, Port Darwin; Whitley (1926): 228; Paradice and Whitley (1927): 78, Pellew Islands; *Taeniura lymnia halgani* Whitley (1931b): 97, pl. 11, Pellew Islands and Port Darwin.

M48-21, Reef, Yirrkalla, 11 August 1948, USNM 174083, 1 female, 610 mm. total length.

Colour in life. From Miller's notes and colour transparencies: disk orange brown, with rather large, chalky blue blotches scattered over it from rostrum to end of pelvic fins; tail greyish brown above, whitish below, with a dorso-lateral chalky blue line on each side from disk to first spine; spines bluish; upper surface of tail, behind spine, and edge of caudal fold blue.

MYLIOBATIDAE

Aetobatus narinari (Euphrasen)

Raja narinari Euphrasen, *K. Svenska VetenskAkad. Nya Handl.* 11: 217, pl. 10 (1790); type locality, St Bartholomieu, West Indies; reference not seen.

Recorded as: *Aetobatus punctatus* Whitley (1940a): 224, figs. 251 and 257, Port Darwin.

MOBULIDAE

Manta alfredi (Krefft)

Deratoptera alfredi Krefft, *Illustrated Sydney News* 5: 3 and 9 (1868); type locality, Sydney Harbour; reference not seen.

Daemomanta alfredi Whitley, *Rec. Aust. Mus.* 18: 328, pl. 37 (1932); Whitley, *The Fishes of Australia*, pt. 1: 226, figs. 260-7 (1940); 'appears in Arnheim Land cave paintings'.

ELOPIDAE

In lieu of further investigation, I follow Weber and de Beaufort (1913) in placing *Megalops* Lacépède in the Elopidae.

The genus *Tarpon* Jordan and Evermann (1896) is probably congeneric with *Megalops*. The position of the dorsal fin relative to the pelvic fin is not a strong character. It should be pointed out, however, that the pelvic rays are constantly i, 8 on both sides and the scales smaller in eight specimens of *Tarpon* from the Florida Keys and Haiti, whereas they are either i,9 or 1,10 and the scales larger in the Australian material of *Megalops*.

Megalops cyprinoides (Broussonet)

Clupea cyprinoides Broussonet, *Ichthyologia*: pl. 9 (1782); type locality, Tanna Islands, New Hebrides.

Megalops setipinnis Richardson, *Ann. Mag. Nat. Hist.* 11: 493 (1843); type locality, Port Essington, and Coburg Peninsula, Australia.

Brisbania staigeri Castelnau, *Proc. Linn. Soc. N.S.W.* 2: 241, pl. 3 (1878); type locality, Brisbane River, Queensland.

Recorded as: *Megalops cyprinoides* Macleay (1878): 364 and (1879a): 383; Klunzinger (1879): 417, Port Darwin.

M48-1, Nightcliff, 19 March 1948, USNM 173575, 1 specimen, 107 mm. standard length.

M48-6, Billabong on Fred Gray's Camp, Little Lagoon, USNM 173576, 14 specimens, 66 to 155 mm. standard length.

M48-14, Emerald River, Groote Eylandt, USNM 173577, 2 specimens, 151 to 245 mm. standard length.

M48-16, Billabongs adjacent to Emerald River, Groote Eylandt, USNM 173578, 11 specimens, 77 to 117 mm. standard length.

M48-18, Rocky pools, Bickerton Island, USNM 173579, 5 specimens, 122 to 161 mm. standard length.

M48-31, Creek, Yirrkalla, USNM 173580, 20 specimens, 99 to 174 mm. standard length.

M48-35, Billabong, Oenpelli, USNM 173581-3, 12 specimens, 156 to 236 mm. standard length.

M48-37, Fish Creek, Oenpelli, USNM 173584, 1 specimen, 169 mm. standard length.

Native names: *Álkúra*, Groote Eylandt; *Garlárlpa*, Oenpelli.

Description. Based on the specimens from Bickerton Island: anal rays v,20(in 1), v,21(3) or v,22(1); dorsal rays v,13(3) or v,14(2); branched caudal rays 9+8=17(5); pelvic rays 1,9(3) or i,10(7); pectoral rays i,13(3), i,14(5) or i.15(2); scales in lateral line 37(1), 38(3) or 39(1); scale rows between dorsal and anal fins 6+1+7 or 8=14(4) or 15(1); scales midline before dorsal fin 19 to 23.

Snout 4·1 to 4·3, posterior end of maxillary to end of snout 1·7 to 1·8, both in head length; predorsal length 1·9 to 2·0, prepelvic length 1·9 to 2·0, head length 3·5 to 3·8, maximum depth of body 3·0 to 3·4, all in standard length.

Last dorsal ray elongated, reaching halfway from postdorsal base to caudal fin; axillary processes present above bases of pelvic and pectoral fins.

Colour in alcohol. Brownish silvery, darker above; lower abdomen without pigment; fins greyish brown.

Colour in life. Side with golden yellow reflections from scales, over olivaceous background; background colour becoming paler downward to dull white below.

Remarks. Megalops is believed to spawn in the billabong at Fred Gray's Camp. It was collected in the Emerald River at a point where the water is always fresh.

*Elops** *hawaiensis* Regan

Elops hawaiensis Regan, *Ann. Mag. Nat. Hist.* ser. 8, 3: 37 and 39 (1909); type locality, Hawaii.

Elops australis Regan, *Ann. Mag. Nat. Hist.* ser. 8, 3: 37 and 39 (1909); type locality, Port Jackson, New South Wales.

M48-2, Nightcliff, 19 March 1948, USNM 173682, 1 specimen, 306 mm. standard length.

M48-19, Creek, Little Lagoon, USNM 173683, 1 specimen, 258 mm. standard length.

Description. Dorsal rays v or vi,19(in 2); anal rays iii or iv,12(2); pelvic rays i,12(1) or i,13(3); pectoral rays i,15(1), i,16(1) or i,17(2); branched caudal rays 9+8=17(2); scales in lateral line about 91(1) or 93(1); scale rows between dorsal and pelvic fins 15+1+12=28(1); gillrakers on first arch 7 above + 2 (possibly one belongs with the count for the upper arch) + 14 below = 23, including a papilla-like rudiment at the anterior end on both the upper and lower limbs of the arch; vertebrae 67(2).

Lower jaw 55 per cent to 60 per cent of head length; lower jaw 1·7 to 1·8, interorbital width 4·6 to 5·0, both in head length; head length 3·8 to 4·1 in standard length.

Colour in alcohol. Generally golden brown; upper surface medium brown; median fins dusky.

Remarks. Regan (loc. cit., 37) described three new species (*Elops senegalensis, E. hawaiensis,* and *E. australis*) with 68 or 69 vertebrae and large scales. I find nothing in the descriptions given to indicate that they are not conspecific.

ALBULIDAE

Albula vulpes (Linnaeus)

Esox vulpes Linnaeus, *Systema Naturae* ed. 10, 1: 313 (1758); type locality, Bahama Islands.

Recorded as: *Albula vulpes neoguinaica* Whitley (1958): 28, Darwin.

* *Gularus* Whitley (1940b): 397; type species, *Elops australis* Regan, by original designation; a synonym of *Elops* Linnaeus.

CLUPEIDAE

I have followed Regan (1917a and b) in the generic allocation of the Arnhem Land species. The small size of my material does not permit an evaluation of the genera *Harengula* and *Sardinella*.

Key to the Arnhem Land Species

1 Anal rays about 30 or more . . . . *Pellona ditchela*
   Anal rays fewer than 27 2
2 Last dorsal ray not elongated; dentary not reflected outward into a sharp edge 3
   Last dorsal ray elongated into a filament; dentary reflected outward, sharp-edged; maxillary expanded posteriorly and curved downward in adults . . 5
3 Total ventral scutes 30 to 32; anal rays iii,19 to 21; gillrakers on lower limb of first arch about 50 or more; posterior portion of scale perforated, its margin ragged . . . . *Sardinella jussieu*
   Total ventral scutes 27 to 30; anal rays iii,16 to 19; gillrakers on lower limb of first arch fewer than 45 4
4 Gillrakers on lower limb of first arch 30 to 32; a row of 8 to 13 blackish spots along upper side between head and caudal peduncle . . . . *Harengula konigsbergeri*
   Gillrakers on lower limb of first arch 39 to 42; about five longitudinal brownish lines along upper side . . . . *Harengula macrolepis*
5 Branched dorsal rays usually 13 or 14, seldom 11 or 12; ventral scutes 29 to 31; fins light yellowish, the anal and pelvic fins, inside of mouth, and most of ramus of lower jaw usually devoid of dark pigment; post cephalic spot obovate; snout conical, the lower jaw included; extension of maxillary behind upper edge of lower jaw very short, about 2 times width of lower lip in adults; pectoral axillary process long, nearly equal to eye diameter . . *Nematalosa come*
   Branched dorsal rays 11 or 12, seldom 10 or 13; ventral scutes 28 to 30; fins dull brownish; all fins, oral valves and rami of lower jaw rather densely sprinkled with dark pigment; post cephalic spot, if present, usually more bar-like; snout obtusely pointed, the lower jaw terminal or subterminal; extension of maxillary behind upper edge of lower jaw longer, about 3 times width of lower lip in adults; pectoral axillary process very short or obscure, its length less than pupil diameter . . . . *Nematalosa erebi*

Sardinella jussieu (Lacépède)

Clupanodon jussieu Lacépède, *Histoire naturelle des poissons* 5: 469 and 474, pl. 11, fig. 2 (1803); type locality, Isle de France.

Fimbriclupea dactylolepis Whitley, *Aust. Zool.* 9: 399, fig. 5 (1940); type locality, Point Sampson, Australia.

Records probably referable: *Clupea tembang* Macleay (1878): 364 and (1879a): 377; Kent (1889): 240, Port Darwin; *Harengula bulan* Paradice and Whitley (1927): 79 and 97 and Munro (1956): 23, Darwin; *Clupalosa bulan* Whitley (1941b): 1, fig. 1, Darwin.

M48-2, Nightcliff, 25 March 1948, USNM 173712, 1 specimen, 56 mm. standard length.

M48-15, Emerald River, Groote Eylandt, USNM 173713, 1 specimen, 41 mm. standard length.

M48-26, Sheltered bay, island, Port Bradshaw, USNM 173714, anterior portion of one fish.

Description. Dorsal rays iv,14(in 1) or iv,15(2); anal rays iii,19(1) or iii,20(1); pelvic rays i,7(6); pectoral rays i,13(2) or i,14(4); branched caudal rays 9+8=17(2); ventral scutes before pelvic base 18 plus 13 behind equal 31(3); scales between head and caudal fin about 40; scales between ventral and predorsal midlines 12(2); gillrakers on lower limb of first arch 44 (in smallest specimen), 54(1) or 56(1) plus about 27 on upper limb of first arch.

In the above specimens, head length 3·7 to 3·9, body depth 3·4 to 3·8, distance from dorsal fin to tip of snout 2·3 to 2·4, distance from pelvic fin to tip of snout 1·8 to 2·0, all in standard length; eye 3·0 to 3·1 in head length.

Scales with rear margin ragged or pectinate, the posterior field with tiny round holes; radii vertical, usually only the posterior one continuous across scale.

Fine teeth present on pterygoid, palatine, hyoid and lower jaw.

Colour in alcohol. Nearly uniformly brown above and golden brown below; head silvery; tip of snout and of lower jaw dark brown; dorsal and pectoral fins yellowish; anterior end of dorsal fin with a dark brown spot, nearly as large as pupil, near base.

Harengula konigsbergeri (Weber and de Beaufort)

Clupea (*Harengula*) *konigsbergeri* Weber and de Beaufort, *Verh. Akad. Wet., Amst.* 17: 14 (1912); type locality, Aru Islands.

Recorded as: *Harengula kanagurta* Paradice and Whitley (1927): 79 and 97, pl. 12, fig. 1, Pellew Islands.

M48-2, Nightcliff, USNM 173710, 3 specimens, 56 to 62 mm. standard length.

M48-7, Mouth of billabong, Little Lagoon, USNM 173711, 3 specimens, 50 to 71 mm. standard length.

Description. Dorsal rays iv,15(in 6); anal rays iii,16(4) or iii,17(2); pelvic rays i,7(12); pectoral rays i,13(6), i,14(5) or i,15(1); branched caudal rays 9+8=17(6); scales between head and caudal fin about 39; scales between predorsal and ventral midlines 12(2); ventral scutes 16 or 17 before pelvic fin base plus 11 to 13 behind equal 28(1) or 29(5); gillrakers on first arch 30(2) or 31(1) on lower limb plus 15 or 16 on upper limb.

In the above specimens, head length 3·3 to 3·5, body depth 2·7 to 3·0, distance from dorsal fin to tip of snout 2·2 to 2·5, distance from pelvic fin to tip of snout 1·7 to 1·9, all in standard length; eye 2·6 to 2·9 in head length.

Teeth present on jaws, hyoid, and palatine. Scales with posterior margins rounded or shallowly crenulated, without perforations, the vertical radii usually continuous across scale.

Colour in alcohol. Back medium brown; side and lower surface yellowish or very light brown; pelvic, pectoral and anal fins yellowish; caudal and dorsal fins greyish; tip of anterior dorsal rays, end of snout and end of lower jaw brownish black; a row of 8 to 13 dark brown spots, about two-thirds pupil diameter, from upper corner of gill opening to upper caudal base.

Harengula macrolepis (Steindachner)

Clupea macrolepis Steindachner, *Denkschr. Akad. Wiss. Wien* 41: 13 (1879); type locality, Cleveland Bay, Townsville, Australia.

M48-15, Emerald River, Groote Eylandt, USNM 173715, 4 specimens, 50 to 56 mm. standard length.

Description. Dorsal rays iv,14(in 4); anal rays iii,16(2) or iii,17(2); pelvic rays i,7(8); pectoral rays i,13(5) or i,14(3); branched caudal rays 9+8=17(4); scales between head and caudal fin 38 to 41; scales between ventral and predorsal midlines 12; ventral scutes before pelvic base 17(4), plus 10 or 11 behind pelvic base equal 27(1) or 28(3); gillrakers on lower limb of first arch 39(1), 41(1) or 42(2), plus 16 to 20 on upper limb.

In the above specimens, body depth 3·1 to 3·3, head length 3·4 to 3·7, distance from dorsal fin to snout tip 2·2 to 2·6, distance from pelvic fin to snout tip 1·7 to 1·8, all in standard length; eye 2·8 to 3·0 in head length.

Strong teeth on jaws and hyoid; teeth weak or absent on palate. Scales with posterior margins rounded or shallowly crenulated, without perforations, the vertical radii usually continuous across scale.

Colour in alcohol. Body medium brown, darker above, the midside with a silvery cast; upper side with about five narrow brown lines along scale borders, from head to caudal peduncle; the interspaces light brown below, darker above; upper head, snout, end and line along lower surface of lower jaw dark brown; lower head silvery; pelvic, pectoral and anal fins yellowish; caudal and dorsal fins greyish, the end of the anterior dorsal rays black.

Pellona ditchela Cuvier and Valenciennes

Pellona ditchela Cuvier and Valenciennes, *Histoire naturelle des poissons* 20: 314 (1847); type locality, Vizagapatam.

Meletta schlegelii Castelnau, *Proc. Zool. Acclim. Soc. Vict.* 2: 93 (1873); type locality, Port Darwin.

Harengula schlegeli Bertin, *Catalogue des types de poissons du Museum National d'Histoire Naturelle* pt. 2: 291 (1940); Port Darwin.

Recorded as: *Neosteus ditchela* Paradice and Whitley (1927): 79 and Whitley (1928a): 214, Pellew Islands; *Clupea schlegelii* Macleay (1879a): 380, Port Darwin; *Neosteus schlegelii* Whitley (1940b): 405, fig. 13, Pellew Islands.

In the genus *Nematalosa** the maxilla fits into a groove over the posterior end of the lower jaw, anterior to which the dentary is reflected outward into a somewhat flattened, sharp-edged surface; maxillary, in adults, projecting little behind rear end of lower jaw and curved downward behind; mouth toothless; symphysis of lower jaw with a small tubercle, fitting into a groove or depression in upper jaw; gillrakers small, numerous, about half as long as filaments, those of epibranchial of first arch directed upward and outward; axillary processes present above pectoral and pelvic fins; belly acute with sharp scutes; posterior dorsal ray greatly elongated.

Nematalosa erebi (Günther)

Chatoëssus erebi Günther, *Catalogue of the Fishes in the British Museum* 7: 407 (1868); type locality, eastern Australia.

Fluvialosa paracome Whitley, *Aust. Zool.* 11: 267 (1948); type locality, Noonkanbah, Fitzroy River, Western Australia.

Fluvialosa bulleri Whitley, *Aust. Zool.* 11: 267 (1948); type locality, Ord River, Western Australia.

Billabong, about 3 miles above Roper River Mission, Northern Territory,

**Fluvialosa* Whitley (1943a): 170; type species, *Chatoessus elongatus* Macleay, by original designation; a generic synonym.

17 August, 1948, Blitner, USNM 173590, 15 specimens, 70 to 103 mm. standard length.

M48-35, Billabong, Oenpelli, USNM 173587-8, 116 specimens, 14 to 332 mm. standard length.

M48-39, Red Lily Lagoon, Oenpelli, USNM 173589, 4 specimens, 151 to 174 mm. standard length.

Native name: *Mahbiba,* Oenpelli.

Description. Dorsal rays iv or possibly v,11(in 8) or 12(11), rarely 10 or 13 in many other random counts of Arnhem Land material, and including the specimens reported by Fowler (1941); anal rays iii or iv,18(2), 19(7), 20(7), 21(2) or 22(1), somewhat more variable than the dorsal; pectoral rays i,13(1), i,14(20) or i,15(16); branched caudal rays 9+8=17(4); pelvic rays i,7(8); scales between head and caudal fin 40 to 45, and between dorsal and anal fins 17 or 18; scales along midline before dorsal fin 17 to 19; scales from in front of upper pectoral base to belly scutes 2, rarely 3; sharp scutes midline of lower surface 16 or 17 before pelvic fins; total ventral scutes 28(6), 29(8) or 30(5).

Body depth 2·2 to 2·6, head length 3·1 to 3·3, pectoral fin 4·2 to 4·4, all in standard length; eye 3·9 to 4·5, snout 5·0 to 5·3, both in head length.

Snout pointed, obtusely angular; lower jaw terminal or subterminal; posterior end of maxilla downcurved, extending behind lower jaw about three times width of lower lip; free membrane of operculum broad, nearly as long as pupil; axillary processes of paired fins rather short, that of pectoral obtuse, barely evident to seldom longer than pupil; last dorsal ray not reaching upper caudal fin, extending to approximately opposite posterior end of anal base.

Colour in alcohol. Body and head medium brown, lighter below; scales outlined in darker brown, and sometimes with indistinct longitudinal brownish stripes along upper scale rows; bar-like to slightly rounded blackish spot usually present behind head, but sometimes indistinct or absent; end of snout and inside of mouth, including area over lower jaw and lower oral valve, usually rather densely covered with dark pigment; rami of lower jaw and side of head suffused with brownish; all fins dusky, heavily dotted with dark brown.

Remarks. This is evidently a freshwater species. I suspect that most of the other freshwater species of this group described from Australia are conspecific, with the possible exception of *Fluvialosa vlaminghi* Munro (1956) which may be a freshwater representative of *N. come* (Russell). Specimens reported by Fowler (1941) from the Mary and Burdekin rivers (USNM 47866 and 47867) agree with Arnhem Land material and do not resemble the figure of *N. erebi* given by Munro (1956). It is possible that his figure was made from a specimen of the more slender southern and eastern form.

In young up to approximately 40 mm. in standard length, the posterior dorsal ray does not extend behind the tip of the depressed anterior rays; above this size, the last ray elongates. In the young, the lower edge of the maxillary, behind the union with the lower jaw, is serrated. At this size, the maxillary is curved, and the extension behind the lower jaw is greater than the anterior length of the upper jaw. With increase in body size, the posterior extension becomes relatively shorter, the maxillary curves downward behind the lower jaw, and the serrae are lost.

Nematalosa come (Russell)

Chatoësus come Russell in Richardson, *Ichthyology of the Voyage of the H.M.S. Erebus and Terror*: 62, pl. 38, figs. 7 to 9—as *Chatoëssus,* (1846); type locality, western Australia and Indian Ocean.

Recorded as: *Nematalosa come* Paradice and Whitley (1927): 79, Darwin; *Chatoëssus erebi* Kent (1889): 240, Port Darwin.

M48-8, Little Lagoon, USNM 173585, 2 specimens, 63 to 74 mm. standard length.

M48-24, Beach, Port Bradshaw, 22 July 1948, USNM 173586, 1 specimen, 115 mm. standard length.

Description. Dorsal rays iv or v,13(in 2) or 14(1), the branched rays 13 or 14, with one count of 11, in other specimens from the Celebes, Borneo, Mindanao and China; anal rays iii or iv,18(2) or 20(1); pectoral rays i,14(4), or i,15(2); branched caudal rays 9+8=17(3); pelvic rays i,7(6); scales between head and caudal fin 40 to 46, between dorsal and anal fins 17 or 18, along midline before dorsal fin 18 or 19, and from before upper pectoral base to belly scutes 2; sharp scutes on midline of lower surface 18 before pelvic fins; total ventral scutes 30(3).

Body depth 2·3 to 2·6, head length 3·3 to 3·6, pectoral fin 4·1 to 4·5, all in standard length; eye 3·0 to 3·3, snout 4·1 to 4·6, both in head length.

Snout rounded; lower jaw included; posterior end of maxilla downcurved, very short, extending little behind lower jaw; free membrane of operculum narrow; axillary processes of paired fins moderately long, that of pectoral about as long as eye diameter; last dorsal ray reaching upper caudal fin.

Colour in life. Bronze to brownish spot just behind head; tip of dorsal slightly blackened.

Colour in alcohol. Body and head light yellowish brown; six or seven brownish stripes along scale rows on upper trunk; rounded black spot, nearly as large as eye, behind head; end of snout, inside of mouth and lower head almost devoid of pigment; fins yellowish, the anal and pelvics without pigment; upper surface of outer pectoral rays brownish; membrane of dorsal and caudal fins lightly pigmented with brownish, the margins dark brown to blackish; elongated last dorsal ray pale yellow.

Remarks. Nematalosa come appears to be a marine species. I have followed apparent current practice in using the name *come* but with some reservation. It has not been clearly demonstrated that Richardson's material was obtained from the sea, and his specimens have not been well described.

In addition, I cannot say whether *Clupea nasus* Bloch and *come* should be regarded as conspecific. There are specimens of *Nematalosa* in the National Museum from the Persian Gulf that strongly resemble the Australian material, but are considerably more elongate. I think the latter are probably the true *nasus.*

In this study, I have examined other material, referable to *come* as here interpreted, from Shanghai, Mindanao, Borneo, and the Celebes. Those from Mindanao and Borneo were reported by Fowler (1941) as *Clupanodon thrissa.*

DUSSUMIERIDAE

Spratelloides delicatulus* (Bennett)

Pl. 1

Clupea delicatula Bennett, *Proc. Zool. Soc. Lond.* 1: 168 (1832); type locality, Mauritius.

M48-7, Mouth of billabong, Little Lagoon, USNM 173692, 7 specimens, 23 to 33 mm. standard length.

M48-8, Little Lagoon, USNM 173693, 117 specimens, 25 to 48 mm. standard length.

* *Stolephorus* Lacépède has been applied by many authors to the 'round herrings', but that generic name has been fixed as a genus of anchovies, Family Engraulidae, which see.

M48-9, Reefs and tide pools, entrance, Little Lagoon, USNM 173694, 21 specimens, 17 to 37 mm. standard length.

M48-17, Reef, Bartalumba Bay, USNM 173695, 20 specimens, 23 to 51 mm. standard length.

M48-21, Reef, Yirrkalla, 12 August 1948, USNM 173696, 1 specimen, 42 mm. standard length.

M48-25, Beach on island, Port Bradshaw, USNM 173697, 4 specimens, 41 to 43 mm. standard length.

M48-26, Sheltered bay, island, Port Bradshaw, USNM 173698, 19 specimens, 39 to 47 mm. standard length.

M48-27, Beach, Port Bradshaw, USNM 173699, 34 specimens, 35 to 44 mm. standard length.

Description. Dorsal rays ii,10(in 5); anal rays ii or iii,7 or 8, total = 10(5); branched caudal rays 8+7=15(5); pelvic rays i,7(10); pectoral rays i,10(1) or i,11(9); scales between head and caudal fin 34 to 36; scales between dorsal and anal fins 7; gillrakers 25 to 31 on lower part of arch, plus 1, plus 9 or 10 on upper = 35(1), 40(3) and 42(1).

In five specimens 40 to 44 mm. in standard length, head length 3·6 to 3·8, body depth 4·6 to 5·0, both in standard length; eye 3·0 to 3·4, snout 3·3 to 3·6, pectoral fin 1·5 to 1·6, pelvic fin 1·9 to 2·0, all in head length.

Dorsal fin origin nearer snout tip than caudal fin; pelvic fin inserted under or behind rear half of dorsal fin; exposed field of scales tall, nearly three times exposed length, the free margin smooth; lower jaw subterminal.

Colour in alcohol. Back and top of head dark brown; side and lower surface lighter brown, becoming yellowish below; no well-defined silvery lateral band delineated by black; caudal and dorsal fins dusky yellowish; caudal with a U-shaped dark brown mark passing through alar scales, the base of the U near caudal base; no specimen has two black marks to each alar scale like those present in the widespread Pacific form; other fins pale yellowish; anterior head yellowish, darker above, with pigment along dorsal part to end of snout and at end of lower jaw; very little pigment on side of head or near nares.

Colour in life. Back emerald green; side bright silvery (Miller's notes).

Remarks. Spratelloides delicatulus (Bennett) is applied to the Arnhem Land material with some doubt, since the original description given by Bennett is inadequate, since the only specimens of this complex of *Spratelloides* in the National Museum from the Indian Ocean area—many from the Persian Gulf—are like the present material, and since it is apparent that at least three more or less uniform and geographically distinct populations exist. The complex is characterized by a short anal fin (9 to 11 rays) and no well defined silver lateral band.

One form which has a moderate number of scales along the body, four longitudinal lines of black pigment—two over each alar scale—on the caudal fin (Bleeker, 1866-72) and a rather heavily pigmented head has been widely accepted as *Spratelloides delicatulus.* We have material in the National Museum of it from many parts of the Pacific Ocean, but none from the Indian Ocean. I suspect that it is not the true *delicatulus.*

Another form found off southern Australia, has finer scales, a lighter coloured head, with little or no pigment surrounding the nares posteriorly, and only one dark line through each alar scale, forming a broad U at the caudal base. This form has been described as *Spratelloides robustus* (Ogilby, 1897).

The present material and specimens from the Persian Gulf (USNM 148037 and 148040) are similar to *Spratelloides robustus* in coloration, but have only about

Dorothea B. Schultz

PLATE 1. *Spratelloides delicatulus*
Port Bradshaw, USNM 173699, 44 mm. standard length

Dorothea B. Schultz

PLATE 2. *Thryssa scratchleyi*
Billabong above Roper River Mission, USNM 173686, 154 mm. standard length

34 to 37 lateral scales. I suspect that these may represent the true *delicatulus*, hence the use of that name here. Possibly all the above are conspecific.

ENGRAULIDAE

In the Arnhem Land material of this family, the caudal and anal fins are separate, the caudal fin is moderately forked, and none of the fin rays is especially produced; teeth in the mouth are small, never canine-like.

Key to the Arnhem Land Genera

1 Dorsal fin preceded by a short, free spine-like process; sharp scutes extending along ventral margin from isthmus to anus; branched anal rays 30 to 40 . . . . genus *Thryssa**

Dorsal fin not preceded by a free spine; sharp scutes along ventral margin 4 to 8, confined to area between pectoral and pelvic fins; branched anal rays 15 to 21 . . . . genus *Stolephorus*

Key to the Arnhem Land Species of *Thryssa*

1 Posterior tip of maxilla elongate, slender, extending to or beyond middle of pelvic fin, usually nearly to anal fin, its posterior portion weakly toothed; branched dorsal rays usually 12, sometimes 11; pectoral rays i,12 or i,13 . . *Thryssa setirostris*

Posterior tip of maxilla not reaching pectoral fin base; branched dorsal rays usually 11 2

2 End of maxilla narrow, pointed behind, reaching to or beyond gill membrane; pectoral rays i,11 . . . . *Thryssa hamiltoni*

End of maxilla obliquely truncated behind, extending only to anterior margin of vertical arm of preopercle; pectoral rays i,12, usually i,13 . . . . *Thryssa scratchleyi*

Thryssa scratchleyi (Ramsay and Ogilby)

Pl. 2

Engraulis scratchleyi Ramsay and Ogilby, *Proc. Linn. Soc. N.S.W.* ser. 2, 1: 18 (1886); type locality, Strickland River, New Guinea.

Billabong, about 3 miles above Roper River Mission, Northern Territory, 17 August 1948, Jerry Blitner, USNM 173686, 2 specimens, 154 to 157 mm. standard length.

Description. Dorsal rays iii,11(in 2), the fin preceded by a short, free, spine-like process; anal rays iv,34(1) or iv,37(1); branched caudal rays 9+8=17(2); pelvic rays i,6(4); pectoral rays i,12(1) or i,13(3); sharp scutes between isthmus and anus 29(2); scales between head and caudal fin 40 to 41 and between anal and dorsal fins 11; gillrakers on first arch 19 above + 21 below = 40(1).

Head length 4·5 to 4·6, body depth 3·9 to 4·0, both in standard length; eye 4·4 to 4·5, snout 6·1 to 6·3, both in head length.

Dorsal origin over middle of body, slightly nearer end of snout than caudal fin; anal fin origin behind base of last dorsal ray; pectoral rays reaching base of pelvic fin; maxilla obliquely truncated behind, ending in an angle of about 70° at or near the anterior margin of the vertical arm of the preopercle; its teeth small and of uniform size except those before tip of lower jaw, where a few teeth are

* For the generic synonymy involving *Thryssa* see Fowler (1941): 669.

about one-fourth larger than those of remainder of maxilla and turned inward; snout short and rounded.

Colour in alcohol. Body and head medium brownish above, lighter below; lower side of body almost devoid of pigment except for a concentration of chromatophores above anterior end of anal fin; back with a median light streak which is bounded by a darker streak on each side; caudal, dorsal, and anterior half of pectoral fins dusky; other fins pale.

Remarks. Engraulis nasutus Castelnau (1879), as figured and described by Munro (1956), is a different species with larger scales and a longer, more pointed snout.

Thryssa hamiltoni (Gray)

Thrissa hamiltonii Gray in Hardwicke, *Illustrations of Indian Zoology* 2: pl. 92, fig. 3 (1832-34); type locality not given.

Recorded as: *Thrissocles hamiltoni* Paradice and Whitley (1927): 79, Darwin.

M48-7, Mouth of billabong, at Little Lagoon, USNM 173684, 2 specimens, 68 and 71 mm. standard length.

Description. Dorsal rays iii,11(in 2), the fin preceded by a short, free, spine-like process; anal rays iv,34(1) or iv,35(1); branched caudal rays 9+8=17(2); pelvic rays i,6(2); pectoral rays i,11(4); sharp scutes from isthmus to anus 28(1) or 31(1); scales between head and caudal fin about 41 to 43, and scales between anal and dorsal fins 11 or 12.

Head length 4·0 to 4·2, depth of body 3·6 to 3·8, both in standard length; eye 4·1 to 4·2, snout 4·9 to 5·2, both in head length.

Dorsal origin midway between tip of snout and caudal fin; anal origin beneath last dorsal ray; pectoral fin reaching middle of pelvic fin; maxilla rather slender and pointed behind, extending slightly behind gill margin, its teeth small and of uniform size; snout short and rounded.

Colour in alcohol. Upper surface of body and head light brownish, becoming yellowish below; a dusky brownish blotch, large as pupil, behind upper rear corner of opercle; midline of back with yellowish line, surrounded on each side by a row of brown chromatophores which more or less continue over head and around end of snout as two brown lines; caudal fin dusky; other fins pale yellowish.

Thryssa setirostris (Broussonet)

Clupea setirostris Broussonet, *Ichthyologia, Sistens Piscium Descriptiones et Icones* 1: pl. 2 (1782); type locality, Pacific Ocean near Tanna Island, Society Group; reference not seen.

M48-2, Nightcliff, 19 March 1948, USNM 173685, 3 specimens, 28 to 30 mm. standard length.

Description. Dorsal rays iii,11(in 1) or iii,12(2); anal rays iv,35(1) or iv,37(2); branched caudal rays 9+8=17(3); pelvic rays i,6(6); pectoral rays i,12(2) or i,13(3); sharp scutes between isthmus and anus 28(1) or 29(2); scales between head and caudal fin about 40 to 43 and between anal and dorsal fins about 11 or 12.

Head length 3·7 to 3·9, body depth 4·6 to 4·9, both in standard length; eye 4·3 to 4·6, snout 4·7 to 4·9, both in head length.

Dorsal origin slightly nearer caudal fin than end of snout; anal fin origin under posterior half of dorsal fin; pectoral rays reaching base of pelvic fin; maxilla extremely elongate, reaching pelvic fin—in the present young extending approximately to middle of pelvic rays—or beyond; its posterior end slender, filamentous, with very fine teeth; snout short and rounded.

Colour in alcohol. Body and head light yellowish brown, with a few scattered brown chromatophores chiefly concentrated on occiput, along back, and at dorsal and anal bases; caudal fin dusky; other fins light yellowish.

Key to the Arnhem Land Species of *Stolephorus*

1 Maxillary extending to or beyond near margin of gill membrane; abdominal scutes 6 to 8; anal fin origin under dorsal fin; branched anal rays 18 to 20; branched pectoral rays 10 to 12 . . . . *Stolephorus commersoni*

Maxillary not extending behind preopercle; abdominal scutes 4 or 5; branched anal rays 15 to 17 2

2 Anal fin origin under middle of dorsal fin; branched dorsal rays 12 or 13, usually 13; branched pectoral rays 13 to 15; maxillary extending backward to preopercle . . . . *Stolephorus indicus*

Anal fin origin under, or behind, posterior end of dorsal fin; branched dorsal rays 11 (10 to 12 in literature); branched pectoral rays 11; maxillary extending backward to middle of vertical arm of preopercle . . . . *Stolephorus heterolobus*

I believe the action of the International Commission on Zoological Nomenclature (Opinion 93) wise in selecting *Stolephorus commersonii* Lacépède as the type species of *Stolephorus** Lacépède. Thus the name *Stolephorus* is removed from the round herrings, family Dussumieridae, where it has sometimes been placed since Jordan and Gilbert (1883) selected *Atherina japonica* Gmelin as the type species.

Stolephorus indicus (van Hasselt)

Engraulis indicus van Hasselt, *Algemeene Konst-en Letter-bode* 1, Nr. 21: 329 (1823); type locality, Java.

M48-7, Mouth of billabong, at Little Lagoon, USNM 173689, 4 specimens, 43 to 73 mm. standard length.

M48-25, Beach, on island, Port Bradshaw, USNM 173690, 10 specimens, 48 to 56 mm. standard length.

M48-26, Sheltered bay, island, Port Bradshaw, USNM 173691, 2 specimens, 59 to 63 mm. standard length.

Description. Dorsal rays iii,12(in 1) or iii,13(15); anal rays iii,15(1), iii,16(3), or iii,17(12); branched caudal rays 9+8=17(16); pelvic rays i,6(32); pectoral rays i,13(16), i,14(14), or i,15(2); abdominal scutes confined to area between pelvic and pectoral fins 4(14) or 5(2); scales between head and caudal fin about 36 to 39, and between anal and dorsal fins about 9 or 10.

In four specimens 43 to 73 mm. in standard length, head length 4·0 to 4·3, body depth 5·5 to 5·8, both in standard length; snout 4·2 to 4·5, eye 3·3 to 3·6, both in head length.

Pectoral rays reaching last or next to last abdominal scute; anal fin inserted approximately under middle of dorsal fin; maxillary bluntly rounded behind, reaching vertical arm of preopercle, its teeth of nearly uniform size; snout short, rounded; tip of lower jaw slightly in advance of middle of snout.

Colour in alcohol. Body yellowish; caudal fin dusky; other fins yellowish; bases of dorsal and anal fins surrounded by dark pigment; top of head dark brownish; dark brown band along midline of yellowish snout.

* *Amentum* Whitley (1940b): 403; type species, *Stolephorus commersoni* Lacépède, by original designation; a synonym.

Stolephorus heterolobus (Rüppell)

Engraulis heterolobus Rüppell, *Neue Wirbelthiere zu der Fauna von Abyssinien gehörig, Fische*: 79, pl. 21, fig. 4 (1835); type locality, Red Sea.
Stolephorus pseudoheterolobus Hardenberg, *Natuurk. Tijdschr. Ned.-Ind.* 93: 261 (1933); type localities, Riouw and Lingga Archipelago and Moluccas.
Amentum devisi Whitley, *Aust. Zool.* 9: 404, fig. 11 (1940); type locality, 'Cape York', but presumably Norman River, since mixed with cotypes of *Engraulis carpentariae* De Vis.

M48-26, Sheltered bay, on island, Port Bradshaw, USNM 173687, 1 specimen, 46 mm. standard length.

Description. Dorsal rays iii,11(in 1); anal rays iii,16(1); branched caudal rays 9+8=17(1), pelvic rays i,6(2); pectoral rays i,11(2); abdominal scutes confined to area between pelvic and pectoral fins 5(1); scales between head and caudal fin about 36.

Snout 4·4, eye 4·2, both in head length; head length 3·8, body depth 6·0, both in standard length.

Pectoral rays reaching slightly beyond last abdominal scute; anal fin origin below last dorsal ray; maxillary bluntly rounded behind, reaching vertical arm of preopercle, its posterior end with a few enlarged teeth which point forward and alternate with series of tiny teeth; snout long, acutely rounded; tip of lower jaw behind middle of snout.

Colour in alcohol. Body yellowish, the scales missing; a narrow brownish band on peduncle; dark pigment outlines a few scale pockets on back and faintly the bases of anal rays; caudal dusky; other fins yellowish; occipital region brownish; snout yellowish with midline brownish.

Stolephorus commersoni Lacépède

Stolephorus commersonnii Lacépède, *Histoire naturelle des poissons* 5: 381-2, pl. 12, fig. 1 (1803); no type locality; based on Commerson.
Engraulis carpentariae De Vis, *Proc. Linn. Soc. N.S.W.* 7: 320 (1883); type locality, Norman River, Queensland.
Amentum carpentariae Whitley, *Aust. Zool.* 9: 403, fig. 10 (1940); types redescribed.
Stolephorus waitei Jordan and Seale, *Bull. Mus. Comp. Zool. Harv.* 67: 379 (1926); type locality, Beddone, Queensland.

Recorded as: *Anchoviella carpentariae* Paradice and Whitley (1927): 79, Pellew Islands.

M48-28, Creek mouth and bay, Port Bradshaw, USNM 173688, 6 specimens, 46 to 55 mm. standard length.

Description. Dorsal rays iii,12(in 1), iii,13(3) or iii,14(2); anal rays iii,18(1), iii,19(4) or iii,20(1); branched caudal rays 9+8=17(5); pelvic rays i,6(12); pectoral rays i,10(1), i,11(7) or i,12(2); abdominal scutes confined to area between pelvic and pectoral fins, 6(4), 7(1) or 8(1); scales from head to caudal fin about 35 to 38; scale rows between dorsal and anal fins about 9.

Body depth 5·0 to 5·4, head length 4·0 to 4·3, both in standard length; snout 4·2 to 4·4, eye 3·7 to 4·1, both in head length.

Pectoral rays reaching to last or next to last abdominal scute; anal fin inserted approximately under middle of dorsal fin; maxillary pointed behind, reaching to rear margin of, or slightly behind opercle, its teeth nearly uniform in size; snout acutely rounded; tip of lower jaw below or slightly in advance of approximate middle of snout.

Colour in alcohol. Body yellowish, the scales missing; dark pigment outlining scale pockets along back and at base of anal fin; caudal dusky with broad basal yellowish area; other fins yellowish; occipital region brownish; snout yellowish, with little or no pigment.

CHANIDAE

Chanos chanos (Forskål)

Mugil chanos Forskål, *Descriptiones Animalium*: 74 (1775); type locality, Djedda, Red Sea.

Previous Arnhem Land records: *Leuciscus (Ptycholepis) salmoneus* Richardson (1843a): 489 and *Lutodeira salmonea* Richardson (1844-8): 58, pl. 36, figs. 1 and 2, near Point Smith, Port Essington; *Chanos salmoneus* Whitley (1940b): 406, fig. 14, Bedwell Point.

M48-7, Mouth of billabong, at Little Lagoon, USNM 173572, 5 specimens, 65 to 168 mm. standard length.

M48-19, Creek, tributary to Little Lagoon, USNM 173591, 16 specimens, 111 to 233 mm. standard length.

M48-28, Creek mouth and bay, Port Bradshaw, USNM 173573, 3 specimens, 144 to 194 mm. standard length.

M48-30, Reef, Yirrkalla, 19 August 1948, 1 specimen (photographed; not preserved), 25 inches standard length, or 8 pounds weight.

M48-31, Creek, Yirrkalla, USNM 173574, 4 specimens, 206 to 220 mm. standard length.

Native name: *Ímrára* (Im rah rah), Groote Eylandt.

Description. Based on five specimens 65 to 168 mm. standard length: Dorsal rays v or probably vi,11(in 2) or 12(3); anal rays iii,7(4) or iii,8(1); pelvic rays 9(1), 10(1) or 11(8); pectoral rays i,15(1) or i,16(9); branched caudal rays 9+8=17(5); scales in lateral line 77 to 85; scale rows between dorsal and anal fins 12 to 14+1+10 or 11=23(1), 24(1), 25(2) or 26(1); scales midline before dorsal fin 32 to 36.

Head length 3·3 to 3·8, body depth 3·6 to 4·4, tip of snout to dorsal fin 1·9 to 2·0, tip of snout to pelvic fin 1·7 to 1·9, all in standard length; eye 3·6 to 4·1 and snout 3·9 to 4·2, both in head length.

Mouth toothless, the opening subterminal; upper jaw with median notch; lower jaw sharp-edged with median ridge which fits into notch of upper jaw; gill membranes broadly connected, free from isthmus; opercular membrane touching base of pectoral fin; large alar scale overlapping 6th and 12th branched caudal rays; dorsal fin with basal sheath of scales, the last ray not elongated; anal fin about three-fourths covered by sheath of scales; axillary scales above pectoral and pelvic fins; eye and side of head covered by transparent membrane.

Colour in alcohol. General colour medium brown above, lighter below; underside of head immaculate; dorsal fin yellowish to greyish, its anterior edge black; caudal and anal fins greyish brown; membranes of pelvic and pectoral fins dark brown above; anterior edge and lower surfaces of paired fins pale yellowish.

Colour in life. Bright silvery.

CHIROCENTRIDAE

Chirocentrus dorab (Forskål)

Clupea dorab Forskål, *Descriptiones Animalium*: xiii and 7 (1775); type localities, Djedda and Mochha, Red Sea.

Neosudis vorax Castelnau, *Proc. Zool. Acclim. Soc. Vict.* 2: 118 (1873); type locality, Noumea, New Caledonia.

Recorded as: *Chirocentrus dorab* Kent (1889): 240, Port Darwin; Paradice and Whitley (1927): 79, Pellew Islands; *Chirocentrus vorax* Whitley (1940b): 398, fig. 3, Sir Edward Pellew Group, Gulf of Carpentaria.

Channel between Groote Eylandt Mission and mouth of Roper River, Gulf of Carpentaria, USNM 173592, 15 August 1948, 1 specimen, 185 mm. standard length, and 8 August 1948, 1 specimen, 348 mm. standard length.

Description. Dorsal rays v,12(in 1) or v,13(1); anal rays iv or v,27(1) or 29(1); pectoral rays i,13(3) or i,14(1); pelvic rays i,6(4); branched caudal rays 9+8=17(1); gillrakers on first arch, count includes both rakers and rudiments, 3 or 4+1+11 to 13=15(1), 16(2) or 18(1).

Head length 5·5, tip of snout to pelvic fin 1·9 to 2·0, tip of snout to dorsal fin 1·4 to 1·5, tip of snout to anal fin 1·4, body depth 6·8 to 6·9, all in standard length; snout 3·1 to 3·2, eye 4·5 and pelvic fin 5·2 in smaller specimen, or eye 5·2 and pelvic fin 6·8 in larger specimen, all in head length.

Upper jaw reaching backward almost to posterior tip of preopercle; scales on body small, deciduous, missing in present specimens; large axillary flap above pectoral fin; tiny teeth in small patch on palate and on branchial and hyoid arches; two rather large canine-like teeth on left and one on right on premaxilla, these directed forward and covered by a horn-like flap; teeth on maxilla in a single series, projecting downward and forward from lower border of bone, and becoming progressively smaller posteriorly, but all visible when mouth is closed; teeth in lower jaw a single series of rather large canines, more or less directed upward and backward; lower jaw projecting; eye covered by an adipose membrane; belly carinate, without scutes.

Colour in alcohol. Generally a medium brown, with a yellowish brown streak along middle of body; fins yellowish, the pectoral and caudal margined with dark brown.

SYNODONTIDAE

Saurida undosquamis (Richardson)

Saurus undosquamis Richardson, *Ichthyology of the Voyage of H.M.S. Erebus and Terror*: 138, pl. 51, figs. 1-6 (1848); type locality, north-west coast of Australia.

Recorded as: *Saurida undosquamis* Whitley (1954): 27, Charles Point and Croker Island, Northern Territory.

OSTEOGLOSSIDAE

Scleropages leichardti Günther

Scleropages leichardti Günther, *Ann. Mag. Nat. Hist.* ser. 3, 14: 196, pl. 7 (1864): type locality, Burdekin River, Queensland.

Recorded as: *Scleropages leichardti jardinii* Whitley (1941a): 265, 'depicted in aboriginal drawings in the Northern Territory'; *Scleropages leichardti* Munro (1958): 28.

The presence of this fish in Arnhem Land is questioned. Miller made special attempts to determine if it was present in the fresh waters collected by him, without success. He showed a drawing of the fish to inhabitants of the area. At Oenpelli, some of the natives recognized the drawing as Gooloóecheer and stated

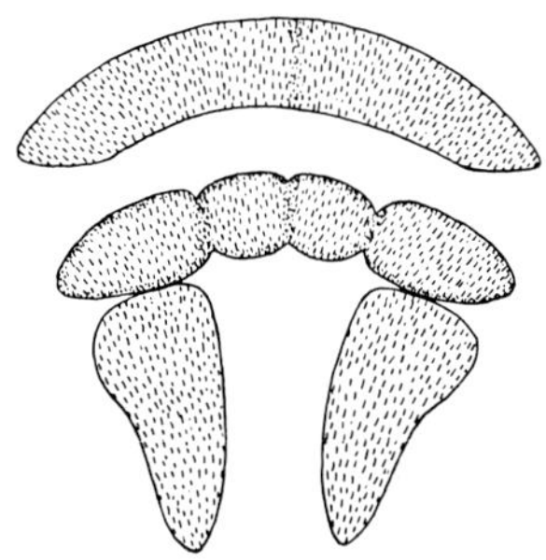

FIG. 1. *Arius thalassinus*
USNM 173567, 353 mm.

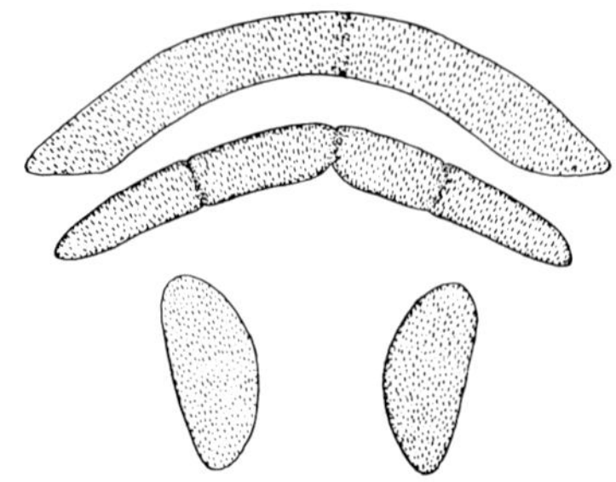

FIG. 2. *Arius mastersi*
USNM 173566, 323 mm.

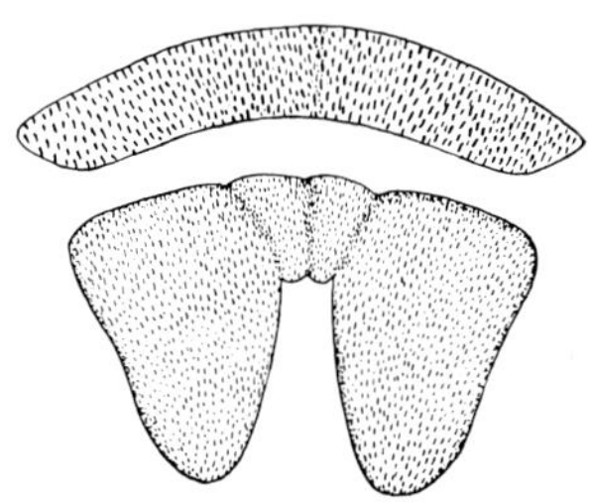

FIG. 3. *Arius proximus*
USNM 173569, 293 mm.

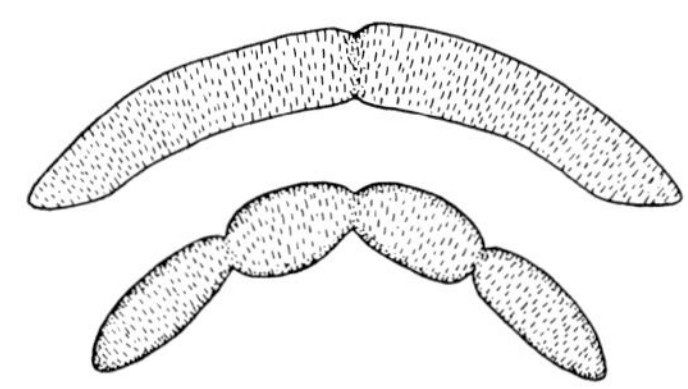

FIG. 4. *Arius australis*
USNM 173571, 242 mm.

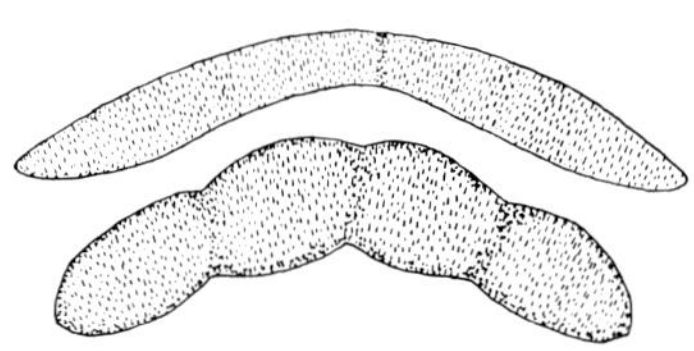

FIG. 5. *Arius leptaspis*
USNM 173565, 334 mm.

Tooth pattern on palate in species of *Arius*

Drawings by Dorothea B. Schultz

that the fish inhabited Red Lily Lagoon. Rotenone sampling in the lagoon did not yield it; and Miller doubts that it was present in the area, indicating in his notes that he believed the natives said it was present because they thought that was what he wished them to say.

ARIIDAE

Tachysurus sinensis Lacépède (1803) based on a Chinese painting does not appear to have been properly identified with any known species of catfish. Thus the name *Tachysurus*, sometimes employed for species of *Arius*, is not used here.

Characteristics of Arnhem Land species of *Arius* are pelvic rays i,5; branched caudal rays $6 + 7 = 13$; pectoral and dorsal fins each with a spine; adipose fin a short fleshy flap, free from back at posterior end; caudal fin deeply forked, not continuous forward with anal fin or as a second dorsal; three pairs of barbels, one maxillary and two mental; sensory canal of head much branched, dendritic; lateral line extending almost to base of caudal rays, thence deflected upward on to caudal fin, its pores opening from small tubules which extend away from lateral line; body with vertical lines of sensory papillae; bony head shield exposed, continuous backward to dorsal spine; anterior and posterior nares adjacent, the posterior more or less covered by a fleshy flap; teeth present on palate, premaxilla and lower jaw; short gillrakers present on anterior faces of all arches, those on rear arches reduced in length; posterior faces of first and second arches nearly smooth, or the second with very fine, scattered papillae; posterior faces of third and fourth arches with gillraker-like structures.

Key to the Arnhem Land Species

1 Granular teeth in two small, oval, widely separated patches, far backward on palate; pectoral rays 1,12 (compiled) . . . . *Arius colcloughi*

Villiform or pointed teeth anteriorly on palate in four or more patches (sometimes appearing as three due to fusion at midline) 2

2 The two or three palatal tooth patches of each side broadly occupying a triangular area, the length of which is greater than the width of single patches; pectoral rays I,9 to I,13, usually I,11 or I,12; midlength of predorsal plate 1·9 to 4·0 in eye length 3

Palatal tooth patches, two on each side, aligned transversely across mouth,their maximum length mostly less than width of outer patch; pectoral rays I,9 or I,10; midlength of predorsal plate 1·0 to 2·3, usually 1·8 or less, in eye length 5

3 Teeth in three groups on each side of palate, the outer anterior patch with long axis transverse; gillrakers and rudiments on first arch 14 or 15 4

Teeth in two groups on each side of palate, the outer patch with its long axis longitudinal; gillrakers and rudiments on first arch 10 to 12; humeral process an acute, moderately granular triangle; predorsal plate short, granular, crescentric . . . . *Arius proximus*

4 Posterior group of palatal teeth adjacent to and directly behind outer anterior patch; humeral process long, relatively broad, without granules, but moderately rugose; dorsal spine long, 1·3 to 1·5 in head; predorsal plate a short, nearly smooth crescent . . . . *Arius thalassinus*

Posterior patch of palatal teeth more mediad, separated rather broadly, by a space more than half its width, from anterior groups; humeral process a broad, strongly granular triangle; dorsal spine shorter, about half head length; predorsal plate long, broad, truncated anteriorly, U-shaped, strongly granular . . . . *Arius mastersi*

5 Total anal rays 16 or 17; vertebrae 46 to 48; predorsal plate short, granular, crescentric; body depth 3·0 to 4·1 in standard length; gillrakers on first arch 14 to 16, usually 14 or 15; humeral process an acute granular triangle; maxillary barbel extending to middle of pectoral fin or beyond . . . . *Arius australis*

Total anal rays 18 to 20, usually 19 or 20; vertebrae 49 to 51; predorsal plate longer, truncated anteriorly, broadly U-shaped; body depth 4·1 to 5·7 in standard length; gillrakers on first arch 14 to 17, usually 16; humeral process an acute granular triangle; maxillary barbel extending to base of pectoral spine or beyond . . . . *Arius leptaspis*

Anal rays 22 or 23; predorsal plate a short granular crescent; body depth 4·6 to 5·0 in standard length; gillrakers on first arch about 16; humeral process a sharp pointed, smooth triangle; maxillary barbel extending beyond tip of pectoral fin (compiled) . . . . *Arius stirlingi*

Arius colcloughi (Ogilby)

Hemipimelodus colcloughi Ogilby, *Proc. Roy. Soc. Qld.* 23: 7 (1910); type locality, Croker Island, Northern Territory.

Cochlefelis colcloughi Whitley, *Aust. Zool.* 10: 172, fig. 4 (1943).

Arius thalassinus (Rüppell)

Pl. 3; Fig. 1

Bagrus thalassinus Rüppell, *Neue Wirbelthiere zu der Fauna von Abyssinien gehörig, Fische*: 75, pl. 20, fig. 2 (1835); type locality, Massaua, Red Sea.

Record questionably referable: *Arius thalassinus* Macleay (1878): 363, Port Darwin.

M48-11, Little Lagoon, 14 April and 20 May 1948, USNM 173567, 2 specimens, 259 and 353 mm. standard length.

Native name: *Yéta*, Groote Eylandt.

Description. Dorsal rays I,6(in 2); pelvic rays i,5(4); pectoral rays I,11(4); branched caudal rays 6+7=13(2); anal rays vii,11(1) or vi,12(1)=18(2); gillrakers first arch 14(2) or 15(2) of which 4 or 5 are on upper limb.

Dorsal spine 1·3 to 1·5, pectoral spine 1·5 to 1·6, eye 6·0 to 6·7, postorbital length of head 2·0 to 2·1, snout 2·6 to 2·7, all in head length; midlength predorsal plate 2·9 to 4·0 in eye; head length 3·4 to 3·9, body depth 4·4 to 5·2, tip of snout to dorsal spine 2·6 to 2·9, tip of snout to pelvic fin 1·7 to 1·9, distance between dorsal and adipose fins 2·8 to 3·0, all in standard length; base of adipose fin 9·5 to 9·6 in distance between dorsal and adipose fins.

Dorsal and pectoral spines compressed, with coarse granules anteriorly and fine, short serrae posteriorly; adipose fin height about equal to eye height; maxillary barbel extending to opercular margin; outer mental barbel almost reaching margin of gill membrane; inner mental barbel to vertical below eye.

Teeth in jaws small, pointed; premaxillary band about five times as broad as long; band on lower jaw divided at midline, each half tapering posteriorly to a point; palatal teeth (Fig. 1) in three groups on each side, the inner anterior groups sub-united at midline, the other groups distinctly separated; the two anterior groups of each side rectangular and together forming a band across mouth; posterior group adjacent to and behind outer anterior group, of about same width, but nearly twice as long, its posterior margin rounded; humeral process flexible, a narrow triangle, without distinct granules, its slight rugosity nearly obscured by

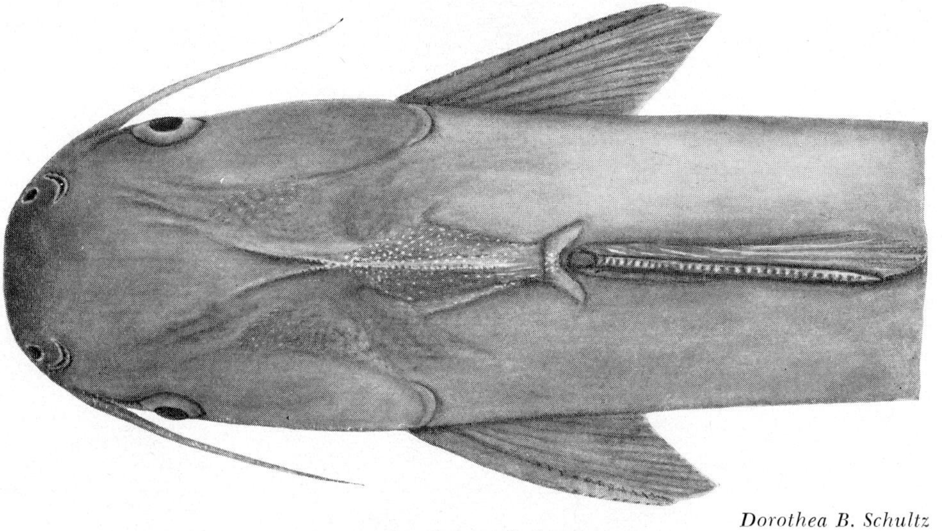

Dorothea B. Schultz

PLATE 3. *Arius thalassinus*
Little Lagoon, USNM 173567, 353 mm. standard length

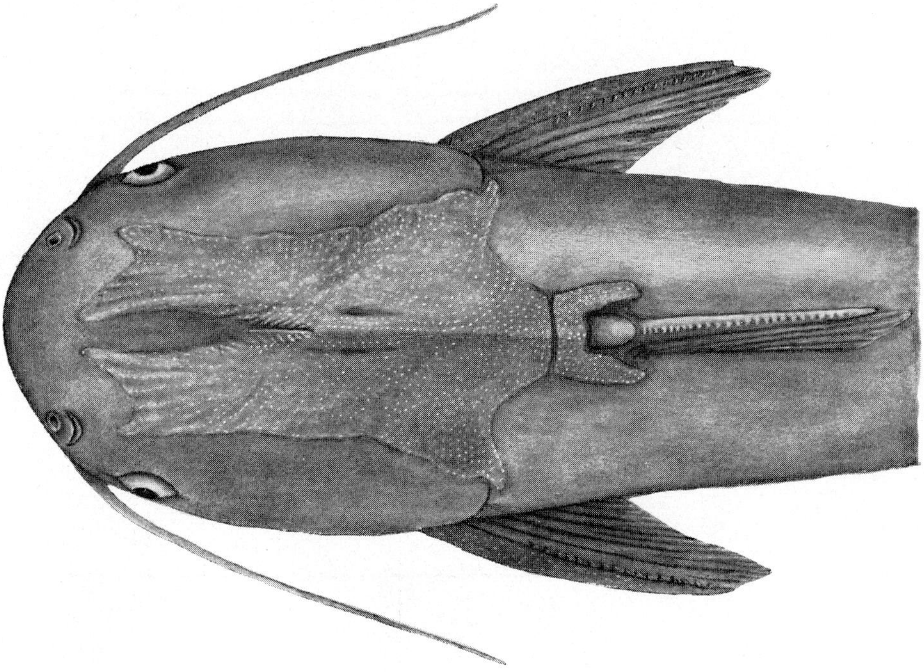

Dorothea B. Schultz

PLATE 4. *Arius mastersi*
Nightcliff, USNM 173566, 323 mm. standard length

skin; head shield nearly smooth anteriorly, moderately rugose or weakly granulated posteriorly, flattened between eyes; occipital process striated or granulated, its midline carinate, with median granules which extend forward to median groove; predorsal plate a short, nearly smooth crescent with granules only near midline.

Colour in alcohol. Body and head light reddish brown above, greyish brown below and on belly; belly covered with fine brown chromatophores; lower head and mental barbels pale yellowish, mostly unpigmented; maxillary barbel dark brown above, pale yellowish below; lips yellowish; pectoral fin brown above, lighter below; pelvic and anal fins brown, the margins light yellowish; caudal and dorsal fins brownish; tips of longer dorsal rays dark brown; adipose fin dark brown; small areas, about half size of pupil, at posterior and anterior margins of orbit pale yellowish.

Arius mastersi Ogilby

Pl. 4, Fig. 2

Arius mastersi Ogilby, *Proc. Linn. Soc. N.S.W.* 23: 34 (1898); type locality, northern Australia, evidently Port Darwin.

Tachysurus (*Pararius*) *godfreyi* Whitley, *Aust. Zool.* 10: 11, fig. 8, Nos. 3 and 4 (1941); type locality, Port Darwin.

Arius sagoroides Hardenberg, *Treubia* 18: 221, fig. 2 (1941); type locality, Merauke, New Guinea.

Recorded as: *Arius gagorides* Macleay (1881): 213, Port Darwin.

M48-2, Nightcliff, 19 March 1948, USNM 173566, 1 specimen, 323 mm. standard length.

Description. Dorsal rays I,6; pelvic rays i,5, both sides; pectoral rays I,11 on left, right side broken, at least I,10; branched caudal rays 6+7=13; anal rays vii,12=19; gillrakers on first arch 14, on both sides, of which 5 or 6 are on upper limb.

Dorsal spine 2·1, pectoral spine about 1·7, eye 7·7, postorbital length of head 1·7, snout 3·0, all in head length; midlength predorsal plate 1·9 in eye; head length 3·2, body depth 5·4, distance between dorsal and adipose fins 3·4, tip of snout to dorsal fin 2·9, tip of snout to pelvic fin 2·0, all in standard length; base of adipose fin 3·6 in distance between dorsal and adipose fins.

Dorsal and pectoral spines slightly compressed, coarsely granular anteriorly, the dorsal spine rugose posteriorly, the pectoral with moderate serrae; adipose fin height about equal to eye length, greater than eye height; maxillary barbel extending to tip of smallest pectoral ray; outer mental barbel reaching gill opening, the inner mental barbel to vertical below posterior margin of orbit.

Teeth in jaws small, pointed; premaxillary band about six times as broad as long; band on lower jaw divided at midline, each half tapering posteriorly; palatal teeth (Fig. 2) in three groups on each side, distinctly separated, the two anterior groups rectangular, forming a band across mouth; posterior group ovoid, behind junction of outer and inner anterior groups, separated from them by about half its width; humeral process stiff, an obtuse triangle, strongly granular; head shield and predorsal plate strongly granulated, rugose anteriorly, flattened between eyes; midline of occipital process and predorsal plate only slightly elevated; predorsal plate long, broadly U-shaped, its lateral wings wide, anteriorly truncated to meet the broad end of occipital process.

Colour in alcohol. Dark brown above, lighter below; lower surface of head and body yellowish, without chromatophores; anal, dorsal and caudal fins light brownish; upper surface of pectoral and pelvic fins medium brown, lower surface lighter; adipose fin dark brown; lips and inner mental barbels yellowish; outer mental barbels dusky; maxillary barbel dark brown, except yellowish lower margin.

Arius proximus Ogilby

Pl. 5; Fig. 3

Arius proximus Ogilby, *Proc. Linn. Soc. N.S.W.* 23: 280 (1898); type locality, Port Darwin.

Arius arafurensis Hardenberg, *Treubia* 19: 409 (1948); type locality, Dobo Market, Aru Islands.

Record questionably referable: *Arius (Tachysurus) graeffei* Paradice and Whitley (1927): 80 and 97, Pellew Islands.

M48-2, Nightcliff, 19 March 1948, USNM 173568, 1 specimen, 289 mm. standard length.

M48-20, Main jetty, Melville Bay, 11 July 1948, USNM 173569, 1 specimen, 293 mm. standard length.

M48-20, Old jetty, Melville Bay, 17 August 1948, USNM 173570, 2 specimens, 316 and 359 mm. standard length.

Description. Dorsal rays I,6(in 4); pectoral rays I,10(1), or I,11(7); pelvic rays i,5(8); branched caudal rays 6+7=13(4); anal rays vi,11(1) or vi,12(3)=17(1) or 18(3); gillrakers first arch 10(4), 11(2) or 12(2) of which 3 to 5 are on upper limb.

In the four specimens, dorsal spine 1·4 to 1·7, pectoral spine 1·7 to 1·9, eye 6·2 to 6·7, postorbital length of head 1·9 to 2·0, snout 2·7 to 3·0, all in head length; midlength predorsal plate 2·6 to 3·6 in eye; head length 3·2 to 3·5, body depth 4·7 to 5·5, distance between dorsal and adipose fins 3·2 to 3·8, tip of snout to dorsal spine 2·6 to 2·7, tip of snout to pelvic fin 1·8 to 1·9, all in standard length; base of adipose fin 4·8 to 5·7 in distance between dorsal and adipose fins.

Dorsal and pectoral spines compressed, with coarse granules anteriorly and rather short serrae posteriorly; adipose fin taller than eye height, nearly equal to eye length; maxillary barbel reaching base pectoral spine; outer mental barbel extending backward to behind gill opening; inner mental barbel to slightly behind vertical from eye.

Teeth in jaws small, pointed; premaxillary band about five times as broad as long; band on lower jaw distinctly divided at midline, each half tapering posteriorly to a point; palatal teeth (Fig. 3) in two adjacent groups on each side, the inner group, close to that of opposite side, nearly square, less than half width of outer patch; outer patch oblong, narrower behind, nearly three times length of inner patch; humeral process rather stiff, acutely triangular, weakly granular; head shield moderately granular behind, the granules anteriorly in more or less radiating series to eye region; head flattened between eyes; occipital process keeled forward to median groove; predorsal plate short, crescentric, moderately granular anteriorly.

Colour in alcohol. Medium to dark brown above; yellowish to yellowish brown below; scattered chromatophores on abdomen of smallest specimen; others with abdomen nearly immaculate; lower head without pigment; adipose fin dark brown; lower surface of paired fins weakly pigmented, light yellowish; vertical fins and upper surface of paired fins medium brown; margin of caudal and tips of longest dorsal rays dark brown or blackish; maxillary barbel dark brown above; its lower surface and mental barbels yellowish.

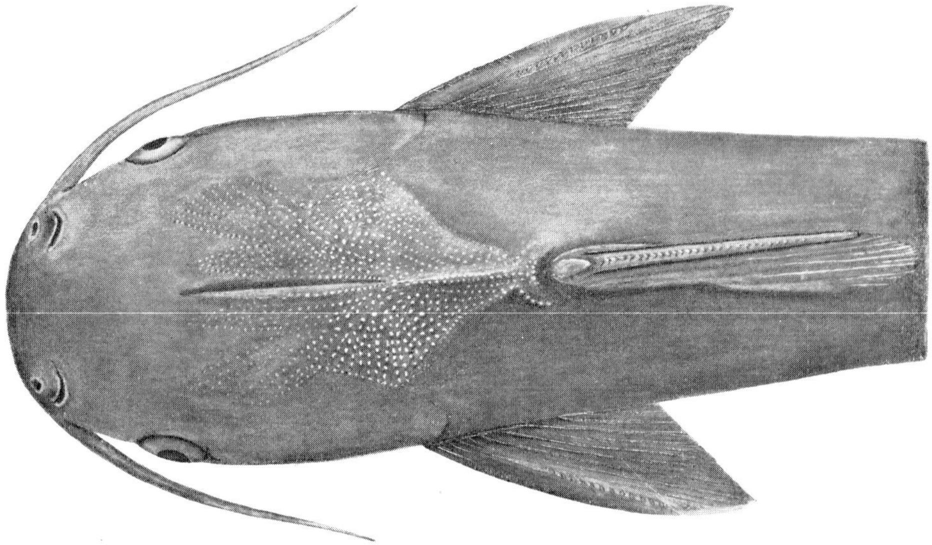

Dorothea B. Schultz

PLATE 5. *Arius proximus*
Melville Bay, USNM 173569, 293 mm. standard length

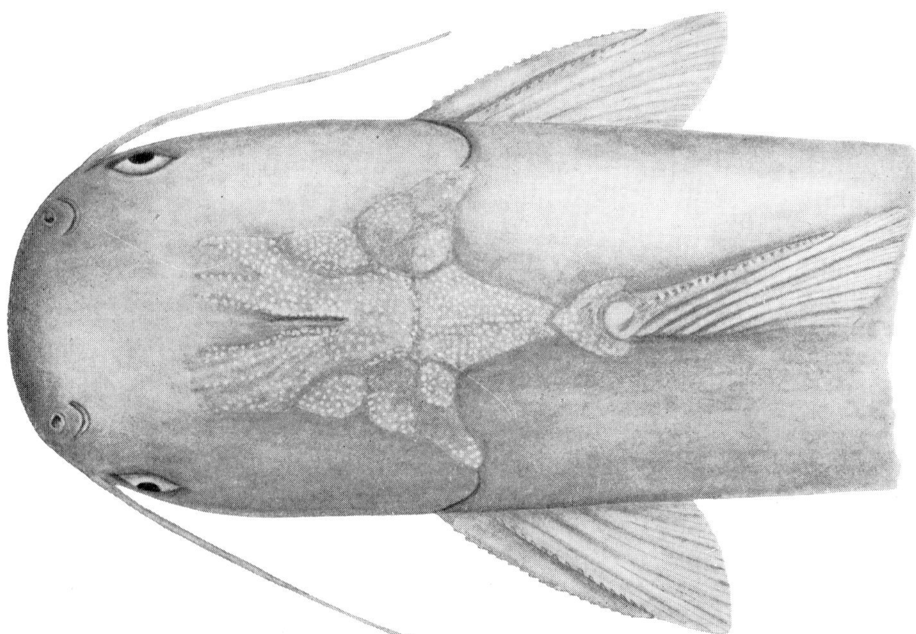

Dorothea B. Schultz

PLATE 6. *Arius australis*
Oenpelli, USNM 173571, 242 mm. standard length

Arius australis Günther

Pl. 6; Fig. 4

Arius australis Günther, *Proc. Zool. Soc. Lond.*: 103 (1867); type locality, Hunter River, near Ash Island, New South Wales.

M48-35, Billabong, Oenpelli, USNM 173571, 12 specimens, 201 to 435 mm. standard length.

Native name: *Mánmákajuári.*

Description. Dorsal rays I,6(in 12); pelvic rays i,5(24); pectoral rays I,9(7) or I,10(15); branched caudal rays 6+7=13(11); anal rays vi or vii,9 to 11=16(3) or 17(9); vertebrae, exclusive of anterior fused vertebrae, 46(2), 47(3) or 48(1); gill-rakers on first arch 14(4), 15(6) or 16(1), of which 5 or 6 are on upper limb.

In 10 specimens, 201 to 435 mm. in standard length, dorsal spine 2·0 to 2·4, pectoral spine 1·9 to 2·2, eye 7·6 to 10·1, postorbital length of head 1·6 to 1·9, snout 2·7 to 3·2, all in head length; midlength of predorsal plate 1·3 to 1·6 in eye; head length 3·0 to 3·5, body depth 3·0 to 4·1, distance between dorsal and adipose fins 3·3 to 4·2, tip of snout to dorsal fin 2·5 to 2·7, tip of snout to pelvic fin 1·6 to 1·7, all in standard length; base of adipose fin 2·6 to 3·3 in distance between dorsal and adipose fins.

Dorsal and pectoral spines only moderately compressed, the dorsal spine granular anteriorly, and rugose or slightly serrate posteriorly; pectoral spine with anterior granular serrae and very short serrae posteriorly; adipose fin height greater than length of small eye, about 0·5 to 0·7 in eye length; maxillary barbel reaching middle of pectoral fin, or in one specimen to below dorsal fin; outer mental barbel extending to base of pectoral spine; inner mental barbel not reaching margin of gill membrane.

Teeth in jaws small, pointed; premaxillary band about eight times as broad as long; band on lower jaw narrow, divided at midline and tapering backward; palatal teeth (Fig. 4) in two groups on each side, each distinctly separated, forming a curved line across palate, the groups subrectangular, wider than long, with round ends; outer group widest; humeral process a narrow stiff triangle, strongly granular; head shield without semblance of median ridge; it and short, concentric predorsal plate strongly granular; shield slightly convex between eyes.

Colour in alcohol. Fins and body dark brown above; lower surfaces of paired fins, caudal peduncle, head and the abdomen yellowish, with little or no pigment; distal portions of fins pale; upper surface of maxillary barbel brown, its lower surface, the lips and mental barbels yellowish.

Arius leptaspis (Bleeker)

Pl. 7; Fig. 5

Hexanematichthys leptaspis Bleeker, *Atlas ichthyologique des Indes Orientales Néêrlandaises* 2: 27, pl. 55, fig. 2 (1862); type locality, south-west New Guinea. Billabong, about 3 miles above Roper River Mission ,17 August 1948 (Blitner), USNM 173564, 1 specimen, 111 mm. standard length.

M48-40, East Alligator River, USNM 173565, 14 specimens, 149 to 406 mm. standard length.

Description. Dorsal rays I,6(in 11); pelvic rays i,5(22); pectoral rays I,10(21); branched caudal rays 6+7=13(11); anal rays vii to x,10 to 13, totals equal 18(1), 19(9) or 20(5); gillrakers on first arch 14(2), 15(3), 16(6) or 17(1) of which 5 to 7 are on upper limb; vertebrae, except anterior fused vertebrae, 49(3), 50(1) or 51(1).

In 11 specimens, 111 to 403 mm. in standard length, dorsal spine 1·6 to 2·4,

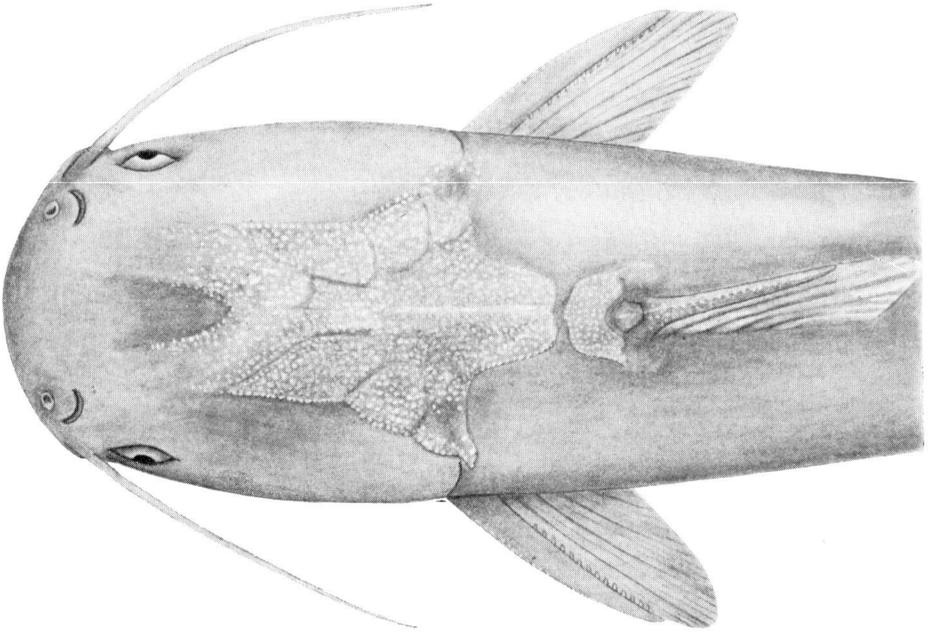

Dorothea B. Schultz

PLATE 7. *Arius leptaspis*

East Alligator River, USNM 173565, 334 mm. standard length

pectoral spine 1·6 to 2·2, eye 6·5 to 9·8, postorbital length of head 1·6 to 1·9, snout 2·9 to 3·2, all in head length; midlength predorsal plate 1·0 to 2·3 in eye; head length 3·1 to 3·8, body depth 4·1 to 5·7, distance between dorsal and adipose fins 3·4 to 4·1, tip of snout to dorsal fin 2·4 to 2·9, tip of snout to pelvic fin 1·7 to 2·0, all in standard length; base of adipose fin 2·3 to 3·4 in distance between dorsal and adipose fins.

Dorsal and pectoral spines only moderately compressed, the dorsal spine granular anteriorly near base, rugose posteriorly, the pectoral spine granular anteriorly and with very short posterior serrae; adipose fin height greater than eye length, about 0·5 to 0·8 in eye; maxillary barbel reaching base of pectoral fin, or to below dorsal fin in small specimen; outer mental barbel extending to middle of shortest pectoral ray; inner mental barbel extending beyond gill opening.

Teeth in jaws small, pointed; premaxillary band about seven times as broad as long; band on lower jaw rather narrow, divided at midline, tapering backward; palatal teeth (Fig. 5) in two groups on each side, the lateral groups distinctly separated from the medial groups, the medial groups separated in the young, but fused across midline in large specimens; palatal teeth forming a curved line across palate, the groups subrectangular, wider than long, with rounded ends, outer group widest; humeral process a stiff, narrow triangle, strongly granular; head shield without median ridge, moderately granular, flattened between eyes; predorsal plate short, sub-crescentric in young, broadly U-shaped in larger specimens, moderately granular.

Colour in alcohol. Fins and body greyish brown to dark brown above; lower surface of paired fins, caudal peduncle, head and the abdomen whitish, mostly without pigment; upper surface of maxillary barbel brown; its lower surface, the lips and mental barbels yellowish white.

Remarks. Arius australis, Arius leptaspis and *Arius stirlingi* are very similar morphologically, differing from each other chiefly in body proportions and in certain meristic characters. Consequently, it is unknown whether they are true species, or each represents population divergence within one species.

Arius stirlingi Ogilby

Arius stirlingi Ogilby, *Proc. Linn. Soc. N.S.W.* 23: 281 (1898); type locality, estuary of Adelaide River, Northern Territory, Australia.

PLOTOSIDAE

Marine and freshwater 'stinging' catfishes with four pairs of barbels—one associated with the posterior nares, one maxillary and two mental—and in some species with an additional barbel-like filament from the upper lip near the maxillary barbel; first dorsal fin and pectoral fins each with a spine; anal fin long, of approximately 70 to 90 rays, continuous with the rounded to pointed caudal fin; 'second dorsal' without interneurals, hence an extension forward of the caudal, of various lengths; humeral pore present; pelvic rays 10 to 15; sensory canals of head, especially the upper ones, opening through a relatively large number of tiny, irregularly placed pores; lateral line ending near base of caudal rays; slender, pointed gillrakers confined to first two arches, those of second arch short, sometimes modified into transverse ridges; anterior face of other arches and posterior face of third and fourth arches with transverse ridges or processes; processes on posterior face of first or second arches may be absent or variously modified, reduced in length or broken into one or more rows of papillae along arch margins.

Because it was not always possible to distinguish between simple and branched rays, they are not differentiated in the counts. Anterior rays in the second dorsal and anal are simple, followed by rays branched near their tip, the branching becoming progressively deeper posteriorly to a maximum depth in the caudal fin. The first vertebra counted is the first one, bearing a rib, behind the fused anterior vertebrae.

Key to the Genera

1 Anal dendritic organ present; distance between origins of first and second dorsal fins less than head length; chiefly marine catfishes, a few freshwater forms . . 2

Anal dendritic organ absent; distance between origins of first and second dorsal fins usually greater than head length (except *Anodontiglanis*); freshwater catfishes 5

2 Anterior nares on end of snout and directed upward or outward 3

Anterior nares near oral border of upper lip and directed downward, surrounded by the papillae extensively covering the lips; gill membranes broadly united, the notch shallow; premaxillary teeth in two triangular or subcircular patches . . . . . . *Paraplotosus albilabris*

3 Premaxillary teeth present 4

Premaxillary teeth absent (compiled) . . . . . . *Oloplotosus mariae*

4 Gill membranes free from each other and from isthmus, except anteriorly, forming a deep notch; lips plicate or moderately papillose; band of teeth on premaxilla nearly continuous across midline, about three times as broad as long . . . . . . genus *Plotosus*

Gill membranes broadly attached to isthmus, or to each other, the notch between them shallow; lips papillose or moderately plicate; teeth on premaxilla in two small triangular or subcircular groups . . genus *Cnidoglanis*

5 Palate and lower jaw edentulous; premaxilla edentulous, or with one or two deciduous teeth on each side; distance between origins of first and second dorsal fins less than head length . . . . . . *Anodontiglanis dahli*

Palate, lower jaw and premaxilla with firmly rooted teeth; distance between origins of first and second dorsal fins equal to or usually greater than head length 6

6 Anterior nares piercing the lower border of the thin upper lip and directed downward; spines smoothly rounded or nearly so, without serrae . . . . . . *Porochilus obbesi*

Anterior nares on end of snout and directed upward or outward; spines, especially the pectoral, serrate . . . . . . genus *Tandanus*

Key to the Species of *Plotosus*

1 Nasal barbel extending well behind eye, nearly to posterior end of head; band of pointed teeth on premaxilla in about three to five irregular rows, the largest teeth anteriorly and mesially; pelvic rays 12 to 14, usually 13 or 14; pectoral spine with numerous small serrae posteriorly; humeral process equal to or longer than pectoral spine width; posterior face of second branchial arch with short, somewhat pendant ridges or papillae, behind which are a few to many tiny supernumerary papillae; brownish, slightly lighter below . . . . . . *Plotosus canius*

Nasal barbel extending well behind eye; band of mostly molariform teeth on premaxilla in about four rows; pelvic rays 12; black, with a superior longitudinal whitish band extending backward from below first dorsal and another, inferior, backward above anal fin from whitish undersurface (compiled) . . . . *Plotosus papuensis*

Nasal barbel extending backward approximately to eye; band of pointed teeth on premaxilla in two or three irregular rows, the lateral teeth nearly as large as medial teeth; posterior edge of pectoral spine with few serrae usually fused into a narrow ridge; humeral process short or obscure, not as long as width of pectoral spine; pelvic rays 11 or 12; posterior face of second branchial arch with short, adnate ridges, without supernumerary papillae behind; brownish; young with three longitudinal whitish stripes; one, superior, backward from upper snout nearly to caudal fin; another backward from below eye and above pectoral fin; the third backward above anal fin from belly . . . . *Plotosus lineatus*

Plotosus lineatus (Thunberg)

Pl. 8

Silurus lineatus Thunberg, *K. Svenska VetenskAkad. Nya Handl.* 12: 191, pl. 6 (1791); type locality, East Indian seas.

Platystacus anguillaris Bloch, *Naturgeschichte der Ausländischen Fische* 8: 61, pl. 373, fig. 1 (1794), not seen; Bloch, *Ichthyologie* 11: 49, pl. 373, figs. 1-2 (1797); type locality not given; in collection of Missionary John.

Plotosus anguillaris Weber and de Beaufort, *Fishes of the Indo-Australian Archipelago* 2: 229 (1913); McCulloch, *Aust. Mus. Mem.* 5: 54 (1929); names in synonymy.

Plotosus flavolineatus Whitley, *Aust. Mus. Mag.* 7: 311, fig. (1941); type locality, Sydney, Australia.

Recorded as: *Plotosus anguillaris* Macleay (1878): 363 and (1881): 203, Port Darwin.

M48-21, Reefs, Yirrkalla, 11 August 1948, USNM 173543, 2 females, slightly more than 155 (tail missing) to 251 mm. standard length.

Description. Rays in first dorsal I,4(in 2); second dorsal-upper caudal 103(1) or more than 97(1) + anal-lower caudal 86(1)=189(1); interhaemals 78(1), each supporting one anal ray; pectoral I,11(4); pelvic 12(4), the anterior ray apparently simple; gillrakers on first arch 6+1+19=26(2), 6+1+21=28(1) or 8+1+21=30(1); branchiostegal rays 10(2), 11(1) or 12(1); precaudal vertebrae 10(1) or 12(1)+ caudal vertebrae 44(1)=54(1), in Arnhem Land specimens. In specimens from Zamboanga, Palawan, Sandaham Bay and Sulu Archipelago, second dorsal-upper caudal 91(1), 93(1), 99(2), 117(1) + anal-lower caudal 70(2), 73(1), 75(1), 80(1)= 161(1), 166(1), 169(1), 174(1), 197(1); interhaemals 62(1), 66(2), 67(1), 72(1), each supporting one anal ray; precaudal vertebrae 10(3) or 12(2) + caudal vertebrae 41(1), 42(2), 43(1), 45(1)=51(1), 52(2), 55(1), 57(1).

In the two Arnhem Land specimens, dorsal spine 2·1 to 2·5, pectoral spine 2·5 to 2·8, snout 2·4 to 2·6, postorbital length of head 2·1 to 2·2, eye to posterior naris 4·5 to 4·7, eye 7·1 to 7·2, origin first dorsal fin to origin second dorsal fin 1·4 to 2·0, all in head length; origin second dorsal fin to tip caudal fin 1·4, origin anal fin to tip caudal fin 1·5, tip of snout to first dorsal 4·1, tip of snout to anal fin 2·4, tip of snout to second dorsal fin 2·9, head 4·5, body depth 6·7, all in standard length.

Lateral line extending nearly to end of caudal peduncle; axillary pore present; dorsal spine with moderate posterior serrae and recurved anterior hooks; pectoral spine with recurved anterior hooks, the posterior serrae somewhat fused into a thin shelf; branchiostegal membranes separate and free from isthmus; premaxilla with 10 to 15 pointed teeth on each side of midline, in two or three rows, the anterior and lateral teeth enlarged; peglike teeth, with rounded crowns, on palate in a broad triangular patch, the largest posteriorly; teeth on lower jaw with truncate crowns, the longest anteriorly, those posteriorly molarlike; lips plicate; maxillary and nasal barbels reaching posterior margin of eye; other barbels not quite reaching gill membrane; sensory canal of head with many, apparently little branched, tubes extending to the surface; posterior faces of second to fourth branchial arches with single series of short adnate, transverse ridges, their length less than width of arch, opposing similar ridges on anterior face of each following arch; posterior face of first gill arch nearly smooth; gillrakers confined to first two arches; anal dendritic organ larger than orbit, much branched; anterior naris just above plicae of upper lip and slightly lateral to nasal barbel, pointing forward as an extremely short, projecting tube; humeral process barely evident, obtuse.

Colour in alcohol. Larger immature female medium brown; lower part of head and belly whitish; three whitish stripes; one as wide as pupil, extending backward from snout tip through posterior naris, above eye, thence just below dorsal fins nearly to end of caudal peduncle; another whitish stripe, nearly as wide as eye, extending from mid-cheek through midside of body nearly to end of caudal peduncle; the third whitish stripe extending backward from the belly, and just above the anal fin, on to caudal peduncle; fins dusky; margins of vertical fins blackish. Small female, with large eggs, nearly uniformly medium brown.

Miller's notes suggest that the longitudinal stripes are white in life.

Cnidoglanis megastoma (Richardson)

Plotosus megastomus Richardson, *Ichthyology of the Voyage of H.M.S. Erebus and Terror*: 31, pl. 21, figs. 1 to 3 (1845); type locality, Sydney Cove, New South Wales.

Recorded as: *Cnidoglanis megastoma* Macleay (1878): 363 and Klunzinger (1879): 410, Port Darwin.

Cnidoglanis microceps (Richardson)

Plotosus microceps Richardson, *Ichthyology of the Voyage of H.M.S. Erebus and Terror*: 31, pl. 21, figs. 4 to 7 (1845); type locality, north-west coast of Australia.
Cnidoglanis microcephalus Günther, *Catalogue of the Fishes in the British Museum* 5: 28 (1864); emendation of *microceps*; Richardson's type redescribed.

Cnidoglanis muelleri Klunzinger

Cnidoglanis mülleri Klunzinger, *S.B. Akad. Wiss. Wien* 80: 411 (1879); type locality, Port Darwin; Macleay, *Proc. Linn. Soc. N.S.W.* 9: 53 (1884).

Cnidoglanis nudiceps Günther

Cnidoglanis nudiceps Günther, *Report of the Voyage of the Challenger, Zoology* 1: 49 (1880); type locality, Arafura Sea; Weber and de Beaufort, *Fishes of the Indo-Australian Archipelago* 2: 232, fig. 92 (1913).

The genus *Cnidoglanis* Günther is not understood because of the insufficient descriptions and lack of material. Some of the above listed forms may be separable to other genera.

Patricia Hogue

PLATE 8. *Plotosus lineatus*
Yirrkalla, USNM 173543, 251 mm. standard length

Patricia Hogue

PLATE 9. *Paraplotosus albilabris*
Yirrkalla, USNM 173544, 208 mm. standard length

*Paraplotosus** *albilabris* (Cuvier and Valenciennes)

Pl. 9

Plotosus albilabris Cuvier and Valenciennes, *Histoire naturelle des poissons* 15: 427 (1840); type locality, Batavia, Java.
Plotosus macropthalmus Bleeker, *Verh. Batavia Genoot.* 21 (reprint): 45 (1847): type locality, Batavia, Java.
Copidoglanis longifilis Macleay, *Proc. Linn. Soc. N.S.W.* 6: 207 (1881); type locality, Long Island, Torres Straits.

Recorded as: *Paraplotosus albilabris* Whitley (1940b): 407, Port Darwin; *Plotosus elongatus* Kent (1889): 240, Cambridge Gulf, north Australia.

M48-3, East Point Reef, USNM 173545, 4 specimens, 95 to 243 mm. standard length.

M48-17, Reef in Bartalumba Bay, USNM 173546, 5 specimens, 86 to 272 mm. standard length.

M48-21, Reefs, Yirrkalla, 14 July 1948, USNM 173544, 14 specimens, 56 to 227 mm. standard length; 29 July 1948, 2 specimens, 168 to 215 mm. standard length; 11 August 1948, 17 specimens, 76 to 210 mm. standard length; 12 August 1948, 19 specimens, 66 to 254 mm. standard length.

Description. Rays in first dorsal I,4(in 12); second dorsal-upper caudal 97(1), 98(2), 99(4), 100(1), 101(1), 102(1), 105(2) + anal-lower caudal 86(1), 89(3), 91(1), 92(1), 94(1), 97(2), 100(2), 104(1) = 187(1), 188(3), 189(2), 195(1), 199(3), 200(1), 209(1); interhaemals 81(1), 82(2), 83(1), 84(1), 86(1), 90(3), 93(1), 95(1), 96(1), each supporting one anal ray; pectoral I,12(6) or I,13(18); pelvic 12(24), the outer ray deeply branched; gillrakers, right side, first arch 5+1+16=22(1), 4+1+17=22(1), 5+1+17=23(1); branchiostegal rays 9(3); precaudal vertebrae 10(2), 11(8), 12(2) + caudal vertebrae 47(1), 49(4), 50(1), 51(5), 53(1) = 58(1), 59(1), 60(3), 61(1), 62(4), 63(1), 64(1).

In eight specimens 93 to 205 mm. in standard length, dorsal spine 1·9 to 2·2, pectoral spine 2·3 to 2·6, snout 2·5 to 3·0, postorbital length of head 2·2 to 2·5, eye to posterior naris 4·8 to 5·5, eye 4·0 to 4·6, origin of first dorsal fin to origin of second dorsal fin 1·5 to 2·4, all in head length; origin second dorsal fin to tip of caudal fin 1·2 to 1·4, origin of anal fin to tip of caudal fin 1·3 to 1·5, tip of snout to first dorsal fin 4·0 to 5·0, tip of snout to anal fin 2·5 to 2·9, tip of snout to second dorsal 3·0 to 3·4, head 4·7 to 5·5, body depth 5·8 to 7·4, all in standard length.

Lateral line extending to or nearly to end of caudal peduncle; axillary pore present; dorsal spine with short posterior serrae and recurved anterior hooks; pectoral spine with anterior hooks and short, irregular to fused posterior serrae; branchiostegal membranes broadly united, the notch between shallow; premaxillary teeth pointed, arranged in two small, triangular to subcircular patches, the largest teeth anteriorly; teeth on palate short, molarlike, with rounded crowns, in a broad triangular patch, the largest teeth postmedial; anterior teeth in lower jaw peglike, with truncate crowns; posteriorly, teeth short, molarlike and in several rows; lips strongly papillated, the lower lip somewhat expanded posteriorly around the inner mental barbel into a papillose flap; nasal barbel long, reaching posterior end of head or often to dorsal fin; maxillary barbel extending a little behind the eye; outer mental barbel reaching base of pectoral fin, the inner mental barbel slightly shorter; sensory canal of head, especially the circumorbital canals, opening

* *Endorrhis* Ogilby (1898): 283; type species, *Copidoglanis longifilis* Macleay, by original designation; a synonym of *Paraplotosus*.

to the surface through numerous tiny uneven size pores; gillrakers confined to first two arches; posterior face of first arch nearly smooth; posterior faces of second to fourth arches and anterior faces of third to fifth arches with opposing transverse appendages which are long, low, narrow, and adnate to arch, each somewhat bulb-like at posterior end; appendages on second arch longer than arch width, projecting behind arch, their length about half that of gill filament; posteriorly, appendages decrease in size, those of last arch not longer than arch width; anal dendritic organ large, half eye diameter, much branched; anterior naris emerging between papillae on lower side of upper lip, pointing downward, with a prominent ovoid flap on aboral side, extending almost a pupil diameter below papillae of lip; humeral process short, acute, its length about half the diameter of pectoral spine.

Colour. A dark species, as recorded by Miller. In alcohol, dark brown to blackish above and on both sides; some specimens slightly greyish mottled; belly and lower head lighter, but densely covered with fine chromatophores; lips pale, slightly yellowish; fins dark brown to greyish brown, the margins usually darker; peritoneum black.

Remarks. According to Miller's notes, this species inflicts a painful wound with its pectoral spine.

Anodontiglanis dahli Rendahl

Pl. 10

Anodontiglanis dahli Rendahl, *Medd. Zool. Mus. Krist.* 5: 169, figs. 2-3 (1922); type locality, Glencoe, Northern Territory, Australia; also recorded from Daly River, Northern Territory.

Billabong, about 3 miles above Roper River Mission, 17 August 1948, Blitner, USNM 173547, 2 specimens, 104 and 115 mm. standard length.

Description. First dorsal rays I,6(in 2); second dorsal-upper caudal 95(1) or 96(1) + anal-lower caudal 77(1) or 83(1) = 173(1) or 178(1); pectoral I,11(4); pelvic 12(3) or 13(1), the first ray simple; gillrakers, right side, first arch 4+1+10=15(1) or 5+1+10=16(1); branchiostegal rays 8(1) or 9(3); precaudal vertebrae 10(2) + caudal vertebrae 42(1) or 44(1)=52(1) or 54(1).

Dorsal spine 1·9 to 2·0, pectoral spine 2·1 to 2·2, snout 2·1, postorbital length of head 2·7 to 2·8, eye to posterior naris 3·0 to 3·1, eye 4·7 to 4·9, origin of first dorsal fin to origin of second dorsal 1·8 to 2·0, all in head length; origin of second dorsal to tip of caudal fin 1·4 to 1·5, origin of anal fin to tip of caudal fin 1·5 to 1·6, tip of snout to first dorsal fin 3·3 to 3·5, tip of snout to anal fin 2·2 to 2·3, tip of snout to second dorsal 2·3 to 2·5, head 3·9 to 4·0, body depth 5·1 to 5·2, all in standard length.

Lateral line extending nearly to end of caudal peduncle; axillary pore present; dorsal spine nearly smooth posteriorly, with hooked steplike processes anteriorly; pectoral spine finely serrate anteriorly, with straight, slightly coarser serrae posteriorly, the distal third of anterior edge with hooked steps; branchiostegal membranes united, the notch shallow, confluent anteriorly with isthmus; mouth edentulous, except one small pointed, movable, probably deciduous tooth on each side of midline of premaxilla in larger specimen and two on each side in smaller individual; pharyngeal teeth numerous, pointed; lips finely plicate; maxillary barbel reaching pupil; nasal barbel extending to eye; outer mental barbel extending to vertical below anterior rim of orbit; inner mental barbel slightly shorter; sensory canals of head opening through a moderate number of pores from short, branched tubes; moderately long, slender gillrakers confined to first arch, those of second arch short, rectangular papillae or ridges similar to structures along anterior half of posterior face of first arch; posterior half of posterior face of first arch smooth or with a few tiny bumps; posterior faces of second to fourth arches and anterior faces of third to fifth arches with transverse, long, low, adnate, opposing ridges, the length of first opposing series about twice that of arch width, that of others equal to or shorter than arch width; no anal dendritic organ; anterior naris on end of snout slightly lateral to nasal barbel, pointing forward as a short projecting tube; humeral process moderate, pointed, about equal to spine width.

Colour in alcohol. Body and head apparently dark brown, becoming lighter below; belly and lower head pale, without chromatophores; fins dusky, dark grey to blackish, lighter toward tips; upper barbels dark brown; mental barbels pigmented only on anterior surface; lower lip brownish; inner margin of upper lip whitish.

*Porochilus** *obbesi* Weber

Pl. 11

Porochilus obbesi Weber, *Nova Guinea* 9: 523, figs. 4 and 5 (1913); type locality, Lorentz River, New Guinea.

Recorded as: *Porochilus obbesi* Whitley (1941b): 7, Yam Creek, Northern Territory.

M48-37, Fish Creek, USNM 173548, 15 specimens, 45 to 66 mm. standard length.

Description. Rays in first dorsal I,4(in 8) or I,5(2); second dorsal-upper caudal 24(1), 25(2), 26(4), 28(2), 29(1) + anal-lower caudal 83(1), 84(1), 85(1), 87(3), 89(1), 91(2), 95(1) = 109(2), 111(2), 112(1), 115(2), 117(1), 120(1), 123(1); pectoral I,7(2), I,8(11), I,9(6); pelvic 10(1), 11(9), 12(10), the anterior ray apparently simple; branchiostegal rays 7(4), 8(6); precaudal vertebrae 6(15) + caudal vertebrae 37(3), 38(8), 39(4) = 43(3), 44(8), 45(4).

In eight specimens 47 to 66 mm. in standard length, dorsal spine 2·1 to 2·7, pectoral spine 2·2 to 2·9, snout 2·7 to 3·1, postorbital length of head 1·9 to 2·3, eye to posterior naris 7·5 to 11·5, eye 4·5 to 5·1, origin of first dorsal fin to origin of second dorsal 0·2 to 0·4, all in head length; origin of second dorsal to tip of caudal fin 3·4 to 4·2, origin of anal fin to tip of caudal fin 1·1 to 1·3, tip of snout to first dorsal fin 3·6 to 4·2, tip of snout to anal fin 3·0 to 3·4, tip of snout to second dorsal 1·1 to 1·2, head 5·3 to 5·6, body depth 5·5 to 6·1, all in standard length.

Lateral line discontinuous, a series of short, isolated tubules, ending under or behind origin of second dorsal; axillary pore present; spines rounded, nearly smooth, with steplike processes only on distal portion of anterior edge and at most a short, thin sheath from posterior edge; branchiostegal membranes broadly united to each other; premaxillary teeth pointed, in two small, ovoid, almost continuous patches; teeth on palate with pointed and rounded crowns, the patch broadly triangular to subcrescentric; anterior teeth in lower jaw pointed, posteriorly molarlike; lips thin, nearly smooth, inner portions slightly papilliated; nasal barbel reaching to or beyond occiput; maxillary barbel extending to behind eye; outer mental barbel extending slightly beyond base of pectoral fin; inner mental barbel reaching pectoral spine; sensory canal of head relatively simple, the pores few in number and in nearly a single series along each canal; moderate, slender

* For *Porochilus* Weber, a monotypic genus, see the species citation.

Patricia Hogue

PLATE 10. *Anodontiglanis dahli*
Billabong above Roper River Mission, USNM 173547, 104 mm. standard length

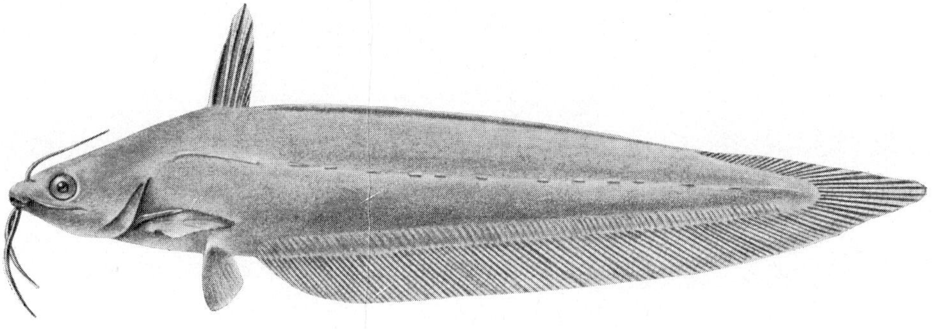

Patricia Hogue

PLATE 11. *Porochilus obbesi*
Fish Creek, USNM 173548, 60 mm. standard length

gillrakers confined to anterior face of first arch; posterior faces of all arches and anterior faces of the second to fifth arches with transverse, low, adnate, opposing ridges, their length about equal to arch width; ridges of first arch expanded at each end, appearing more or less as two rows of papillae along arch; no anal dendritic organ; anterior naris opening downward near corner of mouth through the thin upper lip, without a projecting tube; humeral process acute, longer than diameter of pectoral spine.

Colour in alcohol. Almost uniformly dark brown to blackish; abdomen and lower head densely covered with chromatophores; fin margins slightly darker.

Key to the Arnhem Land Species of *Tandanus*

1 Second dorsal longer, with 39 to 52 rays to middle of caudal fin, its length from origin to tip of caudal fin 2·5 to 3·3 in standard length; head 4·0 to 4·7 in standard length; posterior face of first gill arch with only a single, anterior, row of papillae; pectoral rays I,11 to I,13; pelvic rays 12 to 15, usually 13 or 14; vertebrae 50 to 52; snout longer than postorbital length of head . . . . *Tandanus ater*

Second dorsal shorter, with 24 to 37 rays to middle of caudal fin, its length from origin to tip of caudal fin 3·3 to 5·0 in standard length; head 4·7 to 5·7 in standard length; posterior face of first gill arch with two rows of papillae along arch, the posterior row small; pectoral rays I,9 to I,11; vertebrae 44 to 49 2

2 Snout longer, 2·3 to 3·0 in head, about equal to, usually slightly shorter, sometimes a little longer than, postorbital length of head; pelvic rays 12 to 14, usually 13 or 14; pectoral rays I,10 or I,11; nasal barbel reaching to or slightly behind eye; transverse ridges on posterior face of second gill arch and anterior face of third arch longer than arch width, overlapping base of gill filaments . . . . *Tandanus glencoensis*

Snout shorter, 2·7 to 3·5, usually more than 2·9 in head, shorter than postorbital length of head; pelvic rays 10 to 13, usually 11 or 12; pectoral rays I, 9 or I,10, infrequently I,11; nasal barbel reaching posterior end of head; transverse ridges on posterior face of second gill arch and anterior face third arch about equal to, or slightly shorter than arch width . . . . *Tandanus rendahli*

Tandanus ater* (Perugia)

Pl. 12

Lambertia atra Perugia, *Ann. Mus. Stor. Nat. Genova* ser. 2, 14: 551 (1894); type locality, Inawi, Papua.

M48-14, Emerald River, Groote Eylandt, above and below falls, USNM 173549, 64 specimens, 42 to 302 mm. standard length.

M48-35, Billabong, Oenpelli, USNM 173550, 19 specimens, 169 to 387 mm. standard length.

M48-37, Fish Creek, USNM 173551, 14 specimens, 83 to 129 mm. standard length.

* See McCulloch (1929-30): 55 for generic synonymy, with possible exception of name *Eumeda* Castelnau; other generic synonyms are: *Copidoglanis* Günther (1864): 25, type species, *Plotosus tandanus* Mitchell, by subsequent designation; *Lambertia* Perugia (1894): 550, type species, *Lambertia atra* Perugia, by monotypy; *Lambertichthys* Whitley (1938): 223, substitute name for the preoccupied *Lambertia* Perugia.

M48-38, Oenpelli Creek, USNM 173552, 16 specimens, 80 to 117 mm. standard length.

M48-39, Red Lily Lagoon, USNM 173553, 5 specimens, 147 to 372 mm. standard length.

Native names: *Ángalpía,* Groote Eylandt; *Namardága* and *Náhgool,* Oenpelli.

Description. Rays in first dorsal I,5(in 14), I,6(6), I,7(1); second dorsal-upper caudal 39(1), 40(2), 41(3), 42(1), 43(3), 44(2), 45(5), 48(2), 49(1), 52(1) + anal-lower caudal 84 to 103(21) = 124(1), 129(2), 130(2), 131(2), 132(2), 134(1), 135(2), 137(1), 138(1), 139(3), 140(1), 143(1), 146(1), 155(1); interhaemals 73(1), 75(1), 77(1), 78(2), 79(3), 80(4), 81(1), 82(3), 83(1), 85(1), 86(1), 87(1), 92(1), supporting an additional, small, anal ray at anterior end; pectoral I,11(9), I,12(25), I,13(8); pelvic 12(2), 13(17), 14(17), 15(1), the outer ray simple or very shallowly branched; gillrakers on first arch 24(1), 25(5), 26(8), 27(4), 28(1), 29(1), 30(1), of which 6 to 9 are on upper limb; branchiostegal rays 8(1), 9(8), 10(12); precaudal vertebrae 8(18), 9(3) + caudal vertebrae 42(3), 43(12), 44(6) = 50(2), 51(11), 52(8).

In 15 specimens 78 to 155 mm. in standard length, dorsal spine 1·6 to 2·1, pectoral spine 1·8 to 2·4, distance between origins first and second dorsal fins 0·3 to 0·5, snout 2·1 to 2·4, postorbital length of head (typically shorter than snout) 2·2 to 2·7, eye 4·9 to 6·3, all in head length; origin of second dorsal to tip of caudal fin 2·5 to 3·3, origin of anal fin to tip of caudal fin 1·2 to 1·5, tip of snout to anal fin 2·5 to 2·8, tip of snout to second dorsal 1·1 to 1·4, head 4·0 to 4·7, body depth equal to or less than head length, except in adults, 4·5 to 5·9, tip of snout to first dorsal 3·1 to 3·7, all in standard length.

Lateral line extending to or almost to base of caudal rays; canals of head opening through a moderate number of pores; a cluster of about 5 to 25 temporal pores between eyes; axillary pore present; dorsal spine with short, recumbent posterior serrae and anterior hooked step-like processes; pectoral spine nearly smooth anteriorly, with prominent recurved posterior serrae about three-fourths length of spine width; branchiostegal membranes rather broadly united; premaxilla with a small rectangular patch of tiny, pointed teeth on each side of midline; teeth on palate about four times as large, with rounded crowns, arranged in a large semicircular to broadly triangular, posteriorly truncated, patch; teeth in lower jaw pointed anteriorly, molarlike posteriorly; end of snout somewhat thickened, bulbous, or slightly turned upward; lips plicate; maxillary and nasal barbels reaching to or very slightly behind eye; outer mental barbel longest, extending to or beyond base of pectoral fin; inner mental barbel reaching gill membrane; slender gillrakers present on anterior faces of first and second arches, those of first arch about half length of gill filament, those of second arch short, about as long as arch width; posterior face of first arch with a row of enlarged papillae along anterior edge; the posterior area smooth; posterior faces of second to fourth arches and the anterior faces of the third to fifth arches with transverse, low, adnate, opposing ridges, the first opposing series longer than arch width, free at end, the other series about equal in length to arch width; no anal dendritic organ; anterior naris on end of snout, anterior to and slightly lateral to nasal barbel, pointing forward as a short projecting tube; humeral process acute, about as long as spine width.

Colour in alcohol. Almost uniformly brown to black, sometimes slightly mottled; lighter below, but lower head and abdomen usually sprinkled with large chromatophores; lips brown to yellowish.

Evidently this species is black in life, as judged from Miller's notes.

Tandanus glencoensis (Rendahl)

Pl. 13

Copidoglanis glencoensis Rendahl, *Medd. Zool. Mus. Krist.* 5: 170, fig. 4 (1922); type localities, Daly River and Glencoe, Northern Territory.
Neosilurus mortoni Whitley, *Aust. Zool.* 10: 7, fig. 7 (1941); type locality, Yam Creek, 9 miles from Brocks Creek, Northern Territory.

Large billabong about 3 miles above Roper River Mission, Northern Territory, 17 August 1948, Blitner, USNM 173560, 5 specimens, 94 to 132 mm. standard length.

M48-35, Billabong, Oenpelli, USNM 173561, 28 specimens, 140 to 272 mm. standard length.

M48-37, Fish Creek, USNM 173562, 19 specimens, 77 to 99 mm. standard length.

M48-38, Oenpelli Creek, USNM 173563, 21 specimens, 74 to 100 mm. standard length.

Native name: *Náhgool,* Oenpelli.

Description. Rays in first dorsal I,5(in 11), I,6(9); second dorsal-upper caudal 25(1), 28(1), 30(5), 31(4), 32(5), 33(2), 36(1), 37(1) + anal-lower caudal 84 to 98(20) = 114(2), 116(3), 117(3), 118(1), 119(2), 120(2), 122(3), 123(1), 126(1), 129(1), 131(1); interhaemals 76(1), 77(2), 78(1), 80(3), 85(1), usually supporting an additional small anal ray at anterior end of fin; pectoral I,10(18), I,11(22); pelvic 12(5), 13(20), 14(15), the outer ray simple; gillrakers on first arch 19(1), 20(2), 21(5), 22(2), 23(2) of which 6 to 8 are on the upper limb; branchiostegal rays 8(7), 9(5); precaudal vertebrae 7(4), 8(4) + caudal 38(1), 39(2), 40(4), 41(1) = 45(1), 47(4), 48(3).

In 14 specimens, 94 to 226 mm. in standard length, dorsal spine 1·6 to 2·1; pectoral spine 1·6 to 2·2; distance between origins of first and second dorsal fins 0·2 to 0·4; snout 2·3 to 3·0; postorbital length of head 2·2 to 2·6, about equal to snout length, sometimes shorter, usually slightly longer; eye 4·4 to 6·0, all in head length; distance from origin of second dorsal to tip of caudal fin 3·3 to 4·3; anal fin origin to tip of caudal fin 1·3 to 1·5; tip of snout to anal fin 2·5 to 2·8; tip of snout to second dorsal 1·1 to 1·2; head 4·7 to 5·0; depth of body 4·7 to 5·2, about equal to head length; tip of snout to first dorsal 3·3 to 3·6, all in standard length.

Lateral line reaching to or almost to base of caudal rays; canals of head opening through an intermediate number of pores; cluster of temporal pores about 5 to 15; axillary pore present; both edges of dorsal spine and anterior edge of pectoral spine roughened or with tiny, almost straight serrae; length of distinct, recurved serrae on posterior face of pectoral spine over half width of spine; branchiostegal membranes rather broadly united; premaxilla with a small sub-ovate patch of pointed teeth on each side of midline; teeth on palate only slightly larger with mixed molarlike and pointed ends, arranged in a small triangular patch; teeth in lower jaw pointed anteriorly, molarlike posteriorly; end of snout rounded; lips plicate; nasal barbel reaching eye or slightly beyond; maxillary barbel a little longer, not reaching posterior end of head; mental barbels reaching gill membrane, the outer longest; gillrakers on anterior face of first arch slender; anterior face of second arch with large papillae-like rakers, broad at base, grading above and below into transverse ridges; posterior face of first arch with two rows of papillae, the anterior row notably larger; transverse ridges on posterior face of second arch and anterior face of third arch longer than arch width, slightly overlapping base of gill filaments; transverse ridges on opposing faces of remaining arches equal to or shorter than arch width; no anal dendritic organ; anterior naris on end of snout, anterior to and slightly lateral to nasal barbel, pointing forward,

Patricia Hogue

PLATE 12. *Tandanus ater*
Red Lily Lagoon, USNM 173553, 165 mm. standard length

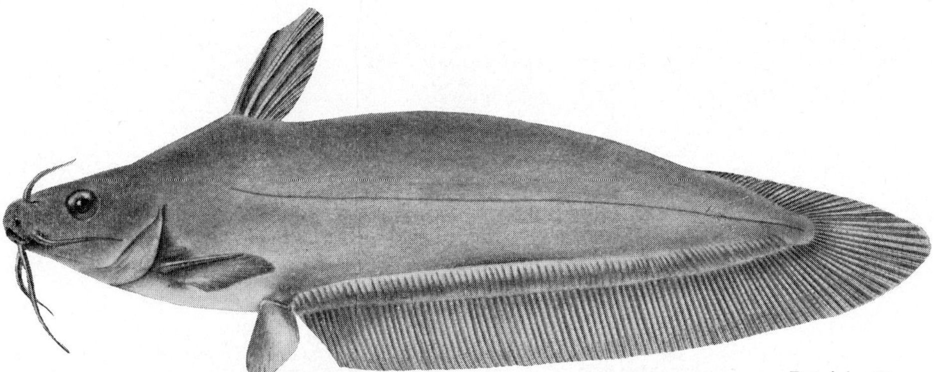

Patricia Hogue

PLATE 13. *Tandanus glencoensis*
Oenpelli, USNM 173561, 158 mm. standard length

Patricia Hogue

PLATE 14. *Tandanus rendahli*
Red Lily Lagoon, USNM 173559, 143 mm. standard length

and opening through a low rim or very short projecting tubule; humeral process acute, about as long as spine width.

Colour in alcohol. Medium brown to greyish, lighter below; lower head and abdomen mostly without pigment.

Tandanus rendahli (Whitley)

Pl. 14

Copidoglanis obscurus Rendahl, *Medd. Zool. Mus. Krist.* 5: 173, fig. 5 (1922); type localities, Glencoe and Hermit Hill, Northern Territory; preoccupied by *Copidoglanis obscurus* Günther, *Catalogue of the Fishes in the British Museum* 5: 26 (1864).

Copidoglanis rendahli Whitley, *Rec. Aust. Mus.* 16: 214 (1928); substitute name for *Copidoglanis obscurus* Rendahl.

Record probably referable: *Tandanus* (*Neosilurus*) *hyrtlii* Paradice and Whitley (1927): 79, Howard River, Darwin.

M48-14, Emerald River, Groote Eylandt, above and below falls, USNM 173554, 203 specimens, 38 to 124 mm. standard length.

Large billabong, about 3 miles above Roper River Mission, Northern Territory, 17 August 1948, Blitner, USNM 173555, 29 specimens, 67 to 127 mm. standard length.

M48-35, Billabong, Oenpelli, USNM 173556, 9 specimens, 69 to 187 mm. standard length.

M48-37, Fish Creek, USNM 173557, 47 specimens, 60 to 98 mm. standard length.

M48-38, Oenpelli Creek, USNM 173558, 1 specimen, 78 mm. standard length.

M48-39, Red Lily Lagoon, USNM 173559, 3 specimens, 135 to 143 mm. standard length.

Native names: *Ángalpía*, Groote Eylandt; *Mároonung* and *Náhgool*, Oenpelli.

Description. Rays in first dorsal I,5(in 15), I,6(14), I,7(1); second dorsal-upper caudal 24(4), 25(5), 26(5), 27(7), 28(2), 29(1), 30(1), 31(1) + anal-lower caudal 79 to 97(27) = 104(1), 106(2), 107(1), 108(4), 109(3), 111(3), 113(1), 114(1), 115(1), 117(3), 118(1), 120(1), 121(1), 122(2), 127(1); interhaemals 69(1), 70(2), 71(1), 72(1), 73(1), 74(3), 75(1), 76(3), 77(1), 78(1), 80(2), 81(1), 82(3), 83(1), 84(2), 85(1), 86(1), 87(1), each equal to or in 20 of 27 cases supporting one additional anal ray at the front of the fin; pectoral I,9(15), I,10(41), I,11(4); pelvic 10(1), 11(9), 12(47), 13(1), the outer ray simple; gillrakers first arch 22(5), 23(4), 24(4), 25(3), 26(2) of which 5 to 7 are on the upper limb; branchiostegal rays 8(2), 9(15), 10(1); precaudal vertebrae 6(27), 7(3) + caudal vertebrae 38(2), 39(8), 40(9), 41(5), 42(4), 43(2) = 44(1), 45(7), 46(11), 47(5), 48(4), 49(2).

In 18 specimens 85 to 123 mm. standard length, dorsal spine 1·5 to 2·0; pectoral spine 1·6 to 1·9; distance between origins of first and second dorsals 0·2 to 0·4; snout 2·7 to 3·5; postorbital length of head (longer than snout) 2·0 to 2·3; eye 4·3 to 5·4, all in head length; origin of second dorsal to tip of caudal fin 3·5 to 5·0; origin of anal fin to tip of caudal fin 1·2 to 1·3; tip of snout to anal fin 2·8 to 3·1; tip of snout to second dorsal 1·1 to 1·3; head length 4·8 to 5·7; body depth (greater than head length) 4·5 to 5·2; tip of snout to first dorsal 3·5 to 3·8, all in standard length.

Lateral line extent to opposite or behind beginning of second dorsal, an eye diameter or more from base of caudal rays; canals of head opening from relatively few pores; temporal pores, between the eyes, few in number, usually one on each side of midline, totalling 2 with seldom as many as 8 in the cluster; axillary pore present; spines slender, smooth in front; dorsal spine roughened behind or with a few short serrae; pectoral spine with straight to slightly recurved posterior serrae, less than half length of spine diameter; branchiostegal membranes rather broadly united; premaxilla with a small rectangular patch of tiny, pointed teeth on each side of midline; teeth on palate much larger, with rounded crowns, arranged in a large semicircular to shallowly crescentric patch, the posterior end of which is truncated to strongly concave; teeth in lower jaw pointed in front, molarlike behind; snout broad, rounded; lips plicate; nasal barbel extending to or beyond posterior end of head; maxillary barbel reaching well behind eye; outer mental barbel reaching almost to, or beyond, base of pectoral fin, the inner mental barbel slightly shorter; gillrakers on anterior face of first arch slender, those of middle of anterior face of second arch short, chubby, broad at base, grading above and below into transverse ridges; posterior face of first arch with two rows of papillae along arch near margins, the anterior row slightly larger; second arch posteriorly with papillae more or less connected transversely across arch into ridges; exposed faces of remaining arches with low, transverse ridges, the length of which is equal to or less than arch width; no anal dendritic organ; anterior naris on end of snout, anterior to and slightly lateral to nasal barbel, without vestige of a projecting tubule; humeral process acute, nearly as long as spine width.

Colour in alcohol. Medium to dark brown above, lighter below; lower head and abdomen without pigment, or pigmented posteriorly and toward sides.

Variation. The specimens from Groote Eylandt differ noticeably from those of the mainland in having fewer gillrakers on the first arch, more vertebrae, a crescentric to shallowly triangular palatal tooth patch which is strongly concave behind versus a rounded semicircular to deeply triangular patch, which is rounded to very slightly concave behind. The specimens from Groote Eylandt are mainly faintly mottled or marbled; those from the mainland show few or no reticulations.

Gillrakers on first arch: mainland 24(3), 25(3), 26(2); Groote Eylandt 22(5), 23(4), 24(1); caudal vertebrae: mainland 39(3), 40(5); Groote Eylandt 41(4), 42(4), 43(2); total vertebrae: mainland 45(1), 46(7); Groote Eylandt 47(4), 48(4), 49(2); the barbels are slightly longer in specimens from Groote Eylandt.

SYNBRANCHIDAE

Synbranchus bengalensis (M'Clelland)

Ophisternon bengalensis M'Clelland, *Calcutta J. Nat. Hist.* 5: 197 and 200, pl. 11, figs. 1 and 2 (1844); type locality, Bengal; reference not seen.

Synbranchus gutturalis Richardson, *Ichthyology of the Voyage of H.M.S. Erebus and Terror*: 49, pl. 30, figs. 14 to 17 (1845); type locality, Dampier Archipelago, Western Australia.

Recorded as: *Symbranchus gutturalis* Macleay (1881): 265, Port Darwin.

M48-35, Billabong, Oenpelli, 16 October 1948, USNM 174018, 1 specimen, 207 mm. total length.

M48-37, Fish Creek, USNM 174019, 14 specimens, 72 to 282 mm. total length.

M48-38, Oenpelli Creek, USNM 174020, 20 specimens, 60 to 234 mm. total length.

Remarks. Further study may separate *Ophisternon* M'Clelland, with *O. bengalensis* as type species (*loc. cit.*), from *Synbranchus* Bloch, partly on the basis of the morphology of the branchial opening. In *Synbranchus marmoratus* Bloch the opening is small and confined to a narrow grooved area of the ventral surface.

The branchial orifice in *S. bengalensis* is broad; the gill membranes are attached laterally just below the side and join medially at an obtuse angle; they are free of the isthmus for a considerable distance anteriorly.

Other characteristics are: no scales; pelvic and pectoral fins absent; median fins a fleshy ridge, without suporting rays; four gill arches with a slit behind each; teeth in jaws pointed, in a single row posteriorly and tiny along side of upper jaw; palatal teeth in a single row along palatines, and in one or more rows on vomer, all somewhat thickened incisors. In preservation, the colour is almost uniformly dark brown.

MURAENESOCIDAE

Muraenesox cinereus (Forskål)

Muraena cinerea Forskål, *Descriptiones Animalium*: x and 22 (1775); type locality, Red Sea.

Muraena arabica Bloch and Schneider, *Systema Ichthyologiae*: 488 (1801); based on Forskål, hence type locality, Red Sea.

Recorded as: *Muroenesox arabicus* Paradice and Whitley (1927): 80, Pellew Islands.

OPHICHTHIDAE

Yirrkala chaselingi* Whitley

Yirrkala chaselingi Whitley, *Aust. Zool.* 9: 410, fig. 17 (1940); type locality, Yirrkalla, near Caledon Bay, Northern Territory.

Ophichthus species

A single juvenile, 141 mm. total length, USNM 174036, apparently referable to the genus *Ophichthus*, was captured at station M48-19, creek, tributary to Little Lagoon, 26 June 1948. The teeth are conical and in two rows along the upper jaw; on the lower jaw and vomer the bands consist of two rows anteriorly, grading to one row behind. The head length is stepped into the pre-anal length about four times; pectoral fins well developed; dorsal and anal fins of about equal height, the dorsal beginning behind the tip of the pectoral fin; snout short and blunt.

Pisodonophis cancrivorus (Richardson)

Ophisurus cancrivorus Richardson, *Ichthyology of the Voyage of H.M.S. Erebus and Terror*: 97, pl. 50, figs. 6 to 9 (1848); type locality, Port Essington.

? *Myrophis chrysogaster* Macleay, *Proc. Linn. Soc. N.S.W.* 6: 271 (1881); type locality, Port Darwin.

Recorded as: *Ophichthys cancrivorus* Macleay (1881): 275, Port Essington; *Pisodonophis cancrivorus* Whitley (1950): 237, fig. 2, Port Darwin and Port Essington. Record probably referable: *Ophicthys* Macleay (1878): 364, Port Darwin.

MORINGUIDAE

Aphthalmichthys intermedius Ogilby

Aphthalmichthys intermedius Ogilby, *Proc. Roy. Soc. Qld.* 20: 9 (1907); type locality, 'North Coast of Australia, probably Port Darwin'.

Record possibly referable: *Moringua* Macleay (1878): 364, Port Darwin.

* *Yirrkala* Whitley (1940b): 410; type species, *Yirrkala chaselingi* Whitley, by original designation.

MURAENIDAE

*Gymnothorax** *fimbriatus* (Bennett)

Muraena fimbriata Bennett, *Proc. Zool. Soc. Lond.*: 168 (1831); type locality, Mauritius.

Recorded as: *Muraena fimbriata* Günther (1870): 108 and Macleay (1881): 283, Port Essington; *Gymnothorax fimbriatus* Ogilby (1907): 11.

Gymnothorax undulatus (Lacépède)

Muraenophis undulata Lacépède, *Histoire naturelle des poissons* 5: 629 and 642, pl. 641, fig. 2 (1803); type locality not given.

Recorded as: *Gymnothorax undulatus* Paradice and Whitley (1927): 80 and 97, Pellew Islands.

M48-3, East Point Reef, USNM 174039, 1 specimen, 350 mm. total length.

Uropterygius concolor Rüppell

Uropterygius concolor Rüppell, *Neue Wirbelthiere zu der Fauna von Abyssinien gehörig . . .*: 83, pl. 20, fig. 4 (1835); type locality, Red Sea.

Recorded as: *Uropterygius concolor* Whitley (1954): 27, fig. 5, Yirrkalla, Northern Territory.

M48-21, Reef, Yirrkalla, 27 and 28 July 1948, USNM 174038, 3 specimens, 408 to 460 mm. total length; August 1948, USNM 174037, 1 specimen, 471 mm. total length.

Evenchelys macrurus (Bleeker)

Muraena macrurus Bleeker, *Natuurk. Tijdschr. Ned.-Ind.* 7: 324 (1854); type locality, Java.

Recorded as: *Evenchelys macrurus* Whitley (1954): 27, Darwin.

BELONIDAE

The Arnhem Land species of the genus *Strongylura* are without gillrakers; anterior dorsal and anal rays elongate, forming lobes; two or three pairs of upper pharyngeals toothed; jaws subequal, normal; body rounded to moderately compressed; branched caudal rays 13 and pelvic rays i,5, on both sides, in two specimens of each species examined.

Key to the Arnhem Land Species

1 Anal rays 25 or more; no keel along side of caudal peduncle; body moderately compressed; head about 3·5 in standard length . . . . . *Strongylura ferox*
Total anal rays 24 or fewer, or if more, side of caudal peduncle with a keel . . 2

2 Opercle scaled; no keel along caudal peduncle 3
Opercle naked, or with one or two vertical rows of scales just behind preopercle; lateral line, forming a strong or a weak keel along caudal peduncle 5

3 Dorsal rays ii,12 or 13; anal rays ii,13 to 15; branched pectoral rays 9 or 10; a round black spot at caudal fin base . . . . *Strongylura strongylura*
Branched pectoral rays 11 or 12; no prominent spot at base of caudal fin . . 4

4 Dorsal rays ii,15 to 17; anal rays ii,17 to 19; origin of pelvic fin about midway between opercle or preopercle and base of caudal fin . . *Strongylura krefftii*

* The type species of *Gymnothorax* Bloch, *G. recticularis* Bloch, was selected by the International Commission on Zoological Nomenclature, Opinion 93.

Dorsal rays ii,17 or 18; anal rays ii,19, usually ii,20 or 21; origin of pelvic fin about midway between eye or rear end of snout and base of caudal fin . . . . *Strongylura incisa*

5 Dorsal rays ii,19 or 20; anal rays ii,18 or 19; branched pectoral rays 12 or 13; keel along caudal peduncle very low, barely evident, yellowish . . . . *Strongylura groeneri*

Dorsal rays ii,20 or 21; anal rays ii,18 or 19; branched pectoral rays 14; keel along caudal peduncle prominent, blackish . . . . . . *Strongylura gigantea*

Strongylura strongylura (van Hasselt)

Belone strongylura van Hasselt, *Algemeene Konst-en Letter-bode* 2, No. 35: 130 (1823); type locality, Batavia, Java; reference not seen.

Strongylura caudimaculata van Hasselt, *Bull. Sci. Nat. Ferussac* 2: 374 (1824); type locality, Vizagapatam; based on Russell, *Description and Figures of Two Hundred Fishes collected at . . . Coromandel*: pl. 176 (1803).

Recorded as: *Tylorsurus strongylurus* Paradice and Whitley (1927): 80, Pellew Islands; *Belone caudimaculata* Günther (1866): 245, Port Essington; Macleay (1878): 363 and (1881): 241, Port Darwin; *Tylosurus caudimacula* Paradice and Whitley (1927): 80 and 97, Pellew Islands.

M48-2, Nightcliff, USNM 173995, 6 specimens, 176 to 305 mm. standard length.

Description. Dorsal rays ii,12(in 5) or ii,13(1); anal rays ii,13(1), ii,14(4), or ii,15(1); pectoral rays ii,9(5) or ii,10(7).

No keel along lateral line; opercle covered with small scales; nostril nearly oval, its projection a broad flap; scales along length of interorbital groove; a distinctive, dark brown or black, oval spot, which nearly as large as pupil, at mid-base of caudal fin.

Strongylura kreffti (Günther)

Belone kreffti Günther, *Catalogue of the Fishes in the British Museum* 6: 250 (1866); type locality, Australia.

Billabong, about 3 miles above Roper River Mission, Northern Territory, 17 August 1948, Blitner, USNM 173996, 5 specimens, 188 to 267 mm. standard length.

M48-35, Billabong, Oenpelli, USNM 173997-9, 27 specimens, 36 to 635 mm. standard length.

M48-39, Red Lily Lagoon, USNM 174000, 10 specimens, 271 to 319 mm. standard length.

Native name: *Goorgabál,* Oenpelli.

Description. In the specimens from the Roper River area and a sample from Oenpelli billabong: dorsal rays ii,15(in 2), ii,16(8), or ii,17(2); anal rays, ii,17(3), ii,18(7), or ii,19(2); pectoral rays ii,11(22) or ii,12(2).

Pelvic fin inserted about midway between vertical arm of preopercle, or opercle, and base of caudal fin; opercle covered with small scales; scales along length of interorbital groove; lateral line pale, without elevation; nostril sub-oval, its projection slender, not greatly expanded distally.

Strongylura incisa (Cuvier and Valenciennes)

Belone incisa Cuvier and Valenciennes, *Histoire naturelle des poissons* 18: 451 (1846); type locality, Indian Ocean.

M48-15, Emerald River, Groote Eylandt, USNM 174005, 8 specimens, 302 to 366 mm. standard length.

M48-26, Sheltered bay, island in Port Bradshaw, USNM 174006, 1 specimen, 244 mm. standard length.

Description. Dorsal rays ii,17(in 6) or ii,18(2); anal rays ii,19(1), ii,20(4), or ii,21(3); pectoral rays ii,11, both sides (8).

Lateral line pale, not forming a keel; opercle covered with small scales; nostril triangular, its projection broad, flap-like; scales along length of interorbital groove; a distinctive dark brown longitudinal dash at upper pore of preopercle.

Strongylura gigantea (Temminck and Schlegel)

Belone gigantea Temminck and Schlegel, *Fauna Japonica, Pisces*: 245 (1846); type locality, Japan.

Recorded as: *Tylosurus giganteus* Paradice and Whitley (1927): 80, Pellew Islands. Records probably referable: *Belone melanotus* Macleay (1878): 363 and (1881): 239; Kent (1889): 240, Port Darwin.

M48-15, Emerald River, Groote Eylandt, USNM 174007, 4 specimens, 417 to 668 mm. standard length.

Description. Dorsal rays ii,20(in 1) or ii,21(1); anal rays ii,18(1) or ii,19(1); pectoral rays ii,14(4).

Lateral line elevated on rear end of caudal peduncle and base of caudal fin, forming a sharp, blackish keel; distal half of rear dorsal rays, middle caudal rays, and most of upper part of pectoral fin dark brown or black; pelvic fin inserted midway between rear end of snout or eye and base of caudal fin; scales not continuous along shallow interorbital groove, with an isolated patch before eye; opercle mostly naked, with a few rows of scales along its anterior margin; nostril subtriangular, its flap broad and plicate.

Strongylura groeneri (Klunzinger)

Belone groeneri Klunzinger, *S.B. Akad. Wiss. Wien* 80: 414 (1879); type locality, Port Darwin; Macleay, *Proc. Linn. Soc. N.S.W.* 9: 57 (1884).

Djulongius groeneri Whitley, *Aust. Zool.* 9: 415, fig. 26 (1940); holotype redescribed.

M48-8, Little Lagoon, USNM 174001, 2 specimens, 134 and 136 mm. standard length.

M48-15, Emerald River, Groote Eylandt, USNM 174002, 3 specimens, 313 to 677 mm. standard length.

M48-28, Creek mouth and bay, Port Bradshaw, USNM 174003, 1 specimen, 260 mm. standard length.

M48-31, Creek, Yirrkalla, USNM 174004, 2 specimens, 376 and 411 mm. standard length.

Description. Dorsal rays ii,19(in 2) or ii,20(3); anal rays ii,18(4) or ii,19(1); pectoral rays ii,12(6) or ii,13(4); lateral line scarcely to slightly elevated on caudal peduncle, pale.

Colour in life. In the two smallest specimens: caudal peduncle bronze, the colour extending on to upper caudal lobe; fins light, except last half of dorsal, which blackened; about 12 brown oval spots on mid-side.

Remarks. This species is similar to *Strongylura gigantea* (Temminck and Schlegel) in morphology and colour pattern, differing chiefly in number of branched pectoral rays and in the degree of elevation of the lateral line. The name here assigned is regarded as tentative, and probably will be replaced by older names given to species from the Indo-Australian area. The two young have about 12 brown bars, nearly as wide as the eye, along the trunk.

Strongylura ferox (Günther)

Belone ferox Günther, *Catalogue of the Fishes in the British Museum* 6: 242 (1866); type locality, Sydney, Australia.

Recorded as: *Tylosurus ferox* Paradice and Whitley (1927): 80, Pellew Islands.

HEMIRAMPHIDAE

Key to the Arnhem Land Species

1 Caudal fin moderately to deeply forked; fin rays of males not modified; dorsal origin above or before anal fin origin; naris with flap, which spatulate or leaf-shaped, extending little beyond nasal pit; pectoral fin shorter than head length; teeth confined to mouth area; triangular area of upper jaw broader than long 2

Caudal fin rounded or truncate behind; naris with a vermiform projection which extends well beyond nasal pit; anal rays and sometimes dorsal rays in males specialized, some of them may be greatly elongate; teeth confined to mouth; lower jaw elongate, about as long as head length; anal fin origin below or behind dorsal fin origin—genus *Zenarchopterus* 5

2 Triangular area of upper jaw without scales 3

Triangular part of upper jaw with small scales 4

3 Lower jaw terminal, but not greatly produced, about equal to or less than snout length in adults, longer in young; pelvic fin inserted midway between caudal fin base and tip of snout or beyond; anal rays ii,12 or ii,13 . . . . . *Arrhamphus sclerolepis*

Lower jaw produced, longer than head length; pelvic fin inserted about midway between base of caudal fin and middle of depressed pectoral fin; anal rays about ii,9 or ii,10 . . . . *Hemiramphus far*

4 Pelvic fin inserted midway between caudal fin base and front border of eye or preorbital; branched pectoral rays 10, seldom 11 . . *Hyporhamphus gaimardi*

Pelvic fin inserted midway between caudal fin base and opercle or rear part of preoperculum; branched pectoral rays 11, seldom 12 . . . . . . *Hyporhamphus quoyi*

Pelvic fin inserted about midway between base of caudal fin and base of pectoral fin; branched pectoral rays 10 or 11 . . . . *Hyporhamphus dussumieri*

5 Triangular area of upper jaw nearly twice as long as broad; snout, in adults, 2·1 to 2·5 in head length; 2 to 4 moderate size scales across upper half of opercle; interorbital pores nearest midline opening into a median pit . . 6

Triangular area of upper jaw nearly as wide as long; snout in adults 2·8 to 3·2 in head length; 1 or 2 scales across upper half of opercle one of which large and covering most of area 7

6 Pectoral fin longer than extension of lower jaw beyond tip of snout, 3·7 to 3·9 in standard length; distance from pelvic fin origin to caudal fin base 3·0 to 3·2 in standard length; male with ii,12 anal rays . . . . . . *Zenarchopterus caudovittatus*

Pectoral fin shorter than prerostral length of lower jaw, 6·5 to 7·2 in standard length; distance from pelvic fin base to caudal fin 3·8 to 3·9 in standard length; male with ii,11 (2 specimens) anal rays and branched pectoral rays 8(3) or 9(1) . . . . *Zenarchopterus ectuntio*

7 Triangular area of upper jaw slightly longer than broad; anal rays ii,6 or ii,7; branched pectoral rays 8 or 9; a tiny blackish dot on base of caudal rays at end of lateral silvery band . . . . . . *Zenarchopterus rasori*

Triangular area of upper jaw not longer than broad; anal rays ii,9 to ii, 12 8

8 Branched pectoral rays 9; top of snout unicolour, brownish; in males, fourth dorsal ray thickened and elongate, and sixth anal ray greatly developed, often reaching tip of caudal fin . . . . *Zenarchopterus gilli*

Branched pectoral rays 10; a dark brown line along midline of snout; dorsal and anal rays of males modified, but not reaching behind base of caudal fin . . . . *Zenarchopterus buffonis*

Arrhamphus sclerolepis Günther

Arrhamphus sclerolepis Günther, *Catalogue of the Fishes in the British Museum* 6: 277 (1866); type locality unknown, believed to be New Zealand.

Recorded as: *Arrhamphus sclerolepis* Klunzinger (1879): 415; Macleay (1881): 248; Kent (1889): 240, Port Darwin; Paradice and Whitley (1927): 80, Pellew Islands and Darwin; *Arrhampus sclerolepis* Macleay (1878): 364, Port Darwin.

M48-2, Nightcliff, USNM 173771, 26 specimens, 34 to 98 mm. standard length.

M48-8, Little Lagoon, USNM 173772, 1 specimen, 85 mm. standard length.

M48-10, Estuary, Thompson Bay, Port Langdon, USNM 173773, 5 specimens, 66 to 106 mm. standard length.

M48-13, South Bay, Bickerton Island, USNM 173774, 1 specimen, 103 mm. standard length.

M48-24, Beach, Port Bradshaw, 26 July 1948, USNM 173775, 11 specimens, 70 to 83 mm. standard length.

M48-28, Creek mouth and bay, Port Bradshaw, USNM 173776, 34 specimens, 85 to 150 mm. standard length.

Native name: *Ályanárra*, Groote Eylandt.

Description. Dorsal rays ii,12(in 3) or ii,13(2); anal rays ii,12(1) or ii,13(4); pelvic rays i,5(10); pectoral rays i,12(10); branched caudal rays 6+7=13(5); pelvic fin inserted slightly in advance of midway between caudal fin base and snout or its tip; lower jaw short, slender in young and about equal to head length in individuals 40 mm. in standard length, but reduced and blunt in adults where about as long as snout length, projecting less than an eye diameter beyond snout tip; pectoral fin shorter than head length; no scales on upper end of snout; in five individuals 102 to 145 mm. standard length, head length 3·9 to 4·2 in standard length.

Colour in alcohol. Caudal fin yellowish distally and greyish basally; other fins yellowish, the dorsal, caudal and anal slightly margined in black; general body colour yellowish; a blackish band from upper base pectoral fin to caudal base; above this band three longitudinal lines, plus one at midline of back, of dusky dashes or tiny spots; a narrow, solid blackish line along each side of midline of back.

Colour in life. Tip of lower jaw orange to reddish orange; lateral band bluish green or emerald green with silvery blue reflections; dorsal, caudal, and anal fins edged with black.

Remarks. Oxyporhamphus brevis (Seale, 1909) is another species of *Arrhamphus,* differing in having ii,13 to ii,15, usually ii,14 dorsal rays; prominent black tips to the upper and lower lobes of the caudal fin and to the anterior lobes of the dorsal and anal fins; the lower jaw is slightly shorter at comparable sizes; and the distance from the pelvic fin origin to the caudal fin when projected forward does not extend beyond the eye. In *A. sclerolepis* this measurement extends beyond the tip of the lower jaw in adults, and to or beyond the upper jaw in young.

Hemiramphus far (Forskål)

Esox far Forskål, *Descriptiones Animalium*: 67 (1775); type locality, presumably Red Sea.

Recorded as: *Hemirhamphus far* Paradice and Whitley (1927): 80, Pellew Islands and Darwin; *Hemirhamphus welsbyi* Paradice and Whitley (1927): 80, pl. 11, fig. 3, Pellew Islands.

M48-10, Estuary, Thompson Bay, Port Langdon, USNM 173777, 1 specimen, about 205 mm. standard length.

Native name: *Ákuréra*, Groote Eylandt.

Description. Dorsal rays ii,11; anal rays ii,9; pelvic rays all branched, 6 both sides; pectoral rays i,11 both sides; branched caudal rays 6+7=13. Pectoral fin shorter than head; pelvic fin inserted about midway between caudal fin and middle of depressed pectoral fin; head approximately 4·1 in standard length; upper end of snout naked.

Colour in alcohol. Dorsal and caudal fins greyish, the anterior dorsal rays blackish; anal, pelvic, and pectoral fins yellowish; a broad, blackish band from upper base of pectoral fin to caudal base; above this scales outlined in brown, which nearly covers scales on dorsum; three faint longitudinal dark lines along middle of back; lower side yellowish; about four dark brown blotches along upper side, above lateral band, those anteriorly irregular and broken but that below anterior half of dorsal fin round and nearly uniformly dark.

Remarks. Paradice and Whitley (loc. cit.) have figured a specimen that they identified as *Hemirhamphus welsbyi* from the Pellew Islands. This illustration agrees with specimens of *H. far* in body proportions, in scale size and fin rays, and in colour, except that the anterior spots are not present. It would appear that the anterior blotches could readily disappear, because they are extensively broken in the present specimen, although they are reported to remain in large individuals. Their illustration does not agree with Ogilby's original description of *Hemiramphus welsbyi* in colour pattern and fin ray counts.

Hyporhamphus gaimardi (Cuvier and Valenciennes)

Hemiramphus gaimardi Cuvier and Valenciennes, *Histoire naturelle des poissons* 19: 36 (1846); type localities, Ambon and New Guinea.

M48-2, Nightcliff, USNM 173790, 11 specimens, 38 to 64 mm. standard length.

M48-13a, South end of Winchelsea Island, 6 June 1948, USNM 173791, 3 specimens, 48 to 116 mm. standard length.

M48-24, Beach, Port Bradshaw, 26 July 1948, USNM 173792, 12 specimens, 45 to 118 mm. standard length.

M48-25, Beach, island, Port Bradshaw, USNM 173793, 2 specimens, 116 to 117 mm. standard length.

Description. Dorsal rays ii,12(in 1) or ii,13(4); anal rays ii,14(2) or ii,15(3); pelvic rays i,4(1) or i,5(9); pectoral rays i,10(9) or i,11(1); branched caudal rays 6+7=13(5). Triangular portion of snout with small scales; pectoral fin shorter than head length; pelvic fin inserted about midway between caudal fin and front border of eye or middle of preorbital; in five specimens, 109 to 117 mm. in standard length, head length 4·3 to 4·7 in standard length.

Colour in alcohol. Caudal fin dusky, faintly bordered by black; other fins yellowish, the end of the anterior dorsal rays black; body yellowish, with a black band from upper pectoral base to caudal fin; above this two anteriorly or three posteriorly lines of brown dashes or spots on upper side; a line of similar dashes along midline of back, outside of which on each side a continuous fine, brown line.

Hyporhamphus quoyi (Cuvier and Valenciennes)

Hemiramphus quoyi Cuvier and Valenciennes, *Histoire naturelle des poissons* 19: 35 (1846); type locality, Port Dorey, New Guinea.

Recorded as: *Hemiramphus quoyi* Whitley (1928a): 214, Port Darwin; *Hemirhamphus quoyi* Kent (1889): 240, Port Darwin; Paradice and Whitley (1927): 80, pl. 11, fig. 2, Pellew Islands; *Belone quoyi* Klunzinger (1879): 415, Port Darwin.

M48-2, Nightcliff, USNM 173779, 9 specimens, 101 to 184 mm. standard length.

M48-10, Estuary, Thompson Bay, Port Langdon, USNM 173780, 3 specimens, 93 to 125 mm. standard length.

M48-13, South Bay, Bickerton Island, USNM 173781, 2 specimens, 71 to 76 mm. standard length.

M48-13a, South end of Winchelsea Island, 6 June 1948, USNM 173782, 2 specimens, 93 to 110 mm. standard length.

M48-15, Emerald River, Groote Eylandt, USNM 173783, 4 specimens, 110 to 149 mm. standard length.

M48-17, Reef, Bartalumba Bay, USNM 173784, 2 specimens, 110 to 114 mm. standard length.

M48-19, Little Lagoon at creek mouth, USNM 173785, 3 specimens, 88 to 117 mm. standard length.

M48-21, Reef, Yirrkalla, USNM 173786, 29 July 1948, 1 specimen, 135 mm. standard length; 11 August 1948, 1 specimen, 145 mm. standard length.

M48-24, Beach, Port Bradshaw, USNM 173787, 32 specimens, 77 to 158 mm. standard length.

M48-25, Beach, island, Port Bradshaw, USNM 173788, 20 specimens, 81 to 120 mm. standard length.

M48-28, Creek mouth and bay, Port Bradshaw, USNM 173789, 4 specimens, 94 to 125 mm. standard length.

Native names: *Ályandárra,* Groote Eylandt; *Mínga,* Yirrkalla.

Description. Dorsal rays ii,13(1), ii,14(3), or ii,15(1); anal rays ii,13(5); pelvic rays i,4(1), or i,5(9); pectoral rays i,11(9) or i,12(1); branched caudal rays 6+7=13(5); pectoral fin shorter than head; pelvic fin inserted about midway between middle of opercle or rear part of preopercle and base of caudal fin; in five specimens, head length 4·2 to 4·6 in standard lengths of 101 to 184 mm.; small scales on triangular part of upper jaw.

Colour in alcohol. Caudal fin light grey, with two faint patches of dusky pigment, one at base of upper and one at base of lower lobe; its edge margined with black; dorsal fin yellow, the anterior edge and distal portions of longest rays sooty; anal, pelvic, and pectoral fins yellow, the pelvic and pectorals with traces of brown at their bases; body generally yellowish; a black band from upper base of pectoral fin to caudal fin; scales above band rather heavily pigmented but forming lines only on dorsum and extreme upper side; midline of back with three fine, solid lines.

Colour in life. Flap at tip and ventral surface of lower jaw blood red, becoming paler and suffusing with black posteriorly; trace of black on dorsal and anal fins; caudal fin with a very narrow black margin; lateral band emerald green.

Hyporhamphus dussumieri (Cuvier and Valenciennes)

Hemiramphus dussumieri Cuvier and Valenciennes, *Histoire naturelle des poissons* 19: 33, pl. 554 (1846); type locality, Seychelles.

M48-21, Reef, Yirrkalla, 29 July 1948, USNM 173778, 2 specimens, 158 to 166 mm. standard length.

Description. Dorsal rays, ii,13 or 14; anal rays ii,13 or 14; pelvic rays i,5, both sides, in both; pectoral rays i,10(in 3) or i,11(2); branched caudal rays 5 or 6+7=12 or 13; pectoral fin shorter than head length; pelvic fin about midway between bases of caudal fin and pectoral fin; head length 4·5 or 4·6 in standard length; small scales on triangular part of upper jaw.

Colour in alcohol. Caudal fin greyish, its tip blackish; at its base a black blotch which about two-thirds as large as eye, more or less divided at middle of fin, and extending as two pointed projections backward on fin; dorsal fin yellow with its anterior margin and ends of long rays black; anal, pelvic, and pectoral fins yellow, the pelvic fin with a cluster, and the pectoral fin with a bar, of black pigment at the base; body yellowish brown; a line of black pigment along each side of lower surface of caudal peduncle behind anal fin; lateral line accentuated by a line of narrow black dashes; black band from upper pectoral fin base to caudal fin base; scales of upper side and back more or less heavily pigmented with dark brown, which tends to form two or three lines; three prominent, narrow, continuous black lines on back, that of midline broadest.

Zenarchopterus ectuntio (Hamilton-Buchanan)

Esox ectuntio Hamilton-Buchanan, *An Account of the Fishes Found in the River Ganges*. . .: 212, 380 (1822); type locality, Gangetic provinces.

Hemiramphus amblyurus Bleeker, *Verh. Batavia Genoot.* 22: 11 (1849); type locality, Madura Island.

Recorded as: *Hemirhampus amblyurus* Macleay (1878): 364, Port Darwin; *Hemirhamphus amblyurus* Macleay (1881): 248, Port Darwin.

Zenarchopterus caudovittatus (Weber)
Pl. 15

Hemiramphus (*Zenarchopterus*) *caudovittatus* Weber, *Nova Guinea* 5: 229, pl. 13, figs. 1a and 1b (1908); type locality, mouth of Merauke River.

M48-40, East Alligator River, USNM 173769, 5 specimens, 44 to 112 mm. standard length.

Description. Total dorsal rays 12(in 1) or 13(4), variously branched, sometimes all simple; anal rays ii,11(1) or ii,12(4); pelvic rays i,5(8); pectoral rays i,9(5) or i,10(3); branched caudal rays 6+7=13(4); pored scales in lateral line, 3 in section below pectoral fin, plus 45(2) or 47(1) from that branch to base of caudal fin; scale rows between head and caudal fin 47(4); two to four moderate scales across upper half of opercle; median-most interorbital pores opening into a groove or pit at midline; on caudal fin, lateral line divided, extending backward between branched rays 7 and 8 and between 10 and 11.

In the four largest specimens, which 92 to 112 mm. standard length, pectoral fin 3·7 to 3·9; head length 3·8 to 4·0; head length plus lower jaw 2·1 to 2·6; distance from pelvic fin origin to caudal fin 3·0 to 3·2, all in standard length; eye diameter 5·8 to 6·3 and snout length 2·3 to 2·5, both in head length.

In the male, dorsal rays 2 to 5 with expanded tips, the rays not greatly elongated; in the male anal fin, first 2 rays simple, the second about as long as rays 3 to 5,

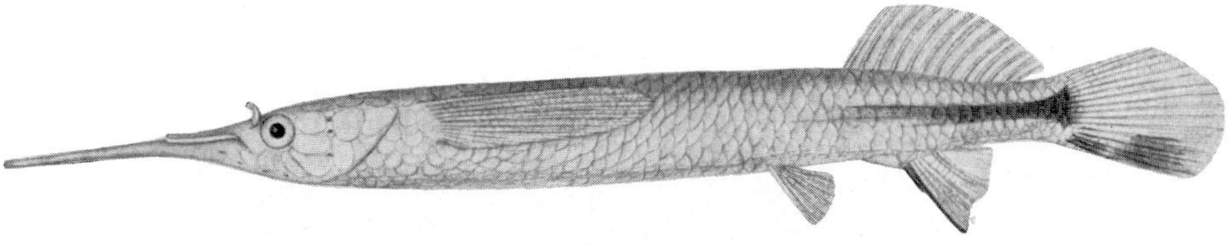

Dorothea B. Schultz

Plate 15. *Zenarchopterus caudovittatus*
East Alligator River, USNM 173769, 99 mm. standard length

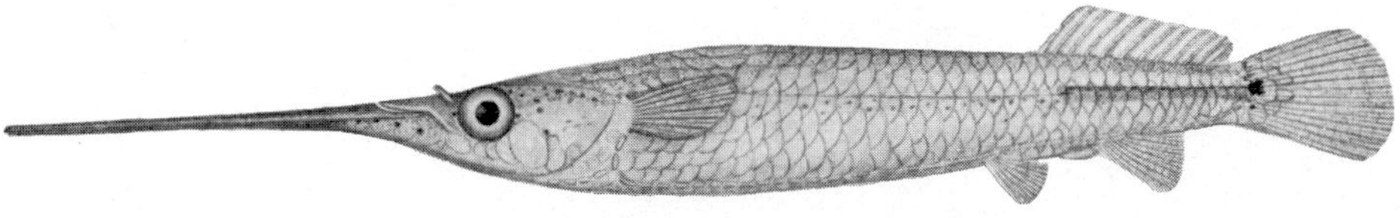

Dorothea B. Schultz

Plate 16. *Zenarchopterus rasori*
Woods Inlet, Delissaville, USNM 173770, 54 mm. standard length

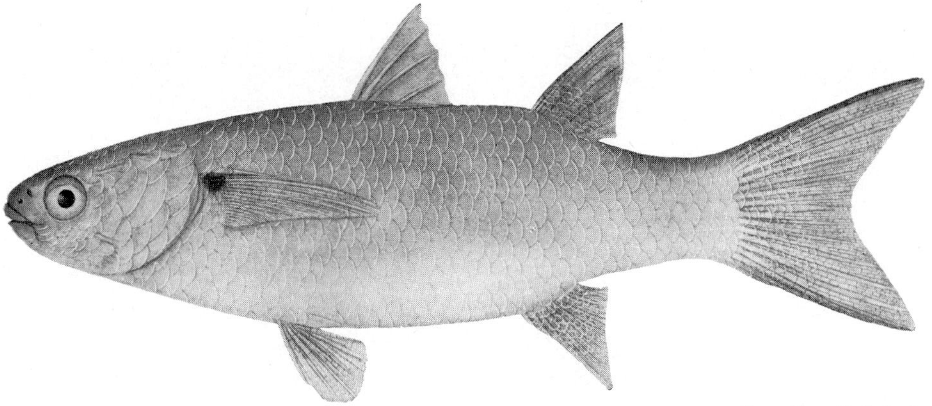

Dorothea B. Schultz

Plate 17. *Chelon strongylocephalus*
Port Bradshaw, USNM 173899, 190 mm. standard length

all of which branched and of nearly normal shape, the fifth shortest; ray 6 of same length with outer half branched; its base broad, about 4 times as wide as adjoining rays, but narrower distally, the anterior branch widest, with 3 to 5 lengthwise grooves or striae along its distal third; at junction of the two branches 4 to 12 tubercles; ray 7 branched only at tip and about as broad as base of ray 5, and as long as ray 6; ray 8 simple, half diameter of ray 7 and slightly shorter; ray 9 shorter and narrower; rays 10 through 13 thin splints, close together and becoming progressively shorter toward rear; last ray branched, of equal diameter to but half the length of ray 7.

Colour in alcohol. Body, head, and fins brown, the fins lighter; the two anterior (simple) rays of anal fin black; lower lobe of caudal fin dusky or with black blotch extending well out on to it; pale line along side of body from pectoral fin to beneath dorsal fin origin where narrow black band begins and extends to large blackish blotch at base of caudal fin.

Zenarchopterus rasori (Popta)

Pl. 16

Hemiramphus rasori Popta, *Notes Leyden Mus.* 34: 192 (1912); type locality, Muna Island.

M48-5, Woods Inlet, Delissaville, 26 to 28 March 1948, USNM 173770, 2 specimens, 54 to 60 mm. standard length.

Description. Total dorsal rays 11 or 12, of which 2 or 3 at rear distinctly branched; anal rays ii,6 or 7; pelvic rays i,5(in 4); pectoral rays i,8(3) or i,9(1); branched caudal rays 5 or 6+7=12 or 13; pored scales in lateral line, 2 or 3 in branch below pectoral fin, plus 38 or 41 from that branch to base of caudal fin; scale rows between head and caudal fin 41 or 42; one large scale across upper part of opercle; median-most interorbital pores opening into shallow valley that extends backward to between eyes; fin rays of the two specimens not modified.

In the two specimens, head length 3·7 and 4·0; pectoral fin 7·1 and 7·5; head length plus lower jaw 1·5 and 1·6; and distance from pelvic fin orgin to base of caudal fin 3·3 and 3·5, all in standard length; eye diameter 4·1 and 4·4 and snout length 2·8 and 2·9, both in head length.

Colour in alcohol. Body, head, and fins yellowish, the fins unicolour; indistinct dark line on middle of abdomen extends backward around anal fin base, thence more prominently to caudal fin; band of scattered dots backward from pectoral fin base to a more solid dark brown or blackish band, below dorsal origin, which continues to caudal fin base; a tiny distinct black spot at end of band on base of caudal rays; scales of back with light brown pigment; three narrow brown lines along midline of back.

Zenarchopterus buffonis (Cuvier and Valenciennes)

Hemiramphus buffonis Cuvier and Valenciennes, *Histoire naturelle des poissons* 19: 48 (1846); type locality, Pulo-Pinan.

M48-2, Nightcliff, 19 March 1948, USNM 173766, 5 specimens, 21 to 93 mm. standard length.

M48-5, Woods Inlet, Delissaville, 26 to 28 March 1948, USNM 173767, 4 specimens, 73 to 111 mm. standard length.

M48-28, Creek mouth and bay, Port Bradshaw, USNM 173768, 30 specimens, 46 to 75 mm. standard length.

Description. Dorsal rays 12 (in 5 adults and 4 young), some or all branched; anal rays ii,10 (2 females) or ii,11 (3 males); pelvic rays i,5(10); pectoral rays i,10(10); branched caudal rays 6+7=13(5); pored scales in lateral line, 3 or 4 in branch below pectoral fin plus 40 to 44 between that point and caudal fin base; scales between head and caudal fin 42(1) or 44(4); interorbital pores opening at head surface, not into a pit or groove; one, sometimes two, large scales across upper part of opercle; in 4 specimens 73 to 111 mm. standard length, snout length 2·9 to 3·2 in head length.

In the male, dorsal rays 4 to 7 elongated and broadened distally, the fifth longest, about twice length of anterior rays; its anal fin with first 5 rays nearly normal, of moderate length forming a broadly rounded lobe; sixth anal ray modified, twice length of fifth, its base broad, and distal end greatly expanded; the end with numerous longitudinal and cross striations, the lengthwise ones running down nearly to a posterior hump at midsection; no tubercles; no longitudinal striae on basal third of ray; on distal, anterior section of ray, striae diagonal; ray 7 branched to middle of its length; its diameter about equal to that of base of 5 and length three-fourths that of 6; ray 8 simple and narrower, nearly as long; rays 9 to 12 simple, very short, the rear ones most slender, crowded due to proximity of distal end of longe simple last ray and ray 8.

Colour in alcohol. Body, head, and fins generally yellowish or yellowish brown; dorsal fin sometimes medium brown or black; anterior anal fin lobe dark brown; a black spot on middle of anal ray 6 in two males; black band, narrow in front, from upper pectoral base to caudal fin; about 3 fine brown lines along back; a characteristic, narrow brown band along midline of snout.

In young, 21 to 22 mm. standard length, body dusky, with about 8 or 9 brown bars or blotches, which nearly as wide as eye, across body; the bars grade into the lateral dusky band in larger individuals.

Colour in life. Tip of lower jaw bright yellow to yellowish green.

Remarks. This fish was noted to move into Woods Inlet creek with the rising tide and out when the creek began flowing with the falling tide.

Zenarchopterus gilli Smith

Hemiramphus brevirostris Günther, *Catalogue of the Fishes in the British Museum* 6: 274 (1866); type locality, East Indian Archipelago; name preoccupied in *Hemiramphus brevirostris* Cuvier.

Zenarchopterus brevirostris Mohr, *Zool. Jb. Abth. Syst.* 52: 250, fig. 15 (1926).

Zenarchopterus gilli Smith, *U.S. Nat. Mus. Bull.* 188: 432 (1945); new name to replace *H. brevirostris* Günther, preoccupied.

M48-3, East Point Reef, USNM 173880, 1 specimen, 26 mm. standard length.

M48-15, Emerald River, Groote Eylandt, USNM 173763, 99 specimens, 26 to 66 mm. standard length.

M48-27, Bay and beach, Port Bradshaw, USNM 173764, 15 specimens, 40 to 72 mm. standard length.

M48-28, Creek mouth and bay, Port Bradshaw, USNM 173765, 6 specimens, 39 to 68 mm. standard length.

Description. Dorsal rays 10 to 12, most or all of them branched, and anal rays ii,8 to ii,12 (see below under *Remarks*); pelvic rays, i,5(in 8); pectoral rays i,9(40); branched caudal rays 5 or 6+6 or 7=11(1), 12(12), or 13(4); in six specimens 61 to 72 mm. standard length, snout length 2·9 to 3·1 in head length; one large scale covering most of upper opercle; median-most interorbital pores opening at surface of head, not into a depression; scale rows between head and caudal fin 39(1); pored scales in lateral line 3 in branch below pectoral fin and 40 in section from there to caudal fin.

In the male, dorsal ray 4 elongated, thickened distally, curving backward as far as base of caudal rays; other dorsal rays normal, the 7th usually next longest. In the male anal fin the first 2 rays simple, the first 5 rays not elongated, normal, but forming a pointed lobe; rays 3 to 5 branched; anal ray 6 greatly elongated, folding against caudal fin, and in large males extending behind tip of longest caudal ray; its base relatively narrow, its distal end blunt, its width half length of anal base or more; its anterior branch prominent, large, broadest basally and narrow distally, with cross striations over most of its length and with a narrow unmarked flap on anterior distal third; from posterior edge of anterior branch numerous, fine branches, the branching extending from its tip to the basal one-sixth; the branches are with minute cross striae, and closely crowded together so that they appear as longitudinal striae; anal rays 7 to 9 branched and nearly normal, the seventh somewhat thickened along its length and about as long as ray 2; remaining rays simple, progressively narrower and shorter, except 13 which has a thickness slightly greater than ray 10; rays 11 and 12 fine splints; no tubercles on fin.

Colour in alcohol. Body and head brown; fins yellowish, the dorsal, anal, and caudal suffused with brown; blackish lateral band faint or absent anteriorly, evident about mid-body and extending to caudal base; lower surface of body, especially abdomen, yellowish; about one narrow, continuous line on back, extending about half distance from head to dorsal fin; top of snout nearly unicolour, brownish.

Remarks. Collected near mangroves in quiet water, the young resembling floating twigs.

I find that the fourth dorsal ray, in agreement with Günther (loc. cit.), rather than the fifth as described by Mohr (loc. cit) and others, is elongate in all twenty adult males in the present collection.

Sexual dimorphism is indicated in fin ray counts in both this species and in *Z. buffonis*. In *buffonis*, the dorsal fin count seems to be constant in both males and females, but the male anal fin appears to average more rays. In *gilli*, the male averages more rays in both the dorsal and anal fins than does the female; variation in other counts was not noted in either species.

The following are my *gilli* data, with total number of dorsal rays followed by total number of anal rays, thence frequence of occurrence in parentheses. In females: 10 and 11(7), 10 and 12(3), or 11 and 12(1). In males: 10 and 13(1) or 11 and 13(9). In unsexed individuals: 10 and 10(1), 10 and 11(9), 11 and 11(5), 11 and 12(4), 10 and 13(1), 11 and 13(20), 12 and 13(1), and 11 and 14(1). In the unsexed individuals, note the modes at 10 and 11, and 11 and 13, suggesting that all or most individuals of the first group are females and those of the last group males, in agreement with the modes for the sexed individuals. It would thus appear that the male normally has 11 or more, seldom 10, while the female has a modal number of 10 dorsal rays, and that the male typically has 13 while the female has a modal number of 11 anal rays.

EXOCOETIDAE

Exocoetus volitans Linnaeus

Exocoetus volitans Linnaeus, *Systema Naturae* ed. 10, 1: 316 (1758); type locality, Atlantic Ocean.

Recorded as: *Exocoetus volitans* Munro (1957): 57, Northern Territory.

Parexocoetus brachypterus (Richardson)

Exocoetus brachypterus Richardson, *Rep. Fifteenth Meet. Brit. Ass. Adv. Sci.*: 265 (1846); based on Solander.

Recorded as: *Parexocoetus brachypterus* Paradice and Whitley (1927): 81, between Cape Wessel and Cape York; Munro (1957): 57.

Cypselurus species

Arafura Sea, about 100 miles east of Goulburn Island and 40 miles north of mouth of Liverpool River, Northern Territory, collected at night by Nick Haritos, 28 July 1948, USNM 174231, 2 specimens, 140 to 142 mm. standard length.

Remarks. Initial distribution records of three species referred to the genus *Cypselurus* were given by Munro (1957): 58, as follows:

Cypselurus heterurus (Rafinesque), Timor Sea, Northern Territory.

Cypselurus bahiensis (Ranzani), off Cape Croker, Northern Territory.

Cypselurus poecilopterus (Valenciennes), Arafura Sea, west of Wessel Island, Northern Territory.

BREGMACEROTIDAE

Bregmaceros nectabanus Whitley

Bregmaceros nectabanus Whitley, *Aust. Zool.* 10: 25, fig. 18 (1941); type locality, Darwin.

Recorded as: *Bregmaceros mcclellandi* McCulloch (1926): 29; Paradice and Whitley (1927): 81 and 97, Darwin; *Bregmaceros nectabanus* Munro (1950): 41, Northern Territory records: Darwin; Marchinbar Island, Wessel Group; and Timor Sea, 30 miles west-north-west of Charles Point.

CENTRISCIDAE

Centriscus scutatus Linnaeus

Centriscus scutatus Linnaeus, *Systema Naturae* ed. 10, 1: 336 (1758); type locality, India.

Recorded as: *Centriscus scutatus* Whitley and Allan (1958): 77, Arafura Sea and Northern Territory.

SYNGNATHIDAE

The method of counting rings employed here corresponds to that described by Herald (1941).

Choeroichthys brachysoma (Bleeker)

Syngnathus brachysoma Bleeker, *Natuurk. Tijdschr. Ned.-Ind.* 8: 327 (1855); type locality, Batu Archipelago.

Recorded as: *Choeroichthys serialis* Whitley (1954): 27, Port Darwin.

M48-9, Reefs and tide pools, entrance to Little Lagoon, USNM 173703, 2 specimens, 45 to 50 mm. standard length.

M48-9a, Reef, entrance to Little Lagoon, USNM 173704, 8 specimens, 40 to 54 mm. standard length.

M48-17, Reef, Bartalumba Bay, USNM 173705, 2 specimens, 38 and 48 mm. standard length.

M48-21, Reef, Yirrkalla, 14 July 1948, USNM 173706, 40 specimens, 37 to 59 mm. standard length; 12 August 1948, USNM 173707, 38 specimens, 36 to

55 mm. standard length; 25 August 1948, USNM 173708, 5 specimens, 40 to 51 mm. standard length.

M48-26, Sheltered bay, island in Port Bradshaw, USNM 173709, 7 specimens, 54 to 61 mm. standard length.

Description. Trunk rings 14(in 2) or 15(7); tail rings 17(1), 18(6), or 19(2); predorsal rings 11(9); subdorsal rings 3+2(1), 4+1(5), or 4+2(3); brood pouch rings 14(3) or 15(3); anal rays iv(9); dorsal rays xix(1), xx(3), xxi(4), or xxii(1); pectoral rays xviii(3), xix(11), xx(3), or xxi(1); caudal rays x(8). In 9 specimens 45 to 61 mm. in standard length, eye 2·9 to 3·7 in snout; snout 0·7 to 1·0 in postorbital length of head.

Brood pouch abdominal, extending virtually the length of the trunk, its membranes scarcely overlapping; superior cristae continuous; median cristae joining inferior cristae in tail rings 1 to 3; no median tail cristae; inferior cristae of trunk and tail interrupted or occasionally continuous; head with numerous finely serrated carinae and rugosities; opercle rugose, with several smaller carinae directed backward and upward or downward from a prominent large longitudinal carina that extends the length of the opercle; no longitudinal keels on body plates.

Colour in alcohol. Medium brown; caudal fin dark brown to blackish; line along superior and inferior cristae of trunk dark brown; dark brown lines on head as follows: one from snout through lower naris and eye on to upper opercle; another along each side of top of head extending from the eye almost to first trunk ring, and one extending backward along each side of lower head from snout to lower half of opercle.

Black dots, each about one-fourth the diameter of the pupil and located at the junctions of the lateral trunk shields, form longitudinal rows. They are in a single line through the upper half of the lower lateral shields, with sometimes a second short line, confined to the area below the dorsal fin, along the middle of the upper lateral shields in males. In females there are two complete lines of dots on the trunk, one as in males; the other passes through the approximate middle of the upper lateral shields. Dorsal shields pale in males; females with a more or less continuous faint black mid-dorsal line.

Colour in life. Uniformly greyish brown to chocolate brown, with a brown band through snout, eye and operculum. Males with one row of small black dots along lower side of trunk; females with an additional row of black dots along upper side of trunk. Eggs deep orange.

Remarks. This is a rock-inhabiting fish. One was found under a large flat rock in daytime on the reef at Yirrkalla.

Relatively few eggs are present in the pouches of the males at hand. Most are empty. A few specimens in the collection taken at Yirrkalla, on 14 July 1948, contain well-developed eggs which are chiefly confined to the posterior end of the pouch.

Micrognathus brevirostris (Rüppell)

Syngnathus brevirostris Rüppell, *Neue Wirbelthiere zu der Fauna von Abyssinien gehörig, Fische*: 144 (1838); type locality, Massowah, Red Sea.

M48-21, Reef, Yirrkalla, 11 August 1948, USNM 173701, 1 male, 45 mm. standard length; 25 August 1948, USNM 173702, 1 male, 44 mm. standard length.

Description. Trunk rings 15(in 1) or 16(1); tail rings 27(1) or 28(1); predorsal rings 14(2); subdorsal rings 1+3(1) or 2+3(1); brood pouch rings 12(2); anal rays at least 3, fin tiny; pectoral rays xii(4); caudal rays x(2); dorsal rays xvii(1) or xviii(1). Snout 1·0 to 1·1 in postorbital head length; eye 1·5 to 1·6 in snout.

Brood pouch subcaudal, extending backward from first tail ring, its membranes overlapping; sac fry filling the pouch in one specimen and eggs with well-developed embryos filling that of the other; superior cristae interrupted below dorsal fin, those of caudal extending forward to first two or three tail rings as a dorso-lateral ridge; median cristae joining inferior cristae of tail at anterior end of brood pouch; inferior cristae of tail and trunk apparently not continuous; head relatively smooth; a rather prominent nonserrated keel along upper midline of posterior half of snout; anterior half of opercle with a faint longitudinal ridge, opercle otherwise smooth; several fleshy projections scattered over head and body; those of head large and vermiform to leaf-like, dark brown, confined chiefly to its upper and lateral posterior surface; body plates not keeled.

Colour in alcohol. Almost uniformly medium brown; three dark brown to blackish blotches, large as eye, on upper lateral trunk; upper snout and tip of caudal fin light yellowish.

Remarks. The specimen, USNM 173702, was obtained from a coral that has been tentatively identified by Dr John W. Wells, Cornell University, as *Porites (Porites)* affinity of *suppressa* Crossland 1952, indicating that this is a coral-inhabiting species. The identification of one of the specimens was confirmed by Dr Earl Herald.

Ichthyocampus annulatus Macleay

Ichthyocampus annulatus Macleay, *Proc. Linn. Soc. N.S.W.* 2: 364, pl. 10, fig. 6 (1878) and $6\frac{1}{2}$; 293 (1881); type locality, Port Darwin; Duncker, 'Pisces. Part I: Syngnathidae' in *Die Fauna Südwest-Australiens* 2: 241 (1909).
Yozia annulata Duncker, *Mitt. Naturh. Mus. Hamburg* 32: 110 (1914).

The status of this nominal species is uncertain. Duncker, in questionably placing it with the genus *Yozia,* indicated that Macleay's description and figure disagree widely.

Stigmatophora argus (Richardson)

Syngnathus argus Richardson, *Proc. Zool. Soc. Lond.* 8: 29 (1840); type locality not given.

Recorded as: *Stigmatophora argus* Klunzinger (1879): 420; Duncker (1909): 239, Port Darwin.

Haliichthys taeniophora Gray

Haliichthys taeniophora Gray, *Proc. Zool. Soc. Lond.* 27: 39, pl. 7 (1859); type locality, Shark's Bay, Western Australia.

Recorded as: *Haliichthys taeniophora* Whitley and Allan (1958): 49 and Munro (1958): 91, Northern Teritory.

Syngnathoides biaculeatus (Bloch)

Syngnathus biaculeatus Bloch, *Naturgeschichte der Ausländischen Fische* 1: 11, pl. 121, fig. 1 (1785); type locality, East Indies.
Recorded as: *Gastrotokeus biaculeatus* Macleay (1881): 300, Port Essington.

HOLOCENTHRIDAE

Key to the Arnhem Land Genera

1 Dorsal spines XI, the last shortest; a large, long spine at angle of preopercle .. *Holocenthrus*

Dorsal spines X-I, the last longer than the preceding one; no prominent spine at angle of preopercle . . . . *Myripristis*

Myripristis macrolepis Bleeker

Myripristis macrolepis Bleeker, *Ned. Tijdschr. Dierk.* 4: 195 (1873); type localities, Batavia, Java; Nias; Macassar, Celebes; Kajeli, Buro; and Ambon; Bleeker, *Atlas icthyologique des Indes Orientales Néêrlandaises* 9: pl. 359, fig. 3 (1878).

M48-21, Reef, Yirrkalla, 11 August 1948, USNM 174045, 2 specimens, 39 to 130 mm. standard length.

Description. In the two specimens, dorsal rays X-I,13 or 14; anal rays IV,12; pelvic rays I,7, both sides; pectoral rays ii,13, both sides; branched caudal rays 9 or 10+8=17 or 18; pored scales in lateral line 27; scales between dorsal and anal fins 3+1+6; gillrakers on first arch 13 or 14+1+24 or 26=39 or 40.

Snout length slightly less than interorbital width; third anal spine about equal to eye diameter; smaller specimen without external teeth; larger specimen with slightly enlarged tubercular teeth in outer row of upper jaw, projecting outward, and with two pairs of clusters of about five to seven teeth on lower jaw near symphysis; one of the pairs on upper side of jaw, the other on lower side; one tubercular tooth on lateral edge of right lower jaw which is extensively perforated by sensory canal pores; a series of five to seven, short, peg-like teeth on lower side of posterior end of maxilla; nostril rugose, irregularly serrated.

Colour in alcohol. Fins all yellowish, except first dorsal in small specimen which is dusted with brown; background colour of body and head yellowish or yellowish brown, the brown more prominent on head and upper surface of body; each scale with a broad brown bar; these unite in larger specimen to form wavy vertical lines.

Colour in life. Body brick red; fins pinkish red.

Remarks. This material appears to be close to *Myripristis australis* Castelnau (1875) but that species is described by Munro (1958) as having a black spot at the angle of the operculum. I also do not think it wise to regard *M. macrolepis* as a synonym of *M. melanostictus* Bleeker as has been done by Weber and de Beaufort (1929), without further study. The latter species has been described as having much black on the body and fins.

Holocenthrus ruber (Forskål)

Sciaena rubra Forskål, *Descriptiones Animalium*: 48 (1775); type locality, Arabia.

Recorded as: *Holocentrum rubrum* Macleay (1878): 354 and (1880): 514; Klunzinger (1879): 359, Port Darwin; *Holocenthrus praslin* Whitley (1941b): 26, Northern Territory.

M48-9a, Reefs and tide pools, entrance to Little Lagoon, USNM 174046, 4 specimens, 44 to 46 mm. standard length.

M48-17, Reef, Bartalumba Bay, USNM 174047, 1 specimen, 78 mm. standard length.

M48-21, Reef, Yirrkalla, 14 July 1948, USNM 174048, 12 specimens, 45 to 99 mm. standard length; 11 and 12 August 1948, USNM 174049, 31 specimens, 44 to 113 mm. standard length.

Description. In 6 specimens, dorsal rays XI,i,12; anal rays IV,9; pelvic rays I,7, both sides; branched caudal rays 9+8=17; pectoral rays ii,11(in 2) or ii,12(10); pored scales in lateral line 33(4) or 34(1).

A large spine extending backward from angle of preopercle; preorbital, edge and ridge of suborbitals, preopercle, opercle, subopercle, and interopercle all spined or serrated; opercle with two large spines, the upper longest; last dorsal spine shortest; three rows of scales between spinous dorsal and lateral line; no spines at nasal opening.

Colour in alcohol. Body and head light brown, the bars, bands, and spots which in life are red become medium reddish brown to brownish black; in young, membrane of first dorsal fin brownish black, this pigment becoming more diffuse in larger specimens until only a trace remains in the largest; middle, and sometimes inner, rays of pelvic fin brownish black, the outer ray and spine yellowish white, in young; in larger specimens, the dark pelvic fin pigment is concentrated terminally and greatly faded proximally.

Colour in life. Miller's colour sketch of this fish shows the second dorsal, anal, and pelvic fins and the inter-radial membranes of the first dorsal fin red; red stripes, spots, and bars on body and head as follows: one from head extending near dorsal fin base to rear base dorsal fin; two others, one from upper head, the other from upper margin of operculum, both converging below soft dorsal into one band which then extends along upper margin caudal fin; two others beginning above pectoral fin, converge below soft dorsal and extend as one band along lower side caudal peduncle thence along lower margin caudal fin; another from pectoral fin to rear base anal fin; two rows of dots or crescents along lower side; a broad reddish bar on opercle; a narrower one along margin of vertical arm of preopercle, and an oblique bar from eye to angle of preopercle, the latter two meeting ventrally (specimens 44 to 46 mm. standard length).

In a colour transparency, the background body colour is silvery white; the second dorsal, anal, and caudal fins are light pinkish white; the anterior border of the second dorsal and the upper and lower margins of the caudal fin are red; a red bar extends to the anal fin margin along and just behind the spines.

Remarks. I do not believe that *Holocenthrus* Scopoli (1777) is a misprint for *Holocentrus,* as interpreted by many recent workers, since the name also appears as *Holocenthrus* in the index to the work. The type species of the genus *Holocenthrus* is here interpreted as *Holocentrus sogo* Bloch, since *Holocentrus maxilla,* as given by Woods (1955), is part of Gronow's polynomial description.

SPHYRAENIDAE

Key to the Arnhem Land Species

1 Rear lower margin of preopercle angular, a membranous flap; first gill arch with a raker at the angle and one on lower limb; pectoral rays ii,12; pored scales in lateral line 80 to 90 . . . . *Sphyraena chrysotaenia*
Rear lower margin of preopercle rounded, not extended as membranous flap; no gillrakers 2

2 Pored scales in lateral line about 75 to 85; short bars along upper side of body inclined slightly backward . . . . *Sphyraena barracuda*
Pored scales along lateral line about 120 to 130; slender bars along side of body nearly vertical . . . . *Sphyraena jello*

Sphyraena barracuda (Walbaum)

Esox barracuda Walbaum, *Petri Artedi sueci Genera Piscium*: 94 (1792); type locality, West Indies.

Sphyraena akerstromi Whitley, *Aust. Zool.* 11: 131 (1947); type locality, Lowendal Island, Western Australia; complex, as certain paratypes appear to be *Sphyraena jello* Cuvier and Valenciennes.

M48-10, Estuary, Thompson Bay, Port Langdon, USNM 173989, 1 specimen, 163 mm. standard length.

M48-15, Emerald River, Groote Eylandt, USNM 173990-1, 8 specimens, 158 to 340 mm. standard length.

M48-24, Beach, Port Bradshaw, USNM 173992, 1 specimen, 302 mm. standard length.

M48-28, Creek mouth and bay, Port Bradshaw, USNM 173993, 4 specimens, 201 to 271 mm. standard length.

M48-31, Creek, Yirrkalla, USNM 173994, 1 specimen, 344 mm. standard length.

Native names: *Áronguárguá*, young, or *Ámatárna*, large individuals, Groote Eylandt; *Wárrooki*, Yirrkalla.

Description. No gillrakers; posterior margin of preopercle rounded; pored scales in lateral line 77(in 1), 78(2), 79(1), 81(1), and 84(1).

Colour in life. Tips of second dorsal and anal fins and of upper and lower lobes of caudal fin milky white; pectoral fin yellowish at base; pelvic fin and base of dorsal fin light yellow.

Colour in alcohol. About a dozen or more slightly oblique bars extending downward and forward approximately to lateral line; round, dark brown spots, about as large as pupil, irregularly along lower side of caudal peduncle; middle parts of dorsal, anal, and pelvic fins dark brown; caudal fin dusky; margin of opercle, chiefly above pectoral fin, dark brown.

Sphyraena jello Cuvier and Valenciennes

Sphyraena jello Cuvier and Valenciennes, *Histoire naturelle des poissons* 3: 349 (1829); type locality, Vizagapatam.

Sphyraena altipinnis Ogilby, *Proc. Roy. Soc. Qld.* 23: 8 (1910); type locality, Aru Islands.

Recorded as: *Sphyraena altipinnis* Paradice and Whitley (1927): 81 and 97, Pellew Islands.

M48-23, Port Bradshaw, USNM 173987, 1 specimen, 560 mm. standard length.

M48-28, Creek mouth and bay, Port Bradshaw, USNM 173988, 2 specimens, 194 and 201 mm. standard length.

Description. No gillrakers; preopercle rounded behind; pored scales in lateral line 123(in 1) or 125(1); 15 to 20 vertical brown bars along side of body; first dorsal fin dark brown or black; second dorsal dusky, or yellowish grey; anal fin yellow in small specimens, with trace of medium brown on middle of anterior rays; in large specimen, anal with large, dark brown blotch.

Sphyraena chrysotaenia Klunzinger

Sphyraena chrysotaenia Klunzinger, *Die Fische des Rothen Meeres* pt. 1: 128, pl. 9, fig. 3 (1884); type locality, Red Sea.

Records probably referable: *Sphyraena langsar* Macleay (1878): 359 and (1881): 35; Klunzinger (1879): 374, Port Darwin.

M48-8, Little Lagoon, USNM 173986, 1 specimen, 65 mm. standard length.

Remarks. The name adopted here follows Williams (1959). The single specimen has two slender gillrakers on the first arch, one at the angle and one along the lower limb; there are 87 scales along the lateral line. The posterior preopercular margin is angular. Although a juvenile, as were cited records of *Sphyraena langsar* from Arnhem Land, body proportions coincide with those given by Williams. In the specimen and in two of this species from China (USNM 130357), the pectoral ray count is ii,12 on both sides. In three specimens of *Sphyraena flavicauda* Rüppell, from Mauritius (USNM 153571), the pectoral fin rays are ii,11 on both sides, in agreement with Williams.

MUGILIDAE

In the study of this family, I have followed Thomson (1954) and here make only comments regarding differences of interpretations of the nomenclature and morphology.

As pointed out by Macleay (1879b), *Agonostoma darwiniense* Macleay from Port Darwin (Macleay, 1878) is a species of gobioid fish rather than a mugilid. Schultz (1946) has indicated that yet another gobioid fish was originally described in the genus *Agonostomus*.

I am unable to identify the Port Darwin records of *Mugil dobula* and *Mugil peronii,* as given by Macleay (1878), with any species. Thomson (1954) has assigned them to *Mugil cephalus* Linnaeus and *Liza* (= *Chelon*) *argentea* (Quoy and Gaimard) respectively, both perhaps in error. He has been followed by Munro (1958).

Key to the Arnhem Land Species

1 Upper lip inferior to fleshy snout, not forming anterior tip of head; eyes dorsolateral; anterior naris about on level with lower border of eye and posterior naris near upper border of eye . . . . *Squalomugil nasutus*

Upper lip terminal, not inferior to remainder of snout; eyes lateral; nares above middle of eye 2

2 Upper lip folded, the inner fold forming two lobes in adult, with about four more lobes at corner of mouth; inner lobes of upper lip and lateral portions of lower lip plicate to subpapilliate; outer lobe of upper lip fimbriated to papilliate . . . . *Plicomugil labiosus*

Upper lip not folded, it and the lower lip smooth or nearly so 3

3 Oral edge of preorbital gently curved, not notched; maxilla hidden when mouth is closed; anal fin origin about opposite to or behind origin of second dorsal fin—genus *Valamugil* 4

Oral edge of preorbital notched at corner of mouth; posterior end of maxilla exposed (not or weakly so in *strongylocephalus*) when mouth is closed; origin of second dorsal fin behind origin of anal fin—genus *Chelon* 5

4 Scales between head and caudal fin 37 to 42 . . . . *Valamugil seheli*

Scales between head and caudal fin 34 to 37; median fins strongly falcate; scale bases on dorso-lateral surface dark, the pigment extending into faint longitudinal lines . . . . *Valamugil buchanani*

5 Anal rays usually III,8; scales between head and caudal fin 25 to 28; dorsal anal, and pectoral fins blackened; dorso-lateral surfaces with dark brown longitudinal lines . . . . *Chelon vaigiensis*

Anal rays usually III,9; fins not black 6

6 Scales between head and caudal fin 33 to 37; pectoral axillary scale prominent; pectoral fin extending to or behind vertical through dorsal fin origin; posterior end of maxilla not visible, or barely so when mouth is closed; tongue margin rounded; silvery with black spot at upper base of pectoral fin . . . . *Chelon strongylocephalus*

Scales between head and caudal fin 28 to 33; pectoral axillary scale vestigial; pectoral fin not reaching vertical through dorsal origin; posterior end of maxilla prominent when mouth is closed; anterior end of tongue notched; no black spot at upper pectoral fin base 7

7 Usually longitudinal dark brown lines along scale rows; adipose eyelid prominent and extending well on to iris in most specimens . . *Chelon dussumieri*

Scale centres light, margined with dark brown, forming a reticulate pattern; adipose tissue little developed and mostly extra orbital . . *Chelon diadema*

Chelon strongylocephalus* (Richardson)

Pl. 17

Mugil strongylocephalus Richardson, *Rep. Fifteenth Meet. Brit. Ass. Adv. Sci.*: 249 (1846); type locality, Hong Kong.

Recorded as: *Liza strongylocephalus* Thomson (1954): 95, Pellier [Pellew] Islands and Adam Bay, Northern Territory; *Mugil longimanus* Paradice and Whitley (1927): 81, Pellew Islands.

M48-1, Nightcliff, USNM 173892, 3 specimens, 55 to 61 mm. standard length.

M48-7, Mouth of billabong at Little Lagoon, USNM 173893, 17 specimens, 30 to 69 mm. standard length.

M48-9, Reefs and tide pools, entrance to Little Lagoon, USNM 173894, 17 specimens, 44 to 72 mm. standard length.

M48-10, Estuary, Thompson Bay, Port Langdon, USNM 173895, 4 specimens, 72 to 80 mm. standard length.

M48-15, Emerald River, Groote Eylandt, USNM 173896, 1 specimen, 107 mm. standard length.

M48-19, Creek, tributary to Little Lagoon, USNM 173897, 33 specimens, 57 to 127 mm. standard length.

M48-21, Reef, Yirrkalla, 12 August 1948, USNM 173898, 1 specimen, 93 mm. standard length.

M48-28, Creek mouth and bay, Port Bradshaw, USNM 173899, 35 specimens, 42 to 190 mm. standard length.

M48-29, Pool near small creek, Yirrkalla, USNM 173901, 1 specimen, 68 mm. standard length.

M48-31, Creek, Yirrkalla, USNM 173900, 28 specimens, 32 to 124 mm. standard length.

Native names: *Yébúa*, Groote Eylandt; *Wáhcoon*, Yirrkalla.

Colour in life. Side uniformly silvery white; upper surface light brownish; orange to bronze gold spot or reflection on middle of opercle; small black spot on upper base of pectoral fin; dorsal fins, caudal and anal fins dusky.

Remarks. I have followed Thomson (1954) in the identification and nomenclature of this species, but note that the adipose eyelid is little developed, forming only a narrow ring around the eye. The posterior end of the maxilla is not visible, or barely so. In my counts there are 33 to 37 scales between the head and caudal fin and III,9 anal rays. The tip of the pectoral fin extends to or behind a vertical through the base of the first dorsal spine.

Chelon dussumieri (Cuvier and Valenciennes)

Mugil dussumieri Cuvier and Valenciennes, *Histoire naturelle des poissons* 11: 147 (1836); type localities, Bombay and Coromandel.

Recorded as: *Liza dussumieri* Munro (1958): 104, fig. 709, Northern Territory; *Mugil compressus* Klunzinger (1879): 395; Macleay (1879b): 421, Port Darwin.

* I agree with Schultz (1946): 391, that *Chelon* Röse, as cited, is available as a generic name.

M48-2, Nightcliff, USNM 173905, 3 specimens, 59 to 123 mm. standard length.

M48-13a, South end of Winchelsea Island, 6 June 1948, USNM 173906, 20 specimens, 11 to 14 mm. standard length.

M48-15, Emerald River, Groote Eylandt, USNM 173907, 12 specimens, 114 to 231 mm. standard length.

M48-19, Creek, tributary to Little Lagoon, USNM 173908, 7 specimens, 68 to 163 mm. standard length.

M48-21, Reef, Yirrkalla, 1 August 1948, USNM 173909, 1 specimen, 14 mm. standard length.

M48-28, Creek mouth and bay, Port Bradshaw, USNM 173910, 10 specimens, 14 to 259 mm. standard length.

M48-31, Creek, Yirrkalla, USNM 173911, 10 specimens, 94 to 144 mm. standard length.

M48-34, Creek, Yirrkalla, USNM 173912, 1 specimen, 25 mm. standard length.

Native names: *Yébúa*, Groote Eylandt; *Wáhcoon*, Yirrkalla.

Remarks. The posterior end of the maxilla in the material of this species is broadly exposed. The fish is usually striped, without a prominent pectoral axillary scale, and with about 28 to 32 scales between the head and caudal fin. I find that the adipose eyelid is well developed in most of the specimens, but in some it is reduced to an orbital ring. The tongue is sub-truncate and slightly notched. These characters disagree somewhat with those given by Thomson (1954). The variations in adipose eyelids and development of lateral stripes lead me to suspect that two similar species may be in the Australian fauna or that the characters used are not constant. Yet, I am unable to divide the Arnhem Land specimens into two kinds on any character or group of characters.

Chelon diadema (Gilchrist and Thompson)

Mugil diadema Gilchrist and Thompson, *Ann. S. Afr. Mus.* 11: 42 (1911); type locality, Natal.

Moolgarda (*Planiliza*) *ordensis* Whitley, *Aust. Zool.* 11: 17, fig. 9 (1945); type locality, Carlton Reach, Ivanhoe Station, Ord River, north-western Australia.

Recorded as: *Liza diadema* Thomson (1954): 106, fig. 10, Melville Island and Ord River, Northern Territory; *Moolgarda* (*Planiliza*) *ordensis* Whitley (1951): 63, Melville Island and Darwin, Northern Territory.

Arnhem Land; no further data; in collection with *Squalomugil nasutus*; USNM 173902, 4 specimens, 225 to 259 mm. standard length.

M48-35, Billabong, Oenpelli, USNM 173903, 6 specimens, 192 to 233 mm. standard length.

M48-39, Red Lily Lagoon, USNM 173904, 13 specimens, 123 to 195 mm. standard length.

Native name: *Guálily*.

Remarks. C. diadema is very similar morphologically to *Chelon dussumieri*. It differs in lacking both longitudinal dark lines and well-developed adipose eyelids. The scale centres are light, with dark margins.

Chelon vaigiensis (Quoy and Gaimard)

Mugil vaigiensis Quoy and Gaimard, *Voyage autour du monde . . . sur . . . l'Uranie et la Physicienne . . . Zoologie*: 337, pl. 59, fig. 2 (1825); type locality, Waigiou.

Recorded as: *Liza vaigiensis* Thomson (1954): 102, fig. 8, Pellew Islands, Darwin, and Melville Island, Northern Territory; *Mugil waigiensis* Macleay (1878):

359 and (1879b): 420; Kent (1889): 240, Port Darwin; *Mugil (Liza) waigiensis* Paradice and Whitley (1927): 81, Pellew Islands; *Ellochelon vaigiensis* Whitley (1953): 123, West Headland, Port Keats.

M48-1, Nightcliff, USNM 173913, 14 specimens, 14 to 66 mm. standard length.

M48-2, Nightcliff, USNM 173914, 5 specimens, 20 to 37 mm. standard length.

M48-7, Mouth of billabong, at Little Lagoon, USNM 173915, 2 specimens, 18 and 22 mm. standard length.

M48-9, Reefs and tide pools, entrance to Little Lagoon, USNM 173916, 3 specimens, 15 to 30 mm. standard length.

M48-10, Estuary, Thompson Bay, Port Langdon, USNM 173917, 1 specimen, 242 mm. standard length.

M48-15, Emerald River, Groote Eylandt, USNM 173918, 2 specimens, 339 to 389 mm. standard length.

M48-17, Reef, Bartalumba Bay, USNM 173919, 2 specimens, 93 to 160 mm. standard length.

M48-21, Reef, Yirrkalla, 12 August 1948, USNM 173920, 2 specimens, 52 to 61 mm. standard length.

M48-24, Beach, Port Bradshaw, USNM 173921, 5 specimens, 110 to 129 mm. standard length.

M48-27, Bay and beach, Port Bradshaw, USNM 173922, 1 specimen, 17 mm. standard length.

M48-28, Creek mouth and bay, Port Bradshaw, USNM 173923, 1 specimen, 360 mm. standard length.

M48-31, Creek, Yirrkalla, USNM 173924, 2 specimens, 32 to 38 mm. standard length.

Native names: *Termálya* or *Termálakáya*, Groote Eylandt; *Kúlroo* (young) and *Chápimílimoo* (large individual), Yirrkalla.

Colour in life. (In young to 66 mm. standard length): Body with (about 6) chocolate brown stripes; caudal fin bright yellow; dorsal, anal, and pectoral fins all blackened; two or three dark brown saddles, separated by light tan interspaces, on back.

Valamugil buchanani* (Bleeker)

Mugil buchanani Bleeker, *Verh. Batavia Genoot.* 25: 99 (1853); type locality, Hooghly River.

? *Moolgarda pura* Whitley, *Aust. Zool.* 11: 16, fig. 8 (1945); holotype, but not cotypes; type locality, Point Cloates, Western Australia.

Record probably referable: *Valamugil puro* Munro 1958): 106, Northern Territory.

M48-15, Emerald River, Groote Eylandt, USNM 173927, 1 specimen, 522 mm. standard length.

Native name: *Birúcha* (Beer-roochah).

Valamugil seheli (Forskål)

Mugil crenilabris seheli Forskål, *Descriptiones Animalium*: 73 (1775); type locality, Lohajae, Red Sea.

Recorded as: *Valamugil seheli* Thomson (1954): 108, 'other material from Darwin'; Munro (1958): 106.

* *Valamugil* Smith (1948): 841; type species, *Mugil scheli* Forskål, by original designation.

Plicomugil labiosus (Cuvier and Valenciennes)

Mugil labiosus Cuvier and Valenciennes, *Histoire naturelle des poissons* 2: 125 (1836); type locality, Red Sea.

? *Oedalechilus kesteveni* Whitley, *Aust. Zool.* 10: 178 (1943); type locality, Port Essington, Northern Territory.

M48-21, Reef, Yirrkalla, 14 July 1948, USNM 173925, 1 specimen, 40 mm. standard length; 12 August 1948, USNM 173926, 1 specimen, 40 mm. standard length.

Colour in life. White (presumably meaning silvery).

Squalomugil nasutus (De Vis)

Mugil nasutus De Vis, *Proc. Linn. Soc. N.S.W.* 7: 621 (1883); type locality, Cardwell, Queensland.

Recorded as: *Squalomugil nasutus* Paradice and Whitley (1927): 96, Adam Bay, near Darwin, Northern Territory; *Rhinomugil nasutus* Thomson (1954): 122, fig. 16, Adam Bay, Northern Territory.

Arnhem Land; no further data; USNM 173928, 16 specimens, 90 to 162 mm. standard length.

M48-40, East Alligator River, USNM 173929, 23 specimens, 27 to 113 mm. standard length.

Remarks. Squalomugil nasutus differs from *Rhinomugil* Gill (1863), to which Thomson (1954) assigned it, in having larger scales and in the following characters. The gently curved oral edge of the preorbital, in *Rhinomugil,* is smooth or nearly so, while the bone's posterio-ventral edge has a few prominent spines; in *Squalomugil* the posterio-ventral preorbital edge is imbedded, with serrae only at the oral corner and forward along the slightly curved oral edge. The anterior and posterior nares in *Rhinomugil* have been displaced ventrad; both are approximately on a level with the lower border of the eye. In *Squalomugil* the anterior naris is considerably below the level of the posterior naris, that is, the anterior naris is about on a level with the lower border of the eye and nearer the postmaxillary groove than to the posterior naris. The rear nostril is about on a level with the upper border of the eye and closer to its anterior edge than to the anterior naris. Primarily because of the difference in position of the nares, I recognize as genera both *Squalomugil* and *Rhinomugil.*

ATHERINIDAE

Key to the Arnhem Land Genera

1 Anal fin origin beneath, or closer to first dorsal fin than to second dorsal fin; air bladder extending backward beyond anal fin origin; depressed pelvic fin extending to or nearly to anal fin base 2

Anal fin origin closer to second dorsal fin than to first dorsal fin; air bladder extending backward to above anal fin origin (character unknown in *Quirichthys*) 3

2 Mouth very small, the upper jaw protractile; second dorsal fin without a spine; second dorsal and anal fin bases short, of relatively few rays . . . . . genus *Pseudomugil*

Mouth of moderate size; upper jaw not protractile or weakly movable; a spine preceding both the second dorsal and anal fins; anal fin base long, with 15 or more soft rays . . . . genus *Melanotaenia*

3 Dorsal spines produced, when depressed reaching second dorsal fin; pelvic fin

length at least two-thirds the distance between pelvic and anal fin bases; first dorsal fin and a narrow band along body black; aspect of *Melanotaenia*, but with about 10 anal rays . . . . *Quirichthys stramineus*

First dorsal spines not nearly reaching second dorsal fin; pelvic fin and anus remote from anal fin, the pelvic rays separated from the anal fin by at least half their length 4

4 Shelf of preopercle with an indentation or concavity along vertical arm; teeth present on vomer 5

Shelf of vertical arm of preopercle truncate or rounded, not indented; no teeth on vomer 6

5 Mandibular ramus elevated posteriorly; vent slightly behind tip of depressed pelvic fin . . . . genus *Hypoatherina*

Mandibular ramus not elevated; vent beneath pelvic rays, or at their tip . . . . genus *Pranesus*

6 Gillrakers on lower limb of first arch fewer than 13; no teeth on palatine or pterygoid, and apparently none on tongue . . . . . . genus *Craterocephalus*

Gillrakers on lower limb of first arch more numerous, 18 to 20; a few teeth present on palatine, pterygoid and tongue . . . . genus *Allanetta*

Key to the Arnhem Land Species of *Melanotaenia*

1 A black lateral band, usually prominent and well defined, extending from head to caudal fin; spines in first dorsal fin IV to VI, usually V; side of caudal peduncle, below lateral band, yellowish, below which a parallel band of dark pigment may extend forward from middle of anal fin 2

A series of narrow longitudinal brownish stripes between scale rows on body, or its midside, with an obscure, irregular, brownish band; spines in first dorsal fin V to VII, usually VI 3

2 Black band, one scale in width, extending from head through upper pectoral fin base on to caudal fin; palatine teeth present, distinct; scales below eye in about one row; pectoral rays ii,10 to ii,13; scales across side of caudal peduncle, midline above to midline below, usually 7, infrequently 8 or 9 *Melanotaenia nigrans*

Black band, about one and one-half scales wide, on one full scale row and the adjacent half or more of next row above, extending from head, around but not on pectoral fin base, to base of caudal fin; palatine teeth absent or irregularly developed; scales below eye usually in two rows; pectoral rays ii, 14 to ii,18; 8 scales across side of caudal peduncle . . *Melanotaenia trifasciata*

3 Stripes, narrow, outlining scale rows, most prominent on rear half of body; scales between head and caudal fin 31 to 34, usually 31 to 33; scale rows across caudal peduncle usually 8, infrequently 7 . . . . *Melanotaenia rubrostriata*

Narrow brown stripes, faint, most prominent posteriorly on caudal peduncle; brownish band more prominent, irregular, often broken up posteriorly by light scale centres, occupying approximately the adjacent three-fourths of two scale rows; scales between head and caudal fin 32 to 35, usually 33 or 34; scale rows across caudal peduncle 7, infrequently 6 or 8; body with golden stripes in life . . . . *Melanotaenia solata*

The four Arnhem Land species of melanotaenids are here grouped in the original genus *Melanotaenia* since I do not find sufficient differences among them to recognize the several genera of many recent authors. In this material, the first dorsal fin has a rather stiff, moderately robust, spine followed by several more slender and flexible spines. The jaws are equal, or nearly so, and the mouth rather large with an abrupt curve or bend in the upper jaw. The teeth are in bands in each jaw, without the described prominent interspaces delimiting separate rows. Those teeth in the lower jaw are largest externally, extending to the outside at the jaw's tip; in the upper jaw, the tooth band is prominently reflected to the outside along the jaw's length and the teeth are largest externally. *M. trifasciata* is most divergent of the Arnhem Land species. In it, the lower jaw is usually slightly included and palatine teeth are not well developed.

In *Melanotaenia,* the premaxilla is bound, and not protractile, or at most slightly movable; its median process is short, about as long as the pupil diameter; the gillrakers are slender, but short, usually less than half as long as the filaments; the mandible has a relatively straight upper surface, without elevation of the ramus; the preopercle is truncate behind; the vent opens nearer the anal fin than to the pelvic fin base; the pelvic fin is close to the anal fin, its rays usually reaching to that fin; the air bladder extends well into the area above the anal fin; the cheek has one or two rows of scales; and scales do not extend on to the dorsal and anal fin bases.

The individual and geographic variation, especially in meristic characters, in the species of this genus is extensive and makes species definition difficult.

Melanotaenia nigrans (Richardson)

Pl. 18

Atherina nigrans Richardson, *Ann. Mag. Nat. Hist.* 11: 180 (1843); type locality, freshwater streams that flow into the harbour, Port Essington, Northern Territory.

Atherinichthys nigrans Günther, *Catalogue of the Fishes in the British Museum* 3: 406 (1861); Kings River, near Victoria, Port Essington.

Zantecla pusilla Castelnau, *Proc. Zool. Acclim. Soc. Vict.* 2: 88 (1873); type locality, Port Darwin.

Nematocentris nigra Günther, *Ann. Mag. Nat. Hist.* ser. 3, 20: 65 (1867); Kings River, Victoria, Port Essington.

Recorded as: *Nematocentris pusilla* Macleay (1881): 44, Port Darwin; *Nematocentris nigrans* Macleay (1881): 44, Port Essington; *Melanotaenia nigrans* Whitley (1953): 126, Bankers Jungle Creek, Koolpinyah Station, about 30 miles inland from Port Darwin, and Bakers Jungle, Northern Territory.

M48-4, Creek, Delissaville, USNM 173743, 61 specimens, 24 to 57 mm. standard length.

M48-6, Billabong, Fred Gray's Camp, USNM 173744, 76 specimens, 21 to 56 mm. standard length.

M48-7, Mouth of billabong, at Little Lagoon, USNM 173745, 1 specimen, 24 mm. standard length.

M48-14, Emerald River, Groote Eylandt, USNM 173746, 129 specimens, 18 to 66 mm. standard length.

M48-35, Billabong, Oenpelli, 16 October 1948, USNM 173747, 31 specimens, 23 to 30 mm. standard length.

M48-37, Fish Creek, USNM 173748, 9 specimens, 25 to 35 mm. standard length.

M48-38, Oenpelli Creek, below waterfall, USNM 173749, 16 specimens, 22 to 28 mm. standard length.

Native name: *Emátchenámura,* Groote Eylandt.

Description. Based on specimens from M48-4, M48-6, and M48-37; rays in first dorsal fin IV(in 2), V(18) or VI(4); rays in second dorsal fin I,8(3), I,9(5), I,10(8), I,11(6), or I,12(2), those from Fish Creek, M48-37, ranging I,8 to I,10, the remainder I,10 to I,12; anal rays I,15(1), I,16(2), I,17(6), I,18(6), I,19(7), I,20(2), those from Fish Creek ranging I,15 to I,17, the remainder I,18 to I,20; pelvic rays I,4(2) or I,5(25). Pectoral rays, M48-6, ii,10(5) or ii,11(5); M48-37, ii,11(9) or ii,12(7); M48-4, ii,11(1), ii,12(14) or ii,13(5). Branched caudal rays 8+7=15(22). Scales between head and caudal fin 32(3), 33(4) or 34(1), in Fish Creek specimens; those from M48-4 and M48-6, 34(4), 35(8) or 36(3). Scale rows between anal fin origin and predorsal midline 11(2), 12(7), 13(4), 14(7) or 15(1); predorsal scales 13 to 19; scale rows around caudal peduncle 12(14), 13(5) or 14(2). In specimens from the seven collections, scale rows across caudal peduncle, from upper to lower midlines 7(78), 8(12) or 9(1). Vertebral counts in specimens from Emerald River and Delissaville: precaudal 18(6), 19(19) or 20(6); caudal 15(2), 16(13) or 17(16); total vertebrae 35(32), 36(34) or 37(1).

Small teeth present in a small cluster on the pterygoid and in a line on the vomer, palatine and tongue; the band on lower jaw reflected to outside of mouth anteriorly; that of upper jaw extensively reflected outward along its length. Scales below eye in one, sometimes in two rows.

Colour in alcohol. A blackish line, nearly as wide as a scale, along midline of back; scales of dorsum and upper side, downward to lateral dark band, outlined in dark brown, their centres yellowish, giving the impression of longitudinal streaks; on sixth scale row downward from predorsal midline a prominent black band, which about as wide as the scale row, perhaps slightly narrowed anteriorly; it extends backward from snout, through eye, is interrupted on preopercle, thence backward across upper part of opercle, prominently through upper base of pectoral fin, backward along the single scale row, thence outward on middle caudal fin rays. Head below band dusky yellowish; two scale rows below band on body yellowish; below the latter area, scale rows yellowish or narrowly outlined in medium brown; pelvic rays, usually the outer ones, black; first dorsal and margins of anal and second dorsal fins black; the remainder of those fins and the caudal dark brown to blackish; pectoral fin light yellowish brown.

Colour in life. Based on specimens from M48-4; black band on side; tail bright yellow; second dorsal and anal fins margined with dark red band.

Based on specimens from M48-6; dorsal, anal, and caudal fins reddish orange, the dorsal and anal with a black border; largest individuals with a bright orange stripe between pectoral and caudal fins, just below middle of side; three or four less distinct stripes below; upper side olive green, with brownish horizontal streaks; edge of belly and pelvic fins red; orange spot on operculum, just below black band.

Remarks. The Arnhem Land records of this species are all from fresh water. Miller observed it amongst mangrove roots on Groote Eylandt at low tide, but in that case there was some inflowing fresh water.

It is essentially a midwater swimmer, but surface feeder. Young were found in shallow, quiet backwaters of large pools, about fallen logs, brush and dense green algae, and the adults in current and in pools of streams. Young were observed on 29 March 1948 at the Delissaville Station.

As is obvious from the accompanying figure this is not the popular fish that is known to aquarists as *Melanotaenia nigrans.*

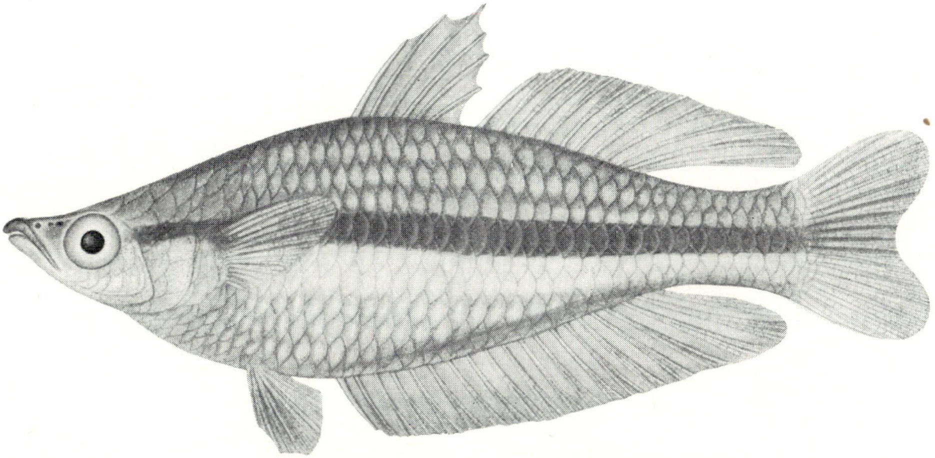

Dorothea B. Schultz

PLATE 18. *Melanotaenia nigrans*
Emerald River, USNM 173746, 62 mm. standard length

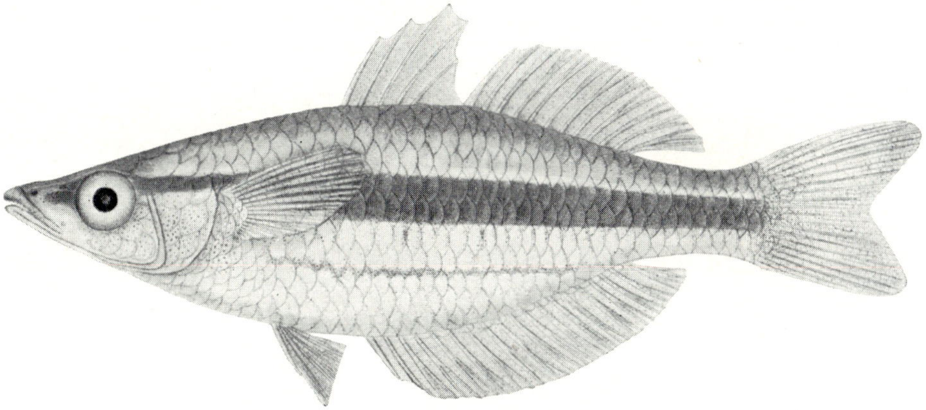

Dorothea B. Schultz

PLATE 19. *Melanotaenia trifasciata*
Yirrkalla, USNM 173750, 51 mm. standard length

Melanotaenia trifasciata (Rendahl)

Pl. 19

Rhombosoma trifasciata Rendahl, *Medd. Zool. Mus. Krist.* 5: 182 (1922); type locality, Mary River, Northern Territory.

M48-34, Creek, Yirrkalla, USNM 173750, 71 specimens, 19 to 56 mm. standard length.

Native names: *Karlpán* or *Karóopa*.

Description. Rays in first dorsal fin IV(in 3), V(7) or VI(1); rays in second dorsal fin I,10(1), I,11(6) or I,12(4); anal rays I,18(4), I,19(2), I,20(4), or I,21(1); pelvic rays I,5(22); pectoral rays ii,14(4), ii,15(14), ii,16(3) and ii,18(1); branched caudal rays 8+6=14(1) or 8+7=15(10); scale count between head and caudal fin 33(1), 34(4), 35(4) or 36(2); scale rows from predorsal midline to anal fin 11(4), 12(4) or 13(3); scales midline, between head and dorsal fin 12 to 16; scale rows around caudal peduncle 14(11); scale rows across caudal peduncle, midline above to midline below 8(30); precaudal vertebrae 19(9), 20(23) or 21(6); caudal vertebrae 15(7), 16(21), 17(9) or 18(1); total vertebrae 35(5), 36(39) or 37(5); scales in one, usually two rows, below eye. Small teeth present on vomer and rarely a few arranged along palatine; they were not observed on the pterygoid and are apparently usually absent from or at least poorly developed on the palatine; a tiny cluster of teeth on tongue; teeth in lower jaw in a band which extends outward at end of jaw where teeth slightly enlarged; those in upper jaw in a band extending outward along length of jaw, the teeth slightly largest externally; mouth slightly protractile.

Colour in alcohol. Dorsal fins dark brown; caudal fin dusky brown; anal fin yellowish basally, becoming darker distally; pelvic rays lightly brownish; pectoral rays light yellowish; midline of back with dark brown line, which nearly as wide as a scale; dorsum and upper side medium brown, the scale bases slightly lighter; posteriorly, just above black lateral band, a narrow yellowish line; the black lateral band extending from snout, through eye, upper preopercle and opercle, thence around, but not on, pectoral fin base, backward to base of caudal fin; the band not wider than pupil on head, but wide as eye on anterior body; it occupies the lower three-fourths of scale row five, as counted from the predorsal midline, and most of scale row six; on the caudal peduncle, it covers about the same portions of scale rows 4 and 5; lower head yellowish or dusky; one or two of scale rows below black band on body yellow, faintly outlined in brown; next rows darker and more intensely outlined in brown, tending to form a lower brown band.

Remarks. The water was 'strictly fresh' where the specimens of *M. trifasciata* were obtained.

Melanotaenia rubrostriata (Ramsay and Ogilby)

Pl. 20

Nematocentris rubrostriatus Ramsay and Ogilby, *Proc. Linn. Soc. N.S.W.* ser. 2, 1: 14 (1886); type locality, Strickland River, New Guinea.

Amneris rubrostriata Whitley, *Rec. Aust. Mus.* 19: 226, fig. 5 (1935); holotype of *N. rubrostriatus* redescribed.

Melanotaenia maculata Weber, *Nova Guinea* 5: 239, pl. 11, fig. 4 (1908); type locality, Merauke River, New Guinea.

Recorded as: *Amneris rubrostriata* Whitley (1956a): 44, fig. 6, Howard Creek Koolpinyah, Northern Territory; Whitley (1954): 27, Yam Creek, Howard Creek, and Mataranka, Northern Territory. Records probably referable: *Melanotaenia nigrans* Rendahl (1922): 177, Port Darwin, Daly River, South Alligator River, and McKinley River, Northern Territory; Paradice and Whitley (1927): 81, Howard River, Darwin.

Billabong, about 3 miles above Roper River Mission, Northern Territory, 17 August 1948 (Blitner), USNM 173751, 13 specimens, 29 to 86 mm. standard length.

M48-35, Billabong, Oenpelli, 29 September 1948, USNM 173752, 34 specimens, 26 to 45 mm. standard length; 16 October 1948, USNM 173753, 78 specimens, 12 to 60 mm. standard length.

M48-37, Fish Creek, USNM 173754, 24 specimens, 40 to 59 mm. standard length.

M48-38, Oenpelli Creek, below waterfall, USNM 173755, 11 specimens, 17 to 62 mm. standard length.

M48-39, Red Lily Lagoon, USNM 173756, 7 specimens, 33 to 95 mm. standard length.

Description. Counts based on 12 specimens from Roper River area and 10 from Oenpelli: rays in first dorsal fin V(in 6), VI(14), or VII(2); rays in second dorsal fin I,9(1), I,10(10), I,11(9), or I,12(2); anal rays I,17(1), I,18(2), I,19(8), I,20(7), or I,21(4); pelvic rays I,5(44); pectoral rays ii,12(16), ii,13(26), or ii,14(1); branched caudal rays 8+7=15(12); scales between head and caudal fin 31(6), 32(9), or 33(6); scale rows predorsal midline to anal fin origin 13(2), 14(8), or 15(6); predorsal scale count 13 to 17; scale rows around caudal peduncle 13(3) or 14(13); scale rows across peduncle, midline above to midline below, based on specimens from the five collections 7(6) or 8(78). Vertebral counts, based on 12 specimens from the Roper River area and 19 from Oenpelli, are: precaudal vertebrae 19(7), 20(19), or 21(5); caudal vertebrae 12(3), 13(21), or 14(7); total vertebrae 32(5), 33(19), or 34(7).

In five specimens 66 to 86 mm. in standard length, head length 3·7 to 4·0 and body depth 2·1 to 2·6, both in standard length; snout length 3·1 to 3·5 and eye diameter 3·6 to 3·9, both in head length.

Premaxillary not protractile, only slightly movable; scales in 1, more often 2 rows below eye; a small cluster of teeth on the pterygoid and a few on tongue, those on palatine and vomer in narrow lines; teeth in bands on jaws; the band reflected outward at end of lower jaw; that of upper jaw reflected outward along its length, with slightly larger teeth externally.

Colour in alcohol. Body in young generally brownish, lighter below, with lengthwise narrow brown lines posteriorly. Fins brownish, sometimes with transluscent spotting.

In largest specimen, first dorsal fin dark brown, blackish posteriorly with transluscent spotting; second dorsal fin, caudal, and anal fins dark brown, with lines of oblong transluscent or whitish (red in life?) spots on membrane between the fin rays; outer pelvic rays brown, the inner rays yellowish; pectoral fin yellowish; dorsum with a broad, dark brown median band; head and anterior side of body nearly uniformly light brown to medium brown; a faint brownish band about as wide as eye, extending backward from head along middle of body; rear part of trunk and caudal peduncle with alternating yellow and brown lines; the yellow lines half as wide as a scale, due to a round yellow spot on each scale centre; the brown lines at junction of scale rows; those lines at midside darkest.

Remarks. Several earlier named species of *Melanotaenia* that have been described from Australia are suspected to be identical, or conspecific, with the present material. Their descriptions, however, are too incomplete to permit evaluation.

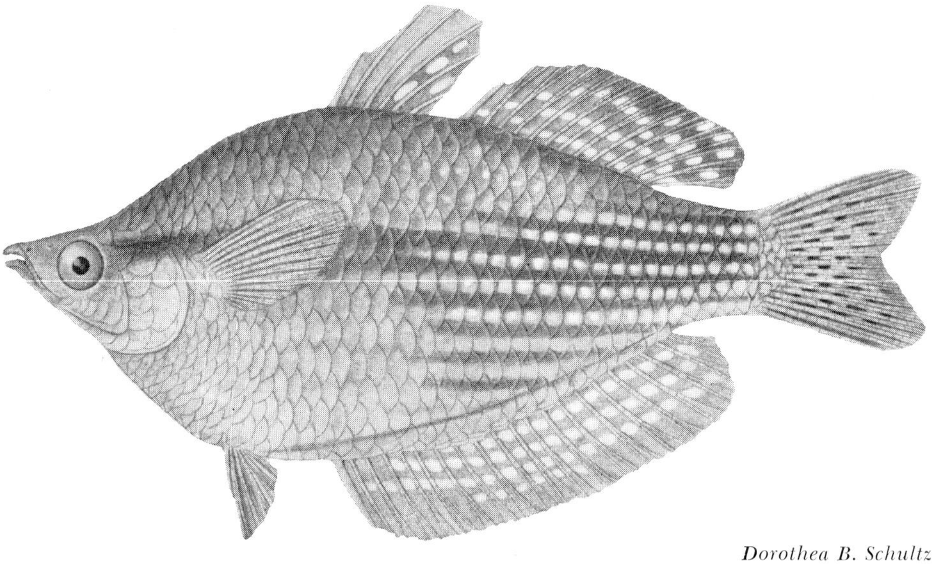

Dorothea B. Schultz

PLATE 20. *Melanotaenia rubrostriata*
Red Lily Lagoon, USNM 173756, 95 mm. standard length

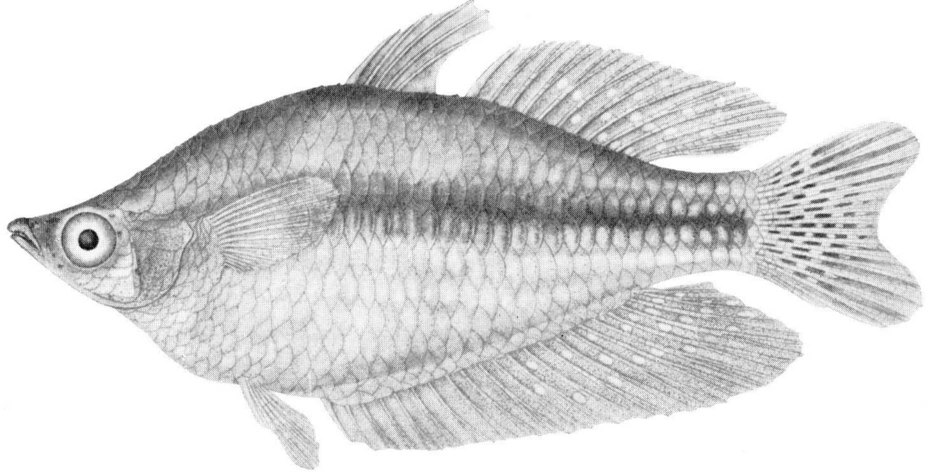

Dorothea B. Schultz

PLATE 21. *Melanotaenia solata* sp. nov., holotype
Emerald River, Australian Museum register no. IB.5236, 100 mm. standard length

Melanotaenia solata sp. nov.

Pl. 21

Holotype: Australian Museum register No. IB.5236, a male 100 mm. in standard length taken from Emerald River (Station M48-14), at old mission site, about 3·5 miles above river mouth, Groote Eylandt, Northern Territory, on 4 June 1948, by Robert R. Miller, Jerry Blitner, Fred Gray and natives.

Paratypes: Emerald River, Groote Eylandt (Station M48-14), taken with holotype and bearing same data, USNM 173758, 95 specimens, 20 to 104 mm. standard length.

Billabongs adjacent to Emerald River (Station M48-16) at old mission site, Groote Eylandt, Northern Territory, 5 June 1948, USNM 173759, 16 specimens, 14 to 76 mm. standard length.

Referred material: M48-18, Freshwater rock pools, tributary to north side of south bay, Bickerton Island, Northern Territory, 8 June 1948, USNM 173760, 1032 specimens, 11 to 56 mm. standard length.

TABLE I

MEASUREMENTS OF *Melanotaenia solata* EXPRESSED IN THOUSANDTHS OF THE STANDARD LENGTH

	Holotype Australian Museum Register No. IB.5236	Paratypes USNM 173758	
	♂	♂	♀
Standard length, mm.	100	79	92
Length of head	255	268	272
Greatest depth of body	404	394	395
Length of snout	80	90	83
Diameter of eye	71	76	74
Postorbital length of head	103	101	109
Interorbital width	101	101	102
Length of caudal peduncle	161	148	163
Least depth of caudal peduncle	126	129	125
Width of head	133	130	134
Distance from snout tip to pelvic fin	368	373	391
Distance from snout tip to anal fin	473	500	516
Distance from snout tip to first dorsal fin	501	506	522
Distance from snout tip to second dorsal fin	644	678	678
Distance from snout tip to pectoral fin	268	276	280
Length pectoral fin	184	191	185
Longest dorsal spine	178	196	167
Longest dorsal ray	218	197	141
Longest anal ray	268	266	189
Length caudal fin	254	237	233

M48-29, Mud pool, near small creek at Yirrkalla, Northern Territory, 30 July 1948, USNM 173761, 31 specimens, 30 to 43 mm. standard length.

M48-34, Creek at Yirrkalla Mission, Northern Territory, 21 August 1948, USNM 173762, 9 specimens, 21 to 38 mm. standard length.

Diagnosis. A species of *Melanotaenia* with a rather slender, compressed body; with complete dentition which poorest developed in upper jaw; with a rather faint brown lateral body band and with numerous, characteristic, golden yellow life stripes through brown band as well as along body.

Description of holotype. Dorsal rays V-I,11; anal rays I,21; pelvic rays I,5, both sides; branched caudal rays 8+7=15; pectoral rays ii,10 on left side and ii,12 on right; scales between head and caudal fin base 34; scale rows between anal fin origin and predorsal midline 13; those between midline of breast and predorsal midline, inclusive, 15; scale rows around caudal peduncle 12; those across caudal peduncle, midline above to midline below 7; scales in two rows below pupil; scales between head and dorsal fin 16; gillrakers on first right arch, short, 5+16=21.

Description. Specimens from Emerald River: rays in first dorsal fin V(in 5), VI(4), or VII(1); rays in second dorsal fin I,10(3) or I,11(7); anal rays I,19(2), I,20(7), or I,21(1); pelvic rays I,5(20); pectoral rays ii,11(3), ii,12(8), ii,13(5), or ii,14(4); branched caudal rays 8+7=15(10); scales between head and caudal fin 33(4), 34(9), or 35(2); scale rows between anal fin origin and predorsal midline 12(1), 13(7), 14(1), or 15(1); scales midline between head and dorsal fin 14 to 16; scale rows around caudal peduncle 10(1), 11(1), 12(3), 13(4), or 14(3); scale rows across side of caudal peduncle, midline above to midline below, 6(4), 7(23) or 8(12).

In referred specimens from Yirrkalla: first dorsal fin rays V(1) or VI(6); second dorsal fin rays I,9(4) or I,10(3); anal rays I,18(5) or I,19(2); pelvic rays I,5(18); pectoral rays ii,12(1), ii,13(10) or ii,14(3); branched caudal rays 7+7=14(1) or 8+7=15(5); scale rows between head and caudal fin 32(1), 33(4) or 34(3); scale rows across body 11(1), 12(4), or 13(2); predorsal scales 13 to 17; scale rows around caudal peduncle 12(6) or 13(1).

Scale rows across caudal peduncle, midline above to midline below, based on specimens from the five collections 6(4), 7(97) or 8(20).

Vertebral counts from Emerald River are: precaudal 20(2) or 21(8); caudal 13(1), 14(8), or 15(1); total 34(2) or 35(8); counts from Bickerton Island specimens are: precaudal 18(1), 19(9), or 20(5); caudal 13(1), 14(4), 15(9), or 16(1); total 32(1), 33(2), 34(9), or 35(3); from Yirrkalla: precaudal 19(4), 20(4), or 21(2); caudal 14(3), 15(5), or 16(2); total 34(4), 35(5), or 36(1).

Additional measurements of typical material are given in Table 1. In six specimens, 75 to 94 mm. in standard length, head length 3·8 to 4·2, body depth 2·6 to 3·1, distance from tip of snout to pelvic fin 2·7 to 2·9, distance from snout tip to dorsal fin 2·0 to 2·2, all in standard length; length of pectoral fin 1·3 to 1·6, least depth of caudal peduncle 2·2 to 2·5, eye diameter 3·3 to 3·6, length of snout 3·1 to 3·5, and postorbital head length 2·5 to 2·8, all in head length.

Teeth are present in a small cluster on both the pterygoid and tongue, and in narrow bands on the vomer and palatine; teeth on lower jaw in a narrow band, broadest anteriorly, where exposed, the outer ones largest; those on upper jaw in a single row of rather large exposed teeth posteriorly and in several rows anteriorly where teeth are smallest, and only the outer ones exposed; jaws about equal, the upper with rear six-tenths abruptly bent downward; premaxilla not protractile, a distinct frenum present; first dorsal fin with a stout spine followed by flexible spines that reach to second dorsal fin; spines in second dorsal and anal fins stout, generally curved backward, the posterior rays of these fins, in males reaching well beyond base of caudal fin and remote from caudal fin in females; pelvic fin bound to abdomen by membrane extending between tip of vent and pelvic fin base; caudal fin posteriorly emarginate; body strongly compressed, rather elongate, the greatest depth below origin of first dorsal fin; upper body profile a gentle curve; lower profile nearly straight above anal fin, thence as a prominent curve below abdomen to isthmus.

Colour in alcohol. First dorsal fin dark brown, black posteriorly; second dorsal and anal fins brown or black with rows of tiny yellowish or transluscent spots on membranes, the distal margin black in young males, dusky in females; caudal fin greyish, the membrane of its central portion black with numerous yellow or transluscent spots which about half as large as black interspaces; pelvic fin dusky, the outer rays darkest; pectoral fin yellowish; body dark brown above and yellowish brown below; dorsum with a broad black band along midline; centre of scales on body yellowish brown, becoming yellow on rear end of body and lower side; the spots smallest above and largest below, tending to form longitudinal yellow lines, which delimited by narrow medium brown lines through scale junctions; an irregular dark brown band on lower three-fourths of scale row five from predorsal midline and upper three-fourths of scale row six; the band broken posteriorly by yellow scale centres; head medium brown, yellowish below.

Colour in life. Large adult topotypes with diffuse dark band and about ten brilliant golden longitudinal stripes on each side; basal half of caudal fin bright yellow; bronze bar behind eye crossing preopercle and diffusing downward on opercle; belly and bases of second dorsal and anal fins pinkish, the inter-radial membranes paler outward; inter-radial membranes of first dorsal fin red.

In specimens from Yirrkalla, the dorsal and anal fins red; caudal fin yellowish orange; pelvic fins deep red; lower side bluish silvery; scale centres on side generally golden bronze; they form about five longitudinal rows, the lowermost of which is approximately on a level with the ventral surface of the caudal peduncle.

Remarks. The name *solata* is from *solatus* Latin, meaning sunburned. It is given to this species of sunfish in reference to the characteristic golden yellow life colours.

*Quirichthys** *stramineus* (Whitley)

Quiris stramineus Whitley, *Rec. Aust. Mus.* 22: 239, fig. 3 (1950); type locality, Katherine River, Northern Territory.

In the material that I have been able to examine of the genus *Pseudomugil*,† the first dorsal fin has four to six slender, flexible spines; the second dorsal fin has no spines; its first ray is segmented and simple. In *Pseudomugil gertrudae* the first anal ray is a short spine and the second is segmented and usually simple; in *Pseudomugil signifer* Kner, and in the new Arnhem Land species, the first anal ray is segmented and simple. The group is characterized by a reduced number of branched caudal rays. I find that they range in number, in the above mentioned species, from 11 to 14, whereas in the other atherinidae studied, there are normally 8+7=15. In the remaining nominal forms, referred to the genus, these characteristics were not described.

In the two Arnhem Land species, teeth in bands on both jaws; mandibular ramus not elevated; premaxilla protractile, its median process short; margin of preopercle entire, not indented; no scales on fins; air bladder ending over anterior half of anal fin; pelvic rays extending to anal fin or nearly so.

Pseudomugil gertrudae Weber

Pseudomugil gertrudae Weber, *Abhandl. Senckenb. Nat. Ges.* 34: 23, figs. 1 and 2 and pl. 1, fig. 4 (1913); type locality, Aru Islands.

* *Quiris* Whitley (1950): 239; type species, *Quiris stramineus* Whitley, by original designation; preoccupied by *Quiris* Pate in Hymenoptera. *Quirichthys* Whitley (1951): 63, new name to replace *Quiris* Whitley.

† *Pseudomugil* Kner (1866): 543; type species, *Pseudomugil signifer* Kner, by monotypy.

M48-13a, South end of Winchelsea Island, USNM 174227, 1 specimen, 14 mm. standard length.

M48-14, Emerald River, Groote Eylandt, USNM 174228, 1 specimen, 11 mm. standard length.

M48-15, Emerald River, Groote Eylandt, USNM 174229, 3 specimens, 10 to 12 mm. standard length.

M48-16, Billabongs adjacent to Emerald River, Groote Eylandt, USNM 174230, 10 specimens, 14 to 19 mm. standard length.

Description. Dorsal rays IV-i,5(in 2), IV-i,6(1), or V-i,6(1); anal rays I,i,8(1), I,i,9(2) or I,i,10(1); pelvic rays I,5(8); pectoral rays ii,8(4) or ii,9(2); scales, midline, between head and dorsal fin 11(4); scale rows across caudal peduncle, midline below to midline above inclusive 7(4); scale rows across body, predorsal midline to abdominal midline inclusive 9(6); branched caudal rays 6+6=12(4); scales between head and caudal fin 27(2) or 28(2).

Vent opening about midway between bases of pelvic and anal fins; scales on cheek in one row; gillrakers slender, less than half as long as filaments.

Colour in alcohol. Body yellowish brown; a black band along upper and lower midline of caudal peduncle, curving laterally on to base of caudal fin; prominent black spots, one per scale, in longitudinal lines on body; the spots absent from caudal peduncle, except its middle, where they form a rather solid band that extends to caudal fin base; pelvic and pectoral fins dusky; remaining fins with prominent black spots similar to those of body.

Remarks. Small specimens, USNM 132603, from Cairns, Queensland, collected 19 August 1942, agree in morphology with the Arnhem Land material, but are not as heavily pigmented. I find in them, dorsal rays IV-i,5(in 1) or V-i,5(1); anal rays I,i,8(2) or I,i,9(1); pelvic rays I,5(2), both sides; pectoral rays ii,8(2), both sides; branched caudal rays 6+6=12(1); scales, midline, between head and dorsal fin 9(1) or 10(1); scale rows across caudal peduncle, midline below to midline above inclusive 7(2); scale rows across body, predorsal to abdominal midlines inclusive 9(4); scales between head and caudal fin 27(1) or 28(1).

Pseudomugil tenellus sp. nov.

Pl. 22

Holotype: Australian Museum register No. IB.5237, 19 mm. in standard length, taken from the large billabong at Oenpelli (Station M48-35), Northern Territory, on 29 September 1948 by Robert R. Miller and party.

Paratypes: Billabong, Oenpelli (Station M48-35), taken with holotype and bearing same data, USNM 174252, 66 specimens, 17 to 22 mm. standard length; same station and locality, 16 October 1948, USNM 174253, 90 specimens, 17 to 21 mm. standard length.

Oenpelli Creek (Station M48-38), below waterfall, about 4 miles south-south-east of Oenpelli, Northern Territory, 4 October 1948, USNM 174254, 2 specimens, 16 to 18 mm. standard length.

Red Lily Lagoon (Station M48-39), about 7 miles west-south-west of Oenpelli, Northern Territory, 10 October 1948, USNM 174255, 28 specimens, 15 to 26 mm. standard length.

Diagnosis. A species of *Pseudomugil* related to *Pseudomugil signifer* Kner by lacking spines in both the second dorsal and anal fins; characterized by a sombre coloration, and short bases to the second dorsal and anal fins, the anal rays usually i,8 or i,9.

Description of holotype. Dorsal rays IV-i,5; anal rays i,8; pelvic rays I,5, both sides; pectoral rays ii,10 on right side, left fin damaged; branched caudal rays 6+7=13; scales between head and caudal fin 28; scales, on midline, between head and dorsal fin 10; scale rows across caudal peduncle, midline below to midline above 7; scale rows across body, abdominal midline to predorsal midline 10; cheek scales in one row; gillrakers very short, stump-like, 8 on lower limb of first arch. Measurements are given in Table 2.

TABLE 2

MEASUREMENTS OF *Pseudomugil tenellus* EXPRESSED IN THOUSANDTHS OF THE STANDARD LENGTH

	Holotype Australian Museum Register No. IB.5237	Paratype USNM 174255
Standard length, mm.	19	23
Length of head	279	291
Greatest depth of body	221	248
Length of snout	63	74
Diameter of eye	89	87
Postorbital length of head	126	130
Interorbital width	100	113
Length caudal peduncle	274	252
Least depth caudal peduncle	105	113
Width of head	147	141
Distance from snout tip to pelvic fin	437	443
Distance from snout tip to anal fin	605	626
Distance from snout tip to first dorsal fin	495	557
Distance from snout tip to second dorsal fin	668	704
Distance from snout tip to pectoral fin	268	278
Length caudal fin	242	209
Length pectoral fin	205	204
Length pelvic fin	147	157

Description. Dorsal rays IV-i,5(in 5), IV-i,6(2) or V-i,5(3); anal rays i,8(9) or i,9(1); pelvic rays I,5(20); pectoral rays ii,10 (11) or ii,11(2); branched caudal rays 6 to 7 in upper lobe, plus 5 to 7 below, equals 11(2), 12(4) or 13(2); scales between head and caudal fin 27(6), 28(3) or 29(1); scales, midline between head and dorsal fin 10 (2) or 11(8); scale rows across caudal peduncle, midline below to midline above inclusive 7(10); scale rows across body, predorsal midline to abdominal midline inclusive 10(10); gillrakers on first arch very short, stump-like, 9(2) on lower limb of arch.

Villiform teeth in narrow bands on jaws; mouth strongly inclined, opening dorsally; lower jaw terminal, its proximal end about on a level with the pupil; snout short, about as long as eye diameter; scales on head deciduous, thus not shown accurately on drawing; they cover opercles and one row extends forward below eye; vent opening beneath rear half of pectoral fin; fin rays not elongated, the pelvic rays barely touching anal fin base.

Colour in alcohol. Head dark brown above, lighter below; body yellowish brown, the scales narrowly outlined in brown, forming a reticulate pattern except over abdomen which is black due to viscera showing through body wall, abdomen otherwise yellowish; scales of back and upper side sprinkled with dark brown chromatophores; a narrow dark brown line on midline from anal fin base to caudal fin base and another, somewhat narrower, along side between epaxial and hypaxial muscles; fins light yellowish to slightly dusky without prominent pattern.

Remarks. The new species of *Pseudomugil* differs from all described species except *gertrudae* in having generally fewer rays in the second dorsal fin and considerably fewer in the anal fin. In contrast to *gertrudae,* it lacks an anal spine and spotted coloration. It has 10 scale rows from the predorsal midline to the abdominal midline, whereas in seven specimens of *P. signifer* Kner from New South Wales and in other specimens of *Pseudomugil* from Queensland that probably are *P. signatus* (Günther) the count is 8; *P. gertrudae* Weber has 9 scales across the body.

The name *tenellus,* Latin, is given to this fish because of its small dainty or delicate appearance.

Allanetta mugiloides (McCulloch)
Pl. 23

Atherinichthys punctatus De Vis, *Proc. Linn. Soc. N.S.W.* 9: 869 (1885); type locality, Cape York.

Atherina mugiloides McCulloch, *Proc. Roy. Soc. Qld.* 24: 47, text fig. 1 (1912); new name for *Atherinichthys punctatus* De Vis, a homonym of *Atherina punctata* Bennett.

M48-2, Nightcliff, USNM 173722, 2 specimens, 30 and 36 mm. standard length.

M48-8, Little Lagoon, USNM 173723, 5 specimens, 33 to 38 mm. standard length.

M48-9, Reefs and tide pools, entrance to Little Lagoon, USNM 173724, 4 specimens, 32 to 38 mm. standard length.

M48-15, Emerald River mouth, Groote Eylandt, USNM 173725, 5 specimens, 20 to 32 mm. standard length.

M48-17, Reef, Bartalumba Bay, USNM 173726, 4 specimens, 27 to 31 mm. standard length.

M48-19, Little Lagoon at creek mouth, USNM 173727, 2 specimens, 38 to 40 mm. standard length.

Description. Rays in first dorsal fin VI(in 2) or VII(2); rays in second dorsal fin I,i,7(4); anal rays I,i,8(3) or 9(1); branched caudal rays 8+7=15(4); pelvic rays I,5(8); pectoral rays ii,12(4); scales between head and caudal fin 34(1) or 35(3); scales, midline between head and dorsal fin 10 to 12, usually 11; scales predorsal midline to abdominal midline inclusive 8(4); gillrakers slender, slightly shorter than gill filaments, on first arch 5 or 6 above, plus 1, plus 18(1), 19(1) or 20(2). No teeth on vomer; a few scattered teeth along palatine and a tiny cluster on both the tongue and pterygoid; teeth on each jaw in a narrow band; rear preopercular margin truncate, not indented; ramus of mandible distinctly elevated posteriorly; vent beneath distal half of pelvic fin; dorsal, anal and pectoral fins naked; premaxilla protractile, its median process about equal to or greater than pupil; scales on cheek in one row; air bladder ending approximately opposite anal fin origin; pelvic fin inserted closer to pectoral fin base than to anal fin.

Colour in alcohol. Body light yellowish; midline of back with a brown line; scale pockets on dorsum and upper side poorly outlined in brown; lateral dark brown band from head to caudal fin, approximately half as wide as a scale, through fourth scale row from midline; pigment of lateral band extending outward between bases of caudal rays, forming a rounded blotch on that fin; a line of small brown chromatophores, each side of anal fin, extends forward to vent; a prominent and characteristic dark brown spot at lower base of pectoral fin.

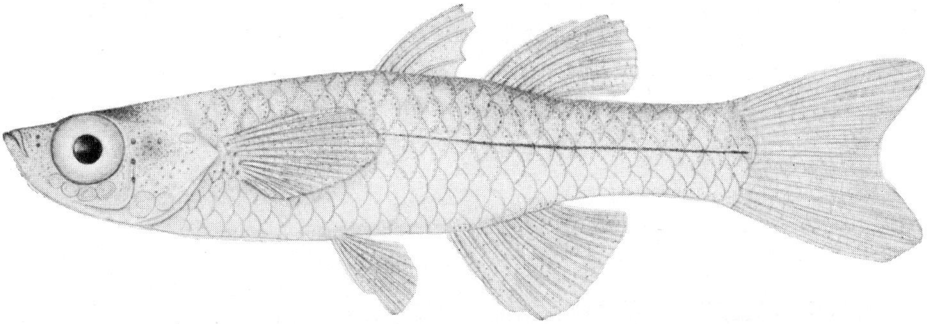

Dorothea B. Schultz

PLATE 22. *Pseudomugil tenellus* sp. nov., holotype
Oenpelli, Australian Museum register no. IB.5237, 19 mm. standard length

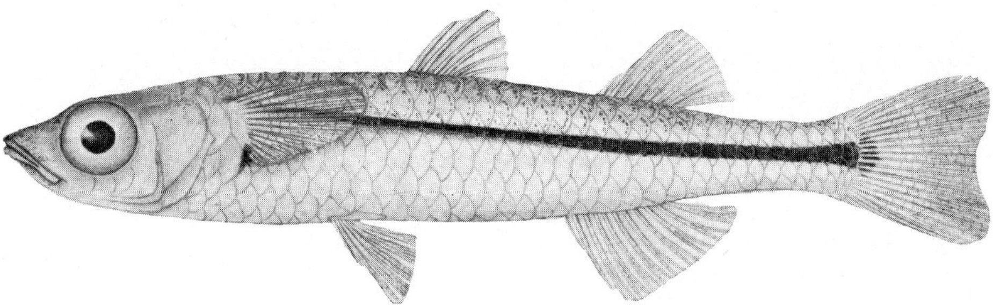

Dorothea B. Schultz

PLATE 23. *Allanetta mugiloides*
Little Lagoon, USNM 173724, 38 mm. standard length

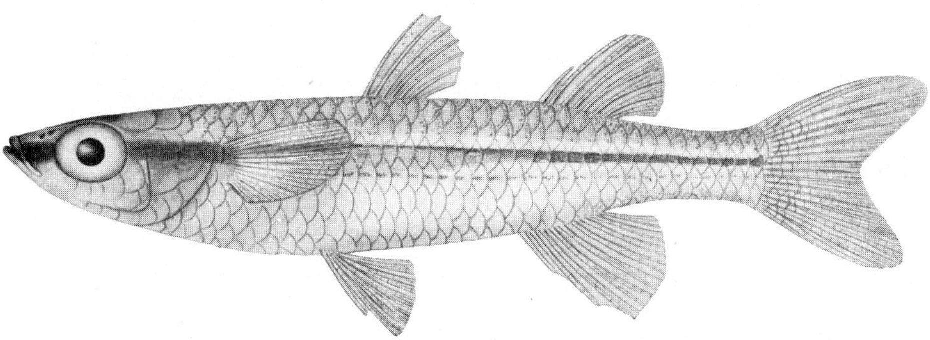

Dorothea B. Schultz

PLATE 24. *Craterocephalus fluviatilis*
Oenpelli, USNM 174238, 33 mm. standard length

Remarks. The Arnhem Land material agrees well with the description of *Atherinichthys punctatus* De Vis, and the description and figure of *Atherina mugiloides* McCulloch, the genotype of *Allanetta.* Neither McCulloch's figure nor the present specimens exhibit the indented preopercle that was described by Schultz (1948) and Schultz *et al.* (1953) for this genus. The cephalic sensory pits that are enlarged in both *Pranesus* and the material assigned to *Allanetta* by Schultz are small in *A. mugiloides.* Consequently, if the present interpretation proves correct, certain species currently assigned to *Allanetta* must be removed.

The relationship of this species seems to be closest to *Craterocephalus* but the dentition is better developed and the mouth larger. As mentioned by De Vis, a characteristic black dot is located at the lower pectoral fin base.

The genus *Craterocephalus* McCulloch has been characterized by a short mouth, not reaching backward to eye; short gillrakers, with generally fewer than 13 on lower limb of first arch; moderately large scales, usually 25 to 35 along side of body; preopercular margin not indented posteriorly; and vent forward, generally beneath pelvic fin. Most species, including the Arnhem Land collection material, have a moderate to strong elevation of the mandibular ramus. However, *Craterocephalus worrelli* Whitley, described from the Roper River, is stated to have this area not elevated, and its gillrakers are not mentioned. It is obvious that the genus requires study, since diverse forms are included.

Craterocephalus worrelli Whitley

Craterocephalus worrelli Whitley, *Rec. Aust. Mus.* 22: 86 (1948); type locality, Mataranka, Roper River system, Northern Territory.

Craterocephalus eyresi (Steindachner)

Atherinichthys eyresii Steindachner, *S.B. Akad. Wiss. Wien* 88: 1075 (1884); type locality, probably Lake Eyre.

Recorded as: *Craterocephalus eyresi* Munro (1958): 102, inland Northern Territory.

Craterocephalus pauciradiatus (Günther)

Atherina pauciradiata Günther, *Catalogue of the Fishes in the British Museum* 3: 401 (1861); type locality, north-west coast of Australia.

Craterocephalus pauciradiatus Whitley, *Proc. Linn. Soc. N.S.W.* 68: 135, fig. 10, no. 1 (1943); type redescribed.

Craterocephalus capreoli Rendahl, *Medd. Zool. Mus. Krist.* 5: 175, fig. 6 (1922); type locality, Roebuck Bay, Australia.

M48-1, Nightcliff, USNM 173716, 3 specimens, 27 to 40 mm. standard length.

M48-7, Mouth of billabong, at Little Lagoon, USNM 173717, 1 specimen, 41 mm. standard length.

M48-17, Reef, Bartalumba Bay, USNM 173718, 1 specimen, 36 mm. standard length.

M48-19, Little Lagoon at mouth of creek, USNM 173719, 2 specimens, 18 and 37 mm. standard length.

M48-21, Reef, Yirrkalla, 14 July 1948, USNM 173720, 3 specimens, 26 to 40 mm. standard length; from crevices in ironstone rocks on reef, 1 August 1948, USNM 173721, 7 specimens, 26 to 38 mm. standard length.

Description. Rays in first dorsal fin IV(in 3) or V(2); rays in second dorsal fin I,i,4(1) or I,i,5(4); anal rays I,8(4) or I,9(1); branched caudal rays 8+7=15(5);

pelvic rays I,5 (10); pectoral rays ii,11(8) or ii,12(2); scales between head and caudal fin 30(1), 31(3), or 32(1); scale rows predorsal midline to abdominal midline inclusive 8(5); scales midline, between head and dorsal fin 12 or 13; gillrakers slender, rather short, much shorter than filaments, 4 or 5 on upper arch plus 10(1), 11(3) or 12(1) on lower arch, equal 14(1), 15 (1), 16(2) or 17(1).

Upper lip expanded, thickened; mouth short, not reaching eye; anal opening between proximal halves of pelvic fin; premaxilla protractile, the median process about as long as pupil diameter; cheek scales in one row; pelvic fin base closer to pectoral fin base than to anal fin; air bladder ending posteriorly above anal fin origin; bases of dorsal, anal and pectoral fins naked; preopercle truncate behind; ramus of mandible distinctly elevated posteriorly; teeth in a narrow band on each jaw, but none on palatine or vomer, and apparently none on tongue or pterygoid; snout length equal to eye diameter.

Colour in alcohol. Body yellowish brown; a brown line along midline of back; scales on dorsum and upper side outlined in brown; lateral silvery band from head to caudal fin about four-tenths of a scale width, on fourth scale row from dorsal midline; lateral band extending outward between bases of caudal rays, forming an obscure, round spot on caudal fin base; lower side pale or with a line of dots along each scale row.

Craterocephalus fluviatilis McCulloch
Pl. 24

Craterocephalus fluviatilis McCulloch, *Proc. Roy. Soc. Qld.* 24: 49, pl. 1, fig. 1 (1912); type locality, North Yanko Creek, Narrandera, New South Wales.

M48-35, Billabong, Oenpelli, 29 September 1948, USNM 174237, 42 specimens, 17 to 33 mm. standard length; 16 October 1948, USNM 174238, 108 specimens, 9 to 37 mm. standard length.

M48-39, Red Lily Lagoon, USNM 174239, 1 specimen, 32 mm. standard length.

Description. Rays in first dorsal fin V(1), VI(4) or VII(5); rays in second dorsal fin I,i,5 (2), I,i,6(7) or I,i,7(1); anal rays I,i,6(2), I,i,7(6) or I,i,8(2); pelvic rays I,5(10); pectoral rays ii,10(3) or ii,11(15); branched caudal rays 8+7=15; scale rows from predorsal to abdominal midlines inclusive 10(10); scales between head and caudal fin 32(4), 33(3) or 34(3); gillrakers very short, stump-like, about as long as width of base, 4 or 5 on upper arch, plus 8(2) or 9(3) on lower arch; anus beneath distal one-third of depressed pelvic fin; upper lip thickened, fleshy, the mouth short, not reaching eye; premaxilla protractile, its median process about equal to pupil diameter; scales in one row on cheek; dorsal, anal, and pectoral fins naked; ramus of mandible elevated posteriorly; preopercle truncate behind; pelvic fin base closer to pectoral fin base than to anal fin; air bladder ending posteriorly above anal fin origin; teeth in one or two rows on upper jaw and in a narrow band on lower; no teeth on palatine, vomer, or pterygoids and apparently none on tongue.

Colour in alcohol. Body and head yellowish brown; a black band, broadest and most prominent anteriorly where wide as pupil, extending from tip of lower jaw through eye to caudal fin base; the band located on the fifth and upper part of the sixth row of scales from upper midline anteriorly and on the fourth row of scales on the caudal peduncle; fins yellowish or dusky; scales on dorsum and upper side dusky, margined with dark brown; lower side pale, but with a line of chromatophores on scale row below lateral band on caudal peduncle, the line joining lateral band anteriorly.

Remarks. This is probably the atherinid referred to by Miller as having a butter-yellow belly.

This species and *pauciradiatus* differ not only in coloration, but also in the number of scale rows across the body, in fin ray counts, and in the branching of the segmented first anal ray.

Key to the Arnhem Land Species of *Pranesus*

1 Scales between head and caudal fin 36 to 39; lateral silvery band slightly narrower than pupil diameter, ending abruptly at base of caudal fin; lines of chromatophores on scales below lateral band, if present, not joining it . . *Pranesus endrachtensis*

Scales between head and caudal fin 41 to 44; lateral silvery band broader than pupil diameter; dark or dusky projections from lateral band extending on to base of caudal fin 2

2 Tip of pectoral fin blackened; line of chromatophores on scale row immediately below lateral band separate from band between head and abdomen, thence joining lateral band from above or before anal fin origin to caudal fin; a blackish blotch on lower side at anterior end of anal fin; gillrakers on lower limb of first arch 20 to 23 . . . . . . *Pranesus maculatus*

Pectoral fin yellowish, with little pigmentation; line of chromatophores on scale row below lateral band separate from it or joining band above or behind middle of anal fin; no blotch at anal fin; gillrakers on lower limb of first arch 18 to 20 . . . . . . *Pranesus vaigiensis*

In the Arnhem Land species of *Pranesus*, the body is robust; the head is large and broad, with enlarged and prominent openings to the sensory canal system; the shelf of the vertical arm of the preopercle is distinctly indented; the mandibular ramus is straight, without elevation; the dorsal, anal, and pectoral fin bases are scaleless; the mouth is large, the jaw extending to below the eye; the lateral premaxillary process is narrow and the median one is short, shorter than the pupil; the premaxillary is protractile; the pelvic fin base is closer to the pectoral fin base than to the anal fin; and the posterior end of the air bladder is approximately over the origin of the anal fin.

Small teeth cover the jaws; a few rows of them reflected on to outside of jaw and visible when mouth is closed; vomerine tooth band moderately long, in a straight line across mouth; teeth on tongue in a broad elongate band; palatine and pterygoid teeth in narrower bands.

The following Northern Territory records probably apply to either *Pranesus vaigiensis* or *P. maculatus*: *Hepsetia pinguis* Paradice and Whitley (1927): 81, Pellew Islands; *Pranesus ogilbyi* Munro (1958): 101.

Pranesus endrachtensis (Quoy and Gaimard)

Atherina endrachtensis Quoy and Gaimard, *Voyage autour du monde sur... l'Uranie et la Physicienne... Zoologie*: 334 (1824); type locality, Sharks Bay, Western Australia.

M48-2, Nightcliff, USNM 173728, 69 specimens, 34 to 47 mm. standard length.

M48-7, Mouth of billabong, Little Lagoon, USNM 173729, 6 specimens, 30 to 38 mm. standard length.

M48-8, Little Lagoon, USNM 173730, 50 specimens, 27 to 51 mm. standard length.

M48-9, Reefs and tide pools, entrance to Little Lagoon, USNM 173731, 9 specimens, 24 to 40 mm. standard length.

M48-13, South bay of Bickerton Island, USNM 173732, 17 specimens, 24 to 42 mm. standard length.

M48-13a, South end of Winchelsea Island, USNM 173733, 7 specimens, 37 to 44 mm. standard length.

M48-15, Emerald River, Groote Eylandt, at mouth, USNM 173734, 4 specimens, 26 to 36 mm. standard length.

M48-17, Reef, Bartalumba Bay, USNM 173735, 77 specimens, 35 to 47 mm. standard length.

M48-19, Mouth of creek and Little Lagoon, USNM 173736, 13 specimens, 23 to 29 mm. standard length.

M48-25, Beach, island, Port Bradshaw, USNM 173737, 45 specimens, 36 to 51 mm. standard length.

M48-27, Bay and beach, Port Bradshaw, USNM 173738, 71 specimens, 28 to 49 mm. standard length.

M48-28, Creek mouth and bay, Port Bradshaw, USNM 173739, 21 specimens, 35 to 45 mm. standard length.

Description. Rays in first dorsal fin IV(in 2), V(6) or VI(2); rays in second dorsal fin I,i,7(1) or 8(9); anal rays I,i,11 (8) or 12(2); pelvic rays I,5(12); pectoral rays ii,13(1), ii,14(7) or ii,15(3); branched caudal rays 8+7=15(6); scales between head and caudal fin 37(2) or 38(4); scales midline between head and dorsal fin 15 to 17, usually 16; scale rows predorsal midline to abdominal midline inclusive 8(6); gillrakers on first arch slender, elongate, longer than filaments, 4 or 5 on upper limb, plus 1, plus 17(1), 18(1), 19(3) or 20(1) below; scales in one row across cheek; vent beneath middle of pelvic fin.

Colour in alcohol. Background body colour generally yellowish; a fine brown line along middle of back; upper two or three scale rows outlined in brown with heaviest pigment near scale centre, tending to form longitudinal lines; lateral silvery band through middle of scale row four, counted from predorsal midline; its width nearly equal to pupil anteriorly, or about four-tenths postorbital head length, narrowest on caudal peduncle; band not extending on to caudal fin, which with rounded area of lines of pigment between rays near base; lower side yellowish or with a longitudinal line of brown dots on each scale row, the upper line remaining distinct from lateral silvery band throughout its length; each side of abdominal midline a brown line on median scale row; each line broadens into a narrow band around anal fin base; a median fine brown line from anal fin base to caudal fin; fins otherwise yellowish, without prominent markings.

Pranesus vaigiensis (Quoy and Gaimard)
Pl. 25

Atherina vaigiensis Quoy and Gaimard, *Voyage autour du monde sur*. . .*l'Uranie et la Physicienne*. . .*Zoologie*: 335 (1824); type locality, Waigeu; Schultz, *U.S. Nat. Mus. Bull.* 202, 1: 310 (1953).

Atherina cylindrica Cuvier and Valenciennes, *Histoire naturelle des poissons* 10: 453 (1835); type locality, Waigeu.

M48-7, Mouth of billabong, Little Lagoon, USNM 174256, 1 specimen, 85 mm. standard length.

M48-8, Little Lagoon, USNM 174257, 2 specimens, 61 and 62 mm. standard length.

M48-17, Reef, Bartalumba Bay, USNM 174240, 1 specimen, 75 mm. standard length.

M48-21, Reef, Yirrkalla, 14 July 1948, USNM 174241, 8 specimens, 73 to 94 mm. standard length.

M48-25, Beach, island, Port Bradshaw, USNM 174242, 12 specimens, 66 to 79 mm. standard length.

M48-27, Bay and beach, Port Bradshaw, USNM 174243, 1 specimen, 82 mm. standard length.

M48-28, Creek mouth and bay, Port Bradshaw, USNM 174244, 20 specimens, 70 to 84 mm. standard length.

Description. Rays in first dorsal fin V(in 4) or VI(2); rays in second dorsal fin I,i,8(6); anal rays I,i,12(5) or 13(1); pelvic rays I,5(12); pectoral rays ii,15(7) or ii,16 (5); branched caudal rays 8+7=15(6); scales between head and caudal fin 41(1), 42(3) or 43(2); scales midline between head and dorsal fin 17 or 18; scale rows predorsal midline to abdominal midline inclusive 8(6); gillrakers on first arch slender, elongate, longer than filaments, 5 or 6 on upper limb, plus 1, plus 18(1), 19(3) or 20(2) on lower limb; scales in one row across cheek; vent opening beneath distal one-fourth or tip of pelvic fin.

In six specimens 70 to 77 mm. standard length, head length 3·4 to 3·6, pectoral fin 4·1 to 4·4, body depth 4·4 to 4·8, all in standard length; eye 2·2 to 2·5 in head length; width of lateral silvery band, above anal fin origin, 1·8 to 2·1 in postorbital length of head.

Colour in life. Lateral band steel grey; orange reflecting spot on middle of opercle.

Colour in alcohol. Background body colour generally yellowish; scales of upper side and dorsum sprinkled with and prominently outlined in brown, with a pigment concentration near scale centre forming longitudinal lines; brown line at midline of back broad and most intense; lateral silvery band wider than pupil, on lower two-thirds of scale row four and adjacent one-third of scale row five, counted from predorsal midline; its width about 50 per cent or more of post-orbital head length; posteriorly it extends on to base of caudal fin as a dusky blotch, of which a small upper area is intensely black; lower side yellowish, each scale often with a brown chromatophore; these arranged in longitudinal lines; upper line of chromatophores fusing with lateral band on rear part of caudal peduncle, to rear of anal fin origin; fine lines of chromatophores along midline of abdomen and caudal peduncle, the bases of the anal rays dusky; except as noted above, all fins generally yellowish.

Remarks. Data from the Arnhem Land specimens agree with those given by Schultz *et al.* (1953) for the type of *Atherina vaigiensis* Quoy and Gaimard. He has placed *vaigiensis* in the synonymy of *Pranesus pinguis* (Lacépède), but since the pectoral fin of the specimens is uniformly yellow and the colour of that fin was not mentioned by either Quoy and Gaimard or Cuvier and Valenciennes, although the latter authors discussed the fin colour in *pinguis*, I have applied the name *vaigiensis* to my material and tentatively recognize it as distinct from *pinguis*.

Pranesus maculatus sp. nov.

Pl. 26

Holotype: Australian Museum register No. IB.5238, 82 mm. standard length, taken on coral reef at Yirrkalla, Northern Territory (M48-21), 14 July 1948, by Robert Rush Miller and native party.

Paratypes: The following specimens are all regarded as paratypes. For complete data see list of localities in introduction.

Dorothea B. Schultz

PLATE 25. *Pranesus vaigiensis*
Yirrkalla, USNM 174241, 87 mm. standard length

Dorothea B. Schultz

PLATE 26. *Pranesus maculatus* sp. nov., holotype
Yirrkalla, Australian Museum register no. IB.5238, 82 mm. standard length

M48-8, Little Lagoon, USNM 174245, 10 specimens, 56 to 67 mm. standard length.

M48-17, Reef, Bartalumba Bay, USNM 174246, 4 specimens, 60 to 71 mm. standard length.

M48-21, Reef, Yirrkalla, 14 July 1948 (taken with holotype and bearing same data), USNM 174247, 88 specimens, 70 to 93 mm. standard length.

M48-25, Beach, island, Port Bradshaw, USNM 174248, 7 specimens, 62 to 77 mm. standard length.

M48-27, Bay and beach, Port Bradshaw, USNM 174249, 7 specimens, 66 to 81 mm. standard length.

M48-28, Creek mouth and bay, Port Bradshaw, USNM 174250, 4 specimens, 70 to 75 mm. standard length.

Diagnosis. A species of *Pranesus*, related to *Pranesus pinguis* (Lacépède), but with a broad, intensely dark lateral band; a black blotch along each side of anal fin base; and the line of chromatophores adjacent to lateral band joining it above or before anal fin origin.

Description of holotype. Dorsal rays VI-I,i,8; anal rays I,i,12; pelvic rays I,5, both sides; pectoral rays ii,16, both sides; branched caudal rays 8+7=15; scales between head and caudal fin 43; scales between head and dorsal fin 19; scale rows predorsal midline to abdominal midline inclusive 8; gillrakers on first right arch 5,1,23. Measurements are given in Table 3.

TABLE 3

MEASUREMENTS OF *Pranesus maculatus* EXPRESSED IN THOUSANDTHS OF THE STANDARD LENGTH

	Holotype Australian Museum Register No. IB.5238	Paratypes USNM 174247	
Standard length, mm.	82	80	79
Length of head	279	280	276
Greatest depth of body	216	224	220
Length of snout	67	70	71
Diameter of eye	100	100	101
Postorbital length of head	110	111	111
Interorbital width	90	100	96
Length of caudal peduncle	173	160	171
Least depth caudal peduncle	76	76	76
Width of head	166	179	178
Distance from snout tip to pelvic fin	422	438	435
Distance from snout tip to anal fin	707	725	709
Distance from snout tip to first dorsal fin	559	550	551
Distance from snout tip to second dorsal fin	727	725	722
Distance from snout tip to pectoral fin	270	275	270
Length longest dorsal ray	104	109	103
Length longest anal ray	126	126	124
Length caudal fin	195	231	192
Length pectoral fin	213	213	216
Length pelvic fin	149	135	143

Description of paratypes. Rays in first dorsal fin V(in 3), VI(12) or VII(1); rays in second dorsal fin I,i,7(2), 8(13) or 9(1); anal rays I,i,11(6), 12(9) or 13(1); pelvic rays I,5(32); pectoral rays ii,14(4), 15(25) or 16(3); branched caudal rays 8+7=15 (16); scales between head and caudal fin 41(1), 42(4), 43(7) or 44(4); scales, midline head to dorsal fin 16 to 19, most frequently 18 or 19; scale rows predorsal midline to abdominal midline inclusive 8(16); gillrakers on first arch slender, as long as filaments, 5 to 7, usually 6 above, plus 1, plus 20(2), 21(8), 22(5) or 23(1) below; scales in one row across cheek; teeth present on jaws, vomer, tongue, pterygoid and palatine; vent opening beneath tip or distal one-fourth of pelvic fin.

In seven specimens 66 to 81 mm. standard length, head length 3·4 to 3·8, pectoral fin 4·7 to 5·1 and body depth 4·5 to 4·9, all in standard length; eye 2·5 to 2·8 and snout length 4·1 to 4·4, both in head length; width of lateral band, above anal fin, 1·8 to 2·0 in postorbital length of head.

Colour in alcohol. Background body colour generally yellowish; scales of upper side and dorsum sprinkled with and outlined in brown, a brown spot on each scale forming longitudinal lines; line of spots along midline of back most intense; lateral band rather intensely blackish, half as wide as eye anteriorly, but constricted on caudal peduncle; it occupies the lower eight-tenths of scale row four and the upper one-third of scale row five as counted from predorsal midline or almost all of scale row four and the upper one-third to half of scale row five on the caudal peduncle; it extends on to the caudal fin base as two obovate or triangular blotches, of which the upper is darkest; lower side yellowish, with a line of chromatophores on scale row just below lateral band, the line joining band anteriorly, over or considerably forward of anal fin origin; an elongate blackish blotch near anterior half of anal fin base; lower surface pale, except a dark line along each side of midline on lower caudal peduncle; pectoral fin yellowish, with a black blotch, two-thirds eye diameter, on upper rays distally followed by a yellow margin; otherwise fins yellowish.

Remarks. This species is easily recognized by the intensely dark, broad lateral band and the black spot or blotch adjacent to the anal fin base, hence the name *maculatus*.

Hypoatherina panatela (Jordan and Richardson)

Atherina panatela Jordan and Richardson, *Bull. U.S. Bur. Fish.* 27: 243, fig. 6 (1908); type locality, Calayan Island, Philippines.

M48-13, Bickerton Island, USNM 174232, 1 specimen, 52 mm. standard length.

M48-13b, Winchelsea Island, USNM 174233, 1 specimen, 54 mm. standard length.

M48-17, Reef, Bartalumba Bay, USNM 174234, 1 specimen, 53 mm. standard length.

M48-21, Reef, Yirrkalla, 14 July 1948, USNM 174235, 1 specimen, 42 mm. standard length.

M48-27, Beach, Port Bradshaw, USNM 174236, 2 specimens, 54 and 64 mm. standard length.

Description. Rays in first dorsal fin V(in 2) or VI(4); rays in second dorsal fin I,i,7(1) or 8(5); anal rays I,i,10(1) or 11(5); pelvic rays I,5(6); pectoral rays ii,15(1), ii,16(4) or ii,17(3); branched caudal rays 8+7=15(3); scales between head and caudal fin 39(2), 40(2) or 41(1); scales, midline between head and dorsal fin, 16 or 17; scale rows, midline of abdomen to and including predorsal midline 8(5); gillrakers slender, the median ones about as long as filaments, 20(2), 21(2) or 22(1) on lower limb of first arch. Anus opening slightly to rear of tip of pelvic fin; mandible ramus distinctly elevated; preopercular ridge indented behind. Teeth present on both jaws, vomer, palatine, tongue, and in several irregular rows on pterygoid; bases of fins scaleless; premaxilla protractile, its median process slightly shorter than pupil; scales on cheek in one row; pelvic fin closer to pectoral fin base than anal fin. Upper median and two upper lateral scale rows with dark brown crescents near rear submarginal scale edge, and each scale with a posterior spot, forming longitudinal lines along body; lateral silvery band about three-fourths as wide as a scale anteriorly, narrower, but nearly as wide as a scale posteriorly, ending on caudal fin as a dark blotch on lower lobe and as a distinct projection on to base of upper lobe. Side below band immaculate.

POLYNEMIDAE

Eleutheronema tetradactylum (Shaw)

Polynemus tetradactylus Shaw, *General Zoology* 5: 155 (1804); type locality, India.
Polynemus caecus Macleay, *Proc. Linn. Soc. N.S.W.* 2: 354, pl. 9, fig. 1 (1878) and 5: 519 (1880); type locality, Port Darwin.

Recorded as: *Polynemus* (*Eleutheronema*) *tetradactylus* Paradice and Whitley (1927): 82, Pellew Islands.

M48-40, East Alligator River, USNM 174050, 5 specimens, 26 to 31 mm. standard length.

Remarks. These small, pale specimens all have four free pectoral appendages on each side. The two free lower rays extend backward beyond the tips of the upper pectoral rays, and the two, free upper rays reach to the rear of the origin of the anal fin. I assume that this elongation is characteristic of juveniles.

AMBASSIDAE

This group of fishes has been placed in the families Centropomidae, Ambassidae, and Chandidae. I separate them from the Centropomidae because of the presence of cycloid scales and the absence of a supramaxilla. The synopsis by Fraser-Brunner (1955) has been followed in allocating the Arnhem Land species to genera, but I believe that further study of the group will considerably realign the arrangement.

Key to the Arnhem Land Species

1 Supraorbital shelf with three to seven spines posteriorly; preorbital with a large, backwardly directed spine covering upper, anterior surface (in addition to serrae-like spines along edge of bone); interorbital ridge ending posteriorly in one or more spines; anal rays III,9; dorsal rays VII-I,8 or 9; membrane between second and third dorsal spines black . . . . *Ambassis dussumieri*

Supraorbital shelf rarely ending posteriorly in more than two spines (not counting spine at upper corner of preopercle); neither a spine at rear end of interorbital ridge nor one directed backward across upper, anterior surface of preorbital (condition in *Ambassis muelleri* not described) 2

2 Supraorbital shelf entire, without a spine at upper rear corner of eye 3

Supraorbital shelf with one or two, rarely three spines at upper rear corner of eye; body without bars or bands (a dark line at junction of epaxial and hypaxial muscles may be present) 4

3 Vertical and horizontal edge of preopercle strongly serrate; dorsal rays VII-I,9 or 10, usually 10; anal rays III,8 or 9, usually 9; four to ten brownish bars on body; gillrakers short and stump-like . . . . *Denariusa bandata*

Vertical edge and vertical ridge of preopercle about equally serrate, the serrae not prominent; dorsal rays VII-I,8; anal rays III,8; body without brown bars, but with about two light bands . . . . *Ambassis muelleri*

4 Membrane between second and third dorsal spines black, the mid-portion of

the first dorsal fin with few chromatophores; pectoral rays ii,9 to ii,11; anal rays III,7 to 9, usually III,8; dorsal rays VII-I,7 to 9, usually VII-I,8 . . *Ambassis agrammus*

Fins dusky, the heaviest concentration of chromatophores on first dorsal at its middle and rear portion, the membrane between the second and third spines lighter; pectoral rays ii,12 or ii,13; dorsal rays VII-I,9 to 11, usually VII-I,10; anal rays III,9 to 11, usually III,10 . . . . *Ambassis macleayi*

Ambassis muelleri Klunzinger

Ambassis mülleri Klunzinger, *S.B. Akad. Wiss. Wien* 80: 346, pl. 1, fig. 3 (1879); type locality, Port Darwin; Macleay *Proc. Linn. Soc. N.S.W.* 9: 7 (1884); Fraser-Brunner, *Bul. Raffles Mus.* 25: 196 (1955).

Ambassis agrammus Günther

Pl. 27

Ambassis agrammus Günther, *Ann. Mag. Nat. Hist.* ser. 3, 20: 57 (1867); type locality, Cape York.

Pseudoambassis elongatus Castelnau, *Proc. Linn. Soc. N.S.W.* 3: 44 (1878); type locality, Norman River, Gulf of Carpentaria; Whitley, *Rec. S. Aust. Mus.* 5: 358, fig. 6 (1935).

? *Ambassis dalyensis* Rendahl, *Medd. Zool. Mus. Krist.* 5: 187 (1922); type locality, Daly River, Northern Territory.

Recorded as: *Blandowskiella dalyensis* Whitley (1953): 124, fig. 1, Northern Territory records; *Blandowskiella agassizi* Whitley (1955a): 47, Northern Territory. Records probably referable: *Ambassis* (*Ambassis*) *nalua* Paradice and Whitley (1927): 84, Darwin; *Blandowskiella castelnaui* Whitley (1945): 12, Coglin Creek, Charlotte Waters, Northern Territory.

Arnhem Land; no further data, but in collection with *Apogon hyalosoma*; USNM 173830, 6 specimens, 17 to 45 mm. standard length.

M48-7, Mouth of billabong at Little Lagoon, USNM 173877, 1 specimen, 26 mm. standard length.

M48-14, Emerald River, Groote Eylandt, USNM 173829, 1 specimen, 26 mm. standard length.

Billabong, about 3 miles above Roper River Mission, Northern Territory, Blitner, 17 August 1948, USNM 173828, 37 specimens, 20 to 49 mm. standard length.

M48-35, Billabong, Oenpelli, 29 September 1948, USNM 173832, 22 specimens, 19 to 35 mm. standard length; 16 October 1948, USNM 173831, 19 specimens, 16 to 35 mm. standard length.

M48-37, Fish Creek, USNM 173825, 64 specimens, 27 to 41 mm. standard length.

M48-38, Oenpelli Creek, USNM 173826, 5 specimens, 20 to 39 mm. standard length.

M48-39, Red Lily Lagoon, USNM 173827, 83 specimens, 15 to 22 mm. standard length.

Goulburn Islands, about 180 miles east-north-east of Darwin, Northern Territory (Nick Haritos), 29 July 1948, USNM 173878, 2 specimens, 20 and 22 mm. standard length.

Description. Rays in first dorsal fin VII(in 12) or VIII(1); second dorsal fin rays I,7(2), I,8(38), I,9(7), or II,7(2), the fin united by a low membrane to first dorsal; anal fin rays III,7(3), III,8(9), or III,9(3); pectoral fin rays ii,9(2), ii,10(2), or

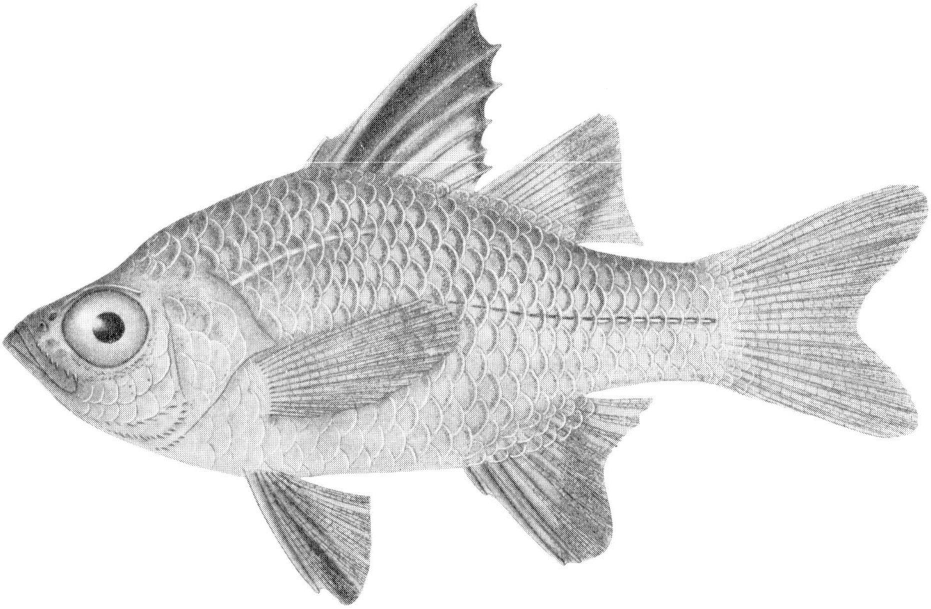

Dorothea B. Schultz

PLATE 27. *Ambassis agrammus*
Oenpelli Creek, USNM 173826, 39 mm. standard length

ii,11(8); branched caudal rays 8+7=15(5); pelvic rays I,5(10); gillrakers, lower limb of first arch, 14(1), 15(7), or 16(4), all slender; scale rows between head and caudal fin 28(1) or 29(1); predorsal scales on midline, 14(1), 15(1), or 16(1); two rows of scales on cheek; lateral line short, with a few scattered tubes along body.

Top of head without spines or serrae, but extensively pitted, the depressions covered by skin; supraorbital shelf smooth, ending posteriorly in one or two spines; suborbitals and ridge of preorbital roughened to variously serrate; preorbital with a shallow groove which passes under a bridge of bone near upper end; its edge with about half a dozen spine-like serrae; vertical ridge of preopercle nearly smooth, its edge variously roughened to serrate; a bluntish spine at upper end of preopercle; horizontal edge and ridge of preopercle usually with spine-like serrae; lower edge of interopercle, posteriorly, with about six small spine-like serrae; small, pointed teeth on jaws, vomer, palatine, and tongue.

Second anal spine only slightly shorter than the third; second dorsal spine longest, 2·4 to 2·9 in standard length, in four specimens 32 to 45 mm. standard length.

Colour in alcohol. Body and head light yellowish brown; scales outlined in medium brown, the concentration heaviest on upper side and back; chin and end of snout dark brown, the lower head lighter behind; pelvic fin with a narrow black band from tip of outer branched ray to base of inner ray; anal fin with black band passing obliquely backward from anterior border to base of last ray; membrane between second and third, and sometimes the fourth, dorsal spines black, the central part of fin lighter; margin of soft dorsal blackish; a black line, showing through scales, along midside of body.

Ambassis macleayi (Castelnau)

Pl. 28

Pseudoambassis macleayi Castelnau, *Proc. Linn. Soc. N.S.W.* 3: 43 (1878); type locality, Norman River, Gulf of Carpentaria; Whitley, *Rec. S. Aust. Mus.* 5: 358 (1935).

Billabong, about 3 miles above Roper River Mission, Northern Territory (Blitner), 17 August 1948, USNM 173815, 89 specimens, 19 to 66 mm. standard length.

M48-35, Billabong, Oenpelli, 29 September 1948, USNM 173816, 126 specimens, 13 to 50 mm. standard length; 16 October 1948, USNM 173817, 57 specimens, 14 to 50 mm. standard length.

Description. Spines in first dorsal fin VII(in 12); rays in second dorsal fin which united with first dorsal fin by a low membrane I,9(3), I,10(63), or I,11(1); anal fin rays III,9(3), III,10(11), or III,11(2); branched caudal rays 8+7=15(5); pelvic rays I,5(10); pectoral rays ii,12(9) or ii,13(3); gillrakers, lower limb of first arch slender, 24(3), 25(3), 26(4), or 27(2); scale rows between head and caudal fin 28(2) or 29(1); predorsal scales, at midline, 12(1), 13(1), or 14(1); two rows of cheek scales; lateral line short, with occasional scattered tubes along body.

Top of head without spines or serrae, but extensively pitted, the depressions covered by skin; supraorbital shelf smooth, ending posteriorly in one or two spines; suborbitals and ridge of preorbital roughened to serrate, usually serrate; preorbital with a shallow groove which passes under a bridge of bone near its upper end; its edge with several spine-like serrae; vertical ridge of preopercle nearly smooth; its edge roughened to prominently serrate; a bluntish spine at upper end of preopercle; horizontal edge and ridge of preopercle usually with

Dorothea B. Schultz

PLATE 28. *Ambassis macleayi*

Billabong above Roper River Mission, USNM 173815, 56 mm. standard length

spine-like serrae; lower edge of interopercle, posteriorly, with several spine-like serrae; small, pointed teeth on jaws, vomer, palatine, and tongue.

Second anal spine shorter than third; second dorsal spine longest, 2·6 to 2·9 in standard length in four specimens 45 to 57 mm. standard length.

Colour in alcohol. Body and head light yellowish brown; scales heavily outlined in brown, forming a prominent reticulate pattern; vertical fins all dusky, the dark pigment heavy on all inter-radial membranes of the dorsal and anal fins; pectoral fin lighter, with a dark brown blotch at the base of its rays; a dark line along body between epaxial and hypaxial muscles; lower surface of head and body dusky brown; membranes between outer rays of pelvic fin dark brown.

Remarks. I base my identification of this species on the description of the type given by Whitley (loc. cit.). The other nominal forms listed by him, with the possible exception of *Ambassis elevatus* Macleay, were wrongly included in the synonymy. The records of *Blandowskiella reticulata* given by Whitley (1956b) may also be of this species.

Ambassis dussumieri Cuvier and Valenciennes

Ambassis dussumieri Cuvier and Valenciennes, *Histoire naturelle des poissons* 2: 181 (1828); type locality, Malabar.

Ambassis telkara Whitley, *Rec. S. Aust. Mus.* 5: 349, fig. 2 (1935); type locality, Bathurst Head, north Queensland.

Recorded as: *Ambassis dussumieri* Klunzinger (1879): 345, Port Darwin. Records probably referable: *Ambassis commersonii* Günther (1859): 223, Port Essington; Macleay (1880): 337, Port Darwin; *Ambassis* (*Priopis*) *gymnocephalus* Paradice and Whitley (1927): 84 and 98, Darwin.

M48-2, Nightcliff, USNM 173818, 16 specimens, 34 to 47 mm. standard length.

M48-7 and 8, Mouth of billabong, at Little Lagoon, and Little Lagoon, USNM 173819, 262 specimens, 19 to 43 mm. standard length (recorded as common at M48-8).

M48-15, Emerald River, Groote Eylandt, USNM 173820, 21 specimens, 16 to 35 mm. standard length.

M48-19, Creek, tributary to Little Lagoon, USNM 173821, 13 specimens, 22 to 38 mm. standard length.

M48-21, Reef, Yirrkalla, 14 July 1948, USNM 173822, 1 specimen, 24 mm. standard length.

M48-27, Bay and beach, Port Bradshaw, USNM 173823, 29 specimens, 23 to 43 mm. standard length.

M48-31, Creek, Yirrkalla, USNM 173824, 1 specimen, 35 mm. standard length.

Description. Dorsal rays VII-I,8(in 3) or VII-I,9(9); anal rays III,9(12); pectoral rays ii,12(3), ii,13(8), or ii,14(1); pelvic rays I,5(10); branched caudal rays 8+7=15 (5); gillrakers slender, those on lower limb of first arch, 20(1), 21(7), 22(2), or 23(2); scale rows between head and caudal fin 29(3); predorsal scales at midline 12(1), 13(1), or 15(1); two rows of cheek scales; lateral line interrupted below dorsal fin, with an anterior curved section and a posterior straight section, both extending over 12 to 15 scales.

Lateral interorbital ridge ending posteriorly, behind eye, in one or more spines; top of head extensively pitted, the depressions covered by skin; supraorbital shelf ending posteriorly in three to five spines; suborbitals nearly smooth; ridge and edge of preorbital with spine-like serrae; shallow groove of preorbital passing beneath a large, backward projecting spine at its upper end; vertical ridge of preopercle nearly smooth, the edge usually distinctly serrate; a bluntish spine at upper end of preopercle; horizontal edge and ridge of preopercle with spine-like serrae; interopercle ending posteriorly in a sharp point or spine; small, pointed teeth on jaws, palatine, vomer, and tongue.

Second anal spine shorter than third; second dorsal spine longest, 4·2 to 4·5 in standard length in four specimens 34 to 41 mm. standard length.

Colour in alcohol. Body and head light yellowish; lower side immaculate, the upper sparsely dotted with chromatophores; chin and snout dusky brown; a few chromatophores, showing through the scales, between the pelvic and anal fins; these continued around base of anal rays and posteriorly along lower midline; a blackish line along midside of body; membrane between second and third dorsal spines black; some black about spine of second dorsal fin and spines of anal fin; otherwise, fins without chromatophores.

Colour in life. Anterior tip of dorsal fin black; caudal fin yellow.

Denariusa bandata Whitley
Pl. 29

Denariusa bandata Whitley, *Rec. Aust. Mus.* 22: 92, fig. 11 (1948); type locality, coastal plain of Arnhem Land, Northern Territory.

Listed as: *Denariusa bandata* Whitley (1953): 126, near Cape Stewart, Northern Territory.

M48-14, Emerald River, Groote Eylandt, USNM 173811, 42 specimens, 15 to 36 mm. standard length.

M48-16, Billabongs, adjacent to Emerald River, Groote Eylandt, USNM 173812, 15 specimens, 15 to 33 mm. standard length.

M48-35, Billabong, Oenpelli, 16 October 1948, USNM 173813, 1 specimen, 20 mm. standard length.

M48-39, Red Lily Lagoon, USNM 173814, 23 specimens, 18 to 25 mm. standard length.

Description. Dorsal rays VII-I,9(in 3) or VII-I,10(10); anal rays III,8(9) or III,9(4); pelvic rays I,5(26); branched caudal rays 8+7=15(13); pectoral rays ii,7(4) or ii,8(9); gillrakers, all short and stump-like, on lower limb of first arch 7(1), 8(7), 9(4), or 10(1); scale rows between head and caudal fin 25(2), 26(1), or 27(1); predorsal scales at midline 15(1), or 17(3); two rows of cheek scales; lateral line very short, with a few scattered tubes along body.

Top of head, interorbital shelf, infraorbitals, ridge of preorbital, and horizontal and vertical ridge of preopercle without spines or serrae; edge of preorbital with spine-like serrae; posterior end of interopercle and horizontal and vertical edge of preopercle serrate; no spine at upper end of preopercle; head deeply pitted and grooved, the groove of the preorbital passing under a bridge of bone at its upper end; small, pointed teeth on jaws and vomer.

Second anal spine about equal to or only slightly shorter than third; second dorsal spine longest, 3·3 to 3·9 in standard length in four specimens 31 to 36 mm. standard length.

Colour in alcohol. Background colour of body and head yellowish brown; scales on side of body with a prominent reticulate pattern of dark brown, this pigment becoming diffuse on dorsal surface, dorso-lateral surface, lower head, and breast; several dark brown bars, wide as pupil, along side of body; scale in axil of pectoral fin and pectoral fin base dark brown to black; caudal and pectoral fins light yellowish brown, the rays outlined in darker brown; other fins dark brown, the

Dorothea B. Schultz

PLATE 29. *Denariusa bandata*

Red Lily Lagoon, USNM 173814, 24 mm. standard length

pelvics almost black; spinous dorsal with a prominent brownish black distal band which becomes most concentrated and blotch-like between last few spines.

Colour in life. Silvery with greyish brown vertical bars.

APOGONIDAE

In the study of this family I have been aided by Dr Ernest A. Lachner who has an extensive revision of the Apogonidae in progress. For his help and aid in the identifications, I am sincerely grateful. Some of the names used here, as in other families considered herein, are regarded as tentative, pending the examination of typical specimens. An abbreviated description of the colour pattern in alcohol will be found in the key; count frequencies are given under the species.

Key to the Arnhem Land Species

1 Jaws with one to six small canine teeth on each side (none at tip) in addition to the band of villiform teeth; edge of preopercle serrated; a black basicaudal spot and six brownish black longitudinal body stripes; dorsal rays VI-I,9; anal rays II,8; pectoral rays ii,10 . . *Paramia quinquelineata*

Jaws without canine teeth; bands of villiform teeth only; pectoral rays ii,11 or more, rarely ii,10 . . . 2

2 Anal rays II,16 to II,18; dorsal rays VI-I, 9; pectoral rays ii,12; edge of preopercle serrated; a broad brownish or brownish black bar across opercle and one downward from eye . . . *Archamia melasma*

Anal rays II,7 to II,9 . . . 3

3 Spines in first dorsal fin VI . . . 4

Spines in first dorsal fin VII; suborbitals, preorbital, and ridge of preopercle entire . . . 9

4 Dorsal rays VI-I,10; anal rays II,9, infrequently II,8; preorbital, suborbitals and preopercle entire, or edge of preopercle with a trace of serrae at angle; pectoral rays ii,11, rarely ii,10; body brownish, with irregular, wavy bars or mottlings . . . *Apogon aprion*

Dorsal rays VI-I,9, infrequently VI-I,8; anal rays II,8; edge of preopercle serrated . . . 5

5 Suborbitals and ridge of preopercle serrated; pectoral rays ii,12 . . . 6

Suborbitals, ridge of preopercle, and preorbital not serrated . . . 7

6 Edge of preorbital fleshy; body and fins yellowish, without prominent marking; palatine teeth in a small, usually rounded, patch on each side; gillrakers on first arch 3,2+1+7,7=20; opercular spine bifid (abnormal?) . . . *Apogon unicolor*

Edge of preorbital serrate; body with dark brown bars, most prominent as follows: downward from dorsal fin origin; downward from end of second dorsal; and a narrow one from eye to angle of preopercle; a black spot at rear base of anal fin and one near caudal fin base; palatine teeth in a band on each side; gillrakers 3,1 or 2+1+12 or 13,2 or 3=20 or 21 . . . *Apogon trimaculatus*

7 Body yellowish brown, with a prominent black spot, larger than pupil, on centre of caudal peduncle near caudal fin base; membrane between second and third dorsal spines blackened; a short black bar extending upward from rear end of anal fin; pectoral rays ii,12 or ii,13 . . . *Apogon hyalosoma*

Body and fins dusky, yellowish or yellowish brown (red in life), without prominent black markings . . . 8

8 Pectoral rays ii,11; gillrakers about 4,1+1+5,5=16 . . . . *Apogon* species
Pectoral rays ii,12; gillrakers about 4,2+1+8,6=21 . . . . *Apogon* species
9 Anal rays II,9, infrequently II,8; edge of preopercle serrated 10
Anal rays II,7 or II,8 11
10 Dorsal rays VII-I,10; pectoral rays ii,14 or ii,15; body, head, and fins brownish; a narrow dark bar from eye to angle of preopercle; soft dorsal fin with a narrow dark band near base; obscure blotches or spots on side; one below soft dorsal, and an indistinct spot at base of caudal fin; opercular area dark opalescent . . . . *Apogon opercularis*
Dorsal rays VII-I,9; pectoral rays ii,12 or ii,13; body, head, and fins yellowish; a narrow black line from eye to angle of preopercle; a dusky band on soft dorsal near base; membrane between third and fourth dorsal spines blackish; a series of black dots (some may be absent) over body as follows: one at rear base of anal fin; about eight along lateral line, followed by a large one at caudal fin base and one on caudal fin; one in area between first and second dorsal fins and one near rear base of second dorsal fin; about three on dorsal surface of caudal peduncle, at and before caudal fin . . . . *Apogon rueppelli*
11 A black spot, larger than pupil, at base of caudal fin; six, more or less prominent, dark brown bands on trunk; edge of preopercle serrated 12
Longitudinal trunk bands and large spot at base of caudal fin absent 13
12 Bands prominent at all sizes; two bands curving around basicaudal blotch on to caudal fin; inner base pectoral fin blackish; pectoral rays ii,13 . . . . . . *Apogon cooki*
Bands prominent in young, but becoming obscure with age, ending at or before base of caudal rays; pectoral fin base dusky; pectoral rays ii,11 or ii,12 . . . . *Apogon angustatus*
13 Edge of preopercle entire; pectoral rays ii,12, infrequently ii,13; caudal fin rounded behind; black or dark brown rounded spot, large as pupil, on opercle; yellowish brown band, bordered by dark brown, from eye to angle of preopercle; body variably punctulated with brown and dorsal, anal, and pelvic fins variably mottled to punctulated with brown . . *Apogon auritus*
Edge of preopercle serrate; pectoral rays ii,13 or ii,14; caudal fin shallowly forked; body and head brownish; a narrow dark brown line from lower border of eye to angle of preopercle and a dark brown punctulation at upper end of vertical arm of preopercle; other obscure brown markings over preopercle and opercle; pelvic fins and membrane between second to fifth dorsal spines dark brownish or black, especially so in young, the pigment becoming concentrated distally on dorsal in adults; bases of second dorsal and anal fins yellowish, the fins dusky, brownish outward, and the dorsal with prominent brown band adjacent to yellowish base . . . . *Apogon darnleyensis*

Apogon auritus Cuvier and Valenciennes

Pl. 30

Apogon auritus Cuvier and Valenciennes, *Histoire naturelle des poissons* 7: 443 (1831); type locality, Mauritius.

Apogonichthys polystigma Bleeker, *Natuurk. Tijdschr. Ned.-Ind.* 6: 484 (1854); type localities, Wahai, Ceram and Priaman, Sumatra.

Recorded as: *Apogon auritus* Klunzinger (1879): 344, Port Darwin; *Apogonichthys polystigma* Macleay (1878): 348 and (1880): 347, Port Darwin.

Dorothea B. Schultz

PLATE 30. *Apogon auritus*
Yirrkalla, USNM 173862, 44 mm. standard length

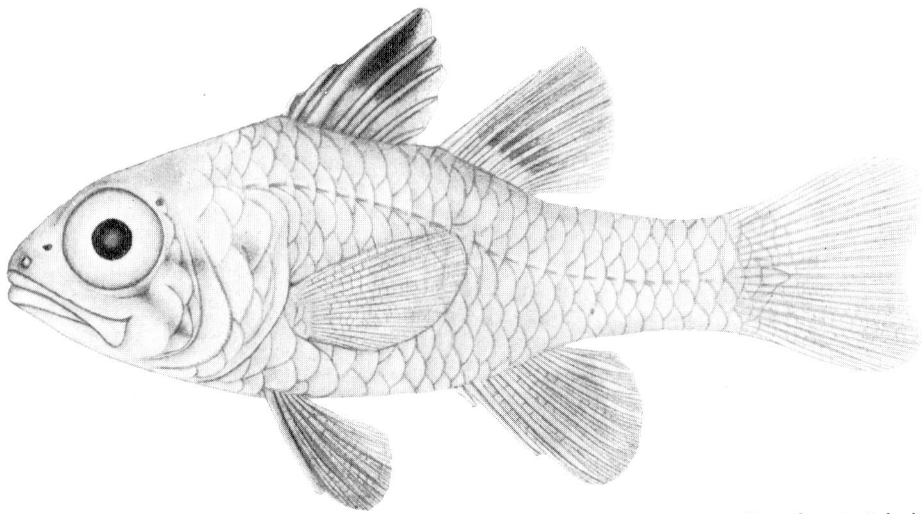

Dorothea B. Schultz

PLATE 31. *Apogon darnleyensis*
Port Bradshaw, USNM 173836, 48 mm. standard length

M48-9a, Reef and tide pools, entrance to Little Lagoon, USNM 173860, 40 specimens, 18 to 45 mm. standard length.

M48-17, Reef, Bartalumba Bay, USNM 173861, 25 specimens, 33 to 46 mm. standard length.

M48-21, Reef, Yirrkalla, 14 July 1948, USNM 173862, 113 specimens, 25 to 68 mm. standard length; 11 and 12 August 1948, USNM 173863, 10 specimens, 35 to 81 mm. standard length.

M48-26, Sheltered bay, island, Port Bradshaw, USNM 173864, 8 specimens, 31 to 37 mm. standard length.

M48-27, Bay and beach, Port Bradshaw, USNM 173865, 3 specimens, 32 to 42 mm. standard length.

Description. Dorsal rays VII-I,9(in 10); anal rays II,7(1) or II,8(9); branched caudal rays 8+7=15(10); pelvic rays, both sides, I,5(10); pectoral rays ii,12(9) or ii,13(1); caudal fin margin rounded; preorbital, suborbitals and preopercle entire.

Colour in life. Body brown or chocolate brown, often with some black punctulations, or uniformly yellowish brown; opercular spot black, narrowly ringed with yellow; fins, of individuals with yellowish brown body, bright red.

Apogon darnleyensis (Alleyne and Macleay)

Pl. 31

Apogonichthys darnleyensis Alleyne and Macleay, *Proc. Linn. Soc. N.S.W.* 1: 268, pl. 5, fig. 3 (1877); type locality, Darnley Island, Queensland.

Apogon darnleyensis Whitley, *Proc. Roy. Zool. Soc. N.S.W.* 1949-50: 64, fig. 9 (1951).

M48-9a, Reef, entrance to Little Lagoon, USNM 173835, 2 specimens, 21 and 23 mm. standard length.

M48-27, Bay and beach, Port Bradshaw, USNM 173836, 1 specimen, 48 mm. standard length.

Description. Dorsal rays VII-I,9(in 3); anal rays II,8(3); branched caudal rays 8+7=15(3); pelvic rays I,5, both sides (3); pectoral rays ii,13(2) or ii,14(4); edge of preopercle serrated, its ridge and the preorbital and suborbitals entire.

Apogon cooki Macleay

Pl. 32

? *Apogon victoriae* Günther, *Catalogue of the Fishes in the British Museum* 1: 243 (1859); type locality, Victoria, Australia [possibly equals Victoria at Port Essington]; Macleay, *Proc. Linn. Soc. N.S.W.* 5: 344 (1880).

Apogon cookii Macleay, *Proc. Linn. Soc. N.S.W.* 5: 344 (1880); type localities, Endeavour River and Darnley Island, Queensland.

Record questionably referable: *Apogon fasciatus* Klunzinger (1879): 344, Port Darwin.

Arnhem Land; no further data; USNM 173849, 1 specimen, 66 mm. standard length.

M48-9a, Reef, entrance to Little Lagoon, USNM 173850, 19 specimens, 43 to 68 mm. standard length.

M48-21, Reef, Yirrkalla, 14 July 1948, USNM 173851, 81 specimens, 41 to 67 mm. standard length; 12 August 1948, USNM 173852, 1 specimen, 58 mm. standard length.

M48-32, Yirrkalla, USNM 173853, 10 specimens, 50 to 82 mm. standard length.

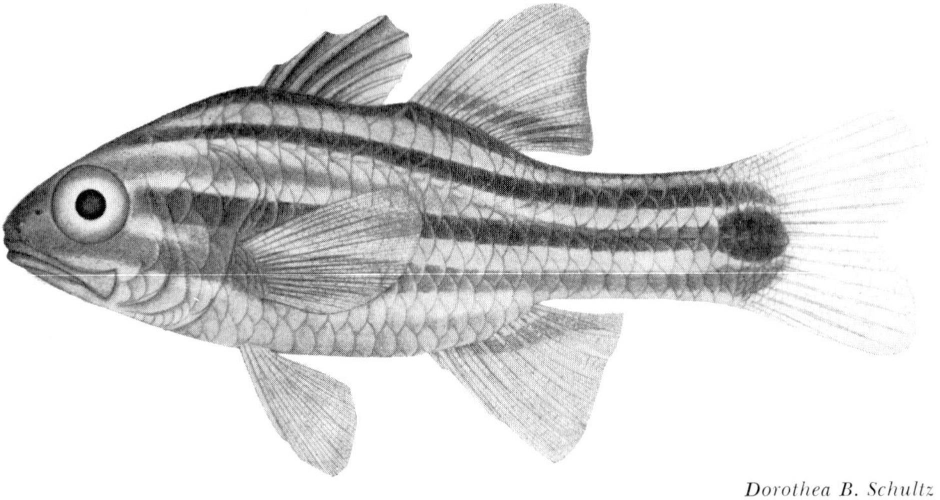

Dorothea B. Schultz

PLATE 32. *Apogon cooki*
Little Lagoon, USNM 173850, 68 mm. standard length

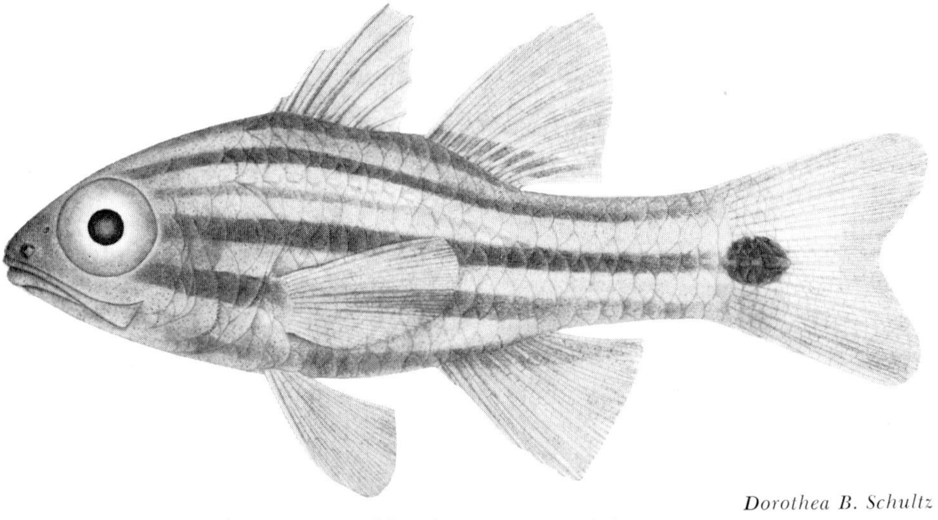

Dorothea B. Schultz

PLATE 33. *Apogon angustatus*
Little Lagoon, USNM 173846, 47 mm. standard length

Description. In ten specimens, dorsal rays VII-I,9; anal rays II,8; branched caudal rays 8+7=15; pelvic rays I,5, on both sides. The pectoral rays were found to be ii,13 in the left fin of 19 fishes. Preorbital, suborbitals and ridge of preopercle entire, the preopercular edge serrate.

Colour in life. The following notes, taken from a coloured drawing in Miller's notebook and his description, may also apply to *Apogon angustatus,* since it and *cooki* are very similar. I interpret the drawing as representing *cooki,* however, since Miller likely would have drawn one of the larger specimens. The preserved adults of *cooki* are sharply marked, whereas those of *angustatus* are drab and the bands diffuse, characteristics which I believe were also true in life in both species. In addition, the drawing shows two of the lateral bands extending on to the caudal fin, true only of *cooki.*

Background colour of body and upper head whitish or dusky white; lower head, snout, and breast light brownish; six dark brown bands on body; spot at base of caudal fin very dark, brownish to black; spinous dorsal fin medium brown; pelvic fin brownish pink; anal, caudal, and second dorsal fins pink; a narrow dark brown band outward from base of second dorsal fin; a red band of similar width along base of anal fin; two of dark brown lateral bands extend around basicaudal spot slightly on to caudal fin.

Apogon angustatus (Smith and Radcliffe)
Pl. 33

Amia angustata Smith and Radcliffe, *Proc. U.S. Nat. Mus.* 41: 253, fig. 1 (1911); type locality, Malanipa Island, east of Zamboanga.

Arnhem Land; no further data; USNM 173845, 2 specimens, 52 to 60 mm. standard length.

M48-9a, Reef, entrance to Little Lagoon, USNM 173846, 26 specimens, 34 to 75 mm. standard length.

M48-17, Reef, Bartalumba Bay, USNM 173847, 14 specimens, 55 to 81 mm. standard length.

M48-21, Reef, Yirrkalla, 14 July 1948, USNM 173848, 26 specimens, 52 to 72 mm. standard length.

Description. In ten specimens, dorsal rays VII-I,9; anal rays II,8; branched caudal rays 8+7=15; pelvic rays I,5, on both sides. The left pectoral fin was found to have ii,12 rays in 39 specimens and ii,11 in 1 specimen. Preorbital, suborbital and ridge of preopercle entire, the edge of the preopercle serrate.

Colour in life. See *Apogon cooki.*

Apogon rueppelli Günther
Pl. 34

Apogon rüppellii Günther, *Catalogue of the Fishes in the British Museum* 1: 236 (1859); type locality, Australia; stated to be Victoria [at Port Essington]; McCulloch, *Aust. Mus. Mem.* 5: 171 (1929).

Recorded as: *Apogon ruppellii* Macleay (1878): 347 and (1880): 346; Paradice and Whitley (1927): 84 and 97, Port Darwin; *Apogon ruppelli* Klunzinger (1879): 344, Port Darwin; Whitley (1953): 124, West Headland, Port Keats.

M48-17, Reef, Bartalumba Bay, USNM 173870, 2 specimens, 27 to 49 mm. standard length.

M48-26, Sheltered bay, island, Port Bradshaw, USNM 173871, 3 specimens, 43 to 57 mm. standard length.

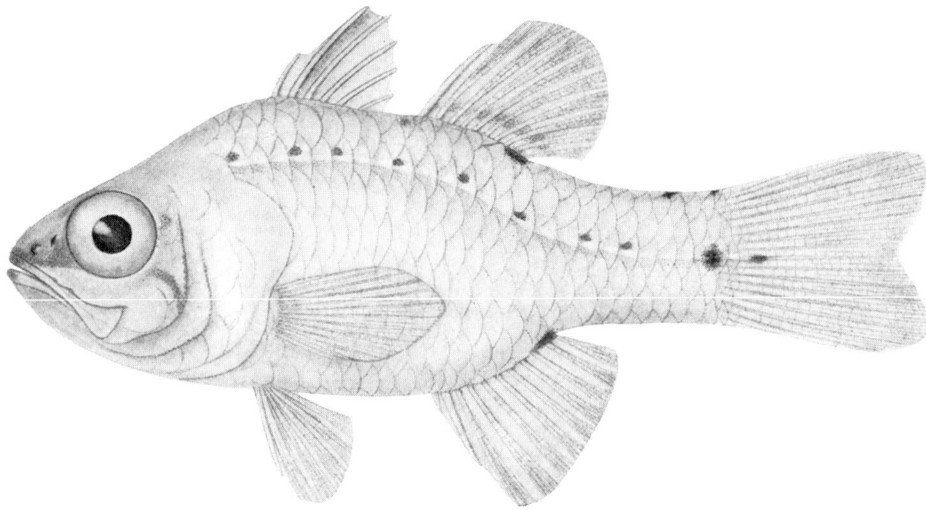

Dorothea B. Schultz

PLATE 34. *Apogon rueppelli*
Port Bradshaw, USNM 173871, 57 mm. standard length

Dorothea B. Schultz

PLATE 35. *Apogon opercularis*
Nightcliff, USNM 173867, 73 mm. standard length

M48-27, Bay and beach, Port Bradshaw, USNM 173872, 7 specimens, 35 to 53 mm. standard length.

Description. In ten specimens, dorsal rays VII-I,9; anal rays II,9; branched caudal rays 8+7=15; pelvic rays I,5, both sides; pectoral rays ii,12(in 6) or ii,13(4); preorbital, suborbitals, and ridge of preopercle entire; edge of preopercle serrate.

Colour in life. Pale yellowish white, with black dots along lateral line and a few scattered elsewhere.

Apogon opercularis Macleay
Pl. 35

Apogon opercularis Macleay, *Proc. Linn. Soc. N.S.W.* 2: 347, pl. 7, fig. 1 (1878) and 5: 346 (1880); type locality, Port Darwin.

Arnhem Land; no further data; USNM 173866, 3 specimens, 57 to 62 mm. standard length.

M48-1, Nightcliff, 19 March 1948, USNM 173867, 1 specimen, 73 mm. standard length.

M48-2, Nightcliff, 19 March 1948, USNM 173868, 2 specimens, 74 and 75 mm. standard length.

M48-21, Reef, Yirrkalla, 14 July 1948, USNM 173869, 24 specimens, 14 to 66 mm. standard length.

Description. Dorsal rays VII-I,10(in 30); anal rays II,8(1) or II,9(29); branched caudal rays 8+7=15(10); pelvic rays I,5(10), both sides; pectoral rays ii,14(11) or ii,15(9); suborbitals, preorbital and ridge of preopercle entire, the edge of preopercle serrated.

Apogon hyalosoma Bleeker

Apogon hyalosoma Bleeker, *Natuurk. Tijdschr. Ned.-Ind.* 3: 63 (1852); type localities, Java, Sumatra, and Sumbawa.

Recorded as: *Apogon hyalosoma* Macleay (1878): 348 and (1880): 345, Port Darwin.

Arnhem Land; no further data; in collection with *Ambassis agrammus*; USNM 173854, 4 specimens, 77 to 100 mm. standard length.

M48-15, Emerald River, Groote Eylandt, USNM 173855, 2 specimens, 83 to 87 mm. standard length.

M48-19, Creek, tributary to Little Lagoon, USNM 173856, 9 specimens, 41 to 80 mm. standard length.

M48-31, Creek, Yirrkalla, USNM 173857, 8 specimens, 53 to 77 mm. standard length.

Description. Dorsal rays VI-I,8(in 1) or VI-I,9(9); anal rays II,8(10); branched caudal rays 8+7=15(10); pelvic rays I,5, both sides (10); pectoral rays ii,12(9) or ii,13(1); edge of preopercle serrated, the preopercular ridge, suborbitals, and preorbital entire.

Colour in life. Dull brown with black spot at base of caudal fin.

Apogon species

M48-21, Reef, Yirrkalla, 14 July 1948, USNM 173873, 1 specimen, 31 mm. standard length.

Description. Dorsal rays VI-I,9; anal rays II,8; branched caudal rays 8+7=15; pelvic rays I,5, both sides; pectoral rays ii,11, both sides; gillrakers on first arch

4,1+1+5,5=16; edge of preopercle serrated, its ridge and the circumorbitals entire in the single specimen.

Colour in life. All of body, head and fins, except eye, bright red.

Remarks. This specimen apparently belongs to the species group of *Apogon* exemplified by *Apogon erythrinus* Snyder, as discussed by Lachner (Schultz, *et al.*, 1953). It is more slender bodied and with longer snout, than in *Apogon erythrinus* or *A. doryssa* Schultz. The descriptions and materials of the other nominal species of this group are inadequate for further comparison.

Apogon species
Pl. 36

M48-21, Reef, Yirrkalla, 11 August 1948, USNM 173876, 1 specimen, 39 mm. standard length.

Description. Dorsal rays VI-I,9; anal rays II,8; branched caudal rays 8+7=15; pelvic rays I,5, both sides; pectoral rays ii,12, both sides; edge of preopercle serrated, its ridge and the circumorbitals entire; gillrakers on first arch 4,2+1+8, 6=21 in the single specimen.

Colour in life. As intimated by Miller's notes, the body, head, and fins were red.

Remarks. This individual may belong to the *Apogon erythrinus* species group, discussed by Lachner (Schultz *et al.,* 1953). It differs from the available Pacific material of that complex in having a more fusiform body, darker upper and lateral surfaces, and more gillrakers.

Apogon trimaculatus Cuvier and Valenciennes

Apogon trimaculatus Cuvier and Valenciennes, *Histoire naturelle des poissons* 2: 156, pl. 22 (1828); type locality, Bouron, Moluccas.

M48-21, Reef, Yirrkalla, 11 August 1948, USNM 173834, 1 specimen, 118 mm. standard length.

Description. In the single specimen, dorsal rays VI-I,9; anal rays II,8; branched caudal rays 8+7=15; pelvic rays I,4 on both sides (abnormal?); pectoral rays ii,12, both sides; preorbital, suborbitals, and edge and ridge of preopercle all serrated; gillrakers on first arch 3,1 or 2+1+12 or 13,2 or 3=20 or 21.

Apogon unicolor Döderlein

Apogon unicolor Döderlein in Jordan and Snyder, *Proc. U.S. Nat. Mus.* 23: 749, pl. 33 (1901); type locality, Yokohama, Japan.

M48-21, Reef, Yirrkalla, 11 August 1948, USNM 173833, 1 specimen, 87 mm. standard length.

Description. In the single specimen, dorsal rays VI-I,9; anal rays II,8; branched caudal rays 9+7=16; pelvic rays I,5, both sides; pectoral rays ii,12, both sides; preorbital with fleshy edge; suborbitals and ridge and edge of preopercle serrated; gillrakers on first arch 3,2+1+7,7=20 on one side.

Remarks. The present specimen is very similar morphologically to the holotype of *Apogon unicolor* Döderlein, USNM 49708, which I have examined, but detailed comparison of colour pattern is not possible. Although the name is regarded as tentative for the specimen, it seems likely that the two are conspecific. The individual differs chiefly in that the opercular spine on each side is bifid, possibly an abnormality, while in typical *unicolor* the spine has a single point.

Dorothea B. Schultz

PLATE 36. *Apogon* species
Yirrkalla, USNM 173876, 39 mm. standard length

Dorothea B. Schultz

PLATE 37. *Apogon aprion*
Red Lily Lagoon, USNM 173844, 104 mm. standard length

Apogon aprion Richardson

Pl. 37

Apogon aprion Richardson, *Ann. Mag. Nat. Hist.* 9: 16 (1842); type locality, King's River, near Victoria, at Port Essington.

Gulliveria ramsayi Macleay, *Proc. Linn. Soc. N.S.W.* 9: 11 (1884); type locality, fresh water inland from Port Darwin.

Recorded as: *Apogonichthys aprion* Günther (1859): 247 and Macleay (1880): 347, Port Essington; *Glossamia aprion* McCulloch (1917): 169, fig. 1, Collects Creek, 50 miles inland from Port Darwin.

M48-14, Emerald River, Groote Eylandt, USNM 173837, 243 specimens, 15 to 177 mm. standard length.

M48-16, Billabong adjacent to Emerald River, Groote Eylandt, USNM 173838, 3 specimens, 68 to 103 mm. standard length.

Billabong, about 3 miles above Roper River Mission, Northern Territory, 17 August 1948, Blitner, USNM 173839, 52 specimens, 26 to 97 mm. standard length.

M48-35, Billabong, Oenpelli, 29 September 1948, USNM 173840, 90 specimens, 30 to 111 mm. standard length; September and October 1948, USNM 173841, 5 specimens, 18 to 113 mm. standard length; 16 October 1948, USNM 173842, 14 specimens, 11 to 57 mm. standard length.

M48-38, Oenpelli Creek, USNM 173843, 19 specimens, 58 to 91 mm. standard length.

M48-39, Red Lily Lagoon, USNM 173844, 17 specimens, 17 to 112 mm. standard length.

Goulburn Islands, Arafura Sea, about 180 miles east-north-east of Darwin, Northern Territory, 29 July 1948 (Nick Haritos), USNM 173879, 1 specimen, 28 mm. standard length.

Description. In ten specimens from each of the following localities: Roper River drainage; Emerald River, Groote Eylandt; and from the vicinity of Oenpelli, the following combined data were obtained: dorsal rays VI-I,10(in 30); anal rays II,8(2) or II,9(28); branched caudal rays 8+7=15(10); pelvic rays I,5, both sides (10); pectoral rays, left side, ii,10(1), or ii,11(29); preorbital, suborbitals and preopercle entire, or a few, poorly developed, serrae on edge of preopercle at its angle; gillrakers on first arch, in a few specimens from the above areas, 2,1+1+5,6 to 8, the variation 6(1), 7(3), or 8(1).

Archamia melasma Lachner and Taylor

Pl. 38

Archamia melasma Lachner and Taylor, *Proc. Biol. Soc. Wash.* 73: 29-34 (1960); type locality, coral reef, Yirrkalla, Northern Territory.

Recorded as: *Archamia lineolata* Paradice and Whitley (1927): 84, Pellew Islands.

M48-21, Reef, Yirrkalla, 11 August 1948, Australian Museum No. IB.4473, holotype, 60 mm. standard length; USNM 173794, 6 paratypes, 55 to 59 mm. standard length.

Pellew Islands, Gulf of Carpentaria, June 1923, Dr W. E. J. Paradice, Australian Museum No. IA. 1499, 1 specimen, 56 mm. standard length.

Description. Dorsal rays VI-I,9(in 8); anal rays II,16(2), II,17(5) or II,18(1); branched caudal rays 8+7=15(8); pelvic rays I,5, both sides (8); pectoral rays ii,12(8); edge of preopercle serrate, its ridge, the suborbitals and preorbitals entire.

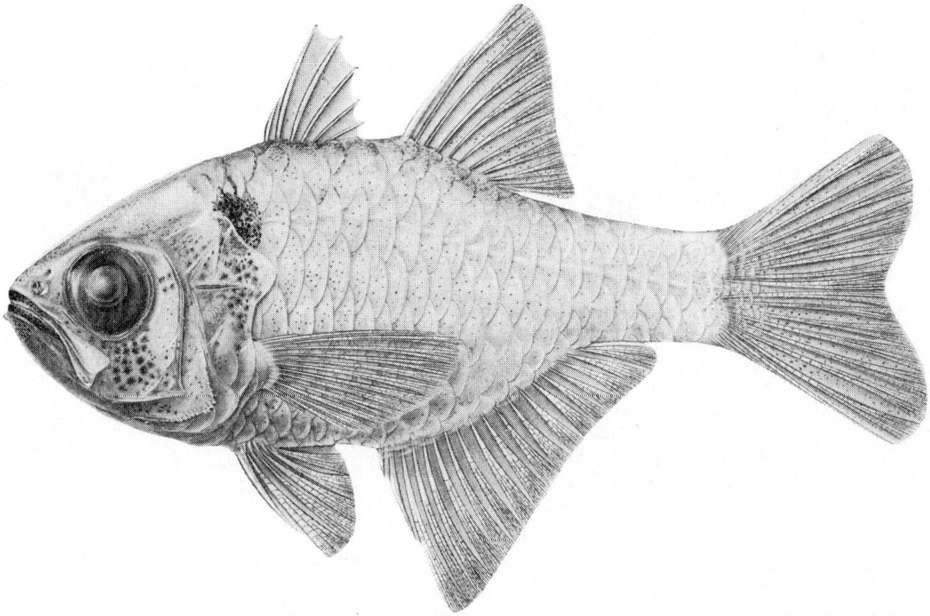

Carolyn B. Lutz

PLATE 38. *Archamia melasma,* paratype
Yirrkalla, USNM 173794, 56 mm. standard length

Colour in life. Silvery white, with curving, deep reddish orange vertical lines along body scale margins; yellowish orange band from tip of snout to eye, and on eye anterior and posterior to pupil; dark brown bar extending from eye obliquely backward to angle of preopercle; black scapular bar; all fins light pink.

Paramia quinquelineata (Cuvier and Valenciennes)

Cheilodipterus quinquelineatus Cuvier and Valenciennes, *Histoire naturelle des poissons* 2: 167 (1828); type locality, Borabora, Society Islands.

M48-21, Reef, Yirrkalla, 14 July 1948, USNM 173874, 2 specimens, 69 to 72 mm. standard length; 11 and 12 August 1948, USNM 173875, 6 specimens, 51 to 83 mm. standard length.

Description. In the eight specimens, dorsal rays VI-I,9; anal rays II,8; branched caudal rays 8+7=15; pelvic rays I,5, both sides; pectoral rays ii,10, left side; edge of preopercle serrated, its ridge, the suborbitals and preorbital entire.

Colour in life. Body silvery white, with black stripes and sky-blue reflections from scales; a large yellow area, surrounding a black spot, at caudal base.

LATIDAE

The two species included here are variously placed in the families Centropomidae, Serranidae, and Latidae. Because their relationships are not known, they are retained in the present family. The material of the genus *Psammoperca* is characterized by an included lower jaw, posterior end of upper jaw below eye, and a single large preopercular spine, which at the angle. *Lates* has the lower jaw terminal, the upper jaw ending behind the eye and 4 large preopercular spines, one of which is at the angle; the other three are along the lower edge.

Günther (1859) described *Psammoperca macroptera,* characterized by III,13 anal rays, and strongly serrated preorbital and suborbitals. His type came from Victoria, which probably was the hamlet at Port Essington, Northern Territory. Whitley (1959) has given a synonymy and figured Günther's type, under the name *Hypopterus macropterus.* I am unable to recognize this form in my material; it may be an abnormal individual of *P. waigiensis.*

Psammoperca waigiensis (Cuvier and Valenciennes)

Labrax waigiensis Cuvier and Valenciennes, *Histoire naturelle des poissons* 2: 83 (1828); type locality, Waigiou.

Recorded as: *Psammoperca waigiensis* Günther (1859): 69, Victoria, Australia [probably equals Port Essington]; Paradice and Whitley (1927): 84, Pellew Islands.

M48-9, Reefs and tide pools, entrance to Little Lagoon, USNM 174178, 2 specimens, 42 and 46 mm. standard length.

M48-9a, Reef, entrance to Little Lagoon, USNM 174179, 6 specimens, 35 to 58 mm. standard length.

M48-15, Emerald River, Groote Eylandt, USNM 174180, 3 specimens, 176 to 237 mm. standard length.

M48-22, Bay of Yirrkalla, 14 and 16 August 1948, USNM 174181, 2 specimens, 199 and 254 mm. standard length.

Native name: *Ijárgaru,* Groote Eylandt.

Colour in life. Young individuals with a broad brown band from end of snout, through eye, to rear end of opercle, and a narrow one along upper jaw, backward to base of pectoral fin; the two bands defining a broad yellow band from snout tip to pectoral fin; a trace of yellow just below lower brown band on breast.

Lates calcarifer (Bloch)

Holocentrus calcarifer Bloch, *Naturgeschichte der Ausländischen Fische* 4: 100, pl. 244 (1790); type locality, Japan; reference not seen.
Lates darwiniensis Macleay, *Proc. Linn. Soc. N.S.W.* 2: 345 (1878) and 5: 306 (1880); type locality, Port Darwin.

Recorded as: *Lates calcarifer* Kent (1889): 239, Port Darwin; Paradice and Whitley (1927): 84, Pellew Islands and Darwin. Record probably referable: *Pseudolates cavifrons* Macleay (1878): 345 and (1880): 307, Port Darwin.

Arnhem Land; no further data; USNM 174171, 5 specimens, 208 to 376 mm. standard length.

M48-2, Nightcliff, USNM 174172, 1 specimen, 439 mm. standard length.

M48-14, Emerald River, Groote Eylandt, USNM 174173, 2 specimens, 245 and 255 mm. standard length.

M48-15, Emerald River, Groote Eylandt, USNM 174174, 3 specimens, 172 to 287 mm. standard length.

M48-19, Creek, tributary to Little Lagoon, USNM 174175, 3 specimens, 231 to 255 mm. standard length.

M48-35, Billabong, Oenpelli, USNM 174176, 5 specimens, 280 to 370 mm. standard length.

M48-39, Red Lily Lagoon, USNM 174177, 3 specimens, 259 to 284 mm. standard length.

Native names: *Yúlkubénda* (Yool-koo-benda), Groote Eylandt; *Malálalk*, Oenpelli.

PSEUDOCHROMIDAE

Key to the Arnhem Land Species

1 Dorsal rays III,22 to 24; anal rays III,12 or 13; body, head and fins nearly uniformly brown, with scattered blue spots, which most concentrated on head; rays of dorsal and anal fins elongated . . . . *Pseudochromis punctatus*

Dorsal rays III,25 to 28; anal rays III,13 or 14; body and head almost uniformly brown, without spots; traces of narrow dark bars on rear part of body; paired fins pale, not elongated; borders of upper and lower lobes of caudal fin yellowish . . . . *Pseudochromis wilsoni*

Pseudochromis punctatus (Richardson)

Assiculus punctatus Richardson in Stokes, *Discoveries in Australia* . . . 1: 494, pl. 2, figs. 1 to 5 (1846); type locality, coast of Australia.
Pseudochromis mülleri Klunzinger, *S.B. Akad. Wiss. Wien* 80: 380 (1879); type locality, Port Darwin.
Cichlops filamentosus Macleay, *Proc. Linn. Soc. N.S.W.* 5: 570 (1880); type locality, Port Darwin.

Recorded as: *Pseudochromis punctatus* Paradice and Whitley (1927): 85 and 98, Pellew Islands and Darwin; *Pseudochromis mülleri* Macleay (1884): 28, Port Darwin.

Groote Eylandt; trawled at 6 fathoms; collected by Fred Gray, 1938, USNM 174251, 1 specimen, 32 mm. standard length.

Remarks. Specimen badly damaged and fins broken; dorsal rays III,24; anal rays III,13; lateral line scales about 34 anteriorly and 9 in rear straight section.

Pseudochromis wilsoni (Whitley)

Leptochromis tapeinosoma wilsoni Whitley, *Rec. Aust. Mus.* 17: 113 (1929); type locality, Port Darwin.

M48-3, East Point Reef, USNM 174040, 7 specimens, 36 to 47 mm. standard length.

M48-21, Reef, Yirrkalla, 14 July 1948, USNM 174041, 20 specimens, 36 to 50 mm. standard length.

Description. Dorsal rays III,25(in 3), III,26(4), or III,28(1); anal rays III,13(2) or III,14(6), the first spine of the dorsal and anal fins very short and slender and easily overlooked; pectoral rays ii,14(1), ii,15(7), or ii,16(8); pelvic rays I,5, the second branched ray generally longest; branched caudal rays 8+8=16(1) or 8+7=15(3); pored scales in lateral line: anterior segment 32(1) or 33(3); posterior straight portion 6(1), 10(2), or 12(1); total 39(1), 43(2), or 44(1); edge of opercle entire; third anal spine about twice as long as second.

Colour in alcohol. Body and head nearly uniformly medium brown, the posterior part of body with traces of darker brown, almost indistinct, bars; pelvic fins yellowish; median fins medium brown with light borders; faint traces of alternating light and dark lines on basal portions of dorsal and anal fins.

Colour in life. Olive green with outer half of dorsal fin bright yellowish orange; border or marginal band along upper and lower lobes of caudal fin yellowish orange, the two areas delimiting a median darker (colour not described) rectangular band; abdomen and breast with yellowish tinge.

Other specimens, which are described as more slender, were noted as lead grey in colour with the yellowish orange on the fins replaced by milky white.

SERRANIDAE

Key to the Arnhem Land Species

1 Dorsal spines X or fewer, infrequently XI 2
Dorsal spines usually XI or XIII; preopercular ridge imbedded in flesh, not exposed; edge of preopercle variously serrate 6

2 Dorsal spines VIII 3
Dorsal spines IX or X; some of teeth depressible; no forward directed spines on lower edge of preopercle 4

3 Edge and posterior ridge of preopercle exposed, both with serrae in adult; short spines on lower edge of preopercle not distinctly directed forward; gill membranes confluent; teeth on jaws villiform, not depressible . . . . . *Diploprion bifasciatum*
Preopercular ridge not exposed, the edge rugose to serrate near corner, with two or three enlarged and forward directed spines below; gill membranes separate; some teeth enlarged, the inner series depressible; a prominent pair of canines near middle of side of lower jaw; dorsal and anal fins rounded posteriorly; small, blue, lenticular shaped spots on head, fins and upper body . . . . . *Plectropomus maculatus*

4 Dorsal spines IX; anal rays III,8 or 9; canine teeth present at front of jaw . . 5
Dorsal spines X; anal rays III,10; no enlarged canine teeth on jaws; brown, covered with rounded blackish spots which approach size of pupil . . . . *Cromileptes altivelis*

5 Body, head and fins brown, with about eight dark brown bars, wide as eye, on body; soft dorsal, anal and caudal fins with white margin; tip of spinous dorsal fin and opercular spot dark brown . . *Cephalopholis boelang*
Head orange, throat vermilion, and body orange brown in life, all turning

brown in preservation; light blue spots about as large as pupil, on body, head and fins, persisting as light or dark spots when preserved; fins narrowly margined with blue; dark opercular spot .. *Cephalopholis argus*

6 Dorsal spines XI; a few canine teeth in outer series on jaw and depressible teeth in inner series; no forward directed spines at lower edge of preopercle 7

Dorsal rays XIII,11; anal rays III,5; teeth villiform, neither enlarged nor depressible; about three enlarged, sharp, forward directed spines at lower edge of preopercle .... *Centrogenys vaigiensis*

7 Palatines toothed 8

Palatines toothless; five or six whitish stripes on body, some extending on to head; small, rusty brown spots on body, head and fins .... *Anyperodon leucogrammicus*

8 Body, head and fins light brown or yellow, without dark or light spots in preservation; six or seven darker brown bars, wide as eye, on body; dorsal fin with black margin .... *Epinephelus fasciatus*

Dark or light spots present; bars present or absent 9

9 Rear margin of opercular flap rounded to truncate, except in young where somewhat pointed; in adult, a line through its upper rear margin nearly intersects anterior base of dorsal fin; middle opercular spine nearer lower spine than upper 10

Rear margin of opercular flap ending in a triangular point; a line through its upper margin passing about midway between dorsal fin and eye .. 11

10 Body, head and vertical fins with numerous, close-set white spots, which mostly smaller than pupil; with age, faint brown bars may appear on body, the white spots diminish in size, mostly disappearing on head and abdomen, and forming wavy or irregular longitudinal light lines, which as broad as dark interspaces, on body and fins .... *Epinephelus summana*

Numerous small brown spots on body, head and fins; body with four or five faint, irregular dark brown bars; five black saddles on back, two of which beneath spinous dorsal fin, two under soft dorsal and one at narrow portion of caudal peduncle; the anterior one, beneath the spinous dorsal, nearly double, with a deep indentation at its side .. *Epinephelus fuscoguttatus*

11 Body and vertical fins with rounded to hexagonal dark brown spots which separated by whitish lines, forming a network; spots on preopercle and cheek tiny, black; no scales on upper jaw .... *Epinephelus gilberti*

Brown spots, if present on body, separated by spaces equal to or greater than their diameter 12

12 Fins with black bars or spots; body and head marbled with brown and with two or three dark brown or black body bars in young; interorbital distance greater than eye diameter, the space increasing with age; pores of lateral line opening through several radiating tubules .. *Epinephelus lanceolatus*

Fins with brown spots or whitish, dark brown to black margined ocelli .. 13

13 Young with black rimmed ocelli which break up into widely spaced oblong dark brown spots as whitish areas disappear; distance between brown spots two to four times their diameter; one or two brown saddles on back and faint brown bars on body appearing with age .. *Epinephelus corallicola*

Young with about five broad brown bars and a few rounded brown spots along lower side of body and head and on lower fins; with age bars become obscure, spots increase in number and extend to upper surface of body and fins; distance between spots about one or two times their diameter .... *Epinephelus tauvina*

Diploprion bifasciatum Kuhl and van Hasselt

Diploprion bifasciatum Kuhl and van Hasselt in Cuvier and Valenciennes, *Histoire naturelle des poissons* 2: 137, pl. 21 (1828); type locality, Java.

Other Arnhem Land records: *Diploprion bifasciatum* Macleay (1878): 346 and (1880): 326; Klunzinger (1879): 340, Port Darwin.

Little Lagoon, Groote Eylandt, found floating 19 June 1948, Jerry Blitner, USNM 174258, 1 specimen, 32 mm. standard length.

Remarks. The second dorsal spine, and probably also the third which is now broken, is elongate and slender; its tip extends to the base of the caudal fin. The specimen, in preservation, is nearly uniformly pale yellowish.

Plectropomus maculatus (Bloch)

Bodianus maculatus Bloch, *Naturgeschichte der Ausländischen Fische* 4: 48, pl. 228 (1790); reference not seen; Bloch, *Ichthyologie* 7: pl. 228 (1797); type locality, Japan.

Recorded as: *Plectropomus maculatus* Paradice and Whitley (1927): 84, Pellew Islands.

M48-21, Reef, Yirrkalla, 29 August 1948, USNM 174287, 1 specimen, 412 mm. standard length; 14 July 1948, USNM 174285, 1 specimen, 147 mm. standard length; 11 August 1948, USNM 174286, 1 specimen, 135 mm. standard length.

Native name: *Dángbi.*

Colour in life. Body and head olive brown; blue oval to lentical shaped spots on body, head, and dorsal, caudal and anal fins; soft dorsal, caudal, anal and pelvic fins reddish distally, yellowish orange medially, becoming olive brown basally.

Remarks. Dorsal spines VIII(in 3); lower margin of preopercle with antrorse spines; one or two large, curved canine teeth at middle of side of lower jaw in addition to those in front.

Cephalopholis boelang (Cuvier and Valenciennes)

Serranus boelang Cuvier and Valenciennes, *Histoire naturelle des poissons* 2: 308 (1828); type locality, East Indies; Quoy and Gaimard, *Voyage de la Corvette l'Astrolabe . . .* 2: pl. 3, fig. 4 (1833).

Epinephelus boelang Bleeker, *Atlas ichthyologique des Indes Orientales Néêrlandaises* 8: pl. 346, fig. 5 (1876-77).

Recorded as: *Serranus pachycentrum* Macleay (1878): 346 and (1880): 314; *Serranus boenak* Klunzinger (1879): 335, Port Darwin; *Cephalopholis pachycentron* Paradice and Whitley (1927): 84 and 98, Pellew Islands.

M48-13b, Hemple Bay, 27 May 1948, USNM 174301, 3 specimens, 124 to 132 mm. standard length.

M48-13c, Bartalumba Bay, USNM 174302, 1 specimen, 151 mm. standard length.

M48-21, Reef, Yirrkalla, 14 July 1948, USNM 174304, 5 specimens, 47 to 116 mm. standard length; 12 August 1948, USNM 174303, 3 specimens, 96 to 106 mm. standard length.

M48-22, Bay of Yirrkalla, 23 August 1948, USNM 174268, 3 specimens, 99 to 118 mm. standard length.

M48-23, Port Bradshaw, USNM 174264, 1 specimen, 158 mm. standard length.

M48-27, Beach, Port Bradshaw, USNM 174305, 4 specimens, 126 to 186 mm. standard length.

Goulburn Island, in Arafura Sea, about 180 miles east-north-east of Darwin, Northern Territory, 29 July 1948, Nick Haritos, USNM 174306, 2 specimens, 127 and 164 mm. standard length.

Colour in life. Brown with about eight or nine darker brown bars across body; soft dorsal, anal, and caudal fins with white margin; tip of spinous dorsal fin and spot between upper and middle opercular spines often dark brown.

Remarks. I doubt that the name *Serranus pachycentron* Cuvier and Valenciennes (op. cit., p. 295), as commonly used for this species, is applicable to it. The figures that I have cited in Bleeker and in Quoy and Gaimard agree in colour pattern with the Arnhem Land material.

Cephalopholis argus Bloch and Schneider

Cephalopholis argus Bloch and Schneider, *Systema Ichthyologiae*: 311, pl. 61 (1801); type locality, East Indies.

M48-21, Reef, Yirrkalla, 14 July 1948, USNM 174322, 1 specimen, 120 mm. standard length; 11-12 August 1948, USNM 174323, 6 specimens, 81 to 181 mm. standard length.

M48-22, Bay of Yirrkalla, 3, 19 and 23 August 1948, USNM 174324, 3 specimens, 117 to 185 mm. standard length.

Native name: *Komuru.*

Colour in life. Body deep orange brown; head orange; throat vermilion; light blue spots on body, head and fins; fins darker than body, the dorsal, anal, caudal and pelvic fins narrowly margined with blue.

Cromileptes altivelis (Cuvier and Valenciennes)

Serranus altivelis Cuvier and Valenciennes, *Histoire naturelle des poissons* 2: 324, pl. 35 (1828); type locality, Java.

Recorded as: *Serranus altivelis* Günther (1859): 152 and Macleay (1880): 319; *Cromileptes altivelis* Boulenger (1895): 271, Port Essington.

*Epinephelus** *fasciatus* (Forskål)

Perca fasciata Forskål *Descriptiones Animalium*: 40 (1775); type locality, Red Sea

Recorded as: *Epinephelus fasciatus* Paradice and Whitley (1927): 84, Pellew Islands.

M48-13e, Chasm Island, 3-4 July 1948, USNM 174297, 1 specimen, 175 mm. standard length.

M48-21, Reef, Yirrkalla, 14 July 1948, USNM 174300, 46 specimens, 40 to 166 mm. standard length; 6 August 1948, USNM 174299, 19 specimens, 36 to 105 mm. standard length.

M48-22, Bay of Yirrkalla, 3 August 1948, USNM 174266, 174267 and 174298, 5 specimens, 124 to 177 mm. standard length.

Native name: *Nyirkala,* Yirrkalla.

Colour in life. Body and head red, yellow or yellowish brown above, the red grading on lower side to lighter red or orange red below; about six deep reddish brown or brown bars on body; fins red with milky white borders; scattered orange to reddish orange splotches on cheek, jaws and throat; edge of spinous dorsal fin and orbital margin dark brown to blackish.

* *Epinephelus marginalis* Bloch was selected as the type species of *Epinephelus* Bloch by the International Commission on Zoological Nomenclature, Opinion 93.

Epinephelus summana (Forskål)

Perca summana Forskål, *Descriptiones Animalium*: xi and 42 (1775); type locality, Arabia.

Epinephelus bataviensis Bleeker, *Atlas ichthyologique des Indes Orientales Néêrlandaises* 7: pl. 282, fig. 2 (1873-76).

Epinephelus alboguttatus Bleeker, *Atlas ichthyologique des Indes Orientales Néêrlandaises* 7: pl. 284, fig. 1 (1873-76).

M48-21, Reef, Yirrkalla, 29 July 1948, USNM 174307, 2 specimens, 267 and 276 mm. standard length; 11 August 1948, USNM 174308, 3 specimens, 82 to 302 mm. standard length.

Colour in life. Brown with irregular stripes and bars; dorsal, anal, caudal and pectoral fins with milky white border; white spots pronounced in young.

Remarks. The small specimen is almost identical in colour pattern to Bleeker's figure of *alboguttatus* (loc. cit.); the larger specimens approach in colour pattern his illustration of *bataviensis*, but are somewhat more blotched or mottled than shown, with lighter and darker areas.

Epinephelus fuscoguttatus (Forskål)

Perca summana fusco-guttata Forskål, *Descriptiones Animalium*: xi and 42 (1775); type locality, Djedda.

Recorded as: *Serranus fuscoguttatus* Günther (1859): 127 and (1867): 57, Port Essington; Macleay (1878): 346, Port Darwin.

M48-21, Reef, Yirrkalla, 29 July 1948, USNM 174319, 2 specimens, 252 and 291 mm. standard length; 6 August 1948, USNM 174320, 1 specimen, 123 mm. standard length; 11 August 1948, USNM 174318, 2 specimens, 324 and 370 mm. standard length.

Colour in life. Olive brown with irregular dark bars and dark spots.

Epinephelus gilberti (Richardson)

Serranus gilberti Richardson, *Ann. Mag. Nat. Hist.* 9: 19 (1842); type locality, Port Essington.

Recorded as: *Serranus gilberti* Günther (1859): 148, Port Essington; *Epinephelus gilberti* Boulenger (1895): 220, Port Essington. Records probably referable: *Serranus merra* Macleay (1878): 346, Port Darwin; *Serranus hexagonatus* Klunzinger (1879): 334 and Kent (1889): 239, Port Darwin; *Epinephelus megachir* Paradice and Whitley (1927): 85, Pellew Islands and Darwin.

M48-9, Reef, entrance to Little Lagoon, USNM 174273, 1 specimen, 70 mm. standard length.

M48-9a, Reef, entrance to Little Lagoon, USNM 174274, 54 specimens, 39 to 134 mm. standard length.

M48-17, Reef, Bartalumba Bay, USNM 174275, 11 specimens, 58 to 180 mm. standard length.

M48-21, Reef, Yirrkalla, 14 July 1948, USNM 174276, 6 specimens, 69 to 266 mm. standard length; 6 August 1948, USNM 174277, 6 specimens, 86 to 133 mm. standard length; 11-12 August 1948, USNM 174278, 2 specimens, 118 and 238 mm. standard length; August 1948, USNM 174279, 2 specimens, 56 and 93 mm. standard length.

M48-22, Bay of Yirrkalla, 3 August 1948, USNM 174280, 1 specimen, 262 mm. standard length.

M48-26, Sheltered bay, island in Port Bradshaw, 23 July 1948, USNM 174281, 4 specimens, 73 to 93 mm. standard length.

Native names: *Tinárra,* Groote Eylandt; *Kúrutoo,* Yirrkalla.

Colour in life. Large brown spots, which round to six-sided, on body and upper head, the spots separated by narrow light interspaces, producing a net-like effect; slightly smaller rounded brown spots on dorsal, caudal, anal and pelvic fins; pectoral fin with suggestion of concentric, wavy, brown bands; several wavy brown bars or large blotches with light interspaces, on breast and lower head; anal and pelvic fin margined with greyish; a few small, widely spaced dark brown or black spots which in a longitudinal row beneath eye and scattered on cheek and preopercle; the spots on opercle larger, approaching size of those on body.

Remarks. Richardson (1842a) recorded *Serranus merra* and originally described *Serranus stellans* on the basis of specimens of unknown origin in the museum at Haslar. He assumed that they originated from Melville Island, Torres Straits on the north coast of Australia—evidently meaning Melville Island, Northern Territory. Boulenger (1895) examined the type specimen of *Serranus stellans* and listed it—incorrectly, as judged from Richardson's description—under *Epinephelus merra.*

Both forms are generally similar in colour pattern to *Epinephelus gilberti,* but differ in detail; especially notable in *gilberti* is the presence of small blackish spots and the absence of chain-like reticulations on the cheek and preopercle. I suspect that neither Richardson's *stellans* nor his *merra* came from Arnhem Land, and have found no evidence, aside from the doubtful literature references given above, that either lives in that region.

Epinephelus corallicola (Kuhl and van Hasselt)

Serranus corallicola Kuhl and van Hasselt in Cuvier and Valenciennes, *Histoire naturelle des poissons* 2: 336 (1828); type locality not given.

Epinephelus altivelioides Bleeker, *Atlas ichthylologique des Indes Orientales Néêrlandaises* 7: pl. 308, fig. 1 (1873-76).

Epinephelus hoevenii Bleeker, *Atlas ichthyologique des Indes Orientales Néêrlandaises* 7: pl. 286, fig. 4 (1873-76).

Recorded as: *Serranus corallicola* Macleay (1878): 346 and (1880): 316 and Klunzinger (1879): 334, Port Darwin.

Arnhem Land; no further data; USNM 174296, 1 specimen, 96 mm. standard length.

M48-1, Nightcliff, 19 March 1948, USNM 174288, 5 specimens, 60 to 85 mm. standard length.

M48-9, Reef, entrance to Little Lagoon, USNM 174269, 2 specimens, 54 and 63 mm. standard length.

M48-9a, Reef, entrance to Little Lagoon, USNM 174291, 15 specimens, 38 to 145 mm. standard length.

M48-15, Emerald River, Groote Eylandt, USNM 174290, 2 specimens, 249 and 254 mm. standard length.

M48-19, Creek, tributary to Little Lagoon, USNM 174295, 19 specimens, 62 to 240 mm. standard length.

M48-21, Reef, Yirrkalla, 14 July 1948, USNM 174293, 2 specimens, 166 and 169 mm. standard length; 6 August 1948, USNM 174292, 9 specimens, 75 to 167 mm. standard length; 11-12 August 1948, USNM 174289, 10 specimens, 70 to 154 mm. standard length.

M48-26, Sheltered bay, island in Port Bradshaw, USNM 174272, 2 specimens, 343 and 402 mm. standard length.

M48-31, Creek, Yirrkalla, USNM 174294, 1 specimen, 160 mm. standard length.

Native names: *Thámapóona*, Groote Eylandt; *Kórmuloo*, Yirrkalla.

Colour in life. Young dark brown with large white spots which surrounded by brownish black or black ring or the ring broken into three or four smaller spots. With age the dark spots separate, becoming more numerous, the white spots disappear, and three or four small, dark brown saddles appear along the base of the dorsal fin.

Remarks. I am unable to determine the earliest name that should be applied to this species since I suspect that the young were first described and have since been widely confused in a complex of white spotted forms; the descriptions are vague.

Small specimens generally resemble the figure of *E. hoevenii* given by Bleeker (loc. cit.). With age and increase in size the black bordered whitish ocelli break up. The white coloration gradually disappears. The black rings, after fragmentation into three or four dark brown or black marks, persist on the body, head and fins in a pattern similar to that illustrated by Bleeker in *E. altivelioides* (loc. cit.); in the two largest specimens the spots are more numerous, and broad, irregular, faint brownish bars can be distinguished on the body.

Epinephelus tauvina (Forskål)

Perca tauvina Forskål, *Descriptiones Animalium*: xi and 39 (1775); type locality, Djedda.

Records probably referable: *Serranus crapao* Richardson (1842a): 25 and Günther (1859): 137, Port Essington; Macleay (1878): 346 and (1880): 317, Port Darwin; *Serranus pantherinus* Klunzinger (1879): 334, Port Darwin; *Epinephelus tauvina* Boulenger (1895): 244, Port Essington; Paradice and Whitley (1927): 85, Pellew Islands.

M48-1, Nightcliff, USNM 174315, 1 specimen, 94 mm. standard length.

M48-9, Reef, entrance to Little Lagoon, USNM 174270, 12 specimens, 34 to 132 mm. standard length.

M48-9a, Reef, entrance to Little Lagoon, USNM 174316, 2 specimens, 61 and 87 mm. standard length.

M48-15, Emerald River, Groote Eylandt, USNM 174314, 2 specimens, 70 and 133 mm. standard length.

M48-17, Reef, Bartalumba Bay, USNM 174312, 1 specimen, 91 mm. standard length.

M48-20, Old jetty, Melville Bay, 17 August 1948, USNM 174317, 1 specimen, 385 mm. standard length.

M48-21, Reef, Yirrkalla, 14 July 1948, USNM 174309, 1 specimen, 154 mm. standard length.

M48-23, Port Bradshaw, USNM 174265, 1 specimen, 485 mm. standard length.

M48-26, Sheltered bay, island in Port Bradshaw, USNM 174271, 4 specimens, 108 to 150 mm. standard length.

M48-27, Beach, Port Bradshaw, USNM 174311, 1 specimen, 223 mm. standard length.

M48-31, Creek, Yirrkalla, USNM 174310, 5 specimens, 75 to 260 mm. standard length.

M48-33, Milingimbi Island, USNM 174313, 1 specimen, 78 mm. standard length.

Native names: *Tinárra,* Groote Eylandt; *Kúrutoo,* Yirrkalla.

Colour in life. Body greyish brown with irregular brown bars and bronze spots.

Epinephelus lanceolatus (Bloch)

Holocentrus lanceolatus Bloch, *Naturgeschichte der Ausländischen Fische* 4: 92, pl. 242, fig. 1 (1790); reference not seen; Bloch, *Ichthyologie* 7: pl. 242, fig. 1 (1797); type locality, East Indies.

M48-15, Emerald River, Groote Eylandt, USNM 174321, 1 specimen, 410 mm. standard length.

Anyperodon leucogrammicus (Cuvier and Valenciennes)

Serranus leucogrammicus Cuvier and Valenciennes, *Histoire naturelle des poissons* 2: 347 (1828); type localities, Moluccas and Seychelles.

M48-21, Reef, Yirrkalla, 14 July 1948, USNM 174283, 1 specimen, 156 mm. standard length; 11-12 August 1948, USNM 174284, 5 specimens, 117 to 157 mm. standard length.

Colour in life. Five or six light yellowish white stripes along body of which three or four extend on to head, the one below eye nearly to tip of snout; numerous rusty brown spots on head, body, dorsal fin and base af caudal fin; soft dorsal, anal and caudal fins olive yellow basally to yellowish distally.

Remarks. Anal rays III,9(in 6); palatines toothless.

Centrogenys vaigiensis (Quoy and Gaimard)

Scorpaena vaigiensis Quoy and Gaimard, *Voyage autour du monde . . . sur . . . l'Uranie et la Physicienne . . . Zoologie* 324, pl. 58, fig. 1 (1824); type locality, Waigeu; reference not seen.

Recorded as: *Centropristes scorpenoides* Richardson (1842a): 120, Port Essington; *Myriodon waigiensis* Günther (1859): 175, Port Essington; Macleay (1878): 347 and (1880): 327, Port Darwin; *Centrogenys waigiensis* Klunzinger (1879): 340; Paradice and Whitley (1927): 84, Port Darwin; *Centrogenys vaigiensis* Boulenger (1895): 147, Port Essington.

M48-17, Reef, Bartalumba Bay, USNM 174282, 2 specimens, 35 and 44 mm. standard length.

M48-21, Reef, Yirrkalla, 12 August 1948, USNM 174259, 1 specimen, 94 mm. standard length.

Remarks. Dorsal rays XIII,11(in 3); anal rays III,5(3); branched caudal rays 6+6=12(2).

SILLAGINIDAE

In the genus *Sillago,* the teeth in the mouth are uniformly small, there are typically 11 spines in the first dorsal fin, and the scales are moderately large.

Key to the Arnhem Land Species

1 Rays in second dorsal fin I,16 to I,18, usually I,17; anal rays II,14 to II,16, usually II,16; pored scales between head and caudal fin 56 to 60 . . . . . *Sillago ciliata*

Rays in second dorsal fin I,18 to I,22, usually more than I,19; pored scales between head and caudal fin 66 to 72 2

2 Dorsal fin rays usually XI-I,20; anal rays II,18 to II,20, usually II,19; lateral line scales 68 to 71; an intense concentration of pigment, usually black, at pectoral fin base . . . . *Sillago maculata*

Dorsal fin rays usually XI-I,21; anal rays II,22, infrequently II,23; lateral line scales 66 to 69; no concentration of dark pigment at pectoral fin base . . . . *Sillago sihama*

Sillago ciliata Cuvier and Valenciennes

Sillago ciliata Cuvier and Valenciennes, *Histoire naturelle des poissons* 3: 415 (1829); type locality, 'Southern Seas'.

Sillago analis Whitley, *Aust. Zool.* 10: 184 (1943); type locality, Sharks Bay, Western Australia.

Recorded as: *Sillago ciliata* Paradice and Whitley (1927): 89, Pellew Islands; *Sillago analis* Whitley (1954): 27, Melville Island, Northern Territory.

M48-2, Nightcliff, USNM 174051, 1 specimen, 75 mm. standard length.

M48-7, Mouth of billabong, Little Lagoon, USNM 174052, 4 specimens, 72 to 142 mm. standard length.

M48-15, Emerald River, Groote Eylandt, USNM 174053, 7 specimens, 93 to 246 mm. standard length.

M48-19, Creek, tributary to Little Lagoon, USNM 174054, 16 specimens, 43 to 154 mm. standard length.

M48-22, Bay of Yirrkalla, July or August 1948, USNM 174055, 1 specimen, 47 mm. standard length.

M48-28, Creek mouth and bay, Port Bradshaw, USNM 174056, 4 specimens, 177 to 243 mm. standard length.

M48-31, Creek, Yirrkalla, USNM 174057, 9 specimens, 66 to 101 mm. standard length.

Native name: *Árnda*, Groote Eylandt.

Description. Dorsal rays X-I,16(in 1), XI-I,16(2), XI-I,17(35), XI-I,18(3), or XII-I,17(1); anal rays II,14(1), II,15(1), or II,16(40); pelvic rays I,5(12); pectoral rays ii,13(1), ii,14(6), or ii,15(5); branched caudal rays 7+7=14(1) or 8+7=15(4); scale rows between dorsal and anal fins 5+1+8(1) or 5+1+9(6); lateral line scales between head and caudal fin base 56(1), 57(2), 58(2), or 59(1).

Colour in alcohol. Larger specimens almost uniformly medium yellowish brown over head, body, and caudal fin; body and head slightly darker above; one series of specimens with dusky on upper and lower lobe of caudal fin; anal, pectoral, and pelvic fins light yellowish; membrane of first dorsal fin dotted with greyish black; second dorsal fin with two or three narrow bands or rows of irregular blackish spots.

Young with interrupted, dark brown, lines along dorsal fin base; otherwise somewhat resembling *Sillago maculata* in colour pattern, but without a concentrated dark spot at pectoral fin base; with six or seven dark brown, oblong spots along midside and dark brown, oblique, bar-like blotches on upper side.

Remarks. Miller's notes indicate that both this species and *Sillago sihama* live in sandy areas.

Sillago maculata Quoy and Gaimard

Sillago maculata Quoy and Gaimard, *Voyage autour du monde . . . sur . . . l'Uranie et la Physicienne . . . Zoologie*: 261, pl. 53, fig. 2 (1824); type locality, Sydney, Port Jackson, Australia.

Sillago burrus Richardson, *Ann. Mag. Nat. Hist.* 9: 128 (1842); type locality, north-west coast of Australia.

Recorded as: *Sillago maculata* Paradice and Whitley (1927): 89, Pellew Islands; *Sillago gracilis* Kent (1889): 239, Port Darwin.

M48-7, Mouth of billabong, Little Lagoon, USNM 174058, 5 specimens, 53 to 75 mm. standard length.

M48-8, Little Lagoon, USNM 174059, 12 specimens, 32 to 95 mm. standard length.

M48-9, Reefs and tide pools, entrance to Little Lagoon, USNM 174060, 7 specimens, 32 to 54 mm. standard length.

M48-10, Estuary, Thompson Bay, Port Langdon, USNM 174061, 3 specimens, 38 to 87 mm. standard length.

M48-24, Beach, Port Bradshaw, USNM 174062, 1 specimen, 49 mm. standard length.

M48-25, Beach, island, Port Bradshaw, USNM 174063, 2 specimens, 37 to 45 mm. standard length.

M48-28, Creek mouth and bay, Port Bradshaw, USNM 174064, 3 specimens, 64 to 100 mm. standard length.

Native name: *Árnda,* Groote Eylandt.

Description. Dorsal rays X-I,21(in 1), XI-I,18(1), XI-I,19(4), XI-I,20(26), or XI-I,21(1); anal rays II,18(2), II,19(28), or II,20(3); pelvic rays I,5(6); pectoral rays ii,14(6); branched caudal rays 8+7=15(3); scale rows between dorsal and anal fins 7+1+12(3); lateral line scales, between head and caudal fin base 68(1), 69(1), or 71(1).

Colour in alcohol. Body and head medium yellowish brown, slightly darker above; caudal fin yellowish brown; pectoral, pelvic, and anal fins yellowish; first dorsal fin dotted with dark brown; second dorsal fin with two rows of spots on fin membrane and a narrow dusky margin; base of pectoral fin with an intense dark brown or black spot; about seven to nine oblong brown blotches along midside of body and one on caudal fin base; a series of narrow brown saddles across back, of which six beneath dorsal fins extend downward and forward toward the lateral blotches and one at the nape extends downward and backward toward the pectoral fin; saddles and mid-lateral blotches not continuous in the Arnhem Land specimens, which are all small.

Colour in life. In specimens 32 to 95 mm. standard length: general colour yellowish white or yellowish brown; abdomen and lower side white; pelvic fin yellow; other fins pale; a greyish black or black spot at base of pectoral fin; brown markings on side of body.

Remarks. Miller's notes indicate that the spotted species of *Sillago,* evidently referring to *S. maculata,* inhabits rocky areas.

Four specimens (USNM 59933) of *Sillago maculata* from Port Jackson, Australia, have XI-I,19(in 3) or XI-I,20(1) dorsal rays and II,19 anal rays. They are large and badly discoloured, but it appears that some of the lateral blotches and the saddles are continuous.

Sillago sihama (Forskål)

Atherina sihama Forskål, *Descriptiones Animalium*: xiii and 70 (1775); type locality, Lohajae, Red Sea.

Recorded as: *Sillago sihama* Paradice and Whitley (1927): 89, Darwin.

M48-2, Nightcliff, USNM 174066, 3 specimens, 33 to 113 mm. standard length. M48-28, Creek mouth and bay, Port Bradshaw, USNM 174065, 23 specimens, 37 to 102 mm. standard length.

Description. Dorsal rays X-I,22 (in 1), XI-I,20(3), or XI-I,21(22); anal rays II,22(11), or II,23(3); pelvic rays 1,5(14); pectoral rays ii, 13(2), ii,14(9), or ii,15 (3); branched caudal rays 8+7=15(7); scale rows between dorsal and anal fins 6+1+10(1), or 6+1+11(3); lateral line scales between head and caudal fin base 66(1), 67(1), 68(2), or 69(1).

Colour in alcohol. Body and head almost uniformly medium brown, slightly darker above; a dark line or trace of oblong dark blotches along midside of caudal peduncle; pelvic, pectoral, and anal fins yellowish; caudal and first dorsal fins dusky; second dorsal fin dusky or with trace of spots on fin membranes; supracleithrum bordered with a narrow, dark brown crescent.

Remarks. An inhabitant of sandy areas as judged from Miller's notes.

CARANGIDAE

Key to the Arnhem Land Species

1 Lateral line without scutes; pectoral fins short, not falcate, not as long as head 2
Lateral line with elevated scutes posteriorly, and/or prominently arched anteriorly; pectoral fins falcate, generally longer than head 7

2 Posterior rays of dorsal and anal fins normal, neither free, broadened nor finlet-like 3
Posterior rays of dorsal and anal fin expanded, finlet-like, but connected by membrane; scales small, oblong-ovate, lanceolate or needle-like, more or less embedded 5
A free two-rayed finlet behind both the dorsal and anal fins; body elongate, the depth about one-fifth standard length; two narrow, dark longitudinal lines along side . . . . . *Elagatis bipinnulatus*

3 Soft dorsal and anal fin bases of about equal length; scales small, rounded; lateral line nearly straight 4
Anal base about half length of dorsal base, with 15 to 17 soft rays; soft dorsal rays 31 to 34; 8 or 9 short gillrakers on first arch; scales minute, rudimentary; lateral line slightly arched anteriorly . . . . *Seriola nigrofasciata*

4 Body without spots, its height about two times or less in standard length; mouth cleft below eye or opposite its lower margin . . *Trachinotus ovatus*
About five spots (4 in Arnhem Land specimen) saddling lateral line; cleft of mouth opposite lower half of eye; body depth about 2·2 or more in standard length . . . . . *Trachinotus bailloni*

5 Upper jaw slender, elongate, extending far behind eye, its posterior end about midway between tip of snout and rear base to middle of depressed pelvic fin; snout short, blunt, its length about equal to eye diameter; gillrakers short, only a little longer than broad, and not prominent when viewed from side of open mouth; body deep; a series of rounded dark spots, about as large as eye, along upper side . . . . . *Scomberoides lysan*
Upper jaw slightly broadened posteriorly, reaching to or behind rear margin of eye, its rear end decidedly closer to tip of snout than to base of pelvic fin; gillrakers elongate, slender and prominently projecting forward into mouth; snout slightly longer than eye diameter 6

6 Body slender, elongate; a large black blotch anteriorly on soft dorsal fin and

a blackish band from occiput to pectoral fin base; a longitudinal series of dark spots above lateral line and presumably a similar one below; gillrakers on first arch 22 (in 1 specimen) . . . . *Scomberoides sanctipetri*

Body deeper, obovate; vertical fins yellowish, without black blotch; a series of brownish blotches along side, those anteriorly bar-like, about as tall as snout length, and becoming rounded posteriorly; slender gillrakers on first arch 12(in 1 specimen) or 13(in 1) . . . . *Scomberoides tala*

7 Six or more free finlets behind both the dorsal and anal fins; high scutes along lateral line, beginning below spinous dorsal fin . . . . *Megalaspis cordyla*

No free finlets behind dorsal or anal fins 8

8 A deep furrow across shoulder girdle at its junction with isthmus and above this a fleshy projection; body rather elongate; dorsal rays normal; eye large, its diameter 2·8 to 3·2 in head length 9

Shoulder girdle normal, smooth 10

9 Anterior part of lateral line little arched, almost as long as straight part which begins below anterior short rays of soft dorsal fin . . . . *Selar crumenophthalmus*

Anterior part of lateral line moderately arched, approximately half as long as straight part, which begins below, or slightly to rear of, spinous dorsal fin . . . . *Selar boops*

10 Dorsal spines rudimentary, not connected by membrane; anterior dorsal and anal rays elongate, about as long as body; eye diameter about one-half of preorbital depth . . . . *Alectis indica*

Dorsal spines developed and connected by membrane 11

11 Gillrakers very long, visible from side of open mouth . . . . *Ulua mandibularis*

Gillrakers normal, not notably projecting into mouth 12

12 Teeth absent or greatly reduced on jaws and palate; 19 to 21 branched dorsal rays . . . . *Caranx speciosus*

Teeth present on jaws and usually on palate 13

13 A row of larger, pointed teeth along jaws exterior to a row or band of villiform teeth 14

Teeth in jaws nearly uniform in size, mostly villiform, in narrow rows to broad bands 15

14 Breast covered with scales; young with dark bars which disappear with age; straight portion of lateral line somewhat longer than curved portion, beginning under lobe of soft dorsal fin; upper jaw extending to below eye; lower jaw teeth in about one row . . . . *Caranx sexfasciatus*

A patch of scales before pelvic fins, surrounded by a naked area . . . . *Caranx ignobilis*

Breast naked in front of a line between pectoral fin base and pelvic fin; straight portion of lateral line beginning under rear part of spinous dorsal fin, the anterior part strongly arched; body usually with numerous small dark spots; lower jaw teeth in about one row . . . . *Caranx bucculentus*

15 Breast scaly 16

Breast naked in front of a line between pelvic and pectoral fins; villiform teeth in bands on each jaw; curved portion of lateral line longer than straight part, ending below middle of soft dorsal fin 17

16 Snout longer than eye; teeth minute, conical, in several rows anteriorly in upper jaw; dorsal and anal rays produced, filamentous . . *Caranx radiatus*

Eye diameter greater than snout length; teeth apparently in single series on each jaw; curved part of lateral line strongly arched, ending under anterior part of soft dorsal, shorter than straight part which weakly armed;

a prominent black opercular spot and dark tips to caudal fin; fin rays not produced . . . . *Caranx parasitus*

17 Body elongate, the depth shorter than soft dorsal base; tip of mouth on level with eye, which has a diameter about one-half of snout length; branched dorsal rays 30(in 1) or more; branched anal rays 25(in 1) or more . . . *Caranx gymnostethoides*

Anterior dorsal and anal rays forming a falcate lobe, which about as long as or longer than fin base; dorsal rays 20 or 21 and anal rays 16 or 17; body deeper, the depth greater than length of soft dorsal base; tip of mouth on level with eye, which about equal to snout length . . . . *Caranx armatus*

Median fin rays not notably elongate; branched dorsal rays 19(in 1) or 20(1); branched anal rays 16(2); eye diameter about half snout length; body depth greater than length of soft dorsal base; mouth distinctly below eye level . . . . *Caranx chrysophrys*

Scomberoides sanctipetri (Cuvier and Valenciennes)

Chorinemus sancti-petri Cuvier and Valenciennes, *Histoire naturelle des poissons* 8: 379, pl. 236 (1831); type locality, Malabar.

Recorded as: *Scomberoides sancti-petri* Paradice and Whitley (1927): 83, Pellew Islands. Record questionably referable: *Scomberoides tolooparah* Whitley (1951): 65, Darwin.

M48-11, Little Lagoon, 26 April 1948, USNM 173955, 1 specimen, 430 mm. standard length.

Scomberoides lysan (Forskål)

Scomber lysan Forskål, *Descriptiones Animalium*: 54 (1775); type localities, Djeddae and Lohajae, Red Sea.

Records questionably referable: *Chorinemus lysan* Günther (1860): 471, Port Essington; Macleay (1878): 355, Port Darwin; Kent (1889): 239; *Chorinemus forsteri* Richardson (1843a): 24, Port Essington.

M48-15, Emerald River, Groote Eylandt, USNM 173954, 7 specimens, 274 to 341 mm. standard length.

Scomberoides tala (Cuvier and Valenciennes)

Chorinemus tala Cuvier and Valenciennes, *Histoire naturelle des poissons* 8: 377 (1831); type locality, Malabar.

M48-11, Little Lagoon, 6 April and 2 May 1948, USNM 173956, 2 specimens, 271 and 300 mm. standard length.

M48-15, Emerald River, Groote Eylandt, USNM 173957, 5 specimens, 296 to 312 mm. standard length.

M48-23, Port Bradshaw, USNM 173958, 1 specimen, 310 mm. standard length.

M48-26, Sheltered bay, island, Port Bradshaw, USNM 173959, 1 specimen, 327 mm. standard length.

Trachinotus bailloni (Lacépède)

Caesiomorus baillonii Lacépède, *Histoire naturelle des poissons* 3: 91 and 93, pl. 3, fig. 1 (1802); type locality not given.

Recorded as: *Trachinotus baillonii* Paradice and Whitley (1927): 83, Pellew Islands.

M48-11, Little Lagoon, 21 May 1948, USNM 173951, 1 specimen, 259 mm. standard length.

Trachinotus ovatus (Linnaeus)

Gasterosteus ovatus Linnaeus, *Systema Naturae* ed. 10, 1: 296 (1758); type locality, Asia.

Recorded as: *Trachinotus ovatus* Paradice and Whitley (1927): 83, Pellew Islands.

*Seriola** *nigrofasciata* (Rüppell)

Nomeus nigrofasciatus Rüppell, *Atlas zu der Reise im Nördlichen Afrika. Fische des Rothen Meeres*: 92, pl. 24, fig. 2 (1828); type locality, Massaua, Red Sea.

Recorded as: *Zonichthys nigrofasciata* Munro (1958): 123, fig. 805, 90 miles south-west of Cape Fourcroy, Bathurst Island, Northern Territory.

Elagatis bipinnulatus (Quoy and Gaimard)

Seriola bipinnulata Quoy and Gaimard, *Voyage autour du monde...sur... l'Uranie et la Physicienne...Zoologie*: 363, pl. 11, fig. 3 (1824); type locality, Iles des Papous.

Recorded as: *Elagatis bipinnulatus* Whitley (1954): 27, Lynedoch Bank, Northern Territory.

Selar boops (Cuvier and Valenciennes)

Caranx boops Cuvier and Valenciennes, *Histoire naturelle des poissons* 9: 46 (1833); type localities, Ambon and Vanicolo.

M48-7, Mouth of billabong, Little Lagoon, USNM 173949, 5 specimens, 45 to 80 mm. standard length.

Selar crumenophthalmus (Bloch)

Scomber crumenophthalmus Bloch, *Naturgeschichte der Ausländischen Fische* 7: 77, pl. 343 (1793); reference not seen; type locality, Acara Bay, Guinea; Bloch, *Ichthyologie* 10: 65, pl. 343 (1797).

M48-7, Mouth of billabong, Little Lagoon, USNM 173948, 2 specimens, 72 and 77 mm. standard length.

Caranx† *speciosus* (Forskål)

Scomber speciosus Forskål, *Descriptiones Animalium*: xii and 54 (1775); type locality, Djeddae, Red Sea.

Caranx speciosus Ogilby, *Mem. Qld. Mus.* 3: 67, pl. 21 (1915).

Recorded as: *Caranx speciosus* Paradice and Whitley (1927): 82, Pellew Islands.

M48-13b, Hemple Bay, Port Langdon, 27 May 1948, USNM 173971, 2 specimens, 304 and 385 mm. standard length.

M48-15, Emerald River, Groote Eylandt, USNM 173972, 1 specimen, 320 mm. standard length.

M48-22, Bay of Yirrkalla, 13 to 16 August 1948, USNM 173973, 3 specimens, 338 to 440 mm. standard length.

Colour in life. All fins, side of head, and abdomen deep yellow.

* *Seriola* Cuvier (1817): 315; type species, *Caranx dumerili* Risso, designated by International Commission on Zoological Nomenclature, Opinion 461.

† The type species of *Caranx* Lacépède is uncertain, but usually considered to be *Scomber speciosus* Forskål, *Scomber carangus* Bloch, or, in error, *Scomber ruber* Bloch.

Caranx radiatus Macleay

Caranx radiatus Macleay, *Proc. Linn. Soc. N.S.W.* 5: 537 (1881); type locality, Rockingham Bay, Queensland.

Recorded as: *Caranx radiatus* Paradice and Whitley (1927): 82, Pellew Islands.

Caranx parasitus Garman

Caranx parasitus Garman, *Bull. Mus. Comp. Zool. Harv.* 39: 232 (1903); type locality, Great Barrier Reef, Cairns.

Recorded as: *Caranx parasitus* Paradice and Whitley (1927): 82, Pellew Islands; *Pantolabus parasitus* Whitley (1931b): 108, pl. 12, fig. 1, Pelican Spit, near Pellew Islands.

Caranx gymnostethoides (Bleeker)

Carangoides gymnostethoides Bleeker, *Natuurk. Tijdschr. Ned.-Ind.* 1: 343 and 364 (1851); type locality, Batavia, Java.

M48-30, Reefs, Yirrkalla, 11 August 1948, USNM 173960, 1 specimen, 285 mm. standard length.

Colour in life. Abdomen yellow.

Caranx armatus (Forskål)

Sciaena armata Forksål, *Descriptiones Animalium*: xii and 53 (1775); type locality, Red Sea.

Recorded as: *Caranx armatus* Paradice and Whitley (1927): 83, Pellew Islands.

Caranx chrysophrys Cuvier and Valenciennes

Caranx chrysophrys Cuvier and Valenciennes, *Histoire naturelle des poissons* 9: 77, pl. 247 (1833); type locality, Seychelles.

M48-15, Emerald River, Groote Eylandt, USNM 173974, 1 specimen, 252 mm. standard length.

M48-27, Bay and beach, Port Bradshaw, USNM 173975, 1 specimen, 340 mm. standard length.

Caranx sexfasciatus Quoy and Gaimard

Caranx sexfasciatus Quoy and Gaimard, *Voyage autour du monde . . . sur . . . l'Uranie et la Physicienne . . . Zoologie*: 358, pl. 65, fig. 4 (1824); type locality, Papous Islands.

Caranx forsteri Cuvier and Valenciennes, *Histoire naturelle des poissons* 9: 107 (1833); type localities, Mauritius, Malabar, Celebes, New Guinea, New Ireland and Vanicolo.

Caranx parapistes Richardson, *Ichthyology of the Voyage of H.M.S. Erebus and Terror*: 136, pl. 58, figs. 6 and 7 (1848); type locality, Port Essington.

Recorded as: *Caranx forsteri* Paradice and Whitley (1927): 82, Pellew Islands; *Caranx hippos* Günther (1860): 449, Port Essington; Macleay (1878): 355, Port Darwin.

M48-7, Mouth of billabong, Little Lagoon, USNM 173961, 5 specimens, 75 to 93 mm. standard length.

M48-10, Estuary, Thompson Bay, Port Langdon, USNM 173962, 6 specimens, 98 to 123 mm. standard length.

M48-19, Creek, tributary to Little Lagoon, USNM 173963, 2 specimens, 87 to 139 mm. standard length.

M48-28, Creek mouth and bay, Port Bradshaw, USNM 173964, 1 specimen, 134 mm. standard length.

M48-31, Creek, Yirrkalla, USNM 173965, 9 specimens, 124 to 180 mm. standard length.

M48-34, Creek, Yirrkalla, USNM 173966, 1 specimen, 132 mm. standard length.

Native names: *Állogúanjíra* (Ahlloo-guan-jeerah), Groote Eylandt; *Tárapa,* Yirrkalla.

Colour in life. Based on specimens 98 to 123 mm. in standard length: side with about 5 faint, broad, grey bars; eye orange, the pupil blue; anal and pelvic fins light yellow, tipped with milky white; soft dorsal and caudal fins yellow with a broad greyish black distal margin; margin of spinous dorsal fin greyish black; a tiny jet black spot at upper posterior corner of opercle which faintly yellow; light yellow bar, almost wide as eye, passing obliquely downward and backward across cheek from eye; and a similar, but slightly narrower band along lower midside of caudal peduncle, running parallel to anal fin base, extending from opposite vent diagonally upward to lateral line scutes opposite rear end of anal fin.

In the specimen 134 mm. in standard length, side below lateral line, most of caudal fin, medial and basal parts of dorsal and anal fins, abdomen and cheek rich orange yellow; eye reddish.

Caranx ignobilis (Forskål)

Scomber ignobilis Forskål, *Descriptiones Animalium*: xii and 55 (1775); type localities, Djeddae and Lohajae, Red Sea.

Recorded as: *Caranx ignobilis* Klunzinger (1879): 377 and Macleay (1884): 24, Port Darwin.

Caranx bucculentus Alleyne and Macleay

Caranx bucculentus Alleyne and Macleay, *Proc. Linn. Soc. N.S.W.* 1: 326, pl. 11, fig. 1 (1877); type locality, Cape Grenville, Queensland.

Carangus bucculentus Ogilby, *Mem. Qld. Mus.* 3: 73, pl. 22 (1915).

Recorded as: *Caranx bucculentus* Paradice and Whitley (1927): 82, Pellew Islands.

M48-13a, South end Winchelsea Island, 2 June 1948, USNM 173967, 5 specimens, 173 to 201 mm. standard length.

M48-20, Main jetty, Melville Bay, 11 July 1948, USNM 173969, 1 specimen, 239 mm. standard length.

M48-20, Old jetty, Melville Bay, 17 August 1948, USNM 173968, 8 specimens, 200 to 232 mm. standard length.

M48-23, Port Bradshaw, USNM 173970, 5 specimens, 191 to 353 mm. standard length.

Ulua mandibularis (Macleay)

Caranx mandibularis Macleay, *Proc. Linn. Soc. N.S.W.* 7: 356 (1882); type locality, near Port Moresby, New Guinea.

Recorded as: *Ulua mandibularis* McCulloch (1924): 75, pl. 14; Paradice and Whitley (1927): 83 and 97, Pellew Islands.

M48-20, Main jetty, Melville Bay, 11 July 1948, USNM 173950, 1 specimen, 262 mm. standard length.

Alectis indica (Rüppell)

Scyris indicus Rüppell, *Atlas zu der Reise im Nördlichen Afrika*: 128, pl. 33, fig. 1 (1828); type locality, Djedda, Red Sea.
Alectis indica Ogilby, *Mem. Qld. Mus.* 3: 83, pl. 26 (1915).

Recorded as: *Alectis indica* Paradice and Whitley (1927): 83, Pellew Islands.

M48-11, Little Lagoon, 23 April and 21 May 1948, USNM 173952, 2 specimens, 261 and 295 mm. standard length.

M48-24, Beach, Port Bradshaw, USNM 173953, 1 specimen, 320 mm. standard length.

Megalaspis cordyla* (Linnaeus)

Scomber cordyla Linnaeus, *Systema Naturae* ed. 10, 1: 298 (1758); type locality, not known.

Recorded as: *Megalaspis cordyla* Whitley (1954): 27, near Peron Islands and from English Company Islands, Northern Territory.

CORYPHAENIDAE

Coryphaena† *hippurus* Linnaeus

Coryphaena hippurus Linnaeus, *Systema Naturae* ed. 10, 1: 261 (1758); type locality, open seas.

Recorded as: *Coryphaena hippurus* Munro (1958): 121, Timor Sea, Northern Territory.

TERAPONIDAE

Key to the Arnhem Land Species

1 Teeth compressed, tricuspidate; five to eight straight longitudinal lines, or bands, from head to caudal peduncle; spinous dorsal black distally, yellowish or silvery below . . . . . *Helotes sexlineatus*
  Teeth conical, simple 2
2 Lower opercular spine greatly developed, extending to the rear of the opercular membrane; spinous dorsal with a prominent large black blotch; three to five longitudinal lines or bands along middle and upper side from head to caudal peduncle; supracleithrum exposed, serrated (subgenus *Terapon* Cuvier) 3
  Opercular spines shorter, not extending to the rear of the opercular membrane; no prominent large black blotch on spinous dorsal fin and body without longitudinal lines 5
3 Scales small, 14 to 19 between lateral line and predorsal midline, or 80 or more between head and caudal fin 4
  Scales larger, 10 to 13 between lateral line and predorsal midline, or fewer than 65 between head and caudal fin; subopercle smooth or nearly so . . . . *Terapon theraps*
4 Longitudinal bands almost straight; subopercle smooth, or nearly so . . . . . *Terapon puta*
  Longitudinal bands strongly curved; subopercle weakly to moderately serrate . . *Terapon jarbua*

* *Megalaspis* Bleeker (1851): 342 and 352; type species, *Caranx rottleri* Bloch.

† *Coryphaena hippurus* Linnaeus was selected as the type species of *Coryphaena* Linnaeus by the International Commission on Zoological Nomenclature, Opinion 77.

5 Supracleithrum exposed; soft dorsal rays 12 or 13; gillrakers on first arch about 9 above plus 17 below; a black basal blotch on the posterior border of the soft dorsal and anal fins (subgenus *Mesopristes* Bleeker) . . . . *Terapon alligatoris*

Supracleithrum concealed by scales; soft dorsal rays 8 to 11; gillrakers on first arch 22 or fewer; subopercle not serrate (subgenus *Leioptherapon* Fowler) 6

6 Scale rows between head and caudal fin, above lateral line 36 to 45; scale rows across cheek, from angle of preopercle to eye 4 or 5; about 5 narrow dark brown bars across body . . . . *Terapon percoides*

Scale rows between head and caudal fin, above lateral line 48 to 62; scale rows across cheek 7 to 10 7

7 Body and fins plain, the body with brown spots that are about half the diameter of the pupil; dorsal spines XI to XIII, usually XII . . . . *Terapon unicolor*

Body rather densely covered with dark brown spots, each about half pupil diameter; black obovate blotches, large as eye, near end of both upper and lower lobes of caudal fin; dorsal spines XII to XIV, usually XIII . . . . . *Terapon caudavittatus*

*Terapon** *theraps* Cuvier and Valenciennes

Therapon theraps Cuvier and Valenciennes, *Histoire naturelle des poissons* 3: 129, pl. 53 (1829); type localities, Java and Mahé.

Therapon rubricatus Richardson, *Ann. Mag. Nat. Hist.* 9: 127 (1842); type locality, north-west coast of Australia.

Recorded as: *Therapon theraps* Richardson (1842a): 126, Port Essington; Macleay (1880): 360, Port Darwin; Ogilby and McCulloch (1916): 102; *Terapon theraps* Paradice and Whitley (1927): 87, Pellew Islands.

Terapon puta Cuvier and Valenciennes

Therapon puta Cuvier and Valenciennes, *Histoire naturelle des poissons* 3: 131 (1829); type localities, Pondicherry and Mahé.

Recorded as: *Terapon puta* Paradice and Whitley (1927): 87, Pellew Islands.

Terapon jarbua (Forskål)

Sciaena jarbua Forskål, *Descriptiones Animalium*. . .: xii and 50 (1775); type locality, Djedda, Red Sea.

Holocentrus servus Bloch, *Naturgeschichte der Ausländischen Fische* 4: 80, pl. 238, fig. 1 (1797); type locality, Japan.

Recorded as: *Therapon jarbua* Klunzinger (1879): 349, Port Darwin; *Therapon servus* Richardson (1842a): 126, Port Essington; Macleay (1878): 348; Ogilby and McCulloch (1916): 104, Port Darwin; *Terapon servus* Paradice and Whitley (1927): 88, Pellew Islands; *Therapon theraps* Castelnau (1873): 83, Port Darwin.

M48-1, Nightcliff, USNM 173656, 1 specimen, 76 mm. standard length.

M48-7, Mouth of billabong, Little Lagoon, USNM 173657, 18 specimens, 18 to 81 mm. standard length.

M48-8, Little Lagoon, USNM 173658, 4 specimens, 64 to 85 mm. standard length.

* Generic synonyms: *Leiopotherapon* Fowler (1931): 328 and 353; type species, *Datnia plumbea* Kner by original designation. *Madigania* Whitley (1945): 10; type species, *Therapon unicolor* Günther, by original designation.

M48-19, Creek, tributary to Little Lagoon, USNM 173659, 13 specimens, 25 to 61 mm. standard length.

M48-29, Pool near small creek, Yirrkalla, USNM 173660, 5 specimens, 60 to 94 mm. standard length.

M48-31, Creek, Yirrkalla, USNM 173661, 5 specimens, 49 to 75 mm. standard length.

M48-34, Creek, Yirrkalla, USNM 173662, 4 specimens, 51 to 96 mm. standard length.

Native names: *Túrmungóoraárnda,* Groote Eylandt; *Móngurándi,* Yirrkalla.

Description. Dorsal rays XI,10(in 2), XI,11(1) or XII,10(6); anal rays III,8(9); pelvic rays I,5(18); pectoral rays ii,10(2), ii,11(15) or ii,12(1); branched caudal rays 8+7=15(9); scale rows between head and caudal fin 85 to 96 above lateral line and 82 to 91 below; scale rows across body 29 to 33 between vent and lateral line and 17 to 19 between lateral line and predorsal midline; scale rows across cheek 9 or 10; gillrakers on first arch, on upper limb 7+12 on lower =19(4); branchiostegal rays 6.

In five specimens, 60 to 94 mm. in standard length, eye 3·9 to 4·3 and snout 3·7 to 4·1, both in head length; head length 2·7 to 3·1, body depth 2·9 to 3·1 and distance from origin of pelvic fin to anal fin 3·2 to 3·3, all in standard length.

Subopercle with weak to moderate serrae; preorbital, preopercle, postcleithral process and exposed supracleithrum distinctly serrated; lower opercular spine extending backward beyond edge of opercular membrane; distance between pelvic and anal fins about equal to or shorter than head length; jaws equal or the lower slightly included; teeth on palatines and vomer small, those on jaws simple, conical, in several series, the anterior or outer row considerably the largest, canine-like, and not close-set; gillrakers moderate, slender; second anal spine equal to or a little longer than the third; distance between anterior and posterior nares considerably less than distance from anterior naris to groove behind premaxilla.

Colour in life. Body silvery white, with prominent brownish grey to chocolate brown lateral stripes; some red on margins of stripes extending on to caudal fin; blotch on inter-radial membranes between dorsal spines IV to VI, two spots on soft dorsal fin, and one on upper lobe of caudal fin black, often margined below with reddish; dorsals, caudal, and anal fins milky white, except for broad orange to reddish bar from base to tip of middle portion of soft anal fin; proximal half of pelvic fin orange, the distal half milky white; medial rays of pectoral fin often reddish.

Colour in alcohol. Body and head yellowish below, medium brown on sides, and darker above; four curving dark brown lines or bands on body as follows: one, very faint, from pectoral base, along lower side to above anal fin; another, half pupil width, curving gently downward from upper gill opening and backward through mid-side of caudal peduncle, thence along shortest caudal ray in upper caudal lobe to its tip; another, wide as pupil, curving slightly downward from mid-nape, and backward to upper surface caudal peduncle—just behind dorsal fin; and one, somewhat wider, slightly curved, extending from anterior end of dorsal fin to anterior half of soft dorsal base and outward on fin, as a black bar, to tip of the middle rays; fins generally yellowish to dusky brownish, the base of the pectoral fin dark brown on its proximal side; dorsal fin with a large distal, black blotch between spines IV to VI and a smaller one, sometimes present, between spines VIII and IX; soft dorsal with ends of first three rays black, in addition to the bar described above; a black bar on each caudal lobe extending obliquely backward across fin; tip of upper caudal lobe black.

Terapon alligatoris Rendahl

Terapon alligatoris Rendahl, *Medd. Zool. Mus. Krist.* 5: 185 (1922); McKinley River, Northern Territory; type locality, South Alligator River, Northern Territory.

Terapon caudavittatus (Richardson)

Pl. 39

Datnia? caudavittata Richardson, *Ichthyology of the voyage of the H.M.S. Erebus and Terror*: 24, pl. 18, figs. 3 to 5 (1845); type locality, Harvey River, Western Australia.

Recorded as: *Therapon caudavittatus* Ogilby and McCulloch (1916): 108, Port Essington and Port Darwin; *Terapon caudavittatus* Paradice and Whitley (1927): 88, Pellew Islands and Darwin; *Therapon caudovittatus* Macleay (1878): 348, Port Darwin.

M48-1, Nightcliff, USNM 173668, 4 specimens, 51 to 62 mm. standard length.

M48-7, Mouth of billabong, Little Lagoon, USNM 173669, 10 specimens, 38 to 97 mm. standard length.

M48-9, Reefs, entrance to Little Lagoon, USNM 173670, 5 specimens, 35 to 44 mm. standard length.

M48-10, Estuary, Thompson Bay, Port Langdon, USNM 173671, 5 specimens, 44 to 79 mm. standard length.

M48-15, Emerald River, Groote Eylandt, USNM 173672, 7 specimens, 56 to 154 mm. standard length.

M48-19, Creek, tributary to Little Lagoon, USNM 173673, 76 specimens, 31 to 155 mm. standard length.

M48-22, Bay of Yirrkalla, USNM 173674, 1 specimen, 48 mm. standard length.

M48-26, Sheltered bay, island, Port Bradshaw, USNM 173675, 2 specimens, 114 to 119 mm. standard length.

M48-27, Beach, Port Bradshaw, USNM 173676, 1 specimen, 157 mm. standard length.

M48-28, Creek mouth and bay, Port Bradshaw, USNM 173677, 25 specimens, 83 to 174 mm. standard length.

M48-29, Pool near small creek, Yirrkalla, USNM 173678, 3 specimens, 49 to 53 mm. standard length.

M48-31, Creek, Yirrkalla, USNM 173679, 35 specimens, 26 to 105 mm. standard length.

M48-34, Creek, Yirrkalla, USNM 173680, 6 specimens, 62 to 174 mm. standard length.

Native names: *Tilyángua,* Groote Eylandt; *Chánook,* Yirrkalla.

Description. Dorsal rays XIII,9(in 8), XIV,9(1), or XIV,10(1); anal rays II,8(1), III,8(8), or III,9(1); pelvic rays I,5(20); pectoral rays ii,12(6), or ii,13 (14); branched caudal rays 7+7=14 (1) or 8+7=15(9); scale rows between head and caudal fin 48 to 54 above lateral line and 44 to 49 below; scale rows across body 17 to 20 between vent and lateral line and 9 to 11 between lateral line and predorsal midline; cheek scales in 8 or 9 rows; gillrakers on first arch 7 to 9 on upper limb plus 12 or 13 on lower limb equals 19(1) or 21(4); branchiostegal rays 6.

In five specimens 49 to 119 mm. standard length, eye 3·1 to 3·8, snout 2·8 to 3·2, both in head length; head length 2·7 to 3·3, body depth 2·6 to 2·9, and distance from pelvic fin origin to anal fin 2·9 to 3·4, all in standard length.

Preorbital, preopercle and postcleithral process serrated; supracleithrum not exposed; lower spine of opercle not extending behind edge of membrane of opercle; distance from pelvic fin base to anal base little longer than or shorter than head length; lower jaw about equal to or slightly terminal to upper; second anal spine longer than third; no teeth on palate, those on jaws simple, not compressed, in several series, the anterior series largest, close-set and slightly curved backward; distance between anterior and posterior nares about equal to or greater than distance from anterior naris to groove behind premaxilla.

Colour in life. Each lobe of caudal fin with a prominent black blotch, these and the black tip of the soft dorsal fin outlined by light to lemon yellow; pectoral fin light yellow; anal, especially the spinous portion, and pelvic fins lemon yellow to deep yellow; spots on body reddish brown.

Colour in alcohol. Body and head yellowish to light brown, darkest above and lightest below; side and upper portion of body and head densely sprinkled with small reddish brown spots; back with a series of five to seven faint brown saddles, wide as eye, that extend downward approximately to lateral line, the first, on nape, most prominent; pelvic, pectoral, and anal fins light yellowish, dusted with brownish; spinous dorsal yellow, suffused or marbled with brown; soft dorsal fin with tips of anterior rays dark brown to black, especially in young, this followed inwardly by a light yellow band, a narrow interrupted brown band, and a light yellow base; caudal fin with reddish brown spots or reticulations on the central portion and base, the upper and lower lobes light yellow, with a prominent black, subrectangular blotch on each, both nearly as wide as eye height, and extending obliquely backward.

Terapon unicolor Günther

Therapon unicolor Günther, *Catalogue of the Fishes in the British Museum* 1: 277 (1859); type locality, head of Mosquito Creek, near Darling Downs, and Gwydir River, New South Wales, the latter indicated as type locality by Ogilby and McCulloch, *Mem. Qld. Mus.* 5: 109-12, pl. 11, fig. 1 (1916).

Therapon longulus Macleay, *Proc. Linn. Soc. N.S.W.* 5: 367 (1880); type locality, fresh water inland from Port Darwin.

Recorded as: *Terapon unicolor* Paradice and Whitley (1927): 88, Howard River, 35 miles east of Darwin; *Madigania unicolor* Whitley (1945): 10, Northern Territory records.

Billabong, about 3 miles above Roper River Mission, Northern Territory, 17 August 1948 (Blitner), USNM 173859, 7 specimens, 68 to 110 mm. standard length.

M48-35, Billabong, Oenpelli, USNM 173653, 16 specimens, 85 to 204 mm. standard length.

M48-37, Fish Creek, USNM 173654, 22 specimens, 39 to 135 mm. standard length.

M48-38, Oenpelli Creek, USNM 173858, 37 specimens, 41 to 119 mm. standard length.

M48-39, Red Lily Lagoon, USNM 173655, 4 specimens, 116 to 184 mm. standard length.

Native name: *Bood,* Oenpelli.

Description. Dorsal rays XI to XIII,10 or 11, the spines XI(in 1), XII(33) or XIII(8), the soft rays 10(7) or 11(2); anal rays III,7(1) or III,8(8); pelvic rays I,5(18); pectoral rays ii,12(1), ii,13(16), or ii,14(1); branched caudal rays 8+6=14(1) or 8+7=15(8); scale rows between head and caudal fin 56 to 64 above lateral line

Dorothea B. Schultz

PLATE 39. *Terapon caudavittatus*
Thompson Bay, USNM 173671, 65 mm. standard length

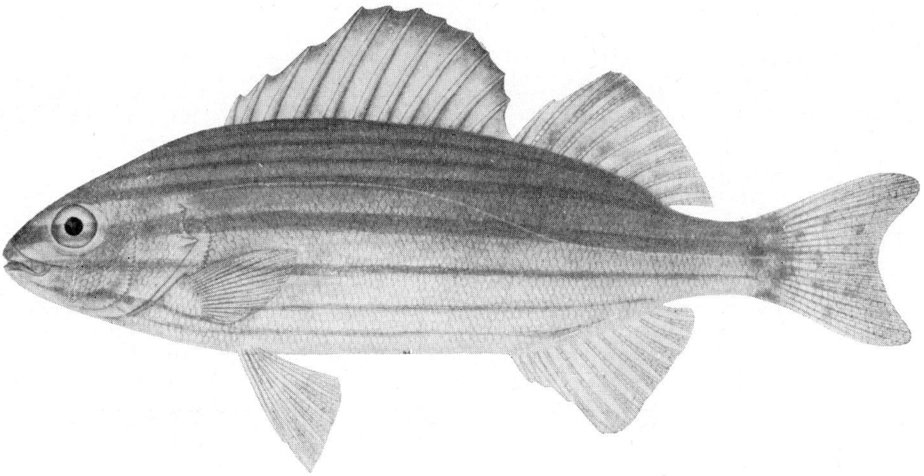

Dorothea B. Schultz

PLATE 40. *Helotes sexlineatus*
Port Bradshaw, USNM 173681, 184 mm. standard length

and 49 to 54 below; scale rows across body 18 to 20 between vent and lateral line and 11 or 12 between lateral line and predorsal midline; cheek scales in 7 to 10 rows; gillrakers on first arch 6 or 7 on upper limb +10 to 12 on lower limb equal 17(4), 18(4), or 19(1); branchiostegal rays 6.

In five specimens 111 to 137 mm. standard length, eye 5·1 to 5·5, snout 3·1 to 3·6, both in head length; head length 2·7 to 3·1, body depth 2·5 to 2·9, distance from pelvic fin origin to anal fin 2·8 to 3·2, all in standard length.

Supracleithrum not exposed; preorbital, preopercle, and postcleithral process serrated; lower spine of opercle not extending behind opercular membrane; distance from pelvic base to anal base shorter than head length; second anal spine longer than the third; jaws about equal or the lower slightly terminal; no teeth on palate, those on jaws simple, not compressed, in several series, the anterior series largest and rather close-set; distance between anterior and posterior nares about equal to or greater than distance from anterior naris to groove behind premaxilla.

Colour in alcohol. Head, body, and fins medium to dark brown above and on sides; lower surface yellow; small dark brown spots, half pupil diameter, most noticeable on side of head, but generally extending backward over trunk; anal spines and, in specimens from Oenpelli, the long lower caudal ray whitish; caudal fin nearly unicolour in specimens from Red Lily Lagoon.

Terapon percoides Günther

Therapon percoides Günther, *Ann. Mag. Nat. Hist.* ser. 3, 14: 374 (1864); type locality, Fitzroy River, near Rockhampton.

Arnhem Land, no further data; USNM 173664, 31 specimens, 36 to 84 mm. standard length.

Billabong, about 3 miles above Roper River Mission, 17 August 1948, Blitner, USNM 173665, 16 specimens, 38 to 71 mm. standard length.

M48-35, Billabong, Oenpelli, 29 September 1948, USNM 173666, 9 specimens, 80 to 99 mm. standard length; September and October 1948, USNM 173667, 30 specimens, 92 to 157 mm. standard length.

Native name: *Mándídi*, Oenpelli.

Description. Dorsal rays XII,10(in 2), XIII,8(2), XIII,9(20), or XIII,10(1); anal rays III,7(4) or III,8(21); pelvic rays I,5(18); pectoral rays ii,11(2), ii,12(34), or ii,13(14); branched caudal rays 8+7=15(9); scale rows between head and caudal fin 39 to 44 above lateral line and 37 to 41 below; scale rows across body 14 to 16 between vent and lateral line and 8 or 9 between lateral line and predorsal midline; scale rows across cheek 4; gillrakers on first arch, 7 on upper limb plus 11 or 12 on lower equal 18(2) or 19(1); branchiostegal rays 6.

In five specimens 80 to 99 mm. standard length, eye 3·4 to 3·7, snout 3·0 to 3·4, both in head length; head length 2·9 to 3·3, body depth 2·3 to 2·5, distance from origin of pelvic fin to anal fin 2·9 to 3·5, all in standard length.

Gillrakers short, almost knob-like; preorbital, preopercle, and postcleithral process serrated; supracleithrum not exposed; jaws equal or the lower slightly included; second anal spine longer than the third; lower opercular spine not extending behind opercular membrane; no teeth on palate, those on jaws simple, not compressed, in several series, the anterior ones slightly enlarged and close-set; distance between pelvic and anal fins about equal to or slightly longer than head length; distance between anterior and posterior nares about equal to or greater than distance from anterior naris to groove behind premaxilla.

Colour in alcohol. Body, head, and fins medium to light brown, the lower head and abdomen yellowish to yellowish brown; longest lower caudal rays dark brown to blackish; a dark brown spot, size of pupil, at base of middle caudal rays in young; body with five vertical dark brown bars, each about a pupil width, as follows: first from just before dorsal fin to postcleithral process; second from dorsal spines V to VII to midside; third from spines XI or XII to midside; fourth from midbase soft dorsal downward to or sometimes on to soft anal fin as a black blotch; and fifth across narrow portion of caudal peduncle; membrane of spinous dorsal light brownish black in adults.

Remarks. A single specimen (USNM 173663) 48 mm. in standard length was sorted from the above series of 31 *Terapon percoides* which bears no locality data. It has XIII,10 dorsal rays; III,8 anal rays; I,5 pelvic rays on both sides; ii,14 pectoral rays, both sides; 8+7=15 branched caudal days; 6 branchiostegal rays; 5+11=16 gillrakers on the first arch; 6 rows of cheek scales; 49 scales above and 44 scales below the lateral line between the head and caudal fin; 17 scales between the vent and the lateral line plus 10 above to the predorsal midline; there are 7 scales between the lateral line and the middle of the dorsal fin. The supracleithrum is not exposed; the preorbital, the exposed postcleithral process, and the preopercle are serrate; the spines of the opercle are short.

The colour in alcohol is medium brown, darker above; five faint brownish bars, nearly as wide as eye, cross the body and peduncle about as in *percoides*; head, side, and upper surface of body with tiny brown spots which form a mottled effect above and rather distinct longitudinal lines below; soft anal fin and lower lobe of caudal fin brownish black.

I suspect that this small specimen is a natural hybrid between *Terapon percoides* Günther and *Terapon unicolor* Günther. Its meristic characters and colour pattern are nearly intermediate between those species.

Helotes sexlineatus (Quoy and Gaimard)

Pl. 40

Terapon sexlineatus Quoy and Gaimard, *Voyage autour du monde . . . sur . . . l'Uranie et la Physicienne . . . Zoologie*: 340, pl. 60, fig. 1 (1825); type locality, Sharks Bay, Western Australia.

Recorded as: *Helotes sexlineatus* Macleay (1878): 348; Macleay (1880): 368, Port Darwin; Paradice and Whitley (1927): 87, Pellew Islands; *Therapon cuvieri* Günther (1859): 282, Port Essington; Macleay (1878): 348, Port Darwin.

M48-28, Creek mouth and bay, Port Bradshaw, USNM 173681, 9 specimens, 157 to 199 mm. standard length.

Description. Dorsal rays XII,10(in 9); anal rays III,10(9); pelvic rays I,5(18); pectoral rays ii,13(16) or ii,14(2); branched caudal rays 8+7=15(8); scale rows between head and caudal fin 82 to 90 above lateral line and 81 to 88 below; scale rows across body 22 to 26 between vent and lateral line and 14 or 15 between lateral line and predorsal midline; cheek scales in 5 or 6 rows; branchiostegal rays 6; gillrakers on first arch 8 on upper limb plus 15 on lower equal 23(1).

In five specimens, 157 to 199 mm. in standard length, eye 3·6 to 4·0 and snout 3·3 to 3·8, both in head length; head length 4·0 to 4·3, body depth 3·4 to 3·8, and distance from base pelvic fin to anal fin 2·6 to 2·8, all in standard length.

Supracleithrum hidden by scales; preorbital, suborbital, and postcleithral process roughened to distinctly serrated; preopercle strongly serrated; lower opercular spine extending backward approximately to edge of opercular membrane; distance between pelvic and anal fins considerably greater than head length; jaws equal or the lower slightly included; third anal spine longest, considerably longer than the second anal spine; teeth in several rows in jaws, compressed, tricuspidate, the middle cusp largest; no teeth on palate; distance between anterior and posterior nares less than distance from anterior naris to groove behind premaxilla.

Colour in alcohol. Body and head medium to dark reddish brown above, lighter to yellowish brown below; approximately 8 straight reddish brown lines or bands from head to caudal peduncle, the band extending backward through the eye to the caudal fin about as wide as pupil, the others narrower; those below, especially on lower side, faint and often discontinuous, and those above tending to be obscured by the dark brown back; pectoral, pelvic, and anal fins yellowish, lightly dusted with reddish brown; caudal fin medium yellowish brown; in a few specimens several faint bar-like rows of spots near basal half of caudal; spinous dorsal with yellowish to silvery base and dark brown to blackish margin; soft dorsal yellowish, heavily dusted with brown, especially near tips of rays.

LUTJANIDAE

Key to the Arnhem Land Species

1 Preopercle entire, not serrated, with at most a shallow notch or concavity along vertical arm near lower border; scales on body in horizontal series 2

Preopercle possibly weakly serrate in young, smooth in adults; dorsal and anal fins naked; anal rays III,8; dorsal rays X,11; pectoral fin short, about as long as snout; rear dorsal and anal rays produced . . . . *Aprion virescens*

Preopercle more or less serrate at all ages, usually with a prominent notch just above angle; dorsal and anal fins scaly; some or all of scale rows on body inclined; gillrakers moderate, with seldom more than 13 developed (except *Lutjanus lineolatus*) on lower limb of first arch 3

2 Anal rays III,9; dorsal rays X,16; soft dorsal fin higher than spines, some of the rays usually produced; dorsal and anal fins without scales; vomer toothless; mouth large . . . . *Paradicichthys venenatus*

Anal rays III,11; dorsal rays X,15; median fin rays not produced but with scales on the membranes; mouth small, somewhat oblique; vomer with short blunt teeth; body depth about 2·2 to 2·4; dorsal fin blackish, other fins nearly unicolour, yellowish; supratemporal bands of scales continuous across top of head . . . . *Caesio cuning*

3 Dorsal spines X, infrequently XI; scale rows below lateral line parallel with body axis 4

Dorsal spines XI, infrequently X, suborbitals naked 10

4 Anal rays III,8, infrequently III,7 or III,9; suborbitals (except in *L. quinquelineatus*) naked, not with small scales 5

Anal rays III,9, infrequently III,8; suborbitals and margin of preopercle covered with series of small scales, which distinct from rows of larger cheek scales; body with faint (not prominent in preservation), broad, light and dark bands; a dark brown spot at upper base of pectoral fin . . . . *Lutjanus carponotatus*

5 Scale rows above lateral line parallel with dorsal profile anteriorly and slightly inclined posteriorly, meeting dorsal fin at or behind spinous dorsal 6

Most of upper scale rows inclined, meeting dorsal base at and behind anterior spinous part 7

6 Upper scale rows nearly parallel with dorsal profile, meeting it at or behind

soft dorsal fin; a large round black spot on lateral line, beneath rear end of spinous dorsal; each scale of body with prominent dark brown spot, these forming longitudinal lines . . . . *Lutjanus johni*

Upper scale rows more inclined, meeting dorsal profile at and behind spinous dorsal; body and head nearly unicolour, each scale often with a dark vertical bar; young often with a pattern of broad dark bars separated by narrow, pale interspaces; two horizontal, dark, wavy lines across cheek, especially in young; pelvic fin and anterior part of anal fin, except spines, black . . . . *Lutjanus argentimaculatus*

7 A large black blotch on lateral line beneath rear end of spinous dorsal; young with narrow brown longitudinal lines which disappear with age; edge of preopercle naked 8

Body with brown lines and one broad brown band or five bluish longitudinal bands; rows of small scales on preopercle below and behind normal band of cheek scales; lateral dark blotch present in *L. quinquelineatus* 9

8 Blotch rounded with more than half its area above lateral line; margin of dorsal fin dusky, blackish; soft dorsal rays typically 14 . . *Lutjanus russelli*

Blotch elongate, oblong, nearly centred on lateral line; dorsal fin unicolour, yellowish; soft dorsal rays typically 13 . . . . *Lutjanus fulviflamma*

9 Five bluish (silvery in preservation) longitudinal bands: one from head to middle of spinous dorsal base; one from eye to anterior base soft dorsal; another from rear of eye to or behind rear end of soft dorsal base; two or three bands across cheek and opercle, of which two above pectoral fin base; thence as one from angle of opercle to mid-base caudal fin; and another from lower base pectoral fin to or behind rear end of anal fin base; a blackish spot on lateral line, beneath rear end of spinous dorsal fin in young, disappearing with age; suborbitals with rows of small scales which separated from the normal rows of cheek scales . . *Lutjanus quinquelineatus*

A prominent, longitudinal, irregular bordered, yet straight, dark brown band from head to base of caudal fin; narrow dark brown lines along scale rows which inclined above lateral line and horizontal below; no large dark spot on body; suborbitals naked . . . . *Lutjanus vitta*

10 Anal rays III,8; brownish, lighter below, marked by a broad, brown band from head to upper base caudal fin; large dots or narrow brown lines along scale rows, which parallel with axis below and inclined above; eye diameter equal to or greater than snout length; margin preopercle with small irregular scales . . . . *Lutjanus lineolatus*

Most scale rows, both above and below lateral line, somewhat inclined, some crossing it; margin of preopercle naked; anal rays III,9 or III,10 11

11 Anal rays usually III,10; three broad, black bars crossing head, trunk and caudal peduncle . . . . *Lutjanus sebae*

Anal rays typically III,9; bars, if present, on head and across caudal peduncle, not extending down over trunk 12

12 A dark or blackish quadrate saddle which confined to dorsal surface of caudal peduncle; pelvic fin pale, whitish; inner margin of rear nostril crenulate, its posterior edge closer to anterior nostril than to eye . . *Lutjanus malabaricus*

Young with margaritaceous bordered blackish saddle or bar on caudal peduncle, which disappears with age; pelvic fin dusky or blackish; inner margin rear nostril entire, its rear border closer to eye than to anterior nostril . . *Lutjanus erythropterus*

TABLE 4

FIN RAY COUNTS IN ARNHEM LAND SPECIES OF LUTJANIDAE

Species	Dorsal Spines		Soft Dorsal Rays						Anal Rays					Pectoral Rays				
	X	XI	12	13	14	15	16	17	III,7	III,8	III,9	III,10	III,11	ii,13	ii,14	ii,15	ii,16	ii,17
Lutjanus																		
argentimaculatus	42			13	29				1	41					5	1		
johni	1				1					1						2		
vitta	3			3						3				1	5			
russelli	33	1		1	25	8				34				1	15			
fulviflamma	12			12						12					6			
quinquelineatus	5				3	2				5					8	2		
carponotatus	31				1	30				1	30				6			
lineolatus		2	1	1						2					2	2		
erythropterus		4			4						4				2	6		
malabaricus		17	1	2	14						17					6		
sebae	1	7				3	4	1		1	1	6			1	11		
Caesio																		
cuning	8					8							6				3	13

Lutjanus argentimaculatus (Forskål)

Sciaena argentimaculata Forskål, *Descriptiones Animalium*: xi and 47 (1775); type locality, Arabia.

Recorded as: *Mesoprion argentimaculatus* Klunzinger (1879): 341, and Macleay (1884): 6, Port Darwin. Records probably referable: *Mesoprion yapilli* Richardson (1842a): 26, and Günther (1859): 194, Tamar Rock, Port Essington; *Mesoprion roseogaster* Kent (1889): 239, Port Darwin; *Lutjanus coatesi* Whitley (1954): 28, Sahul Bank; Lynedoch Bank; Evans Shoal; and Flinders Reef, Northern Territory.

M48-10, Estuary, Thompson Bay, Port Langdon, USNM 174161, 1 specimen, 92 mm. standard length.

M48-14, Emerald River, Groote Eylandt, USNM 174162, 4 specimens, 72 to 88 mm. standard length.

M48-15, Emerald River, Groote Eylandt, USNM 174163, 7 specimens, 69 to 292 mm. standard length.

M48-19, Creek, tributary to Little Lagoon, USNM 174164, 12 specimens, 54 to 256 mm. standard length.

M48-21, Reef, Yirrkalla, 12 August 1948, USNM 174165, 1 specimen, 82 mm. standard length.

M48-23, Port Bradshaw, USNM 174166, 1 specimen, 300 mm. standard length.

M48-27, Bay and beach, Port Bradshaw, USNM 174167, 3 specimens, 195 to 222 mm. standard length.

M48-29, Pool near small creek, Yirrkalla, USNM 174168, 1 specimen, 57 mm. standard length.

M48-34, Creek, Yirrkalla, USNM 174169, 12 specimens, 57 to 226 mm. standard length.

Native name: *Chírtfalá*, Yirrkalla.

Description. Certain fin ray counts are given in Table 4. Pelvic rays I,5(in 6); branched caudal rays 8+7=15(3); pored scales in lateral line 45(1) or 46(2); scale rows between lateral line and spinous dorsal fin 5(2) or 6(1); scale rows across cheek 7(2) or 8(1); gillrakers, including rudiments, on first arch 15 to 18 of which 1 or 2+1+6 or 7 developed; vomerine tooth patch shallowly triangular or slightly concave posteriorly, developing a slight median rear extension; that of tongue circular; edge of preopercle and suborbitals naked; colour in alcohol as described in the key.

Colour in life. Pelvic fins blood red, bordered by velvet black; dorsal, caudal, and anal fins reddish orange.

Lutjanus johni (Bloch)

Pl. 41

Anthias johnii Bloch, *Naturgeschichte der Ausländischen Fische* 6: 113, pl. 318 (1792); type locality, Suratta; reference not seen.

M48-2, Nightcliff, 19 March 1948, USNM 174122, 1 specimen, 180 mm. standard length.

Description. Certain fin ray counts are given in Table 4. Pelvic rays I,5, both sides; branched caudal rays 8+7=15; pored scales in lateral line 48; scale rows between lateral line and spinous dorsal fin 7; those across cheek 7; gillrakers, including rudiments, on first arch 20, of which 9 or 10 are developed; vomerine tooth patch a crescentric band; teeth present in a subcircular patch on tongue; suborbitals and edge of preopercle naked; colour in alcohol as described in the key.

Lutjanus vitta (Quoy and Gaimard)

Serranus vitta Quoy and Gaimard, *Voyage autour du monde. . .sur. . .l'Uranie et la Physicienne. . .Zoologie*: 315, pl. 58, fig. 3 (1824); type locality, Waigiu.

Recorded as: *Mesoprion vitta* Günther (1859): 208, Port Essington.

M48-20, Melville Bay, 31 July 1948, USNM 174123, 1 specimen, 133 mm. standard length.

M48-22, Bay, Yirrkalla, 18 August 1948, USNM 174124, 2 specimens, 163 and 191 mm. standard length.

Description. Certain fin ray counts given in Table 4. Pelvic rays I,5(in 6); branched caudal rays 8+7=15(3); pored scales in lateral line 47, 48, and 50; scale rows between lateral line and spinous dorsal fin 6(1) or 7(2); scale rows across cheek 9(3) with small ones nearly extending to preopercle edge; gill-rakers, including rudiments, on first arch 16 or 17; teeth on vomer in a triangular patch with a median posterior extension; those on tongue in a large rounded patch; colour in alcohol as described in the key; fins yellowish.

Colour in life. After having been partly dried: longitudinal stripe chocolate brown; caudal fin deep yellow; general coloration reddish.

Lutjanus russelli (Bleeker)

Mesoprion russellii Bleeker, *Verh. Batavia Genoot.* 22: 41 (1849); type locality, Vizagapatam.

Records probably referable: *Mesoprion johnii* Günther (1859): 201, Port Essington; Macleay (1878): 347 and (1880): 329; Kent (1889): 239, Port Darwin; *Lethrinus johnii* Castelnau (1873): 84, Port Darwin; *Genyoroge notata* Kent (1889): 239, Port Darwin; *Lutianus russelli* Paradice and Whitley (1927): 85, Pellew Islands.

M48-1, Nightcliff, 19 March 1948, USNM 174144, 1 specimen, 23 mm. standard length.

M48-7, Mouth of billabong, Little Lagoon, USNM 174145, 5 specimens, 58 to 96 mm. standard length.

M48-9, Reefs, entrance to Little Lagoon, USNM 174146, 3 specimens, 43 to 56 mm. standard length.

M48-9a, Reefs, entrance to Little Lagoon, USNM 174147, 2 specimens, 100 and 115 mm. standard length.

M48-15, Emerald River, Groote Eylandt, USNM 174148, 1 specimen, 66 mm. standard length.

M48-19, Creek, tributary to Little Lagoon, USNM 174149, 7 specimens, 60 to 99 mm. standard length.

M48-21, Reef, Yirrkalla, 14 July 1948, USNM 174150, 1 specimen, 84 mm. standard length; 12 August 1948, USNM 174151, 2 specimens, 107 to 121 mm. standard length.

M48-23, Port Bradshaw, USNM 174152, 1 specimen, 279 mm. standard length.

M48-24, Beach, Port Bradshaw, USNM 174153, 1 specimen, 83 mm. standard length.

M48-26, Sheltered bay, island, Port Bradshaw, USNM 174154, 2 specimens, 77 to 94 mm. standard length.

M48-31, Creek, Yirrkalla, USNM 174155, 8 specimens, 68 to 142 mm. standard length.

Native name: *Mánguchiláranqáni,* Yirrkalla.

Description. Certain fin ray counts are given in Table 4. Pelvic rays I,4(in 1) or I,5(15); branched caudal rays 8+7=15(8); pored scales in lateral line 47(4) or 48(3); scale rows between lateral line and spinous dorsal fin 8(2), 9(5) or 10(1); scales on interopercle in 1, seldom 2 rows; scale rows across cheek in 6(3) or 7(5) rows, the suborbitals and edge of preopercle naked; gillrakers, including rudiments on first arch 16 to 19, of which 8 or 9 are developed; vomerine teeth in a triangular patch with a median backward extension; a few teeth on tongue. In alcohol, the colour is a medium brown, with a slight dark brown or blackish margin to the dorsal fin; the dark brown or blackish blotch below the anterior part of the soft dorsal fin is subcircular, with about three-fourths of its area above the lateral line. In young, narrow brown lines extend backward from the head, those above extending to the midline and dorsal fin base, and horizontal below; they are not evident in large specimens.

Colour in life. In young 58 to 96 mm. standard length from M48-7: bronze horizontal and oblique stripes over a greenish background; a large oval black spot below soft dorsal fin; dorsal, caudal, and anal fins milky, with red borders, but anal fin with a white tip; pectoral fin pale; pelvic fin orange with a white border.

In others, 43 to 56 mm. standard length, from M48-9, in which collection *Lutjanus fulviflamma* was also taken, the stripes are indicated as yellow or orange; the dorsal and caudal fins are light yellow, with a red margin along the upper and lower caudal lobes and on the dorsal fin; the anal fin is light yellow, suffused with pink.

In other collections, the stripes are described as orange or yellow in colour, but in some cases *L. fulviflamma* was also present, and it is not clear whether the yellow or orange was present only in *russelli* or in both species; their colour pattern in alcohol is strikingly similar.

Lutjanus fulviflamma (Forskål)

Sciaena fulviflamma Forskål, *Descriptiones Animalium*: xi and 45 (1775); type locality, Arabia.

Recorded as: *Lutianus fulviflamma* Paradice and Whitley (1927): 85, Pellew Islands.

M48-9, Reefs, entrance to Little Lagoon, USNM 174156, 6 specimens, 44 to 54 mm. standard length.

M48-17, Reef, Bartalumba Bay, USNM 174157, 1 specimen, 94 mm. standard length.

M48-26, Sheltered bay, island, Port Bradshaw, USNM 174158, 1 specimen, 98 mm. standard length.

M48-27, Bay and beach, Port Bradshaw, USNM 174159, 3 specimens, 53 to 97 mm. standard length.

M48-31, Creek, Yirrkalla, USNM 174160, 1 specimen, 115 mm. standard length.

Description. Certain fin ray counts are given in Table 4. Pelvic rays I,5(in 6); branched caudal rays 8+7=15(3); pored scales in lateral line 46(1), 47(1), or 48(1); scale rows between lateral line and spinous dorsal fin 6(2), 7(6) or 8(1); those on interopercle in about 2 rows, sometimes 1 or 3; those across cheek in 5(3) or 6(6) rows, the suborbitals and preopercular edge naked; gillrakers, including rudiments, 18 or 19, of which 8 or 9 are developed; a few teeth on tongue and in a triangular patch, with a median backward extension, on vomer.

Remarks. Life colours of this species may have been confused with those of *Lutjanus russelli*, which see. The colour in preservation of the two forms is very similar, *L. fulviflamma* differing in having a light yellowish margin to the dorsal fin and the oblong lateral dark brown to blackish blotch approximately centred on the lateral line. It typically has 13 soft dorsal rays, whereas *russelli* has 14.

Lutjanus quinquelineatus (Bloch)

Pl. 42

Holocentrus quinquelineatus Bloch, *Naturgeschichte der Ausländischen Fische* 4: 94 (1790); type locality, Japan; reference not seen; Bloch, *Ichthyologie*... 7: 65 (1797).

Holocentrus quinquelinearis Bloch, *Ichthyologie*... 7: pl. 239 (1797); as spelled on plate.

M48-22, Bay of Yirrkalla, 19 August 1948, USNM 174134, 5 specimens, 168 to 191 mm. standard length.

Description. Other fin ray counts are given in Table 4. Pelvic rays I,5(in 10); branched caudal rays 8+7=15(5); pored scales in lateral line 47(2) or 48(3); scale rows between lateral line and spinous dorsal 6(5); scale count across cheek 13 or 14, those scales on the suborbitals and preopercular margin small and in irregular rows; scales in 2 or 3 rows on interopercle and usually in a small cluster at rear end of lower jaw; gillrakers, including rudiments, on first arch 18 or 19 of which 9 to 11 are developed; tongue smooth; vomerine tooth patch crescentric; colour in alcohol as described in the key.

Colour in life. Orange, with longitudinal light blue stripes; fins deep orange; snout and areas above and below eyes maroon.

Remarks. The complex in *Lutjanus* to which this species belongs requires further clarification. I find that there are several species in this group, but their nomenclature and identification are beyond the scope of this problem. In general, all have a deep notch near the base of the vertical arm of the preopercle into which a knob from the subopercle fits. One section of the group has several series of small scales along the suborbitals that extend on to the preorbital; this series is slightly separated from the rows of large, nearly uniform size, scales on the cheek. The other section lacks the suborbital and preorbital scales. The longitudinal bluish stripes on the head and body appear to be constant, and their number and position characteristic for each species, although they sometimes fade.

I have been able to examine some of the material recorded by Fowler (1931) as *Lutjanus spilurus,* presumably a synonym of *L. quinquelineatus,* and find that at least three species of the group and other members of the genus *Lutjanus* are included. The following, using his original numbers, are *Lutjanus kasmira* (Forskål), a distinct species with four bluish stripes along the upper and midside of the body: 16140; 14985; 14986; 14988; 15798; 18558 to 18561; 18869; 21291; 8346; 13791; 13792; 15537; 11046; 13687; 13688; 16215 to 16217; 18141; 19796; 19797; 12471; 8355; and USNM 32715.

The following are *L. quinquelineatus* (Bloch): 6450; 16139; 16141; 8345 (24 specimens); 8296; 8298; 8299; 8300; 17774; 13608; 4329; USNM 56064; USNM 71978 (larger specimen only, smaller specimen not of this group; both are from Okinawa rather than Japan); USNM 30598 and 30580 (both from New Guinea); and USNM 87982 (Sumatra), apparently not recorded by Fowler; number 17775 is *L. quinquelineatus,* but was recorded by Fowler as *L. kasmira* (loc. cit. p. 162).

The five specimens from Mauritius (USNM 19988) have scales on the suborbitals and preorbital, but possess five or six, probably six, stripes that cross the area between the nape and the midside of the caudal peduncle and thus cannot be

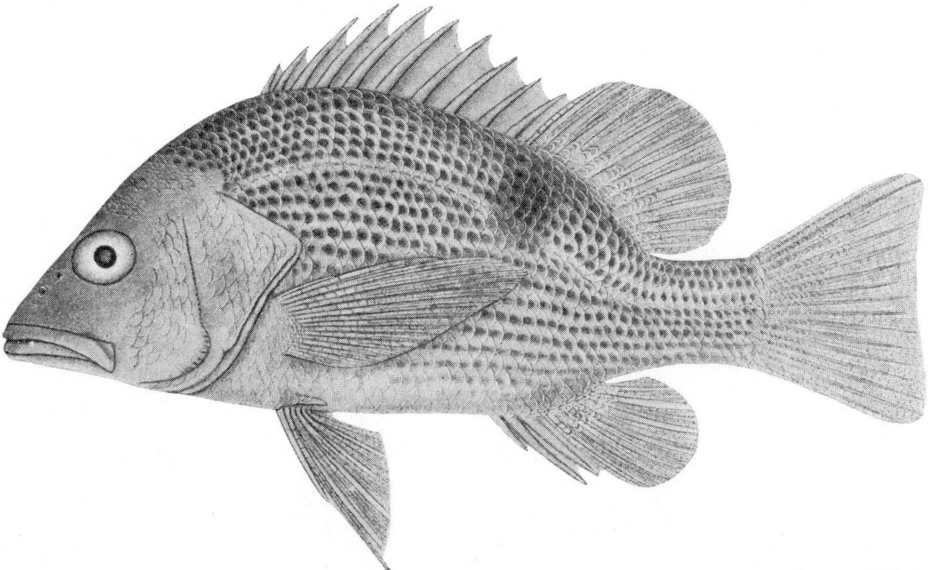

Dorothea B. Schultz

PLATE 41. *Lutjanus johni*
Nightcliff, USNM 174122, 180 mm. standard length

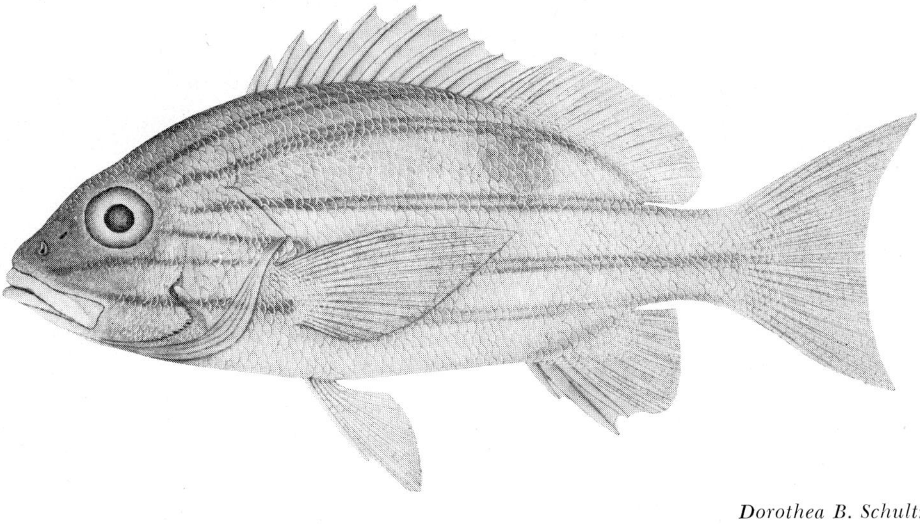

Dorothea B. Schultz

PLATE 42. *Lutjanus quinquelineatus*
Yirrkalla, USNM 174134, 189 mm. standard length

identified as either *L. quinquelineatus* or *L. kasmira*; in USNM 19938, also from Mauritius, six specimens are of this form and six identifiable as *L. kasmira*. USNM 72275 does not belong to this section of the genus.

In all of this material, I find that *L. quinquelineatus* has the five body stripes, although sometimes, especially the stripe along the lower side of the body, faint in young or poorly preserved material. It has scales along the suborbitals and on the preorbital; in larger specimens a small patch of scales is observed on the lower jaw at its rear end while the scales on top of the head extend forward to the anterior margin of the eyes. In *L. kasmira*, there are never more than four light body stripes although a series of fine dusky stripes may be observed along the lower side; the suborbitals, preorbital, and lower jaw are naked, and the scales on top of the head typically do not extend forward to the anterior part of the eye.

Lutjanus carponotatus (Richardson)

Mesoprion carponotatus Richardson, *Ann. Mag. Nat. Hist.* 9: 28 (1842); type locality, Port Essington; Günther, *Catalogue of the Fishes in the British Museum* 1: 190 (1859).

Recorded as: *Mesoprion carponotatus* Macleay (1878): 347 and Klunzinger (1879): 342, Port Darwin. Records probably referable: *Mesoprion waigiensis* Günther (1859): 191, and Macleay (1880): 329, Port Essington; *Lutianus waigiensis* Paradice and Whitley (1927): 86, text fig. 1, Pellew Islands.

M48-2, Nightcliff, USNM 174135, 6 specimens, 122 to 195 mm. standard length.

M48-9a, Reef, entrance to Little Lagoon, USNM 174136, 4 specimens, 54 to 82 mm. standard length.

M48-13a, South end Winchelsea Island, 2 June 1948, USNM 174137, 1 specimen, 209 mm. standard length.

M48-17, Reef, Bartalumba Bay, USNM 174138, 8 specimens, 67 to 258 mm. standard length.

M48-20, Main jetty, Melville Bay, 11 July 1948, USNM 174139, 2 specimens, 163 to 202 mm. standard length.

M48-21, Reef, Yirrkala, 14 July 1948, USNM 174140, 2 specimens, 62 and 85 mm. standard length; 12 August 1948, USNM 174141, 7 specimens, 68 to 108 mm. standard length.

M48-26, Sheltered bay, island, Port Bradshaw, USNM 174142, 1 specimen, 91 mm. standard length.

M48-27, Bay and beach, Port Bradshaw, USNM 174143, 1 specimen, 254 mm. standard length.

Native names: *Géllaéra* (Gell-la-air-ah), Groote Eylandt; *Ílanginámi*, Yirrkalla.

Description. Other fin ray counts are given in Table 4. Pelvic rays I,5(in 6); branched caudal rays 8+7=15(3); pored scales in lateral line 49(2) or 52(1); scale rows between lateral line and spinous dorsal fin 9(4), 10(6), or 11(1); scale count across cheek 11 to 14, of which a series of small scales near preopercular margin and another along suborbitals, both usually slightly separated and distinct from the larger cheek scales; gillrakers on first arch, including rudiments, 15 to 17 of which about 9 are developed; teeth in a round patch on tongue and in a triangular patch with a median posterior extension on vomer; body medium brown, with a pattern of alternating lighter and darker brown bands; fins yellowish; outer pectoral fin base defined with a black vertical line which broadens and becomes a prominent black spot at upper base.

Colour in life. Six or seven yellowish orange to orange horizontal bands; all of caudal fin brilliant yellowish orange; tips of dorsal spines and terminal (subterminal?) band on both the soft dorsal and anal fins orange; outer rays of pectoral and pelvic fins light yellow; outer margin of dorsal and anal fins milky white; pectoral fin base black.

Miller's drawing of a small individual and a colour transparency show seven orange bands as follows: one from nape to base of anterior dorsal spines; another from rear upper margin of eye, curving along back to mid-base of soft dorsal fin; another from rear margin of eye, divides at upper end of gill opening; one branch, above lateral line, curves backward to rear base of soft dorsal; the other branch extends straight backward to the last soft dorsal ray and upper part of caudal peduncle; a band from anterior half of upper jaw extends backward to the eye; it is possibly contiguous with one extending from lower, posterior margin of eye, through angle of opercle, thence straight backward to base of caudal fin just above lateral line; another backward from rear end of upper jaw and just above pectoral fin base, ends at base of caudal fin, below lateral line; and another, slightly curved downward, extends from interopercle (or gill opening?) through lower base pectoral fin, to last ray of anal fin.

Remarks. This is probably the species currently identified as *Lutjanus chrysotaenia* (Bleeker). In examining numerous specimens reported by Fowler (1931) I find that his *chrysotaenia* and the present specimens are very similar morphologically. Among the outstanding characters, the young of both have similar patterns of faint lighter and darker brown stripes that seem to disappear with age or preservation; both nominal forms have dark brown or black at the base of the pectoral fin and small scales on the suborbitals and near the preopercular edge.

Lutjanus lineolatus (Rüppell)
Pl. 43

Diacope lineolata Rüppell, *Atlas zu der Reise im Nördlichen Afrika*: 76, pl. 19, fig. 3 (1829); type locality, Massaua.

M48-21, Reef, Yirrkalla, 11 August 1948, USNM 174129, 2 specimens, 127 and 135 mm. standard length.

Description. Other fin ray counts are given in Table 4. Pelvic rays I,5(in 4); branched caudal rays 8+7=15(2); pored scales in lateral line 45 and 49; scale rows between lateral line and spinous dorsal fin 5 or 6; those across cheek 8(2); gill-rakers, including rudiments, on first arch 24 or 25 of which 22 or 23 are longer than their basal diameter; eye diameter longer than snout length; vomerine tooth patch crescentric or shallowly triangular, with a long median posterior extension; teeth on tongue in a large oblong patch; a band of smaller scales, behind larger rows of cheek scales, encroaching on margin of preopercle; suborbitals naked; fins yellowish; body and head yellowish brown below and medium brown above; a brown band from eye to upper base of caudal fin delimits the upper brown and lower yellowish areas; brown lines along body correspond to scale rows, those below the lateral line longitudinal and above it diagonal.

Lutjanus erythropterus Bloch
Pl. 44

Lutjanus erythropterus Bloch, *Naturgeschichte der Ausländischen Fische* 4: 115, pl. 249 (1790); type locality, Japan; reference not seen; Bloch, *Ichthyologie* 7: pl. 249 (1797).

Lutianus erythropterus Day, *The Fishes of India*: 32, pl. 10, figs. 1 and 2 (1878); Bloch's types re-examined.

Recorded as: *Lutianus erythropterus* Paradice and Whitley (1927): 85, Pellew Islands; *Mesoprion annularis* Klunzinger (1879): 342, Port Darwin.

M48-11, Little Lagoon, 8 July 1948, USNM 174133, 2 specimens, 156 and 179 mm. standard length.

M48-20, Old jetty, Melville Bay, 17 August 1948, USNM 174132, 2 specimens, 221 and 235 mm. standard length.

Description. Other fin ray counts are given in Table 4. Pelvic rays I,5(in 7); branched caudal rays 8+7=15(4); pored scales in lateral line 49(1), 50(2) or 52(1); scale rows between lateral line and spinous dorsal fin 9(2) or 10(2); those across cheek 6, 7, or 8(2), the margin of the preopercle and the suborbitals naked; gillrakers, including rudiments, on first arch 18 to 21 of which 10 to 13 developed; nostril membranes without incised margins, smooth; no teeth on tongue, the vomerine patch crescentric, without a median rear extension.

In preservation, body light brown with faint brown lines corresponding to the scale rows; pelvic fin dark brown or blackish; other fins yellowish, the dorsal, anal and caudal with brown along margin; faint trace of a brown band from eye to dorsal fin origin, thence backward along its base; colour of upper caudal peduncle similar to that of body, but with a small margaritaceous spot in one specimen.

Colour in life. Light red with narrow darker red horizontal and oblique stripes.

Lutjanus malabaricus (Bloch and Schneider)

Sparus malabaricus Bloch and Schneider, *Systema Ichthyologiae*: 278 (1801); type locality, Indian seas about Coromandel.

Lutianus malabaricus Day, *The Fishes of India*: 31, pl. 9, fig. 4 (1878); Bloch and Schneider's type re-examined.

M48-13c, Off reef in Bartalumba Bay, 7 June 1948, USNM 174130, 14 specimens, 200 to 255 mm. standard length.

M48-20, Old jetty, Melville Bay, 17 August 1948, USNM 174131, 3 specimens, 273 to 333 mm. standard length.

Description. Other fin ray counts are given in Table 4. Pelvic rays I,5(in 6); branched caudal rays 8+7=15(3); pored scales in lateral line 49(1) or 50(2); scale rows between lateral line and spinous dorsal fin 7(1) or 8(2); those across cheek 7(3), the suborbitals and preopercular margin naked; gillrakers, including rudiments, on first arch 21(1) or 22(2), with 10 developed; vomerine tooth patch crescentric, without rear extension; no teeth on tongue; membrane of rear nostril distinctly crenulate.

In preservation, the body colour is yellowish or silvery, with a prominent, dark brown quadrate saddle on caudal peduncle; fins all pale, yellowish, the dorsal fin margin brownish; faint trace of a broad brown band from eye to dorsal fin, thence backward along fin base.

Colour in life. Deep red.

Lutjanus sebae (Cuvier and Valenciennes)

Diacope sebae Cuvier and Valenciennes, *Histoire naturelle des poissons* 2: 411 (1828); type localities, Waigiu, Pondicherry, and Java.

Recorded as: *Lutianus sebae* Paradice and Whitley (1927): 86, Pellew Islands.

M48-11, Little Lagoon, 14 April 1948 and 21 May 1948, USNM 174125, 2 specimens, 215 and 219 mm. standard length.

M48-13b, Hemple Bay, Port Langdon, 27 May 1948, USNM 174126, 2 specimens, 221 and 231 mm. standard length.

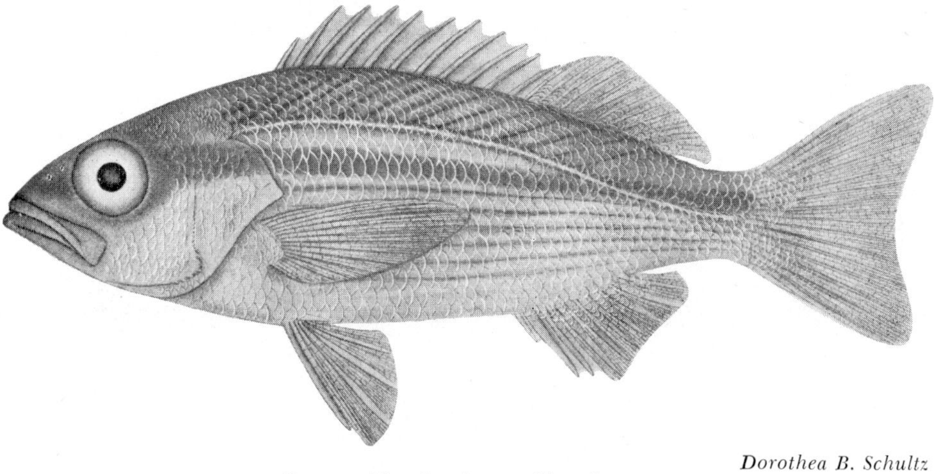

Dorothea B. Schultz

PLATE 43. *Lutjanus lineolatus*
Yirrkalla, USNM 174129, 135 mm. standard length

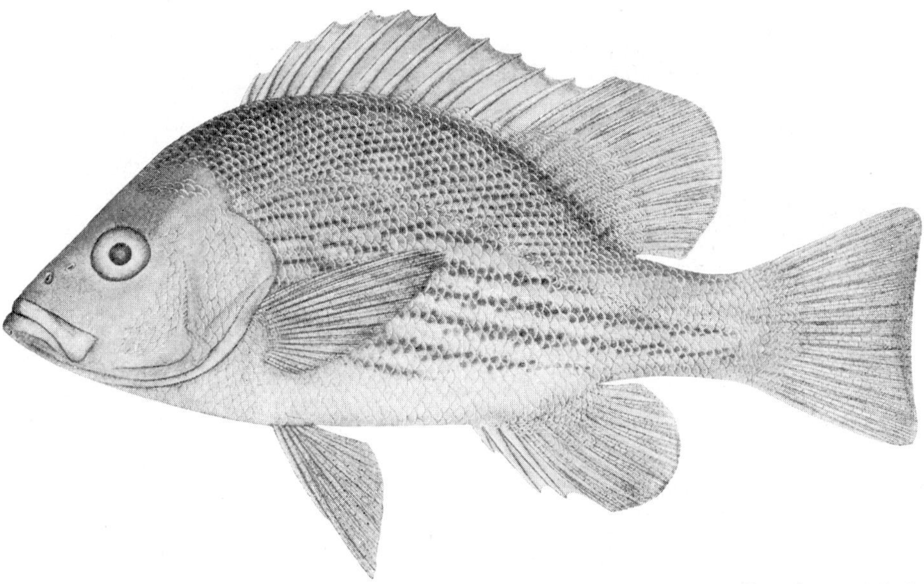

Dorothea B. Schultz

PLATE 44. *Lutjanus erythropterus*
Little Lagoon, USNM 174133, 156 mm. standard length

M48-23, Port Bradshaw, USNM 174127, 3 specimens, 195 to 231 mm. standard length.

M48-27, Bay and beach, Port Bradshaw, USNM 174128, 1 specimen, 247 mm. standard length.

Native name: *Ímáóra*, Groote Eylandt.

Description. Other fin ray counts are given in Table 4. Pelvic rays I,5(in 4); branched caudal rays 8+7=15(2); pored scales in lateral line 47(1) or 49(1); scale rows between lateral line and spinous dorsal fin 6(1) or 7(1); those across cheek in 6(2) rows; the suborbitals and preopercular margin naked; gillrakers, including rudiments, on first arch 17(2) of which 6 or 7 developed; tongue smooth; vomerine tooth patch crescentric.

In preservation, body, head and fins yellowish, except as follows: black band, wide as eye, along side of snout from its tip, through eye to midline of back just before dorsal fin; anterior half and distal two-thirds of rear half spinous dorsal fin, as well as band along front half soft dorsal fin to and including long rays, black; a black bar, wide as snout length, extending downward from anterior part spinous dorsal to lower side, through and behind pectoral fin base, out over pelvic fin and abdomen, and backward to anterior part of anal fin where reaching to tip of longest ray; another bar, narrow above and broad below, passes obliquely backward from rear base spinous dorsal to lower half caudal fin base, thence outward along lower caudal rays; a black blotch distally on upper caudal rays. From a colour transparency, the general body colour appears to be bright red, with dark brown or black bars.

*Paradicichthys** *venenatus* Whitley

Paradicichthys venenatus Whitley, *Mem. Qld. Mus.* 10: 13 (1930); type locality, Townsville, Queensland.

Recorded as: *Paradicichthys venenatus* Whitley (1954): 27, Evans Shoal, Northern Territory.

Caesio cuning (Bloch)

Sparus cuning Bloch, *Naturgeschichte der Ausländischen Fische* 5: 31, pl. 263, fig. 1 (1791); type locality, East Indies; reference not seen.

Caesio cuning Cuvier and Valenciennes, *Histoire naturelle des poissons* 6: 444 (1830); Bloch's type described.

Caesio erythrogaster Cuvier and Valenciennes, *Histoire naturelle des poissons* 6: 442 (1830); type locality, Java.

Recorded as: *Caesio erythrogaster* Klunzinger (1879): 352, Port Darwin.

M48-22, Bay of Yirrkalla, 17 to 19 August 1948, USNM 174120, 4 specimens, 211 to 245 mm. standard length.

M48-23, Port Bradshaw, USNM 174121, 4 specimens, 227 to 266 mm. standard length.

Description. Certain fin ray counts are given in Table 4. Pored scales in lateral line 49(in 1), 51(2) or 52(1); scale rows across cheek 4(4); branched caudal rays 8+6=14(1) or 8+7=15(3); pelvic rays I,5(8); supratemporal bands of scales continuous across midline; gillrakers on first arch 8, 9, or 10+1+24=33 to 35, all developed, slender. In three specimens 211 to 245 mm. standard length, body depth 2·2 to 2·4 in standard length. Colour in alcohol nearly uniform greyish

* *Paradicichthys* Whitley (1930): 13; type species, *Paradicichihys venenatus* Whitley, by monotypy.

brown; traces of bands on cheek and scaleless areas of head medium brown; dorsal fin dusky, blackish; caudal, anal, and pectoral fins yellowish with a blackish blotch in axil of pectoral.

Colour in life. Back bluish grey to steel blue, grading to whitish on side and red on underparts; orange stripes along scale rows of side; operculum, below lower border of orbit, with red; head and nape with purplish blue stripes; caudal fin deep yellow; outer two-thirds of dorsal fin red, with milky border; anal fin orange or red, with milky border; pelvic and pectoral fins reddish orange.

Aprion virescens Cuvier and Valenciennes

Aprion virescens Cuvier and Valenciennes, *Histoire naturelle des poissons* 6: 544 (1830); type locality, Seychelles.

Aprion virescens placidus Whitley, *Mem. Qld. Mus.* 11: 137 (1937); type locality, Rib Reef, Townsville, Australia.

Recorded as: *Aprion virescens placidus* Whitley (1954): 28, Timor Sea, Northern Territory.

POMADASYIDAE

In the Arnhem Land members of this family the anal fin is short, consisting of three spines and 7 or 8 rays; the teeth are small on the jaws and absent from the palate; the scales of the head are highly irregular in shape, mostly ctenoid, some are imbedded; they extend forward on to the cheek, usually to the preorbital and on top of the head to the snout; preopercle serrate; suborbitals not forming a shelf.

Key to the Arnhem Land Species

1 A median longitudinal groove on chin; lips normal, not notably thickened; body with bar-like rows of spots . . . . *Pomadasys nageb*
No prominent median groove on chin; three of sensory canal pores, each side of chin, usually notably enlarged; lips somewhat to greatly thickened . . 2

2 Dorsal rays XIV,14 to 16; third anal spine weak, shorter than second, when depressed not reaching tip of second spine; lips very thick; body dark brown; opercular margin, dorsal, anal and pelvic fins blackish, without spots . . . . *Plectorhinchus nigrus*
Dorsal spines (IX?) X to XIII 3

3 Second and third anal spines of about equal length, or the third slightly shorter 4
Third anal spine longest, extending well beyond tip of depressed second anal spine 5

4 Dorsal rays XIII,21 or 22; colour pattern light brown with a darker reticulation or network on head and upper half of body consisting of narrow irregular stripes and connecting bars . . . . *Plectorhinchus reticulatus*
Dorsal rays XII,20 or 21 (to 22); colour pattern of about eight dark brown longitudinal bands alternating with similar width yellowish brown bands on head and body; fins yellowish, not spotted . . *Plectorhinchus polytaenia*

5 Dorsal rays X (IX to XI), infrequently XII,22 or 23 (to 26); vertical fins variably covered with small spots; body nearly uniformly light brown or faintly spotted (adults) . . . . *Plectorhinchus pictus*
Dorsal spines XII; fins uniform, not spotted 6

6 Body, head, and caudal fin nearly uniformly dark brown; pectoral fin brownish

black; other fins and opercular margin black; pores on chin not greatly enlarged . . . . *Plectorhinchus unicolor*

Body, head and fins yellowish, the pelvic fin dusky; numerous, fine lines about as wide as a scale on head and body; those on head longitudinal, whitish, margined with dark brown; about 20 on body, dark brown with narrow white centres, directed backward and upward to dorsal fin . . . . *Plectorhinchus multivittatus*

Pomadasys nageb (Rüppell)

Pristipoma nageb Rüppell, *Neue Wirbelthiere zu der Fauna von Abyssienien gehörig*. . .: 124, pl. 30, fig. 2 (1835); type locality, Djetta, Red Sea.

Recorded as: *Pomadasys nageb* Paradice and Whitley (1927): 87 and 98, Darwin. Records probably referable: *Pristipoma* sp. Kent (1889): 239, Port Darwin; *Pristipoma hasta* Macleay (1878): 348; Rendahl (1922): 165, Port Darwin.

M48-20, Main jetty, Melville Bay, 11 July 1948, USNM 174097, 1 specimen, 320 mm. standard length.

Description. Dorsal rays XII,14; anal rays III,7; pelvic rays I,5, both sides; pectoral rays ii,14, both sides; branched caudal rays 8+7=15; lateral line scales 49; scale rows between dorsal and anal fins 7+1+12; cheek scales moderate, imbedded; those on top of head irregular, extending forward to anterior margin of eye; a single row of scales along interopercle.

The dorsal fin has about three longitudinal lines of dark brown spots; other fins pale, yellowish; body yellowish with dark brown spots clustered in about seven bar-like groups along upper side; each group about as wide as eye.

Remarks. This specimen compares favourably with the figure given by Rüppell; the colour pattern does not agree with that given for *Lutjanus hasta* by Bloch (1797). I suspect, however without strong evidence, that all may be of the same species. I am unable to understand the reasons cited by Weber and de Beaufort (1936) for uniting material of Bleeker's *Pomadasys hasta* and *P. nageb*, yet recognizing Rüppell's *nageb* as another form.

Plectorhinchus nigrus (Cuvier and Valenciennes)

Pristipoma nigrum Cuvier and Valenciennes, *Histoire naturelle des poissons* 5: 258 (1830); type locality, Manila.

M48-2, Nightcliff, USNM 174091, 1 specimen, 203 mm. standard length.

M48-19, Creek, tributary to Little Lagoon, USNM 174092, 2 specimens, 148 and 230 mm. standard length.

M48-30, Reefs, Yirrkalla, 25 August 1948, USNM 174087, 1 specimen, 344 mm. standard length.

M48-31, Creek, Yirrkalla, USNM 174088, 1 specimen, 89 mm. standard length.

Description. Dorsal spines XIV(in 5), the rays 14(1), 15(2), or 16(2); anal rays III,7(5); pelvic rays I,5(2); branched caudal rays 8+7=15(1); pectoral rays ii,14(2) or ii,15(4); pored scales in lateral line 48(1) or 51(1); scale rows between lateral line and dorsal fin origin 12(3); preorbital not heavily scaled.

Colour in life. Specimen 89 mm. standard length—dark brown; border of second dorsal fin and outer three-fourths of caudal fin milky white.

Plectorhinchus reticulatus (Günther)

Diagramma reticulatum Günther, *Catalogue of the Fishes in the British Museum* 1: 334 (1859); type locality, China.

Plectorhynchus reticulatus McCulloch, *Biological Results . . . of the . . . 'Endeavour'* 4: 185, pl. 53 (1916).

Recorded as: *Plectorhynchus reticulatus* Paradice and Whitley (1927): 87, Pellew Islands.

Plectorhinchus polytaenia (Bleeker)

Diagramma polytaenia Bleeker, *Natuurk. Tijdschr. Ned.-Ind.* 3: 755 (1852); type locality, Macassar.

Recorded as: *Diagramma polytaenia* Günther (1859): 332; Macleay (1880): 371, Port Essington.

M48-9, Reefs and tide pools, entrance to Little Lagoon, USNM 174085, 1 specimen, 43 mm. standard length.

M48-17, Reef, Bartalumba Bay, USNM 174086, 2 specimens, 66 and 80 mm. standard length.

Description. Dorsal rays XII,20(in 2) or 21(1); anal rays III,7(3); pelvic rays I,5(2); branched caudal rays 8+7=15(1); pectoral rays ii,14(6); scales between lateral line and dorsal origin 14, 16, and 17; preorbital densely scaled.

Colour in life. About 7 or 8 orange bands across head and along body; pelvic, anal, caudal, and dorsal fins yellowish.

Plectorhinchus multivittatus (Macleay)
Pl. 45

Diagramma multivittatum Macleay, *Proc. Linn. Soc. N.S.W.* 2: 349, pl. 7, fig. 2 (1878) and 5: 371 (1880); type locality, Port Darwin.

Recorded as: *Plectorhynchus multivittatus* Ogilby (1915): 120, Port Darwin at Spalding, Northern Territory.

M48-23, Port Bradshaw, USNM 174094, 2 specimens, 360 and 363 mm. standard length.

Description. Dorsal rays XII,19 and 20; anal rays III,7(in 2); pectoral rays ii,15(4); pelvic rays I,5(2); branched caudal rays 8+7=15(1); scales between lateral line and dorsal fin origin 15 and 16; preorbital scaled.

Remarks. This species was erroneously placed in the synonymy of *Plectorhinchus goldmanni* (Bleeker) by Fowler (1931). The stripes are narrower and more numerous than in *goldmanni* and inclined on the body; and in addition, spots are not present on the fins and there are twelve instead of thirteen dorsal spines.

Plectorhinchus unicolor (Macleay)

Diagramma unicolor Macleay, *Proc. Linn. Soc. N.S.W.* 8: 261 (1883); type locality, China Straits, New Guinea.

M48-22, Bay of Yirrkalla, 3 August 1948, USNM 174096, 1 specimen, 344 mm. standard length.

M48-23, Port Bradshaw, USNM 174095, 1 specimen, 350 mm. standard length.

Description. Dorsal rays XII,19(in 1) or 20(1); anal rays III,8(2); pelvic rays I,5(2); branched caudal rays 8+7=15(1); pectoral rays ii,15(4); scales between lateral line and dorsal fin origin 14 and 17.

Colour in life. From a colour transparency. Body and head nearly uniformly dark greyish blue; fins bluish black.

Remarks. Placing *Diagramma unicolor* Macleay in the synonymy of *Plectorhinchus nigrus* by Fowler (1931) is unjustified, because the dorsal fin ray counts of the two are considerably different; the colour pattern and degree of lip thickening of both species are somewhat similar.

Plectorhinchus pictus (Thunberg)

Perca picta Thunberg, *K. Svenska VetenskAkad. Nya Handl.* 13: 143 (1792); type locality, Japan; reference not seen.

Recorded as: *Plectorhynchus pictus* Paradice and Whitley (1927): 87, Pellew Islands.

M48-20, Old jetty, Melville Bay, 17 August 1948, USNM 174089, 1 specimen, 315 mm. standard length.

M48-23, Port Bradshaw, USNM 174093, 1 specimen, 258 mm. standard length.

M48-27, Bay and beach, Port Bradshaw, USNM 174090, 1 specimen, 330 mm. standard length.

Description. Dorsal rays X,23(in 2) or XII,22(1); two of spines in latter specimen malformed and stunted of which the most posterior is very slender and flexible; anal rays III,7(2); pelvic rays I,5(2); branched caudal rays 8+7=15(1); pectoral rays ii,15(3) or ii,16(3); scales between dorsal fin origin and lateral line 20(1) or 21(2); preorbital scaled.

This is the only species in the Arnhem Land material with spots on both the fins and body; they are tiny, irregular in development, and almost indistinct in large specimens.

NEMIPTERIDAE

In the Arnhem Land members of this family, the preorbital is naked and is distinctly elevated; preopercle serrate, or the serrae obsolete in large specimens of *Pentapodus setosus*; palate toothless; several rows of scales across cheek; subocular shelf narrow or absent, when present often serrate.

Key to the Arnhem Land Species

1 Posterior end of preorbital serrate, or bearing one or more distinct, backward projecting spines immediately below eye; teeth on jaws small, none enlarged 2

Posterior end of preorbital not serrate, the edge fleshy; preopercular edge weakly serrate in young, fleshy in adult; enlarged canine-like teeth in both jaws in addition to small, villiform teeth; 3 rows of scales between lateral line and middle of dorsal fin; upper caudal lobe elongated into a filament; banded; a small dark spot at base of caudal fin . . . . *Pentapodus setosus*

2 Scale rows between lateral line and middle of spinous dorsal fin 5 or 6 3

Scale rows between lateral line and middle of dorsal fin 3 or 4 4

3 Preorbital spine not prominent; a dark band beneath dorsal fin and a silvery horizontal stripe on preorbital . . . . *Scolopsis affinis*

Preorbital spine prominent, almost as long as pupil; scales closely encircling eye and extending forward on top of head to opposite anterior margin of orbit; upper lobe of caudal fin somewhat elongate; three dark lines across snout, one at anterior margin scaly area, another extending around snout through nares, and one from snout tip to lower margin of eye; from cheek and eye two dark lines extending backward; a narrow brown line along pectoral base and an elongate brownish area on back just before dorsal fin . . . . *Scolopsis regina*

4 Preorbital spine prominent; body depth about 2·3; a broad light bar extending from nape across opercle; upper base of pectoral and opercular margin brownish . . . . *Scolopsis vosmeri*

Preorbital spine not enlarged, little longer than the serrae; body depth 3·1 to 3·5 in standard length; body and head with brown bands, which often indistinct as are the two black spots at base of caudal fin . . *Scolopsis nicanor*

Scolopsis nicanor (Whitley)

? *Scolopsis longulus* Richardson, *Ann. Mag. Nat. Hist.* 9: 389 (1842); type locality, evidently Torres Straits; recorded from Port Essington.

? *Maenoides? auro-frenatus* Richardson, *Icones Piscium* . . .: 8, pl. 5, fig. 2 (1843); type locality, Talc Bay [probably equals Talc Head, Northern Territory].

*Scaevius** *nicanor* Whitley, *Aust. Zool.* 11: 142 (1947); Port Darwin and Pellew Islands; type locality, Gregory, Cape Peron, Shark's Bay, Western Australia.

Scaevius milii Whitley, *Rec. Aust. Mus.* 22: 240, fig. 4 (1950).

Records possibly referable: *Scolopsis longulus* Günther (1859): 363; Macleay (1878): 349 and (1880): 381; Klunzinger (1879): 351, Port Darwin; *Pentapus cyaneotaeniatus* Paradice and Whitley (1927): 88 and 98, Pellew Islands.

M48-9, Reef and tide pools, entrance to Little Lagoon, USNM 174106, 4 specimens, 43 to 57 mm. standard length.

M48-9a, Reef, entrance to Little Lagoon, USNM 174107, 2 specimens, 30 to 53 mm. standard length.

M48-15, Emerald River, Groote Eylandt, USNM 174108, 1 specimen, 102 mm. standard length.

M48-17, Bartalumba Basin, Bartalumba Bay, USNM 174109, 24 specimens, 60 to 165 mm. standard length.

M48-21, Reef, Yirrkalla, 11 and 12 August 1948, USNM 174110, 8 specimens, 75 to 113 mm. standard length.

M48-26, Sheltered bay, island, Port Bradshaw, USNM 174111, 5 specimens, 63 to 189 mm. standard length.

M48-27, Bay and beach, Port Bradshaw, USNM 174112, 5 specimens, 60 to 189 mm. standard length.

Description. Dorsal rays X,9(in 9); anal rays III,7(9); pelvic rays I,5(8); pectoral rays ii,14(1) or ii,15(7); branched caudal rays 8+7=15(4); gillrakers short, on first arch 5+5=10(3) or 6+5=11(1); scale rows across cheek 5(1), 6(4), or 7(1); scales between lateral line and middle of spinous dorsal fin 3(9); scales in lateral line 41(1), 42(3), 43(2), 44(2), or 45(1). Two or three rows of scales across interopercle; preopercle margin naked; suborbitals variously (usually somewhat) serrate, the serrae tiny and between the scales; no forward directed serrae; preorbital serrated on its rear edge, the upper serration largest; scales on top of head extend forward to rear margin of eye; upper lobe of caudal fin longest, but not extended as a filament.

Colour in life. In specimens 43 to 57 mm. standard length, top of head, back, and upper side light olive green; a narrow, nearly straight, brown band through green area from eye to rear end of dorsal fin; two slightly wider, straight, brown bands, separated by yellow or yellowish orange; one from snout through eye to black spot at base of upper caudal lobe; the other from lower margin of eye to black spot at base of lower caudal lobe; beneath latter band a broad area of brown on body with a trace of yellowish brown on cheek and lower part of opercle.

* *Scaevius* proposed as a new genus.

Remarks. Whitley (loc. cit., 1950) placed his *Scaevius nicanor* in the synonymy of *Cantharus milii* Bory de Saint-Vincent. The description of *C. milii** was based on a coloured drawing as were those of *Maenoides auro-frenatus* and *Maenoides cyaneo-taeniatus* by Richardson (1843c). Of the nominal species cited only the descriptions of *Scolopsis longulus* Richardson and *Scaevius nicanor* Whitley were based on actual specimens. Since I am unable to judge the variation in this species, or what was meant by each author, I have used Whitley's name, which was later illustrated by a drawing that is comparable with some of my specimens.

There appears to be variation in colour pattern in this species, especially in preservation. The lateral bands are variously present or faded; the round black spot at the base of the upper caudal lobe is most consistently present but the elongate one below tends to disappear. In some specimens, the four bands or oblique bars on the snout are prominent; in others, they are faded and difficult to distinguish. Since Miller did not illustrate the snout bands in his drawing of the small specimen, and since they are barely discernible in some of the preserved material, it must be concluded that they either develop with age or the degree of intensity is changeable, most likely the latter.

Scolopsis affinis Peters

Scolopsis affinis Peters, *Mber. K. Preuss. Akad. Wiss.*: 832 (1877); type locality, New Britain.

Recorded as: *Scolopsis affinis* Klunzinger (1879): 351; Macleay (1884): 13, Port Darwin.

Scolopsis vosmeri (Bloch)

Anthias vosmeri Bloch, *Naturgeschichte der Ausländischen Fische* 6: 120, pl. 321 (1792); type locality, Japan Sea; reference not seen; Bloch, *Ichthyologie* 9: pl. 321 (1797).

M48-30, Reef, Yirrkalla, 28 August 1948, USNM 174113, 1 specimen, 161 mm. standard length.

Description. Dorsal rays X,9; anal rays III,7; pelvic rays I,5, both sides; pectoral rays ii,17, both sides; branched caudal rays 8+7=15; gillrakers short, on first arch 7+5=12; scale rows across cheek 7; scales between lateral line and middle of spinous dorsal fin 4; pored scales in lateral line 41.

Preorbital with rear margin somewhat concave, serrated, with a large spine below eye; suborbitals forming a continuous naked ridge below eye, with a spine at both ends, one directed forward and one backward; rear margin of preopercle concave, prominently serrated; ridge along anterior margin of opercle forming a thin, irregularly serrated shelf; scales on head closely approaching eye and extending forward to and before a line joining anterior nares; upper caudal lobe longest but not extended.

Colour in life. Body pinkish red; dorsal, anal, and pelvic fins reddish orange; pectoral and caudal fins deep orange.

Remarks. In preservation, this specimen bears a colour pattern characteristic of *Scolopsis torquatus* Cuvier and Valenciennes, but most authors believe such to be a variant of *vosmeri* as noted by Weber and de Beaufort (1936).

It has a broad, faint, light bar extending from the nape across the opercle; upper margin of opercular flap, a scale just below anterior end of lateral line, upper base of pectoral fin, and extremity of scaly dorsal sheath dark brown;

* Bory de Saint-Vincent (1823): 160, pl. 90, fig. 3.

otherwise, body and head uniformly light brown; no longitudinal light band on body.

Scolopsis regina Whitley

Scolopsis regina Whitley, *Mem. Qld. Mus.* 11: 137 (1937); type locality, Rib Reef, Townsville, Queensland.

Scolopsis temporalis McCulloch, *Proc. Linn. Soc. N.S.W.* 46: 469, pl. 40, fig. 3 (1921).

Recorded as: *Scolopsis temporalis* Paradice and Whitley (1927): 87, fig. 2, Pellew Islands. Records probably referable: *Scolopsis personatus* Macleay (1878): 349 and (1880): 382, Port Darwin.

M48-20, Main jetty, Melville Bay, 11 July 1948, USNM 174114, 1 specimen, 154 mm. standard length.
M48-22, Bay of Yirrkalla, 20 July and 3 to 6, and 23 August 1948, USNM 174115, 7 specimens, 125 to 237 mm. standard length.
M48-23, Port Bradshaw, 22 to 26 July 1948, USNM 174116, 1 specimen, 242 mm. standard length.
M48-26, Sheltered bay, island, Port Bradshaw, USNM 174117, 1 specimen, 82 mm. standard length.

Description. Dorsal rays IX,10(in 1) or X,9(9); anal rays III,7(10); pelvic rays I,5(6); pectoral rays ii,16(18) or ii,17(2); branched caudal rays 8+7=15(3); gill-rakers short, on first arch 5+5=10(3); scale rows across cheek 11(2) and 12(1); scales between lateral line and middle of spinous dorsal fin 5(3) or 6(1); scales in lateral line 47(2), 48(1), or 49(1).

Preorbital serrated, with a prominent spine below eye; suborbitals with serrae covered by scales, and without forward directed spine; preopercle with rear margin somewhat concave, serrated; scales on head closely approaching eye and extending forward above to opposite anterior margin of eye; no naked area at rear upper corner of eye; upper caudal lobe prolonged, somewhat filamentous.

Colour in life. Body greyish white; light blue lines on snout, around eye, and on opercles; an orange band above the light blue band on opercles; dorsal fin deep orange with a light purple submarginal band; upper caudal lobe and most of caudal fin orange, but fork of caudal broadly margined by purplish grey; outer margin of lower caudal lobe purplish grey; anal, pelvic, and pectoral fins watery yellow—almost clear; upper and lower jaws, inside of mouth, pinkish.

Remarks. The Arnhem Land material agrees with the figure of *Scolopsis temporalis* given by McCulloch (loc. cit.), except that the bands on the head are dark brown, in agreement with Whitley. *S. regina* is similar in colour pattern arrangement to *Scolopsis temporalis* (Cuvier and Valenciennes), but the supraorbital and 'temporal' region in *temporalis* is naked. In *regina* that area is densely scaled. A dark brown bar is present at the pectoral fin base in *regina*, but absent in *temporalis*. I have examined one specimen of *S. temporalis* (from Great Tobea Island, USNM 146707) which was recorded by Fowler (1931) as *Scolopsis phaeops*.

Pentapodus setosus (Cuvier and Valenciennes)

Pentapus setosus Cuvier and Valenciennes, *Histoire naturelle des poissons* 6: 270 (1830); type locality, Batavia, Java.

Recorded as: *Pentapus setosus* Paradice and Whitley (1927): 88, Pellew Islands.

SPARIDAE

I have followed the 'Revision of Australian Silver Breams . . .' by Munro (1949) in the identification of this material. If the name *Mylio* Lacépède is not accepted for this genus, the next available name is *Caeso** Gistel, rather than *Acanthopagrus* Peters.

Key to the Arnhem Land Species of *Caeso*

1 Dark dashes or short bars usually through scale centres, forming interrupted longitudinal dark streaks; caudal fin yellowish with prominent black rear margin; pelvic fins uniformly yellowish . . . . *Caeso latus*

Scale centres dusky or pale, sometimes forming light longitudinal lines; caudal fin dusky, with only a slightly darker margin; pelvic fin dark brown, especially on distal half of rays . . . . *Caeso berda*

In two specimens of each of the two forms the following data were obtained: dorsal rays XI,11; anal rays III,8; branched caudal rays 8+7=15; pelvic rays I,5, both sides; pored scales in lateral line 43 or 44; scale rows counted obliquely backward between upper and lower midlines of trunk 6 or 7+1+11 to 13; scale rows across cheek 6 or 7; pectoral rays in *latus*, ii,13 on both sides (3 specimens) and in *berda*, ii,13(in 5 counts) or ii,14(3).

Caeso latus (Houttuyn)

Sparus latus Houttuyn, *Natuurk. Verh. Holland. Maatsch. Wet.* 20: 322 (1782); type locality, Japan.

Recorded as: *Chrysophrys australis* Günther (1859): 494, Port Essington; Rendahl (1922): 165, Port Darwin; *Sparus berda* Paradice and Whitley (1927): 88, Pellew Islands, in part; *Mylio latus* Munro (1949): 201, pl. 19, Pellew Islands.

M48-2, Nightcliff, USNM 173944, 2 specimens, 154 and 168 mm. standard length.

M48-19, Creek, tributary to Little Lagoon, USNM 173945, 1 specimen, 161 mm. standard length.

Caeso berda (Forskål)

Sparus berda Forskål, *Descriptiones Animalium*: 32 (1775); type locality, Red Sea.

Recorded as: *Sparus berda* Paradice and Whitley (1927): 88, Pellew Islands, in part, and Port Darwin; *Chrysophrys hasta* Macleay (1878): 351, and (1880): 418, Port Darwin; *Mylio berda* Munro (1949): 195, pl. 18, Darwin and Pellew Islands, Northern Territory.

M48-15, Emerald River, Groote Eylandt, USNM 173942, 7 specimens, 121 to 235 mm. standard length.

M48-20, Main jetty, Melville Bay, 11 July and 3 September 1948, USNM 173943, 3 specimens, 172 and 225 mm. standard length.

M48-20, Old jetty, Melville Bay, 17 August 1948, USNM 173947, 2 specimens, 163 and 176 mm. standard length.

Native names: *Inuabejágpa* (In-ooah-bejagpah), Groote Eylandt; *Mitebóonah*, Melville Bay.

LETHRINIDAE

In the Arnhem Land members of the genus *Lethrinus*, the cheek and most of the preopercle are naked, the latter usually with a small cluster of scales on its upper limb near rear of eye; palate toothless; gill membranes broadly united.

* *Caeso* Gistel (1848): viii; type species, *Sparus berda* Forskål, designated by Jordan (1919): 235.

Counts were made on the two specimens of *Lethrinus ornatus* and three specimens each of *lentjan* and *punctulatus*; no variation was found in the following: dorsal rays X,9; anal rays III,8; pelvic rays I,5, both sides; pectoral rays ii,11, both sides; branched caudal rays 8+7=15; scale rows between lateral line and middle of spinous dorsal fin $5\frac{1}{2}$ (equals 6).

Lethrinus ornatus Cuvier and Valenciennes

Lethrinus ornatus Cuvier and Valenciennes, *Histoire naturelle des poissons* 6: 310 (1830); type locality, Java.

M48-22, Bay of Yirrkalla, 3 August 1948, USNM 174100, 2 specimens, 190 and 194 mm. standard length.

Description. Pored scales in lateral line 46 and 47; gillrakers very short, 5+5=10 or 5+6=11; total number of scales in cluster behind eye 11 to 14, usually 13; teeth as follows: in upper jaw 2 moderate canines across front plus 4 or 5 smaller pointed teeth in a row, followed by 2 or 3 blunt teeth or molars, total each side 8 to 10; in lower jaw with similar arrangement and shapes as follows, front to rear, 2 plus 3 or 4 plus 3; totals each side 8 or 9.

In the specimens, head length 3·1 or 3·2 and body depth 2·3 or 2·4, both in standard length; snout length 1·9, eye diameter 3·6 or 3·7 and caudal peduncle depth 2·7 or 2·8, all in head length.

Posterior naris closer to free edge of membrane encircling eye than to anterior naris; snout blunt, steep, its profile an angle of about 75°; mouth extending backward to below rear nostril or front of eye; colour in alcohol nearly uniform yellowish; scales of body broadly outlined in brown; about half a dozen faint, oblique bars or mottlings along lower side; a faint, narrow brown line along base of pectoral fin.

Colour in life. From a colour transparency and field notes: dorsal and caudal fins red; a red bar along rear edge of opercle and another along rear edge of preopercle; pectoral fin yellowish orange; anal fin reddish; body and head silvery, the scales prominently outlined in darker; bands along side of body reddish orange, of which I can discern four in the transparency, each band as wide as a scale row, as follows: one anteriorly along, or on scale row beneath, lateral line, posterior extension not visible; the remaining reddish orange bands below, each regularly separated from the above and each other by two rows of scales.

Lethrinus lentjan (Lacépède)

Bodianus lentjan Lacépède, *Histoire naturelle des poissons* 4: 281, 293 (1802); type locality not given.

Lethrinus lentjan Smith, *Rhodes Univ. Ichthyol. Bull.* 17: 291, pl. 21, figs. A and B (1959); Lacépède's type re-examined and stated to be from Java.

Lethrinus fusciceps Macleay, *Proc. Linn. Soc. N.S.W.* 2: 350, pl. 8, fig. 1 (1878) and 5: 414 (1880); type locality, Port Darwin.

Recorded as: *Lethrinus nebulosus* Klunzinger (1879): 356; Macleay (1884): 16, Port Darwin. Records probably referable: *Lethrinus cocosenis* Kent (1889): 239, Port Darwin; *Lethrinus glyphodon* Paradice and Whitley (1927): 88, Pellew Islands.

M48-22, Bay of Yirrkalla, 20 July and 3 August 1948, USNM 174098, 3 specimens, 209 to 262 mm. standard length.

M48-26, Sheltered bay, island, Port Bradshaw, USNM 174099, 3 specimens, 77 to 88 mm. standard length.

Description. Pored scales in lateral line 46(in 2) or 47(1); gillrakers short and thick, 5+5=10(1) or 6+5=11(2) on first arch; total scales in postorbital cluster 10 to 14, usually 13 or 14; teeth (2 specimens) as follows: in upper jaw 2 larger canines across front, plus a longitudinal row of 4 to 6, usually 6, smaller canines plus 2 to 4 blunt molars; total teeth each side 10 or 11; in lower jaw (with similar shapes and arrangement), 1 or 2 plus 5 to 7 plus 3 to 5, totals each side 10 to 13.

In three specimens 209 to 262 mm. standard length, head length 2·7 to 2·9 and body depth 2·3 to 2·5, both in standard length; snout length 1·8 to 1·9, eye diameter 4·0 to 4·3 and caudal peduncle depth 3·2 to 3·4, all in head length; distance between anterior and posterior nares less than distance from posterior naris to edge of orbital membrane, but greater in specimens 77 to 88 mm. standard length; snout moderately pointed, forming an angle of about 60° with lower surface of head; mouth extending backward to opposite rear nostril; in alcohol, the specimens are nearly uniformly light brownish, with prominent light scale centres on body.

Colour in life. In the large specimens dorsal and anal fins margined with bright red; dorsal rays, from fifth spine backward to end of fin, fade to pinkish basally; bright red spot at base of pectoral fin and at upper angle of opercle; caudal fin deep blood red on lobes, fading basally; gill cavity blood red, the red extending forward to mouth, opposite posterior end of lower jaw; pectoral and pelvic fins yellowish orange; scale pattern on body prominent, whitish or silvery, outlined in bluish or grey. The small specimens with indication of orange horizontal stripes along side.

Remarks. Smith (loc. cit.) has figured this species in colour and maintains that Lacépède's type justifies the application of the name as used here. His figures agree remarkably with a colour transparency of one of the Arnhem Land specimens, especially in the disposition of the red colour at the base of the pectoral fin and at the rear end of the opercle.

Lethrinus punctulatus Macleay

Lethrinus punctulatus Macleay, *Proc. Linn. Soc. N.S.W.* 2: 351, pl. 8, fig. 2 (1878) and 5: 414 (1880); type locality, Port Darwin.

? *Lethrinus fletus* Whitley, *Proc. Linn. Soc. N.S.W.* 48: 138 (1943); type locality, Pellew Islands.

Recorded as: *Lethrinus richardsoni* Klunzinger (1879): 357, Port Darwin; *Lethrinus reticulatus* Macleay (1884): 17, Port Darwin, error for *richardsoni*. Record probably referable: *Lethrinus haematopterus* Paradice and Whitley (1927): 88, Pellew Islands.

M48-9a, Reef, entrance, Little Lagoon, USNM 174101, 1 specimen, 112 mm. standard length.

M48-13a, South end of Winchelsea Island, 2 June 1948, USNM 174102, 1 specimen, 216 mm. standard length.

M48-13b, Hemple Bay, Port Langdon, 23 and 27 May 1948, USNM 174103, 2 specimens, 176 and 205 mm. standard length.

M48-13d, Bay of unnamed island, about 5 miles east of north end of Chasm Island, 8 June 1948, USNM 174104, 1 specimen, 306 mm. standard length.

M48-20, Old jetty, Melville Bay, 17 and 18 August 1948, USNM 174105, 2 specimens, 166 and 173 mm. standard length.

Native names: *Ámóngu*, Winchelsea Island; *Yálgiangóora* (Yahlgeeangoorah), Groote Eylandt.

Description. Pored scales in lateral line 46(in 1) or 47(2); gillrakers on first arch short, thick, 5+5=10(2) or 5+6=11(1); scales in cluster just behind eye 10 to 13; teeth all pointed or nearly so, no molars, those in upper jaw of 2 enlarged canines in front plus 8 to 11 smaller teeth in a longitudinal row behind; in lower jaw 2 larger canines followed by a row of 10 to 13 smaller teeth.

In two specimens, 166 to 173 mm. standard length, head length 2·7 to 3·0 and body depth 2·3 to 2·6, both in standard length; snout length 1·8 to 2·0, eye 4·0 to 4·1 and depth of caudal peduncle 3·0 to 3·1, all in head length.

Snout moderately pointed forming an angle of about 65° with lower surface of head; mouth extending backward to below (or slightly to rear of) anterior naris.

In alcohol, body and head medium brown with darker brown mottlings and irregular spotting on body, these tending to form bars; a dark brown spot, large as eye, below lateral line, above basal half of extended pectoral fin; head faintly spotted, with two narrow brown bars enclosing a lighter brown area; one bar along vertical arm of preopercle, and the other from eye obliquely downward toward rear end of mouth; pelvic fin blackish; dorsal and anal fins yellowish, mottled or marbled with brown.

Colour in life. In a 112 mm. specimen, brown bars as follows: one, wide as pupil from eye to rear end of mouth opening; another, not as wide, on vertical arm of preopercle, extending across interopercle; rear edge of opercle brown; about ten narrow brown bars, which are broken into blotches anteriorly, across body and caudal peduncle; pectoral fin pink; dorsal and anal fins light yellowish with two or three longitudinal rows of small brown spots, the fin margins red; rear end of caudal peduncle and basal portion caudal fin bright yellow, the yellow grading to red marginally.

In larger specimens, inside of mouth red or with reddish orange blotches.

Remarks. Although older names may be available I have used Macleay's name for this species, since the forms in this difficult group have not been adequately defined.

GERRIDAE

Key to the Arnhem Land Species

1 Dorsal rays normally IX,10; anal rays usually III,7; top of head smooth; scales moderate, 33 to 45 in lateral line . . . 2

Dorsal rays VIII,15 or 16 and anal rays III,14; scales usually small, deciduous, 30 to 60 tubes in lateral line; top of head rugose, with bony ridges visible under skin . . . 5

2 Scales on top of head covering premaxillary groove and extending forward nearly to nares; faint, short vertical bars across midside of body . . . *Gerres subfasciatus*

Premaxillary groove naked or nearly so between eyes, with scales only along its side . . . 3

3 Lateral line scale rows 34 to 38; body depth about 1·9 to 2·2 in standard length; yellowish, with narrow brown longitudinal streaks along scale rows of upper side . . . *Gerres abbreviatus*

Lateral line scale rows 38 to 45; body depth 2·5 to 3·0 in standard length . . 4

4 Body somewhat elongate, its depth 2·6 to 3·0 in standard length; lateral line scale rows 38 to 42; second dorsal spine not reaching posteriorly beyond base of spinous dorsal fin; five or six oblique brown bars extending downward and forward above lateral line; short vertical bars below lateral line . . . *Gerres oyena*

Body shorter and deeper, the depth stepped about 2·5 to 2·8 times in standard length; lateral line with 41 to 45 scale rows; second spine of dorsal fin elongate, often reaching to or beyond rear half of soft dorsal fin; about ten narrow brown bars extending down across side of body, those posteriorly often faint or obscured . . . . *Gerres filamentosus*

5 Teeth in jaws rather large, sharp, pointed, canine-like, the largest pairs near symphysis of upper jaw and laterally in lower jaw; upper half of side of body marbled with brown; lower half dusky yellowish . . . . *Gazza minuta*

Teeth in jaws nearly uniformly minute, setiform 6

6 Mouth tiny, directed upward when protracted; body depth stepped about 1·7 times in standard length; a black line from eye to lower jaw; a black spot on opercle and rear lower part of preopercle . . . . *Leiognathus ruconius*

Mouth small, directed forward or downward when protracted 7

7 Lateral line extending to caudal fin; breast appearing naked; preopercle distinctly serrated; cheek in young unpigmented, naked; irregular brown bars on upper side of body . . . . *Leiognathus equulus*

Lateral line not reaching caudal fin, usually ending beneath dorsal fin; breast scaled 8

8 Dorsal fin black distally; upper side of body with irregular brown blotches; lower side nearly uniformly dusky except posteriorly where unpigmented; supraorbital serrate . . . . *Leiognathus splendens*

Dorsal fin yellowish, without black blotch; upper side of body with curving dark bars; a longitudinal band along middle of side; supraorbital entire to rugose . . . . *Leiognathus moretoniensis*

In the genus *Gerres*, the edge of the preopercle is smooth and there are normally IX,10 dorsal rays. The Arnhem Land species, except *G. subfasciatus* Cuvier and Valenciennes, have been characterized by Weber and de Beaufort (1931).

Gerres abbreviatus Bleeker

Gerres abbreviatus Bleeker, *Natuurk. Tijdschr. Ned.-Ind.* 1: 103 (1850); type locality, Batavia.

Gerres profundus Macleay, *Proc. Linn. Soc. N.S.W.* 2: 350, pl. 7, fig. 3 (1878) and 5: 376 (1880); type locality, Port Darwin.

Recorded as: *Xystaema abbreviatum* Paradice and Whitley (1927): 89, Pellew Islands; *Parochusus cheverti* Whitley (1955b): 157, Darwin. Record probably referable: *Gerres ovatus* Klunzinger (1879): 354, Port Darwin.

M48-7, Mouth of billabong, Little Lagoon, USNM 174194, 7 specimens, 28 to 50 mm. standard length.

M48-9, Reefs and tide pools, entrance to Little Lagoon, USNM 174195, 3 specimens, 33 to 45 mm. standard length.

M48-10, Estuary, Thompson Bay, Port Langdon, USNM 174196, 15 specimens, 23 to 58 mm. standard length.

M48-15, Emerald River, Groote Eylandt, USNM 174197, 4 specimens, 53 to 108 mm. standard length.

M48-19, Creek, tributary to Little Lagoon, USNM 174198, 5 specimens, 41 to 50 mm. standard length.

M48-24, Beach, Port Bradshaw, USNM 174199, 3 specimens, 125 to 141 mm. standard length.

M48-25, Beach, island, Port Bradshaw, USNM 174200, 6 specimens, 30 to 70 mm. standard length.

M48-28, Creek mouth and bay, Port Bradshaw, USNM 174201, 84 specimens, 33 to 97 mm. standard length.

M48-31, Creek, Yirrkalla, USNM 174202, 7 specimens, 50 to 76 mm. standard length.

Native names: *Állogórnbóona*, Groote Eylandt; *Kúrnoo*, Yirrkalla.

Colour in life. Dorsal fin black tipped; anal and pelvic fins yellow; body silvery.

Gerres filamentosus Cuvier

Gerres filamentosus Cuvier, *Règne animal* ed. 2, 2: 188 (1829); type locality, Vizagapatam.

Recorded as: *Gerres filamentosus* Richardson (1842b): 25, Port Essington; *Gerres punctatus* Macleay (1878): 350 and (1880): 376, Port Darwin; *Xystaema punctatum* Paradice and Whitley (1927): 89, Pellew Islands.

M48-8, Little Lagoon, USNM 174203, 10 specimens, 43 to 55 mm. standard length.

M48-14, Emerald River, Groote Eylandt, just below falls, 4 June 1948, USNM 174204, 2 specimens, 50 and 55 mm. standard length.

M48-15, Emerald River, Groote Eylandt, USNM 174205, 5 specimens, 60 to 84 mm. standard length.

M48-19, Creek, tributary to Little Lagoon, USNM 174206, 30 specimens, 22 to 80 mm. standard length.

M48-29, Pool, near small creek, Yirrkalla, USNM 174207, 2 specimens, 35 and 37 mm. standard length.

M48-31, Creek, Yirrkalla, USNM 174208, 28 specimens, 39 to 69 mm. standard length.

M48-34, Creek, Yirrkalla, USNM 174209, 4 specimens, 54 to 77 mm. standard length.

Native name: *Kúrnoo*, Yirrkalla.

Colour in life. Elongated 2nd and 3rd dorsal spines blackened at tips; about 8 narrow grey bars on side; nape metallic blue.

Gerres oyena (Forskål)

Labrus oyena Forskål, *Descriptiones Animalium*: 35 (1775); type localities, Suez and Djeddae, Red Sea.

Recorded as: *Gerres oyena* Klunzinger (1879): 353, Port Darwin; *Xystaema carinatum* Paradice and Whitley (1927): 89, Pellew Islands.

M48-7, Mouth of billabong, Little Lagoon, USNM 174210, 7 specimens, 54 to 73 mm. standard length.

M48-8, Little Lagoon, USNM 174211, 17 specimens, 36 to 75 mm. standard length.

M48-9, Reef and tide pools, entrance to Little Lagoon, USNM 174212, 21 specimens, 20 to 40 mm. standard length.

M48-9a, Reef, entrance to Little Lagoon, USNM 174213, 8 specimens, 27 to 77 mm. standard length.

M48-10, Estuary, Thompson Bay, Port Langdon, USNM 174214, 22 specimens, 27 to 78 mm. standard length.

M48-17, Reef, Bartalumba Bay, USNM 174215, 7 specimens, 22 to 96 mm. standard length.

M48-21, Reef, Yirrkalla, 12 August 1948, USNM 174216, 2 specimens, 49 to 70 mm. standard length.

M48-25, Beach, island, Port Bradshaw, USNM 174217, 44 specimens, 25 to 108 mm. standard length.

M48-26, Sheltered bay, island, Port Bradshaw, USNM 174218, 9 specimens, 55 to 100 mm. standard length.

M48-27, Bay and beach, Port Bradshaw, USNM 174219, 1 specimen, 81 mm. standard length.

M48-28, Creek mouth and bay, Port Bradshaw, USNM 174220, 36 specimens, 50 to 94 mm. standard length.

Native names: *Állogórnbóona*, Groote Eylandt; *Yóongala*, Yirrkalla.

Colour in life. Spinous dorsal fin with black tip; irregular dark bars on side of body; pelvic fin yellow toward base; other fins clear.

Gerres subfasciatus Cuvier and Valenciennes

Gerres subfasciatus Cuvier and Valenciennes, *Histoire naturelle des poissons* 6: 477 (1830); type locality, Port Jackson.

M48-7, Mouth of billabong, Little Lagoon, USNM 174221, 1 specimen, 30 mm. standard length.

M48-8, Little Lagoon, USNM 174222, 1 specimen, 53 mm. standard length.

M48-15, Emerald River, Groote Eylandt, USNM 174223, 3 specimens, 36 to 60 mm. standard length.

M48-19, Creek, tributary to Little Lagoon, USNM 174224, 4 specimens, 37 to 52 mm. standard length.

M48-28, Creek mouth and bay, Port Bradshaw, USNM 174225, 1 specimen, 56 mm. standard length.

M48-31, Creek, Yirrkalla, USNM 174226, 5 specimens, 48 to 88 mm. standard length.

Remarks. In this species, the premaxillary groove is covered by scales which extend almost as far forward as the nares; in the other Arnhem Land species of *Gerres*, the groove is scaleless or nearly so. In most of the specimens the scales, chiefly those of the body, have been irregularly rubbed off, indicating that they are not strongly attached in contrast to the condition in the other species in the present collection with generally adhering scales. The body is medium brown in colour, darker above, with faint brown bars across the midside; fins yellowish; spinous dorsal fin broadly tipped with black.

Fowler (1933) has indicated that both *Gerres ovatus* Günther and *Gerres subfasciatus* have the premaxillary groove completely scaled in front. This seems to be in disagreement with the original description, as Günther (1859 and 1862) in describing *Gerres ovatus* states that the premaxillary groove is short and 'there are no scales between that groove and the orbit'. Cuvier and Valenciennes' description of *Gerres subfasciatus* states that it has 6 or 7 bars on the flanks and no triangular groove between the eyes. I identify as *Gerres subfasciatus* the above material and USNM 59936, which was reported by Fowler as *Gerres ovatus* (loc. cit.), on the basis of the description by Cuvier and Valenciennes. USNM 42030, also listed by Fowler as *Gerres ovatus* (loc. cit.), is probably correctly identified; it is a more elongate fish, yellowish, without bands or bars, and with a naked premaxillary groove.

Gazza minuta (Bloch)

Scomber minutus Bloch, *Ichthyologie* 12: 110, pl. 429, fig. 2 (1797); type locality, Malabar.

Recorded as: *Gazza equuliformis* Paradice and Whitley (1927): 84, Pellew Islands.

M48-7, Mouth of billabong, at Little Lagoon, USNM 174183, 2 specimens, 23 and 47 mm. standard length.

M48-8, Little Lagoon, USNM 174184, 33 specimens, 21 to 45 mm. standard length.

M48-26, Sheltered bay, island, Port Bradshaw, USNM 174185, 2 specimens, 43 and 45 mm. standard length.

Remarks. This species is characterized by a series of prominent, sharp or pointed, teeth in each jaw, tiny scales, and a serrated lower preopercular margin. The specimens were easily sorted from the species of *Leiognathus* because of the nearly complete and uniform distribution of dark pigment cells over the cheek.

The Arnhem Land collection members of the genus *Leiognathus* are all small; consequently the identifications are tentative. They are characterized by a serrated lower margin to the preopercle, tiny scales, and fine, villiform teeth in each jaw.

Leiognathus ruconius (Hamilton-Buchanan)

Chanda ruconius Hamilton-Buchanan, *Fishes of the River Ganges*: 106 and 371, pl. 12, fig. 35 (1822); type locality, estuaries of Ganges River.

Recorded as: *Equula interrupta* Günther (1860): 504; Macleay (1880): 550, Port Essington.

M48-10, Estuary, Thompson Bay, Port Langdon, USNM 174186, 2 specimens, 13 to 14 mm. standard length.

Remarks. The mouth in the two juveniles protracts diagonally upward; the body depth is stepped into the standard length about 1·7 or 1·8 times; the fins are broken, but only the third dorsal spine has anterior serrae; body and head yellow with a line of dark brown chromatophores from eye to lower jaw, a patch of them over opercle and rear lower part of preopercle, some on top of head, a cluster on breast, one at tip of each interhaemal spine, and two short longitudinal lines along midside; upper side and base of dorsal fin with light brown blotches.

Leiognathus splendens (Cuvier)

Equula splendens Cuvier, *Règne animal* ed. 2, 2: 212 (1829); type locality, Vizagapatam.

Recorded as: *Leiognathus splendens* Paradice and Whitley (1927): 84, Pellew Islands and Darwin.

M48-7, Mouth of billabong, at Little Lagoon, USNM 174187, 7 specimens, 20 to 34 mm. standard length.

M48-10, Estuary, Thompson Bay, Port Langdon, USNM 174188, 5 specimens, 15 to 23 mm. standard length.

M48-25, Beach, island, Port Bradshaw, USNM 174189, 22 specimens, 23 to 28 mm. standard length.

Description. Mouth extending slightly downward when protracted; lateral line incomplete, ending below rear half of dorsal fin; breast covered with thin scales; supraorbital slightly serrate; lower edge, and sometimes adjacent ridge, of preopercle serrated; anterior serrae on third and fourth dorsal and third anal spines, the front of the other spines smooth; cheek naked, yellowish, with 2 distinct horizontal rows of chromatophores in smaller individuals; tip of spinous dorsal fin with small dusky or black blotch; a narrow brown band around end of snout; upper side of body with irregular brown blotches; lower side with numerous fine dusky chromatophores except on rear part of caudal peduncle.

Leiognathus moretoniensis Ogilby

Leiognathus moretoniensis Ogilby, *Mem. Qld. Mus.* 1: 59 (1912); type locality, Bulwer, Moreton Bay, Queensland.

Recorded as: *Equulites moretoniensis* Whitley (1932a): 108, Port Darwin and Pellew Islands; *Leiognathus mortoniensis* Paradice and Whitley (1927): 84, Pellew Islands.

Leiognathus equulus (Forskål)

Scomber equula Forskål, *Descriptiones Animalium*: 58 (1775); type locality, Red Sea.

Recorded as: *Equula edentula* Castelnau (1873): 84; Macleay (1884): 25, Port Darwin; *Equula caballa* Klunzinger (1879): 379, Port Darwin.

M48-7, Mouth of billabong, at Little Lagoon, USNM 174190, 19 specimens, 34 to 46 mm. standard length.

M48-10, Estuary, Thompson Bay, Port Langdon, USNM 174191, 25 specimens, 13 to 22 mm. standard length.

M48-25, Beach, island, Port Bradshaw, USNM 174192, 135 specimens, 18 to 37 mm. standard length.

M48-28, Creek mouth and bay, Port Bradshaw, USNM 174193, 4 specimens, 34 to 50 mm. standard length.

Description. Lateral line ending at base of caudal fin; breast naked; supraorbital rugose to serrate; lower edge of preopercle strongly and the adjacent ridge roughly serrated; second to fourth dorsal and second and third anal spines serrated, the serrae of the second spine of both fins fine in young, becoming obsolete in the large individuals; small specimens frequently with fine serrae on pelvic and first spines of dorsal and anal fins; end of snout crossed with a broad brownish band; cheek unpigmented, naked; dorsal fin dusky; a series of narrow, irregular brown bars on upper side, which not prominent in small specimens; lower side, except rear part, dotted with brown.

SCIAENIDAE

*Sciaena** *diacanthus* (Lacépède)

Pl. 46

Lutjanus diacanthus Lacépède, *Histoire naturelle des poissons* 4: 195 and 240 (1802); type locality not given.

Pseudosciaena diacanthus Bleeker, *Atlas ichthyologique des Indes Orientales Néêrlandaises* 9: pl. 388, fig. 2 (1877).

Sciaena antarctica rex Whitley, *Aust. Zool.* 11: 26 (1945) and 11: 141, fig. 2 (1947); type locality, Onslow, Western Australia.

Near flying boat buoys, south-east part Little Lagoon, Groote Eylandt, Northern Territory, 24 July 1948 (collected by natives), USNM 174119, 1 specimen, 308 mm. standard length.

Native name: *Bol-u-ji-lar,* Groote Eylandt.

Description. Dorsal rays X,I,25; anal rays II,7, the second spine length about twice eye diameter; pelvic rays I,5, both sides; branched caudal rays 8+7=15; pectoral rays ii,19 (left) or ii,21 (right); pored scales in lateral line about 48; scale rows between origin dorsal fin and highest part lateral line 8; five enlarged

* *Sciaena umbra* Linnaeus was selected as the type species of *Sciaena* Linnaeus by the International Commission on Zoological Nomenclature, Opinion 93.

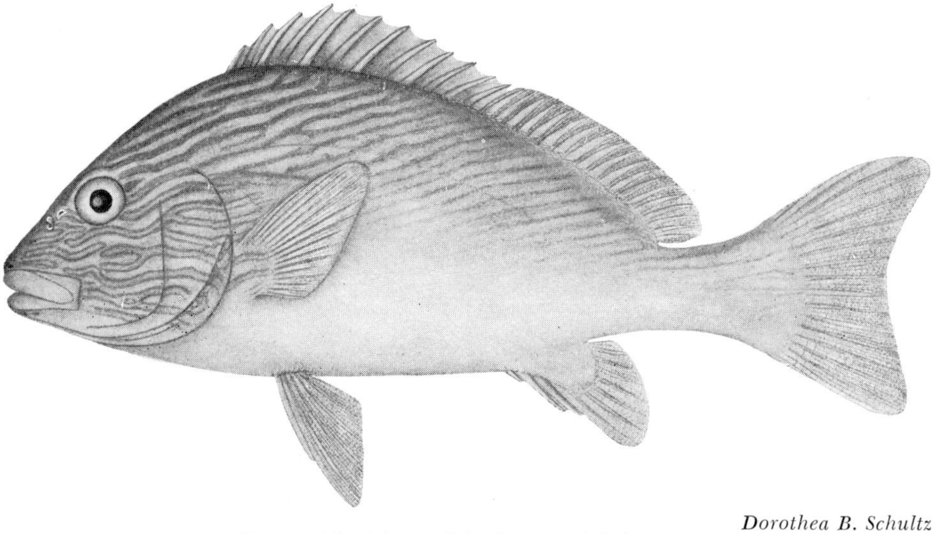

Dorothea B. Schultz

PLATE 45. *Plectorhinchus multivittatus*
Port Bradshaw, USNM 174094, 363 mm. standard length

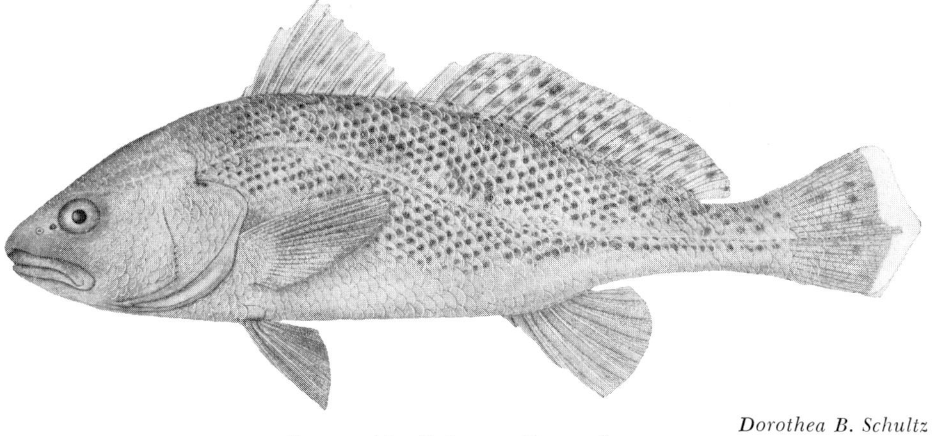

Dorothea B. Schultz

PLATE 46. *Sciaena diacanthus*
Little Lagoon, USNM 174119, 308 mm. standard length

pores anteriorly on chin, one terminal, medium, transverse; the other four paired on each jaw; no barbels; lower jaw included; teeth on upper jaw a band of villiform ones outside of which a larger row of moderate canines; in lower jaw, chiefly villiform, in broad band at front, with some teeth slightly longer toward rear; pseudobranchae well developed; lateral line extending well on to caudal fin, which moderately scaled; scales on body and head ctenoid, except cheek where smooth; gillrakers on first arch, short, longest about 3 times its basal width, 5+1+7=13 on right, including rudiments. Body depth 3·8 and head length 3·4, both in standard length; pelvic fin 1·8, pectoral fins broken, about 1·6 or less, snout 4·0, and eye 5·8, all in head length.

Body, head, and upper fins yellowish; pelvic and anal fins dark brown; large, prominent dark brown spots on upper body, dorsal, and caudal fins; those on body beginning about midside and arranged more or less in inclined rows; those on dorsal fin membranes in about three or four longitudinal rows.

MULLIDAE

Upeneus tragula Richardson

Upeneus tragula Richardson, *Rep. Fifteenth Meet. Brit. Ass. Adv. Sci.*: 220 (1846); type locality, Canton.

Recorded as: *Upeneus tragula* Kent (1889): 239, Port Darwin; *Upeneoides tragula* Macleay (1878): 350 and (1880): 402; Klunzinger (1879): 354, Port Darwin.

*Parupeneus** *jeffi* (Ogilby)?

Pseudupeneus jeffi Ogilby, *Proc. Roy. Soc. Qld.* 21: 19 (1908); type locality, Brisbane River.

Recorded as: *Pseudupeneus jeffi* Paradice and Whitley (1927): 89, Pellew Islands; *Upeneoides jeffi* Whitley (1929a): 123, Pellew Islands; identifications questioned.

MONODACTYLIDAE

Monodactylus argenteus (Linnaeus)

Chaetodon argenteus Linnaeus, *Systema Naturae* ed. 10, 1: 272 (1758); type locality, 'Habitat in Indiis'.

Recorded as: *Psettus argenteus* Macleay (1878): 353; Klunzinger (1879): 379, Port Darwin; *Monodactylus argenteus* Paradice and Whitley (1927): 89, Pellew Islands.

M48-1 or 2, Nightcliff, 19 March 1948, USNM 173504, 5 specimens, 107 to 128 mm. standard length.

M48-7, Mouth of billabong, Little Lagoon, USNM 173505, 1 specimen, 61 mm. standard length.

M48-15, Emerald River, Groote Eylandt, USNM 173506, 7 specimens, 41 to 100 mm. standard length.

M48-19, Creek, tributary to Little Lagoon, USNM 173507, 8 specimens, 17 to 72 mm. standard length.

M48-26, Sheltered bay, island in Port Bradshaw, USNM 173508, 7 specimens, 90 to 119 mm. standard length.

M48-31, Creek, Yirrkalla, USNM 173509, 8 specimens, 33 to 50 mm. standard length.

* *Parupeneus* Bleeker (1863): 234; type species, *Mullus trifasciatus* Lacépède, designated by Fowler (1933): 271.

Description. Dorsal rays VIII,27(in 1), VIII,28(5), VIII,29(1), or VIII,30(1); anal rays III,27(1), III,28(4), or III,29(2); pectoral rays ii,14(6) or ii,15(8); branched caudal rays 8+6=14(1), 8+7=15(5), or 8+8=16(1); pelvic rays, in specimens 33 to 40 mm. in standard length, I,3(1), I,4(1), or I,5(8); scales in lateral line 52 to 57; above lateral line 18 to 20; below lateral line 43 to 48.

In seven specimens 38 to 72 mm. in standard length, snout 4·5 to 4·9, eye 2·5 to 2·7, third anal spine 1·1 to 1·4, all in head length; head length 2·6 to 2·8, pectoral fin 3·9 to 4·4, depth of body 1·1 to 1·3, all in standard length.

Body deep, strongly compressed; pelvic fins greatly reduced in size, the rays all small, proportionately larger in young; teeth present on vomer, palatines and jaws; those in jaws simple; lateral line arched, continuous, extending on to basal portion of caudal fin; mouth terminal, the jaws equal; upper jaw protractile; anterior dorsal and anal rays elongated, the posterior rays short; caudal fin slightly to deeply concave posteriorly; upper pectoral rays longest, the fin margin rounded.

Colour in alcohol. In young and adults, background colour light brown; fine brown lines along side of body corresponding to each scale row; dark brown bar, wide as pupil, from occiput through eye to isthmus; another from nape through opercle and pectoral fin base to pelvic fin and belly, thence continued along anterior rays anal fin, including first two or three soft rays to their tip; dorsal fin and soft rays anal brown anteriorly and at base, yellowish toward margin; caudal and pectoral fin yellowish.

PEMPHERIDAE

Pempheris adjustus Bleeker

Pempheris adjustus Bleeker, *Arch. Néerl. Sci. Nat.* 12: 50 (1877); type locality, Amboina; Bleeker, *Atlas ichthyologique des Indes Orientales Néêrlandaises* 9: 7, pl. 383, fig. 1 (1878).

M48-21, Reef, Yirrkalla, 14 July 1948, USNM 173533, 50 specimens, 69 to 112 mm. standard length.

Description. Dorsal rays VI,9(in 9), or VI,10(1); anal rays III,40(2), III,42(5), III,43(2), or III,45(1); pelvic rays I,5(19); pectoral rays ii,15(1), or ii,16(19); branched caudal rays 8+7=15(10); pored scales in lateral line 53(1), 54(1), 56(2), 57(3), 58(1), or 59(2); scales above lateral line 5(9), or 6(1); scales below lateral line 12(4), 13(5), or 14(1). Gillrakers, first arch on right side, specimens 69 to 83 mm. standard length, 8+1+19=28(2) or 8+1+20=29(1); specimens 101 to 108 mm. standard length, 9+1+20=30(2), 8+1+21=30(1), or 9+1+21=31(4). In the same specimens: snout 5·3 to 6·6, tip of snout to posterior end of maxilla 1·9 to 2·1, eye 2·1 to 2·5, all in head length; pectoral fin 3·6 to 4·0, head 3·1 to 3·6, body depth 2·2 to 2·6, distance dorsal origin to tip of snout 2·5 to 2·7, all in standard length.

Lateral line continuous from upper corner of gill cavity to tip of middle caudal rays; eye very large; snout short, obtuse; first three branched pectoral rays longest; scales large, irregular, and somewhat deciduous, with small auxiliary scales which remain when large scales fall off; regular scales on side of body cycloid; those on lower head, circumorbitals, area above operculum, narrow portion of caudal peduncle, on back behind origin of dorsal fin and on breast mostly ctenoid; very small teeth in V-shaped band on vomer and in an elongate band on palatine, those on jaws larger; outer teeth of upper jaw in some individuals slightly the largest and directed outward; tongue pale yellowish, its tip forming an acute angle when viewed from above; posterior margin caudal fin deeply concave; anterior two or three anal rays longest forming a shallowly concave lower margin; third anal spine a little more than half as long as longest anal ray and about equal to pupil; dorsal spines graded, the first shorter than pupil, the last nearly as long as first dorsal ray; first three or four dorsal rays prominently the longest. Preopercle with three or four irregular, blunt spinelike processes from lower portion of posterior margin.

Colour in life. 'Body golden brown with bronze reflections from scales'. Based on a colour transparency: Body and head light coppery brown, with silvery and brassy reflections; eye olive green, pupil dark blue; caudal fin dirty red, its margin dark grey to black; other fins without trace of pink.

Colour in alcohol. Body and head a copper brown; head lighter above, grading to brownish yellow on snout; belly and breast lighter; often a purplish patch on breast below pectoral fin; pectoral fin light yellow, its base dusky to prominently blackened; pelvic fin light yellowish with scattered brownish chromatophores, its base dark brown; anal fin dusted with brown, its margin light yellowish and base dark brown; dorsal fin anteriorly dusky grey, lighter posteriorly, the anterior margin and blotch on distal one-third of longest rays dusky grey to blackish; caudal fin dusky brown, its posterior end narrowly or in some specimens broadly margined with black.

Remarks. The variation in pigmentation exhibited by these specimens suggests that certain colour characters, such as presence or absence of a black blotch at the inner and outer base of the pectoral fin, are of little value in distinguishing the forms. In this sample, the pectoral blotch is present and jet black in some individuals, dusky in others, and in some the area is yellowish, with only a small darkened line near the base of the pectoral rays. Likewise the colour of the dorsal fin varies. The anterior margin and apex may be uniformly black, or the anterior margin may be greyish black and the apex jet black.

This form is somewhat similar in colour and scale arrangement to the central Pacific *Pempheris oualensis* Cuvier and Valenciennes, but differs markedly in having coarser scales, a larger eye, and a shorter, more obtuse, snout. No trace of black was noted on the anal fin.

TOXOTIDAE

Key to the Arnhem Land Species

1 Lateral line relatively straight, without an inflection, with about 45 scales between head and caudal fin; head and body without prominent markings; dorsal rays V,13 to 14 . . . . *Toxotes lorentzi*

Lateral line with an inflection in its anterior part, with 28 to 38 scales between head and caudal fin; head and body with dark blotches or bars; dorsal rays V, seldom IV or VI,12 to 15 . . . . *Toxotes chatareus*

*Toxotes** *chatareus* (Hamilton-Buchanan)

Coius chatareus Hamilton-Buchanan, *An Account of the Fishes Found in the River Ganges and Its Branches*: 101 and 370, pl. 14, fig. 34 (1822); type locality, 'mouthes of the Ganges'.

Toxotes carpentariensis Castelnau, *Proc. Linn. Soc. N.S.W.* 3: 47 (1878); type locality, Norman River, Queensland.

* *Toxotes* Bosc (1816): 442; species references vernacular; type species, *Toxotes jaculator* equals *Labrus jaculator* Shaw, by monotypy in Cloquet (1816): 116. *Protoxotes* Whitley (1950): 244; type species, *Toxotes lorentzi* Weber, by original designation; here regarded as a generic synonym.

Toxotes dorsalis Whitley, *Rec. Aust. Mus.* 22: 242, fig. 5 (1950); type locality, Flinders River near Hughenden and Richmond, Queensland; other localities: Mataranka, Roper River, Northern Territory; Katherine River; and the neighbourhood of Darwin.

Toxotes ulysses Whitley, *Rec. Aust. Mus.* 22: 243, pl. 17, fig. 3 (1950); type locality, Townsville, Queensland.

Recorded as: *Toxotes jaculator* from Port Darwin by Castelnau (1873): 84, and Macleay (1880): 399.

M48-5, Woods Inlet, Delissaville, USNM 173501, 2 specimens, 18 and 31 mm. standard length.

M48-35, Large billabong, Oenpelli, 29 September and 1 October 1948, USNM 173502, 16 specimens, 18 to 191 mm. standard length.

Freshwater billabong, about 3 miles above Roper River Mission, Northern Territory, 17 August 1948 (Jerry Blitner), USNM 173503, 12 specimens, 43 to 68 mm. standard length.

Native name: *Nárlgan,* Oenpelli.

Description. Delissaville area: Dorsal rays V,12(in 2); anal rays III,16(2); pelvic rays I,5(4); pectoral rays ii,11(2) or ii,12(2); branched caudal rays 8+7=15(1); pored lateral line scales 29(2); scales above lateral line 4(1); scales below lateral line 12(1). Oenpelli area: Dorsal rays V,12(2), V,13(7), V,14(4), or VI,13(1); anal rays III,16(12) or III,17(2); pelvic rays I,5(20); pectoral rays ii,8(1), ii,9(2), ii,10(3), ii,11(8), or ii,12(4); branched caudal rays 8+6=14(1), 8+7=15(6), 8+8=16(1), or 9+7=16(1); pored lateral line scales 31(1), 32(2), 33(4), 34(4), or 36(1); scales above lateral line 4(7) or 5(5); scales below lateral line 11(7) or 12(4). Roper River area: Dorsal rays V,12(1), V,13(9), V,14(1), or V,15(1); anal rays III,15(1), III,16(5), or III,17(6); pelvic rays I,5(24); pectoral rays ii,10(1) or ii,11(23); branched caudal rays 7+7=14(1) or 8+7=15(9); pored scales in lateral line 33(1), 34(1), 36(1), 37(1), or 38(1); scales above lateral line 4(2) or 5(3); scales below lateral line 11(1), 12(3), or 13(1).

In five specimens, 160 to 191 mm. standard length, eye 4·4 to 5·0, snout 3·4 to 3·7, tip of snout to posterior end of maxilla 2·1 to 2·3, all in head length; pectoral fin 3·8 to 4·4, head length 3·0 to 3·3, depth of body 2·0 to 2·3, distance from tip of snout to first dorsal spine 1·7 to 1·8, distance from tip of snout to first anal spine 1·5 to 1·7, all in standard length.

Numerous short rounded teeth on tongue, palatines, pterygoids, vomer, and jaws; preopercle smooth posteriorly, but preopercle and preorbital strongly serrated along lower margins; a few serrae on each side along inside lower border of anterior end of lower jaw; lower jaw projecting; premaxilla protractile; a long axillary flap present above pelvic spine; inner pelvic ray bound for most of length to abdomen; anterior dorsal and anal spines short, their length about equal to or shorter than eye; margins of dorsal and anal fins rounded; caudal fin notched or shallowly forked in large specimens, and rounded in young; pectoral fin falcate, the first branched ray longest; body and head covered with large scales; soft dorsal and anal fins covered with fine scales; lateral line continuous between gill opening and base of caudal fin, but a slight arch is present below the anterior end of dorsal fin.

Colour in alcohol. Background body colour brown or yellowish brown, the scales on trunk and caudal peduncle outlined by darker brown; head with dark brown spot at upper corners of opercle and preopercle; young with four dark brown bars on body, each nearly as wide as eye; first bar from nape through base of pectoral fin; second from base of spinous dorsal to belly; third from middle of soft dorsal fin to middle of soft anal fin; fourth across caudal peduncle at base of caudal rays; in larger specimens bars disappear, becoming large irregular spots which alternate with smaller spots on the upper lateral surface; soft pelvic rays black in young, paler in adults; caudal fin yellowish in young, the inter-radial membranes dark brown in adult; dorsal and anal fins dark brown in adult; anal, in young, greyish brown, the outer one-third black; dorsal fin in young with a median basal dark brown blotch and a distal black area near posterior end.

Remarks. Variation in lateral line scale count, given above, is greater than that indicated in the literature. The dorsal spines, however, seem to remain relatively constant in both this species and in *Toxotes jaculator* (Pallas); those in material listed at *T. jaculator* by Fowler and Bean (1929) are III(in 1), IV(66), and V(1). Two specimens of *T. jaculator*, from New Guinea (USNM 30544 and 30612), each have four dorsal spines; they are probably the individuals reported as *T. oligolepis* by Fowler (1928).

Due to the variation in scale count, fin ray count, and change in colour with age, I regard the nominal species *Toxotes dorsalis* Whitley, *Toxotes ulysses* Whitley, and *Toxotes carpentariensis* Castelnau as identical to *T. chatareus*.

Toxotes lorentzi Weber

Toxotes lorentzi Weber, *Notes Leyden Mus.* 32: 232 (1911); type locality, Merauke, New Guinea; Weber and de Beaufort, *Fishes of the Indo-Australian Archipelago* 7: 198, fig. 52 (1936).

Recorded as: *Protoxotes lorentzi* by Whitley (1950): 244, from Yam Creek, Northern Territory.

PLATACIDAE

The genera *Zabidius* and *Platax* have tricuspid teeth, whereas *Chaetodipterus* and *Ephippus* have simple teeth on the jaws and are placed in the family Ephippidae.

Key to the Genera

1 Dorsal spines IX, the anterior spines exposed; large, irregular scales across top of head extending forward to the posterior naris . . . . . genus *Zabidius*

Dorsal spines IV to VII, the anterior ones concealed or nearly so; small scales extending forward only to above the eyes; midline of head, anterior to tip of occipital process, almost free of scales . . . . genus *Platax*

*Zabidius** *novemaculatus* (McCulloch)

Platax novemaculatus McCulloch, 'Report on the Fishes' in *Biological Results of the . . . Endeavour* 4: 188, pl. 55, fig. 1 (1916); type locality, north-east of Gloucester Head, Queensland.

Recorded as: *Platax novemaculatus* Paradice and Whitley (1927): 90 and 95, Pellew Islands; Geranium Bay, Pellew Islands.

M48-20, Old jetty, Melville Bay, 17 August 1948, USNM 173519, 1 specimen, 252 mm. standard length.

M48-20, Main jetty, Melville Bay, 3 September 1948, USNM 173520, 1 specimen, 251 mm. standard length.

* *Zabidius* Whitley (1930): 17; type species, *Platax novemaculatus* McCulloch, by original designation.

Native name: *Úruroo,* Melville Bay.

Description. Dorsal rays IX,28(in 2); anal rays III,21(2); pelvic rays I,5(4); pectoral rays ii,18(2) or ii,19(2); branched caudal rays 8+7=15(2); scales between head and caudal fin 53 and 54; lateral line pores 46 and 48; scales above lateral line 18 or 19, and below lateral line 31 and 33; vertebrae 10+14=24(1); ribs 8(1), the last rudimentary; mandibular pores 5(both sides in 2); free interneurals anterior to dorsal fin 3(1).

Snout 2·4 and 2·5, eye 3·7 and 3·8, fourth dorsal spine 4·7 or 5·0, ninth dorsal spine 2·1 or 2·5, all in head length; head length 3·3 or 3·4, body depth 1·3 or 1·4, pectoral fin 4·6 or 4·7, all in standard length.

Lateral line slightly arched, continuous to base caudal fin; teeth in jaws tricuspid, the median cusp only slightly longer than the lateral cusps; teeth arranged in four or five rows in the upper jaw and four in the lower, the rows separated by spaces broader than the width of the crown of a tooth; length of rows decreasing posteriorly; a few teeth on midline of palate; pelvic fin extending posteriorly of anus; snout prominent, blunt.

Colour in alcohol. General colour brown, lighter below; anterior snout and upper lip dark brown; diffuse dark brown bar across head, through eye to lower margin preopercle, thence continued rather indistinctly to breast; another from nape across operculum and base of pectoral fin to belly; outer pelvic fin black, inner ray whitish; median fins dusky with narrow black margin.

Examination of the specimens of *Platax** in the United States National Museum indicates that there are four species, in agreement with Weber and de Beaufort (1936). The following key is based on this material; proportions apply best to specimens larger than 60 mm. in standard length.

Key to the Species of *Platax*

1 Anterior profile, from dorsal to pelvic fin, evenly rounded; length from eye to anterior margin preorbital in least depth caudal peduncle 2·1 to 2·8; pelvic fin inserted ventral to preopercle in small specimens; middle tooth cusp equal to or only slightly longer than lateral cusps; dorsal rays V, seldom VI,30 to 34; mandibular pores usually 5 on both sides (total=9 or 10 in 14 specimens); branched pectoral rays 14 to 16, usually 15 .. *Platax teira*

Anterior profile more angular; length from eye to anterior margin preorbital in least depth caudal peduncle 1·2 to 2·4; pelvic fin inserted ventral to opercle; middle tooth cusp notably longer than lateral cusps 2

2 Dorsal rays VI or VII,28 to 32; anal rays III,19 to 23; body depth 1·3 to 1·7 in length; middle tooth cusp broad; lateral cusps about one-fourth its size; on both sides, mandibular pores 5 and branched pectoral rays 18 (in 2 specimens) . . . . *Platax batavianus*

Dorsal rays V, seldom IV or VI,34 to 39; anal rays III,24 to 29; body depth 0·8 to 1·2 in length; middle tooth cusp more slender 3

3 Snout not concave before eye; mandibular pores 5 on both sides (total=9, usually 10 in 32 specimens); branched pectoral rays 15 or 16, rarely 17; distance from base last dorsal spine to lateral line 0·8 to 1·1 in head length . . . . *Platax orbicularis*

* The type species of *Platax* Cuvier is *Chaetodon vespertilio* Bloch, designated by Guichenot (1839): 136.

Patricia Hogue

PLATE 47. *Platax teira*
Off Skirmish Point, USNM 173521, 82 mm. standard length

Snout with a slight concavity before eye; mandibular pores usually 4 on both sides (total=6 to 8 in 8 specimens); branched pectoral rays 16 to 18; distance from base last dorsal spine to lateral line 1·1 to 1·4 in head length . . . . . *Platax pinnatus*

Platax teira (Forskål)

Pl. 47

Chaetodon teira Forskål, *Descriptiones Animalium* . . .: 60 (1775); type locality, Red Sea.

Platax teira Klunzinger, *Synopsis der Fische des Rothen Meeres*: 791 (1870), description.

Record probably referable: *Platax teira* Paradice and Whitley (1927): 89, Pellew Islands.

Reef, off Skirmish Point, near Liverpool River, Northern Territory, 4 May 1948, David H. Johnson, D.H.J. No. 5466, USNM 173521, 1 specimen, 82 mm. standard length.

Description. Dorsal rays V,32; anal rays III,24; pelvic rays I,5, both sides; pectoral rays ii,15, both sides; branched caudal rays 8+7=15; pores in lateral line 63; scale rows between head and caudal fin 66; above lateral line 34; below lateral line 49; mandibular pores 5, both sides.

Snout 2·6, eye 3·6, base last dorsal spine to lateral line 0·9, all in head length; head length 2·8, pectoral fin 3·9, body depth 0·9, all in standard length; snout 1·1 in postorbital length of head or 1·3 in least depth caudal peduncle; distance eye to anterior margin preorbital 2·3 in least depth caudal peduncle.

Lateral line strongly arched, continuous to base of caudal fin; teeth present in a small patch on midline of palate and jaws; those of jaws in about 4 rows, the anterior teeth tricuspid; middle tooth cusp about same size as lateral cusps; pelvic fin extending beyond rear of body; margins of dorsal and anal fins nearly straight; longest dorsal ray 0·6 and longest anal ray 0·7 in standard length; middle caudal rays about as long as outer rays, the upper and lower lobes of the caudal fin slightly concave posteriorly; anterior profile rounded.

In an X-ray of a specimen (USNM 122245, Batananan Island, Philippines), vertebrae 10+14=24; ribs 8, the last very small; three free interneurals anterior to dorsal fin, their dorsal plates not greatly expanded, separated by spaces nearly equal to their length.

Colour in alcohol. Background colour medium brown, lighter anteriorly; dark brown bar, nearly as wide as eye, from occipital process through eye to breast and pelvic fin; another, 1·5 times as wide as eye, from spinous dorsal and tip first two segmented rays through posterior end of opercle and pectoral fin base to belly; another bar, as wide as head length, from the tip of anterior dorsal rays through middle of dorsal fin, across posterior half of body through base anal fin, to tip of anterior anal rays; caudal, pectoral, and posterior halves of dorsal and anal fins yellowish; pelvic fins dark brownish black.

Platax orbicularis (Forskål)

Pl. 48

Chaetodon orbicularis Forskål, *Descriptiones Animalium*: 59 (1775); type locality, Red Sea.

Records probably referable: *Platax orbicularis* Macleay (1878): 353, and (1880): 547, Port Darwin.

Dorothea B. Schultz

PLATE 48. *Platax orbicularis*
Little Lagoon, USNM 173522, 58 mm. standard length

M48-8, Little Lagoon, USNM 173522, 1 specimen, 58 mm. standard length.

Description. Specimen badly damaged; dorsal rays V,37; anal rays III,26; pelvic rays I,5, both sides; pectoral rays ii,16 on left and ii,17 on right; branched caudal rays 8+7=15; lateral line pores 49; scale rows between head and caudal fin 54; above lateral line 24; below lateral line 43; mandibular pores 5, both sides.

Snout 2·5, eye 3·1, base last dorsal spine to lateral line 0·9, all in head length; head length 2·7, pectoral fin 3·7, body depth 0·9, all in standard length; snout 1·0 in both postorbital length of head and least depth caudal peduncle; distance eye to anterior margin preorbital 1·8 in least depth caudal peduncle.

Lateral line strongly arched, continuous to base caudal fin; a few teeth on midline of palate; teeth in jaws tricuspid, the middle cusp about twice as long as the lateral cusps; longest pelvic rays reaching posterior half of anal fin; anterior profile an angle of about 90°. In an X-ray of a specimen (USNM 122244, Leyte Island, Philippines) vertebrae 10+14=24; ribs 8; 3 free interneurals anterior to dorsal spine, their plates small.

Colour in alcohol. Body colour dark brown; indistinct brownish black bar, half width of eye, from occiput through eye to base of pelvic fin; semblance of another bar from spinous dorsal through pectoral fin base to belly; dorsal and anal fins brownish black, the margin of their shortest rays whitish; brownish black bar, wide as pupil, near base of caudal rays; remainder of caudal and pectoral fins whitish; pelvic fin black.

Platax pinnatus (Linnaeus)

Chaetodon pinnatus Linnaeus, *Systema Naturae* ed. 10, 1: 272 (1758); type locality, 'Habitat in Indiis'.

Recorded as: *Platax orbicularis* Richardson (1842b): 32, harbour of Port Essington.

Platax batavianus Cuvier and Valenciennes
Pl. 49

Platax batavianus Cuvier and Valenciennes, *Histoire naturelle des poissons* 7: 225 (1831); type locality, Batavia.

Record probably referable: *Platax arthriticus* Kent (1889): 239, Port Darwin.

M48-11, Little Lagoon, 21 June 1948, USNM 174067, 1 specimen, 284 mm. standard length.

M48-22, Bay of Yirrkalla, 4 August 1948, USNM 173523, 1 specimen, 193 mm. standard length.

Description. Dorsal rays VII,29(in 1), the large specimen with 28 soft rays; anal rays III,19(1) or III,21(1); pelvic rays I,5, both sides(1); pectoral rays ii,18, both sides(2); branched caudal rays 8+7=15(1); lateral line pores 65(1); scale rows between head and caudal fin 84(1); above lateral line 26(1); below lateral line 41(1); mandibular pores 5, both sides(2).

In the small specimen, snout 2·2, eye 3·8, base last dorsal spine to lateral line 1·4, all in head length; head length 3·1, pectoral fin 4·4, body depth 1·3, all in standard length; snout 0·9 in postorbital length of head or 0·8 in least depth of caudal peduncle; preorbital 1·3 in least depth caudal peduncle.

Lateral line slightly arched, continuous to base of caudal fin; teeth present in a large patch on palate and on jaws; those of the jaws tricuspid and arranged in about 6 rows which are separated by spaces about equal to the width of the anterior teeth; middle tooth cusp broad and about 4 times as long as lateral

Patricia Hogue

PLATE 49. *Platax batavianus*
Yirrkalla, USNM 173523, 193 mm. standard length

cusps; pelvic fin reaching base first segmented anal ray; margins of dorsal and anal fins slightly convex; longest dorsal ray 2·1 and longest anal ray 2·6 in standard length; caudal fin concave posteriorly, each lobe again slightly concave; anterior profile decidedly angular; snout bluntly pointed; large specimen with hump on forehead above nostrils.

From an X-ray of the small specimen, vertebrae 10+14=24; ribs 6, the last three very short; two free interneurals anterior to dorsal fin, their dorsal plates greatly expanded and forming a continuous bridge from base of dorsal fin to the large occipital process.

Colour in life. Based on colour transparencies; general body colour rich brown; dark brown bar, three-fourths width of eye, from top of head through eye to breast and base pelvic fin; another, wider than eye, from spinous dorsal through posterior third of opercle and base pectoral fin to posterior base pelvic fin and belly; outer third dorsal fin dark brown, inner portion medium brown; caudal fin yellowish brown; anal fin blackish anteriorly, becoming dark brown posteriorly; pelvic fin black; pectoral fin brown, becoming yellowish brown distally; eye black.

DREPANEIDAE*

Drepane punctata (Linnaeus)

Chaetodon punctatus Linnaeus, *Systema Naturae* ed. 10, 1: 273 (1878); type locality, Asia.

Recorded as: *Drepane punctata* Macleay (1878): 353; Klunzinger (1879): 363; Kent (1889): 239, Port Darwin; Paradice and Whitley (1927): 90 and 95, Pellew Islands; Geranium Bay, Pellew Islands.

SCATOPHAGIDAE

Key to the Arnhem Land Species

1 Dorsal spines normally XI; soft anal rays usually 13 or 14, sometimes 15; gill membranes forming a free, fleshy fold across isthmus .. *Scatophagus argus*
Dorsal spines normally XII; soft anal rays usually 16, sometimes 15 or 17; gill membranes not forming a free, fleshy fold across isthmus .. .. *Selenotoca multifasciata*

Scatophagus semi-striatus was listed by Kent, *Proc. Roy. Soc. Qld.* 6: 239 (1889), from Port Darwin. The name is a *nomen nudum.*

Scatophagus argus (Linnaeus)

Chaetodon argus Linnaeus, *Systema Naturae* ed. 12, 1: 464 (1766); type locality, 'Habitat in India'; Bloch, *Ichthyologie* 6: 63, pl. 204, fig. 1 (1788).

? *Scatophagus altermans* Castelnau, *Proc. Linn. Soc. N.S.W.* 3: 47 (1878); type locality, Norman River, Australia; Whitley, *Aust. Zool.* 9: pl. 31, fig. 39 (1940); type figured.

? *Scatophagus alternans* Castelnau, *Proc. Linn. Soc. N.S.W.* 3: 376 (1878); correction of misprint *altermans.*

Scatophagus argus var *ocellata* Klunzinger, *S.B. Akad. Wiss. Wien* 80: 363 (1879); type locality, Port Darwin.

Scatophagus quadranus De Vis, *Proc. Linn. Soc. N.S.W.* 9: 455 (1884); type locality,

* Corrected spelling to replace Drepanidae, which now fixed as a family of insects; see Opinion 610, International Commission on Zoological Nomenclature.

Queensland coast; McCulloch, *Aust. Mus. Mem.* 5: 241 (1929); 'originally described in the Proceedings of the Queensland Philosophical Society in a Brisbane newspaper about 1882'.

Scatophagus quadratus Kent, *The Great Barrier Reef of Australia . . .*: 369 (1893); misspelling of *quadranus*.

Other Arnhem Land records: *Scatophagus tetracanthus* Macleay (1878): 353 and (1880): 396, Port Darwin; *Desmoprenes tetracanthus* Whitley (1940b): 424, Northern Territory; *Prenes ornatus* Whitley (1956b): 252, Groote Eylandt and Darwin.

M48-7, Mouth of billabong, at Little Lagoon, USNM 173515, 2 specimens, 47 and 50 mm. standard length; M48-7, mouth brackish creek, Little Lagoon, 1 July 1948, USNM 173516, 1 specimen, 29 mm. standard length.

M48-14, Emerald River, Groote Eylandt, just below falls, USNM 173517, 1 specimen, 45 mm. standard length.

M48-15, Emerald River, Groote Eylandt, USNM 173518, 6 specimens, 141 to 197 mm. standard length.

Native names: *Woorarúomba* and Bull Skin, Groote Eylandt; *Ranooga*, Roper River.

Description. Dorsal rays X,16(in 1), XI,16(8), or XI,17(1); anal rays IV,14(9), or IV,15(1); pelvic rays I,5(20); pectoral rays ii,15(6), ii,16(13), or ii,17(1); branched caudal rays 7+7=14(9), or 7+8=15(1). Pored scales in lateral line approximately 108 to 131; scales above lateral line 27 to 33; below lateral line 53 to 62.

In six specimens 141 to 197 mm. in standard length: eye 4·4 to 4·9, snout 2·9 to 3·2, both in head length; head 3·0 to 3·5, depth of body 1·6 to 1·8, pectoral fin 5·2 to 6·4, all in standard length. In four specimens 29 to 50 mm. in standard length: eye 2·8 to 3·7, snout 3·3 to 3·7, both in head length; head 2·5 to 2·9, depth of body 1·5 to 1·6, pectoral fin 4·6 to 5·2, all in standard length.

Teeth on jaws tricuspid, in about six irregular rows; lateral tooth cusps about one-third the height of middle cusp; dorsal and anal spines heterocanth; lateral line arched, complete to base of caudal fin, the arch parallel with the contour of the back; caudal fin in young convex and in adult truncate posteriorly; scales covering body, head except lips, and fins except membranes between spines; small calcareous plates on concealed portion of eye; preopercle and preorbital finely serrated; occular margin circumorbital bones slightly projecting, the edge serrated; first interneural of dorsal fin with a forward projecting sharp spine.

Colour in life. From a coloured drawing of a young specimen 29 mm. in standard length in Miller's notes: background colour of body and head yellow; a deep chocolate brown, almost black bar, wide as eye, from top of head through eye to lower margin preopercle; another from nape through opercle to breast; three others, nearly as wide, from spinous dorsal to belly; another from midbase of soft dorsal fin to midbase of soft anal fin, and one, nearly as wide as eye, at base of caudal fin; spinous portion dorsal and anal fins red; basal one-fourth of soft dorsal and soft anal fins deep chocolate brown; outer three-fourths soft anal and soft dorsal fins, caudal fin, and pectoral fin yellow; pelvic fin dark brown, tinged with red; small red areas at upper base of caudal fin, at midbase of soft dorsal fin, on back at posterior base of dorsal fin, at junction of spinous and soft dorsal fins, and at dorsal origin; two yellow bars on head grade to red at top of head; tip of snout red; upper lip brown laterally; lower lip and margin of opercle orange.

Colour in alcohol. In specimens 45 to 50 mm. in standard length: colour pattern similar to the 29 mm. specimen; body and head very dark brown; body bars more numerous and tending to obscure the adjacent narrow light brown areas which were red, yellow, or orange in the small specimen; bars in largest specimen tending to break into vertical rows of dark spots; bars in the two smaller individuals somewhat obscured by small, irregular lighter areas or reticulations which are narrowly margined by dark brown; pelvic fin and spinous portions dorsal and anal fins dark brown, almost black; caudal fin and soft portions dorsal and anal fins, except base, cream coloured.

In specimens 141 to 197 mm. in standard length: body and head dark brown, becoming yellowish brown on lower surface; numerous spots, some as large as eye, on body; spots dark brown to pale yellowish, ringed with brownish black; pectoral fin yellowish brown; other fin rays brownish, the fin membranes dark brown to black.

Remarks. The small specimens from this collection are distinctly barred, whereas those of comparable size in the National Museum from India, Burma, Java, and the Philippine Islands are prominently spotted. These differences do not appear to be present in larger specimens. This feature of the young probably accounts for the identification of Port Darwin specimens as *Scatophagus tetracanthus* (Lacépède), an African species, by Macleay and others (loc. cit.).

*Selenotoca** *multifasciata* (Richardson)

Scatophagus multifasciatus Richardson, *Ichthyology of the Voyage of H.M.S. Erebus and Terror*. . .: 57, pl. 35, figs. 4-6 (1846); type locality, King George's Sound, Australia.

Scatophagus aetate-varians De Vis, *Proc. Linn. Soc. N.S.W.* 9: 456 (1885); type locality, Queensland coast; Kent, *Great Barrier Reef of Australia*. . .: pl. 47, fig. 4 (1893).

Other Arnhem Land records: *Scatophagus multifasciatus* Macleay (1878): 353 and Klunzinger (1879): 363, Port Darwin; *Scatophagus aetate-varians* Paradice and Whitley (1927): 90, Pellew Islands.

M48-7, Mouth of billabong, Little Lagoon, USNM 173510, 4 specimens, 35 to 75 mm. standard length.

M48-10, Estuary, Thompson Bay, Port Langdon, USNM 173511, 1 specimen, 111 mm. standard length.

M48-15, Emerald River, Groote Eylandt, USNM 173512, 1 specimen, 215 mm. standard length.

M48-19, Creek, tributary to Little Lagoon, USNM 173541, 5 specimens, 52 to 112 mm. standard length.

M48-28, Creek mouth and bay, Port Bradshaw, USNM 173513, 6 specimens, 93 to 111 mm. standard length.

M48-31, Creek, Yirrkalla, USNM 173514, 21 specimens, 82 to 143 mm. standard length.

M48-34, Creek, Yirrkalla, USNM 173542, 1 specimen, 73 mm. standard length.

Native names: *Tee Ján Belóoa,* Groote Eylandt; *Kápila,* young, or *Bírcoocoo,* large individuals, Yirrkalla.

Description. Dorsal rays XII,15(in 1), XII,16(3), XII,17(7), or XII,18(1); anal rays IV,15(2), IV,16(8), or IV,17(2); pectoral rays ii,14(1), ii,15(8), ii,16(14), or ii,17(1); pelvic rays I,4(3) or I,5(21); branched caudal rays 7+6=13(1), or 7+7=14 (11).

* *Selenotoca* Myers (1936): 84; type species, *Scatophagus multifasciatus* Richardson, by original designation.

In ten specimens 42 to 202 mm. in standard length: eye 3·4 to 4·3, snout 2·9 to 3·4, longest dorsal spine 1·0 to 1·7, all in head length; head 2·8 to 3·6, depth of body 1·6 to 1·9, pectoral fin length 5·7 to 7·5, all in standard length.

In 12 specimens: number of bars touching base dorsal fin 5 to 9; bars crossing lateral line 5 to 10; number of spots 23 to 72; third dorsal spine longest in three, or fourth dorsal spine longest in seven.

Teeth in jaws tricuspid, in about 6 irregular rows; middle tooth cusp equal to or slightly longer than lateral cusp; dorsal and anal spines heterocanth; lateral line arched, complete to base of caudal fin; the arch parallel to the contour of the back; posterior margin of caudal fin in specimens to 45 mm. slightly convex; in larger specimens truncate to slightly concave; scales covering body, head except lips, and fins except membranes between spines; no small calcareous plates on eye; preopercle and preorbital finely serrated; occular margin of circumorbitals slightly projecting, the edge serrated; first interneural of the dorsal fin with a forward projecting, sharp spine; scales very small, rows along lateral line about 170; pectoral fin short and rounded.

Colour in life. From a specimen (M48-10), 111 mm. in standard length: 'body bright silvery; bars and spots lead grey'. From a notebook drawing: pectoral and pelvic fins, membrane of operculum, top of head except darkened areas, dorsal and anal spines, basal one-fourth soft dorsal and anal fins, and basal three-fourths caudal fin yellow.

Colour in alcohol. Body and head light brown, lower surface light yellowish to silvery; lips black, grading to pale yellow around mouth; brown bar, wide as pupil, from eye across cheek; narrower brown vertical bars as follows: between anterior nares; two or three on top of head; one from nape to opercle; several on upper half of body and caudal peduncle; brown spots nearly as large as pupil on side of body and peduncle; fins all yellowish except black inter-radial membranes of dorsal and anal fins and dark brown posterior margin of caudal.

Habits. Miller states: These fish travel in close-knit schools of five or six, swimming in unison near mangroves and along sandy stretches, coming into shallow water in protected embayments.

Remarks. Selenotoca papuensis Fraser-Brunner (1938) cannot be distinguished from *S. multifasciata* on the basis of the relative length of the third and fourth dorsal spines.

CHAETODONTIDAE

Key to the Arnhem Land Species

1 Preopercle serrate or entire, without a strong, prominent, spine at the angle; pelvic axillary scale present 2

Preopercle serrate, with a strong, prominent, spine at the angle; no pelvic axillary scale 5

2 Snout notably produced, tube-like, 2·0 to 2·3 in head length; dorsal rays IX to X,29 to 31 . . . . . *Chelmon rostratus*

Snout 2·2 or more times in head length, conic or rounded, not notably produced as a tube 3

3 Fourth dorsal spine light yellowish, extended into a flexible filament as long as body; two broad dark bars on body, first from front of spinous dorsal through opercle and anterior trunk to abdomen, second across rear of spinous dorsal and caudal peduncle to posterior half of anal fin . . . . . *Heniochus acuminatus*

Dorsal spines graduated, none notably elongated 4

4 Dorsal rays VI,28 to 30; five brown bars on body and head; black ocellus at anterior base of soft dorsal in fourth brown band . . *Parachaetodon ocellatus*

Dorsal rays X or XI,20 to 22; brown bar across caudal peduncle in young; two yellow bars anteriorly, first from nape through eye to breast, second from below spinous dorsal through opercle and pectoral base to abdomen . . . . *Chaetodon aureofasciatus*

5 Scales between head and caudal fin about 110; interoperculum large; dorsal rays XI,21 to 22; usually many wavy light lines across body; a broad yellow bar from front of spinous dorsal to pectoral base and abdomen; a similar band from middle of spinous dorsal backward along base of dorsal fin to caudal peduncle . . . . *Chaetodontoplus duboulayi*

Scales between head and caudal fin fewer than 65, about 45 to 60 6

6 Interoperculum large, without spines; lateral line ending at base of caudal fin; posterior margin of preorbital not free . . . . genus *Euxiphipops*

Interoperculum small, serrated or with spines posteriorly; lateral line terminating below end of soft dorsal fin; hind margin of preorbital free, serrated or with strong spines . . . . genus *Centropyge*

Chaetodon aureofasciatus Macleay

Chaetodon aureofasciatus Macleay, *Proc. Linn. Soc. N.S.W.* 2: 351, pl. 8, fig. 3 (1878); type locality, Port Darwin.

Recorded as: *Chaetodon aureofasciatus* Macleay (1880): 388; Klunzinger (1879): 360; Kent (1889): 239; McCulloch (1923): 6, pl. 2, fig. 2; Ahl (1923): 84, Port Darwin; and Paradice and Whitley (1927): 90, Pellew Islands.

M48-21, Reefs, Yirrkalla, 14 July 1948, USNM 173526, 18 specimens, 54 to 95 mm. standard length; 11 August 1948, USNM 173527, 1 specimen, 64 mm. standard length; 12 August 1948, USNM 173528, 8 specimens, 46 to 96 mm. standard length.

Description. Dorsal rays X,21(in 1), XI,20(I), XI,21(4), or XI,22(2); anal rays III,17(4) or III,18(4); pelvic rays I,5(16); pectoral rays ii,13(12), or ii(14(4); branched caudal rays 8+7=15(8); pored scales in lateral line 33 to 38; scale rows from upper edge of gill opening to caudal fin base 43 to 46, above lateral line 13 to 16, below lateral line 25 to 27 (five counts).

Based on five specimens 67 to 96 mm. standard length: greatest depth 1·1 to 1·2, head length 3·0 to 3·2, longest pectoral ray 3·3 to 3·7, all in standard length; snout 3·0 to 3·2, tip of snout to posterior end of maxilla 5·7 to 6·0, tip of snout to rictus 6·3 to 7·0, all in head length.

Cleithrum, supracleithrum and preorbital weakly serrated to smooth along free margins; preoperculum finely serrate; orbit serrated above eye; lateral line ending at the oblique scale row at base of 4th to 7th from last dorsal ray.

Colour in alcohol. General background colour of body and fins yellowish brown; margins of scale pockets dark brown; head and breast light purple; narrow yellow (or orange) mid-dorsal line, margined with dark purple, from occiput to tip of snout; two similar coloured bars, as wide as pupil, curving from first dorsal spine through eye, preoperculum, and suboperculum to breast and from just above lateral line through opercular membrane and base of pelvic fin to tip of axillary scale; an almost obscure brown bar on caudal peduncle at base of caudal rays; tips of caudal rays black.

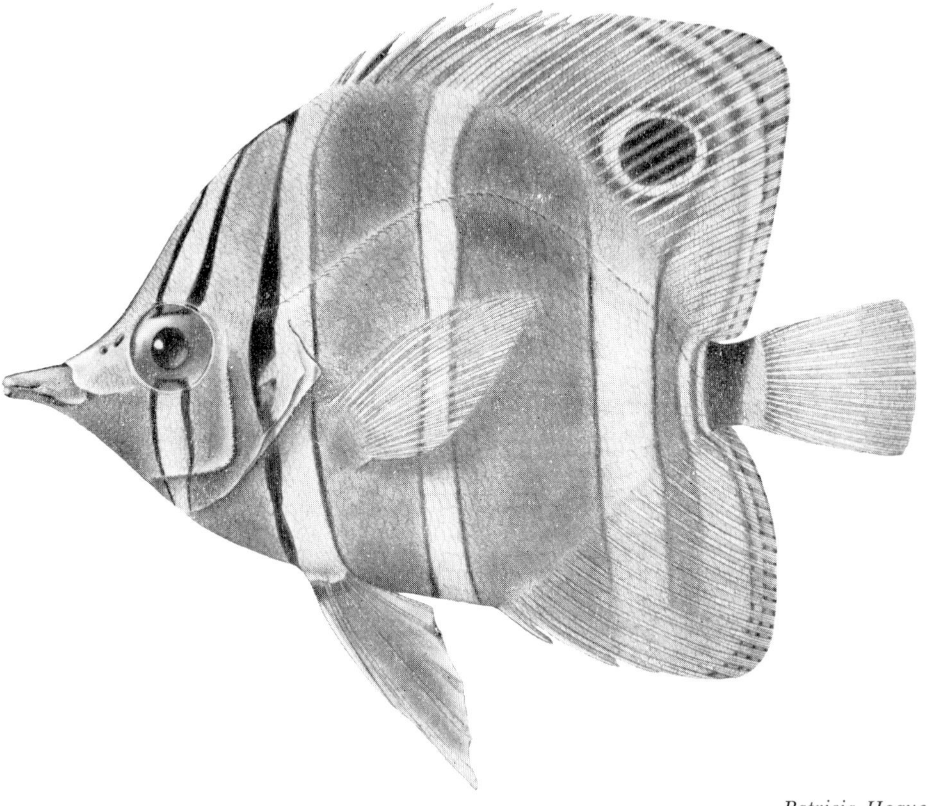

Patricia Hogue

PLATE 50A. *Chelmon rostratus*
Yirrkalla, USNM 173532, 31 mm. standard iength

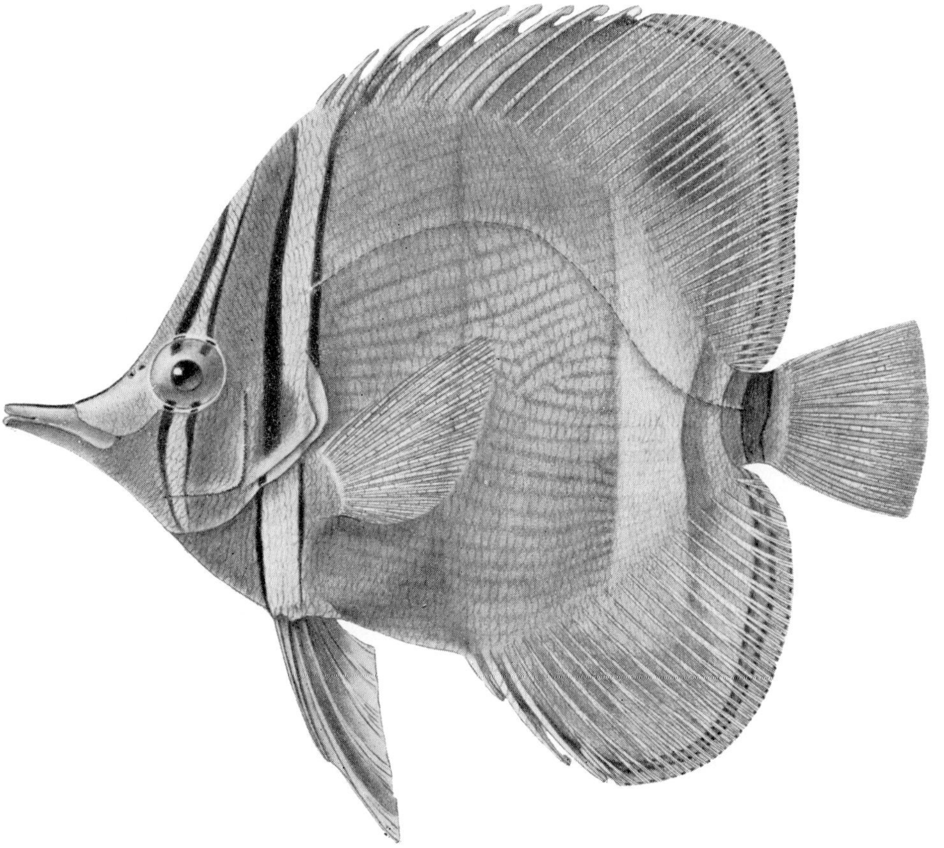

Patricia Hogue

PLATE 50B. *Chelmon rostratus*
Yirrkalla, USNM 173532, 57 mm. standard length

*Chelmon** *rostratus* (Linnaeus)

Pl. 50A and 50B

Chaetodon rostratus Linnaeus, *Systema Naturae* ed. 10, 1: 273 (1758); type locality, 'Habitat in Indiis'.

Chelmon marginalis Richardson, *Ann. Mag. Nat. Hist.* 10: 29 (1842); type locality, Port Essington; Richardson, Descriptions of Six Fish, in Stokes, *Discoveries in Australia*. . . 1: 491, pl. 4 (1846).

Chelmo tricinctus Castelnau, Researches on the Fishes of Australia, in *Philadelphia Centennial Exhibition 1876, Melbourne. Intercolonial Exhibition Essays, 1875-6,* no. 2: 14 (1875); type locality, Port Darwin, Australia; Macleay, *Proc. Linn. Soc. N.S.W.* 5: 392 (1880).

Chelmo mülleri Klunzinger, *S.B. Akad. Wiss. Wien* 80: 361 (1879); type locality, Australia.

Other Arnhem Land records: *Chelmo rostratus* Macleay (1878): 352 and Kent (1889): 239, Port Darwin; Macleay (1880): 391, Port Essington and Port Darwin; *Chelmon rostratus* Whitley (1955b): 158, fig. 5, drawing on shield, Raffles Bay, Northern Territory; *Chelmo marginalis* Macleay (1878): 352 and Klunzinger (1879): 361, Port Darwin; *Chelmon mülleri* Ahl (1923): 14, Port Darwin; and questionably as: *Chelmo truncatus* Kent (1889): 239, Port Darwin.

M48-3, East Point Reef, USNM 173529, 1 specimen, 86 mm. standard length.

M48-9a, Reef, entrance to Little Lagoon, USNM 173530, 1 specimen, 41·5 mm. standard length.

M48-17, Reef, Bartalumba Bay, USNM 173531, 2 specimens, 93 and 98 mm. standard length.

M48-21, Reef, Yirrkalla, 14 July 1948, USNM 173532, 7 specimens, 15 to 86 mm. standard length.

Native name: *Bócuroo,* Yirrkalla.

Description. Dorsal rays IX,29(in 2), IX,30(5), IX,31(2), or X,29(1); anal rays III,18(1), III,19(1), III,20(1), III,21(5), or III,22(2); pelvic rays I,5(18); pectoral rays ii,13(5), ii,14(12), or ii,15(1); branched caudal rays 8+7=15(10). Pores in lateral line 46 to 49; scales from upper edge gill opening to caudal fin base 48 to 50, above lateral line 11 to 13, below lateral line 24 to 26 (five counts).

Based on five specimens 31 to 93 mm. in standard length: greatest depth 1·4 to 1·6, head length 2·2 to 2·4, both in standard length; snout 2·0 to 2·3, tip of snout to posterior end of maxilla 2·6 to 3·1, tip of snout to rictus 6·4 to 7·0, all in head length.

Cleithrum, supracleithrum, and preoperculum strongly serrated on free edges; circumorbitals variably serrated; preorbital weakly serrated along lower margin; lateral line continuous to base of caudal fin.

Colour in alcohol. The 31 mm. specimen (Pl. 50A) has prominent markings: background body colour light brown, the scale pockets lighter; scales margined above and below by darker brown, forming longitudinal streaks; narrow cream coloured streak margined by black on mid-dorsal line from snout tip to occiput; two cream-coloured vertical bars on head, both margined by black, one from nape through eye and subopercle, the other from first dorsal spine through opercle,

* As noted in Herre (1953): 494 and Fowler and Bean (1929): 42, both the generic names *Chelmon* Cloquet and *Chelmo* Oken were established in 1817 with the subsequent designation of *Chaetodon rostratus* Linnaeus as type species of both. Determination of priority in the names has not been done at this date. The generic name employed here is based on estimated frequency of usage in later years, since I have found, in 32 references, *Chelmon* used 18 times and *Chelmo* 14.

thence anterior to pectoral fin base to pelvic base; third cream-coloured bar, margined with dark brown, extends from dorsal spines VI and VII through midside to anus; fourth light bar, margined with brown, from soft dorsal fin to anal fin; prominent black spot in soft dorsal fin ringed with creamy white, margined with dark brown; dorsal and soft anal fins margined with black; black bar as wide as pupil on caudal peduncle at base of fin rays; pelvic fin unpigmented except middle rays blackish.

Colour changes with increase in size; the 41·5 mm. specimen has third bar diffuse brown, and spot in dorsal fin black, grading into background fin colour. Largest specimens (57 to 98 mm., Pl. 50B) have third bar faint and scarcely discernible or absent; spot in dorsal fin greyish, faint, or indistinct; bar at caudal fin base cream coloured, margined with black; largest specimens have the caudal bar margined with grey.

Colour in life. Miller's colour sketch of an individual 41·5 mm. standard length shows tube-like snout yellow; two anterior bars orange, outlined by black; third bar on midbody, dark brown or grey, lightly outlined by black; posterior broad orange bar bordered anteriorly by broad brown line extending from orange anal fin to posterior base spinous dorsal, with a black spot at base of soft dorsal fin; an elongate narrow light bar near base caudal fin; posteriorly solid black bar at base caudal rays; pelvic fin orange.

Remarks. Arnhem Land and Philippine specimens (Herre and Montalban, 1927) differ when alive and when preserved in alcohol as follows: in Philippine specimens up to 139 mm., third body bar orange in life remaining prominent in preservation; in Arnhem Land specimens, this body bar prominent, probably orange at 31 mm. (Pl. 50A), brown at 41·5 mm., and scarcely evident above 57 mm. (Pl. 50B). Fifth bar, on caudal peduncle, solid black in Philippine specimens up to 139 mm., but in Arnhem Land material, solid black to approximately 41 mm., thence becoming a cream-coloured bar, margined with black or grey in all specimens 57 to 98 mm.

Philippine specimens have a prominent ocellus in dorsal fin up to 139 mm.; Arnhem Land specimens have a prominent ocellus at 31 mm., the light and dark borders of the ocellus disappear at about 41 mm., the black fades at 57 mm., and the ocellus is scarcely discernible in larger specimens.

These differences may indicate racial divergence between the two regions but not enough material is available to solve the problem at present.

Heniochus acuminatus (Linnaeus)

Pl. 51

Chaetodon acuminatus Linnaeus, *Systema Naturae* ed. 10, 1: 272 (1758); type locality, 'Habitat in Indiis'.

Recorded as: *Heniochus macrolepidotus* Macleay (1878): 352 and (1880): 394, Port Darwin.

M48-26, Sheltered bay, island in Port Bradshaw, USNM 173525, 1 specimen, 79 mm. standard length.

Description. Dorsal rays XI,25; anal rays III,18; pelvic rays I,5, both sides; pectoral rays ii,16, both sides; branched caudal rays 8+7=15; pored scales in lateral line 51; scale rows from upper edge gill opening to caudal fin base 46; scales above lateral line 12; scales below lateral line 26.

Greatest depth 1·2, head length 2·7, and longest pectoral ray 3·0, all in standard length; snout 2·7, tip of snout to posterior end of maxilla 4·1, and tip of snout to rictus 7·5, all in head length.

Dorothea B. Schultz

PLATE 51. *Heniochus acuminatus*
Port Bradshaw, USNM 173525, 79 mm. standard length

Cleithrum rugose, nearly smooth posteriorly; supracleithrum, preoperculum, and circumorbitals serrated along free margin; lower edge preorbital smooth; lateral line continuous to base of caudal fin; fourth dorsal spine long, flexible, nearly as long as fish, 0·9 in standard length.

Colour in alcohol. Background body colour brown anteriorly, yellowish brown posteriorly; dorsal part of snout, anterior to nostrils, brownish black; interorbital with black band; broad black bar from tips of dorsal spines I to III and anterior margin spine IV, through opercle and trunk, thence to abdomen, between anus and pelvic insertion; pectoral base black; broad black bar extending diagonally from tip of dorsal spine V across base of spine XI to posterior half anal fin; remainder anal fin margined with black; pelvic fin black.

Parachaetodon ocellatus (Cuvier and Valenciennes)

Platax ocellatus Cuvier and Valenciennes, *Histoire naturelle des poissons* 7: 229 (1831); type locality not given.

Recorded as: *Chaetodon oligacanthus* Macleay (1878): 351 and (1880): 388; Klunzinger (1879): 360, Port Darwin; *Parachaetodon ocellatus* Ahl (1923): 18, Port Darwin; Paradice and Whitley (1927): 90, Pellew Islands.

Euxiphipops sexstriatus* (Cuvier and Valenciennes)

Holacanthus sexstriatus Cuvier and Valenciennes, *Histoire naturelle des poissons* 7: 194 (1831); type locality, Java.

Recorded as: *Holacanthus sexstriatus* Macleay (1878): 352 and (1880): 395, Port Darwin; *Heteropyge sexstriatus* Fraser-Brunner (1933): 570, Pellew Islands.

Euxiphipops xanthometopon (Bleeker)

Holacanthus xanthometopon Bleeker, *Natuurk. Tijdschr. Ned.-Ind.* 4: 258 (1853); type locality, Telokbetong, Sumatra.

Recorded as: *Heteropyge xanthometopon* Fraser-Brunner (1933): 571, Pellew Islands.

Centropyge bicolor (Bloch)

Chaetodon bicolor Bloch, *Naturgeschichte der Ausländischen Fische* 3: 94, pl. 206, fig. 1 (1787); type localities, both Indies.

Recorded as: *Centropyge bicolor* Fraser-Brunner (1933): 590, Pellew Islands.

Centropyge tibicen (Cuvier and Valenciennes)

Holacanthus tibicen Cuvier and Valenciennes, *Histoire naturelle des poissons* 7: 173 (1831); type locality not given.

Recorded as: *Centropyge tibicen* Fraser-Brunner (1933): 591, Pellew Islands.

Chaetodontoplus duboulayi (Günther)

Holacanthus duboulayi Günther, *Ann. Mag. Nat. Hist.* ser. 3, 20: 67 (1867); type locality, north-west coast of Australia.

Holacanthus darwiniensis Kent, *Proc. Roy. Soc. Qld.* 6: 235 and 239 (1889); type locality, Port Darwin.

* *Euxiphipops* Fraser-Brunner (1934): 192, was proposed to replace *Heteropyge* Fraser-Brunner, a preoccupied name.

Recorded as: *Holacanthus duboulayi* Macleay (1878): 352 and (1880): 395; McCulloch (1913): 360, pl. 14, Port Darwin; *Holacanthus douboulayi* Klunzinger (1879): 361, Port Darwin; *Holacanthus (Chaetodontoplus) duboulayi* Paradice and Whitley (1927): 96, Darwin; *Chaetodontoplus duboulayi* Whitley (1940b): 423, pl. 31, fig. 38, Port Darwin.

M48-22, Bay, Yirrkalla, 16 August 1948, USNM 173524, 1 specimen, 216 mm. standard length.

Description. Dorsal rays XI,22; anal rays III,19; pelvic rays I,5, both sides; pectoral rays ii,15, both sides; branched caudal rays 8+7=15; scales small, irregular, usually with an auxiliary scale; those from upper edge gill opening to caudal fin base approximately 110; scales below lateral line 55; scales above lateral line 22; lateral line not prominent, interrupted near base last dorsal ray, thence continued along middle of caudal peduncle to base caudal fin.

Snout 2·3, tip of snout to posterior end of maxilla 4·7, both in head length; tip of snout to end preopercular spine 3·3, pectoral fin 5·1, body depth 1·9, all in standard length.

Posterior naris slightly larger than anterior naris; preoperculum irregularly and coarsely serrated, its spine shallowly grooved; posterior edge of cleithrum smooth; preorbital smooth, not free posteriorly; interoperculum large, without serrae; upper and lower caudal rays elongated; margin of middle caudal rays convex; pelvic rays reaching anus.

Colour in life. From Miller's notes and colour transparencies: anterior snout, lips, isthmus, and chin yellow; broad greyish blue bar from nape through eye to lower margin preoperculum where interrupted and then continued across breast; eye bluish margined with yellowish; preopercular spine, operculum and area above and below pectoral fin whitish; upper and lower borders caudal fin, pectoral and pelvic fins, and bar from dorsal spine II and nape to belly and posterior breast orange; posteriorly, body deep purplish blue, crossed by approximately 40 longitudinal white lines; lighter purplish blue dorsal fin with submarginal maroon band and white margin; broad, rich orange band along back and base of dorsal fin; bluish anal fin with maroon submarginal band and white margin; tail and caudal peduncle brilliant yellow flecked with orange.

Remarks. The white lines in this specimen are longitudinal whereas those described for the type of *duboulayi* are vertical.

POMACENTRIDAE

Marine, often brightly coloured fishes with a single pair of nares and I,5 pelvic rays; teeth in mouth small, restricted to one or two rows in each jaw; scales ctenoid; tubed lateral line ending high on side below soft dorsal fin; a row of scales along midside of caudal peduncle, each with a small round pit; spinous and soft dorsal fins continuous; anal fin normally with two spines.

Key to the Arnhem Land Genera

1 Opercle and subopercle with numerous radiating ridges or striae, ending posteriorly in narrow spine-like serrae; suborbital and preopercle serrated; dorsal spines IX to XII, usually X or XI; teeth conical, not compressed; tubed scales in lateral line 30 to 45; scales in lateral series 50 to 60 . . . . *Amphiprion*

Opercle and subopercle smooth or relatively so, only the opercle with one or two broad based short spine-like projections along its upper posterior margin; dorsal spines XII to XIV; teeth compressed, incisor-like; tubed scales in lateral line 15 to 23; scales in lateral series 25 to 31 2

2 Free margins of preopercle and suborbitals smooth, or the margin of the preopercle rough, wavy or crenulated, never serrate . . . . *Abudefduf*

Posterior margin of preopercle serrated; anterior suborbital serrated in adults 3

3 Lips normal, nearly smooth . . . . *Pomacentrus*

Lips expanded, plicate and fimbriate, curled back over end of snout; preopercle and suborbital finely serrated . . . . *Cheiloprion*

Key to the Arnhem Land Species of *Abudefduf*

1 Preorbital and anterior suborbitals naked, usually scaled forward to below pupil; scales on snout extending forward approximately to nares; tubed scales in lateral line 19 to 23; anterior teeth in upper jaw uniserial 2

Anterior suborbitals and at least lower portion of preorbital scaled; scales on snout extending forward nearly to upper lip (except in *A. bankieri*) 4

2 Body with distinct dark bars; no prominent black spot on upper side of caudal peduncle 3

Body without distinct dark brown bars, in young a trace of possibly as many as three or four bar-like blotches on upper half; a prominent blackish patch on upper side of caudal peduncle; dorsal rays XII or XIII,14 to 16; anal rays II,14 or 15 . . . . *Abudefduf sordidus*

3 Seven bars, the first downward from nape and the last near base of caudal rays; dorsal rays XIII,14 or 15; anal rays II,13 or 14 . . . . *Abudefduf palmeri*

Five bars, the first near dorsal fin origin and the last across caudal peduncle; dorsal rays XIII,13 or 14; anal rays II,12 or 13 . . . . *Abudefduf saxatilis*

4 Body with five prominent dark brown bars, the first downward from just before dorsal fin, the last across caudal peduncle; both upper and lower lobes of caudal fin with a dark brown band near margin; anterior teeth in upper jaw uniserial; tubed scales in lateral line 18 to 21 . . *Abudefduf sexfasciatus*

Body without prominent dark bars; tubed scales in lateral line 15 to 19; anterior teeth in upper jaw biserial, i.e. larger incisor-like teeth in front more or less alternating with small, more posterior, pointed teeth (the latter tiny and often deciduous in *A. bankieri* and *A. aureus*) 5

5 Dorsal rays XIII,10 to 12 6

Dorsal rays usually XIII,13 to 15; anal rays usually II,13 or 14 7

6 Anal rays II,10 to 12; pectoral rays ii,14 to 16, usually ii,15 or 16; dorsal, anal, and caudal lobes long and pointed, filamentous; posterior part of dorsal and anal fins, and caudal peduncle yellowish, lighter than remainder of body and fins; upper preorbital scaleless . . . . *Abudefduf bankieri*

Anal rays usually II,13; pectoral rays usually ii,13; median fins rounded to slightly pointed behind, not filamentous; posterior trunk and fins not notably lighter than remainder of body; preorbital completely scaled . . . . *Abudefduf aureus*

7 Body and median fins nearly uniformly dark brownish black; pelvic fins black; pectoral rays ii,16 or 17 . . . . *Abudefduf melas*

Belly, upper head, back above lateral line, spinous dorsal fin and anterior soft dorsal fin light yellowish brown (yellow in life); pelvic fin and anterior anal rays black; remainder of body and median fins medium brown; pectoral rays ii,15 to 17, usually ii,16 . . . . *Abudefduf xanthonotus*

A black bar at rear margin of preopercle and black saddle over pectoral fin base; body dark brown; median fins blackish brown; pelvic fins black; pectoral rays ii,14 to 16 . . . . *Abudefduf behni*

Abudefduf palmeri (Ogilby)

Glyphisodon palmeri Ogilby, *Mem. Qld. Mus.* 2: 87, pl. 22, fig. 2 (1913); type locality, Bulwer, Moreton Bay, Australia; McCulloch, *Mem. Qld. Mus.* 7: 172, pl. 10, fig. 1 (1921).

Recorded as: *Glyphisodon palmeri* Paradice and Whitley (1927): 92, Pellew Islands.

M48-1, Nightcliff, USNM 173593, 7 specimens, 31 to 80 mm. standard length.

M48-3, East Point Reef, USNM 173594, 1 specimen, 99 mm. standard length.

M48-9, Reefs, entrance to Little Lagoon, USNM 173595, 3 specimens, 17 to 52 mm. standard length.

M48-9a, Reef, entrance to Little Lagoon, USNM 173596, 22 specimens, 15 to 48 mm. standard length.

M48-17, Reef, Bartalumba Bay, USNM 173597, 10 specimens, 44 to 90 mm. standard length.

M48-21, Reef, Yirrkalla, 14 July 1948, USNM 173598, 14 specimens, 42 to 78 mm. standard length; 1 August 1948, USNM 173600, 5 specimens, 49 to 65 mm. standard length; 11 August 1948, USNM 173599, 12 specimens, 50 to 83 mm. standard length; 12 August 1948, USNM 173601, 31 specimens, 34 to 88 mm. standard length.

M48-26, Sheltered bay, island, Port Bradshaw, USNM 173602, 1 specimen, 55 mm. standard length.

M48-32, Rock pools, Yirrkalla, USNM 173603, 1 specimen, 56 mm. standard length.

Native name: *Mílika*, Yirrkalla.

Description. Variations in dorsal, anal, and pectoral ray counts are given in Table 5; pectoral rays ii,16 to 18; branched caudal rays 7+6=13(in 12); pelvic rays I,5(24); pored scales in lateral line 20(2), 21(3), or 22(7); scales between head and caudal fin 28(3), 29(8), or 30(1), scale rows between dorsal and anal fins 5(6) or 6(4) above lateral line and 12(7) or 13(3) below; scale rows across cheek 4(7) or 5(5); no scales on preorbital, but suborbitals scaled forward to rear margin of eye.

In five specimens 49 to 65 mm. standard length, head length 2·7 to 3·0, pectoral fin 2·7 to 2·9, pelvic fin 2·5 to 2·9, body depth 1·6 to 1·9, all in standard length; eye 2·7 to 3·1, snout 3·2 to 3·8, both in head length.

Scales on top of head extending forward to nares; preorbital and suborbitals anterior to rear margin of eye naked; gular region and lower jaw naked; no serrae on head; teeth in upper jaw uniserial, moderately compressed, the tips truncate, or at most very slightly bicuspidate.

Colour in life. Vertical bars black [either six or seven noted by Miller; the bar at posterior end of head apparently absent, merely greyish or distinctly present in living material; it is evident, but frequently faint in all of the preserved specimens]; body chocolate brown to yellowish, the scale centres yellowish white.

At night, reflections from scale centres over body pale purplish blue to bluish green; spot in pectoral axil vivid Prussian blue; yellowish green overcast to body and fins, except the pectorals and pelvics.

Colour in alcohol. Background colour of body and head medium reddish brown, the scale outlines darker than the centres, lighter on belly; seven dark brown bars approximately half as wide as eye, narrower than the interspaces, on body as follows: first from nape vertically downward through anterior end of lateral line; second from base first dorsal spine to just behind pectoral fin; third from fifth dorsal spine to mid-length of pelvic fin; fourth from ninth dorsal spine almost to anus; fifth from last dorsal spine to anterior anal soft rays; sixth from lower middle of soft dorsal fin to base of posterior half of anal fin; seventh around narrowest part of caudal peduncle, behind fin bases; pectoral fin yellowish, with a dark brown or blackish subtriangular blotch, about half as large as pupil, on upper outer portion of base; other fins medium to dark brown, slightly lighter posteriorly.

Abudefduf saxatilis (Linnaeus)
Pl. 52

Chaetodon saxatilis Linnaeus, *Systema Naturae* ed. 10, 1: 276 (1758); type locality, India.

M48-21, Reef, Yirrkalla, 11 August 1948, USNM 173615, 1 specimen, 52 mm. standard length; 12 August 1948, USNM 173616, 5 specimens, 50 to 94 mm. standard length.

M48-32, Rockpools, Yirrkalla, USNM 173617, 4 specimens, 45 to 73 mm. standard length.

Description. Dorsal, anal, and pectoral ray counts are given in Table 5; pectoral rays ii,16 to 18; branched caudal rays 7+6=13(in 10); pelvic rays I,5(20); tubed scales in lateral line 20(2), 21(5), or 22(3); scales between head and caudal fin 28(3) or 29(7); scale rows between dorsal and anal fins 5(10) above lateral line and 11(4) or 12(6) below; cheek scales in 3(3) or 4(6) rows; no scales on anterior suborbitals; posterior suborbitals scaled to just behind pupil.

In the ten specimens, head length 2·8 to 3·2, pectoral fin 2·8 to 3·2, pelvic fin 2·6 to 3·3, body depth 1·7 to 2·0, all in standard length; eye 2·5 to 3·3, snout 3·4 to 4·3, both in head length.

Scales on top of head extending forward to anterior edge of nares; preorbital, gular region, and lower jaw naked; no serrae on head; teeth in upper jaw uniserial anteriorly, each with two short cusps.

Colour in alcohol. Body, head, and vertical fins generally light reddish brown in larger specimens; fins light yellowish in smaller specimens; scales outlined by darker reddish brown; five dark brown bars about as wide as snout length as follows: one from posterior nape and dorsal origin to pectoral fin base; one from dorsal sheath, below fifth and sixth dorsal spines, to abdomen near middle of depressed pelvic fin; one from sheath below tenth and eleventh dorsal spines to between anus and anal fin; one from sheath below second to fifth soft dorsal rays to midbase of anal fin; and one across caudal peduncle, behind dorsal and anal fins; pectoral fin light yellowish with a small dark spot at outer upper base, the spot larger and more prominent on axillary side; three outer pelvic rays reddish brown, inner rays light yellowish.

Remarks. The background life colour of *Abudefduf sexfasciatus* and this species appears to have been similar as Miller did not distinguish *A. saxatilis* in his notes.

Weber and de Beaufort (1940) have pointed out the confusion pertaining to the type locality of *Chaetodon saxatilis* Linnaeus. I regard that given in the tenth edition of Linnaeus (loc. cit.) as correct.

Characters that distinguish the American and Indo-Pacific populations of *saxatilis* have not yet been worked out. Those given by Weber and de Beaufort (1940) indicate closely related forms.

TABLE 5

FIN RAY COUNTS IN ARNHEM LAND SPECIES OF POMACENTRIDAE

Species	Total Pectoral Rays						Dorsal Spines					Soft Dorsal Rays							Anal Rays				
	15	16	17	18	19	20	X	XI	XII	XIII	XIV	11	12	13	14	15	16	17	II,11	II,12	II,13	II,14	II,15
Abudefduf																							
palmeri				4	18	2				12					10	2					2	10	
sexfasciatus				13	9					11			4	7					1	9	1		
saxatilis				2	16	2				10				9	1					9	1		
sordidus					21	3			3	9					1	6	5					6	6
bankieri				10						5		5							5				
xanthonotus				15	1					8					8						4	4	
behni		1	2	1						2					2						1	1	
aureus	2									1		1									1		
melas					2					1					1						1		
Pomacentrus																							
nigromanus			4							2				1	1							2	
apicalis					5	7				6						6					6		
littoralis		1	4	19						10	2			1	11								12
milleri			2	16						6	101			1	8						1	1	7
darwiniensis			1	9	14					11	1		1		10	1						11	1
Cheiloprion																							
labiatus	1		17	4						11			1	10							7	4	
Amphiprion																							
percula		2	2	2				3							3				1	2			
ephippium				4	15	1	8	2									5	5				9	1
polymnus					2			1							1						1		

Dorothea B. Schultz

PLATE 52. *Abudefduf saxatilis*
Yirrkalla, USNM 173617, 73 mm. standard length

Abudefduf sordidus (Forskål)

Chaetodon sordidus Forskål, *Descriptiones Animalium*: 62 (1775); type locality, Red Sea.

M48-13e, Off Chasm Island, USNM 173633, 2 specimens, 156 and 173 mm. standard length.

Arnhem Land, no further data, USNM 173632, 4 specimens, 51 to 90 mm. standard length.

M48-21, Reef, Yirrkalla, 6 August 1948, USNM 173634, 1 specimen, 63 mm. standard length; 12 August 1948, USNM 173635, 1 specimen, 43 mm. standard length.

M48-32, Point about three-fourths mile east of Yirrkalla, USNM 173636, 66 specimens, 34 to 156 mm. standard length.

Description. Dorsal, anal, and pectoral ray counts are given in Table 5; pectoral rays ii,17 or 18; branched caudal rays 7+6=13(in 12); pelvic rays I,5(24); tubed scales in lateral line 20(1), 21(6), or 22(5); scales between head and caudal fin 28(8) or 29(4); scales between dorsal and anal fins, 5(12) above lateral line and 13(2) or 14(10) below; scale rows across cheek 3(5) or 4(7); no scales on suborbitals. In five specimens 43 to 90 mm. standard length, head length 2·6 to 3·0, pectoral fin 2·8 to 3·3, pelvic fin 2·5 to 2·9, body depth 1·5 to 1·8, all in standard length; eye 2·8 to 3·9, snout 2·8 to 3·5, both in head length.

Scales on top of head extending forward to anterior margin of eye, almost to the nares; preorbital and all but posterior suborbitals naked; lower surface of head, anterior to isthmus, naked; no serrae on head; teeth in upper jaw uniserial, compressed, strongly bicuspid.

Colour in life. Adults dark brown [at 156 mm. standard length], or grey; prominent black spot on upper side of caudal peduncle; about six dark vertical bands.

Colour in alcohol. In small specimens to about 50 mm. standard length, anterior half of spinous dorsal with blackish blotch which sometimes extends on to upper side as one or more irregular blotches; otherwise, body, head, and median fins medium brown, the fins lighter distally; a black saddle, about three-fourths as large as eye, over caudal peduncle and a tiny black spot on upper, outer base of pectoral rays; pectoral rays yellowish; pelvic rays dusky brown to blackish; faint dusky saddle, about one scale in length and extending downward over three or four scales, across nape.

Remarks. One specimen, 65 mm. in standard length from M48-32, is abnormal in having a blackish band along the lowest branched caudal ray.

Abudefduf sexfasciatus (Lacépède)

Pl. 53

Labrus sexfasciatus Lacépède, *Histoire naturelle des poissons* 3: 430 and 477, pl. 19, fig. 2 (1802); type locality, the great gulf of India.

Glyphisodon coelestinus Cuvier and Valenciennes, *Histoire naturelle des poissons* 5: 464, pl. 135 (1830); type locality, Sea of the Indies.

Recorded as: *Glyphidon coelestinus* Macleay (1878): 361, Port Darwin; *Glyphidodon coelestinus* Macleay (1881): 66, Port Darwin.

M48-21, Reef, Yirrkalla, 14 July 1948, USNM 173628, 2 specimens, 42 and 45 mm. standard length; 11 August 1948, USNM 173629, 13 specimens, 39 to 60 mm. standard length; 12 August 1948, USNM 173630, 11 specimens, 39 to 62 mm. standard length.

Dorothea B. Schultz

PLATE 53. *Abudefduf sexfasciatus*
Yirrkalla, USNM 173629, 49 mm. standard length

Description. Dorsal, anal, and pectoral fin ray counts are given in Table 5; pectoral rays ii,16 or ii,17; branched caudal rays 7+6=13(in 11); pelvic rays I,5(22); tubed scales in lateral line 18(1), 19(3), 20(5), or 21(2); scales between head and caudal fin 26(1), 27(4), 28(3), or 29(3); scale rows between dorsal and anal fins 4(7) or 5(4) above lateral line and 10(1), 11(7), or 12(3) below; cheek scales in 5(10) or 6(1) rows, of which one row is on the suborbital; suborbital very narrow, free at its lower edge.

In five specimens, 43 to 60 mm. in standard length, head length 2·7 to 3·1, pectoral fin 2·7 to 3·0, pelvic fin 2·7 to 2·9, and body depth 1·8 to 2·1, all in standard length; eye 2·6 to 2·9, and snout 4·2 to 4·6, both in head length.

Scales on top of head extending forward nearly to the upper lip; preorbital and suborbitals scaled; lower jaw scaled forward nearly to symphysis; gular region naked, or the anterior part with a few scales; no serrae on head; teeth in upper jaw uniserial, those anteriorly each with two short cusps.

Colour in life. Body light brown above, silvery on lower sides and abdomen; head light brown above, silvery tinged with light brown below; vertical bars dark; horizontal bands on upper and lower caudal lobes jet black (Miller's notes and drawing).

Colour in alcohol. Body and head medium to dark brown, lighter below; five dark brown bars, first from just before dorsal fin downward just behind opercle to pectoral base; second from dorsal spines IV to VI downward; third from near posterior base of spinous dorsal downward; next bar from middle of soft dorsal downward and last bar across narrow part of caudal peduncle; dorsal fin otherwise medium brown; anal fin yellowish brown; caudal fin yellowish with darkened lobes, the membrane of the two or three outer branched rays of each lobe black; pelvic fin dusky yellowish; pectoral fin yellowish with a tiny dark saddle across upper base; trace of a dark bar below eye.

Remarks. I have some doubt regarding the application of the trivial name *sexfasciatus* of Lacépède to this species, but follow current practice. I note that he described this fish as having six bands or bars, but on the figure that he gave there are only five which agrees with the present material. His figure and description, however, do not indicate the prominent bands in the caudal, one to each lobe.

Abudefduf melas (Kuhl and van Hasselt)

Glyphisodon melas Kuhl and van Hasselt in Cuvier and Valenciennes, *Histoire naturelle des poissons* 5: 472 (1830); type locality, Java.

M48-21, Reef, Yirrkalla, 11 August 1948, USNM 173631, 1 specimen, 75 mm. standard length.

Description. Dorsal rays XIII,14; anal rays II,13; pectoral rays ii,17; branched caudal rays 7+6=13; pelvic rays I,5, both sides; tubed scales in lateral line 18; scales between head and caudal fin 27; scale rows between dorsal and anal fins, 4 above lateral line plus 12 below lateral line; scale rows on cheek 4 plus 1 on suborbitals.

Head length 3·1, pectoral fin 3·2, pelvic fin 2·7, body depth 1·7, all in standard length; eye 3·4, and snout 3·3, both in head length.

Preorbital and suborbitals covered with scales; scales on top of head and on lower jaw extending forward to lips, those anteriorly very irregular in shape; teeth in upper jaw compressed, an anterior large row alternating with a posterior row of tiny teeth; bones of head without serrae.

Colour in alcohol. Pelvic fin black; pectoral fin light yellowish brown; anal fin almost black anteriorly and near base posteriorly; remainder of fish dark brownish.

Abudefduf xanthonotus (Bleeker)

Pl. 54

Glyphisodon xanthonotus Bleeker, *Natuurk. Tijdschr. Ned.-Ind.* 18: 357 (1859); type locality, Sanlsapura, Bawean.

Paraglyphidodon xanthonotus Bleeker, *Atlas Ichthyologique des Indes Orientales Néêrlandaises*... 9: pl. 404, fig. 3 (1877); Bleeker, *Natuurk. Verh. Holland. Maatsch. Wet.* ser. 3, 2: 118 and 126 (1877).

Chrysiptera xanthonota Aoyagi, *Coral Fishes*: 136, text fig. 39 (1943).

M48-21, Reef, Yirrkalla, 11 August 1948, USNM 173640, 5 specimens, 36 to 51 mm. standard length; 12 August 1948, USNM 173641, 3 specimens, 45 to 47 mm. standard length.

Description. Dorsal, anal, and pectoral ray counts are given in Table 5; pectoral rays ii,16 or 17; branched caudal rays 7+6=13(in 8); pelvic rays I,5(16); tubed scales in lateral line 16(1), 17(3), or 18(4); scales between head and caudal fin base 26(2), 27(4), or 28(2); scales between dorsal and anal fins 4(8) above lateral line plus 10(1) or 11(7) below; scales across cheek 3 to 5, usually 4, plus 1 row on suborbitals.

In five specimens 36 to 51 mm. in standard length, head length 2·9 to 3·1, pectoral fin 3·1 to 3·8, pelvic fin 2·7 to 2·8, body depth 1·8 to 1·9, all in standard length; eye 2·8 to 3·1 and snout 3·4 to 3·8, both in head length.

Scales on top of head extending forward nearly to the lip; preorbital and suborbitals scaled; scales on lower jaw extending forward to symphysis; gular area naked; bones of head without serrae; teeth in upper jaw compressed, close set, in two alternating rows.

Colour in life. Body and head purplish blue; pelvic and anal fins deep blue, the anal lighter blue posteriorly; upper margin of caudal fin, tip of lower caudal lobe, upper back, posterior part of head downward to a level with top of eye, nape, spinous dorsal fin, and anterior portion of soft dorsal mustard yellow; base of pectoral fin yellow (Miller's notes and drawing).

Colour in alcohol. Lower head, lower body, and caudal peduncle medium reddish brown; upper head and area on trunk above lateral line, extending backward and upward to approximate base of last dorsal spine, yellowish to yellowish brown; belly yellowish; outer pelvic fin black, the two inner rays dusky; anterior half and posterior base of anal fin black, the posterior half dusky distally; caudal and pectoral fins transluscent to very light dusky brown; no blotch on pectoral fin base; spinous dorsal and distal half of first two or three soft dorsal rays yellowish; remainder of soft dorsal fin dusky brownish, darker basally and lighter posteriorly.

Remarks. Smith (1960) has figured this species in colour and applied the name *Abudefduf melanopus* (Bleeker) to it. In my opinion, his synonymizing of *Glyphisodon xanthonotus* Bleeker and *G. melanopus* is not justified, since Bleeker (1877a) figured the latter with a light blue band along the dorsal fin base.

Abudefduf behni (Bleeker)

Pl. 55

Glyphisodon behnii Bleeker, *Verh. Batavia Genoot,* 21: 25 (1847); type locality, Batavia, Java.

Glyphisodon xanthurus Bleeker, *Natuurk. Tijdschr. Ned-Ind.* 5: 345 (1853); type locality, Amboina.

M48-21, Reef, Yirrkalla, 14 July 1948, USNM 173637, 1 specimen, 44 mm. standard length; 11 August 1948, USNM 173638, 1 specimen, 42 mm. standard length.

Description. Dorsal rays XIII,14; anal rays II,13 or 14; pectoral rays ii,14 to 16; branched caudal rays 7+6=13(in 2); pelvic rays I,5(4); pored scales in lateral line 17(1); scales between head and caudal fin 26(1); scale rows between dorsal and anal fins, 4(1) above lateral line and 10(1) below; scale rows on cheek 3(2) plus 1 on suborbitals.

Head length 2·9 to 3·0, pectoral fin 3·0 to 3·1, pelvic fin 2·4 to 2·5, body depth 1·9 to 2·0, all in standard length; eye 2·8 to 2·9 and snout 3·8, both in head length.

Upper head and lower jaw scaled forward nearly to lips; suborbitals and lower part of preorbital scaled, but scales not crowded; upper part of preorbital naked; bones of head not serrate; teeth in upper jaw compressed, in two rows, a larger anterior row alternating with the smaller posterior teeth.

Colour in life. Body dark chocolate brown; dark vertical bar along anterior edge of opercle and a conspicuous dark spot at upper pectoral base.

Colour in alcohol. Black bar, half as wide as pupil, along opercle, behind and partly beneath preopercle; pelvic fins and anterior three-fourths of anal fin black; black blotch covering upper half of outer pectoral base; axillary portion of pectoral base dark brown; pectoral fin yellowish; filaments between dorsal spines and possibly distal portion of soft dorsal fin black; remainder of spinous dorsal with a distinct yellowish brown line followed inward by brown similar to that of body; cheek, breast, and lower snout yellowish brown, with a dark brown bar, as wide as eye, downward from eye across cheek; remainder of body and fins dark brown, lighter and with faint stripes along scale rows posteriorly; caudal fin dark brown, the ends of its rays and short streaks midway along its rays blackish.

Abudefduf aureus (Kuhl and van Hasselt)
Pl. 56

Glyphisodon aureus Kuhl and van Hasselt in Cuvier and Valenciennes, *Histoire naturelle des poissons* 5: 472 (1830); type locality, Java.

M48-21, Reef, Yirrkalla, 14 July 1948, USNM 173642, 1 specimen, 46 mm. standard length.

Description. Dorsal rays XIII,11; anal rays II,13; pectoral rays ii,13; branched caudal rays 7+6=13; pelvic rays I,5; tubed scales in lateral line 16; scales between head and caudal fin 27; scale rows between dorsal and anal fins 4 above lateral line plus 9 below; scale rows across cheek 4, plus 1 on suborbitals.

Head length 3·1, pectoral fin 3·1, pelvic fin 2·7, body depth 1·8, all in standard length; eye 2·7 and snout 3·4, both in head length.

Scales on upper head extending forward nearly to the lip, those near the midline and anterior to the nares tiny; preorbital and suborbitals densely scaled anteriorly, the scales less crowded on the posterior suborbitals; gular region naked; scales on lower jaw extending forward to the lip; second anal spine about equal in length to distance from tip of snout to posterior margin of eye; last two dorsal spines about equal in length, the fifth or sixth spine longest; teeth in upper jaw biserial, the larger compressed anterior teeth alternating with tiny rounded posterior teeth; bones of head without serrae.

Dorothea B. Schultz

PLATE 54. *Abudefduf xanthonotus*
Yirrkalla, USNM 173641, 46 mm. standard length

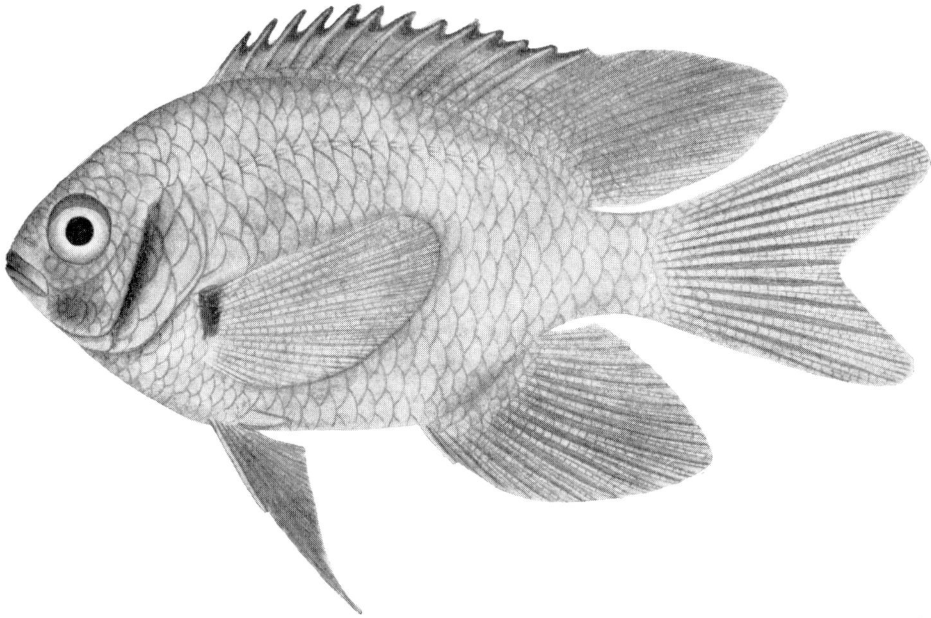

Dorothea B. Schultz

PLATE 55. *Abudefduf behni*
Yirrkalla, USNM 173637, 44 mm. standard length

Colour in life. (Apparently referring to this species.) 'Body deep, almost round, yellowish, with greyish blue scale centres in area behind and above pectoral fin' [middle portion of body and posterior part of head]; basal portions of dorsal, anal, and caudal fins yellowish.

Colour in alcohol. Without prominent bars; medium brown anteriorly on upper parts of side and head; light brownish below and posteriorly; each scale centre on anterior body with a narrow vertical bar; a rather prominent, silvery, triangular mark on lower side between pelvic and anal fins, tending to divide pigment of side into bars; upper surface of caudal peduncle with a medium brown saddle; scales of upper head with faint light centres; dorsal, pectoral, and caudal fins yellowish, the margin of the dorsal fin dusky; pelvic rays dusky; first six soft rays of anal fin dusky, the fin yellowish posteriorly.

Abudefduf bankieri (Richardson)
Pl. 57

Glyphisodon bankieri Richardson, *Rep. Fifteenth Meet. Brit. Ass. Adv. Sci., Ichthyology of China and Japan*: 253 (1846); type locality, Hong Kong.
Glyphisodon nemurus Bleeker, *Natuurk. Tijdschr. Ned.-Ind.* 13: 19, in reprint, (1857); type localities, Kajeli and Amboina, in sea.

M48-21, Reef, Yirrkalla, 14 July 1948, USNM 173639, 5 specimens, 38 to 42 mm. standard length.

Description. Frequencies of dorsal, anal, and pectoral ray counts are given in Table 5; pectoral rays ii,16; branched caudal rays 7+6=13(in 5); pelvic rays I,5(10); gillrakers on first arch 6 or 7 on upper limb, plus 1, plus 13 below; tubed scales in lateral line 15(1), 16(2), or 17(1); scales between head and caudal fin 26(1), 27(1), or 28(2); scale rows between dorsal and anal fins 3 above lateral line and 8 or 9 below; scale rows on cheek 3(3), plus 1 on suborbital; suborbital more or less continuous with cheek, free at its lower edge in one of the specimens.

In the five specimens, head length 3·0 to 3·4, pectoral fin 3·8 to 4·1, pelvic fin 3·3 to 3·7, body depth 2·7 to 3·1, all in standard length; eye 3·1 to 3·6, and snout 4·4 to 4·9 in head length.

Scales on top of head extending forward approximately to the nares; preorbital naked above; suborbitals scaled; lower jaw scaled forward nearly to the symphysis; no serrae on head; the posterior edge of the preopercle slightly wavy or irregular; teeth in upper jaw biserial; those in the posterior row tiny; those of the anterior row truncate, perhaps slightly bicuspidate, large, and close-set; lobes of dorsal, anal, and caudal fins pointed, almost filamentous.

Colour in life. From Miller's notes and drawings. Anterior two-thirds of body dark bluish grey, overlaid by chalky blue spots. Soft dorsal, soft anal, and caudal peduncle from base of soft dorsal to tip of caudal fin bright yellowish orange; peacock blue spots in longitudinal rows along yellow part and on lower side forward to base of pectoral fin; spinous dorsal shown as dark grey.

In a fifth specimen (see Pl. 57) anterior two-thirds of body brown; posteriorly orange rather than yellowish orange; otherwise similar to the other four.

Colour in alcohol. Anterior caudal peduncle, trunk, and head almost uniformly rich medium brown; pectoral fin slightly lighter brown, with a tiny black saddle across upper base; pelvic fin sooty brown, the middle rays blackish distally, the elongated outer filament yellowish distally; anterior two-thirds and posterior base of anal fin medium brown, remainder yellowish; spinous dorsal fin and anterior half and base of soft dorsal fin medium brown, the remainder yellowish; caudal peduncle and caudal fin mostly yellowish, the peduncle with brown streaks along scale centres; one individual with brownish black margin along upper lobe of caudal fin.

Remarks. I find little to differentiate between material identified as *Abudefduf bankieri, A. azysron,* and *A. melanocarpus* by Fowler and Bean (1928) and the present material. In their *bankieri*: dorsal rays XIII,10(in 1), XIII,11(5), and XIV,10(1); anal rays II,10(1) or II,11(7); pectoral rays ii,14(4) or ii,15(12). In the *azysron*: dorsal rays XIII,10(1) or XIII,11(1); anal rays II,11(2); pectoral rays ii,15(2) or ii,16(2). In the *melanocarpus*: dorsal rays XIII,11(3); anal rays II,11(2) or II,12(1); pectoral rays ii,15(1) or ii,16(5). The teeth in the anterior upper jaw are biserial in each of the reported forms. All the specimens have a prominent brownish blotch of pigment in the axil of the pectoral fin; that on the outer side of the fin varies somewhat, and is most prominent in the specimens called *melanocarpus* and only somewhat less so in the specimens reported as *azysron*. The specimens of *bankieri* have round pearly-white marks on the scales; they show no consistent pattern. The fin ray count in this material does not agree with the description of *azysron* by Bleeker; thus that name does not appear to be available for this possible complex. Until further study can be made, I deem it best to refer my material to *Abudefduf bankieri* (Richardson).

It has been observed by me that some material of this complex has been reported by Fowler and Bean (1928) as *Pomacentrus violescens* (Bleeker).

Key to the Arnhem Land Species of *Pomacentrus**

1 Suborbitals and lower part of preorbital scaled; teeth uniserial; dorsal fin with a black spot between second and fourth dorsal spines; body medium brown, with darker oblique streaks through posterior scale margins . . . . *Pomacentrus apicalis*

Suborbitals naked (condition not known in *P. bilineatus*); no black spot near anterior end of dorsal fin 2

2 Suborbitals smooth in young, not or weakly serrated in adults, the anterior serrae at most not better developed than those of preopercle; preorbital scaleless; teeth biserial; anal rays II,14, seldom II,15 3

Suborbitals strongly serrated, the posterior serrae nearly as large as those of preopercle; the two or three anterior serrae enlarged, often fused at base, and separated by a notch from the posterior ones, considerably larger than the serrae of the preopercle; anal rays II,13 to 16, usually II,15 4

3 Anterior suborbital deep and rounded below, the distance from eye to its lower margin greater than from its margin to lower edge of preopercle; fins dusky; a small black saddle on upper base of pectoral fin and a black postcephalic scale (adults) . . . . *Pomacentrus darwiniensis*

Anterior suborbital narrower than its distance from lower edge of preopercle; posterior fins light yellowish; anterior anal rays and broad patch covering outer pectoral base black . . . . *Pomacentrus nigromanus*

4 Fins, especially the caudal, yellowish; dorsal rays XIII,14 or 15; apparently the preorbital naked and teeth uniserial . . . . *Pomacentrus bilineatus*

Teeth biserial; median fins brownish 5

5 A small isolated cluster of scales on the preorbital; body dark brownish black, with a tiny dark saddle on upper pectoral base, a blackish postcephalic scale, and a small cluster of blackish scales on body below anterior end of lateral line; dorsal rays XIII,13 or 14 . . . . *Pomacentrus littoralis*

*Guichenot (1839): 302, designated *Chaetodon pavo* Linnaeus as type species of *Pomacentrus* Lacépède.

Dorothea B. Schultz

PLATE 56. *Abudefduf aureus*
Yirrkalla, USNM 173642, 46 mm. standard length

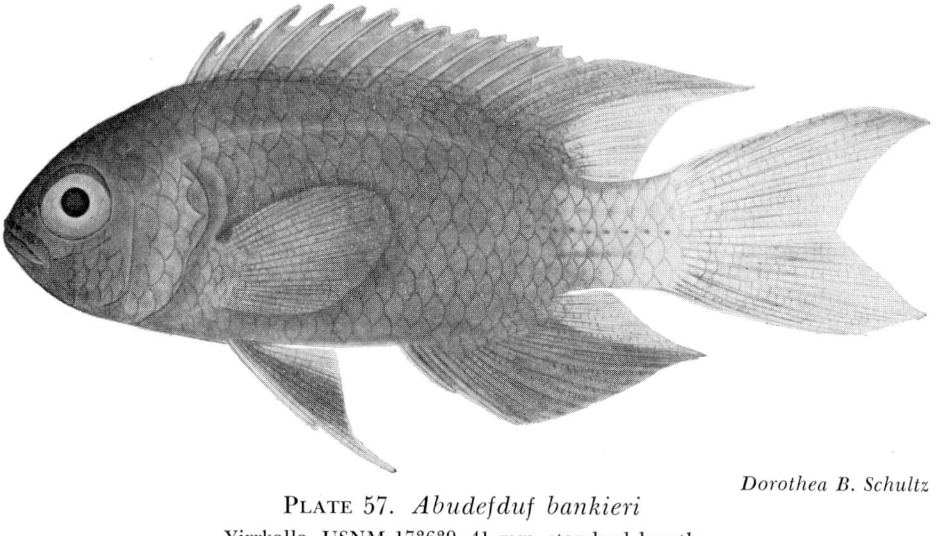

Dorothea B. Schultz

PLATE 57. *Abudefduf bankieri*
Yirrkalla, USNM 173639, 41 mm. standard length

Preorbital scaleless; body medium brown with lighter scale pockets which form faint longitudinal streaks that are most prominent on caudal peduncle; black postcephalic scale; dorsal rays XIV, seldom XIII,13 or 14 . . . . *Pomacentrus milleri*

Indeterminate Northern Territory species records assigned to the genus *Pomacentrus* are: *Pomacentrus bankanensis* Macleay (1878): 361 and (1881): 64, Port Darwin; *Pomacentrus chrysurus* Paradice and Whitley (1927): 96, Melville Island; *Pomacentrus flavicauda* Whitley (1928b): 297, Melville Island.

Pomacentrus nigromanus Weber

Pl. 58

Pomacentrus analis Macleay, *Proc. Linn. Soc. N.S.W.* 7: 364 (1882); type locality, New Guinea; preoccupied in *Pomacentrus analis* Poey.

Pomacentrus insolitus Whitley, *Rec. Aust. Mus.* 16: 221 (1928); substitute name for *Pomacentrus analis* Macleay.

Pomacentrus nigromanus Weber, *Siboga Expedition, Fische* 57: 338, fig. 72 (1913); type locality, Amboina.

M48-21, Reef, Yirrkalla, 14 July 1948, USNM 173646, 2 specimens, 33 to 38 mm. standard length.

Description. Dorsal rays XIII,13(in 1) or XIII,14(1); anal rays II,14(2); pelvic rays I,5(4); branched caudal rays 7+6=13(2); pectoral rays ii,15(4); tubed scales in lateral line 16(1) or 17(1); scale rows between head and caudal fin 27(1) or 28(1); scale rows between dorsal and anal fins 4(2) above lateral line and 10(1) or 11(1) below; scale rows across cheek 3(1); no scales on suborbitals or preorbital; suborbitals very narrow, smooth except anteriorly where rugose, no distinct serrae; rear margin preopercle with moderate serrae; compressed teeth in two alternating rows on jaws.

In the two specimens, head length 3·2 to 3·3, pectoral fin 3·3 to 3·6, pelvic fin 2·8 to 3·3, and body depth 2·3 to 2·4, all in standard length; eye 2·5 to 2·6, and snout 4·6 to 5·0, both in head length.

Colour in alcohol. Head, anterior half of body, and anterior half of spinous dorsal medium brown; caudal fin, rear portion of dorsal fin and posterior half of body light yellowish brown; pelvics dusky, the first ray blackish; pectoral fin yellowish with large obovate blackish blotch over most of outer base; anterior two-thirds of anal fin brownish black, grading to jet black on distal third of rays; posterior third of anal fin yellowish brown.

Colour in life. From Miller's notes: A conspicuous black spot on base of pectoral fin; light yellow on posterior body and dorsal fin forward to about its middle; anterior one-half to two-thirds of anal fin velvet black.

Remarks. This species is close to *Pomacentrus melanopterus* Bleeker, and possibly conspecific, since the coloration of that species seems to vary considerably.

Pomacentrus bilineatus Castelnau

Pomacentrus bilineatus Castelnau, *Proc. Zool. Acclim. Soc. Vict.* 2: 89 (1873); type locality, Port Darwin; Macleay, *Proc. Linn. Soc. N.S.W.* 6: 64 (1881).

Pomacentrus apicalis De Vis

Pl. 59

Pomacentrus apicalis De Vis, *Proc. Linn. Soc. N.S.W.* 9: 874 (1885); type locality, Great Barrier Reef; McCulloch, *Mem. Qld. Mus.* 6: 91, pl. 27 (1918).

Records probably referable: *Pomacentrus scolopsis* Macleay (1878): 361 and (1881): 63, Port Darwin.

Dorothea B. Schultz

PLATE 58. *Pomacentrus nigromanus*
Yirrkalla, USNM 173646, 38 mm. standard length

Dorothea B. Schultz

PLATE 59. *Pomacentrus apicalis*
Yirrkalla, USNM 173645, 85 mm. standard length

M48-21, Reef, Yirrkalla, 14 July 1948, USNM 173643, 4 specimens, 85 to 92 mm. standard length; 11 August 1948, USNM 173644, 1 specimen, 88 mm. standard length; 12 August 1948, USNM 173645, 1 specimen, 85 mm. standard length.

Description. Dorsal rays XIII,15(in 6); anal rays, II,13(6); branched caudal rays 7+6=13(6); pelvic rays I,5(12); pectoral rays ii,17(5), or ii,18(7); tubed scales in lateral line 19(2), or 20(4); scales between head and caudal fin 28(6); scale rows between dorsal and anal fins 4(6) above lateral line and 12(5), or 13(1) below; scale rows across cheek 3(1), or 4(5) plus 1 row on suborbitals, the scales extending on to lower part of preorbital; preorbital naked above; suborbitals with serrae about as large as those of preopercle, slightly larger anteriorly, the bones broad, moderately deep behind and about as wide as a pupil diameter anteriorly; compressed teeth in a single row on each jaw.

In five specimens, head length 3·0 to 3·4, pectoral fin 3·5 to 3·9, pelvic fin 2·8 to 3·0, and body depth 1·8 to 2·0, all in standard length; eye 3·7 to 4·1 and snout 3·1 to 3·6, both in head length.

Colour in alcohol. Body and head generally medium to dark brown; the trunk and caudal peduncle with narrow vertical bluish black lines through posterior margin of each scale row; vertical fins darker brown, their soft portions becoming pale toward margin; spinous dorsal with black tipped filaments and with a black blotch, nearly as large as eye, between the first to fourth spines; pelvic fin nearly black; pectoral fin light greyish brown, its upper base with a small blackish saddle which is barely evident on outer side and only slightly more prominent on axial side; upper head, snout and middle of opercle with brownish black cast.

Remarks. Similarity of colour pattern of *apicalis* and *Pomacentrus scolopsis* probably accounts for the record of the latter species from Port Darwin, although the two differ in colour details and prominently in the number of dorsal spines.

The Arnhem Land material agrees well with the description of *apicalis* as given by McCulloch. There remains some uncertainty about its belonging to the form originally described by De Vis.

The relationship of *apicalis* seems to be with *Pomacentrus jenkinsi* Jordan and Evermann. That species differs by averaging one more soft dorsal ray, by having fewer diagonal scale rows across the cheek, by the presence of a dark margin to the dorsal fin, and by generally lacking a dark spot between the second to fourth dorsal spines.

The name *Pomacentrus obreptus* Whitley (1948) may also apply to this species, but the colour description is insufficient. As noted above, *Pomacentrus scolopsis* Günther is another species.

Pomacentrus littoralis Kuhl and van Hasselt

Pomacentrus littoralis Kuhl and van Hasselt in Cuvier and Valenciennes, *Histoire naturelle des poissons* 5: 425 (1830); type locality, Java.

Recorded as: *Pomacentrus littoralis* Günther (1862): 32, Port Essington; Klunzinger (1879): 397, Port Darwin; Macleay (1878): 361 and (1881): 62.

M48-3, East Point Reef, USNM 173604, 8 specimens, 37 to 79 mm. standard length.

M48-9a, Reef, entrance to Little Lagoon, USNM 173605, 7 specimens, 16 to 49 mm. standard length.

M48-17, Reef, Bartalumba Bay, USNM 173606, 2 specimens, 53 and 54 mm. standard length.

M48-21, Reef, Yirrkalla, 14 July 1948, USNM 173607, 34 specimens, 34 to 79 mm. standard length; 6 August 1948, USNM 173608, 10 specimens, 41 to 73 mm. standard length; 11 August 1948, USNM 173609, 2 specimens, 71 and 76 mm. standard length; 12 August 1948, USNM 173610, 21 specimens, 36 to 82 mm. standard length.

Native name: *Mílika,* Yirrkalla.

Description. Variation in dorsal, anal, and pectoral rays is given in Table 5; pectoral rays ii,15 or ii,16; dorsal spines XIII(in 14) or XIV(1); branched caudal rays 7+6=13(9) or 7+7=14(1); pelvic rays I,5(20); tubed scales in lateral line 16(1), 17(1), 18(4), or 19(4); scale rows between head and caudal fin 26(1), 27(8), or 28(1); scale rows between dorsal and anal fins, 4(10) above lateral line, plus 9(1) or 10(9) below; scale rows across cheek 3(5) or 4(5); infraorbital scaleless except to rear of eye; a small, isolated patch of scales on preorbital in all adults, which characteristic of this species; teeth biserial, compressed, the outer slightly larger teeth alternating with more slender, adjacent, inner teeth; posterior margin of preopercle and lower margin of narrow suborbitals with prominent serrae, those of the suborbital series largest anteriorly where one is greatly enlarged and spinelike, sometimes divided, its tip pointing backward and upward.

In five specimens 51 to 77 mm. standard length, head length 2·9 to 3·3, pectoral fin 3·2 to 3·6, pelvic fin 2·8 to 3·2, and body depth 1·7 to 1·9, all in standard length; eye 3·3 to 3·7 and snout length 3·4 to 3·7, both in head length.

Colour in alcohol. Almost uniformly dark brownish black, with a metallic cast; scale centres indistinctly dark reddish brown; tiny saddle at upper base of pectoral fin, centre of scale at upper rear corner of opercle and a cluster of one to five scales below anterior end of lateral line blackish, darker than body colour; distal half of spinous dorsal fin with a narrow faint longitudinal light line, followed outwardly by dark brown or black on tips of dorsal filaments; upper oral valve and tip of tongue dusky to black.

Colour in life. From Miller's notes: All of body jet black, with turquoise centre to scale at apex of opercle and deep blue reflection from about three scales behind that one; young like adult, but with very faint light blue margin to dorsal, anal and pelvic fins.

Remarks. The tiny specimens, 16 to 18 mm. in standard length, in collection M48-9a, are tentatively identified as *Pomacentrus littoralis.* They are similar in body shape and agree in fin ray counts with *littoralis*; however, the suborbital has not yet become serrated and the patch of preorbital scales is not evident. Presumably, both characters are not yet developed. The colour in alcohol is of a medium dark brown, lighter posteriorly; the caudal fin is yellowish; the spinous dorsal is narrowly edged with black; at the junction of the spinous and soft dorsal fins there is a black spot as large as the eye. Miller's notes and colour sketch indicate that in life, the entire body was brilliant peacock blue; metallic cobalt blue bands the black dorsal spot, forms two continuous blue lines from the snout through the eye, and lines of small blue spots along the body, head and spinous dorsal fin.

Pomacentrus milleri sp. nov.

Pl. 60

Previous published records, probably of this species: *Pomacentrus trilineatus* Klunzinger (1879): 396 and Macleay (1884): 44, Port Darwin; *Pomacentrus wardi* Whitley (1927a): 301 and (1927b): 18, Cape Wessel; Paradice and Whitley (1927): 92 and 98, Cape Wessel and Darwin.

Holotype: Australian Museum register No. IB.5235, 71·0 mm. in standard length, taken at station M48-21, on the coral reef at Yirrkalla, near Cape Arnhem, Northern Territory, 14 July 1948, by Robert R. Miller and party.

Paratypes: M48-3, East Point Reef, about 6 miles north-north-west of Darwin, Northern Territory, 26 March 1948, USNM 173623, 17 specimens, 33 to 66 mm. standard length.

M48-21, Coral reef at Yirrkalla (topotypical), near Cape Arnhem, Northern Territory, 14 July 1948, taken with holotype, USNM 173624, 94 specimens, 36 to 73 mm. standard length; 6 August 1948, USNM 173625, 9 specimens, 48 to 68 mm. standard length; 11 August 1948, USNM 173626, 2 specimens, 54 and 64 mm. standard length; 12 August 1948, USNM 173627, 52 specimens, 37 to 73 mm. standard length.

Diagnosis. A species of *Pomacentrus* with normally XIV,13 or 14 dorsal rays and a relatively plain coloration; scale centres light in preservation, and apparently blue or purple in life, usually forming faint longitudinal streaks; suborbitals relatively narrow, serrated, the anterior one with a large backward directed, generally sharp, spine; teeth biserial.

Description of holotype. Dorsal rays XIV,13; anal rays II,15; pelvic rays I,5, both sides; pectoral rays ii,16, both sides; branched caudal rays 7+6=13; tubed scales in lateral line 19; scales between head and caudal fin 27; a line of 8 scales along caudal peduncle, each with a pit, but the last on body without a pit; scale rows across cheek 4; scale rows across body 3 above lateral line and 11 below. Measurements of the holotype and of paratypes are given in Table 6.

TABLE 6

MEASUREMENTS OF *Pomacentrus milleri* EXPRESSED IN THOUSANDTHS OF THE STANDARD LENGTH

	Holotype Australian Museum Register No. IB.5235	Paratypes USNM 173624	
Standard length, mm.	71·0	72·5	73·0
Depth of body	500	484	503
Length of head	320	320	319
Length of snout	103	102	99
Length of eye	93	88	90
Width of preorbital	42	46	47
Length of upper jaw	90	98	95
Postorbital length of head	162	161	158
Interorbital width	104	113	107
Depth of caudal peduncle	145	143	144
Length of pectoral fin	280	277	290
Length of pelvic fin	331	310	318
Length of caudal peduncle	142	153	155
Length of upper caudal rays	331	313	316
Length of lower caudal rays	285	259	274

Description. Variation in dorsal, anal, and pectoral rays is given in Table 5; pectoral rays ii,15 or ii,16; dorsal spines, including specimens from both localities, XIII(in 6) or XIV(101); branched caudal rays 7+6=13(9); pelvic rays I,5(18); tubed scales in lateral line 18(1) or 19(8); scale rows between head and caudal fin 27(7) or 28(2); scale rows between dorsal and anal fins 3(4) or 4(5) above lateral line plus 10(1) or 11(8) below; scale rows across cheek 4(8) or 5(1); no scales on preorbital or suborbitals in front of posterior margin of eye; suborbitals moderately narrow posteriorly, broader in front, the rear ones irregularly serrated; some of serrae large as those of preopercle, the front ones forming a large backward projection which sometimes blunt, but usually resembling a broad, upturned spine; teeth compressed, in two adjacent rows on each jaw, the outer teeth slightly larger than the inner ones and alternating with them.

In six specimens, including the holotype, 60 to 73 mm. in standard length, head length 3·1 to 3·4, pectoral fin 3·3 to 3·8, pelvic fin 3·1 to 3·5, and body depth 2·0 to 2·2, all in standard length; eye 3·2 to 3·8 and snout 3·0 to 3·5, both in head length.

Colour in alcohol. Almost entirely medium brown to dark reddish brown; never nearly black as in *littoralis*; scale margins on body outlined in darker brown, usually giving the effect of faint longitudinal streaks; fins perhaps slightly darker than body; ends of filaments betwen dorsal spines sooty brown, inward from which is a narrow subterminal yellowish brown streak; no dark spot at base of pectoral; scale at upper rear corner of opercle with black centre; inside of mouth and tongue with little or no pigment. Small specimens to about 50 mm. in standard length often with black ocellus, as large as pupil, narrowly margined with light brown, near middle of soft dorsal fin.

Colour in life. From Miller's notes: Body brownish to reddish brown, with peacock blue to purplish spots in longitudinal rows on scale centres along side below lateral line and behind pectoral fin, the spots running out on to anal fin base; anal and pelvic fins margined with chalky blue; black spot at apex of opercle and in axil of pectoral fins; two rows of peacock blue spots between eye and mouth, two rows along outline of opercle and a row on each side of top of head. Young with sky blue, narrow lines as follows: two on top of head, two through eye, two below eye, and two broadly spaced rows of pale blue spots on opercle; spot at opercle black; middle rays of soft dorsal fin with velvet black spot, ringed with light blue.

Remarks. On the basis of the characteristic dorsal ray counts given by Whitley, I identify his specimens of *Pomacentrus wardi* from Cape Wessel and Darwin as *Pomacentrus milleri.*

This species is named *milleri* in honour of Dr Robert Rush Miller who made the Arnhem Land fish collections.

Pomacentrus darwiniensis Whitley

Pl. 61

Dascyllus fasciatus Macleay, *Proc. Linn. Soc. N.S.W.* 2: 361, pl. 10, fig. 2 (1878) and 6: 62 (1881); type locality, Port Darwin; name preoccupied in *Pomacentrus fasciatus* Cuvier and Valenciennes.

Pomacentrus darwiniensis Whitley, *Rec. Aust. Mus.* 16: 297 (1928); substitute name for *Dascyllus fasciatus* Macleay.

Records probably referable: *Pomacentrus cyanospilus* Klunzinger (1879): 397 and Macleay (1884): 44, Port Darwin; *Pomacentrus fasciatus* Klunzinger (1879): 397 and Macleay (1884): 45, Port Darwin.

M48-3, East Point Reef, USNM 173618, 5 specimens, 52 to 76 mm. standard length.

M48-9a, Reef, entrance to Little Lagoon, USNM 173619, 1 specimen, 35 mm. standard length.

M48-17, Reef, Bartalumba Bay, USNM 173620, 27 specimens, 30 to 86 mm. standard length.

M48-21, Reef, Yirrkalla, 14 July 1948, USNM 173621, 10 specimens, 23 to 80 mm. standard length; 12 August 1948, USNM 173622, 10 specimens, 39 to 83 mm. standard length.

Description. Variation in dorsal, anal, and pectoral ray counts is given in Table 5; pectoral rays ii,15 to ii,17; branched caudal rays 7+6=13(in 11), or 7+7=14(1); pelvic rays I,5(24); tubed scales in lateral line 17(7) or 18(5); scales between head and caudal fin 27(10) or 28(2); scale rows between dorsal and anal fins 4(5) or 5(7) above lateral line plus 10(4) or 11(8) below; scale rows across cheek 4(2), 5(9), or 6(1); no scales on suborbitals or preorbital; preopercle with moderate serrae; suborbitals smooth in young and with small irregular serrae in adults, the chain of bones with almost straight lower or posterior margin, moderately deep posteriorly, and much broader anteriorly, where the width approaches the eye diameter; teeth compressed, in two close-set alternating rows.

In five specimens 52 to 76 mm. standard length, head length 3·1 to 3·4, pectoral fin 3·4 to 3·8, pelvic fin 3·0 to 3·5 and body depth 2·0 to 2·3, all in standard length; eye 3·2 to 3·7 and snout 2·8 to 3·5, both in head length.

Colour in alcohol. In larger specimens, body, head, and paired fins medium to dark brown; scales light centred, forming faint longitudinal streaks; anal fin slightly darker with an arcuate faint line through its middle and a very narrow blackish margin; caudal fin dark brown to brownish black; spinous dorsal fin and anterior soft dorsal fin medium brown with narrow light line just below blackish margin; soft dorsal dark brown, its centre blackish with a distinct black blotch upward from base of posterior rays in smaller specimens; tiny saddle over upper base of pectoral fin and scale centre at upper rear corner of opercle black; traces of about three light lines from anterior and lower margin of eye across suborbitals and preorbital to upper lip; anal region black; smaller individuals with faint bars, intermediate between the adult colour pattern and that of the young.

In young, 30 to 37 mm. in standard length, body and head with light yellowish to reddish brown bars alternating with distinct, broader, dark brown bars; light bars as follows: first, about three-fourths eye width, from anterior nape across opercle and preopercle; another, about half eye width, from anterior dorsal spine to side behind pectoral fin; another, slightly wider, from rear edge of first black blotch in dorsal fin to anterior half of anal fin; another beginning around posterior black dorsal blotch and crossing caudal peduncle to rear of anal fin; last dark bar, almost as wide as eye, abruptly terminated near base of caudal rays; caudal fin light yellowish brown; two black blotches in dorsal fin: first, size of eye, between seventh and twelfth dorsal spines, ringed by light yellow; second bar-like, wide as pupil and height of eye, surrounded by light yellow, extending from middle of soft dorsal, through posterior dorsal base on to caudal peduncle; general pigment pattern otherwise similar to that in adults, the fins perhaps lighter, more yellowish to reddish brown.

Colour in life. From Miller's notes and drawings: In 35 mm. specimen. Black blotch, large as eye, at middle of spinous dorsal fin and another, bar-like and nearly as large, at posterior end of soft dorsal fin, both bordered anteriorly by bright yellow; upper margin of pectoral base and scale at upper corner of opercle black; pectoral fin yellow; bluish line from eye to middle of lip, and another just

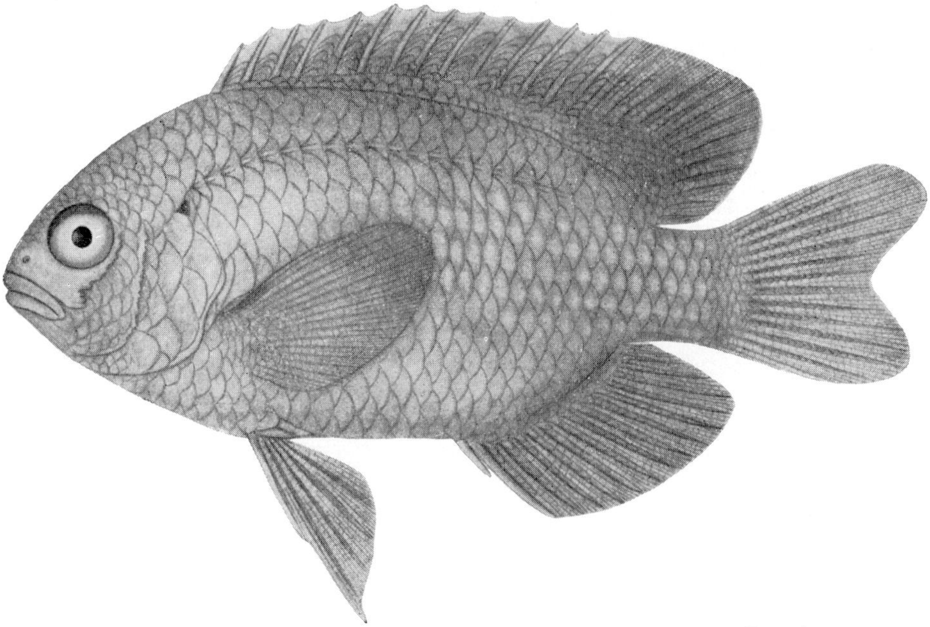

Dorothea B. Schultz

PLATE 60. *Pomacentrus milleri* sp. nov., holotype
Yirrkalla, Australian Museum register no. IB.5235, 71 mm. standard length

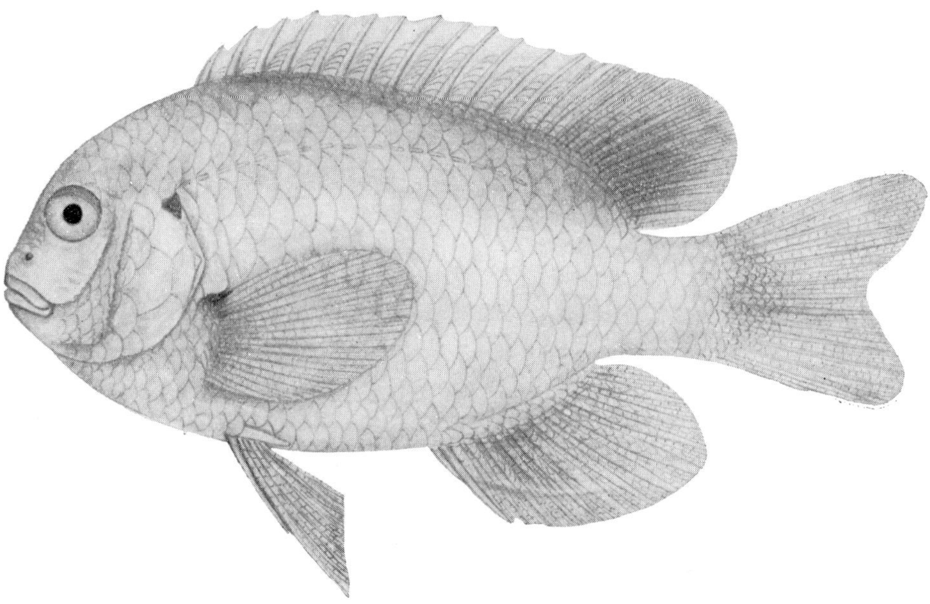

Dorothea B. Schultz

PLATE 61. *Pomacentrus darwiniensis*
Yirrkalla, USNM 173622, 76 mm. standard length

below eye; lower head, area above and before pectoral fin, and lower midside dotted with bluish.

In larger specimens, body yellowish or olive, grading to grey posteriorly; black spot at upper corner of opercle and at upper margin of pectoral fin base; two peacock blue stripes from eye to upper jaw and dots of same colour on opercle, around base of pectoral fin, and on side above origin of anal fin; basal half of pectoral fin yellowish to faintly olive green, greyish to white distally; borders of dorsal, anal and caudal fins olive green, the fins greyish basally.

Remarks. The fish here described belongs to the group regarded by some as the genus *Dischistodus* Gill, characterized by a deep, broad preorbital, which in the young is shallow and has the lower margin smooth. In large specimens, a few serrae are present.

Originally described as a species of *Dascyllus, darwiniensis* has been regarded as a *Pomacentrus* by Whitley. Colour pattern in the present young and counts of some specimens agree with the descriptions of the type given by Macleay and Whitley. As is obvious from the series available for this study, the dark barred colour pattern of the young changes to a more uniform yellowish pattern in adults.

It is likely that the Port Darwin records of *Pomacentrus cyanospilus* and *Pomacentrus fasciatus* given by Klunzinger are colour stages of this species; the counts listed by Macleay for them are compiled and do not agree with those of Klunzinger.

Cheiloprion labiatus (Day)

Pl. 62

Pomacentrus labiatus Day, *Fishes of India*: 384, pl. 81, fig. 2 (1877); type localities, Andamans and Nicobars.

M48-21, Reef, Yirrkalla, 14 July 1948, USNM 173611, 11 specimens, 32 to 72 mm. standard length; 6 August 1948, USNM 173612, 2 specimens, 43 and 61 mm. standard length; 11 August 1948, USNM 173613, 21 specimens, 40 to 83 mm. standard length; 12 August 1948, USNM 173614, 4 specimens, 31 to 66 mm. standard length.

Description. Variation in dorsal, anal, and pectoral fin ray counts is given in Table 5; pectoral rays ii,13 to 16; branched caudal rays 7+6=13(in 8) or 7+7=14(3); pelvic rays I,5(22); tubed scales in lateral line 17(4) or 18(6); scales between head and caudal fin 26(2), 27(7), or 28(1); scales between dorsal fin and lateral line 3 or 4; scales between lateral line and anal fin 10; scales across cheek 3 to 5, usually 4; no scales on suborbital.

In five specimens, 60 to 72 mm. standard length, head length 2·8 to 3·1, pectoral fin 3·5 to 3·8, pelvic fin 3·5 to 3·9, body depth 1·7 to 1·9, all in standard length; eye 3·4 to 4·1 and snout length 3·4 to 3·8, both in head length.

Scales forming rather dense sheaths on vertical fins; those on top of head extending forward to anterior margin of nares, and those on lower jaw nearly to lip; preorbital naked; suborbital and posterior margin of preopercle weakly serrate; teeth fine, compressed, closely-set, in two alternating rows in both jaws; lips coarsely plicate, fimbriate, greatly expanded and folded back over snout.

Colour in life. Body deep brown anteriorly, grading to blackish posteriorly; bluish reflections on scale centres on sides; a faint black crescent at top of opercle; belly reddish brown. Young with a row of peacock blue spots between eye and mid-dorsal base on each side of body; otherwise like adults (from Miller's notes).

Colour in alcohol. Almost uniformly medium reddish brown to dark brown; scales outlined by darker or lighter brown on body, and by dark brown on caudal peduncle where faint longitudinal light lines are evident in small individuals; pectoral fin and posterior ends of vertical fins light yellowish brown; pelvic fins blackish; lips greyish; scale at upper posterior corner of opercle black centred.

Key to the Arnhem Land Species of *Amphiprion*

1 Breast, lower belly and side, and fins light yellowish, the fins at most narrowly margined with black; body and head dark brownish black, crossed by three pale bars in young and but one on head in adults; dorsal rays X or XI,16 to 18; anal rays II,13 to 15; total pectoral rays 18 to 20 . . . . *Amphiprion ephippium*

Dark pigment of body extending on to median fins; soft dorsal, anal, and caudal fins light margined or the caudal very narrowly margined with brown outside of the whitish area in *polymnus*; distal half or more of pelvic fins black; dorsal rays X, usually XI (in *percula* usually IX or X in eastern portion of range),13 to 16; anal rays II,10 to 13 2

2 Body and head dark brown with two light bars, one downward from nape through opercle and preopercle; the other from junction of dorsal fins nearly to anus; dark brown of caudal peduncle extending on caudal fin, nearly to end, as a lanceolate shaped patch; total pectoral rays 19 . . . . *Amphiprion polymnus*

Body and head medium to light brown with three light bars, each bordered with black; first bar through head; second bar, with an anterior projection beneath pectoral fin, from junction of dorsal fins to anus; third bar across caudal peduncle; caudal and pectoral fins with an obovate, black bordered, light to medium brownish patch; posterior margins of both fins pale; total pectoral rays 16 to 18 . . . . *Amphiprion percula*

In the fishes of the genus *Amphiprion* the preorbital, suborbital, preopercle, subopercle, and opercle are serrated. In the Arnhem Land material studied, often only one of the upper, outer pectoral rays is simple, and there are 13 to 15 branched caudal rays. In specimens studied from other areas, the branched caudal ray count is often 7+7=14 and frequently 8+7=15, suggesting considerable variation, and indicating further differences between this genus and the other Arnhem Land pomacentrids, all of which normally possess 7+6=13 branched caudal rays.

Amphiprion ephippium (Bloch)

Lutjanus ephippium Bloch, *Naturegschichte der Ausländischen Fische* 4: 121 (1790); reference not seen; Bloch, *Ichthyologie* 7: 98, pl. 250, fig. 2 (1797); type locality, East Indies.

Amphiprion tricolor Günther, *Catalogue of the Fishes in the British Museum* 4: 8 (1862); type locality, Port Essington, Northern Territory, Australia.

Amphiprion ruppellii Castelnau, *Proc. Zool. Acclim. Soc. Vict.* 2: 91 (1873); type locality, Port Darwin, Australia; Macleay, *Proc. Linn. Soc. N.S.W.* 6: 59 (1881).

Recorded as: *Amphiprion ephippium* Schultz (1953): 199, pl. 10, figs. B and C, incorrectly listed from 'Woods Inlet, west of Darwin, Australia'; see below; *Amphiprion tricolor* Macleay (1878): 360; Klunzinger (1879): 396; Macleay (1881): 59; Kent (1889): 240; Paradice and Whitley (1927): 92 and 98, pl. 13, Port Darwin.

M48-3, East Point Reef, USNM 173649, 10 specimens, 20 to 54 mm. standard length.

Locality unknown; probably East Point Reef (M48-3) according to letter to me from R. R. Miller; incorrectly published, see synonymy, as Woods Inlet west of

Darwin, Australia; 1948, USNM 152777, 55 specimens, 17 to 89 mm. standard length.

M48-9, Reef, entrance to Little Lagoon, USNM 174182, 1 specimen, 14 mm. standard length.

M48-9a, Reef, entrance to Little Lagoon, USNM 173650, 2 specimens, 51 and 57 mm. standard length.

M48-21, Reef, Yirrkalla, 11 August 1948, USNM 173651, 3 specimens, 35 to 36 mm. standard length; 12 August 1948, USNM 173652, 2 specimens, 36 and 57 mm. standard length.

Native name: *Tíndoróo,* Yirrkalla.

Description. Frequencies of dorsal, anal, and pectoral ray counts are given in Table 5; pectoral rays ii,16 to 18; branched caudal rays 7+6=13(in 1), 7+7=14(7), or 8+7=15(2); pelvic rays I,5(20); tubed scales in lateral line 32 to 39, usually 37 or 38; scales between head and caudal base 50 to 59; scale rows between dorsal fin and lateral line 7, usually 8; scale rows between lateral line and anal fin 19 to 22; scale rows across cheek 4, usually 5 or 6; no scales on suborbital.

In six specimens 55 to 89 mm. standard length, head length 3·0 to 3·3, pectoral fin 3·6 to 4·1, pelvic fin 3·2 to 4·1, body depth 1·7 to 1·9, all in standard length; eye 3·1 to 4·2 and snout 3·0 to 3·5, both in head length.

Teeth in a single row in upper and lower jaw, large, bluntly pointed and rounded, largest forward; scales on top of head extending forward to middle of eye; mandible naked; preorbital with a few blunt spine-like serrae; suborbitals, preopercle, subopercle and opercle with prominent serrae.

Colour in life. From Miller's notes and colour transparencies. In larger specimens, a single chalky blue bar from dorsal origin through opercle to base of opercle; black behind to caudal base, including side of body, lower portion of dorsal fin, anterior margin of pelvic fin and upper head; end of snout greyish orange; lips, lower head, breast, belly and remainder of fins deep orange; fins grading to yellowish orange toward margin.

In young, a pale blue bar extending from dorsal fin to midside, and often a blue saddle on caudal peduncle; otherwise like larger specimens.

Colour in alcohol. Head and body dark brownish to blackish, both becoming lighter below and on caudal peduncle where yellowish; extreme base of dorsal fin blackish; fins otherwise yellowish, sometimes with faint narrow black margins; area over pelvic spine prominently black in adults; a light brown bar wider than eye, from just before dorsal fin passing downward behind eye to subopercle; in young an additional one or two similar bars, first from region of dorsal fin to that of anus, and second, in smallest specimens, across caudal peduncle.

Remarks. As pointed out by Schultz (1953) the young of this species have two or three bars that appear to be lost with age, the colour pattern of the young resembling that of the adult of *Amphiprion bicinctus* Rüppell. In the latter species, however, the caudal fin is emarginate or deeply concave posteriorly, the lobes tending to be pointed, with the longest rays sometimes filamentous; the snout is yellowish; and the scales of the nape extend forward only to the rear portion of the eye. In *ephippium,* the caudal fin is convex, becoming slightly concave in the middle in adults, but the lobes remain rounded; the scales of the nape extend forward to above the middle of the eye; and the upper snout is dusky to brownish.

The records listed by Macleay (1878 and 1881) from Port Darwin as *Amphiprion clarkii, Amphiprion bicinctus* and *Amphiprion melanopus* are probably based on colour changes in this species, due to age, as described above.

Amphiprion polymnus (Linnaeus)

Perca polymna Linnaeus, *Systema Naturae* ed. 10, 1: 291 (1758); type locality, 'Habitat in Indiis'.

Recorded as: *Amphiprion polymnus* Schultz (1953): 194, present record.

Submerged canoe on sand bar, Little Lagoon, north-east end of Groote Eylandt, Northern Territory, 16 June 1948, USNM 165534, 1 specimen, 86 mm. standard length.

Description. Dorsal rays XI,14; anal rays II,13; branched caudal rays 7+7=14; pelvic rays I,5, both sides; pectoral rays ii,17, both sides; tubed scales in lateral line 41; scales between head and caudal fin 54; scales between dorsal fin and lateral line 8; scales from lateral line to anal fin 20; scale rows across cheek 5; no scales on suborbital.

Head length 3·2, pectoral fin 4·6, pelvic fin 4·0, body depth 2·2, all in standard length; eye 4·2 and snout 3·2, both in head length.

Teeth in a single row in upper and lower jaw, large bluntly pointed to rounded, largest forward; scales above extending forward only to occipital region; lower jaw naked; preorbital with large blunt serrae; preopercle with small serrae; rather large serrae on suborbitals, subopercle and opercle.

Colour in life. Body velvet black; bars coral pink with narrow emerald borders.

Colour in alcohol. Body and head deep chocolate brown with two light yellow bars, the chocolate brown extending on to caudal fin as lance-shaped area; first light bar about two times width of eye, from nape through preopercle and anterior portion of opercle to subopercle; second bar more than two times width of eye above, extending from rear portion spinous dorsal and anterior half soft dorsal downward, but not quite reaching anus; all of pelvic fins, all but rear portion spinous dorsal fin, basal half of posterior part of soft dorsal, and basal two-thirds of anal fin chocolate brown, the remainder of these fins yellowish; basal third of pectoral fin chocolate brown, grading to yellowish distally; trace of brown along upper and lower margin of caudal lobes, otherwise that fin yellowish except for the chocolate brown central area.

Amphiprion percula (Lacépède)

Lutjanus percula Lacépède, *Histoire naturelle des poissons* 4: 194, 239 and 240 (1802); type locality, Port Praslin, New Britain Island.

Amphiprion bicolor Castelnau, *Proc. Zool. Acclim. Soc. Vict.* 2: 92 (1873); type locality, Port Darwin, Australia; Macleay, *Proc. Linn. Soc. N.S.W.* 2: 361 (1878) and 6: 60 (1881).

Recorded as: *Amphiprion percula* Klunzinger (1879): 396; Whitley (1927b): 24, Port Darwin; *Actinicola bicolor* Whitley (1929b): 214-16, pl. 27, fig. 2, Port Darwin.

North coast of Groote Eylandt, April 1948, USNM 165535, 1 specimen, 54 mm. standard length.

M48-21, Reef, Yirrkalla, 11 August 1948, USNM 173647, 1 specimen, 55 mm. standard length; 12 August 1948, USNM 173648, 1 specimen, 49 mm. standard length.

Description. Frequencies of dorsal, anal, and pectoral ray counts are given in Table 5; pectoral rays i or ii,14 to 16; branched caudal rays 7+6=13(in 1) or 7+7=14(1); pelvic rays I,5(6); tubed scales in lateral line 33 to 38; scales between head and caudal fin 56 to 58; scale rows between dorsal fin and lateral line 7 or 8;

scales between anal fin and lateral line 24 to 26; scale rows across cheek 4 or 5; no scales on suborbital.

Head length 2·9 to 3·1, pectoral fin 3·6 to 4·1, pelvic fin 4·3 to 4·7 and body depth 2·1 to 2·4, all in standard length; eye 4·2 to 5·0 and snout 3·1 to 3·2, both in head length.

Large, bluntly pointed, slightly compressed teeth in a single row in each jaw; scales above extending forward only to occipital region; lower jaw naked; serrae small and blunt on preopercle; those of preorbital, suborbitals, subopercle and opercle large and pointed.

Colour in life. From Miller's notes and colour sketch. Three white bars as follows: on head, behind pectoral fin, and across caudal peduncle; soft dorsal, caudal, anal and pectoral fins with a distal jet black band, margined with white; remainder of dorsal and caudal fins orange (or orange brown ?); anal and pelvic fins deep orange basally, the pelvic jet black distally; head deep orange; body deep orange brown behind first white band.

Colour in alcohol. Breast and lower part of head yellowish; upper head, body, and proximal portion of median fins light brown; three light yellow bars bordered by black, each about twice width of eye as follows: first from nape through opercle to subopercle; second including posterior part of spinous dorsal fin to anus and through first anal spine, with a V-shaped anterior projection toward upper base of pectoral fin; third across posterior part of caudal peduncle; margin of median fins white, bordered inwardly by black; pelvic fins dark brown to black, basally yellowish; centre of pectoral fin yellowish, surrounded by black, the fin margin whitish.

Remarks. Amphiprion percula as understood in recent works is probably divisible into two forms, as the following data suggest. One, the true *percula* appears to inhabit the reefs of the south central Pacific from New Guinea and eastern Australia eastward. For those in the western Pacific, the name *bicolor* is available. Material from the intermediate area, New Guinea and north Australia, however, is insufficient to interpret the specific or subspecific status of these populations; consequently I have retained the name *percula* for the Arnhem Land specimens.

In general, specimens from the western portion of the range have a modal number of XI dorsal spines, 17 total pectoral rays and the body between the alternate bars brownish. Those from eastward have a modal number of X dorsal spines, 16 total pectoral rays, and the areas between the alternate bars light centred or the first area is broadly invaded by black.

The most notable break is in number of dorsal spines as follows (frequency of counts from literature and United States National Museum specimens given in parentheses): Solomon Islands X(10); New Britain Island, type locality of *percula*, X (from the original description); Great Barrier Reef, Australia X (Whitley, 1929b); Admiralty Islands IX(1) or X(1); Schouten Islands IX(1) or X(1); Arnhem Land, Australia XI(3), also XI indicated for material of *bicolor* Whitley (1929b); Indo-Australian Archipelago X(4), XI(50) or XII(4) according to Weber and de Beaufort (1940); Bleeker (1877b) has indicated that those from the East Indies typically have XI spines, infrequently X and rarely IX; Ceram Sea XI(1); Mollucas Islands X(1); Celebes XI(3); Philippine Islands X(2) or XI(10); Calamianes X(1) or XI(1); Palawan X(1), XI(3), or XII(1). Counts of total pectoral rays (branched plus simple rays) are as follows: Solomon Islands 16(12); Admiralty Islands 16(2) or 17(2); Schouten Islands 16(2) or 17(2); Arnhem Land 16(2), 17(2), or 18(2); Philippine Islands 17(5) or 18(1); Palawan Island 17(5) or 18(1).

SCARIDAE

Schultz (1958) has studied the members of this family. In his work he identified the Arnhem Land parrotfishes and made special reference to certain species. I have compiled a list of this material below, giving his identification as made at the time of the study and have included previous references to Arnhem Land forms.

It should also be noted that Smith (1959) maintains *Scarus psittacus* Forskål cannot be identified in the absence of a type specimen and recommends the rejection of the generic name *Scarus*. I have accepted *Scarus* for the genus and as the stem of the family name. If it is not available, the next name for these usages appears to me *Callyodon* Scopoli (1777); see both Schultz (1958) and Smith (1959) for reference to the name and the older synonymy.

*Scarops** *rubroviolaceus* (Bleeker)

Scarus rubroviolaceus Bleeker, *Verh. Batavia Genoot.* 22: 52 (1849); type locality, Batavia, Java.

M48-13e, Chasm Island, 3 and 4 July 1948, USNM 114937, 1 specimen.

M48-30, Reefs, Yirrkalla, 21 and 23 August 1948, USNM 114936, 2 specimens.

Scarus fasciatus Cuvier and Valenciennes

Scarus fasciatus Cuvier and Valenciennes, *Histoire naturelle des poissons* 14: 222 (1839); type locality, Moluccas; Schultz, *U.S. Nat. Mus. Bull.* 214: 82 (1958).

M48-13b, Hemple Bay, Port Langdon, 27 May 1948, USNM 165561, 1 specimen.

M48-13e, Chasm Island, 3 and 4 July 1948, USNM 114939, 1 specimen.

M48-21, Reef, Yirrkalla, 11 August 1948, USNM 114620, 1 specimen.

M48-30, Reef, Yirrkalla, 5 August 1948, USNM 114938, 2 specimens.

Colour in life. From Miller's notes at M48-13b: teeth white; body blue to emerald; all fins bright bluish green with wavy longitudinal, darker blue, bands on anal fin and red band on dorsal fin; opercle, above pectoral base, orange; bright emerald green wavy markings and spots on snout, cheek, and throat. At M48-21, he describes the fish as having rust red scale centres and lower part of head mostly whitish, covered with irregular emerald markings.

Scarus aeruginosus Cuvier and Valenciennes

Scarus aeruginosus Cuvier and Valenciennes, *Histoire naturelle des poissons* 14: 257 (1839); type locality, Red Sea.

M48-9a, Reef, entrance to Little Lagoon, USNM 114622, 2 young, identification not certain.

M48-21, Reef, Yirrkalla, 14 and 29 July and 11 and 12 August 1948, USNM 114621, 21 specimens.

Remarks. The natives on Groote Eylandt gave the name *Wóorabárja* to the specimens from M48-9a. Miller's notes indicate that the life colours of these specimens are: scale centres coppery, forming longitudinal rows along the side; his colour drawing indicates a faint blue area beneath lower jaw connected with a similar coloured band extending backward from mouth to eye, thence along lower margin of eye; a patch on pelvic fin and upper, outer, rays of pectoral fin yellow; borders of dorsal and anal fins greyish blue; narrow basal part of anal fin

* *Scarops* Schultz (1958): 18; type species, *Scarus rubroviolaceus* Bleeker, by original designation.

grey; remainder of these fins orange, the orange grading to yellowish orange posteriorly on dorsal fin; caudal fin faintly yellowish orange, with a blue distal margin.

Scarus ghobban Forskål

Scarus ghobban Forskål, *Descriptiones Animalium*: 28 (1775); type locality, Djedda, Red Sea.

Scarus pyrrostethus australianus Paradice, *Mem. Qld. Mus.* 9: 92, 98, 103, pl. 14 (1927); type locality, Cape Wessel, Northern Territory.

Recorded as: *Scarus pyrrostethus* Paradice and Whitley (1927): 92 and 105, Pellew Islands.

TEUTHIDIDAE

Key to the Arnhem Land Species of *Teuthis**

1 Area between anterior processes of pelvic girdle naked; a free flap on rear part of rim of anterior nostril; anterior profile slightly concave above eye; cheek naked or sparsely scaled only on its lower half; first and last dorsal spines of approximately equal length; colour brownish, with large, irregular dark brown spots scattered over body; a black blotch on lateral line just behind head . . . . . . *Teuthis nebulosa*

Lower surface of body covered with scales; rim of anterior nostril low, without a prominent flap; first dorsal spine shorter than last spine, about 1·4 or more in the latter's length; head not notably concave above eye 2

2 Narrow, dark brown or black bars crossing upper midline from snout to dorsal fin; narrow, brown or blackish bars on side of body, sometimes broken into small rounded spots; fins nearly uniformly light yellowish brown . . . 3

No brown bars crossing upper midline; narrow, light to medium brown stripes, separated by yellowish interspaces, along lower and mid-side of body, the stripes on upper side of body broken and usually forming a reticulated network; a large, light or brownish spot usually evident below rear dorsal rays; caudal, soft dorsal and anal fins dotted with dark brown . . *Teuthis lineata*

3 Numerous narrow dark brown wavy bars on head and body which more or less bent backward or broken into spots on rear half of body; few spots anteriorly; generally longitudinal lines or network on opercle, but only a portion of one line visible on one side in present specimen; lower side silvery as is background colour of upper side; about three diagonal brownish lines from mouth and chin to eye; body depth about 2·4; first dorsal spine 2·2 in last dorsal spine; cheek scales chiefly confined to its lower rear corner, scattered elsewhere . . . . . . *Teuthis doliata*

Fine, dark brown bars on middle of body which become broader and broken into spots dorsally and anteriorly and with age; these bars anteriorly extending downward from similar diagonal bars which run from dorsal fin to head and pectoral fin; spaces between diagonal bars lighter, brown or yellowish and often containing numerous small dark spots or dashes; diagonal rows of dark brown spots or lines from chin and mouth to eye; body depth 1·8 to 2·2 in standard length; first dorsal spine 1·4 to 1·8 in last spine; scales confined chiefly to rear half of cheek . . . . . . *Teuthis virgata*

* In Opinion 93, International Commission on Zoological Nomenclature, *Teuthis javus* Linnaeus was designated as type species of *Teuthis* Linnaeus. *Teuthis* thus replaces *Siganus* Forskål.

The members of this family are identified chiefly on the basis of colour pattern which shows considerable variation, at least in preservation. The presence or absence of scales between the anterior processes of the pelvic girdle apparently delimits certain broad groups of species, the relationships of which are not well defined. Other important characters, which may prove of value in defining the species are the position and density of scalation on the cheek and the gillraker arrangement. It is to be noted, however, that the scalation of the cheek appears to increase with size or age, as the small specimen of *T. nebulosa* in the present collection has no cheek scales, the next largest has several on the lower half of the right cheek and one on the left, while the largest specimens are moderately well scaled.

Fin ray counts in *Teuthis* are relatively constant. A member of each of the species gave the following data: Dorsal spines XIII; anal spines VII; pelvic rays I,3,I; branched caudal rays 8+7=15; segmented dorsal rays 10 and segmented anal rays 9, but an additional specimen of *T. virgata* had 11 soft dorsal and 10 soft anal rays. Pectoral rays are variable, but not useful in identification, as follows: ii,13(in 1), ii,14(12) or ii,15(3) in *T. virgata*; ii,14(5) or ii,15(5) in *T. nebulosa*; ii,14(8) or ii,15(8) in *T. lineata*; and ii,15(2) in *T. doliata*. The number of scales between the lateral line and middle of dorsal fin varies from 19 to 24.

Teuthis nebulosa (Quoy and Gaimard)

Amphacanthus nebulosus Quoy and Gaimard, *Voyage autour du monde . . . sur l'Uranie et la Physicienne . . . Zoologie*: 369 (1824); type locality, Port Jackson.
Siganus nebulosus Stead, *The Edible Fishes of New South Wales . . .*: pl. 49 (1908).
Amphacanthus gymnopareius Richardson, *Ann. Mag. Nat. Hist.* 11: 174 (1843); type locality, bays of the harbour, Port Essington.
Siganus consobrinus Ogilby, *Mem. Qld. Mus.* 1: 54, pl. 13 (1912); type locality, Moreton Bay, Queensland.
Siganus concavocephalus Paradice in Paradice and Whitley, *Mem. Qld. Mus.* 9: 90, 98, 99, pl. 12, fig. 2 (1927); type locality, Pellew Islands, Northern Territory.

Recorded as: *Amphacanthus nebulosus* Whitley (1928a): 232, Pellew Islands; *Teuthis nebulosa* Günther (1861): 321, Coral Bay, Port Essington; Macleay (1880): 444; *Siganus nebulosus* Paradice and Whitley (1927): 91, Pellew Islands; *Teuthis sutor* Klunzinger (1879): 393; Macleay (1884): 20, Port Darwin.

Record probably referable: *T. guttatus* Kent (1889): 239, Port Darwin.

M48-9, Reefs and tide pools, entrance to Little Lagoon, USNM 173932, 1 specimen, 63 mm. standard length.

M48-22, Bay of Yirrkalla, 4 and 6 August 1948, USNM 173933, 3 specimens, 219 to 223 mm. standard length.

M48-27, Bay and beach, Port Bradshaw, USNM 173934, 1 specimen, 138 mm. standard length.

Colour in life. Based on a colour transparency; dark reddish brown on body and head, with dark brown blotches, large as pupil, scattered over body; tiny dark brown spots forming faint, very irregular rows or lines along body, or pigment irregularly arranged in a reticulated network of brown bordered, small whitish blotches; a rounded, dark brown blotch at rear of head, covering lateral line; dorsal and anal fins marbled brownish, the spines brown and the soft rays annulated with dark brown; upper and lower lobes of yellowish olive caudal fin bordered by dark brown which more or less concentrated into blotches that tend to form bars extending on to fin.

Teuthis lineata (Cuvier and Valenciennes)

Amphacanthus lineatus Cuvier and Valenciennes, *Histoire naturelle des poissons* 10: 130, pl. 286 (1835); type localities, Vanicolo and New Guinea.

Siganus aurolineatus Ogilby, *Mem. Qld. Mus.* 1: 56 (1912); type locality, Somerset, Queensland.

Siganus lineatus Marshall in Ogilby, *The Commercial Fishes and Fisheries of Queensland*: 92, fig. 113 (1954).

Recorded as: *Amphacanthus lineatus* Whitley (1928a): 231, Pellew Islands and Cape Wessel, Northern Territory; *Siganus lineatus* Paradice and Whitley (1927): 90.

M48-15, Emerald River, Groote Eylandt, USNM 173935, 1 specimen, 109 mm. standard length.

M48-19, Creek, tributary to Little Lagoon, USNM 173936, 3 specimens, 109 to 125 mm. standard length.

M48-26, Sheltered bay, island, Port Bradshaw, USNM 173937, 1 specimen, 187 mm. standard length.

M48-27, Bay and beach, Port Bradshaw, USNM 173938, 2 specimens, 88 and 102 mm. standard length.

M48-28, Creek mouth and bay, Port Bradshaw, USNM 173939, 5 specimens, 74 to 111 mm. standard length.

M48-30, Reefs, Yirrkalla, 5 August 1948, USNM 173940, 2 specimens, 218 and 224 mm. standard length.

M48-31, Creek, Yirrkalla, USNM 173941, 1 specimen, 155 mm. standard length.

Colour in life. Based on a colour transparency of the largest specimen: general background colour medium brown to greyish above, becoming whitish below; as many as 18 narrow, orange brown or reddish brown, stripes from head, pectoral fin, and middle of lower side to caudal peduncle, where broken into spots posteriorly and below; breast without obvious stripes or spots; fins medium brown, with dark brown spots similar in size to those on caudal peduncle; three or four light bluish lines, intermingled with several orange brown lines, extending diagonally upward from lower part of head to its rear: one blue line from upper jaw to eye; another from rear end of upper jaw, around eye to opercle; a third from chin to opercle, and possibly a fourth along lower part of opercular series; a yellowish orange spot, large as eye, just below rear part of dorsal fin.

Miller's notes indicate that the stripes in a slightly smaller specimen were orange.

In preservation, the stripes are medium brown, yellowish, silvery or obscured. The upper side, which is not visible in the colour photograph, exhibits a network or reticulation, formed from the brownish lines.

Teuthis virgata (Cuvier and Valenciennes)

Amphacanthus virgatus Cuvier and Valenciennes, *Histoire naturelle des poissons* 10: 133 (1835); type locality, Java.

Amphacanthus notostictus Richardson, *Ann. Mag. Nat. Hist.* 11: 172 (1843); type locality, Port Essington.

Recorded as: *Teuthis notosticta* Günther (1861): 320, Coral Bay, Port Essington; Macleay (1880): 443, Port Darwin; *Teuthis notostictus* Macleay (1878): 353; *Amphacanthus virgatus* Whitley (1929a): 126, Port Darwin.

M48-13e, Chasm Island, 3 and 4 July 1948 (Jerry Blitner), USNM 173930, 1 specimen, 171 mm. standard length.

M48-21, Reef, Yirrkalla, 11 and 12 August 1948, USNM 173931, 7 specimens, 80 to 95 mm. standard length.

Colour in life. See comments under *Teuthis doliata.* An abbreviated description of the preserved colour pattern is found in the key.

Teuthis doliata (Cuvier)

Pl. 63

Siganus doliatus Cuvier in Guerin, *Iconographie du règne animal. . . poissons*: pl. 35, fig. 1 (1830-1844).

Amphacanthus doliatus Cuvier and Valenciennes, *Histoire naturelle des poissons* 10: 132 (1835); reference to figure in Guerin; type localities, Bourou and Vanicolo; Cuvier, *Règne animal,* Disciples Edition: pl. 71, fig. 1 (about 1836).

M48-21, Reef, Yirrkala, 11 August 1948, USNM 173946, 1 specimen, 184 mm. standard length.

Colour in life. Possibly confused with *Teuthis virgata.* Body greyish; caudal fin yellowish orange; reddish brown band from eye to corner of mouth; narrow blue stripes (bars) on head, nape, and upper side of body; four orange bars extending obliquely forward from base of pectoral fin; back, caudal peduncle and second dorsal fin yellowish orange.

ACANTHURIDAE

I have followed Randall (1955 and 1956) in the identification and nomenclature of the material in this family. He has studied and recorded some of the Arnhem Land specimens.

Acanthurus grammoptilus Richardson

Acanthurus grammoptilus Richardson, *Ann. Mag. Nat. Hist.* 11: 176 (1843); type locality, harbour of Port Essington; Günther, *Catalogue of the Fishes in the British Museum* 3: 335 (1861).

Other Arnhem Land records: *Acanthurus grammoptilus* Macleay (1878): 354 and (1880): 528; Klunzinger (1879): 393, Port Darwin; *Teuthis grammoptilus* McCulloch (1918): 92, Port Darwin.

M48-9a, Reefs and tidepools, entrance to Little Lagoon, USNM 173742, 2 specimens, 54 and 58 mm. standard length.

M48-11, Little Lagoon, 2 May 1948, USNM 164288, 1 specimen, 242 mm. standard length.

M48-17, Reef, Bartalumba Bay, USNM 173740, 1 specimen, 64 mm. standard length.

M48-21, Reef, Yirrkalla, 14 July 1948, USNM 164287, 1 specimen, 86 mm. standard length; 11 August 1948, USNM 173741, 1 specimen, 74 mm. standard length; 12 August 1948, USNM 164286, 3 specimens, 85 to 92 mm. standard length.

M48-22, Island, near point, east of Yirrkalla, 18 August 1948, USNM 164289, 1 specimen, 204 mm. standard length.

Native names: *Kokwál* and *Góocoodl,* Yirrkalla; *Ilábrangóla,* Groote Eylandt.

Description. Dorsal rays VIII,i,24(in 2), IX,i,21(1), IX,i,23(3), IX,i,24(3), or IX,25(1); anal rays II,i,22(1), III,i,21(1), III,i,22(6), or III,i,23(2); pelvic rays I,5(20); pectoral rays ii,14(14) or ii,15(5); branched caudal rays 7+7=14(10); teeth

14 to 17 in upper jaw and 14 to 19 in lower in specimens below 100 mm. standard length, or 18 to 20 in upper and 20 to 23 in lower in those over 200 mm. standard length.

In the above specimens, least depth caudal peduncle 2·8 to 3·2, snout 1·3 to 1·5 and eye 3·3 in small specimens to 5·3 in adults, all in head length; longest dorsal ray 5·2 to 6·7, head length 3·2 to 3·6, pectoral fin 3·2 to 3·6, and body depth 1·7 in small specimens to 2·3 in adults, all in standard length.

A single spine on each side of caudal peduncle which folds into a groove; dorsal and anal spines moderately slender; scales small, numerous, covering body and head; teeth fixed, broad, compressed, each with dentations on its lateral and medial margins; posterior margin caudal fin prominently concave, the upper caudal lobe longest.

Colour in alcohol. Body and head medium brown to dark reddish brown; body with or without many faint, wavy, longitudinal lines; head usually with about three wavy lines extending forward from eye and numerous small light brown spots on cheek and upper surface; opercular membrane dark brown or black; dorsal and anal fin colours similar to those of body, but both with series of (not prominent) alternating dark and medium brown lines or bands; caudal fin brown to dark brown with white bar (narrow and obscure in the 58 mm. individual from M48-9a), about as wide as eye, at its base and a narrow light margin; spine groove on caudal peduncle narrowly margined, most prominent anteriorly, in dark brown or black.

Colour in life. Body and head greyish brown; rust coloured spots, blotches, and wavy lines on anterior part of head and over nape; pectoral fin yellow; anal fin with pale blue border; caudal fin with white bar across base and a narrow white or yellowish margin.

Remarks. The record of *Acanthurus annularis* Kent (1889): 239, from Port Darwin, is probably referable to this species. No description was given.

OPISTOGNATHIDAE

Gnathypops papuensis (Bleeker)

Opisthognathus papuensis Bleeker, *Versl. Akad. Amst.* ser. 2, 2: 333 (1868); type locality, Waigeu; Boeseman, *Meded. Akad. Wet. Amst.* ser. C, 57: 271, fig. (1954).

Opisthognathus maculatus Alleyne and Macleay, *Proc. Linn. Soc. N.S.W.* 1: 280, pl. 9, fig. 3 (1877); type locality, Palm Island, Queensland.

Batrachus punctatulus Ramsay, *Proc. Linn. Soc. N.S.W.* 8: 177 (1883); type locality, Torres Strait.

Gnathypops maculata Ogilby, *Mem. Qld. Mus.* 7: 27, pl. 3 (1922).

Recorded as: *Opisthognathus maculatus* Kent (1889): 239, Port Darwin.

Goulburn Island, in Arafura Sea, about 180 miles east-north-east of Darwin, Northern Territory, 29 July 1948 (Nick Haritos), USNM 174044, 1 specimen, 208 mm. standard length.

Description. Dorsal rays XII,16 and anal rays II,14, the spines slender and flexible. Upper jaw extending backward to vertical arm of preopercle, truncate behind; gillrakers on first right arch 9+1+16; pelvic rays I,ii,3, both sides; lateral line high on upper side, extending backward to below middle of soft dorsal base; no lower line observed.

In alcohol, body and fins yellowish brown with small dark brown spots; the spots on head, tiny and nearly round; those on trunk of irregular shapes, some

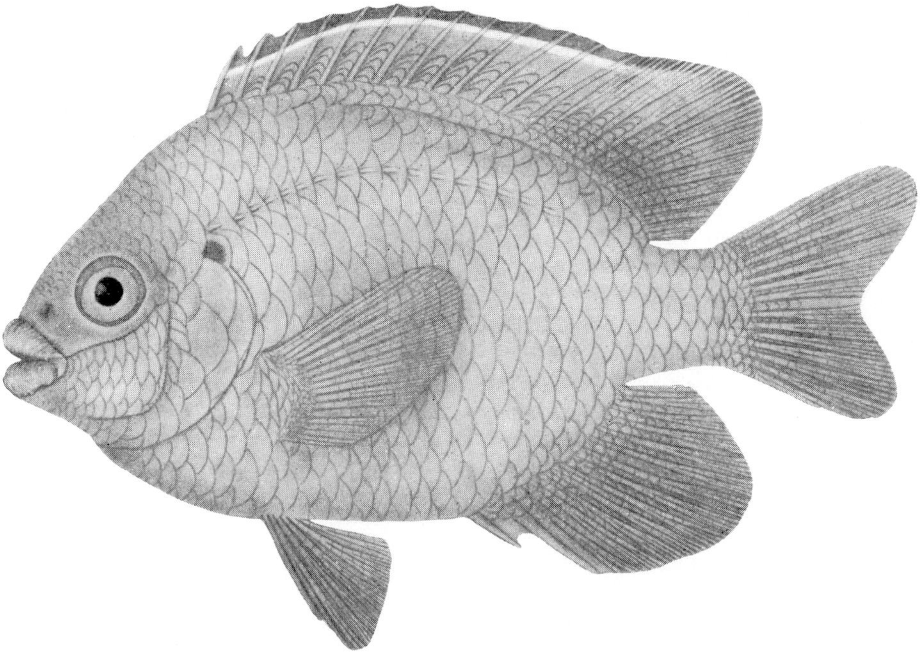

Dorothea B. Schultz

PLATE 62. *Cheiloprion labiatus*
Yirrkalla, USNM 173614, 61 mm. standard length

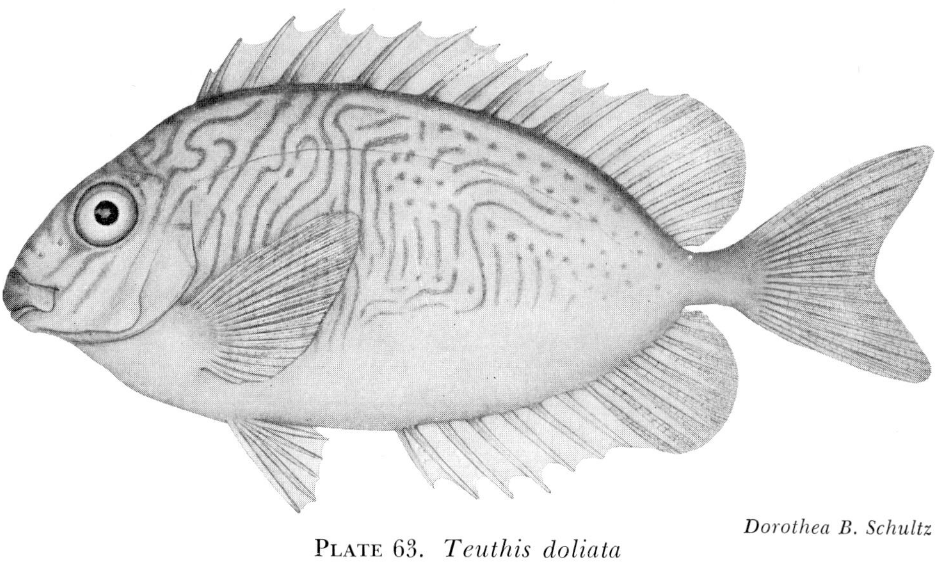

Dorothea B. Schultz

PLATE 63. *Teuthis doliata*
Yirrkalla, USNM 173946, 184 mm. standard length

large and blotch-like, others small and rounded; spots least numerous on rear part of body and median fins.

Gnathypops darwiniensis (Macleay)
Pl. 64

Opisthognathus darwiniensis Macleay, *Proc. Linn. Soc. N.S.W.* 2: 355, pl. 9, fig. 3 (1878); type locality, Port Darwin.

Recorded as: *Opisthognathus darwiniensis* Macleay (1880): 569 and Klunzinger (1879): 371, Port Darwin.

M48-3, East Point Reef, USNM 174043, 1 specimen, 67 mm. standard length.

M48-21, Reef, Yirrkalla, 12 August 1948, USNM 174042, 8 specimens, 53 to 106 mm. standard length.

Description. Dorsal rays X,17(in 1), X,18(2), or XI,17(6); anal rays II,12(1), II,13(1), or II,14(7); dorsal and anal spines slender, flexible; upper jaw extending backward to vertical arm of preopercle; pelvic rays I,ii,3 (both sides, 1); lateral line high on side, ending under rear half of soft dorsal fin; a second, irregularly developed, line of double pores present along lower side of caudal peduncle and abdomen; it appears to begin behind and below the pectoral fin and extends backward to above the anterior half of anal base; one or sometimes two branches extend vertically downward from this line toward the vent, or across the abdomen.

Body and head medium brown, covered with dark brown spots; spots on head and nape tiny and close-set; those on trunk larger, blotch-like and more irregular posteriorly; pectoral fins yellowish with rows of tiny brown spots; all other fins with irregular, oblique alternating dark brown and whitish bars; an obovate or elongate jet black blotch on spinous dorsal fin between spines III and V or VI.

Colour in life. A black rectangular spot on dorsal fin; body mostly yellow with brown markings; oblique white bars on all fins except pectoral.

CREEDIIDAE

Squamicreedia obtusa Rendahl

Squamicreedia obtusa Rendahl, *K. Svenska VetenskAkad. Handl.* 61: No. 9, 20, figs. 5a and 5b (1921); type locality, 45 miles west-south-west Cape Jaubert, Western Australia.

Recorded as: *Squamicreedia obtusa* Whitley (1956b): 260, Charles Point, Northern Territory.

TRICHONOTIDAE

Trichonotus setiger Bloch and Schneider

Trichonotus setiger Bloch and Schneider, *Systema Ichthyologiae*: 179 (1801); type locality, eastern India.

Recorded as: *Trichonotus setigerus* Macleay (1878): 359; Klunzinger (1879): 396, Port Darwin; *Trichonotus blochii* Macleay (1881): 2, Port Darwin.

CONGROGADIDAE

Congrogadus subducens (Richardson)

Machaerium subducens Richardson, *Ann. Mag. Nat. Hist.* 12: 175, pl. 6 (1843); Richardson, Notices and Abstracts of Communications... in *Rep. Twelfth Meet. Brit. Ass.*: 69 (1943); type locality, bays of Port Essington; Richardson,

Ichthyology of the Voyage of the H.M.S. Erebus and Terror...: 72, pl. 44, figs. 1 to 6 (1846).

Recorded as: *Congrogadus subducens* Günther (1862): 388, Port Essington; Macleay (1878): 362; Macleay (1881): 120; Klunzinger (1879): 406; Paradice and Whitley (1927): 93; Whitley (1929a): 108, pl. 30, fig. 1, Port Darwin.

M48-1, Nightcliff, 19 March 1948, USNM 173800, 1 specimen, 289 mm. standard length.

M48-3, East Point Reef, USNM 173801, 2 specimens, 148 and 249 mm. standard length.

M48-9, Reefs and tide pools, entrance to Little Lagoon, USNM 173802, 1 specimen, 172 mm. standard length.

M48-9a, Reef, entrance to Little Lagoon, USNM 173803, 2 specimens, 46 and 317 mm. standard length.

M48-17, Reef, in Bartalumba Bay, USNM 173804, 11 specimens, 144 to 296 mm. standard length.

M48-21, Reef, Yirrkalla, 14 July 1948, USNM 173805, 20 specimens, 114 to 340 mm. standard length; 29 July 1948, USNM 173806, 1 specimen, 273 mm. standard length; 6 August 1948, USNM 173807, 1 specimen, 254 mm. standard length; 6 to 11 August 1948, USNM 173808, 1 specimen, 329 mm. standard length; 11 August 1948, USNM 173809, 6 specimens, 188 to 245 mm. standard length.

M48-33, Milingimbi Island, USNM 173810, 1 specimen, 243 mm. standard length.

Native name: *Dúray*, Yirrkalla.

Description. Rays in vertical fins all branched; no spines; pelvic fins absent; dorsal rays 69(in 2), 70(1), 71(1), and 74(1); anal rays 59(3), 60(1), or 63(1); caudal rays 5+4=9(1), or 5+5=10(4), the fin continuous with the dorsal and anal by membranes; pectoral rays i,7(1), or i,9(9); scales small; pored scales in lateral line, which ends anterior to anal fin, 50(1), 53(1), 54(1), 56(1), and 61(1).

In five specimens 143 to 317 mm. in standard length, head length 6·0 to 7·0, distance from snout tip to anal fin 2·8 to 3·3, and distance from end of snout to dorsal fin 4·5 to 5·2, all in standard length; body depth 1·4 to 1·8, eye 4·9 to 6·7, and snout 3·1 to 3·7, all in head length.

Body eel-like, moderately compressed; small scales covering body, upper opercle, preopercle, and cheek; a prominent ridge or crest on snout, extending backward to between eyes; lips thickened and expanded, folded back over snout and chin; upper lip divided near midline; gill membranes broadly connected to isthmus, with a small, free fold; gill opening extending upward to above pectoral base; small, conical teeth in a single row in each jaw, each row V-shaped; a fleshy fringe on each jaw, lateral to and adjacent to row of teeth.

Colour in alcohol. Body and head medium to dark brown, usually lighter below; markings variable; larger specimens generally with light blotches or reticulations and small specimens usually with small dark spots which tend to form longitudinal rows; a row of dark spots, about as large as pupil, along base of dorsal fin; lips yellowish; anterior naris, which is tiny, encircled in black; blackish blotch, large as eye, on opercle below scaly area; lower jaw, and frequently abdomen, with a series of discrete, dark brown blotches; pectoral fin yellowish; vertical fins brown to dark brown, with variable reticulations or mottling; young yellowish brown with brown circles, large as pupil, on caudal peduncle.

Colour in life. One specimen is described in Miller's notes as dark green; another as having the body orange brown, while an accompanying incomplete sketch shows the eye and upper jaw green.

NOTOGRAPTIDAE

Notograptus guttatus Günther

Notograptus guttatus Günther, *Ann. Mag. Nat. Hist.* ser. 3, 20: 64 (1867); type locality, Cape York.

Blanchardia maculata Castelnau, Researches on the Fishes of Australia, in *Official Record of the Philadelphia Centennial Exhibition of 1876*: 47 (1875); type locality, Queensland.

Recorded as: *Notograptus guttatus* Macleay (1878): 359 and (1881): 30, Port Darwin.

M48-1, Nightcliff, 19 March 1948, USNM 173795, 1 specimen, 84 mm. standard length.

M48-9, Reefs and tide pools, entrance to Little Lagoon, USNM 173796, 3 specimens, 79 to 157 mm. standard length.

M48-17, Reef, Bartalumba Bay, USNM 173797, 6 specimens, 46 to 185 mm. standard length.

M48-21, Reef, Yirrkalla, 14 July 1948, USNM 173798, 15 specimens, 82 to 172 mm. standard length.

M48-33, Milingimbi Island, USNM 173799, 1 specimen, 146 mm. standard length.

Native name: *Garnal,* Yirrkalla.

Description. Dorsal, anal, and caudal fins continuous; dorsal spines 66(in 3), 67(1), or 68(1), the spines tiny and flexible anteriorly and stiff posteriorly; anal spines 39(1), 40(1), 41(1), 42(1), or 43(1), similarly weakest anteriorly; the spines not extending to margin of the fins, the margin supported by flexible, often branched but unsegmented, ray-like filaments that extend outward from the spine; segmented dorsal rays 2(26), both branched in larger specimens; segmented anal rays 1(1) or 2(25), similarly frequently branched; the segmented dorsal and anal rays interpreted as such rather than caudal rays because of their wide interspaces and non-association with the hypural plate; total caudal rays 6+4=10(1), 6+5=11(9), or 6+6=12(1), the branching of the uppermost and lowermost rays variable, of which the lower ray is usually, but not always, simple; pectoral rays i,16,i(2), or i,17,i(8); pelvic fin small, jugular, the rays I,ii(10), the spine tiny; lateral line reflected upward below dorsal fin origin, thence continued along dorsal base, ending about half a head length from caudal fin, with 92(1) pores; scales small, covering body below and behind lateral line; nape and head naked; narrow area between lateral line and dorsal fin with scattered scales. Lower jaw slightly included; lips fleshy, somewhat reflected; end of upper lip leaf-like or triangular, turned back over end of snout; a barbel-like projection at symphysis of lower jaw; two pairs of nares, the anterior tubular and about a pupil diameter from the posterior pair; gill membrane confluent with isthmus, the gill opening extending to upper part of pectoral base; dorsal origin over rear part of head; body eel-like, slightly compressed; broad patches of small teeth on palatine and jaws; snout length about equal to eye diameter.

In three specimens 79 to 157 mm. in standard length, eye 6·4 to 7·4 and body depth 1·3 to 1·9, both in head length; head length 6·4 to 7·1 and distance from end of snout to anal fin 2·1 to 2·3, both in standard length.

Colour in alcohol. Body, head, and fins medium brown to light yellowish brown; dark brown, rounded spots, faintly outlined in yellowish, some nearly as large as eye, on head, body, and vertical fins; spots on body and fins variable, generally forming irregular, longitudinal rows, and absent or disappearing with age on the posterior parts; those on head largest and remaining most prominent.

Colour in life. Body yellowish pink to orange; head pale bluish green; four dark (or orange) spots on maxillary; a row of similar spots under eye; others on top of head and nape and often extending backward along body; vertical fins edged in milky white; pectoral fin greenish at base; belly white.

CARAPIDAE

Material recorded as: *Fierasfer* Kent (1889): 240, Port Darwin.

BROTULIDAE

Approximately 245 specimens of free tailed brotulids were obtained at seven stations during the Arnhem Land survey. This material is now being studied by Dr Boyd W. Walker, University of California, Los Angeles, California. He states that four species are represented; one belongs to the genus *Dinematichthys* Bleeker and the others to a single undetermined genus.

CALLIONYMIDAE

Callionymus calcaratus* Macleay

Callionymus calcaratus Macleay, *Proc. Linn. Soc. N.S.W.* 5: 628 (1881); type locality, Port Jackson.

Recorded as: *Repomucenus calcaratus* Whitley (1948): 275, 12° 12′ S. lat. and 130° 36′ E. long., Northern Territory.

TRICHIURIDAE

Trichiurus savala Cuvier

Trichiurus savala Cuvier, *Règne animal* ed. 2, 2: 219 (1829); reference not seen; type locality, Indies.

Recorded as: *Trichiurus savala* Macleay (1878): 354, and (1880): 524, Port Darwin.

SCOMBRIDAE

Rastrelliger kanagurta (Cuvier)

Scomber kanagurta Cuvier, *Règne animal* ed. 1, 2: 313 (1817); type locality, Vizagapatam.

Recorded as: *Rastrelliger canagurta* Munro (1958): 111, Darwin harbour.

Kishinoella tonggol (Bleeker)

Thynnus tonggol Bleeker, *Natuurk. Tijdschr. Ned.-Ind.* 1: 356 (1851); type locality, Singapore.

Recorded as: *Kishinoella tonggol* Munro (1958): 111, Northern Territory.

Katsuwonus pelamis (Linnaeus)

Scomber pelamis Linnaeus, *Systema Naturae* ed. 10, 1: 297 (1758); type locality, tropics.

Recorded as: *Katsuwonus pelamis* Whitley (1954): 28, near Sahul Bank, Northern Territory.

* *Callionymus lyra* Linnaeus was selected as type species of *Callionymus* Linnaeus by the International Commission on Zoological Nomenclature, Opinion 77.

Euthynnus alletteratus (Rafinesque)?

Scomber alletteratus Rafinesque, *Caratteri di alcuni Nuovi Generi e Nuovi Specie...*: 46 (1810); type locality, Sicily.

Recorded as: *Euthynnus alletteratus* Whitley (1954): 28, Timor Sea, Northern Territory.

The Australian members of the genus *Scomberomorus* were revised by Munro (1943). I have used his study in the identification of my material.

Scomberomorus niphonius (Cuvier and Valenciennes)

Cybium niphonium Cuvier and Valenciennes, *Histoire naturelle des poissons* 8: 180 (1831); type locality not given; based on a Japanese painting.

Recorded as: *Sawara niphonia* Munro (1958): 112, Northern Territory.

Scomberomorus commersoni (Lacépède)

Scomber commerson Lacépède, *Histoire naturelle des poissons* 2: 598 and 600, pl. 20, fig. 1 (1800); type locality not given.

Recorded as: *Scomberomorus commerson* Paradice and Whitley (1927): 82, Pellew Islands.

Scomberomorus queenslandicus Munro

Scomberomorus (*Cybium*) *queenslandicus* Munro, *Mem. Qld. Mus.* 12: 82 (1943); type locality, Cape Cleveland, Queensland.

Recorded as: *Cybium queenslandicum* Whitley (1954): 27, near Parry Shoal, Northern Territory.

Scomberomorus semifasciatus (Macleay)

Cybium semifasciatum Macleay, *Proc. Linn. Soc. N.S.W.* 8: 205 (1883); type locality, lower Burdekin River, Queensland.

Recorded as: *Scomberomorus (Indocybium) semifasciatus* Munro (1943): 82 and 91, Port Darwin; *Indocybium semifasciatum* Whitley (1954): 27, between Darwin and Point Charles, Northern Territory.

M48-2, Nightcliff, USNM 174170, 3 specimens, 132 to 155 mm. standard length.

Description. A single lateral line, which moderately straight for this genus, passing through a keel on caudal peduncle; a small ridge above and below keel; gillrakers on first arch very short, on right side 2 above, plus 1 at angle, plus 6(in 1) or 8(2); dorsal fin black, with large transparent area on its rear half; about 20 dark brown, nearly vertical, bars on upper side of trunk and caudal peduncle; the bars wider than pupil and often Y-shaped above.

ISTIOPHORIDAE

A specimen recorded as *Tetrapturus brevirostris* Whitley (1955a): 52 and (1955c): 296, Arafura Sea north of Melville Island, 11° 12′ S. lat. and 131° 50′ E. long.; *Makaira audax* Munro (1958): 115, same specimen.

The nomenclature of this family is complex and uncertain; therefore the identification of the record is questioned.

SCORPAENIDAE

Key to the Arnhem Land Species

1 Dorsal fin beginning above eye, of XIII,7 rays; pelvic rays I,4; no palatine teeth; scales cycloid; no dermal appendages; branched caudal fin rays 10; head, nape and area along base of dorsal fin naked . . . . *Liocranium praepositum*

Dorsal fin beginning at or behind rear part of head, considerably behind eye; pelvic rays I,5; dorsal spines XII or XIII; no free pectoral appendage . . 2

2 Body and head scaleless 3

Body scaled; head scaled or nearly scaleless 4

3 Mouth terminal; body compressed, the skin smooth, marmorated; head relatively smooth, with two sharp spines; one preopercular, at rear end of suborbital stay; the other preorbital, which prominent, projecting outward and backward; its free length nearly equal to eye diameter; a tiny blunt spine at its base anteriorly; palatine teeth present . . *Gymnapistes marmorata*

Mouth vertical, preceded by the lower jaw; body obovate in cross-section, the skin roughened by numerous wart-like papillae; body colour nearly uniformly dark brown, the fins banded or variegated; head rugose, without sharp spines, but with a deep depression or saddle behind eye and a prominent pit below eye; eye small, elevated above general head profile; palatine teeth absent . . . . *Synanceia horrida*

4 Branched caudal fin rays normally 10 (5+5); dorsal and pectoral rays greatly produced; pectoral and some dorsal rays more than half as long as fish body; pectoral rays simple, 14 or 15; dorsal rays XIII,11 or 12; black and white barred; pelvic fin black with tiny white spots . . . . *Pterois volitans*

Branched caudal fin rays 11 (6+5); dorsal and pectoral rays short, all less than half as long as trunk; dorsal spines XII; some upper pectoral rays branched; suborbital chain attached to vertical arm of preopercle 5

5 Palatine teeth present; scales cycloid; no dermal appendages from lower jaw; chin and lower jaw marmorated . . . . *Scorpaena bynoensis*

Palatine teeth absent; scales ctenoid; dermal appendages from lower jaw as well as other parts of body; interorbital width equal to or greater than eye . . 6

6 Cranial spines branching with age, the nasals each with two or more points; body notably elevated and expanded at anterior base dorsal fin; interorbital ridges obscure or absent . . . . *Scorpaenopsis gibbosa*

Cranial spines mostly simple; body nearly fusiform, not notably broadened at anterior dorsal fin base; three interorbital ridges evident, one each side near eye and an anterior median ridge extending backward between eyes . . . . *Scorpaenopsis novaguinea*

*Synanceia** *horrida* (Linnaeus)

Scorpaena horrida Linnaeus, *Systema Naturae* ed. 12, 1: 453 (1766); type locality, eastern India; reference not seen.

Synanceia trachynis Richardson, *Ann. Mag. Nat. Hist.* 9: 385 (1842); type locality, Port Essington.

Recorded as: *Synancidium horridum* Günther (1860): 144, Port Essington; Klunzinger (1879): 367; Macleay (1880): 442; Kent (1889): 239, Port Darwin; *Synanceja trachynis* Whitley (1930): 25, Port Darwin.

* *Synanceia* Bloch and Schneider (1801): xxxvii and 573; type species *Scorpaena horrida* Linnaeus; *Synanceja* Bloch and Schneider (1801): 194 is apparently a misspelling.

Port Darwin; from Australian Museum; USNM 47875, 1 specimen, 178 mm. standard length.

M48-1, Nightcliff, USNM 174012, 1 specimen, 146 mm. standard length.

M48-7 or 8, Mouth of billabong or Little Lagoon, USNM 174013, 2 specimens, 43 and 78 mm. standard length.

M48-9, Reef, entrance to Little Lagoon, USNM 174014, 5 specimens, 64 to 79 mm. standard length.

M48-9a, Reef, entrance to Little Lagoon, USNM 174015, 1 specimen, 70 mm. standard length.

M48-21, Reef, Yirrkalla, 6 August 1948, USNM 174016, 1 specimen, 102 mm. standard length.

Colour in life. Body dirty green; orange bar, which not prominent, across body between soft dorsal and anal fins; abdomen with diffused orange; soft dorsal fin light yellow, with broad black border; caudal fin with a narrow yellow bar at base, this followed outward by a broad dark bar, a wider yellow bar, and a broad black terminal bar; anal fin orange, diffused with dusky, and with a broad black border; pectoral fins dull green basally, which grades to yellow orange at mid-fin, the terminal part broadly margined with black; pelvic fin orange, with blackened membranes.

Remarks. An inhabitant of rocky pools. Whitley (loc. cit.) regards *S. horrida* (Linnaeus) and *S. trachynis* Richardson as distinct species on the basis of several characters interpreted from descriptions and figures. In comparing specimens from Arnhem Land and Singapore (USNM 142944), I fail to observe the differences mentioned except that the dermal papillae are more numerous and close-set on the Australian specimens, which, judging from my material, is a highly variable character.

*Liocranium** *praepositum* Ogilby

Liocranium praepositum Ogilby, *Proc. Roy. Soc. Qld.* 18: 25 (1903); type locality, coast of Queensland.

Recorded as: *Abcichthys praepositus* Whitley (1954): 28, Port Darwin.

Pterois volitans (Linnaeus)

Gasterosteus volitans Linnaeus, *Systema Naturae* ed. 10, 1: 296 (1758); type locality, Amboina.

Recorded as: *Pterois volitans* Kent (1889): 239, Port Darwin.

Gymnapistes marmorata (Cuvier and Valenciennes)

Apistus marmoratus Cuvier and Valenciennes, *Histoire naturelle des poissons* 4: 416 (1829); type locality, Timor?

Recorded as: *Pentaroge marmorata* Klunzinger (1879): 366, Port Darwin.

Scorpaenopsis gibbosa (Bloch and Schneider)

Scorpaena gibbosa Bloch and Schneider, *Systema Ichthyologiae*: 192, pl. 44 (1801); type locality, America; error?

Scorpaena diabolus Cuvier and Valenciennes, *Histoire naturelle des poissons* 4: 312 (1829); type locality, 'mer du Croisic'.

* *Abcichthys* Whitley (1927a): 304; type species, *Liocranium praepositum* Ogilby, by original designation; a synonym, proposed to replace *Liocranium* Ogilby, incorrectly, believed to be a homonym of *Liocranum* Koch.

Recorded as: *Scorpaena diabolus* Macleay (1878): 353 and (1880): 433, Port Darwin.

Scorpaenopsis novaguinea (Cuvier and Valenciennes)

Scorpaena nova-guinea Cuvier and Valenciennes, *Histoire naturelle des poissons* 4: 320 (1829); type locality, New Guinea.

M48-22, Bay of Yirrkalla, 14 August 1948, USNM 174017, 1 specimen, 170 mm. standard length.

Description. Dorsal rays XII,9; anal rays III,5; pelvic rays, I,5, both sides; pectoral rays i,5,xi, both sides; branched caudal rays 6+5=11.

Body and head with numerous fleshy appendages, those from lower jaw and anterior gular region prominent; large, branched fleshy appendage from lower arm of preorbital, the concealed spine turned backward—upper spine of preorbital inclined backward—nearly horizontal; anterior spine sharp and directed forward; a ridge of sharp spines consisting of 3 suborbital and 2 close together on preopercle forming continuous chain with preorbital spine, all directed backward; nasal spines simple; three spines around upper edge of eye and a tiny cluster of two (on left) or three spines (right) immediately behind eye; at lower rear corner of eye, two tiny spines both sides; interorbital ridges prominent, those next to eye extending from before eye to saddle or depression behind eye; median ridge eye, two tiny spines, both sides; interorbital ridges prominent, those next to eye posteriorly, supraorbital shelf ending in two or three points, separated from rear cranial spines by a cross groove or depression; other spines on operculum and rear part of head; ctenoid scales present over body, and on upper part of opercle, nape, and cheek behind eye; colour light yellowish, irregularly blotched or mottled with brown; fins more or less banded with brown and light yellow; a light yellow saddle, long as eye, over anterior nape, just behind cross depression.

Miller's notes refer to this specimen as a red scorpaenid.

*Scorpaena** *bynoensis* Richardson

Scorpaena bynoensis Richardson, *Ichthyology of the Voyage of H.M.S. Erebus and Terror* . . .: 22, pl. 14, figs. 3 to 5 (1845); type locality, north-west coast of Australia.

Recorded as: *Scorpaena bynoensis* Macleay (1878): 353; Klunzinger (1879): 366, Port Darwin; Paradice and Whitley (1927): 91, Cape Wessel, Northern Territory; McCulloch (1926): 36; *Scorpaena bynaensis* Macleay (1880): 431, Port Darwin.

M48-1, Nightcliff, USNM 174008, 1 specimen, 75 mm. standard length.

M48-3, East Point Reef, USNM 174009, 2 specimens, 96 and 108 mm. standard length.

M48-17, Reef, Bartalumba Bay, USNM 174010, 1 specimen, 80 mm. standard length.

M48-21, Reef, Yirrkalla, 11 and 12 August 1948, USNM 174011, 19 specimens, 38 to 108 mm. standard length.

Description. Dorsal rays XII,9(in 2) or XIII,9(1); anal rays III,5(3); pelvic rays I,5(6); branched caudal rays 6+5=11(3); pectoral rays, i,5,xi(2) or i,4,xii(4); scale rows between head and caudal fin 40(1) or 42(2).

Scales cycloid on body, nape, upper part of opercle and behind eye; interorbital

* *Scorpaena porcus* Linnaeus was selected as the type species of *Scorpaena* Linnaeus by the International Commission on Zoological Nomenclature, Opinion 77.

ridges low, those of opposite sides curving toward, and joining at midline behind eye; dermal appendages scattered over body and head, but none noted on lower surface; spines of head about as figured by Richardson (loc. cit.), the lower spine of preorbital directed downward or curving slightly forward instead of backward as described by Matsubara (1943).

In colour, body mottled; undersurface of head, anteriorly, barred or blotched; mid-length of pelvic fins with large uniformly black area, not barred or blotched.

Colour in life. Generally brick red; abdomen, throat, and bases of pelvic and pectoral fins blood red.

Remarks. Matsubara (1943) has united *Sebastapristes* Gill, in which *S. bynoensis* is often placed, with *Scorpaena* Linnaeus, an action with which I am in tentative agreement. The present species is not the same as that reported from Japan or the mid-Pacific islands as *S. bynoensis,* since the scales are cycloid rather than ctenoid (the latter supposedly a character of *Sebastapristes*) and because of lesser differences in colour pattern and head spinulation.

PLATYCEPHALIDAE

I have referred, in part, to Matsubara and Ochiai (1955) for the definition of genera employed here, but suspect that *Cymbacephalus* belongs with the Inegociinae rather than the Onigociinae, since the side of the head is moderately bicarinate and in other characters, as given in the following key, it approaches *Inegocia.*

Key to the Arnhem Land Species

1 Vomerine teeth in a transverse crescentric band; palatine teeth uniserial; teeth of inner row in lower jaw slightly enlarged, larger than those in adjacent or outer rows; head moderately smooth; two acute preopercular spines, the lower one longest; iris lappet not cirrose; dorsal rays I,VIII-13; anal rays 13; three black bands on caudal fin . . . . . *Platycephalus endrachtensis*

Vomerine teeth in two parallel longitudinal bands; palatine teeth in broad bands; teeth in lower jaw subequal in size, those in the inner row not notably larger than those in adjacent rows; head moderately rugose; two preopercular spines, of which the upper is slightly longer and acute, the lower more obtuse; iris lappet cirrose; head and nape scaly; scales on body normal, not enlarged or spined except a few spined anteriorly along lateral line; side of head bicarinate; anterior nostril with a flap; distal margin of pectoral fin rounded; second dorsal spine nearly as long as third; lateral line scale duct opening through a single downward directed tube; basal tubercles on first gill arch uniserial 2

2 No prominent pit behind eye; no dermal flaps on eye; interorbital space about equal to eye height; interopercular membrane broad, its edge scalloped, not developed into a tongue-like projection; 12 rays in both the second dorsal and anal fins . . . . *Levanaora bosschei*

A large, moderately deep pit behind each eye; interopercular flap prominent, tongue-like; interorbital width about half eye length; second dorsal and anal rays both 10 or 11—genus *Cymbacephalus* 3

3 Several dermal flaps on upper half of eye; inner pelvic ray simple . . . . *Cymbacephalus nematophthalmus*

No dermal flaps on eye; inner pelvic ray branched . . *Cymbacephalus parilis*

Platycephalus endrachtensis Quoy and Gaimard

Pl. 65

Platycephalus endrachtensis Quoy and Gaimard, *Voyage autour du monde . . . sur . . . l'Uranie et la Physicienne . . . Zoologie*: 353 (1824); type locality, Sharks Bay, Western Australia; Cuvier and Valenciennes, *Histoire naturelle des poissons* 4: 240 (1829).

Record probably referable: *Platycephalus arenarius* Paradice and Whitley (1927): 91, Pellew Islands.

M48-8, Little Lagoon, USNM 173881, 2 specimens, 145 and 164 mm. standard length.

M48-10, Estuary, Thompson Bay, Port Langdon, USNM 173882, 1 specimen, 216 mm. standard length.

M48-15, Emerald River, Groote Eylandt, USNM 173883, 9 specimens, 50 to 243 mm. standard length.

M48-19, Creek, tributary to Little Lagoon, USNM 173884, 3 specimens, 81 to 114 mm. standard length.

Description. Dorsal rays I,VIII-13(in 15); anal rays 13(15); pelvic rays I,5(30); pectoral rays i or ii,11 to 13,v to vii, totals equal 19(8), 20(13), or 21(1); branched caudal rays 6+6=12(10).

Colour in alcohol. Medium brown above, with numerous small, rather indistinct, brown spots or flecks; yellowish white below; caudal fin light brownish with three black bands; one from lower procurrent rays through middle of lower half of fin to tip; another curving downward through basal half of upper lobe, thence along middle of fin to tip; and another, blotch like, on distal half of upper part of fin; dorsal and paired fins dotted with medium brown; anal fin immaculate.

Colour in life. Body and head uniformly brown above with darker brown spotting; ventral surface and lower side white; caudal fin milky white with upper lobe bright orange, fading to yellow proximally; three jet black bands on caudal fin; pupil cobalt blue.

Remarks. The name *endrachtensis* is tentatively applied to this species, which is close to *Platycephalus indicus* (*auct.*) and *Platycephalus insidiator* (*auct.*); probably all three names have been widely confused and variously interchanged. The present species appears to be the same as those known under the above names in the Australian literature, as I have examined a specimen from Lindeman Island, Queensland (USNM 177130), and believe the species to be common along the Australian coast.

Japan (USNM 62338) and China (USNM 6533) specimens identified as *Platycephalus indicus* in our collection have two small spines in front of the first dorsal fin, in agreement with counts given by Matsubara and Ochiai (1955), and the black caudal fin bands differ in shape and position from those in my material. In specimens from Travancore, the caudal fin colour approaches that of the Arnhem Land specimens and there appears to be only one small spine in front of the dorsal fin, but the preopercular spines are about equal in length.

In the original description of *P. endrachtensis,* counts of 14 soft dorsal and 14 anal rays are given, but Cuvier and Valenciennes obtained counts of 13 for both fins from the same specimen.

The genus *Cymbacephalus* is characterized by a moderately deep pit or depression at the medial, posterior side of the eye.

Cymbacephalus nematophthalmus* (Günther)

Platycephalus nematophthalmus Günther, *Catalogue of the Fishes in the British Museum* 2: 184 (1860); type localities, Port Essington and Victoria.

Recorded as: *Platycephalus nematophthalmus* Macleay (1878): 356 and (1880): 586, Port Darwin; *Insidiator nematophthalmus* Paradice and Whitley (1927): 92, Pellew Islands.

M48-9, Reefs and tide pools, entrance to Little Lagoon, USNM 173886, 6 specimens, 33 to 74 mm. standard length.

M48-19, Creek, tributary to Little Lagoon, USNM 173887, 1 specimen, 71 mm. standard length.

M48-21, Reef, Yirrkalla, 12 August 1948, USNM 173888, 1 specimen, 164 mm. standard length.

M48-28, Creek mouth and bay, Port Bradshaw, USNM 173889, 2 specimens, 127 and 146 mm. standard length.

Native names: *Áhmayrerra*, Groote Eylandt; *Máhlear*, Yirrkalla.

Description. Dorsal rays I,VIII-10(in 1), I,VIII-11(8), or I,IX-11(1); anal rays 11(10); pelvic rays I,4,i(20); pectoral rays ii,8 to 12,vii to x, totals equal 19(1), 20(9) or 21(2); branched caudal rays 6+6=12(4); cirri on eye 3 in small specimens to 7 or 8 visible in larger specimens, of which one over centre of eye is branched.

Colour in alcohol. Dark brown above and whitish below; a series of about 6 or 8 brownish black saddles on back, about 4 or 5 of which extend down over side as bars; a brownish black blotch at base of caudal fin; anal fin yellowish with a distal row of dark brown spots; other fins dusted, mottled, or spotted with dark brown, the dorsal and caudal fins with fine brownish black lines; head dark brown above, the lower lip with alternate dark brown and light bars.

Colour in life. Brown bands or rows of spots on first dorsal fin; four deep chocolate brown bars on body below dorsal fins and another similarly coloured blotch on caudal peduncle from caudal fin to a vertical through rear end of dorsal fin; a brownish black or black blotch on pelvic fin and a similarly coloured row of spots on anal fin; posterior end of caudal fin yellow (evidently based on small individuals).

Cymbacephalus parilis (McCulloch)

Insidiator parilis McCulloch, *Biological Results of the Fishing Experiments . . . by the F.I.S. 'Endeavour'* 2: 151, fig. 14, pl. 31, fig. 2 (1914); type localities, Port Darwin and Mapoon, Gulf of Carpentaria.

Records probably referable: *Platycephalus japonicus* Macleay (1880): 585 and Kent (1889): 239, Port Darwin.

M48-2, Nightcliff, USNM 173885, 1 specimen, 357 mm. standard length.

Description. In the single specimen, dorsal rays I,VIII-10; anal rays 10; pelvic rays I,5, both sides; pectoral rays ii,12,vi or vii; branched caudal rays 6+6=12.

Colour in alcohol. Dark brown above; yellowish below; fins heavily mottled or spotted with dark brown, the anal with a row of brown spots, each nearly as large as pupil; first dorsal dusky brown with a basal, light, triangular area between first and fourth long spines; brown spots, large as pupil, on caudal, forming zig-zag vertical rows; a series of brownish blotches (projection downward of the upper brown colour) at junction of upper brown and lower yellowish colour on body and head; indistinct dark brown bar below eye; lower lip blotched with brown.

* *Cymbacephalus* Fowler (1938): 90; type species, *Platycephalus nematophthalmus* Günther, by original designation.

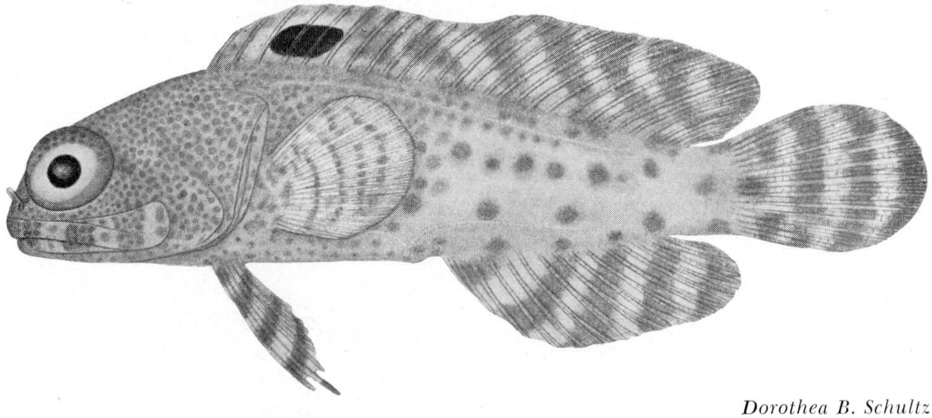

Dorothea B. Schultz

PLATE 64. *Gnathypops darwiniensis*
Yirrkalla, USNM 174042, 106 mm. standard length

Dorothea B. Schultz

PLATE 65. *Platycephalus endrachtensis*
Emerald River, USNM 173833, 161 mm. standard length

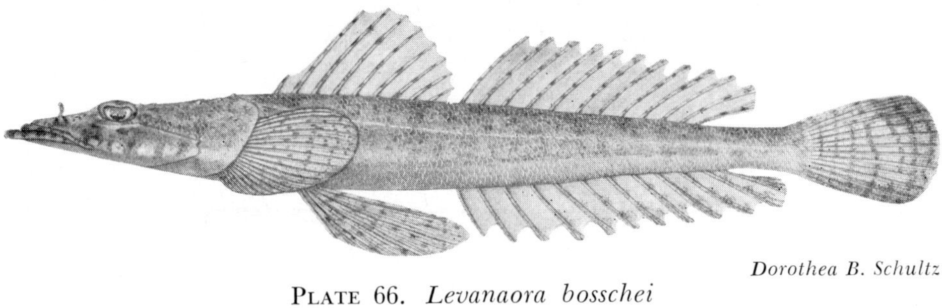

Dorothea B. Schultz

PLATE 66. *Levanaora bosschei*
Little Lagoon, USNM 173890, 136 mm. standard length

Remarks. McCulloch (loc. cit.) gave second dorsal and anal ray counts for this species as 11, whereas my counts for these fins are 10.

Levanaora bosschei* (Bleeker)

Pl. 66

Platycephalus bosscheis Bleeker, *Natuurk. Tijdschr. Ned.-Ind.* 21: 140 (1860); type locality, Muntok.

Platycephalus bosschei Bleeker, *Atlas ichthyologique des Indes Orientales Néêrlandaises* 9: pl. 419, figs. 3 and 3a (1878).

Recorded as: *Insidiator bosschei* McCulloch (1914): 139, Port Darwin; *Platycephalus isacanthus* Macleay (1880): 585, Port Darwin.

Record probably referable: *Platycephalus inops* Macleay (1878): 356, Port Darwin.

M48-9a, Reef, entrance, Little Lagoon, USNM 173890, 2 specimens, 130 and 136 mm. standard length.

M48-17, Reef, Bartalumba Bay, USNM 173891, 2 specimens, 103 and 140 mm. standard length.

Description. Dorsal rays I,VIII-12(in 3) or I,IX-12(1); anal rays 12(4); pelvic rays I,4,i(8); pectoral rays ii,9 to 11,vi to viii, totals equal 17(1), 19(1) or 20(6); branched caudal rays 5+4=9(1) or 6+5=11(3).

Colour in alcohol. Medium to dark brown above, covered with small brown spots, and with several rounded, saddle-like blotches along dorsal fin base; upper head mottled; lower lip with about four dark brown bars, as wide as the interspaces, on each side; lower surface of head and body yellowish brown; caudal, dorsal, pelvic and pectoral fins with small dark brown punctulations, which tend to form rows; an indefinite, almost obscure, blackish blotch on distal third of upper part of caudal fin and a similarly coloured band along lower part of caudal fin; rear half of anal fin with dark punctulations.

Remarks. Levanaora is close to *Inegocia* Jordan and Thompson, as pointed out by Matsubara and Ochiai (1955). My specimens differ from their definition of *Inegocia* in that the interopercular flap is a broad membrane with scalloped edge and the perforating duct of the lateral line scales communicates with the exterior through a single, downward directed tube. From their definition of *Levanaora*, they differ in that the iris lappet is cirrose.

PEGASIDAE

Pegasus volitans Linnaeus

Pegasus volitans Linnaeus, *Systema Naturae* ed. 10, 1: 338 (1758); type locality, Amboina.

Recorded as: *Pegasus volitans* Whitley and Allan (1958): 79, Northern Territory; *Parapegasus natans* Munro (1958): 95, Northern Territory.

BOTHIDAE

Arnoglossus aspilos praeteritus Whitley?

Arnoglossus aspilos praeteritus Whitley, *Proc. Roy. Zool. Soc. N.S.W.* 1948-49: 32, fig. 1 (1950); type locality, between Cape Jaubert and Wallal, Western Australia; recorded from off Darwin, Northern Territory.

* *Levanaora* Whitley (1933): 95; type species, *Platycephalus isacanthus* Cuvier and Valenciennes, by original designation.

Pseudorhombus arsius (Hamilton-Buchanan)

Pleuronectes arsius Hamilton-Buchanan, *An Account of the Fishes Found in the River Ganges*: 128 (1822); type locality, Ganges River, below Calcutta.
Rhombus lentiginosus Richardson, *Ann. Mag. Nat. Hist.* 11: 495 (1843); type locality, Port Essington 'and whole coast of Cobourg Peninsula'.

Recorded as: *Pseudorhombus arsius* Norman (1934): 101, Port Essington; *Pseudorhombus russellii* Günther (1862): 424, Port Essington; Macleay (1878): 362, Port Darwin; Macleay (1881): 124, Port Essington; and questionably as: *Pseudorhombus multimaculatus* Paradice and Whitley (1927); 91, Pellew Islands and Darwin.

M48-15, Emerald River, Groote Eylandt, USNM 174034, 1 specimen, 46 mm. standard length.

M48-28, Creek mouth and bay, Port Bradshaw, USNM 174035, 2 specimens, 71 and 102 mm. standard length.

Remarks. Dorsal rays 72, 73, and 74; scales along lateral line between head and caudal fin 83(in 1); gillrakers on lower limb of first arch, 9, 10, and 11; upper eye beginning slightly in advance of lower eye, the two adjacent, separated by a ridge; dorsal fin beginning over or slightly behind anterior naris of blind side, a little in advance of eye.

Colour pattern not prominent; apparently series of ocelli on body and two or three dark brown blotches along lateral line, the largest and most anterior at rear end of lateral line arch.

SOLEIDAE

Eyes on right side of body; dorsal fin extending forward on head to or beyond eye; fins without spines; preoperculum imbedded, hidden by scales of head.

Key to the Arnhem Land Species

1 Pectoral fins absent; a large pore near base of each ray of dorsal and anal fins 2
Pectoral fins present, variously developed; no large pore on fin rays 3
2 Body depth about 2·4 (2·1 to 2·4) in standard length; body colour medium brown with numerous yellowish, usually circular, patches which outlined by dark brown . . . . . *Pardachirus pavoninus*
Body depth about 2·6 to 2·8 in standard length; body reddish brown with three lengthwise rows of transverse dark blotches, one near dorsal fin base, one near anal fin base and one along mid-body . . . . . *Pardachirus rautheri*
3 Pectoral fin rays usually branched in part; operculum not connected with base of pectoral fin 4
Pectoral fin rays simple; opercular membrane joined to base of upper rays of pectoral fin 6
4 Gill openings restricted, widely separated from each other . . *Paradicula setifer*
Gill openings normal, confluent in front of pelvic fins; caudal, dorsal and anal fins broadly connected; rear nostril of coloured side opening from a tube anterior to lower eye; interorbital area scaled; right pectoral fin longer than eye diameter 5
5 Scales of blind side cycloid; 95 to 115 scales in a longitudinal series . . . . . *Synaptura aspilos*
Scales of blind side ctenoid; 65 to 80 scales in a longitudinal series . . . . . *Synaptura orientalis*

6 Pelvic fin free from anal fin; dorsal and anal fins nearly free from caudal fin, connected only at base; dorsal rays normal, not modified or elongated; head and body crossed by numerous dark bars which irregularly light centred or broken . . . . *Soleichthys heterorhinos*

Pelvic fin more or less united with the anal fin; dorsal, anal and caudal fins broadly united; body brownish, without bars, faintly and irregularly spotted *Phyllichthys sclerolepis*

Synaptura aspilos Bleeker

Synaptura aspilos Bleeker, *Natuurk. Tijdschr. Ned.-Ind.* 3: 74 (1852); type locality, Singapore.

Recorded as: *Brachirus aspilos* Ogilby (1910): 36, Croker Island, Northern Territory; Norman (1926a): 294.

Synaptura orientalis (Bloch and Schneider)

Pleuronectes orientalis Bloch and Schneider, *Systema Ichthyologiae*: 157 (1801); locality, Tranquebar.

Brachirus orientalis Norman, *Biological Results of the Fishing Experiments . . . by the F.I.S. 'Endeavour'* 5: 293 (1926).

Recorded as: *Synaptura nigra* Kent (1889): 240, Port Darwin.

Paradicula setifer* (Paradice)

Synaptura setifer Paradice in Paradice and Whitley, *Mem. Qld. Mus.* 9: 91, 98, and 101, text fig. 3 (1927); type locality, Port Darwin.

Soleichthys heterorhinos (Bleeker)

Solea heterorhinos Bleeker, *Acta Soc. Sci. Indo-Neerl.* 1: 64 (1856); type locality, Amboina.

Recorded as: *Aesopia heterorhinos* Whitley (1951): 62, Yirrkalla, Northern Territory.

M48-3, East Point Reef, USNM 174032, 1 specimen, 70 mm. standard length.

M48-21, Reef, Yirrkalla, 12 August 1948, USNM 174033, 1 specimen, 58 mm. standard length.

Remarks. Anal fin rays 79 and 83; body and head crossed by dark brown wavy, narrow bars which often with light centres; the bars wide as interspaces, which about equal to snout length. The long anterior nostril tube was described as prominently red in life.

Phyllichthys sclerolepis (Macleay)

Synaptura sclerolepis Macleay, *Proc. Linn. Soc. N.S.W.* 2: 363, pl. 10, fig. 4 (1878) and 6: 137 (1881); type locality, Port Darwin.

Phyllichthys sclerolepis McCulloch, *Mem. Qld. Mus.* 5: 66, pl. 9, fig. 2, text fig. 4 (1916); Norman, *Biological Results of the Fishing Experiments . . . by the F.I.S. 'Endeavour'*, 5: 297 (1926); Port Darwin.

M48-21, Reef, Yirrkalla, 12 August 1948, USNM 174031, 1 specimen, 136 mm. standard length (see below).

* *Paradicula* Whitley (1931a): 322; new name to replace the preoccupied *Whitleyia* Chabanaud; type species, *Synaptura setifer* Paradice, by original designation.

Remarks. Left pelvic fin free; right fin connected by broad membrane to anal fin; dorsal rays 87; anal rays 75; caudal rays 8+8=16; the dorsal, anal, and caudal fins continuous.

In preservation, the upper side is medium brown with numerous, faint, light brownish or yellowish spots or blotches.

Miller's notes record a sole, described as brown coloured with irregular white spots and blotches, from M48-3, East Point Reef, but do not note this large specimen which bears a label indicating that it is from M48-21. Since this is the only specimen available that could be described in that manner, I think it necessary to note the possibility of locality error.

Pardachirus pavoninus (Lacépède)

Achirus pavoninus Lacépède, *Histoire naturelle des poissons* 4: 658 and 661 (1802); type locality not given.

Recorded as: *Pardachirus pavoninus* Whitley (1951): 63, Melville Bay and Cape Arnhem, Northern Territory.

M48-21, Reef, Yirrkalla, 12 August 1948, USNM 174030, 1 specimen, 40 mm. standard length.

Pardachirus rautheri (Chabanaud)

Achirus rautheri Chabanaud, *Zool. Anz.* 93: 95, figs. 1-10 (1931); type locality, Port Darwin.

Recorded as: *Solea* (*Achirus*) *poropterus* Klunzinger (1879): 408, Port Darwin; *Solea* (*Achirus*) *poroptera* Macleay (1884): 51, Port Darwin; *Pardachirus klunzingeri* Weber (1908): 250, Port Darwin, but not types; Norman (1926a): 288; Whitley (1958): 42.

CYNOGLOSSIDAE

Eyes on the left side of body; dorsal fin extending forward on to head at least to eye; fins without spines; preopercle imbedded, hidden by scales of head.

Key to the Arnhem Land Species

1 Lips with short, stout, generally unbranched fringes; three lateral lines on coloured side of body . . . . . . *Paraplagusia guttata*

Lips smooth, without fringes 2

2 Three lateral lines on coloured side of body; blind side without lateral line; rear corner of mouth behind eye, closer to gill opening than to snout tip; 13 to 15 scale rows between upper and middle lateral lines, counted above branchial opening; scales weakly ctenoid on blind side, appearing smooth to the touch and strongly ctenoid on coloured side; rear nostril between eyes . . . . . . *Cynoglossus heterolepis*

Two lateral lines on coloured side of body 3

3 Two lateral lines on blind side of body; 19 or 20 rows of scales between lateral lines on coloured side . . . . . . *Cynoglossus sindensis*

No lateral line on blind side of body; 11 rows of scales between lateral lines on coloured side, counted slightly to rear of gill opening; scales ctenoid on both sides of body; interorbital distance small, much less than half diameter of eye; rear end of mouth nearer snout tip than gill opening; rear nostril in interorbital space . . . . . . *Cynoglossus maculipinnis*

Paraplagusia guttata (Macleay)

Plagusia guttata Macleay, *Proc. Linn. Soc. N.S.W.* 2: 362, pl. 10, fig. 3 (1878) and 6: 137 (1881); type locality, Port Darwin.
Paraplagusia guttata Norman, *Biological Results of the Fishing Experiments . . . by the F.I.S. 'Endeavour' 1909-14*, 5: 299 (1926).

Recorded as: *Rhinoplagusia guttata* Paradice and Whitley (1927): 91, Pellew Islands; *Plagusia japonica* Klunzinger (1879): 409, Port Darwin; *Rhinoplagusia japonica* McCulloch and Whitley (1925): 350, fig. 3, Port Darwin.

Cynoglossus sindensis Day

Cynoglossus sindensis Day, *Fishes of India*: 434, pl. 90, fig. 6 (1877); type locality, Sind and seas of India.

Recorded as: *Cynoglossus sindensis* Ogilby (1910): 37, Croker Island, Northern Territory.

Cynoglossus maculipinnis Rendahl

Cynoglossus maculipinnis Rendahl, *K. Svenska VetenskAkad. Handl.* 61: No. 9, 17 (1921); type locality, 45 mi. west-south-west of Cape Jaubert, Western Australia.

M48-28, Creek mouth and bay, Port Bradshaw, USNM 174029, 1 specimen, 42 mm. standard length.

Remarks. This species appears to be closely related to *Cynoglossus sibogae* Weber (1913) and may prove to be identical.

Cynoglossus heterolepis Weber

Cynoglossus heterolepis Weber, *Notes Leyden Mus.* 32: 237 (1910); type locality, Lorentz River, New Guinea; Weber, *Nova Guinea* 9: 590, pl. 14, figs. 3a to 3d (1913).

M48-40, East Alligator River, USNM 174028, 6 specimens, 63 to 132 mm. standard length.

Remarks. The specimens are dark brown in preservation and show only faint traces of darker brown cross bars—similar to, but slightly narrower than those illustrated by Weber (loc. cit., fig. 3b).

ECHENEIDAE

Echeneis naucrates* Linnaeus

Echeneis neucrates Linnaeus, *Systema Naturae* ed. 10, 1: 261 (1758); type locality, Indian Ocean.

Recorded as: *Echeneis naucrates* Richardson (1843a): 498; Günther (1860): 385, Port Essington; Paradice and Whitley (1927): 93, Pellew Islands and Port Darwin, on tiger shark.

M48-11, Little Lagoon, 14 and 15 April 1948, USNM 174118, 3 specimens, 470 to 540 mm. standard length.

* *Echeneis naucrates* Linnaeus was designated as type species of *Echeneis* Linnaeus by the International Commission on Zoological Nomenclature, Opinion 92.

Remora remora* (Linnaeus)

Echeneis remora Linnaeus, *Systema Naturae* ed. 10, 1: 260 (1758); type locality, Indian Ocean.

Recorded as: *Echeneis remora* Günther (1860): 379; Macleay (1880): 561, Port Essington.

TRIACANTHODIDAE

Triacanthus biaculeatus (Bloch)

Balistes biaculeatus Bloch, *Naturgeschichte der Ausländischen Fische* 2: 17, pl. 148, fig. 2 (1786); reference not seen; type localities, Japan and East Indies.

Recorded as: *Triacanthus biaculeatus* Günther (1870): 210 and Macleay (1881): 308, Port Essington; Kent (1889): 240, Port Darwin; Paradice and Whitley (1927): 94, Pellew Islands and Darwin.

Triacanthus brevirostris Temminck and Schlegel

Triacanthus brevirostris Temminck and Schlegel, *Fauna Japonica, Pisces*: 294, pl. 129, fig. 2 (1842); type locality, Japan.

Recorded as: *Triacanthus brevirostris* Klunzinger (1879): 421, Port Darwin.

BALISTIDAE

Abalistes† *stellaris* (Bloch and Schneider)

Balistes stellaris Bloch and Schneider, *Systema Ichthyologiae*: 476 (1801); type locality, Indian Ocean.

Balistes phaleratus Richardson in Stokes, *Discoveries in Australia*... 1: 484, pl. 1, figs. 4 and 5 (1846); type locality, western coasts of Australia.

M48-22, Bay of Yirrkalla, 4, 6, and 18 August 1948, USNM 173976-7, 3 specimens, 251 to 272 mm. standard length.

M48-30, Reefs, Yirrkalla, 28 August 1948, USNM 173978, 1 specimen, 282 mm. standard length.

Colour in life. Spots on body and fins yellow or light yellowish brown.

Remarks. This species, placed in a monotypic genus, is characterized by a depressed caudal peduncle.

MONACANTHIDAE

Monacanthus chinensis (Bloch)?

Balistes chinensis Bloch, *Naturgeschichte der Ausländischen Fische* 2: 29, pl. 112, fig. 1 (1786); reference not seen; Bloch, *Ichthyologie* 5: 26, pl. 112, fig. 1 (1787); type locality, China.

Monacanthus megalourus Richardson, *Icones Piscium*: 5, pl. 1, fig. 3 (1843); type locality, Houtmans Abrolhos, Western Australia.

Recorded as: *Monacanthus chinensis* Macleay (1878): 365, Port Darwin; Paradice and Whitley (1927): 94, Pellew Islands; *Monocanthus megalurus* (error for *megalourus*) Klunzinger (1879): 424, Port Darwin.

* *Echeneis remora* Linnaeus was designated as the type species of *Remora* Gill by the International Commission on Zoological Nomenclature, Opinion 242, but the generic name apparently dates from earlier authors.

† Herre (1953): 822, listed *Leiurus* Swainson in the synonymy of *Abalistes* Jordan and Seale, but failed to indicate that *Leiurus* Swainson, as used for the present species, is preoccupied in the same work by *Leiurus* Swainson, a genus of Gasterosteidae.

M48-9, Reefs and tide pools, entrance to Little Lagoon, USNM 173979, 1 specimen, 22 mm. standard length.

Remarks. The identification of this small individual is tentative since I do not have adequate material to determine the change with age of the pelvic flap. In the specimen, it ends before the tips of the flexible pelvic spines. There are 29 anal rays and 31 dorsal rays.

In *Monacanthus* there is a single row of curved serrae on each side of the dorsal spine posteriorly, the serrae directed backward and downward. *Meuschenia* Whitley, which may not be distinct from *Pseudomonacanthus* Bleeker, differs in possessing both the posterior row and an anterior row of serrae on each side; no flexible pelvic spine is present.

Meuschenia trachylepis (Günther)

Monacanthus trachylepis Günther, *Catalogue of the Fishes in the British Museum* 8: 248 (1870); type locality, Australia.

Recorded as: *Monacanthus trachylepis* Klunzinger (1879): 422, Port Darwin.

Anacanthus barbatus Gray

Anacanthus barbatus Gray, *Zoological Miscellany* 1: 8 (1831); type locality, Singapore.

Recorded as: *Psilocephalus barbatus* Paradice and Whitley (1927): 94, Pellew Islands.

Remarks. Use of the name *Anacanthus* in Dasyatidae or rays is apparently in error, since the frequently quoted reference: Ehrenberg in Cuvier, *Règne Animal* ed. 2, 2: 400 (1829) is to the French vernacular, 'Les Anacanthes'.

OSTRACIIDAE

Two species of *Aracana* Gray have been recorded from Port Darwin. Since the genus is believed confined in Australia to southern latitudes, the records are questioned and thought to be based on locality misstatements (McCulloch and Waite, 1915). They are: *Aracana aurita* Klunzinger (1879): 424 and *Aracana ornata* Klunzinger (1879): 424.

Ostracion tuberculatus Linnaeus

Ostracion tuberculatus Linnaeus, *Systema Naturae* ed. 10, 1: 331 (1758); type locality, India.

Ostracion cubicus Linnaeus, *Systema Naturae* ed. 10, 1: 332 (1758); type locality, India.

Recorded as: *Ostracion cubicus* Macleay (1878): 365 and (1881): 332, Port Darwin.

Ostracion rhinorhynchus Bleeker

Ostracion rhinorhynchus Bleeker, *Verh. Batavia Genoot.* 24: 34, pl. 6, fig. 12 (1852); type locality, East Indies.

Recorded as: *Ostracion rhinorhynchus* Klunzinger (1879): 424 and Paradice and Whitley (1927): 94 and 98, Port Darwin.

Lactoria cornuta* (Linnaeus)

Ostracion cornutus Linnaeus, *Systema Naturae* ed. 10, 1: 331 (1758); type locality, India.

Recorded as: *Ostracion cornutus* Kent (1889): 240, Port Darwin.

TETRAODONTIDAE

Tetraodon immaculatus Bloch and Schneider

Tetrodon immaculatus Bloch and Schneider, *Systema Ichthyologiae*: 507 (1801); type locality not given; based on a drawing by Commerson.

Tetrodon manillensis Procé, *Bull. Soc. Philomathique*: 130 (1822); type locality, Manila, Philippine Islands.

Tetrodon virgatus Richardson, *Ichthyology of the Voyage of H.M.S. Erebus and Terror*. . .: 62, pl. 39, figs. 8 and 9 (1846); type locality, Port Jackson.

Recorded as: *Tetrodon virgatus* Macleay (1878): 365 and (1881): 339, Port Darwin; *Tetraodon immaculatus* var. *manillensis* Paradice and Whitley (1927): 94, 96, and 99, Pellew Islands; poisoning.

Off Nightcliff headland, 7 miles north of Darwin, Northern Territory, 3 April 1948, USNM 173985, 1 specimen, 132 mm. standard length.

Remarks. I have followed usage in identifying this specimen as *immaculatus*, although I am not certain that is correct, since *immaculatus* Bloch and Schneider may be distinct from *manillensis* Procé (= *virgatus* Richardson). The specimen has imperforate nares, with two tentacles at each organ. Between the predorsal midline and the midline of the abdomen, there are approximately nine narrow, dark brown longitudinal lines, similar to those shown by Richardson. The margin of the caudal fin is dark brown or black.

Tetraodon nigropunctatus Bloch and Schneider

Tetrodon nigro-punctatus Bloch and Schneider, *Systema Ichthyologiae*: 507 (1801); type locality, Tranquebar.

Recorded as: *Tetraodon nigropunctatus* Paradice and Whitley (1927): 95, Pellew Islands.

Sphoeroides pleurostictus (Günther)

Tetrodon pleurostictus Günther, *Proc. Zool. Soc. Lond.* 1871: 674, pl. 69, fig. a (1872); type localities, Port Bower and Port Mackay, Queensland.

Tetrodon fasciatus Macleay, *Proc. Linn. Soc. N.S.W.* 2: 365, pl. 10, fig. 5 (1878) and 6: 340 (1881); type locality, Port Darwin.

M48-2, Nightcliff, sandy beach, USNM 173983, 1 specimen, 68 mm. standard length.

M48-5, Woods Inlet, Delissaville, USNM 173984, 1 specimen, 99 mm. standard length.

Remarks. The black bar between the eyes, that one ending behind the pectoral fin, and the one below the dorsal fin, all are prominent. The caudal peduncle, behind the dorsal fin, has no prominent coloration. Each nasal organ is bulb-like, with an anterior and a posterior opening.

* *Lactoria* Jordan and Fowler (1902): 278; type species, *Ostracion cornutus* Linnaeus, by original designation.

Sphoeroides whitleyi Paradice

Spheroides whitleyi Paradice, *Mem. Qld. Mus.* 9: 94, 99, 105, Pl. 15 (1927); Pellew Islands; type locality, Port Denison, Queensland.

Sphoeroides lunaris (Bloch and Schneider)

Tetrodon lunaris Bloch and Schneider, *Systema Ichthyologiae*: 505 (1801); type locality, Malabar.

Tetrodon darwinii Castelnau, *Proc. Zool. Acclim. Soc. Vict.* 2: 94 (1873); type locality, Port Darwin; Macleay, *Proc. Linn. Soc. N.S.W.* 6: 341 (1881).

Recorded as: *Spheroides lunaris* Paradice and Whitley (1927): 94, Pellew Islands.

Sphoeroides richei (Freminville)

Tetrodon richei Freminville, *Nouv. Bul. Soc. Philom. Paris* 3: 250, pl. 4, fig. 2 (1813); type locality, 'terre de Nyts'.

Recorded as: *Tetrodon richei* Klunzinger (1879): 425, Port Darwin.

Sphoeroides hamiltoni (Richardson)

Tetrodon hamiltoni Richardson, *Ichthyology of the Voyage of H.M.S. Erebus and Terror* . . .: 63, pl. 39, figs. 10 and 11 (1846); type locality, Port Jackson.

Recorded as: *Tetrodon hamiltonii* Klunzinger (1879): 425, Port Darwin.

Chelonodon patoca (Hamilton-Buchanan)

Tetrodon patoca Hamilton-Buchanan, *An Account of the Fishes found in the River Ganges* . . .: 7 and 362, pl. 18, fig. 2 (1822); type locality, estuaries of Ganges River; Day, *The Fishes of India*: 703, pl. 182, fig. 4 (1878).

Chelonodon dapsilis Whitley, *Proc. Linn. Soc. N.S.W.* 68: 143 (1943); type locality, near Mackenzie Island, Fitzroy River estuary, Queensland.

Recorded as: *Tetrodon patoca* Macleay (1878): 365 and (1881): 339, Port Darwin; *Leiodon patoca* Paradice and Whitley (1927): 94, 99, Port Darwin.

M48-2, Nightcliff, USNM 173980, 1 specimen, 13 mm. standard length.

M48-5, Woods Inlet, Delissaville, USNM 173981, 1 specimen, 38 mm. standard length.

M48-28, Creek mouth and bay, Port Bradshaw, USNM 173982, 1 specimen, 149 mm. standard length.

Description. Nasal organ a shallow pit with an anterior and a posterior fringe, without an opening. Dark brown above, with faint bars or saddles as follows: between the eyes, above and downward to rear of pectoral fin, below dorsal fin, and on upper caudal peduncle at its narrowest part; remainder of upper surface and upper side a reticulated dark brown network or a series of round to oblong yellowish white spots over the brown background; lower surface and fins, except caudal, yellowish white; caudal fin dusky basally and dark brown distally.

DIODONTIDAE

? *Diodon** *holocanthus* Linnaeus

Diodon holocanthus Linnaeus, *Systema Naturae* ed. 10, 1: 335 (1758); type locality, India.

Recorded as: *Diodon maculatus* Macleay (1878): 366 and Klunzinger (1879): 424, Port Darwin.

* *Diodon hystrix* Linnaeus was selected as the type species of *Diodon* Linnaeus by the International Commission on Zoological Nomenclature, Opinion 77.

BATRACHOIDIDAE

*Halophryne** *diemensis* (LeSueur)

Pl. 67

Batrachoides diemensis LeSueur, *J. Acad. Nat. Sci. Philad.* 3: 402 (1824); type locality, Van Diemen's Land.

Batrachus mülleri Klunzinger, *S.B. Akad. Wiss. Wien* 80: 387, pl. 9, fig. 1 (1879); type locality, Port Darwin; Macleay, *Proc. Linn. Soc. N.S.W.* 9: 29 (1884).

Recorded as: *Batrachus diemensis* Richardson (1843a): 352 and (1844-8): 17, pl. 8, figs. 1 and 2; Günther (1861): 170 and (1867): 61, Port Essington; Macleay (1878): 355 and (1880): 573, Port Darwin; *Coryzichthys diemensis* Ogilby (1908): 51, Port Essington and Port Darwin; McCulloch (1926): 39; Paradice and Whitley (1927): 94 and 96, Darwin.

M48-1, Nightcliff, USNM 174021, 2 specimens, 84 and 139 mm. standard length.

M48-3, East Point Reef, USNM 174022, 1 specimen, 113 mm. standard length.

M48-17, Reef, Bartalumba Bay, USNM 174023, 1 specimen, 109 mm. standard length.

M48-21, Reef, Yirrkalla, 12 August 1948, USNM 174024, 2 specimens, 55 and 83 mm. standard length.

M48-33, Beach and mangroves, at mission, south-east side Milingimbi Island, USNM 174025, 1 specimen, 144 mm. standard length.

Description. Dorsal rays III-20(in 2) or III-21(1); anal rays 17(3); no axillary pore; two subopercular spines, the lower one rather small; teeth small, pointed, on jaws and in an irregularly developed band on vomer.

Batrachomoeus trispinosus (Günther)

Pl. 68

Batrachus trispinosus Günther, *Catalogue of the Fishes in the British Museum* 3: 169 (1861); type localities, Sea of Pinang and Singapore.

Pseudobatrachus dahli Rendahl, *Medd. Zool. Mus. Krist.* 5: 195 (1922); type locality, Roebuck Bay, Western Australia.

Recorded as: *Batrachomoeus dahli* Whitley (1954): 28, Cape Bedwell and Melville Island, Northern Territory.

M48-21, Reef, Yirrkalla, 12 August 1948, USNM 174026, 1 specimen, 97 mm. standard length.

M48-33, Beach and mangroves, at mission, south-east side Milingimbi Island, USNM 174027, 2 specimens, 135 and 138 mm. standard length.

Description. Dorsal rays III-22(in 1) or III-23(2); anal rays 19(3); axillary pore present, small; a single subopercular spine; teeth bluntly conical, those in lower jaw in about 2 rows anteriorly and one posteriorly; in upper jaw in two anterior rows, which reduce to one posteriorly; vomerine teeth in a single row.

This species appears to be close to *Batrachomoeus broadbenti* Ogilby (1908) but differs chiefly in lacking the second subopercular spine.

ANTENNARIIDAE

Schultz (1957) has described the genera and the members of this family. I have compiled below his identifications of the two Arnhem Land specimens. The specific assignment of others from the literature is highly tentative.

* *Halophryne* Gill (1863): 170; type species, *Batrachoides diemensis* LeSueur, by original designation; a synonym of which is *Coryzichthys* Ogilby; for references to latter see McCulloch (1929-30): 358.

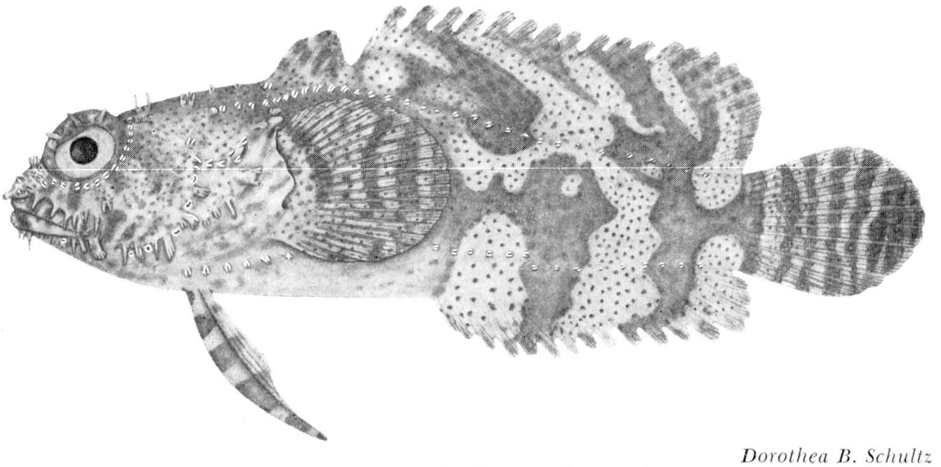

Dorothea B. Schultz

PLATE 67. *Halophryne diemensis*
Nightcliff, USNM 174021, 84 mm. standard length

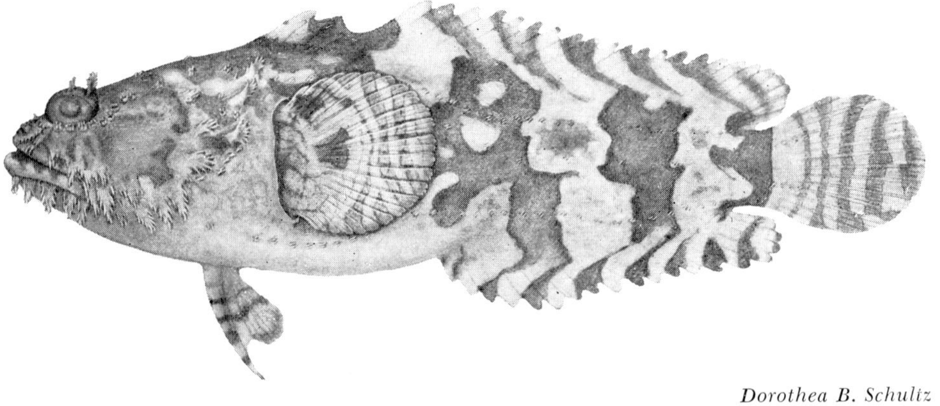

Dorothea B. Schultz

PLATE 68. *Batrachomoeus trispinosus*
Milingimbi Island, USNM 174027, 138 mm. standard length

Antennarius asper Macleay

Antennarius asper Macleay, *Proc. Linn. Soc. N.S.W.* 5: 580 (1880); type locality, Darnley Island, Queensland.

Recorded as: *Antennarius asper* Whitley (1954): 28, Darwin; Schultz (1957): 89, USNM 164194, locality apparently in error, see below. Records probably referable: *Antennaries urophthalmus* Macleay (1878): 356, Port Darwin; *Antennarius urophthalmus* Klunzinger (1879): 388; Macleay (1880): 578; Kent (1889): 239; Paradice and Whitley (1927): 94 and 98, Port Darwin.

Schultz (loc. cit.) has recorded a specimen of this species from Woods Inlet, west of Darwin, collected by Robert R. Miller. Miller made two collections in this area, both in fresh and brackish water. His notes do not indicate that an antennarid was captured there, but record taking one 'adult angler fish . . . from outer edge of reef' at station M48-3 (East Point Reef, about 6 miles north-north-west of Darwin, 26 March 1948). I feel that the latter is probably the correct locality both for this specimen and for the misplaced specimens of *Amphiprion ephippium,* which see.

I am unable to judge the systematic status of the material recorded from the Darwin area. It is to be noted that Macleay recorded *urophthalmus* from Darwin and described *asper* in the same paper, and as above indicated, a specimen was later identified as *asper* from this region. Both are characterized by large white ocelli on the caudal fin.

Antennarius chironectes Lacépède ?

Antennarius chironectes Lacépède, *Histoire naturelle des poissons* 1: 325, pl. 14, fig. 2 (1798); type locality not given.

Recorded as: *Antennarius commersonii* Macleay (1878): 356 and (1880): 578; Klunzinger (1879): 387, Port Darwin.

Tathicarpus butleri Ogilby

Tathicarpus butleri Ogilby, *Proc. Roy. Soc. Qld.* 20: 20 (1907); type locality, Port Curtis, Queensland.

Recorded as: *Tathicarpus butleri* Schultz (1957): 54 and 64, USNM 164195, Jarape (Jarapa), Groote Eylandt, Northern Territory.

REFERENCES

Ahl, E. (1923). Zur Kenntnis der Knochenfischfamilie Chaetodontidae insbesondere der Unterfamilie Chaetodontinae. *Arch. Naturgesch.* 89, abth. A. heft 5: 1-205, 2 pls.

Bleeker, P. (1851). Over eenige nieuwe geslachten en soorten van Makreelachtige visschen van der Indischen Archipel. *Natuurk. Tijdschr. Ned.-Ind.* 1: 341-72.

——— (1863). Onzième Notice sur la Faune ichthyologique de l'Ile de Ternate. *Ned. Tijdschr. Dierk.* 1: 228-38.

——— (1866-72). *Atlas ichthyologique des Indes Orientales Néêrlandaises.* 6: 1-170, pls. 241-78.

——— (1877a). Ibid. 9: 1-80, pls. 363-420.

——— (1877b). Mémoire sur les Chromides marins ou Pomacentroides de l'Inde Archipelagique. *Natuurk. Verh. Holland. Maatsch. Wet.* ser. 3, 2: 1-166.

Bloch, M. E. (1797). *Ichthyologie . . .* 7: 1-104, pls. 217-52.

——— and Schneider, J. G. (1801). *Systema Ichthyologiae.*

Bory de Saint-Vincent, J. B. G. M. (1823). *Dictionnaire Classique d'Histoire naturelle*. 3: 160, pl. 90.
Bosc, L. A. G. (1816). *Nouveau Dictionnaire d'Histoire naturelle* ed. 2, 2: 442.
Boulenger, G. A. (1895). *Catalogue of the Perciform Fishes in the British Museum*. 1: i-xix, 1-394, pls. 1-15.
Castelnau, Count F. L. (1873). Notes on Fishes from North Australia. *Proc. Zool. Acclim. Soc. Vict.* 2: 83-97.
——— (1875). Researches on the Fishes of Australia, in *Intercolonial Exhibition Essays. (Philadelphia Centennial Exhibition of 1876, Official Record)*: 1-52.
——— (1879). Notes on the Fishes of the Norman River. *Proc. Linn. Soc. N.S.W.* 3: 41-51.
Cloquet, H. (1816). *Dictionnaire des Sciences naturelles*. 2, supplement: 116.
Cuvier, le Baron G. (1817). *Règne Animal* ed. 1, 2: 1-532.
——— and Valenciennes, A. (1828). *Histoire naturelle des poissons*. 2: 1-490.
Duncker, G. (1909). Pisces, Part I Syngnathidae, in *Die Fauna Südwest-Australiens*. . . 2: 233-50.
Fowler, H. W. (1928). The Fishes of Oceania. *Mem. Bishop Mus.* 10.
——— (1931). The Fishes of the Families Pseudochromidae, Lobotidae, Pempheridae, Priacanthidae, Lutjanidae, Pomadasyidae, and Teraponidae, collected by the United States Bureau of Fisheries Steamer *Albatross*, chiefly in Philippine Seas and Adjacent Waters. *U.S. Nat. Mus. Bull.* 100 (11).
——— (1933). The Fishes of the Families Banjosidae, Lethrinidae, Sparidae, Girellidae, Kyphosidae, Oplegnathidae, Gerridae, Mullidae, Emmelichthyidae, Sciaenidae, Sillaginidae, Arripidae, and Enoplosidae, collected by the United States Bureau of Fisheries Steamer *Albatross*, chiefly in Philippine Seas and Adjacent Waters. Ibid. 100 (12).
——— (1938). Descriptions of new fishes obtained by the United States Bureau of Fisheries Steamer *Albatross*, chiefly in Philippine Seas and Adjacent Waters. *Proc. U.S. Nat. Mus.* 85: 31-135, figs. 1-61.
——— (1941). The Fishes of the Groups Elasmobranchii, Holocephali, Isospondyli, and Ostarophysi obtained by the United States Bureau of Fisheries Steamer *Albatross*, in 1907 to 1910, chiefly in the Philippine Islands and Adjacent Seas. *U.S. Nat. Mus. Bull.* 100 (13).
——— and Bean, B. A. (1928). The Fishes of the Families Pomacentridae, Labridae, and Callyodontidae, collected by the United States Bureau of Fisheries Steamer *Albatross*, chiefly in Philippine Seas and Adjacent Waters. Ibid. 100 (7).
——— (1929). The Fishes of the Series Capriformes, Ephippiformes, and Squamipennes, collected by the United States Bureau of Fisheries Steamer, *Albatross*, chiefly in Philippine Seas and Adjacent Waters. Ibid. 100 (8).
——— (1930). The Fishes of the Families Amiidae, Chandidae, Duleidae, and Serranidae, obtained by the United States Bureau of Fisheries Steamer *Albatross*, in 1907 and 1910, chiefly in the Philippine Islands and Adjacent Seas. Ibid. 100 (10).
Fraser-Brunner, A. (1933). A Revision of the Chaetodont Fishes of the Subfamily Pomacanthinae. *Proc. Zool. Soc. Lond.* 1933: 543-99, 1 pl., 29 text figs.
——— (1934). Substitute Name for *Heteropyge* Fraser-Brunner, a genus of Chaetodont fishes. *Copeia* 1934: 192.
——— (1938). Notes on the Scatophagid Fishes. *Aquarist and Pondkeeper* 8: 72-5, 3 figs.
——— (1955). A Synopsis of the Centropomid Fishes of the Subfamily Chan-

dinae, with Descriptions of a New Genus and two New Species. *Bull. Raffles Mus.* 25: 185-213, 4 figs.

Garrick, J. A. F. (1962). Reasons in favour of retaining the generic Name *Carcharhinus* Blainville, and a Proposal for identifying its Type Species as the Indo-Pacific black-tipped shark, *C. melanopterus*. *Proc. Biol. Soc. Wash.* 75: 89-96.

Gill, T. N. (1863). Descriptive Enumeration of a Collection of Fishes from the Western Coast of Central America, presented to the Smithsonian Institution by Capt. John M. Dow. *Proc. Acad. Nat. Sci. Philad.* 15: 162-80.

Gistel, J. (1848). *Naturgeschichte des Thierreichs fur höhere Schulen*: i-xvi, 1-216, pls. 1-32.

Guichenot, Alphonse (1839). *Dictionnaire pittoresque d'Histoire naturelle et des Phénomènes de la Nature* . . . 8: 135-6 and 301-2.

Günther, A. (1859). *Catalogue of the Fishes in the British Museum.* Vol. 1.

——— (1860). Ibid. Vol. 2.

——— (1861). Ibid. Vol. 3.

——— (1862). Ibid. Vol. 4.

——— (1864). Ibid. Vol. 5.

——— (1866). Ibid. Vol. 6.

——— (1867). Additions to the knowledge of Australian Reptiles and Fishes. *Ann. Mag. Nat. Hist.* ser. 3, 20: 45-68.

——— (1870). *Catalogue of the Fishes in the British Museum.* Vol. 8.

Hamilton-Buchanan, F. (1822). *An Account of the Fishes found in the River Ganges and its Branches.*

Herald, E. S. (1941). A Systematic Analysis of Variation in the Western American Pipefish, *Syngnathus californiensis*. *Stanf. Ichthyol. Bull.* 2: 49-73, 12 figs.

Herre, A. W. (1953). Check List of Philippine Fishes. *U.S. Fish and Wildlife Service Research Report.* 20: 1-977.

——— and Montalban, H. R. (1927). The Philippine Butterfly Fishes and their Allies. *Philippine J. Sci.* 34: 1-113, pls. 1-24.

Iredale, T. and Whitley, G. P. (1938). The Fluvifaunulae of Australia. *S. Aust. Nat.* 18: 64-8, 1 map.

Jordan, D. S. (1919). *Genera of Fishes* pt. 2: 163-284.

——— and Evermann, B. W. (1896). The Fishes of North and Middle America. *U.S. Nat. Mus. Bull.* 47 (1).

——— and Fowler, H. W. (1902). A Review of the Trigger-Fishes, File-Fishes and Trunk-Fishes of Japan. *Proc. U.S. Nat. Mus.* 25: 251-86, figs. 1-6.

——— and Gilbert, C. H. (1883). Synopsis of the Fishes of North America. *U.S. Nat. Mus. Bull.* 16.

Kent, W. S. (1889). Preliminary Observations on a Natural History Collection made in connection with the Surveying Cruise of H.M.S. *Myrmidon*, at the Port Darwin and Cambridge Gulf—September to November, 1888. *Proc. Roy. Soc. Qld.* 6: 219-42, pl. 13.

Klunzinger, C. B. (1879). Die v. Müllerische Sammlung Australischer Fische in Stuttgart. *S.B. Akad. Wiss. Wien* 80: 325-430, pls. 1-9.

Kner, R. (1866). Specielles Verzeichniss der während der Reise der kaiserl. Fregatte 'Novara' gesammelten Fische. Ibid. 53: 543-50.

Lacépède, Comte B. G. (1803). *Histoire naturelle des poissons* . . . Vol. 5.

Macleay, W. (1878). The fishes of Port Darwin. *Proc. Linn. Soc. N.S.W.* 2: 344-67, pls. 7-10.

——— (1879a). On the Clupeidae of Australia. Ibid. 4: 363-85.

———— (1879b). On the Mugilidae of Australia. Ibid. 4: 410-27.

———— (1880). Descriptive Catalogue of the Fishes of Australia. Ibid. 5: 302-444 and 510-629, pls. 13-14.

———— (1881). Ibid. Ibid. 6: 1-138 and 202-387, pls. 1-2.

———— (1884). Supplement to the Descriptive Catalogue of the Fishes of Australia. Ibid. 9: 2-64.

Matsubara, K. (1943). Studies on the Scorpaenoid Fishes of Japan (1). *Trans. Sigenk. Kenk.* 1: 1-486, pls. 1-4, 155 figs.

———— and Ochiai, A. (1955). A Revision of the Japanese Fishes of the Family Platycephalidae (The Flatheads). *Mem. Coll. Agric., Kyoto Univ.* 68 (Fishery Ser. 4): 1-109, pls. 1-3, 33 figs.

McCulloch, A. R. (1910). In: Notes and Exhibits. *Proc. Linn. Soc. N.S.W.* 35: 688.

———— (1913). Studies in Australian Fishes. *Rec. Aust. Mus.* 9: 355-89, 9 pls., 2 figs.

———— (1914). Report on some Fishes obtained by the F.I.S. *Endeavour* on the coasts of Queensland, New South Wales, Victoria, Tasmania, South and South-Western Australia. Part 2. *Biological Results of the Fishing Experiments . . . by the F.I.S. 'Endeavour'*, 1909-14. 2: 77-165, pls. 13-34.

———— (1917). Studies in Australian Fishes. No. 4. *Rec. Aust. Mus.* 11: 163-88, pls. 29-31.

———— (1918). Four Queensland Fishes. *Mem. Qld. Mus.* 6: 91-6, 4 pls.

———— (1920). Studies in Australian Fishes No. 6, with a description of a new *Girellops* from the Kermadec Islands. *Rec. Aust. Mus.* 13: 41-71, pls. 10-14.

———— (1923). Notes on Fishes from Australia and Lord Howe Island. Ibid. 14: 1-7, 3 pls.

———— (1924). Ichthyological Items, No. 2. *Mem. Qld. Mus.* 8: 61-76, pls. 11-14, 6 text figs.

———— (1926). Studies in Australian Fishes. *Rec. Aust. Mus.* 15: 28-39.

———— (1929-30). A check-list of the fishes recorded from Australia. *Aust. Mus. Mem.* 5: i-x, 1-534.

———— and Waite, E. R. (1915). A Revision of the genus *Aracana* and its allies. *Trans. Roy. Soc. S.Aust.* 39: 477-93, pls. 16-25.

———— and Whitley, G. P. (1925). Some Little Known Australian Flatfishes. *Rec. Aust. Mus.* 14: 342-54, 1 pl., 4 figs.

Munro, I. S. R. (1943). Revision of Australian Species of *Scomberomorus*. *Mem. Qld. Mus.* 12: 65-95, pls. 6-8, 4 text figs.

———— (1949). Revision of Australian Silver Breams, *Mylio* and *Rhabdosargus*. Ibid. 12: 182-223, pls. 16-23, 5 figs.

———— (1950). Revision of *Bregmaceros* with Descriptions of Larval Stages from Australasia. *Proc. Roy. Soc. Qld.* 61: 37-53, text figs. 1-9.

———— (1956). *Handbook of Australian Fishes.* (Canberra). Nos. 1-6: 1-28, figs. 1-198.

———— (1957). Ibid. Nos. 7-18: 29-76, figs. 199-532.

———— (1958). Ibid. Nos. 19-30: 77-124, figs. 533-812.

Myers, G. S. (1936). On the Indo-Australian Fishes of the Genus *Scatophagus*, with Description of a New Genus, *Selenotoca*. *Proc. Biol. Soc. Wash.* 49: 83-6.

Norman, J. R. (1926a). A Report on the Flatfishes (Heterosomata) collected by the F.I.S. *Endeavour*, with a Synopsis of the Flatfishes of Australia and a Revision of the Subfamily Rhombosoleinae. *Biological Results of the Fishing Experiments . . . by the F.I.S. 'Endeavour'*, 1909-14. 5: 219-308, 15 figs.

———— (1926b). A Synposis of the Rays of the Family Rhinobatidae, with a Revision of the Genus *Rhinobatus*. *Proc. Zool. Soc. Lond.* 1926: 941-82, 30 text figs.

——— (1934). *A Systematic Monograph of the Flatfishes (Heterosomata) Vol. 1, Psettodidae, Bothidae, Pleuronectidae.* British Museum.

Ogilby, J. D. (1888a). *Catalogue of the Fishes in the Collection of the Australian Museum.* Part I: 1-28.

——— (1888b). List of the Australian Palaeichthyes with Notes on their Synonymy and Distribution. *Proc. Linn. Soc. N.S.W.* ser. 2, 3: 1765-72.

——— (1889). Ibid. Ibid. ser. 2, 4: 178-86.

——— (1897). New Genera and Species of Australian Fishes. Ibid. 22: 62-5.

——— (1898). New Genera and Species of Fishes. Ibid. 23: 280-99.

——— (1907). Symbranchiate and Apodal Fishes New to Australia. *Proc. Roy. Soc. Qld.* 20: 1-15.

——— (1908). Revision of the Batrachoididae of Queensland. *Ann. Qld. Mus.* 9: 43-55.

——— (1910). On New or Insufficiently Described Fishes. *Proc. Roy. Soc. Qld.* 23: 1-55.

——— (1915). On Some New or Little-Known Australian Fishes. *Mem. Qld. Mus.* 3: 117-29, 2 pls.

——— and McCulloch, Allan R. (1909). A Revision of the Australian Orectolobidae. *Proc. Roy. Soc. N.S.W.* 42: 264-99, 2 pls., 1 text fig.

——— (1916). A Revision of the Australian Therapons with notes on some Papuan Species. *Mem. Qld. Mus.* 5: 99-126, pls. 10-13, 1 text fig.

Paradice, W. E. J. and Whitley, G. P. (1927). Northern Territory Fishes. Ibid. 9: 76-106, pls. 11-15, text figs. 1-3.

Perugia, A. (1894). Viaggio di Lamberto Loria nella Papuasia Orientale XIII. Pesci d'Acque Dolce. *Ann. Mus. Stor. Nat. Genova* ser. 2, 14: 546-53.

Randall, J. E. (1955). Analysis of the Genera of Surgeon Fishes (Family Acanthuridae). *Pacific Sci.* 9: 359-67.

——— (1956). A Revision of the Surgeon Fish Genus *Acanthurus.* Ibid. 10: 159-235, 3 pls., 23 figs.

Regan, C. T. (1917a). A revision of the Clupeoid Fishes of the Genera *Pomolobus, Brevoortia* and *Dorosoma* and their Allies. *Ann. Mag. Nat. Hist.* ser. 8, 19: 297-316.

——— (1917b). A Revision of the Clupeoid Fishes of the Genera *Sardinella, Harengula,* . . . Ibid. ser. 8, 19: 377-95.

Rendahl, H. (1922). A contribution to the Ichthyology of North-West Australia. *Medd. Zool. Mus. Krist.* 5: 163-97, 6 figs.

Richardson, J. (1842a). Contributions to the Ichthyology of Australia. *Ann. Mag. Nat. Hist.* 9: 15-31, 120-31, 207-18, 384-93.

——— (1842b). Ibid. Ibid. 10: 25-34.

——— (1843a). Ibid. Ibid. 11: 22-8, 169-82, 352-9, 422-8, 489-98.

——— (1843b). Description of the Lurking Machete (*Machaerium subducens*) from the Northern Coast of New Holland. Ibid. 12: 175-8, pl. 6.

——— (1843c). *Icones Piscium, or Plates of Rare Fishes.*

——— (1844-8). *Ichthyology of the Voyage of H.M.S. 'Erebus and Terror', under the Command of Captain Sir James Clark Ross, R.N., F.R.S.*

Schultz, L. P. (1946). A Revision of the Genera of Mullets, Fishes of the Family Mugilidae, with Descriptions of Three New Genera. *Proc. U.S. Nat. Mus.* 96: 377-95, 5 figs.

——— (1948). A Revision of Six Subfamilies of Atherine Fishes, with Descriptions of New Genera and Species. Ibid. 98: 1-48, 2 pls., 9 figs.

——— (1953). Review of the Indo-Pacific Anemone Fishes Genus *Amphiprion,* with Descriptions of two New Species. Ibid. 103: 187-201, 2 pls.

——— (1957). The Frogfishes of the Family Antennariidae. Ibid. 107: 47-105, 14 pls., 8 figs.

——— (1958). Review of the Parrotfishes Family Scaridae. *U.S. Nat. Mus. Bull.* 214.

———, Herald, E. S., Lachner, E. A., Welander, A. D. and Woods, L. P. (1953). Fishes of the Marshall and Marianas Islands. Ibid. 202 (1).

———, Chapman, W. M., Lachner, E. A. and Woods, L. P. (1960). Ibid. Ibid. 202 (2).

Scopoli, J. A. (1777). *Introductio ad historiam naturalem sistens genera Lapidum, Plantarum et Animalium* . . . Prague.

Seale, A. (1909). New Species of Philippine Fishes. *Philippine J. Sci.* 4: 491-543, 13 pls.

Smith, J. L. B. (1948). A Generic Revision of the Mugilid Fishes of South Africa. *Ann. Mag. Nat. Hist.* ser. 11, 14: 833-43, figs. 1-15.

——— (1959). The Identity of *Scarus gibbus* Rüppell, 1828 and of other Parrotfishes of the Family Callyodontidae from the Red Sea and the Western Indian Ocean. *Ichthyol. Bull., Rhodes University* 16: 265-82, pls. 41-5, 10 figs.

——— (1960). Coral Fishes of the Family Pomacentridae from the Western Indian Ocean and the Red Sea. Ibid. 19: 317-49, pls. 26-33, 6 figs.

Thomson, J. M. (1954). The Mugilidae of Australia and Adjacent Seas. *Aust. J. Marine Freshw. Res.* 5: 70-131, 2 pls., 16 figs.

Weber, M. (1908). Süsswasserfische von Neu-Guinea, ein Beitrag zur Frag nach dem Früheren Zusammenhang von Neu Guinea und Australien. *Nova Guinea* 5, Zoologie 2: 201-67, 3 pls.

——— (1913). Die Fische der Siboga-Expedition. *Siboga Expeditie.* Monog. lvii, Livr. 65.

——— and de Beaufort, L. F. (1913). *The Fishes of the Indo-Australian Archipelago* Vol. 2.

——— (1929). Ibid. Vol. 5.

——— (1931). Ibid. Vol. 6.

——— (1936). Ibid. Vol. 7.

——— (1940). Ibid. Vol. 8.

Whitley, G. P. (1926). Fishes. Biology of the North-West Islet, Capricorn Group. *Aust. Zool.* 4: 227-36, pls. 33-4, 1 text fig.

——— (1927a). Studies in Ichthyology, no. 1. *Rec. Aust. Mus.* 15: 289-304, pls. 24-5, 1 text fig.

——— (1927b). The Fishes of Michaelmas Cay, North Queensland. Ibid. 16: 1-32, 1 pl., 1 text fig.

——— (1928a). Studies in Ichthyology no. 2. Ibid. 16: 211-39, pls. 16-18, 2 text figs.

——— (1928b). Fishes from the Great Barrier Reef collected by Mr. Melbourne Ward. Ibid. 16: 294-304, 2 figs.

——— (1929a). Studies in Ichthyology no. 3. Ibid. 17: 101-43, pls. 30-4, 5 text figs.

——— (1929b). Some Fishes of the Order Amphiprioniformes. *Mem. Qld. Mus.* 9: 207-46, pls. 27-8, 4 text figs.

——— (1930). Ichthyological Miscellanea. Ibid. 10: 8-31, 3 figs.

——— (1931a). New names for Australian fishes. *Aust. Zool.* 6: 310-34, pls. 25-7.

——— (1931b). Studies in Ichthyology no. 4. *Rec. Aust. Mus.* 18: 96-133, pls. 11-16, 2 text figs.

——— (1932a). Some Fishes of the Family Leiognathidae. *Mem. Qld. Mus.* 10: 99-116, pls. 13-14, 1 text fig.

——— (1932b). Studies in Ichthyology no. 6. *Rec. Aust. Mus.* 18: 321-48, pls. 36-9, 3 text figs.

——— (1933). Studies in Ichthyology no. 7. Ibid. 19: 60-111, pl. 11-15, figs. 1-4.

——— (1938). Descriptions of some New Guinea fishes. Ibid. 20: 223-33, 1 fig.

——— (1940a). *The Fishes of Australia. Part 1. The Sharks, Rays, Devilfish and other Primitive Fishes of Australia and New Zealand.* Royal Zoological Society of New South Wales.

——— (1940b). Illustrations of Some Australian Fishes. *Aust. Zool.* 9: 397-428, 45 figs.

——— (1941a). Burramundi. *Aust. Mus. Mag.* 7: 264-8, 8 figs.

——— (1941b). Ichthyological Notes and Illustrations. *Aust. Zool.* 10: 1-50, 32 figs.

——— (1943a). Ibid. Ibid. 10: 167-86, 10 figs.

——— (1943b). Ichthyological Descriptions and Notes. *Proc. Linn. Soc. N.S.W.* 68: 114-44, 12 figs.

——— (1945). The Simpson Desert Expedition, 1939, Scientific Reports: No. 5 Biology—Fishes. *Trans. Roy. Soc. S.Aust.* 69: 10-13.

——— (1947). The Fluvifaunulae of Australia with particular reference to Freshwater Fishes in Western Australia. *W.A. Nat.* 1: 49-53, 2 figs.

——— (1948). New Sharks and Fishes from Western Australia. *Aust. Zool.* 11: 259-76, 2 pls., 7 figs.

——— (1950). Studies in Ichthyology, no. 14. *Rec. Aust. Mus.* 22: 234-245, 1 pl., 5 figs.

——— (1951). New Fish Names and Records. *Proc. Roy. Zool. Soc. N.S.W.*, 1949-50: 61-8, figs. 8-10.

——— (1953). Fishes collected by the Australian Museum Expedition, 1952. *Rec. Aust. Mus.* 23: 123-31, 4 figs.

——— (1954). New Locality Records for some Australian Fishes. *Proc. Roy. Zool. N.S.W.* 1952-3: 23-30, 5 figs.

——— (1955a). Taxonomic Notes on Fishes. Ibid. 1953-4: 44-57, 8 figs.

——— (1955b). Opus C C C. *Aust. Zool.* 12: 154-9, 7 figs.

——— (1955c) The Australian Museum's Marlins. *Aust. Mus. Mag.* 11: 292-7, 4 figs.

——— (1956a). List of the Native Freshwater Fishes of Australia. *Proc. Roy. Zool. Soc. N.S.W.* 1954-5: 39-47, 16 figs.

——— (1956b). Ichthyological Notes. *Aust. Zool.* 12: 251-61, 7 figs.

——— (1958). Descriptions and Records of Fishes. *Proc. Roy. Zool. Soc. N.S.W.* 1956-7: 28-51, 12 figs.

——— (1959). Ichthyological Snippets. *Aust. Zool.* 12: 310-23, 3 figs.

——— and Allan, Joyce (1958). *The Sea-horse and its Relatives.*

Williams, F. (1959). The Barracudas (Genus *Sphyraena*) in British East African Waters. *Ann. Mag. Nat. Hist.* ser. 13, 2: 92-128, 2 pls.

Woods, L. P. (1955). Western Atlantic Species of the Genus Holocentrus. *Fieldiana Zool.* 37: 91-119, 5 figs.

(*Manuscript received, August 1961*)

4

REPTILES AND AMPHIBIANS OF ARNHEM LAND

by Francis J. Mitchell*

Under the field curatorship of the ichthyologist, Dr R. R. Miller, the expedition made an official collection of 729 specimens. These have been divided between the United States National Museum, Washington, and the South Australian Museum, Adelaide, as indicated by the abbreviations USNM and SAM respectively, which precede the reference numbers entered herein. All holotype specimens have been deposited in the latter institution. In addition, a series of 179 specimens comprising the bulk of a collection obtained by the expedition transport officer, Mr J. E. Bray, and presented to the Australian Museum, Sydney, has also been made available for inclusion in this report. Such specimens are indicated by the abbreviation AM. In all, 908 specimens representing 63 species and subspecies.

The present study was undertaken at the South Australian Museum and extensive use was made of its research collection for purposes of reference and comparison.

The titles of various Australian political areas have been abbreviated as follows: N.T., for Northern Territory; Qld., for Queensland; N.S.W., for New South Wales; Vict., for Victoria; S. Aust., for South Australia and W. Aust., for Western Australia. The more important findings in this report were published in a 'Preliminary Account of the Reptilia and Amphibia collected by the National Geographic Society–Commonwealth Government–Smithsonian Institution Expedition to Arnhem Land'—Mitchell, *Rec. S. Aust. Mus.* 11: 373-408 (1955).

ACKNOWLEDGMENTS

I wish to acknowledge the co-operation of Dr Remington Kellogg and Dr Doris M. Cochran of the United States National Museum, Washington, and Dr A. B. Walkom and Mr J. R. Kinghorn of the Australian Museum, Sydney, in facilitating my examination of the total reptile and amphibian collection. Their assistance is much appreciated. I am also indebted to Mr Arthur Loveridge, Museum of Comparative Zoology, Harvard, Mr G. Mack, Queensland Museum, Brisbane, and Messrs J. Henry and S. J. Copland of the Macleay Museum, University of Sydney, who forwarded type and other material housed in their respective institutions for my examination.

* South Australian Museum, Adelaide.

REPTILIA

TESTUDINATA

CHELONIIDAE

Eretmochelys imbricata (Linnaeus)

Testudo imbricata Linnaeus, *Syst. Nat.* (ed. 12) 1: 350 (1766)
Eretmochelys squamata Agassiz, *Contr. Nat. Hist. U.S.* 1: 382 (1857)

Black Rock Point, Cobourg Peninsula, N.T.: USNM 128527

A shell measuring 390 mm. and a partial skeleton. The cranium is missing. Malcolm A. Smith (1931), followed by Loveridge (1934 and 1948), has referred the alleged Pacific race *squamata* to the synonymy of the Atlantic *imbricata*.

CROCODYLIA

CROCODYLIDAE

Crocodylus porosus Schneider

Crocodylus porosus Schneider, *Hist. Amphib.* 2: 169 (1801)

Black Rock Point, Cobourg Peninsula, N.T.: USNM 128526
South end of Melville Bay, west of Yirrkalla, N.T.: USNM 128685

The Cobourg Peninsula specimen is an almost complete skull, while that from Melville Bay comprises both lower jaws with only a few teeth remaining.

SERPENTES

BOIDAE

Liasis childreni childreni Gray

Liasis childreni Gray, *Zool. Misc.*: 44 (1842)

Port Langdon, Groote Eylandt, N.T.: USNM 128469

Midbody scales in 43 rows; subcaudals in 48 pairs; 290 ventral scales. Thirteen upper and fifteen lower labials, four of the lower labials being deeply pitted.

Measurements: 1084 (989 + 95) mm.

Liasis fuscus fuscus Peters

Liasis fuscus Peters, *Mber. Akad. Wiss. Berl.*: 607 (1873)

Oenpelli, N.T.: USNM 128768

Midbody scales in 47-9 rows; 79 subcaudals, mostly paired; eleven upper labials, the first two pitted, and sixteen lower labials, the 10th-13th being pitted. Anal undivided.

Morelia spilotes variegata Gray

Morelia variegata Gray, *Zool. Misc.*: 43 (1842)

Lee Point, 9 miles north-east of Darwin, N.T.: USNM 128235 (head only)
Near Umbakumba, Groote Eylandt, N.T.: USNM 128442

Midbody scales in 43 rows; 265 ventrals; 83 pairs of subcaudals; anal entire. The rostral and the 1st, 2nd and 3rd of 12 upper labials deeply pitted; six or seven of the lower labials also deeply pitted; 4 or 5 loreal shields. The colouring is characteristic of *variegata*, USNM 128235 being almost topotypical.

As indicated by Loveridge (1934: 270), present data does not suggest any correlation between the distribution of the *variegata* colour phase and a particular geographical region or environment. However, no one worker has yet had the opportunity of examining a large Australia-wide series of this species, and as pure populations of both phases undoubtedly occur, *variegata* is herein retained. (See also Mitchell (1951: 545).)

COLUBRIDAE

Achrochordus javanicus Hornstedt

Achrochordus javanicus Hornstedt, *Abh. Ac. Stockh.* 8: 307, pl. 12 (1787)

Oenpelli, N.T.: USNM 128765-7, SAM R2844 (2 specimens)

Five subadult specimens from fresh water lagoons in the vicinity of Oenpelli, Northern Territory.

The Kakadu native name for this species is *gejebi* (R. R. Miller).

Achrochordus granulatus granulatus (Schneider)

Hydras granulatus Schneider, *Hist. Amphib.* 1: 243 (1799)

Milingimbi Island, N.T.: USNM 128511

A single specimen of this widely distributed species was taken in a marine inlet. M. A. Smith (1943: 131) considers *Chersydrus* to be synonymous with *Achrochordus.*

Fordonia leucobalia (Schlegel)

Homalopsis leucobalia Schlegel, *Phys. Serp.* 2: 345, pl. 13, fig. 8-9 (1837)

Inlet at dock for Delissaville, Cox Peninsula, N.T.: USNM 128234

Midbody scales in 23 rows; 148 ventrals; 39 subcaudals, all but six of which are single; anal sutured. Temporals 2 + 3. The basic dorsal colouring is light brown with an irregular black marbling.

Enhydris polylepis (Fischer)

Hypsirhina polylepis Fischer, *Abh. Naturw. Hamburg*: 14 (1886)

Port Langdon, N.T.: USNM 128474

Near Umbakumba, Groote Eylandt, N.T.: USNM 128450

Midbody scales in 21 rows; 23 rows can be counted at several points forward of midbody in the Groote Eylandt specimen. Ventral scales 140 and 146; 36 and 40 pairs of subcaudals; anal sutured. Generally eight upper labials with the fifth and sixth subocular, but only seven with the fourth and fifth subocular on one side of the head of USNM 128450.

Measurements of the larger specimen, USNM 128450, 431 (368 + 63) mm.

Kinghorn (1929: 90) appears to be in error when he records single subcaudals for *macleayi* Ogilby, a species which Loveridge (1934: 273) places in the synonymy of *polylepis.*

Ahaetulla punctulata punctulata (Gray)

Leptophis punctulatus Gray, in King's *Voy. Aust.* 2: 432, pl. A (1827)

Near Umbakumba, Groote Eylandt, N.T.: USNM 128445-7

Oenpelli, N.T.: USNM 128769-71, SAM R2845

Yirrkalla, N.T.: USNM 128683-4

Port Langdon, Groote Eylandt, N.T.: USNM 128470, 128472

West Point, 5 miles west of Darwin, N.T.: USNM 128233

Midbody scales constantly in thirteen rows; but eleven can be counted at several points along the middle third of the body in some specimens; ventrals 186-215; subcaudals, 110-34 pairs. One pre- and two post-oculars; 4th and 5th upper labials subocular; loreal usually present, but absent in USNM 128445.

The longitudinal dark line under the tail, stated to be characteristic of the New Guinea race *lineolata* (Guichenot) is present in about half of this series, being particularly prominent in the three specimens from Groote Eylandt. As recorded by Loveridge (1948: 385) the horizontal diameter of the eye relative to the other head dimensions varies considerably with age in this species.

The Kakadu native name for this species is *garnbo* and the Murngin name is *karanunka* (R.R.M.).

Natrix mairii mairii (Gray)

Tropidonotus mairii Gray, in Grey's *Journ. Exped. Discov.* 2: 442, pl. 8, fig. 2 (1841)

Near Umbakumba, Groote Eylandt, N.T.: USNM 128443-4
Oenpelli, N.T.: USNM 128772

Midbody scales constantly in 15 rows, strongly keeled; 149-53 ventrals and 60-72 pairs of subcaudals; anal divided. Nasals partly or completely divided; internasals equal in length to the prefrontals; four lower labials contacting the anterior chin shields; temporals 1 + 2. Horizontal diameter of the eye not longer than its distance from the nostril; three post- and one or two pre-oculars; 3rd, 4th and 5th of eight upper labials entering the eye. USNM 128772 is a gravid female containing twelve ova. It measures 740 (580 + 160) mm.

In addition, two South Australian Museum specimens, one from the Adelaide River, Northern Territory and one from Rocky River, Atherton, Queensland, were examined. None of this series can be considered as belonging to *picturatus* (Schlegel), which species is doubtfully recorded from Fergusson Island, Northern Australia by De Rooij (1917: 77).

Cerberus rynchops australis (Gray)

Homalopsis australis Gray, *Zool. Misc.*: 65 (1842)

Nightcliff, near Darwin, N.T.: USNM 128264

A single South Australian Museum specimen from Melville Island, Northern Territory, was also examined.

Ventral scales 146, 148; subcaudals in 48 (tail incomplete) and 56 pairs; anal divided. Midbody scales, keeled, in 23 and 25 rows. A suture through the nostril to the first labial on one side of the head and to the second labial on the other in both specimens; loreal contacting the 1st, 2nd, 3rd and 4th, or 2nd, 3rd and 4th upper labials. A single preocular scale extends around to a point well under the eye; one or two sub- and two post-oculars.

The designation of these two specimens to *australis* is based mainly on the locality, it being evident from the above data that they do not conform satisfactorily to any of the three subspecies as keyed by Loveridge (1948: 389), his key being based on a series of 70 *r. rynchops*, 5 *r. novaeguinea*, 4 *r. australis* and presumably data published by Boulenger (1896, 1897) and De Rooij (1917).

If the tails of these specimens were complete, both would possess more than the maximum of 51 pairs of subcaudals indicated in Loveridge's key and thus would be identifiable as *r. rynchops*. On the other hand, if the diagnosis is based on the nasal cleft there would be equal grounds for placing these specimens with either *novaeguinea* or *australis*. It is possible that the circumocular scalation will be found to provide a better key character for the separation of the three subspecies. Boulenger (1896: pl. 2, fig. 1, 1a) in figuring *australis* shows the nasal cleft joining the second upper labial on one side and the preocular on the other, with only the second and third upper labials contacting the loreal, the fourth being widely separated from it. De Rooij (1917: fig. 70) figures *rynchops* and shows a smaller preocular and two suboculars.

Although it seems probable that *australis* and *novaeguinea* can be satisfactorily distinguished from the type race, many more specimens will have to be examined and their variation tabled before a practicable key can be devised.

Boiga fusca (Gray)

Dendrophis fusca Gray, *Zool. Misc.*: 54 (1842)

Port Langdon, Groote Eylandt, N.T.: USNM 128473

Midbody scales in 19 rows; 266 ventrals; 104 pairs of subcaudals, anal entire. Preocular almost separating the supraocular from the prefrontal; 3rd, 4th and 5th upper labials subocular. The anterior palatine teeth are subequal. Body with 74 distinct black cross bars.

This specimen measures 1254 (1012 + 242) mm.

ELAPIDAE

Aspidomorphus diadema (Schlegel)

Calamaria diadema Schlegel, *Phys. Serp.* 2: 32 (1837)

Nightcliff, near Darwin, N.T.: USNM 128261

Midbody scales in 15 rows; 176 ventrals; 54 pairs of subcaudals; anal divided. Preocular almost separating the supraocular from the prefrontal. Nostril breaking the upper edge of the nasal shield.

The patterning of this specimen very closely resembles that of *christieanus* Fry, with which this species coexists in the vicinity of Darwin. The head and neck are black with a broad lighter band across the nape replacing the more usual small crescent-shaped marking. Upper labials and ventral surfaces white; dorsal surfaces pale yellow, each scale being edged with brown, giving a net-work effect.

Aspidomorphus christieanus (Fry)

Pseudelaps christieanus Fry, *Proc. Roy. Soc. Qld.* 27: 91, fig. 6 (1915)

Nightcliff, near Darwin, N.T.: USNM 128262

Midbody scales in 17 rows; 187 ventrals; 49 pairs of subcaudals; anal divided. The nasal is widely separated from the preocular which joins the frontal.

An additional specimen from Nightcliff was presented to the South Australian Museum in 1949 and a note forwarded with it suggests that the species is quite common within a limited distribution centred near Darwin. It is registered SAM R2995, and possesses 17 midbody scales; 185 ventrals; 43 pairs of subcaudals and a divided anal. It is the larger of the two and measures 375 (319 + 56) mm.

The tails of both of these specimens appear to be complete, indicating that the subcaudal count is variable. It has been recorded as 38 (type specimen), 49 and 43 (the present specimens), 56 (Loveridge, 1934: 276) and 56 (Longman, 1916: 48).

The general ground colour is almost white tending yellow middorsally; each scale is edged with purple, giving the dorsal surface a reticulate appearance. A light band across the nape separates a dark purple neck band from the still darker head.

The distributions of *diadema* and *christienaus* appear to meet in the vicinity of Darwin, and the close superficial resemblance of the two snakes in this district, together with the geographical segregation of *christienaus* in suitable districts of north-western Australia, suggests that it may only be a subspecies of *diadema*, distinguishable by its greater midbody count and several minor differences in coloration. Both snakes display a wide variation in ventral and subcaudal scale counts.

Demansia carinata (Longman)

Diemenia carinata Longman, *Mem. Qld. Mus.* 3: 31 (1915)

Port Langdon, Groote Eylandt, N.T.: USNM 128471

Midbody scales in 17 rows; 203 bi-keeled ventrals; 62 pairs of subcaudals; anal scale divided. Temporals 1 + 3, the lower posterior being partially enclosed within the enlarged 6th upper labial. Only one of the postoculars contacting the anterior lateral edge of the parietal, the second being narrowly separated from it; lower half of preocular deeply sunken, giving a 'brow' effect. Frontal widest anteriorly, mean width a little greater than that of the supraoculars.

Uniform blue-grey dorsally, without darker cross banding; lighter ventrally.

Measurements approximately 1040 (870 + 170) mm.—body set in close spiral.

The record of this specimen from Port Langdon is interesting as it considerably extends the known range of this arboreal *Demansia*, the type locality being Cane Grass Station via Charleville in Western Queensland.

Demansia psammophis psammophis (Schlegel)

Elaps psammophis Schlegel, *Phys. Serp.* 2: 455 (1837)

Near Umbakumba, Groote Eylandt, N.T.: USNM 128448

Nightcliff, near Darwin, N.T.: USNM 128263

Midbody scales in 15 rows; 174-81 ventrals; 79-81 pairs of subcaudals; anal divided. Nostril completely dividing the nasal shield which contacts the preocular. Frontal straight sided, almost three times longer than wide; wider than the supraocular region.

The general patterning on the head complies with that figured for *olivacea* by Boulenger (1896: pl. 18, fig. 2), but definite indications of the circumorbital markings are evident in the larger specimen. The tendency is toward light-edged rather than dark-edged *reticulata* type dorsal scales.

The larger specimen, USNM 128448, measures 471 (417 + 54) mm.

Loveridge (1934: 276; 1949: 209) and Kinghorn (1942: 118) have determined between them that *maculiceps* Boettger, *torquata* Günther, *olivacea* Gray, and *ornaticeps* Macleay are conspecific with *psammophis* Schlegel. Loveridge (1949: 209) suggests that *olivacea* may be retained as a north-eastern subspecies on the grounds of possessing a greater number of subcaudals and more detailed study may reveal grounds for resurrecting other names from the synonomy to cover distinctly marked populations of this variable species. Of twelve South Australian Museum specimens examined, nine possess a colour pattern characteristic of *psammophis psammophis* and show a variation of 70-87 subcaudals and 183-98 ventrals, the counts being disposed as follows: two specimens from Grange, South Australia, (80, 198) (81, 188) Fowler's Bay, South Australia, (73, body damaged); Minnie Downs via Marree, South Australia, (87, 187); Tower Hill, Andamooka Ranges, South Australia, (70, 193); also two South Australian specimens without explicit locality data, (83, 189) (79, 191). A specimen from Darwin, Northern Territory, measuring 665 (494 + 171) mm. and possessing a typical *olivacea* colour pattern has (81, 175), while two specimens of *psammophis reticulata* from 'between Ashburton and Gascoyne Rivers, Western Australia' possess (71, 190) and (77, 191).

Pseudechis australis (Gray)

Naja australis Gray, *Zool. Misc.*: 55 (1842)

Near Umbakumba, Groote Eylandt, N.T.: USNM 128451; SAM R2843 (2 specimens)

Midbody scales in 17 rows; ventrals 202-10; subcaudals 66-8, all but 3-5 single; anal sutured. Uniform light grey-brown dorsally, white ventrally.

These three specimens were compared with eight South Australian specimens preserved in the South Australian Museum collection. This comparison resulted in the complete confirmation of the conclusion of Thomson (1930: 133) that *darwiniensis* is synonymous with *australis*. Except for the fact that the Groote Eylandt specimens seem a little more slender, the variation displayed by the South Australian series completely encloses that shown above, the noted variation being: midbody scales 17; ventrals 194-212, subcaudals 57-69, the greater majority being single—in one specimen all single and all but 4-13 single in the remainder. Proctor (1923: 1073) records an aberrant specimen from Thomby Station, central South Queensland, which possesses 19 midbody scales.

The largest specimen examined, SAM R773 from Sheringa, South Australia, measured 1860 (1670 + 190) mm. It has rather unusual coloration, the dorsum being uniform dark-chocolate brown and the ventral surface pink anteriorly fading to cream posteriorly, speckled and flecked with brown. This specimen is extremely stout, measuring 38 mm. across the widest part of the head.

None of these specimens show any resemblance to the large *Pseudechis* described and figured by Thomson (1933: 859) as *P. platycephalus* sp. nov. It was originally described from the head and neck of a specimen taken at East Alligator River, Arnhem Land, Northern Territory; redescribed and figured by Glauert (1950: 162) from the head of a specimen taken on the Adelaide River Flat, Northern Territory. Proctor's supposedly aberrant specimen from Thomby Station (op. cit.) may be referable to *platycephalus*.

Oxyuranus scutellatus (Peters)

Pseudechis scutellatus Peters, *Mber. Akad. Wiss. Berl.*: 710 (1867)
Oxyuranus maclennani Kinghorn, *Rec. Aust. Mus.* 14: 42-5 (1923)

Twenty miles north-east of Oenpelli, N.T.: USNM 128773 (head and neck only)

Thirty-one to thirty-three convex, but smooth, scales around the neck; prefrontals partially divided by longitudinal sutures; nasal contacting the preocular on one side and separated from it on the other. Six upper labials, the 3rd and 4th contacting the eye with the 5th separated from it by the lower postocular; seven or eight lower labials; temporals 3 + 3.

The keeling of the dorsal scales figured so prominently by Thomson (1933: pl. 1) is not apparent on the nuchal scaling of this specimen, but the elongate scales are strongly convex and no doubt keeling would be evident further along the body.

Unfortunately, the total length of this specimen could not be recorded as the specimen was badly damaged in capture.

Two additional specimens were examined, the skin of a large specimen from Oenpelli, Northern Territory (SAM R1346) and a smaller specimen from Coen,

Queensland (SAM R358). The latter specimen possesses irregular midbody scaling, there being 23 rows at midbody, but 21 at several points thereabouts. Ventral scales 239; 70 pairs of subcaudals; anal single; nasal well separated from the preocular.

Acanthophis antarcticus antarcticus (Shaw)

Boa antarctica Shaw, *Nat. Misc.*: pl. mxxxv (1794)

Port Langdon, Groote Eylandt, N.T.: USNM 128476-7, SAM R3226
Near Umbakumba, Groote Eylandt, N.T.: USNM 128449

Midbody scales in 21 rows; ventrals 127-32, subcaudals 47-9, the last 17-21 beneath the compressed extremity being divided; anal entire. Lower temporal extending down between the 5th and 6th upper labials almost separating them—completely separating them in USNM 128449 on one side of the head, thus becoming a labial.

Head shields multikeeled, one specimen approaching the *palpebrosa* condition. General coloration typical of the species, these specimens possessing a smooth-scaled, strongly compressed, white extremity to the tail.

A comparison of this Arnhem Land material together with other Northern Australian specimens with Southern Australian examples indicates that the colour variation of the caudal extremity is largely if not completely correlated with locality. Also, there would seem to be some variation in its form, the southern specimens possessing an acutely keeled extremity with strongly imbricate scales (Fig. 1).

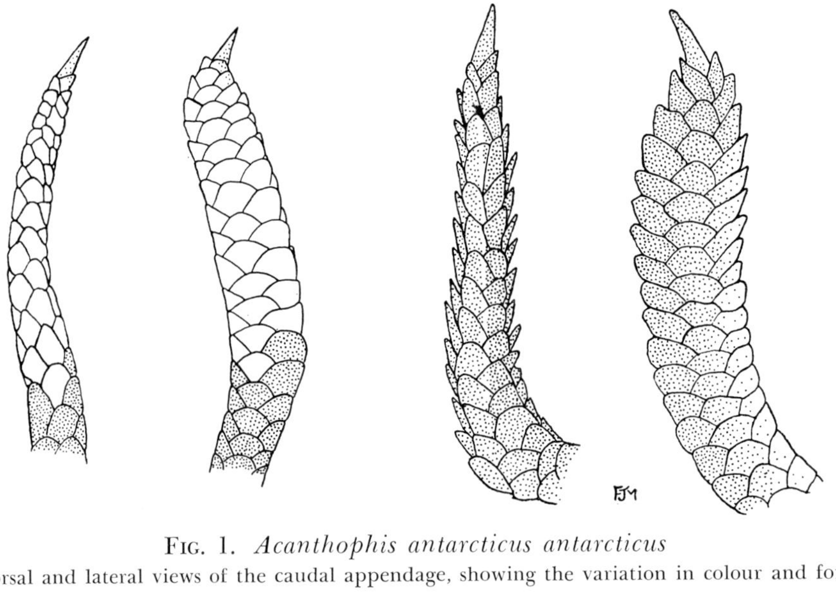

FIG. 1. *Acanthophis antarcticus antarcticus*

Dorsal and lateral views of the caudal appendage, showing the variation in colour and form

Although I have been unable to find any other character correlated with this caudal variation, its constancy within the material examined and apparent geographic segregation indicate that it is more than an irregular variant, and the possibility of the two groups being subspecifically distinct is worthy of further consideration. In dealing with Indo-Australian specimens, De Rooij (1917: 273) states 'end of tail yellow or black'. Whether this statement is based on the Indo-Australian material examined or merely following Boulenger (1896: 355) is not clear. The specimen figured by De Rooij (op. cit. fig. 3) has a light caudal extremity.

HYDROPHIIDAE

Hydrophis elegans (Gray)

Aturia elegans Gray, *Zool. Misc.*: 61 (1842)

Port Langdon, Groote Eylandt, N.T.: USNM 128473

The skin of a large specimen measuring approximately 2380 mm. Forty-two scale rows at midbody; anals and ventrals slightly enlarged.

SAURIA

GEKKONIDAE

Heteronota binoei Gray

Heteronota binoei Gray, *Catalogue of Lizards in the British Museum London*: 174 (1845)

Yirrkalla, N.T.: USNM 128539-56, SAM R2854 (28 specimens)

Umbakumba, Groote Eylandt, N.T.: USNM 128282-91, 128463, AM R12473 (3 specimens), R13474 (3 specimens)

Milingimbi Island, N.T.: USNM 128495-503, SAM R2853 (12 specimens)

Nightcliff, near Darwin, N.T.: USNM 128240-6

Rocky Beach, Cape Arnhem, N.T.: USNM 128518

Black Rock Point, Cobourg Peninsula, N.T.: USNM 128525

Cape Don, Cobourg Peninsula, N.T.: USNM 128521-4

Cape Arnhem, N.T.: AM R12651 (7 specimens)

A comparison of this large series with other material contained in the South Australian Museum collection shows clinal variation in the number of preanal pores. They vary from 2-4 in the extreme north to 6-8 in the south. Of the total number of specimens examined, 94 were males and possessed preanal pores in accordance with the following list.

Preanal pores of male specimens taken from localities north of latitude 28°S.: Arnhem Land, N.T., 46 specimens with 4 pores, 2 specimens with 2 pores; Alice Springs, N.T., 1 specimen with 4 pores; Hermannsburg, N.T., 4 specimens with 4 pores, 1 specimen with 5 pores; Daly Waters, N.T., 1 specimen with 4 pores; Macdonnell Ranges, N.T., 2 specimens with 4 pores; Tennant Creek, N.T., 1 specimen with 4 pores; Cooper's Creek, S. Aust., 2 specimens with 6 pores; Musgrave Ranges, S. Aust., 2 specimens with 4 pores.

Preanal pores of male specimens taken from localities south of latitude 28°S.: Kingoonya, S. Aust., 2 specimens with 6 pores; Hawker, S. Aust., 2 specimens with 7 pores; Tarcoola, S. Aust., 1 specimen with 6 pores; Eucla, S. Aust., 1 specimen with 4 pores; Broken Hill, N.S.W., 2 specimens with 6 pores; Mount Serle, Flinders Ranges, S. Aust., 1 specimen with 6 pores; Moolooloo, S. Aust., 1 specimen with 7 pores; Flinders Ranges, S. Aust., 4 specimens with 6 pores, 2 specimens with 8 pores; Andamooka Ranges, S. Aust., 3 specimens with 6 pores, 2 specimens with 8 pores; Gawler Ranges, S. Aust., 2 specimens with 6 pores; Lake Mulligan (Callabonna), S. Aust., 4 specimens with 4 pores; Ooldea, S. Aust., 2 specimens with 6 pores; Kalgoorlie, W. Aust., 1 specimen with 4 pores.

From the above data it can be seen that the bulk of the male specimens taken at localities north of the latitude 28°S. possess 5 or less preanal pores, while those from localities to the south, 5 or more. The type locality, Houtman's Abrolhos, Western Australia, is approximately midway along the cline, and as might be expected from the present data the holotype is described as possessing 5 preanal pores (see Boulenger 1885: 75).

As there do not appear to be any other variable features it seems best to retain the whole population under the one name, while acknowledging the existence and direction of this variation.

A single specimen (SAM R1807) taken at an unknown locality in Queensland is interesting. The dorsal tubercles are of a rounded, conical form and the ventral scales are much smaller than those of any other specimens examined. It would be interesting to compare this specimen with the types of *eboracensis* Macleay, although Loveridge (1934: 301) had the opportunity of comparing a near topotypic series of that species from Coen, Queensland, with a topotypic *binoei* specimen from West Wallaby Island, Western Australia, and was unable to separate the two.

Hemidactylus frenatus Duméril and Bibron

Hemidactylus frenatus (Schleg.) Duméril and Bibron, *Erp. Gen.* 1-9: 366 (1836)

Milingimbi Island, Crocodile Islands, N.T.: USNM 128494, SAM R2872 (2 specimens)

This widely distributed species has previously been recorded from the Murray Islands by Boulenger (1885: 121) and Barbour (1914: 203), and from the mainland at Port Essington by Boulenger (op. cit.); Zietz (1920: 190) records it from the Monte Bello Islands, Western Australia, but Montague (1914) did not list it.

Of these geckoes, USNM 128494 is a male, and possesses 30 elongate femoral and preanal pores, which form a continuous series. Nostril pierced between the rostral, 1st upper labial, internasal, and two post-nasals; internasals separated by two small scales or forming a median suture.

Oedura rhombifer Gray

Oedura rhombifer Gray, *Zool. Erebus and Terror, Rept.*: pl. 16, fig. 6 (1844)

Yirrkalla, N.T.: USNM 128531-4, SAM R2859 (3 specimens)
Central Hill, Groote Eylandt, N.T.: USNM 128460-1
Umbakumba, Groote Eylandt, N.T.: USNM 128293-9, AM R12627 (5 specimens), R13518 (2 specimens), R13517 (2 specimens)
Oenpelli, N.T.: USNM 128728
Birlouja Creek, N.T.: USNM 128516-7
Cape Arnhem, N.T.: AM R13559 (2 specimens)
Gove Air Base, 8 miles south-west of Yirrkalla, N.T.: SAM R2870 (3 specimens)

The number of femoral and/or preanal pores is very variable. In the present series, five males possess 4 + 4 preanal pores, one 4 + 5, one 6 + 5 femoral and preanal, and one 8 + 8 mostly femoral pores, the preanal interruption being abnormally wide in this last male. Loveridge (1949: 210) records a specimen from near Darwin, N.T., with 12 + 12 pores and an abnormally large preanal interruption. In the specimens with only 4 + 4 preanal pores, the two series are separated by a single scale. Supranasals generally separated, but contacting anteriorly in two specimens, a median cleft in the rostral does not reach the buccal border; ear opening of variable size and shape. The coloration is also variable, but the majority show the usual zig-zag dorso-lateral lines tending to enclose a middorsal series of light quadrangular spots.

The present material supports the findings of Kinghorn (1942: 120), the tail being distinctly depressed at the base in all specimens examined. The only character preventing the relegation of *rhombifer* to the synonymy of the closely allied *lesueurii* is the supposed absence of femoral and preanal pores in males of the latter species. A single specimen from the Bunda Plateau, inland from Fowler's Bay, South Australia (SAM R1926), agrees with the Arnhem Land series in all respects.

Oedura marmorata Gray

Oedura marmorata Gray, *Zool. Misc.*: 52 (1842)

Yirrkalla, N.T.: USNM 128557-65, SAM R2864
Oenpelli, N.T.: USNM 128729
Purlonkoimi Creek, N.T.: USNM 128515
Cape Arnhem, N.T.: AM R13551-3

This species is very variable both in scalation and coloration. The present series contains six males possessing an interrupted series of 13-15 femoral and preanal pores. Except for two Yirrkalla specimens the anterior lower labials are separated behind the mental. Median cleft of the rostral does not reach the buccal border in any of this material. In two specimens a small rhomboidal scale is enclosed between the supranasals and the rostral; AM R13551 has the supranasals separated by a large hexagonal scale.

The colour pattern of most of these specimens is close to that originally described for *marmorata,* although two possess dorsal ocelli corresponding to *ocellata* Boulenger and another has a pattern resembling that described for *fracticolor* De Vis.

While Loveridge (1934: 309) records 18-22 average 20 for seven males mostly from localities in Queensland, and Garman (1901: 3) records 20 (18 in one specimen—see Loveridge op. cit.: 310) for his type series of *mayeri* from Cooktown, Queensland, the present material possesses only 13-15. Two additional males, one from Eucla, South Australia, and one from the Bunda Plateau, inland from Fowler's Bay, South Australia, were examined and found to possess 22 and 16 pores respectively.

Diplodactylus strophurus ciliaris Boulenger

Diplodactylus ciliaris Boulenger, *Catalogue of Lizards in the British Museum, London* 1-3: 98, pl. 8, fig. 2 (1885)

Nightcliff, near Darwin, N.T.: USNM 128247

In the 'Preliminary Account' (Mitchell, 1955) specimens preserved in the South Australian Museum collection were examined with a view to linking the work of Brazenor (1951) and Glauert (1952). A general zoogeographical study was undertaken in which *strophurus* was considered a polytypic species. The subspecies *spinigerus, ciliaris* and *aberrans* were satisfactorily distinguished from the type race; *intermedius* Ogilby was provisionally recognized.

This specimen is almost topotypic.

Supraciliary and caudal spines fully developed; dorsal tubercles distributed uniformly in two longitudinal series.

Gehyra variegata australis Gray

Gehyra australis Gray, *Catalogue of Lizards in the British Museum, London*: 163 (1845)

Umbakumba, Groote Eylandt, N.T.: USNM 128292-300
Yirrkalla, N.T.: USNM 128535-8, SAM R2861
East Alligator River, N.T.: AM R13638 (2 specimens)
Cape Arnhem, N.T.: AM R13556
Oenpelli, N.T.: SAM R2863 (2 specimens)
Milingimbi Island, Crocodile Islands, N.T.: SAM R2862

Subdigital scansors undivided anteriorly but with a median groove posteriorly; a slight rudiment of a claw is evident on the fifth toe in several specimens and the rudiment of a web at the base of the toe seems to be more prominent than in *variegata variegata*. A cutaneous fold present along the back of the thigh; in three specimens a dorso-lateral fold is also present, but this may be the result of dehydration by strong preservative. Nine to eleven enlarged upper, and seven to nine enlarged lower labials. Two or three post-nasals; internasals forming a median suture or separated by one or two small granules. Five of this series are males and they possess 15-29, averaging 22, preanal pores.

Contrary to the findings in other species, the Arnhem Land specimens of this species seem to possess a greater average number of preanal pores. Loveridge (1934: 313) records an average of 15 with a range of 13-19 for five Queensland males, while an examination of the males in the South Australian Museum collection gave the following data. *G. v. australis*—six males possessing 13-21, average 16; *G. v. punctata*—four males possessing 11-13, average 12 [Loveridge (op. cit.) also records 11-13 for *punctata*]; *G. v. variegata*—thirty-three males possessing 9-21, average 15.

The colour varies from uniform light blue to light brown with indistinct darker variegations. When analysed under a microscope the light blue colour is shown to be a translucent white minutely punctate with black.

The position and status of *punctata* Fry have been verified by the examination of eleven specimens in the South Australian Museum collection. The best diagnostic feature for its separation appears to be the shape of the mental and chinshields (Fig. 2), the characters used by Fry in his type description being both variable and difficult to define.

On examining the type description of *variegata* (Duméril and Bibron 1836: 353) it was noted that one of the four type specimens is referable to *punctata*, being from 'baie des Chiens Marins' or Shark Bay, Western Australia. The remaining three specimens, which are mentioned first whenever the variation warrants a

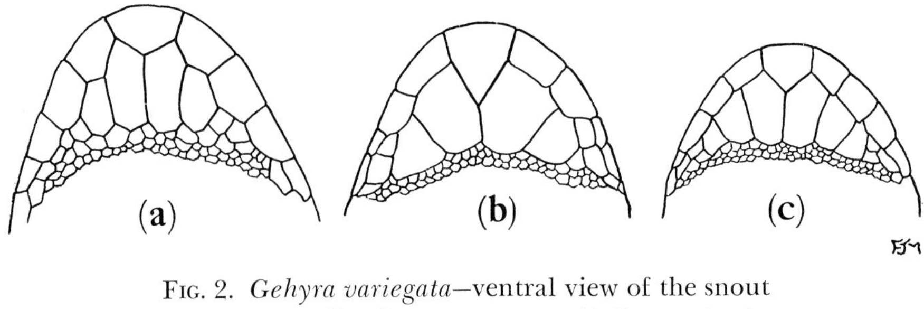

FIG. 2. *Gehyra variegata*—ventral view of the snout
(a) *G. v. australis* (b) *G. v. punctata* (c) *G. v. variegata*

comparison, and are therefore accepted as representing the type race, were reputedly taken in Tasmania ('la terre de Vandieman'). The species has not been subsequently recorded as occurring on the island and the last person to list the lizards of Tasmania, Hewer (1948), completely omits the species.

AGAMIDAE

Chlamydosaurus kingii Gray

Chlamydosaurus kingii Gray, in King's *Voy. Aust.* 2: 425, pl. A (1827)

Milingimbi Island, Crocodile Islands, N.T.: USNM 128480
Umbakumba, Groote Eylandt, N.T.: USNM 128452-3
Yirrkalla, N.T.: USNM 128568, SAM R3228
Oenpelli, N.T.: USNM 128752

Three to five femoral and one or two preanal pores on each side. Several of these specimens were badly infested with ticks (*Ixodes* sp.) about the body, and mites [*Leeuwenhoekia* (*Acomatacarus*) sp.] on the underside of the frill.

Amphibolurus caudicinctus caudicinctus (Günther)

Grammatophora caudicincta Günther, *Zool. Erebus and Terror, Rept.*: 19 (1844)

Four and a half miles SSE. of Oenpelli, N.T.: USNM 128751, SAM R3229

Dorsal scales keeled, the keels being directed sharply toward the midline; a middorsal series of enlarged, strongly keeled scales present; caudal scales strongly keeled. A low spinous nuchal crest. Several small conical tubercles are present about the neck which is strongly plicate. Lateral scales much smaller than the dorsals, tending granular with slight keels directed backward and downward at right angles to those on the dorsal surface. Ventrals smooth or faintly keeled; 29-30 femoral and preanal pores, forming an almost straight series, interrupted in the preanal region; 33-5 bicarinate subdigital lamellae beneath the 4th toe. Nostril in a swollen canthus rostralis, directed slightly upward, nearer the eye than the tip of the snout; 7-8 scales separating the nostril from the upper labials.

The larger specimen, USNM 128751, measures 352 (35 + 66 + 251) mm. A field note attached to these two specimens states that they were 'shot together on sandstone boulders at western edge of Arnhem Land plateau, about 4½ miles SSE. of Oenpelli, and some 400 feet higher than Oenpelli. 30th September (R.R.M.)'.

Loveridge (1934: 319) suggests that *rufescens* Stirling and Zeitz and *imbricatus* Peters may prove to be races of *caudicinctus* Günther. Preliminary work on a revision of the genus *Amphibolurus* has been undertaken by the author and the above possibility investigated. The types (SAM R1424 and R1425) and four other specimens of *rufescens* have been examined and compared with *caudicinctus*. Although they do resemble one another in colour and general form, they differ markedly in structural detail, and until more extensive work on the distribution and variation of these lizards can be undertaken, no advantage is to be gained by linking them under a single specific name, although this may ultimately prove warranted. The following characters serve to distinguish the two lizards:

	A. caudicinctus Gunther	*A. rufescens* S. & Z.
1.	Femoral and preanal pores, 29-35.	Femoral and preanal pores 56-60.
2.	Canthus rostralis swollen and rounded, with nostril in the rostralis, directed upward.	Canthus rostralis obtuse, but angulate, with nostril below the rostralis, directed outward.

3. A longitudinal series of enlarged middorsal scales and low spinous nuchal crest.	No longitudinal series of enlarged middorsal scales, and a slightly enlarged, non spinous series of mid-nuchal scales.

After examining eleven specimens from the vicinity of Marree and Finniss Springs, South Australia, the author is of the opinion that *imbricatus* Peters is specifically distinct from *caudicinctus* Gunther, although confirmatory specimens from other parts of central South Australia are required.

Amphibolurus barbatus barbatus (Cuvier)

Agama barbatus Cuvier, *Régne Animal* (2nd ed.) 2: 35 (1829)

Horse Shoe Bend, Finke River, S. Aust.: USNM 128529

Although most characters of this specimen agree with those of the type race, the 'beard' is not developed to the same extent as that of specimens taken further south. It shows a trend toward the subspecies *minor* Sternfield, which was originally taken at Hermannsburg, Northern Territory. This race apparently occurs in the Macdonnell Ranges and westward, as specimens examined from localities to the south, east and north all show closer affinities with the type race. Loveridge (1934: 325) records a specimen of *minor* from the north-west coast at Broome, Western Australia.

Diporiphora bilineata Gray

Diporiphora bilineata Gray, *Zool. Misc.*: 54 (1842); Loveridge, *Bull. Mus. Comp. Zool. Harv.* 77, No. 6; 327 (1934) (Syn.)

Umbakumba, Groote Eylandt, N.T.: USNM 128301, 129542-61, SAM R2848 (29 specimens)

Groote Eylandt, N.T.: AM R13621 (7 specimens), R13613 (6 specimens), R13617 (6 specimens)

Oenpelli, N.T.: USNM 128734-49, SAM R2847

Nightcliff, near Darwin, N.T.: USNM 128248-9

South end of Lake Hubert, Groote Eylandt, N.T.: USNM 128455

East Alligator River, N.T.: AM R13644 (4 specimens)

Yirrkalla, N.T.: SAM R2858

Preanal pores present in the male, 1 + 1 or 2 + 2; gular fold constantly absent. This latter character serves to readily distinguish *bilineata* from *Physignathus gilberti* (Gray), juvenile specimens of which have a close superficial resemblance.

The inconstancy of the dorsal and ventral scale keeling, and of the colour pattern indicated by this large series fully supports the synonymizing of the names enumerated by Loveridge (1934: 327).

The two specimens from Nightcliff can be considered topotypic. An examination of the Oenpelli series indicates that the colour variation is not correlated with age or locality, there being both adults and juveniles of all recognizable colour phases.

Physignathus gilberti gilberti (Gray)

Lophognathus gilberti Gray, *Zool. Misc.*: 53 (1842)

Oenpelli, N.T.: USNM 128730, 128732-3, SAM R2941 (3 specimens)

Two miles SSE. of Oenpelli, N.T.: USNM 128731

Yirrkalla, N.T.: USNM 128566

West side of Lake Hubert, Groote Eylandt, N.T.: USNM 128456
South end of Lake Hubert, Groote Eylandt, N.T.: USNM 128454
Between Emerald River and Old Mission, Groote Eylandt, N.T.: USNM 128464
East Alligator River, N.T.: AM R13639

One, sometimes two rows of enlarged keeled scales on each side of the dorsal crest, the keels being continuous and parallel to the crest; dorso-lateral scales with majority of keels directed obliquely toward centre; many of the scales between the outer enlarged series and the crest are also directed obliquely to it. Ventral scales larger than the laterals and dorso-laterals but smaller than the middorsals. In the adults the nostril is markedly nearer the tip of the snout than the anterior corner of the eye; horizontal diameter of ear greater than the vertical, which is approximately equal to the length of the orbit. One, occasionally two preanal pores, two femoral pores. The length of the hind limb from the groin to the tip of the longest toe reaches the anterior corner of the eye when measured along the body from the level of the vent.

The Kakadu native name for this species is *majarndemet* (R.R.M.). In the Arnhem Land specimens the greater majority of dorso-lateral and many of the dorsal scales are directed obliquely toward the midline, the only series of scales with continuous keels parallel to the midline being the two slightly enlarged series adjacent to the dorsal crest. However, in this material, and in additional specimens in the South Australian Museum collection, there is considerable variation in the number of scales directed toward the midline and in the degree of inclination, certain extreme specimens closely resembling *temporalis* in that they possess the greater majority of the scales, including those adjacent to the dorsal crest, slightly inclined toward it. The keeling in these extreme variants is not strictly continuous. A population of lizards occurring in New Guinea agrees closely with Boulenger's redescription and figure of *temporalis* (1885: 398, pl. 31, fig. 3), but no additional record of the species occurring on the Australian mainland appears to have been made since Boulenger's 1885 reference to three specimens, including the type, from Port Essington, Northern Territory. Port Essington is also the type locality for *gilberti*, and recent records indicate that the species is still particularly common in this area.

VARANIDAE

Varanus (*Varanus*) *gouldii* (Gray)

Hydrosaurus gouldii Gray, *Ann. Mag. Nat. Hist.* (3) 20: 394 (1838)

Yirrkalla, N.T.: USNM 128571-3
Four miles south-west of Umbakumba, Groote Eylandt, N.T.: USNM 128384
Port Langdon, Groote Eylandt, N.T.: USNM 128479
Red Lily Lagoon, 7 miles west-south-west of Oenpelli, N.T.: USNM 128753
Oenpelli, N.T.: USNM 128754

The scalation and basic coloration of these monitors agrees closely with that of desert adapted specimens from central and southern Australia. Specimens from more humid parts of South Australia, and in particular from Kangaroo and Thistle Islands, possess a distinctive dark colour pattern, and may prove to be subspecifically distinct.

The largest specimen, USNM 128572, measures 1205 (490 + 715) mm. This and several other specimens were badly damaged in capture and are in a poor state of preservation. The above measurements are therefore only approximate.

Varanus (*Varanus*) *varius* (Shaw)

Lacerta varia Shaw, in White's *Journ. Voy. N.S.W.*: 246, pl. 3, fig. 2 (1790)

Near swamp at Lake Hubert, Groote Eylandt, N.T.: USNM 128386
Five miles west of Oenpelli, N.T.: SAM R3230

These two specimens are unusual in that they possess slender digits with almost straight claws, and have the supraocular and interorbital scales larger than the temporals and almost as large as the upper head scales, there being only 15-17 between the supraciliary ridges.

The colouring is uniform black above, with an occasional yellow scale in the dorso-lateral regions of the body and on the limbs.

The larger specimen, USNM 128386, measures 780 (224 + 556) mm.

Dr Miller made an interesting field note on the larger specimen. He referred to it as a 'Water Iguana. Shot by F. M. Setzler in swamp at the south end of Lake Hubert on 24 May.'

Note: Since the manuscript of this paper was originally prepared, noted authority on *Varanus*, Dr Robert Mertens, has examined SAM R3230 and is arranging to study the United States National Museum specimen, as he believes that they may prove to be specifically distinct from *varius*.

Varanus (*Odatria*) *glebopalma* Mitchell

Varanus (*Odatria*) *glebopalma* Mitchell, *Rec. S. Aust. Mus.* 11, pt. 4: 389 (1955)

Holotype: SAM R3222, an adult male taken at the south end of Lake Hubert, Groote Eylandt, N.T.

Diagnosis: This *Varanus* is a typical member of the *Odatria* group and is most nearly approached by the subspecies *timorensis tristis* (Schlegel). It differs from that lizard in possessing a much longer tail, different coloration and characteristic dorsal and caudal scalations. The best diagnostic feature is the presence of highly polished black pads on the soles of all four feet (Fig. 3).

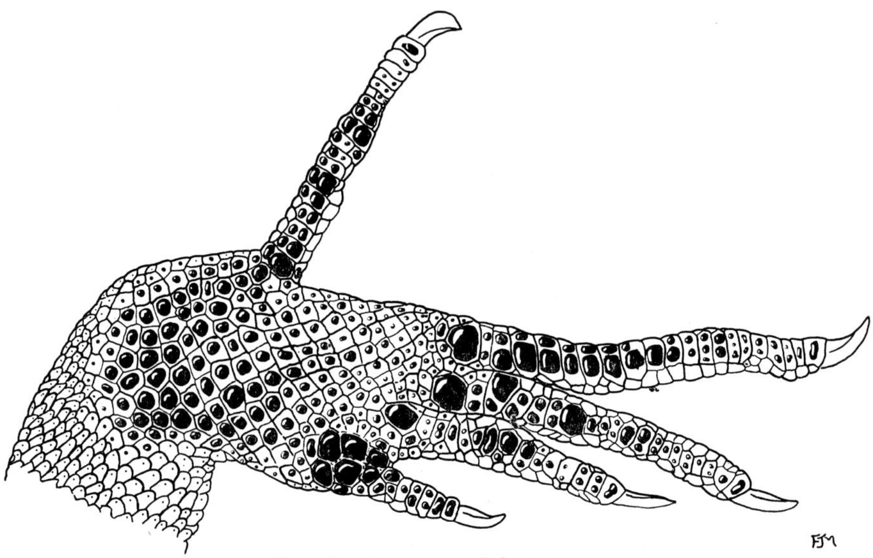

FIG. 3. *Varanus glebopalma*
Sole of the hind foot, showing positions and size of the polished black pads

Description: Teeth acute, slightly compressed. Canthus rostralis acute; nostril oval, below the rostralis, nearer the tip of the snout than the anterior corner of the eye, the measurements from the centre of the nostril being 9 mm. and 12 mm. respectively. Upper head scales moderate, slightly larger than the temporals but smaller than the interorbitals; supraocular scales minute, readily distinguishable from the interorbitals, 37-40 between the supraciliary ridges. Two or three rows of scales at the supraciliary border larger than the supraoculars. Ear opening triangular, slightly oblique; its vertical diameter is approximately equal to the horizontal diameter of the eye. Dorsal and caudal scales of characteristic forms not approached by the variation recorded for *timorensis* (Pl. 1). Abdominal scales smooth, in one hundred and twenty-six transverse rows. Tail slightly depressed basally, compressed posteriorly; covered with uniform obtusely keeled scales which do not rise to a spine. The spines often to be found on each side of the vent in males of *Odatria* species are absent, being replaced by several slightly prominent scales. Scales on the limbs with an obtuse central keel. The soles of both fore and hind limbs are covered with highly polished black pads, the largest pads being at the base of each digit (Fig. 3). The tail is very long, measuring more than twice the combined lengths of the head and body.

The basic colour pattern of the body is light grey-brown with numerous black scales distributed irregularly over the dorsal surface, giving it a punctate appearance (Pl. 2B). On the anterior half of the tail the black scales become dominant, while toward its tip the basal colouring lightens and the black pigmentation disappears. Underside of the lower jaw white with four black transverse bands; remainder of gular region and chest with a black and white reticulate patterning (Pl. 2C).

Measurements: 821 (262 + 559) mm. The distance from the tip of snout to the gular fold is 112 mm., while 75 mm. and 107 mm. are the lengths of the fore and hind limbs respectively.

Comments: The type specimen is unique and was 'shot by R. R. Miller and F. M. Setzler in a crevice in a sandstone boulder at the base of a sandstone escarpment near the south end of Lake Hubert (R.R.M.)'.

Varanus (*Odatria*) *timorensis orientalis* Fry

Varanus punctatus var. *orientalis* Fry, *Rec. Aust. Mus.* 10: 18, fig. 7-10 (1913)

Milingimbi Island, Crocodile Islands, N.T.: USNM 128481-5, 128487-93, SAM R3227

Yirrkalla, N.T.: USNM 128569-70, 128574

Umbakumba, Groote Eylandt, N.T.: USNM 128387, SAM R2851

East Alligator River, N.T.: AM R13637, R13646

Cape Arnhem, N.T.: AM R12550

The following South Australian Museum specimens were also examined: *V. t. orientalis*—R137, R182-3, Eidsvold, Qld. (Type locality); R351-3, Stewart River, Qld.; R362-5, Melville Island, N.T.; R1938, Flinders Island, Qld.

V. t. tristis—R329, Macdonnell Ranges, N.T.: R2058, between Mt Singleton and Treur Range, S. Aust. (4 specimens); R2186, Kairi, Qld.; R3085, Frazer Range, W. Aust.; R3223, Innamincka, S. Aust.; R3224, Tennant Creek, N.T. (5 specimens).

The dorsal and caudal scalation of the material listed above shows considerable variation. This indicates that the scalation characters enumerated and figured by Fry (1913: 18-19, figs. 7-10) as distinguishing features are inadequate to satisfactorily characterize this subspecies. However, several more easily discernible and constant features distinguishing it from the other subspecies are described by

Mertens (1942: 298-307). One of these, the supposed longer tail of *tristis* is not confirmed by the present material. In the Arnhem Land specimens the ratio of the tail length to combined head and body length varies from 1·5-1·7 and in the South Australian Museum specimens of *orientalis* from 1·4-1·7; in the *tristis* material the variation is 1·6-1·8. Except in cases where the difference is extreme, the taxonomic value of this character is limited because of the difficulties in accurately measuring the body lengths of many preserved specimens, and in determining whether or not the last few centimetres of tail are present.

The measurements and the ratios of the tail length to the length of head plus body of the adult specimens taken by the expedition:

Specimen Number	Sex	Total Length	Head and Body Length	Tail Length	Tail Length / Head and Body Length
USNM		mm.	mm.	mm.	
128484	♂	520	194	326	1·7
128485	♂	499	187	312	1·7
128486	♂	487	182	305	1·7
128487	♀	465	185	280	1·5
128488	♂	497	185	312	1·7
128489	♂	451	172	279	1·6
128490	♀	405	160	245	1·5
128492	♂	350	130	220	1·7
128387	♀	440	165	275	1·7
128574	♀	480	190	290	1·5

Several additional adults obviously possess incomplete tails.

Although comparatively constant in specimens from the one locality, the body scalation of this species varies markedly from one locality to another. The scales vary both in shape and relative density. The latter variation can be illustrated by counting the number of micro-scales. In the Arnhem Land specimens the dorsal scales are separated by as many as five rows of micro-scales, while in others, notably those from Tennant Creek, by only one or two rows. Fry (1913: fig. 7) figures 1-2 rows for his Eidsvold specimens, while the three South Australian Museum specimens from the type locality possess 2-3. Further, the relative size of these micro-scales varies. The number bordering each dorsal scale varies from an average of eight for the Stewart River specimens to fourteen for those from Tennant Creek; the average is twelve in the Arnhem Land series. Fry (op. cit., fig. 8) figures eighteen for his *punctatus* var. *typica*. In some specimens there is a tendency for the micro-scales to become fused, forming larger elongate scales along the lateral borders of the dorsals. Working on a limited number of specimens this variation does not appear to be significant, although the examination of a larger series may reveal grounds for the subdivision of the material at present referred to *tristis*. Plate 1 contains photographs illustrating this variation in both *tristis* and *orientalis*.

The number of abdominal scales averages 84 in a range of 79-94 in the Arnhem Land specimens and 90 in a range of 85-104 in the South Australian Museum series of *tristis*.

Two gravid females were opened, one USNM 128574, contained twelve eggs, while the other, USNM 128287, contained eight.

The largest specimen examined, SAM R3085, a male from the Frazer Range, Western Australia, originally recorded by Stirling and Zietz (1893: 170) measures 700 (265 + 435) mm., while the largest Arnhem Land specimen, USNM 128483, also a male, measures 465 (210 + 255 +) mm.—tail incomplete.

PYGOPODIDAE

Delma fraseri fraseri Gray

Delma fraseri Gray, *Zool. Misc.*: 14 (1831)
Delma plebeia De Vis, *Proc. Linn. Soc. N.S.W.* 2: 825 (1888)

Yirrkalla, N.T.: USNM 128679-82
Nightcliff, near Darwin, N.T.: USNM 128260
Cape Arnhem, N.T.: AM R13648 (2 specimens), R13569 (2 specimens), R13570 (2 specimens)
Groote Eylandt, N.T.: AM R13471

The anal scalation of this material is interesting as it confirms that *plebeia* De Vis is a variant of *fraseri*, as was suggested by Longman (1916: 51) and Kinghorn (1926: 53). Although all of the present series possess three anal scales, the size and position of the median scale varies considerably; the outer anals accordingly vary from widely separated by the median anal to forming a median suture thereby excluding the median anal (Fig. 4). Probably the median anal is present in the type of *plebeia*, but its position and size make it of little apparent significance.

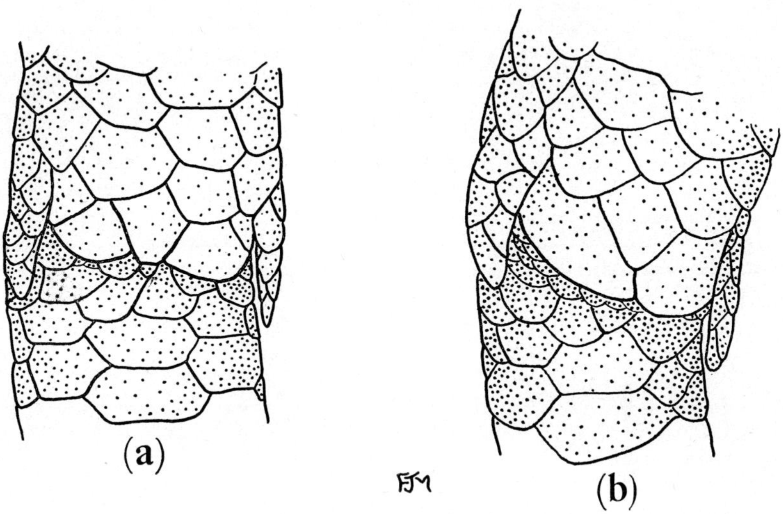

FIG. 4. *Delma fraseri fraseri*—anal region
(a) var. *typica* condition (b) var. *plebeia* condition, Yirrkalla, N.T.

Lialis burtonis Gray

Lialis burtonis Gray, *Proc. Zool. Soc. London*: 134 (1834)

Umbakumba, Groote Eylandt, N.T.: USNM 128379-83, 128441, AM R13610 (3 specimens)
Port Langdon, Groote Eylandt, N.T.: USNM 128467-8
Nightcliff, near Darwin, N.T.: USNM 128259
Yirrkalla, N.T.: USNM 128576
Milingimbi Island, Crocodile Islands, N.T.: USNM 128504
Kuraiturumur, N.T.: USNM 128514
Cape Arnhem, N.T.: AM R13548 (2 specimens), R13647
Oenpelli, N.T.: SAM R2846

Midbody scales in 20 or 21 rows including the ventrals; the four preanal pores are sometimes indistinguishable. As indicated by Kinghorn (1924: 186) the general scalation and colour of this widely distributed species is very variable, the variation being independent of habitat.

SCINCIDAE

Tiliqua scincoides intermedia Mitchell

Tiliqua scincoides intermedia Mitchell, *Rec. S. Aust. Mus.* 11, pt. 4: 393 (1955)

Holotype: Yirrkalla, N.T.: SAM R3095, an adult male

Paratypes: Yirrkalla, N.T.: USNM 128636-7, 128639
Near Umbakumba, Groote Eylandt, N.T.: SAM R3225
Port Langdon, Groote Eylandt, N.T.: USNM 128478

Holotype scalation: Midbody scales in 33 rows; prefrontals meeting at a point on the midline; nasals widely separated. Fifth, sixth and seventh of nine upper labials subocular; four supraoculars and six supraciliaries; three auricular lobules.

Coloration: Somewhat difficult to describe and best conveyed by the figure of a juvenile paratype, SAM R3225 (Pl. 2A). In the holotype the black dorso-lateral cross-bars are more prominent and the light bars less prominent, being partially enclosed by the dark bars at their dorsal ends.

Measurements: 447 (334 + 113 +) mm.—tail incomplete.

Variation: Midbody scales 32-5, irregular in the dorso-lateral region. Prefrontals separated in all paratype specimens; nasals invariably separated. Nine or ten upper labials with the 5th, 6th and 7th or 6th, 7th and 8th subocular; 6-7 supraciliaries; 2-3 auricular lobules.

The most distinctive feature of this race is the colour pattern. The light and dark dorso-lateral bars alternate on each side of the body, a dark bar on one side facing a light bar on the other. The general middorsal colour is darker than the light bars, there being a light blotch the same shade as the light bars at the dorsal end of each dark bar (Pl. 2A). This basic colour pattern is quite constant in the present series and also in the four South Australian Museum specimens recorded by Mitchell (1950: 295). The number of darker bars and bands on the body and tail is the same as that recorded for the type race (Mitchell, op. cit.).

In *intermedia* these bars do not extend to or across the dorsal midline. The dark temporal streak is not as prominent as in the type race.

Another character evident from a comparison of this material with South Australian specimens is the larger size of *intermedia*. Of twenty-one South Australian specimens examined, the largest measured 293 mm. from the tip of the snout to the vent. Of the Arnhem Land series all specimens except the juvenile (USNM 128388) possess body lengths in excess of 300 mm., the largest being the holotype. With the increased length is a marked increase in bulk.

Lygosoma (Sphenomorphus) taeniolata (Shaw)

Lacerta taeniolata Shaw, in White's *Journ. Voy. N.S.W.*: 245, pl. 32, fig. 1 (1790)

Oenpelli, N.T.: USNM 128758, SAM R2868
East Alligator River, N.T.: USNM 128756

Midbody scales in 24 or 26 rows; 22-4 keeled subdigital lamellae beneath 4th toe. In USNM 128756, the prefrontals form a long median suture; USNM 128758 possesses only three upper labials anterior to the subocular on one side of the head. Otherwise these specimens correspond closely with Boulenger's catalogue description of this species (1887: 227). The black vertebral stripe has no light edge in the largest specimen, which measures 145 + (105 + 40 +) mm.—tail incomplete.

Lygosoma (*Sphenomorphus*) *essingtonii* (Gray)

Tiliqua essingtonii Gray, *Zool. Misc.*: 51 (1842)
Nightcliff, near Darwin, N.T.: USNM 128251-6
Yirrkalla, N.T.: USNM 128603-7, 128609-10
Umbakumba, Groote Eylandt, N.T.: USNM 128402-7, AM R13466 (2 specimens), R13467 (3 specimens), SAM R2866 (7 specimens)

Loveridge (1934: 346) raises the question of the affinities of *T. essingtonii* Gray and suggests that it may be more nearly allied to *leonhardii* Sternfeld than to *taeniolata* Shaw. Although varying a little, the coloration of this series of 31 specimens agrees well with Gray's type and Boulenger's catalogue description (1887: 228) of this species. An examination of the scalation supports Boulenger's statement (op. cit.) that it differs in no way from that of *taeniolata* (Shaw).

A summary of the dermal variation shown by these specimens is as follows: Midbody scales in 26 rows (21 specimens) or 28 rows (10 specimens); 21-8 sharply keeled subdigital lamellae beneath the 4th toe. Ear opening a little smaller than the eye opening, with two or three large rounded lobules anteriorly; three or four pairs of laterally expanded nuchal scales. Prefrontals invariably separated, the frontal and fronto-nasal forming a variable length median suture; nasals usually separated, but in several specimens they make point contact on the midline; a subnarial groove is faintly present in many specimens. Four labials anterior to the subocular in all specimens except USNM 128255 which has five on one side and four on the other.

The colour and pattern are distinctive. A broad dorsal stripe varies from light brown in the Nightcliff specimens through light bronze to almost cream in some of the Groote Eylandt material. This is bordered on each side by a wide, light edged, dark dorso-lateral stripe, which breaks up into a series of dark spots on the tail. The light upper edge of this stripe is defined on the nape and anterior half of the body by a further dark line, particularly prominent in the Nightcliff specimens. In the Yirrkalla and Groote Eylandt specimens a series of 16-24 bronze blotches are enclosed within the dark dorso-lateral stripe. In the Nightcliff series there are only occasional small lighter spots in the anterior half. Limbs bronze, spotted or striated with black. Upper labials and ventral surfaces white.

Three specimens from Melville Island, Northern Territory, contained in the South Australian Museum collection have the black line which defines the light upper edge of the dark dorso-lateral stripe much wider and continuing along the full length of the body. Otherwise they are identical with the Nightcliff specimens.

Largest specimen, USNM 128603, measures 159 (56 + 103) mm.—tail reproduced.

Lygosoma (*Sphenomorphus*) *spaldingi* (Macleay)

Hinulia spaldingi Macleay, *Proc. Linn. Soc. N.S.W.* 2: 63 (1877)
Yirrkalla, N.T.: USNM 128594-5, 128548, 128600, 128585
Umbakumba, Groote Eylandt, N.T.: USNM 128391, SAM R2851
Cape Arnhem, N.T.: AM R13576 (part)

Midbody scales in 26 rows (1 specimen), 28 rows (4 specimens), 30 rows (3 specimens); 21-4 rounded subdigital lamellae beneath 4th toe; two or three pairs of enlarged nuchals. The number of upper labials anterior to the subocular is variable, usually four, but USNM 128594 possesses five, while USNM 128391 and 128585 have five on one side and four on the other. Prefrontals forming a median suture in seven of this series. Except for an aberrant specimen, USNM 128595, in which both prefrontals and nasals form median sutures, the frontonasal being absent, no specimen has the nasals contacting on the midline.

The largest specimen, AM R13876, measures 295 (83 + 212) mm.

Copland (1947: 136-44) has redescribed this species from the type material in the Macleay Museum, Sydney, and discusses its variation in detail.

Lygosoma (*Sphenomorphus*) *leonhardii* (Sternfeld)

Lygosoma (*Hinulia*) *taeniolatum* var. *maculata* Rosen, *Ann. Mag. Nat. Hist.* (7) 16: 140 (1905)

Lygosoma (*Hinulia*) *leonhardii* Sternfeld, *Mitt. Senckenb. Naturf. Ges.* 1: 79 (1919)

Yirrkalla, N.T.: USNM 128577-84, 128586-93, 128596-7, 128599, 128601, 128608, 128618, SAM R2860

Near Umbakumba, Groote Eylandt, N.T.: USNM 128389-90, 128392-401, SAM R2869, AM R13605 (3 specimens), R13602 (3 specimens), R13611 (1 specimen), R13465 (1 specimen), R13514 (2 specimens), R 13603 (3 specimens), R13503 (2 specimens), R13604 (3 specimens), R13501 (2 specimens), R13598 (3 specimens), R13508 (2 specimens), R13597 (3 specimens)

Oenpelli, N.T.: USNM 128757, SAM R2867

Cape Arnhem, N.T.: AM R13655 (2 specimens), R13577 (2 specimens), R13576 (1 specimen)

The variation displayed by this large series is as follows: Midbody scales in 28 rows (16 specimens) or 30 rows (24 specimens)—only the adults were counted. Nasals separated behind the rostral in all specimens; prefrontals forming a variable length median suture; ear opening oval, relatively smaller in the adult than in the juvenile; 3-5 auricular lobules, the shape of each varying from obtuse to very acute. A subnarial suture is present in many specimens. Two or three pairs of enlarged nuchals. All specimens possess four supraoculars except USNM 128582 which has only three, there being indications of the fourth suture at the supraciliary border. The second anterior supraocular is invariably the largest. Except for two specimens which possess six on one side and five on the other, all specimens

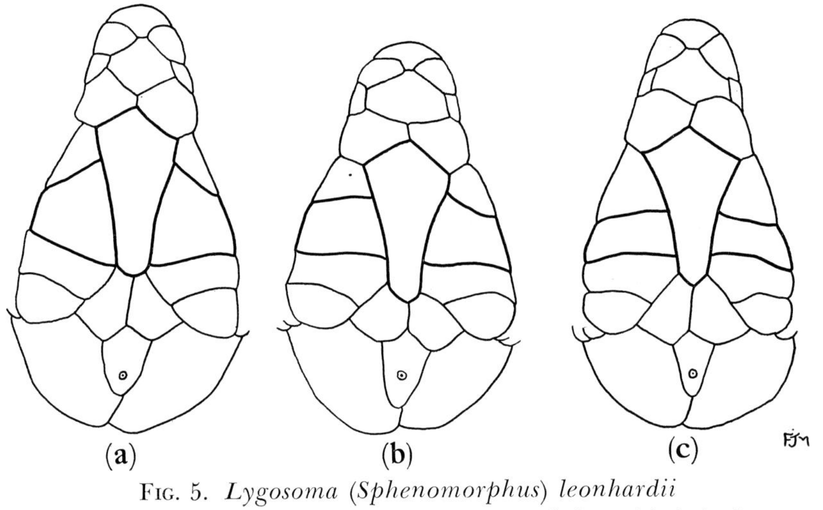

FIG. 5. *Lygosoma* (*Sphenomorphus*) *leonhardii*

Variation in supraocular scalation and its correlation with latitude (a) USNM 128608, Yirrkalla, N.T. (b) SAM R1532, Hermannsburg Mission, N.T. (type locality) (c) SAM R3177, Mern Merna, S. Aust.

have five upper labials anterior to the first subocular. The length of the snout seems to be very variable with consequent variation in the shape and size of the loreals and proportions of the prefrontals and internasal.

The colour and pattern also show some variation. Many adult specimens are without a black vertebral stripe, and when some vestige of this line is evident, it is often without the white border characteristic of specimens taken further south.

A comparison of this material with a large series of over a hundred southern and central Australian specimens preserved in the South Australian Museum has revealed some interesting variation in the size and shape of the supraoculars. Although the size of the second supraocular is variable in all populations examined, there appears to be a south to north gradation tending toward an enlargement of this supraocular, its average size in the Arnhem Land specimens being markedly greater than that of South Australian skinks examined (Fig. 5). In addition, this study revealed considerable random variation in the midbody scales, disposition of the nasals and prefrontals, and in the numbers of subdigital lamellae. The length of the 4th toe in central Australian specimens is particularly variable, the number of subdigital lamellae varying from 22-35.

Lygosoma (*Sphenomorphus*) *isolepis isolepis* Boulenger

Lygosoma isolepis Boulenger, *Catalogue of Lizards in the British Museum, London* 1-3: 234, pl. 15, fig. 1 (1887)

Milingimbi Island, Crocodile Islands, N.T.: USNM 128506, SAM R3249
Nightcliff, near Darwin, N.T.: USNM 128250
Larporkuru, N.T.: USNM 128512
Groote Eylandt, N.T.: AM R13606 (2 specimens), R13468 (3 specimens)
Arnhem Land, N.T.: AM R13549
Cape Arnhem, N.T.: AM R13654
East Alligator River, N.T.: AM R13641

Midbody scales in 26 rows (2 specimens), 27 rows (2 specimens), 28 rows (5 specimens) or 30 rows (3 specimens); 16-24 lamellae beneath the 4th toe. Upper labials irregular, 5th and 6th subocular in eight specimens, 4th and 5th subocular in four specimens, and one possessing 4th and 5th on one side and 5th and 6th on the other. The nuchal scalation is also irregular, varying from no enlarged nuchals to two pairs. Seven of this series possess three enlarged nuchal scales, one pair and an odd scale. The ear opening is variable both in size and shape; no auricular lobules.

The coloration agrees closely with that described by Boulenger (1887: 234). In AM R13641 the dorsal colour is much darker brown, the black speckling being less prominent and the white dorso-lateral spots more prominent.

Lygosoma (*Lygosoma*) *crassicauda* Duméril

Lygosoma crassicaudum A. Duméril, *Cat. Method. Coll. Rept. Paris*: 172 (1851)

Yirrkalla, N.T.: USNM 128611

Midbody scales in 20 longitudinal rows; four pairs of enlarged nuchals, contracting in size to a double series of slightly enlarged vertebrals. Auricular opening round, without lobules; approximately one and three-quarter times longer than that from tip of snout to forelimb. The hind limb measures 14 mm., a length equal to the distance between the centre of the eye and the forelimb. Fifth upper labial centrally subocular; frontonasal forming broad sutures with both rostral and frontal, the suture in each case being a little more than half the frontonasal length. The maximum combined lengths of the frontoparietals and interparietal is a little greater than that of the frontal. Subdigital lamellae smooth, 5, 7-8, 12-13, 18-19, 8-9 being the formula for the toes and 5, 8, 8, 6, 4 the formula for the fingers, the number of lamellae being in direct proportion to the length of the digit in each case.

Dorsal surfaces dark brown, minutely punctate with black; the black spots becoming more prominent as the basic colour lightens ventrolaterally. Upper labials black with a light vertical bar in the centre of each labial; lower labials and underside of tail also punctate with black. Remaining ventral surfaces uniform white. The limbs have longitudinal rows of black spots.

This specimen measures 78 + (56 + 22 +) mm.—tail incomplete.

Although it possesses the long frontonasal-frontal and frontonasal-rostral sutures stated to be characteristic of *mjobergi* Lönnberg and Andersson (1915: 6), this specimen has the larger number of subdigital lamellae of *crassicauda*. While confidently claiming on the evidence of an examination of a specimen from Ravenshoe, Queensland, referred to *mjobergi* by Proctor (1923: 1073), and another without locality data, that these two names refer to distinct species, Loveridge (1934: 367) refers the latter specimen, which possesses only twelve lamellae beneath the 4th toe, to *crassicauda*. If correct, this would greatly reduce the value of the hind limb lamellae formula as a taxonomic character for the separation of these two lizards. Similarly, the validity of the relative lengths of the frontonasal sutures as a distinguishing feature is doubtful if this specimen is referred to *crassicauda*.

A comparison of the present specimen with literature also shows the following characters to vary. The Arnhem Land specimen and the type material, as figured by Hombron and Jacquinot (1842-53: pl. 4, fig. 1), have the fifth upper labial centrally subocular, while Lönnberg and Andersson record the fourth as being 'below the centre of the eye' in the cotypes of *mjobergi*; Boulenger (1887: 325), after examining specimens from Fly River, New Guinea, Murray and Cornwallis Islands records both conditions in his material. A suggestion of variation in the relative proportions of the frontonasal and in the number of supraoculars contacting the frontal is dependent on the accuracy of the Hombron and Jacquinot figure (op. cit.), which shows the frontonasal to be much longer than wide and the 1st, 2nd and 3rd supraoculars contacting the frontal. In coloration *mjobergi* is intermediate between that of the Arnhem Land specimen described above and that of *crassicauda*. No evidence of a dark dorso-lateral stripe, sharply defined at its dorsal edge is to be found in the present specimen, while some reference is made to such in all other relevant literature, including Macleay's *Lygosoma ornatum* (1878: 64).

Loveridge (op. cit.) mentions that *mjobergi* is a much larger species, a suggestion supported by the measurements of the larger cotype.

Leiolopisma

Because of the confused nomenclature in certain sections of this genus, it was found necessary to make a brief revision of the Australian lizards in the 'Heteropus' species group, which is characterized within the genus by the possession of 4 + 5 digits and an undivided frontoparietal, to enable three of the species taken by the expedition to be satisfactorily identified. This revision has been published, *Rec. S. Aust. Mus.* 11: 57-90 (1955), and the synonomies quoted below are discussed in it.

Since the publication of the 'Preliminary Account' (1955), the author's attention has been drawn to a paper by Mittleman (1952). He has made a courageous endeavour to simplify and stabilize the generic classification of the Lygosominae, and has resurrected the first available name for this group, *Carlia* Gray, and given it generic status. For the sake of continuity of reference the name has not been employed in this report, although Mittleman's action in separating this group generically would seem to be well founded.

Leiolopisma fuscum fuscum (Duméril and Bibron)

Heteropus fuscus Duméril and Bibron, *Erp. Gen.* 1-9: 759 (1839)
Heteropus schmeltzii Peters, *Mber. Akad. Wiss. Berl.*: 23 (1867)
Heteropus tricarinatus Meyer, *Mber. Akad. Wiss. Berl.*: 133 (1874)
Heteropus longipes Macleay, *Proc. Linn. Soc. N.S.W.* 2: 66 (1877)
Heteropus sexdentatus Macleay, *Proc. Linn. Soc. N.S.W.* 2: 67 (1877)
Heteropus maculatus De Vis, *Proc. Roy. Soc. Qld.* 1: 169 (1885)
Heteropus rubricatus De Vis, *Proc. Roy. Soc. Qld.* 1: 170 (1885)
Heteropus rostralis De Vis, *Proc. Roy. Soc. Qld.* 1: 171 (1885)

Yirrkalla, N.T.: USNM 128612-17, 128519
Cape Arnhem, N.T.: AM R13583-4, R13656 (3 specimens)

Midbody scales in 32 rows (1 specimen); 34 rows (7 specimens); 36 rows (4 specimens). Dorsal scales weakly tricarinate, almost smooth; occasional quinquecarinate scales towards the nape. Four upper labials (five in AM R13584) anterior to the subocular. Ear opening vertically oval, of similar size and shape to the horizontally oval palpebral disc; with 3-5 short, acute lobules on the anterior border, and several short obtuse lobules on the posterior border; 29-32 subdigital lamellae beneath the 4th toe.

Uniform dark brown to olive above; without any sign of dorso-lateral markings in juvenile or adult specimens. Lighter ventrally.

Leiolopisma vivax (De Vis)

Heteropus peronii Duméril and Bibron, *Erp. Gen.* 1-9: 760 (1839)
Myophila vivax De Vis, *Proc. Roy. Soc. Qld.* 1: 77 (1884)
Heteropus blackmanni De Vis, *Proc. Roy. Soc. Qld.* 1: 168 (1885)

Milingimbi Island, Crocodile Islands, N.T.: USNM 128507-10
Nightcliff, near Darwin, N.T.: USNM 128257
Cape Arnhem, N.T.: AM R13585 (4 specimens), R13586 (4 specimens)
Umbakumba, Groote Eylandt, N.T.: USNM 128439-40, 128412, AM R13464 (2 specimens), R13470 (2 specimens), R13607 (2 specimens), SAM R2857 (3 specimens)
Yirrkalla, N.T.: SAM R2865

Dorsal scales sharply bicarinate, becoming smooth or obtusely tri- or quadricarinate on the nape. Midbody scales in 28 rows (2 specimens), 30 rows (6 specimens) or 32 rows (4 specimens)—only USNM and SAM specimens counted. Prefrontals narrowly separated or making point contact in the majority of specimens, but forming a distinct median suture in two of the Groote Eylandt skinks. Interparietal very small; a single pair of enlarged nuchals. Seven upper labials, fifth subocular. Ear opening variable in size, but usually only about half of the horizontal diameter of the transparent palpebral disc; several short lobules anteriorly. The position of the posterior suture of the second supraocular is variable; in several specimens it makes point contact with the frontal. The frontal is fused with the frontoparietal in USNM 128440. There are 22-6 lamellae beneath the 4th toe.

Metallic green to bronze dorsally with numerous irregularly distributed black edged ocelli. Lateral surfaces uniform light bronze or reticulate with black; ventral surfaces white to pale blue with a dense black reticulate patterning under the throat in several male specimens.

Leiolopisma pectoralis (De Vis)

Carlia melanopogon Gray, *Zool. Erebus and Terror, Rept.*: pl. 7, fig. 1 (1844)
Heteropus lateralis De Vis, *Proc. Roy. Soc. Qld.* 1: 168 (1885)
Heteropus pectoralis De Vis, *Proc. Roy. Soc. Qld.* 1: 168 (1885)
Heteropus mundus De Vis, *Proc. Roy. Soc. Qld.* 1: 172 (1885)
Lygosoma devisii Boulenger, *Proc. Zool. Soc. Lond.*: 79 (1890) (n.n. for *lateralis* De Vis as preoccupied in *Lygosoma*)
? *Lygisaurus foliorum* De Vis, *Proc. Roy. Soc. Qld.* 1: 77 (1884)

Oenpelli, N.T.: USNM 128764
Port Essington, N.T.: USNM 128528

Dorsal scales tricarinate, almost smooth in the Port Essington specimen; 26 (USNM 128764) or 30 scales at midbody. Ear opening smaller than the transparent palpebral disc; auricular lobules extremely small. Prefrontals well separated; lamellae beneath the 4th toe, 22 and 29 respectively.

Uniform dark grey to olive dorsally; lighter ventrally. The Port Essington specimen possesses a pronounced silvery-grey stripe along the upper lip through the ear to the shoulder.

Leiolopisma guichenoti delicata (De Vis)

Mocoa delicata De Vis, *Proc. Linn. Soc. N.S.W.* 2nd ser., 2: 820 (1888)

Groote Eylandt, N.T.: AM R13461

Midbody scales in 26 rows; no auricular lobules. General scalation similar to that of South Australian and Victorian specimens of *guichenoti*, but with a much wider frontonasal-frontal suture. Its general form is more slender and the limbs weaker. The colour corresponds accurately with that described by De Vis, and if constant would readily distinguish it from the type race. On the above evidence *delicata* is retained as a subspecies.

Loveridge (1934: 359) doubtfully referred *delicata* to the synonomy of *guichenoti*, commenting on the fact that all specimens of *guichenoti* possessed enlarged preanals, and that this feature was not characteristic of *delicata*. This is confirmed by material examined.

Rhodona stylis Mitchell

Rhodona stylis Mitchell, *Rec. S. Aust. Mus.* 11: 400 (1955)
Holotype: Yirrkalla, N.T.: SAM R3094
Paratypes: Yirrkalla, N.T.: USNM 128640-78, SAM R2856 (5 specimens)

Umbakumba, Groote Eylandt, N.T.: USNM 128409-11, SAM R2855 (2 specimens)
Rocky Beach, Cape Arnhem, N.T.: USNM 128520
Cape Arnhem, N.T.: AM R13566 (4 specimens), R13567 (5 specimens), R13568 (5 specimens), R13657 (8 specimens), R13658 (8 specimens)

Diagnosis: This species most nearly approaches *Rhodona lineata* (Gray) in the degree of reduction of the limbs, but is immediately distinguished from it by the presence of frontoparietals and at least three supraoculars. The scalation closely resembles that of *Rhodona wilkinsi* Parker, with which it agrees in all essential points including the midbody scale count. However, the presence of a styliform hind limb immediately separates it from that species.

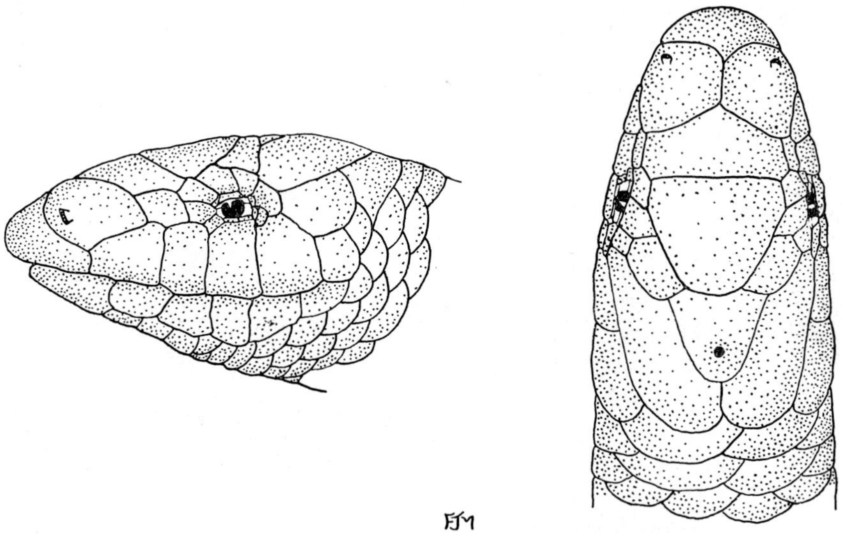

FIG. 6. *Rhodona stylis*
Lateral and dorsal views of the head of the holotype, x 10 approx.

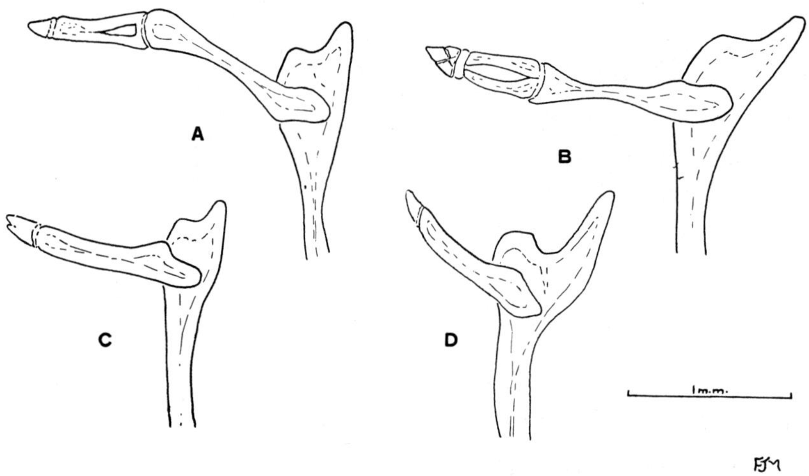

FIG. 7. *Rhodona stylis*—bones dissected from within the styliform hind limb

A, B Specimens from Groote Eylandt, N.T.
C, D Specimens from Yirrkalla, N.T. (type locality)
Body lengths of the specimens 60 ± 2 mm.

Type description: Body elongate; forelimb absent; hind limb minute, styliform, variable in size (Fig. 7), but of equivalent length to two of the adjacent scales in the holotype. Snout cuneiform with angularly projecting labial edge; eye small, lower eyelid with a large transparent disc. Three supraoculars; frontoparietals and interparietal distinct, the former being widely separated and less than half the size of the latter; parietals forming a suture behind the interparietal. A loreal and one or two preoculars; two postoculars. Five upper and five lower labials, the third upper being subocular; two pairs of enlarged nuchals and a pair of slightly enlarged anal plates present. Eighteen smooth scales at midbody. Ear opening minute.

Coloration: Light grey dorsally with a black dorsolateral stripe extending from behind each eye to the tip of the tail. On the dorsal surface a longitudinal line of fine black dots extends down each side of the vertebral line, the two rows coalescing on the tail. Ventral surfaces uniform white.

Measurements: The holotype measures 113 (60 + 53) mm.—tail regenerating.

Variation: The head scalation is very constant within the type series of 82 specimens, the main variation being in the size and number of the small ocular and supraciliary scales. In several of the Groote Eylandt specimens the larger postocular has been forced up into line with the supraoculars by the enlargement of several small scales at the posterior border of the eye, and might therefore be considered an additional supraocular. The number of scales at midbody is usually eighteen, but the lateral scales are often irregular and in such cases twenty can often be counted, particularly in the Yirrkalla material. The basic midbody count for the Groote Eylandt skinks is sixteen, with eighteen occurring in two specimens with irregular lateral scaling. Accepting midbody as being the middle point of the distance between the ear opening and hind limb, fifty specimens were counted with the following result: eighteen specimens were found to possess 20 midbody scales, twenty-seven possessed 18 and five possessed 16. Many specimens have only four lower labials.

Although no point would seem to be gained by separating them, the Groote Eylandt specimens have several characters which distinguish them from the Yirrkalla series. Apart from the small scalation differences indicated above they differ in possessing less degenerate hind limbs, some evidence of the tibia, fibula and tarsals being present. There is only a short rudiment of the tibia-fibula present in the Yirrkalla lizards (Fig. 7).

The uniform degeneration of the limbs and digits in the genus *Rhodona* has been of considerable interest to the author, and limb dissections of the nine species available in the South Australian Museum collection have been made and studied. Although by no means conclusive, the variation in tarsal (carpal) and metatarsal (metacarpal) bones, particularly among specimens of the same species, but from different localities, as is indicated in the present case, suggests that the loss of digits and digital bones may not be always indicative of specific variation. The genus may be found to contain polytypic species within which the number of digits varies.

Ablepharus taeniopleurus Peters

Ablepharus (*Morethia*) *taeniopleurus* Peters, *Mber. Akad. Wiss. Berl.*: 375 (1874)
Red Lily Lagoon, 7 miles WSW. of Oenpelli, N.T.: USNM 128759
Yirrkalla, N.T.: USNM 128635

Two subadult specimens. Twenty-five or twenty-six midbody scales; three supraoculars; four or five large supraciliaries. Prefrontals narrowly separated or joined by a small azygous scale. Supranasals present in USNM 128635, separated; absent in USNM 128759, the nasals making point contact on the midline. The fifth of seven upper labials enlarged, subocular. One pair of enlarged nuchals present in USNM 128759; 20 subdigital lamellae beneath the 4th toe in both specimens.

The markings of both specimens are typical of the species, one specimen possessing the dark ground colour and rufous tail of *ruficaudus* Lucas and Frost.

Ablepharus orientalis (De Vis)

Miculia orientalis De Vis, *Proc. Roy. Soc. Qld.* 5: 160 (1888)

Near Umbakumba, Groote Eylandt, N.T.: USNM 128408

Except for the possession of 20 instead of 18 midbody scales, this specimen corresponds accurately with De Vis's type description. Nasals forming a short suture on the midline; prefrontals widely separated; frontoparietals distinct, approximately equal in area to the interparietal; two pairs of nuchals, the second pair being fused into a single shield. Eye completely surrounded by small rectangular scales; 4th upper labial subocular, but 3rd, 4th and 5th upper labials all enter the proximity of the eye. Digits 4 + 4, relatively long; 12-13 large lamellae beneath the 4th toe. Colour as in the type, but apparently not as dark on the flanks. Tail, translucent pink resembling that of *ruficaudus* Lucas and Frost (= *taeniopleurus* Peters).

Type locality, Barkley (Barkly) Tablelands, Queensland.

Zietz (1920: 222) gives the distribution of *elegans* as Western Australia, South Australia, Central Australia, New South Wales and Victoria. Except for Western Australia, I have been unable to find grounds for suggesting that the species occurs in any of the states mentioned. The South Australian Museum collection contained 18-20 specimens which had been referred to *elegans*, but on examination all proved to have been wrongly identified, most being referable to *A. greyi* Gray or *A. timidus* De Vis. The occurrence of *elegans* in South Australia is therefore doubtful. The present specimen differs from the Western Australian *elegans* in possessing a higher number of midbody scales, the presence of only three upper labials anterior to the subocular and in the complete encirclement of the eye by smaller granular scales.

Ablepharus boutonii metallicus Boulenger

Ablepharus boutonii var. *metallicus* Boulenger, *Catalogue of Lizards in the British Museum, London* 1-3: 347 (1887)

Yirrkalla, N.T.: USNM 128619-34

Near Umbakumba, Groote Eylandt, N.T.: USNM 128413-38, AM R13469 (3 specimens), R13592 (10 specimens), SAM R2852 (3 specimens)

Two miles SSE. of Oenpelli, N.T.: USNM 128760-3, SAM A2850 (3 specimens)

Nightcliff, near Darwin, N.T.: USNM 128282, 128258

Cape Arnhem, N.T.: AM R13582 (5 specimens), R13650 (7 specimens)

This large series agrees for the most part with *metallicus*, although several of the brightly metallic specimens show brown keel-like striations on the mid-dorsal scales suggestive of *virgatus* Garman. Many specimens show a melanic tendency, the characteristic patterning being very dark and non-metallic, or in some, completely indiscernible, the dorsal colour being uniform dark grey to black; ventral surfaces also much darker than in the metallic specimens, where they are pearl white. The two colour phases occur at all of the localities quoted above, the dark phase outnumbering the metallic phase.

The midbody count varies from 22-6—average 23 for 25 specimens counted. Four labials anterior to the subocular; prefrontals forming a median suture of variable length.

Mertens (1931) gives a detailed account and diagnosis of the races of *A. boutonii* Desj.

AMPHIBIA

SALIENTIA

LEPTODACTYLIDAE

Limnodynastes ornatus (Gray)

Discloglossus ornatus Gray, *Zool. Misc.*: 56 (1842)
Central Hill, Groote Eylandt, N.T.: USNM 128457-8
Umbakumba, Groote Eylandt, N.T.: USNM 128274-5, SAM R3250
Port Langdon, Groote Eylandt, N.T.: USNM 128465-6

Only one specimen of this series possesses the ornate colour pattern which can be considered typical of this species. The others are referable to the 'well marked colour variant' mentioned by Parker (1940: 49) or possess markings at some stage intermediate between the two, the cantho-temporal, the labial, and a reduced portion of the middorsal H-shaped markings being persistent.

The largest specimen measures 41 mm. from snout to vent and has a hind limb of 51 mm.

Cyclorana australis (Gray)

Alytes australis Gray, *Zool. Misc.*: 56 (1842)
Nightcliff, near Darwin, N.T.: USNM 128236-9

This species is separable from its more slender ally, *C. alboguttatus* (Günther) by the strong sculpturing of the upper surfaces of the maxillae, premaxillae and zygomatic processes of the squamosals. The occurrence of both species at Alexandra Station, Northern Territory (Loveridge 1935: 13; 1949: 213 quoting Parker 1940: 20) and Port Denison, Queensland (Parker 1940: 19-20) indicates that they coexist over a large range.

Glauertia orientalis Parker

Glauertia orientalis Parker, *Novit. Zool.* 42: 67 (1940)
Umbakumba, Groote Eylandt, N.T.: USNM 128276-7

These two interesting frogs support Parker's type description in all respects, one paratype female having been taken on Groote Eylandt.

Vomerine and maxillary teeth absent; two large compressed metatarsal tubercles, directed transversely. Tarso-metatarsal articulation reaching the posterior corner of the eye. Webbing between the 3rd and 4th toes not quite reaching the proximal tubercle of the 4th toe.

Both specimens measure 20 mm. from snout to vent with hind limbs measuring 25-6 mm.

HYLIDAE

Hyla caerulea (Shaw)

Rana caerulea Shaw, in White's *Journ. Voy. N.S.W.*: 248 (1790)
Umbakumba, Groote Eylandt, N.T.: USNM 128267
Yirrkalla, N.T.: USNM 128530

The largest specimen, USNM 128267, measures 73 mm. from snout to vent.

Hyla rubella Gray

Hyla rubella Gray, *Zool. Misc.*: 57 (1842)

Umbakumba, Groote Eylandt, N.T.: USNM 128281

A single subadult specimen. Vomerine teeth in two small groups posterior to the choanae; tibio-tarsal articulation marks the tympanum which is only half the orbital diameter.

Hyla lesueurii Duméril and Bibron

Hyla lesueurii Duméril and Bibron, *Erp. Gen.* 1-9: 595 (1841)

Oenpelli, N.T.: USNM 128719

This specimen agrees well with Günther's type description (1867: 56) and Boulenger's redescription and figure (1882: 413, pl. 26, fig. 2) of the type of *nigrofrenata,* which name was placed in the synonomy of *lesuerurii* by Loveridge (1935: 51).

First finger a little longer than the second; prominent inner and outer metatarsal tubercles, the outer small, white. When the hind limb is adpressed the tip of the snout is at a point ⅔ to ¾ way along the tibia. The coloration is very similar to that figured by Boulenger, but possessing a quadrangular dark blotch between the hinder half of the eyes and extending a short distance beyond.

This specimen measures 22 mm. from snout to vent.

Hyla peroni (Tschudi)

Dendrohyas peroni Tschudi, *Classif. Batr.*: 75 (1838)

Oenpelli, N.T.: USNM 128715-18

Vomerine teeth in two transverse series of 3-4 teeth between the choanae; tibio-tarsal articulation reaching the tip of the snout; toes three-quarters webbed, the webbing extending as a fringe to the pad on the 4th toe. Tympanum distinct, ⅔ to ¾ of the orbital diameter; a well defined cutaneous fold from behind the eye over the tympanum to the shoulder. This fold has a distinct dark edge in the majority of specimens.

Hyla adelaidensis Gray

Hyla adelaidensis Gray, in Grey's *Journ. Exped. Discov.* 2: 447, pl. 8, fig. 2 (1841)

Oenpelli, N.T.: USNM 128720-5, SAM R3252

Vomerine teeth in two ill-developed oblique series between the choanae; snout acuminate, but not markedly so. Canthus rostralis distinct but rounded; loreal region almost vertical; ear well defined, ⅔ to ¾ of the orbital diameter. Fingers without web, toes approximately ¾ webbed, the webbing extending to the discs of the 3rd and 5th toes. Inner metatarsal tubercle present and a strong subarticular tubercle at the base of the 4th toe. When the hind limb is carried forward along the body the tibio-tarsal articulation reaches the nostril or the tip of the snout. Dorsal surfaces with a few scattered rugosities; under surfaces of body and thighs granulate; a dermal fold across the chest and another along the inner edge of the tarsus.

Dark brown above with a few ill-defined darker blotches. A darker patch extends from the nostril to the supraocular region and above the tympanum in most specimens; a light pink streak joins the eye to the tympanum; alternate light and dark vertical barring on the upper labials. Both anterior and posterior faces of the hind limbs with an irregular brown and pink mottling; the pink colouring has been bleached by the preservative in two specimens. Throat and chest finely punctate, otherwise immaculate beneath.

This series measure 15-18 mm. from snout to vent.

Loveridge (1935: 48) gives some interesting notes on the live coloration of some specimens from Claremont, Western Australia.

Hyla nasuta (Gray)

Pelodytes nasutus Gray, *Zool. Misc.*: 56 (1842)

Umbakumba, Groote Eylandt, N.T.: USNM 128268-71, SAM R3253
Central Hill, Groote Eylandt, N.T.: USNM 128459
Near Old Mission, Groote Eylandt, N.T.: USNM 128462

Vomerine teeth in two slightly oblique series on a level with the anterior edge of the choanae; snout a little shorter than twice the orbital diameter; tympanum round, equal to or slightly smaller than the eye. First finger not webbed, slightly longer than the second; tibio-tarsal articulation reaching well beyond the end of the snout.

Colour markings variable, usually a dark vertebral stripe bordered on each side by a narrow light streak. The longitudinally disposed variegations on the thighs and lateral regions vary considerably. Dorsal surfaces usually smooth, but distinctly tuberculate in the dorso-lateral regions of USNM 128269, the rugosities being irregularly distributed; middorsal skin surface smooth. The prominence of the longitudinal folds varies with the state of dehydration of the specimen.

The largest specimen (USNM 128273) measures 47 mm. from snout to vent and has a hind limb measuring 102 mm.

Hyla bicolor (Gray)

Eucnemis bicolor Gray, *Zool. Misc.*: 57 (1842)

Oenpelli, N.T.: USNH 128712-14
Cahill's Landing, East Alligator River, N.T.: USNM 128686-703, SAM R3251
Red Lily Lagoon, near Oenpelli, N.T.: USNM 128710-11

The latter two localities are within a mile or so of each other, approximately 6 miles west of Oenpelli, Northern Territory.

The specimens in the series agree with the characters described by Loveridge (1948: 399) from a series of eight specimens taken at Knuckey's Lagoon, 9 miles south of Darwin, Northern Territory.

Vomerine teeth absent; tibio-tarsal articulation reaching a point between the anterior corner of the eye and the nostril. The colour of the Oenpelli specimens corresponds with that described by Boulenger (1882: 421), while that of the Cahill's Landing series is much darker, the base colour obscuring the other markings.

These specimens measure 18-21 mm. from snout to vent.

The characters of three tadpoles taken at a billabong on Groote Eylandt and registered under USNM 128266 suggest that they may be referable to this species. Two of the tadpoles possess both fore and hind limbs and partially absorbed tails.

PLATE 1. Form and density of the middorsal (right) and basal third caudal scales (left)

Varanus glebopalma Holotype
Varanus timorensis tristis Fraser Range, W. Aust.
Varanus timorensis tristis Tennant Creek, N.T.
Varanus timorensis orientalis Eidsvold, Qld.
Varanus timorensis orientalis Groote Eylandt, N.T.

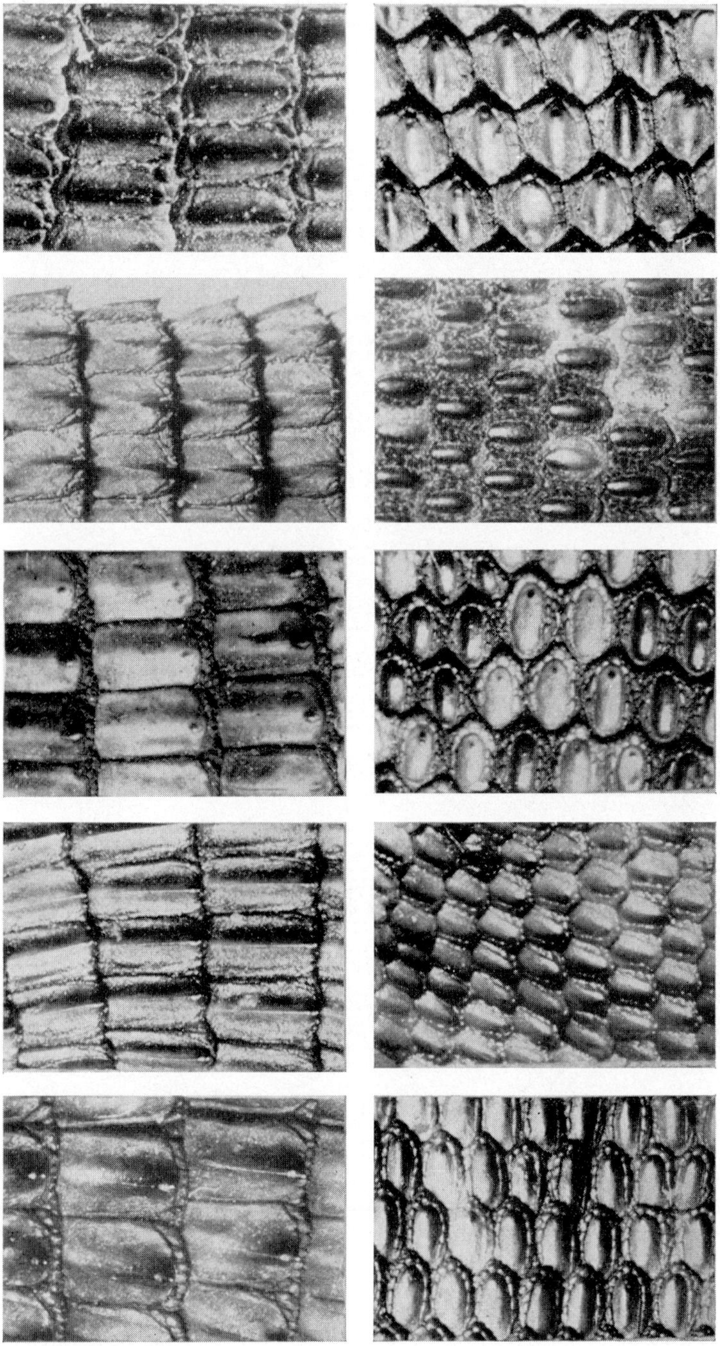

PLATE 1

REFERENCES

Barbour, T. (1914). On some Australasian reptiles. *Proc. Biol. Soc. Wash.* 27: 201-6.

Boulenger, G. A. (1882). *Catalogue of Batrachia Salentia in the British Museum.*

——— (1885-1887). *Catalogue of lizards in the British Museum,* vol. 1-3.

——— (1893-1896). *Catalogue of snakes in the British Museum,* vol. 1-3.

——— (1897). A list of the reptiles and batrachians collected by Mr Alfred Everett in Lombok, Flores, Sumba and Savu with descriptions of new species. *Ann. Mag. Nat. Hist.* 6th ser., 19: 507.

Brazenor, C. W. (1951). On the Victorian species of tuberculated *Diplodactylus. Mem. Nat. Mus. Vict.* 17: 215-22.

Copland, S. J. (1947). Catalogue of reptiles in the Macleay Museum. Part II. *Sphenomorphus spaldingi* (Macleay). *Proc. Linn. Soc. N.S.W.* 71: 136-44.

De Rooij, N. (1917). *Indo-Australian Reptiles* 2: 77.

Duméril, A., and Bibron, G. (1834-1854). *Erpétologie générale,* 9 vols.

Fry, D. B. (1913). On a *Varanus* and a frog from Burnett River, Queensland, and a revision of the variations in *Limnodynastes dorsalis,* Gray. *Rec. Aust. Mus.* 10: 17-34.

Garman, S. (1901). Some reptiles and batrachians from Australasia. *Bull. Mus. Comp. Zool. Harv.* 39: 1-14.

Glauert, L. (1950). Thomson's mulga snake. *W. Aust. Nat.* 2: 162-4.

——— (1952). Herpetological miscellanea. *W. Aust. Nat.* 3: 166-70.

Günther, A. (1867). Additions to the knowledge of Australian reptiles and fishes. *Ann. Mag. Nat. Hist.* 3rd ser., 20: 45-68.

Hewer, A. M. (1948). Tasmanian lizards. *Tasmanian Naturalist* N.S. 1 (Pt. 3): 8-11.

Hombron, J. B., and Jacquinot, C. H. (1842-1853). Voyage au Pôle Sud . . . sur les corvettes l'Astrolabe et la Zélée . . . 1837-1840, sous le commandement de M. J. Dumont d'Urville. *Zoologie* (Reptiles et Poissons par Jacquinot et Guichenot) 16-17, pl. 4, fig. 1.

Kinghorn, J. R. (1924). A short review of the lizards belonging to the genus *Lialis* Gray. *Rec. Aust. Mus.* 14: 184-8.

——— (1926). A brief review of the family Pygopodidae. *Rec. Aust. Mus.* 15: 40-64.

——— (1942). Herpetological notes No. 4. *Rec. Aust. Mus.* 21: 118-21.

Longman, H. A. (1916). Snakes and lizards from Queensland and the Northern Territory. *Mem. Qld. Mus.* 5: 46-51.

Lönnberg, E., and Andersson, L. G. (1915). Results of Dr E. Mjoberg's Swedish Scientific Expeditions to Australia 1910-13. VII. Reptiles collected in Northern Queensland. *K. Svenska VetenskAkad. Handl.,* 4th ser., 52, No. 7: 1-9.

Loveridge, A. (1934). Australian reptiles in the Museum of Comparative Zoology, Cambridge, Massachusetts. *Bull. Mus. Comp. Zool. Harv.* 77 (6): 243-383.

——— (1935). Australian amphibia in the Museum of Comparative Zoology, Cambridge, Massachusetts. *Bull. Mus. Comp. Zool. Harv.* 78 (1): 1-60.

——— (1948). New Guinea reptiles and amphibians in the Museum of Comparative Zoology and United States National Museum. *Bull. Mus. Comp. Zool. Harv.* 101 (2): 305-430.

——— (1949). On some reptiles and amphibians from the Northern Territory. *Trans. Roy Soc. S. Aust.* 72: 208-15.

Mertens, R. (1931). *Ablepharus boutonii* (Desjardin) und seine geographische Variation. *Zool. Jahrb.* 61: 63-210.

H. J. Bowshall

PLATE 2

A *Tiliqua scincoides intermedia*

Dorsal view of the paratype, SAM R3225

B, C *Varanus* (*Odatria*) *glebopalma*

Dorsal and lateral views of the head of the holotype specimen, SAM R3222

——— (1942). Die Familie der Warane (Varanidate). *Abh. Senckenb. Naturf. Ges.* No. 446: 237-391.

Mitchell, F. J. (1950). The scincid genera *Egernia* and *Tiliqua* (Lacertilia). *Rec. S. Aust. Mus.* 9: 275-308.

——— (1951). The South Australian reptile fauna. Part I. Ophidia. *Rec. S. Aust. Mus.* 9: 545-57.

——— (1955). Preliminary account of the Reptilia and Amphibia collected by the National Geographical Society-Commonwealth Government-Smithsonian Institution Expedition to Arnhem Land (April to November 1948). *Rec. S. Aust. Mus.* 11: 373-408.

Mittleman, M. B. (1952). A generic synopsis of the Lizards of the subfamily Lygosominae. *Smithsonian Misc. Coll.* 117 (17): 1-35.

Montague, P. D. (1914). A report on the fauna of the Monte Bello Islands. *Proc. Zool. Soc. Lond.* 1914: 625-52.

Parker, H. W. (1940). Australasian frogs of the family Leptodactylidae. *Novit. Zool.* 42: 1-106.

Proctor, Joan B. (1923). On new and rare reptiles and batrachians from the Australian region. *Proc. Zool. Soc. Lond.* 1923: 1069-77.

Smith, M. A. (1931). *Fauna of British India,* Reptilia and Amphibia, 1: 67.

——— (1943). *Fauna of British India,* Reptilia and Amphibia, 3.

Stirling, E. C., and Zietz, A. (1893). Vertebrata. *Trans. Roy. Soc. S. Aust.* 16: 154-76.

Thomson, D. F. (1930). Observations on the venom of a large Australian snake, *Pseudechis australis* (Gray). 1. Synonymy. *Aust. J. Exp. Biol. Med. Sci.* 7: 125-33.

——— (1933). Notes on Australian snakes of the genera *Pseudechis* and *Oxyuranus. Proc. Zool. Soc. Lond.* 1933: 855-60.

Zietz, F. R. (1920). Catalogue of Australian lizards. *Rec. S. Aust. Mus.* 1: 181-228.

(Manuscript received, May 1956)

5

BIRDS OF THE ARNHEM LAND EXPEDITION

by Herbert G. Deignan*

In the following pages will be found notes upon the 191 forms of Territorian birds collected by members and guests of the Commonwealth of Australia-National Geographic Society-Smithsonian Institution Expedition to Arnhem Land in the Northern Territory of Australia, and sight records for 31 others, not represented in the collection, but considered certainly identifiable to species.

In consequence of the extensive Territorian collections made by such skilled collectors as Gilbert, McLennan, Dahl, Woodward, and May, studied by Gould, H. L. White, Hartert, Collett, Mathews, and others, our knowledge of the extent of the Territorian avifauna has been for many years virtually complete, and, as might have been expected, no startling novelties were discovered by the expedition. Indeed, after the potshot naming of Territorian races over many years by Gregory Mathews, it may be thought remarkable that a frogmouth, a nightjar, and two babblers were found apparently worthy of subspecific separation, yet had no names available in the literature.

The most valuable results obtained from the collection lie in the field of new locality records, extensions of range of a number of forms, notes on moults and breeding condition, and information on migrations.

With the exception of a limited number of specimens taken at intermediate points, my own collection was made at four localities, none of which lay at any great distance from salt water, in 1948: Darwin (19 March to 18 April); Umbakumba, Groote Eylandt (17 May to 6 July); Yirrkalla (7 July to 1 September); Oenpelli (21 September to 6 November).

The first station, offering a great variety of habitats within a limited area, showed the most diversified avifauna. Within a short walk from the tourist hostel at Nightcliff could be found rocky foreshores and sandy beaches, extensive stands of mangrove, monsoonal forest, open eucalypt forest, grassland, cultivation, and paperbark swamps.

Umbakumba sat beside a placid inner bay, edged by sandy beaches that rose into high dunes, partly clothed with wattle scrub. Behind these lay many miles of open eucalypt forest, with, here and there, a pocket of monsoonal forest.

Yirrkalla, on the Cape Arnhem Peninsula, another seaside camp, had sandy beaches, with occasional rocky foreshores, an extensive reach of short grass (annually burned off), an expanse of freshwater marsh (almost birdless), a very small stand of monsoonal forest, and a seemingly endless open eucalypt forest.

* Smithsonian Institution, Washington, D.C.

Oenpelli, away from the sea and near the headwaters of the East Alligator River, lay at the edge of a vast grass-covered plain, during our stay largely burned off, but possessing moist oases in which grew dense stands of pandanus and other moisture-loving plants. From this camp the Arnhem Land Plateau was accessible to us, but the almost complete absence of water at the top of that interesting formation precluded the existence of much bird life.

The localities visited all lie well within the Torresian Sub-region (as demarcated by Serventy and Whittell, *Birds of Western Australia*: 44, 1948), but nevertheless an extraordinary number of characteristic Torresian species were nowhere seen. This resulted in part from the fact that the ranges of these Torresian forms by no means coincide with the accepted Torresian boundaries; they are found, rather, in or at the edge of stands of mangrove and/or monsoonal forest (accordingly, mainly near the coast), and in the cases of some, only in very extensive stands of such growth. Except in the Darwin area, no such important stands were found near the expedition's camps, and even at Darwin they had been seriously reduced by war-time activities of the military forces, while post-war settlement is continuing the process. A fine stand of monsoonal forest, considerably less than one square mile in area, just behind the hostel at Nightcliff, was being cleared during my stay; one may expect that its place will soon be taken by the open eucalypt forest which already surrounded it on its three landward sides. At the same time, the truly Torresian birds inhabiting it will give place to birds common to the greater part of coastal Australia. At other camps, monsoonal forest and mangrove had been originally much more limited in extent, and owing to human occupation were now either wholly or partly destroyed. Doubtless a longer period of collecting near Darwin would eventually have turned up many of the missing species, but it seems certain that birds so dependent upon a definite ecotone must shortly disappear from many places where they have in the past been common.

Although earlier collectors have handed down horrendous stories of the mosquitoes that infested the Territorian mangrove swamps, I nowhere in such growth found any mosquitoes at all. It may be suspected that the efficiency with which the sanitation corps, since 1941, have cleared the swamps of their insect populations is not unrelated to their current dearth of bird life.

ACKNOWLEDGMENTS

It is not easy to give due credit to all who have lent me assistance in the course of my study of the collection. Needless to say, I am highly indebted to the authorities of the American Museum of Natural History (New York), who kindly gave me full access to the important Mathews Collection in their charge. For permission to examine particular specimens, I owe thanks to the officials of the Academy of Natural Sciences (Philadelphia), the South Australian Museum (Adelaide), the National Museum of Victoria (Melbourne), and the Australian Museum (Sydney). Correspondence in connection with various problems has been carried on with Ernest Mayr and Dean Amadon of New York, Rodolphe Meyer de Schauensee of Philadelphia, Herbert T. Condon of Adelaide, Bruce Amos, Jack

Jones, and Charles E. Bryant of Melbourne, Arnold R. McGill and Keith A. Hindwood of Sydney, and D. L. Serventy and H. M. Whittell of Perth. Mr W. Rainbow, Librarian of the Australian Museum, has located for me certain obscure type localities. Finally, I am grateful to all my colleagues and the guests of the Arnhem Land Expedition, who, without exception, missed no opportunity to collect for me specimens that seemed to them of interest for my purposes.

DROMICEIDAE

Dromiceius novaehollandiae (Latham)

[*Casuarius*] *N*[*ovæ*] *Hollandiæ* Latham, *Index ornithologicus* 2: 665 (1790), New Holland = Sydney, New South Wales.

The emu was never seen by me, but other members of our party reported having found its tracks in the interior of the Cape Arnhem Peninsula, and a nest with nine eggs was discovered by aborigines a few miles from Yirrkalla in the middle of August. Since all the eggs were summarily devoured, it was made impossible for me to visit the site for the purpose of making observations on the incubating bird.

PODICIPEDIDAE

Podiceps novaehollandiae novaehollandiae Stephens

Podiceps novæ hollandiæ Stephens, in Shaw, *General zoology* 13 (1): 18 (1826). New Holland = New South Wales.

Oenpelli: 1 ♂, 2 ♀, 30 September, 6, 20 October.

The grebe was found only on the large Oenpelli Lagoon, where it was numerous between 21 September and 6 November.

The two specimens of earlier date are white-throated birds, probably immature; the third, although showing no evidence of moult, differs in having the throat evenly stippled light and dark, to give a grey tone to the area, and in having the rufous on the side of the neck deeper in tone and more clearly defined from the adjacent feather tracts. Although Mayr (*Emu* 43: 5, 1943) has suggested that Australian populations have no eclipse plumage, I am inclined to believe that such a plumage does indeed exist and that the grey-throated example exhibits it. It is noteworthy that, among the many grebes seen at Oenpelli, *none* seemed to be wearing the nuptial dress. Breeding in that district could, of course, not take place until the monsoonal rains, beginning in November, had greatly increased the areas covered with water, and no gonadal activity appeared in any of these three specimens.

I can find no way to distinguish my Territorian birds from a series of six from New South Wales (four of which are stipple-throated, two white-throated), and am convinced that *parryi*, described by Mathews from north-western Australia, will prove to be unrecognizable. Measurements given by Mayr (loc. cit.) for northern and southern populations show too much overlap to justify division into two races.

PELECANIDAE

Pelecanus conspicillatus Temminck

Pelecanus conspicillatus Temminck, in Temminck and Laugier, *Nouveau recueil de planches coloriées d'oiseaux* 5, livr. 47: pl. 276 and text (1824). Terres Australes = New South Wales.

Oenpelli, 1 ♂, 8 October.

The pelican was first seen on 24 April at an islet off Cape Cockburn, asserted by aborigines to be a regular breeding ground. During the third week of August, a flock of some 50 individuals were observed high above our camp at Yirrkalla, soaring in wide circles but simultaneously moving southward along the coast. The species was otherwise found only at Oenpelli, where, throughout our stay, a single band of eight to ten were noted almost daily at one or other of the larger lagoons.

My specimen, with inactive gonads, had the irides brown; the edges of the eyelids plumbeous; the orbital skin dead white, pale plumbeous just at the feather line; the bill light plumbeous, with the tip and edges of the commissure on the apical half yellow; the gular skin dead white; the feet and toes pale plumbeous.

Mathews's *Pelecanus conspicillatus westralis,* named from Perth, W.A., and based upon an inconsequential difference in size, is scarcely worthy of notice.

SULIDAE

Sula leucogaster (Boddaert)

Pelecanus leucogaster Boddaert, *Table des planches enluminéez d'histoire naturelle*: 57 (1783). Cayenne, *ex* D'Aubenton and Buffon.

The brown gannet was seen in some numbers off the mouth of the King River on 27 April; the species probably breeds on one of the islets in this neighbourhood.

PHALACROCORACIDAE

Phalacrocorax sulcirostris sulcirostris (Brandt)

Carbo sulcirostris Brandt, *Bull. Acad. Sci. St-Pétersb.* 3 (4): 56 (16 Nov. 1837). Terrae australes (Südsee) = New South Wales.

Oenpelli: 2 ♂, 7, 19 October.

This was a fairly common species on the lagoons about Oenpelli, where it flocked with *Anhinga* and the pied cormorants.

The brown-flecked neck and upper breast of my specimens indicate that they are sub-adult. The bird of 7 October had the irides deep olive-grey; the bill plumbeous, with culmen black; the gular skin plumbeous flesh; the feet, toes, and claws black.

Mathews (*Birds of Australia* 4 (2): 176, 1915) has set up the name *Mesocarbo ater territori* for a specimen from Hermit Hill, N.T. Since he mistakenly supposed the type of the nominate race to have come from so near a locality as Shark's Bay, W.A., one might have expected him to have indicated how *territori* was thought to differ from *ater*; on the contrary, no comparison at all was made, and the fact that Mathews permitted the description of the type of *territori* to serve for birds of all Australia and Tasmania implies that, in truth, he failed to find any difference whatsoever. My own skins are in no way distinguishable from others taken in New South Wales, and my use of the trinomial is due to the existence of a presumably valid race, *purpuragula,* in New Zealand.

Phalacrocorax melanoleucos (Vieillot)

Hydrocorax melanoleucos Vieillot, *Nouveau dictionnaire d'histoire naturelle,* nouv. éd., 8: 88 (Mar. 1817). Australasie = New South Wales.

The little pied cormorant was not uncommon on the great lagoon at Oenpelli, where it was usually seen in company with *Ph. sulcirostris.*

ANHINGIDAE

Anhinga rufa novae-hollandiae (Gould)

Plotus novæ-hollandiæ Gould, *Proc. Zool. Soc. Lond.* 15: 34 (29 Apr. 1847). Rivers of the whole of the southern coast of Australia = New South Wales, *fide* Stone, *Aust. Avian Rec.* 1: 144 (1913).

Oenpelli: 1 ♀, 6 October.

The darter occurred in small numbers, associated with cormorants, on all the larger lagoons about Oenpelli.

My specimen, believed to be adult, had the irides, lores, bill, gular skin, feet, toes, and claws yellow.

Mathews (*Aust. Avian Rec.* 1: 74, 1912) has named *Plotus novae-hollandiae derbyi* from north-western Australia, because one specimen differed from typical *novae-hollandiae* in its larger wing measurement, 'viz 364 mm.'; in *Birds of Australia*, 4: 196, 1915, however, a male topotype of *derbyi* is said to have the wing length only 347 mm., while a female topotype has it 357 mm. My female from Oenpelli has a wing measuring 353 mm., while two males before me from New South Wales have wing lengths of 358 and 361 mm. In brief, northern birds are *not* larger than southern ones, and the name *derbyi* is a mere synonym of *novae-hollandiae.*

FREGATIDAE

Fregata minor (Gmelin)

[*Pelecanus*] *minor* Gmelin, *Systema naturae* 1 (2): 572 (1789). No locality given; designated as 'the eastern half of the Indian Ocean', by Rothschild, *Novit. Zool.* 22: 145 (1915); restricted to Christmas Island, Indian Ocean at lat. 10° 31′ S., long. 105° 34′ E., by Lowe, *Novit. Zool.* 31: 306 (1924).

Birds of the genus *Fregata* were seen at several points along the Territorian coast, but the only one I was able to identify to species was an example of *minor* off Croker Island on 22 April.

ARDEIDAE

Ardea sumatrana Raffles

Ardea sumatrana Raffles, *Trans. Linn. Soc. Lond.* 13 (2): 325 (not earlier than Nov. 1822). Sumatra.

One was seen on 19 March, flying across a bay near Nightcliff.

Ardea pacifica Latham

A[*rdea*] *pacifica* Latham, *Supplementum indicis ornithologici*: lxv (1801). New Holland = New South Wales.

A single bird, flying at a great height toward the East Alligator River, was seen near Oenpelli on or about 10 October.

Notophoyx novaehollandiae novaehollandiae (Latham)

[*Ardea*] *Novæ Hollandiæ* Latham, *Index ornithologicus* 2: 701 (1709). New South Wales.

Groote Eylandt (Umbakumba): 1 ♂, 15 June.

Oenpelli (Tableland): 1 ♂, 3 October.

The white-faced heron was not seen at Darwin, but was noted in small numbers at most other coastal areas visited, as well as at Oenpelli, where a specimen was taken at a pool, teeming with a small fish, on the Tableland.

Near Umbakumba this bird was several times observed perched upon one or another of the rather low poles that had carried a military telephone line along the beach, despite the obvious difficulty it found in balancing itself against the force of the trade wind.

Mathews's *parryi* from the Kimberley Division is but another of that author's imaginary races. My use of a trinomial name is in recognition of *nana,* described by Amadon in 1942 from New Caledonia.

Notophoyx picata (Gould)

Ardea (Herodias) picata Gould, *Proc. Zool. Soc. Lond.* 13: 62 (Sept. 1845). Port Essington, Northern Territory.

Oenpelli: 1 ♀, 10 October.

The strikingly marked pied heron was seen in flocks of as many as 100 individuals, flying in beautiful formation over the East Alligator River or perched on the mangroves along its banks, often in company with other members of its family; similarly large aggregations frequently visited the larger lagoons about Oenpelli, at periods when the swiftly rising tide was covering the river mud flats. My specimen, and other older ones in the museum collection, have the dark portions of the plumage blackish slate, but I noted that these same parts appeared, in the field, a rich slaty blue; it is possible that a bloom on the feathers is lost after death.

Butorides striatus (Linnaeus)

[*Ardea*] *striata* Linnaeus, *Systema naturae,* ed. 10, 1: 144 (1758). Surinam.

This little heron was generally distributed, although in small numbers, on all the foreshores and islands visited by our party, and occurred also along the banks of the East Alligator River to the limit of mangrove.

Bubulcus ibis coromandus (Boddaert)

Cancroma coromanda 'mihi' Boddaert, *Table des planches enluminéez d'histoire naturelle*: 54 (1783). Côte de Coromandel, *ex* D'Aubenton, pl. 910.

Oenpelli: 1 ♂, 1 ♀, 23 September, 29 October.

The cattle-egret was found only on the grassy plains about Oenpelli, but there proved to be the most numerous of all the herons, occurring at times in aggregations of more than 100 individuals.

The habits of this Australian population differed in no way from those of its relatives in Asia; the birds regularly travelled with the mobs of buffalo and with the domesticated horses and cattle belonging to the mission station. Since the association between bird and mammal was a close one, it was not possible for me to shoot specimens without risking either personal jeopardy in the case of the buffalo or injury to private property in the case of the mission's herds, and I owe my two skins to the skilful handling of a stock-whip by aboriginal cattle tailers.

Neither of these specimens showed gonadal activity, and it is probable that the breeding season had ended before we arrived at Oenpelli, inasmuch as the great majority of those observed wore the pure white plumage of non-breeding individuals. My October bird still carries the buffy-orange nuptial plumes, but new white feathers are appearing on its neck, and the change is much farther advanced in the one collected more than a month earlier.

They had the irides yellow; the orbital region and lores orange; the bill chrome yellow; the exposed portion of the tibiae yellowish to brownish; the tarsi blackish brown; the toes and claws black.

According to Serventy and Whittell (*A handbook of the birds of Western Australia*: 170, 1948), 'The cattle Egret . . . was introduced into the Kimberley Division in 1933. In May of that year the Pastoralists' Association imported 20 individuals from Calcutta and 18 were released but apparently the bird has not succeeded in establishing itself.' Tarr (*Emu* 49: 191-2, 1950), discussing the distributions of foreign birds in Australia, has been unable to trace any other recorded introduction.

It seems to me quite improbable that a flock of 18 indivduals of indeterminable sex should, *within the span of 15 years,* have presumably vanished altogether from the point of introduction and yet have multiplied into a population of hundreds, if not thousands, at a place some 400 miles east of that point, especially when much of the intervening territory, like that between Oenpelli and Darwin, is thoroughly unsuited to the requirements of the species.

That Australian specimens seem to be inseparable from the Asiatic *coromandus* is of little moment in a species so non-plastic as this, but that the cattle-egret's Australian range should be so restricted in comparison with those of other herons common to Australia and Asia seems to point to a relatively recent arrival in the former continent.

The following hypotheses are to be considered:

1. *Bubulcus ibis* may have been imported by the settlers of Port Essington in the same period as the buffalo, the sambhar or rusa, the banteng, the Malayan swine, the Timor pony, and the donkey, all of which have persisted to this day on the Cobourg Peninsula and/or the Oenpelli plains just south and east of that peninsula.

2. It may have been transported to Australia from the Celebes, where it is common, by the Macassar traders who regularly visited the Territorian coasts until the early years of our century. Horsfield (*Zoological researches in Java,* 1824) has observed: 'Our bird, if taken young, is easily domesticated, and is not unfrequently seen in this state, in the villages situated on the banks of rivers and lakes.' One may suppose that it is similarly tamed in other islands than Java. It is, in any case, certain that Austronesian navigators have assisted in the spread of a number of plants and animals, and that they have been responsible for the presence of the tamarind (*Tamarindus indica*) on the northern Australian shores.

3. The cattle-egret may have introduced itself into Australia. The species ranges almost continuously from Portugal and Austria southward to the Cape Colony and Madagascar, north-eastward to Ussuriland and Japan, south-eastward to the Philippines, the Celebes, and Java. As a wanderer and potential settler it has been found in England, the Madeira and Canary Islands, the Comoro Islands and the Seychelles, Sumbra, Timor, the Moluccas, western New Guinea, and the Palau Islands. That so adventurous a bird may have invaded northern Australia and, finding an abundance of the large grazing animals with which it habitually consorts, have prospered there, must be taken into serious account.

Egretta alba (Linnaeus)

[*Ardea*] *alba* Linnaeus, *Systema naturae,* ed. 10, 1: 144 (1758). Europe.

This heron was definitely identified only at Oenpelli, where from September to November it occurred in small numbers at the larger lagoons, associated with the plumed egret.

Egretta intermedia plumifera (Gould)

Herodias plumiferus Gould, *Proc. Zool. Soc. Lond.* 15: 221 (29 Mar. 1848). New South Wales.

Oenpelli: 1 ♂, 1 ♀, 22 September, 22 October.

At Oenpelli, where alone I saw members of the genus *Egretta* in some numbers, the overwhelming majority seemed to be of this species. They were always to be seen there wading near the margins of the larger lagoons.

The bird of 22 October, although with the gonads inactive, is wearing full nuptial plumes, no sign of which had yet appeared in the specimen of 22 September.

The earlier-taken example had the irides yellow; the orbital region, lores, and bill chrome yellow; the inner side of the upper tibiae and the hinder side of the tibiotarsal joint yellowish; the feet, toes, and claws otherwise black. The other differed mainly in having the lores pale yellowish green, and the inner side of the tibiae varying from brown to yellow.

Egretta garzetta nigripes (Temminck)

Ardea nigripes Temminck, *Manuel d'ornithologie,* ed. 2, 4: 376 (1840). l'Archipel des Indes = Sunda Islands.

Darwin (Lee Point): 1 ♂, 18 April.

The little egret is evenly distributed along the Territorian foreshores, whether these are rocky or of sand, feeding at the margins of tidal pools and even on occasion entering the surf, where small fishes are pursued by running with fluttering wings. The species was not observed at all along the muddy tidal rivers or on the Oenpelli lagoons.

The specimen collected had the irides, orbital region, and lores yellow; the bill black, with the extreme base of the maxilla and the basal half of the mandible yellow; the tibiae, tarsi, toes, and claws black; the under side of the toes greenish yellow.

I find no character by which this example, with gonads inactive but wearing full-grown nuptial plumes, may be separated from *nigripes* of New Guinea and the East Indies. From Javanese skins it differs only by lacking the two white occipital plumes, but that *some* Australian individuals wear them is indicated by Gould's report of such a specimen from near Brisbane (*Handbook to the birds of Australia* 2: 305, 1865).

The whole question of moults and presence or absence of occipital plumes in this species requires investigation. Indo-Chinese birds, commonly identified as *E. g. garzetta,* have been taken, wearing apparently full nuptial dress, in every month of the year, and amongst them only a small proportion show plumes from the occiput (three out of ten from Szechwan, one out of eight from Siam and Cambodia).

Demigretta sacra sacra (Gmelin)

[*Ardea*] *sacra* Gmelin, *Systema naturae* 1 (2): 640 (1789). Tahiti.

Croker Island: 1 ♀, 23 April.
Groote Eylandt (Umbakumba): 1 ♂, 24 June.
Arnhem Peninsula (Yirrkalla): 1 ♀, 20 July.

The reef heron was seen, solitary or paired, along all the rocky foreshores visited and on many outlying reefs.

During our entire stay in the Territory, I observed only one individual (at

Yirrkalla) in the white phase of plumage. The three specimens taken are in the grey phase.

For a careful study of the species, see Mayr and Amadon, *Amer. Mus. Novit.* No. 1144, 1941.

Nycticorax caledonicus (Gmelin)

[*Ardea*] *caledonica* Gmelin, *Systema naturae* 1 (2): 626 (1789). New Caledonia.

In the Territory I recorded the night-heron (with one exception) only along the East Alligator River, where at certain places it was very numerous. A single immature individual seen near the mission station at Oenpelli was evidently a wanderer from the river shores.

Dupetor flavicollis (Latham)

[*Ardea*] *flavicollis* Latham, *Index ornithologicus* 2: 701 (1790). India.

A single example of this heron was flushed on 12 April from a dense stand of sedge and fern at the margin of an extensive paperbark swamp near Nightcliff.

CICONIIDAE

Xenorhynchus asiaticus (Latham)

[*Mycteria*] *asiatica* Latham, *Index ornithologicus* 2: 670 (1790). India.

A jabiru was seen on 22 August at an extensive mud flat, exposed by the receding tide, on the coast of the Cape Arnhem Peninsula not far from Melville Bay. Between September and November solitary individuals were often noted at wet areas on the Oenpelli plain, and one haunted the great pool formed by the waterfall that drops over the Escarpment and on occasion even followed the stream up on to the Tableland, doubtless attracted by the abundance of small fish in the potholes.

PLATALEIDAE

Threskiornis moluccus (Latham)

[*Tantalus*] *æthiopicus* Latham, *Index ornithologicus* 2: 706 (1790). Æthiopia = north-eastern Africa.

This ibis occurred in small numbers along the East Alligator River on 19 September, and two or three were sometimes noted at the great lagoon of Oenpelli during the subsequent months.

Threskiornis spinicollis (Jameson)

Ibis spinicollis Jameson, *Edin. New Philos. J.* 19: 213 (July 1835). Murray River, New South Wales.

Oenpelli: 1 ♀, 26 October.

Upon the expedition's return to Darwin at the beginning of September, the straw-necked ibis was found to have made an appearance there, a small flock haunting the lawns of the Botanic Gardens, and two or three individuals an unoccupied block of land opposite the offices of the Department of Native Affairs. These birds were perfectly tame, paying no attention to passing traffic. At Oenpelli the species was seen in small numbers at moist places on the short-grass plains, often in company with the cattle, but here it was very shy.

My specimen is immature, with the whole head and neck, excepting only the orbital region, lores, and chin, clothed in short feathers. It had the irides brown; the bill and naked skin of the face black; the tibiae salmon, changing to dark brown towards the tibiotarsal joint; the tarsi dark brown; the toes and claws black.

Mathews has named *Carphibis spinicollis fitzroyi* from north-western Australia; it is alleged to have a shorter wing, but, to judge from *Birds of Australia* 3: 391, 1914, the author did not distinguish in his measurements between adult and immature birds. It seems improbable that so nomadic a species will show subspecific variation, even when breeding adults alone are considered.

The first use of the name *Ibis spinicollis* Jameson is that given above, if Mathews is correct in his statement that it appeared in July 1835; otherwise. the earliest description was published in *L'Institut* 3 (125): 316, 30 Sept. 1835.

Plegadis falcinellus peregrinus (Bonaparte)

[*Falcinellus*] *peregrinus* Bonaparte, *Conspectus generum avium* 2: 159 (1855). Java, Celebes.

Oenpelli: 2 ♂, 1 ♀, 1, 1, 1 November.

The glossy ibis occurred at a number of the larger lagoons near Oenpelli, in flocks of as many as 50 individuals. Although certain birds were wearing nuptial dress as late as the beginning of November, my three specimens are all in eclipse plumage.

Mathews has named a bird from the Kimberley Division *Plegadis falcinellus rogersi,* but since he carefully refrained from indicating how it was thought to differ from other races, it would be a waste of time to pay further attention to it. In any case, if Australian birds be proved distinct from *peregrinus,* the name *humeralis* De Vis, 1898, given to a specimen from Papua, must probably be used.

Platalea leucorodia regia Gould

Platalea regia Gould, *Synopsis of the birds of Australia* 4: 7 (Apr. 1838). East coast of New South Wales = Moreton Bay, Queensland, *ex* Gould, *Handbook to the birds of Australia* 2: 287 (1865).

Groote Eylandt (Umbakumba): 1 ♂, 19 June.
Oenpelli:: 1 ♀, 30 October.

The spoonbill occurred in small numbers at the edges of the larger lagoons near Oenpelli, and was found once on Groote Eylandt, but was otherwise not noted in the Territory.

My specimens, uncrested birds with inactive gonads, had the irides dark brown; the bill and naked skin of the face slaty; the feet, toes, and claws black.

ANATIDAE

Dendrocygna arcuata australis Reichenbach

Dendrocygna arcuata (*australis*) Reichenbach, *Novitiae ad synopsin avium,* no. 4: 4 (1850). Port Essington, Northern Territory.

Oenpelli: 2 ♀, 27, 27 October.

The whistling tree-duck was common at Oenpelli, where it visited the larger lagoons in flocks of as many as several hundred individuals. It was most active after dark, spending the daylight hours in rafts far from shore, so that it proved to be excessively difficult to collect. On several occasions a wounded bird dived under the surface and was never seen again.

Mayr (*Amer. Mus. Novit.* no. 1294: 2-3, 1945) has shown that this species is divisible into three races, and that Mathews's *Dendrocygna 'javanica' peroni*, named from north-western Australia, is a synonym of *australis*.

Tadorna radjah rufitergum Hartert

Tadorna radjah rufitergum Hartert, *Novit. Zool.* 12 (1): 205 (11 Feb. 1905). South Alligator River, Northern Territory.

Oenpelli: 1 ♀, 23 September.

The Burdekin duck was common on the muddy shoals of the East Alligator River at times of low tide and, when the waters rose, visited all the lagoons and sloughs in the neighbourhood of Oenpelli.

My specimen had the irides pale yellow; the eye ring brownish yellow; the bill livid white, suffused with rose and violet (the nares rimmed with black); the feet and toes livid white; the claws horny brown. The bill and feet have become, in the dried skin, a yellow-brown, as if they might have been orange in life.

Anas poecilorhyncha rogersi Mathews

Anas superciliosa rogersi Mathews, *Aust. Avian Rec.* 1 (2): 33 (2 Apr. 1912). Augusta, Western Australia.

Oenpelli: 1 ♀, 8 October.

The black duck was seen only on the Oenpelli plain, where, however, it was common in flocks at all the larger lagoons.

I am in complete agreement with the view of Delacour and Mayr (*Wilson Bull.* 57: 21, 29, 1945) that *Anas 'superciliosa'* and *A. poecilorhyncha* are conspecific.

Anas gibberifrons gracilis Buller

Anas gracilis Buller, *Ibis*, ser. 2, 5: 41 (Jan. 1869). In the Orona Stream, near its junction with the Manawatu River, in the Province of Wellington, North Island, New Zealand.

Oenpelli: 1 ♀, 4 November.

Alice Springs (Wigley Waterhole): 1 ♂, 20 November.

The grey teal was found in small numbers at Oenpelli; what is supposed to have been the same flock of about a dozen individuals was observed frequently in October and November at several of the larger lagoons, in company with other ducks. The specimen from Alice Springs was shot by H. L. Hillier for D. H. Johnson.

Ripley (*Auk* 59: 90-9, 1942) has attempted to revive an old theory as to the conspecificity of *Anas gibberifrons* with *A. castanea*. His views have been unanimously opposed by Australian students (see Jones, *Emu* 41: 309-10, 1942) and I myself have seen wild grey and chestnut teals together on the lake in the Botanic Gardens at Melbourne. It should be noted that, if Ripley found New Zealand specimens inseparable from the Australian grey teal, as stated on page 94 of his paper, the name *gracilis* has 54 years' priority over *mathewsi*!

Nettapus pulchellus Gould

Nettapus pulchellus Gould, *Birds of Australia* pt. 6 [vol. 7]: pl. 4 (1842). Port Essington, Northern Territory.

Arnhem Peninsula (Yirrkalla): 2 ♂, 2♀, 16, 16, 16, 16 July.

Oenpelli: 1 ♂, 9 October.

The four pygmy geese from Yirrkalla, the only ones recorded at that locality, alit upon the sea just beyond the surf early in the morning of 16 July; all are sub-adult and possibly were members of a single brood. Later, at Oenpelli, the species proved to be exceedingly common on the large lagoon beside our camp.

This little duck, seen paddling among the lily pads in pairs or small parties, is extraordinarily reminiscent of *Chenonetta jubata*. At Oenpelli, where the semipalmated goose offers a more attractive target to the aboriginal 'sportsman', *Nettapus* scarcely bothered to fly at one's approach, preferring rather to swim farther from the shore; when truly alarmed, as by a near shot, it was as likely to dive as to take to wing.

Anseranas semipalmata (Latham)

Anas semipalmata Latham, *Trans. Linn. Soc. Lond.* 4: 103 (1798). Near Hawkesbury River, New South Wales.

Oenpelli: 1 ♂, 4 October.

The pied goose was only seen at Oenpelli, where, from September to November, it occurred in many tens, if not hundreds, of thousands of individuals. During this period of increasing aridity, the species was concentrated at the lagoons and on the moister portions of the grassy plains; on occasion, the browsing geese were so densely flocked as to cover the ground completely over hundreds of acres. In the hottest hours of the day they rested on the water or lay concealed in the marsh vegetation, but they became very active from late in the afternoon until morning: throughout the night could be heard the resonant, honking calls of gaggles passing overhead from one lagoon to another, and the splashing and cackling of birds on the water, disturbed by the new arrivals. When one approached the resting or feeding myriads, the nearer individuals would fly above the others to the farther edge of the mob, and the unceasing roar of great wings, together with the excited trumpeting of the alerted birds, their necks erected to espy the danger, made an ornithological experience more easily to be imagined than described in words.

There was no evidence whatsoever of breeding activity at this season, but nesting probably takes place throughout the rainy season, and even afterward: the aborigines at Milingimbi Island were eating eggs newly collected on the adjacent mainland as late as 10 May.

Mathews's *Anseranas semipalmata hamiltoni*, named from north-western Australia, scarcely deserves consideration: at the first description (*Aust. Avian Rec.* 1: 85, 1912), it is said to be larger than the bird of New South Wales; in *Birds of Australia* 4: 31, 1914 it is admitted to be smaller; in *Systema Avium Australasianarum* 1: 209, 1927, it is recognized without explanation! The size of the knob on the head varies, of course, with the age and sex of specimens.

ACCIPITRIDAE

Elanus notatus Gould

Elanus notatus Gould, *Synopsis of the birds of Australia* pt. 4: 1 app. (Apr. 1838). New South Wales.

A black-shouldered kite was noted near the bank of the East Alligator River on 20 September, and a pair occurred regularly between 28 September and the beginning of November at one particular area between Oenpelli and the river. This is apparently another species of the interior that approaches the coastal districts during the dry weather.

Mathews (*Birds of Australia,* 5 (2): 200-2, 1916) has convincingly shown that *F*[*alco*]. *axillaris* Latham, 1801, must apply to the letter-winged kite, 'but in view of the lack of mention of axillaries in the English description', prefers to consider the name indeterminable. In fact, whatever Mathews's preference may have been, the name is, by his own showing, determinable, since in the Latin description are found the words 'axillaribus nigris'. In the circumstances, *Elanus scriptus* Gould, 1842, must give way to *F*[*alco*]. *axillaris* Latham, 1801.

Milvus migrans affinis Gould

Milvus affinis Gould, *Synopsis of the birds of Australia* 3: pl. [10], fig. 1, and text (Apr. 1838). Australia.

Darwin (Nightcliff): 1 ♀, 16 April.
Oenpelli: 1 ♂, 14 October.

Fork-tailed kites were seen in some numbers over the aerodrome at Alice Springs on 18 March, but proved to be rather uncommon at Darwin in March and April, apparently absent from Umbakumba in May and June, rare at Yirrkalla (first noted there in the third week of August), and, finally, very numerous at Oenpelli from 20 September to 7 November.

The species was breeding at Oenpelli in October and November; my specimen from there had the gonads greatly enlarged.

Haliastur indus (Boddaert)

Falco indus Boddaert, *Table des planches enluminéez d'histoire naturelle*: 25 (1783). Pondicherry, southern India, *ex* D'Aubenton and Buffon.

This miscalled kite was common at all coastal localities visited by us, but was not noted at Oenpelli. It was seen both over the foreshores and in the mangrove.

Haliastur sphenurus (Vieillot)

Milvus sphenurus Vieillot, *Nouveau dictionnaire d'histoire naturelle*, nouv. éd., 20: 564 (May 1818). Australasie = New South Wales.

Skirmish Point: 1 ♀, 4 May.
Groote Eylandt (Umbakumba): 1 ♀, 25 June.
Oenpelli: 1 ♂, 1 ♀, 1 unsexed, 25 September, 18, 27 October.

The whistling eagle was common at every locality visited by the expedition, but especially so at Oenpelli, where it was, during our stay, the second most numerous raptor.

The breeding season is evidently of long duration, for specimens with the gonads enlarged were collected on 25 June and 25 September, and during October and November many were nesting in the trees about Oenpelli. Moreover, while the bird of 27 October is a sub-adult bird of the year, the one of 18 October is a chick only a day or two old.

The feathers of this hawk quickly become faded and worn, and the differences between fresh and old plumages are very striking.

Accipiter novaehollandiae (Gmelin)

[*Falco*] *novae hollandiae* Gmelin, *Systema naturae* 1 (1): 264 (1788). Nova Hollandia = Tasmania.

An example of this species in the white phase of plumage was seen on a number of occasions at various points between the great lagoon at Oenpelli and the foot of the Escarpment, from September to November.

Accipiter fasciatus didimus (Mathews)

Astur fasciatus didimus Mathews, *Aust. Avian Rec.* 1 (2): 33 (2 Apr. 1912). Melville Island, Northern Territory.

Arnhem Peninsula (Yirrkalla): 1 ♀, 27 August.
Oenpelli: 1 ♂, 25 September.

Accipitrine hawks were seen at all our camps, but were difficult to shoot and thus usually not certainly identifiable. At Umbakumba they were frequently observed in the vicinity of the poultry runs, and were considered very destructive to the fowls, so it may be well to note that the female from Yirrkalla had fed upon a skink.

My specimens, apparently fully grown, but wearing one of the immature plumages, had the irides golden yellow; the cere olive-plumbeous or greenish yellow; the maxilla black, plumbeous at the base; the mandible and interior of the mouth plumbeous; the feet and toes yellow; the claws horny black.

Three males from the Northern Territory and the Kimberley Division now in the United States National Museum have the wing length from 221 to 235 mm., while two females from the same regions have it 265 and 270 mm. The examples listed above, with wing lengths of 233 mm. (male) and 261 mm. (female), may therefore properly be placed with the resident race (*didimus*).

Accipiter fasciatus fasciatus (Vigors and Horsfield)

[*Astur*] *fasciatus* Vigors and Horsfield, *Trans. Linn. Soc. Lond.* 15 (1): 181 (1827). Australia = New South Wales.

Groote Eylandt (Umbakumba): 1 ♀, 4 June.

This bird, a fresh-plumaged, very old female, is so large that, until my Australian hawks had been studied in the museum, I had erroneously assumed my specimens of *A. f. didimus* to represent the collared sparrowhawk. With a wing length of 306 mm., it cannot be considered an example of *didimus* (which see), but is inseparable from a series of seven females from New South Wales (wing length: 282-304 mm.).

It had the irides golden yellow; the cere olive-plumbeous; the maxilla black; the mandible plumbeous; the feet and toes yellow; the claws black. In the stomach was an unidentifiable bird (probably a weaver finch).

Since *didimus* is almost certain to be the form resident on Groote Eylandt, it may be assumed that this individual was a winter wanderer from some more southern region of Australia.

Accipiter radiatus (Latham)

F[*alco*]. *radiatus* Latham, *Supplementum indicis ornithologici*: xii (1801). New Holland = New South Wales.

Cobourg Peninsula (Port Essington): 1 ♂, 28 September.

I saw an apparently mated pair of the red goshawk at Cape Cockburn on 25 April, and a solitary bird between Yirrkalla and Melville Bay at the beginning of August, but owe my single specimen to David H. Johnson, who shot it in the act of seizing a 'quail'. It is an old adult, in rather worn plumage.

Mathews (*Birds of Australia* 5 (1): 91, 5 Nov. 1915) has written: 'I am quite unable to even appreciate subspecific differences', but nevertheless five months later named *Erythrotriorchis radiatus katherine* from the Northern Territory (*Aust. Avian Rec.* 3 (3): 57, 7 Apr. 1916) without any indication of how it was supposed to differ from topotypical *radiatus*, or from *Erythrotriorchis rufotibia*

Campbell (north-western Australia). With Mathews's earlier view I am in complete agreement.

Peters (*Check-list of birds of the world* 1: 225, 1931) has surely erred in combining *Accipiter radiatus* and *Megatriorchis doriae* in *Erythrotriorchis*. While *Erythrotriorchis* is hardly to be separated from *Accipiter, Megatriorchis* seems to be a perfectly valid genus.

Uroaetus audax (Latham)

V[ultur]. audax Latham, *Supplementum indicis ornithologici*: ii (1801). New Holland = New South Wales.

This huge bird was seen at least once near Nightcliff, on 13 April. Otherwise, I found it in the Territory only along the Escarpment near the great waterfall of Oenpelli, where a pair haunted the neighbourhood of an enormous nest, apparently then (September, October) unused, at the summit of a lofty pinnacle.

Haliaeetus leucogaster (Gmelin)

[Falco] leucogaster Gmelin, *Systema naturae* 1 (1): 257 (1788). No locality given = Princes Island, off the westernmost cape of Java, *fide* Stresseman, *Auk* 67: 81-2 (1950).

Bremer Island: 1 ♂, 8 August.

The white-breasted sea-eagle was common on all the coasts and islands visited by the expedition.

This specimen, removed from a nest by an aboriginal woman, is probably between two and three months old; it has fully half the body still clothed in the natal down.

Circus approximans gouldi Bonaparte

Circus gouldi Bonaparte, *Conspectus generum avium* 1: 34 (1850). Australia = New South Wales.

Oenpelli: 1 ♀, 27 October.

This harrier occurred in small numbers over the marshy plains along the East Alligator, but was almost impossible to collect.

My specimen, in one of the immature plumages, is so worn and decolourized as to possess very little taxonomic value.

Amadon (*Emu* 40: 370, 1941), finding that the type of '*Circus approximans inexpectatus*' is in fact an example of *Circus assimilis,* has erroneously assumed that the swamp harrier does not reach northern Australia at all!

PANDIONIDAE

Pandion haliaetus melvillensis Mathews

Pandion haliaëtus melvillensis Mathews, *Aust. Avian Rec.* 1 (2): 34 (2 Apr. 1912). Melville Island, Northern Territory.

Groote Eylandt (Umbakumba): 2 ♂, 1 ♀, 6, 6, 24 June.
Bremer Island: 2 ♀, 19, 19 August.

The osprey was evenly distributed along all the Territorian foreshores visited by the expedition. In voice, habits, and general appearance, this small tropical race differs in no consequential manner from its close relatives throughout the world.

The birds taken on 6 June at their nest were a mated pair; the male had the gonads enlarged, while the female carried an egg in the oviduct. When the nest was revisited on 14 June, it proved to be in the possession of a second pair, which were spared. The male of 24 June had the gonads only slightly enlarged. The specimens from Bremer Island are juveniles in postnatal moult, taken from a single nest; one of these is probably a full week older than the other.

I have noted that the colour of the irides ranged between yellow and golden orange, without relationship to age or sex.

FALCONIDAE

Falco longipennis murchisonianus Mathews

Falco lunulatus murchisonianus Mathews, *Novit. Zool.* 18 (3): 252 (31 Jan. 1912). East Murchison River, Western Australia.

Darwin (Nightcliff): 1 ♀, 17 April.
Arnhem Peninsula (Gove [O.B. Camp]): 1 ♂, 15 July.
Oenpelli: 2 ♀, 14, 24 October.

The little falcon was taken at all our stations except Groote Eylandt, and even there, on 27 June, one was seen between Umbakumba and Central Hill. So many were present at Oenpelli during October, that it seems probable that I was witness of the return flight of migrants from islands to the north of Australia.

This was a species of the open eucalypt forest, where, perched upon a dead limb or a leafless tree top, it proved to be extraordinarily trustful. The specimen of 15 July had the remains of an unidentifiable bird in its stomach.

A female had the irides brown; the orbital region yellowish; the bill plumbeous, with the cere and base of the mandible olive-yellow; the feet and toes brilliant yellow; the claws black.

Mathews has named *Falco lunulatus apsleyi* from Melville Island, N.T., but the distinctions adduced by him are simply those of an adult compared with an immature plumage.

Falco cenchroides cenchroides Vigors and Horsfield

[*Falco*] *cenchroides* Vigors and Horsfield, *Trans. Linn. Soc. Lond.* 15 (1): 182-3 (1827). Australia = Parramatta, New South Wales.

Groote Eylandt (Umbakumba): 4 ♀, 4, 10, 11, 22 June.
Arnhem Peninsula (Yirrkalla): 1 ♀, 12 July.
Oenpelli, 1 ♂, 6 October.

Although Condon (*S. Aust. Nat.* 23: 2, 1945) states that the kestrel is rare in the Territory above Tennant Creek, I found it common enough at Darwin and at every one of the expedition's camps, as well as at numerous intermediate localities along the coast between Darwin and Groote Eylandt.

It is a bird of the open country, and especially of the clearings about settlements, where I observed it feeding upon lizards, centipedes, and insects.

All of this series of six seem to be adult, but none of them shows any gonadal activity. It is possible that the species, like a number of others, is merely a dry-season visitor to the northern coasts from some region of lesser rainfall.

The trinominal form of the name is employed, inasmuch as Rand has described, in 1940, *F. c. baru* from New Guinea.

Mathews's *Cerchneis cenchroides milligani* from north-western Australia is *not* smaller than the bird of New South Wales, nor is it darker in colour. It is improbable that more than one race occurs in Australia.

Falco berigora melvillensis (Mathews)

Ieracidea berigora melvillensis Mathews, *Aust. Avian Rec.* 1 (2): 34 (2 Apr. 1912). Melville Island, Northern Territory.

Darwin (Nightcliff): 1 ♂, 23 March; (Delissaville): 1 ♀, 28 March.
Arnhem Peninsula (Yirrkalla): 3 ♀, 11 July, 26, 31 August.
Oenpelli: 1 ♀, 23 October.

The brown hawk was seen in small numbers on Groote Eylandt, was common at Yirrkalla and Darwin, and was exceedingly abundant on the burned-over plains of Oenpelli in September and October. At the last-named locality, birds of this species were always to be seen in the sky, on the watch for smoke and flame, and they gathered from every direction at the first hint of a blaze, to swoop low and harry the insects and small birds driven from the protective grass. This is in no sense a 'noble falcon', for its weak feet compel it to prey mainly on beetles and grasshoppers, but it is able to seize quails and hemipodes when these are sufficiently frightened and confused, and once a brown hawk snatched a wounded land rail from the ground, to fly with it a short distance before letting it fall.

All of my specimens had the irides brown; the orbital region plumbeous; the bill plumbeous, tipped blackish; the feet and toes pale plumbeous.

With six examples of this nigrescent population before me, I must agree with Condon (*Emu* 50 (3): 152-74, 30 Jan. 1951) that *melvillensis* is a recognizable race. The length of the flattened wing of my male is 355 mm.; that of the five females ranges from 344 to 375 mm.

Falco berigora centralia (Mathews)

Ieracidea berigora centralia Mathews, *Birds of Australia* 5 (3): 277 (23 May 1916). Finke River, Northern Territory.

Arnhem Peninsula (Yirrkalla): 1 ♂, 22 August.

After careful study of Condon's revision of the brown hawks (*Emu* 50 (3): 152-74, 30 Jan. 1951), I can no longer doubt that this small, pale, rufescent bird is an example of *centralia* that has wandered into the breeding range of *melvillensis*. Condon (p. 162) has suggested that 'whilst there is nothing to show that [the brown hawk] undertakes regular migrations, nomadic movements, often coinciding with seasonal changes, probably occur on a minor scale as with other Australian birds', and it seems to me certain that this species is but one more of the many that move from the interior toward the northern coasts with the increasing heat and aridity of the Territorian dry season. Such nomadism would account for the extraordinary abundance of the brown hawk at Oenpelli in September and October, although, as it happens, my single skin from that locality is an unusually dark specimen of the resident form.

The length of the flattened wing of this bird is 332 mm.

MEGAPODIIDAE

Megapodius freycinet tumulus Gould

Megapodius tumulus Gould, *Birds of Australia* pt. 6 [vol. 5]: pl. 79 (1842). Cobourg Peninsula, Northern Territory.

Darwin (West Point): 1 ♀, 30 March.
Groote Eylandt (Umbakumba): 1 ♂, 9 June.
Arnhem Peninsula (Yirrkalla): 1 ♂, 11 July.

The scrub-fowl was a common bird at all our sea-side camps; it seems to leave the coasts only along some of the tidal rivers. I found it almost restricted to the densest monsoonal forests and mangrove swamps, only occasionally entering the immediately adjacent wattle scrub or venturing onto the naked dunes.

The extraordinary nest mounds of this species have been described by many travellers. I saw no evidence of active breeding at any examined by me, but a freshly laid egg was brought me by an aborigine at Yirrkalla on 10 August.

A dove-sized bird shot by me on 9 June at Umbakumba from the top of a fairly high, vine-covered tree, about which it was actively clambering, proved, to my surprise, to be a chick of the scrub-fowl; it differs from the adults only in size and by having the centre of the breast and abdomen clothed with soft grey and rusty feathers.

PHASIANIDAE

Coturnix ypsilophorus cervina (Gould)

Synoicus cervinus Gould, *Handbook to the birds of Australia* 2: 195 (1865). Port Essington, Northern Territory.

Oenpelli: 2 ♀, 26, 29 October.

The swamp quail was found only at Oenpelli, where a covey of about a dozen individuals haunted each of several stands of the naturalized Kupang grass growing at the margins of wet places near our camp. In such a habitat, the species was almost impossible to collect without dogs, for, rising one or two at a time from one's very feet, they dived almost instantly back into cover and then ran rapidly away amongst the tussocks.

These specimens had the irides garnet; the bill plumbeous, with tip and culmen black; the feet and toes yellow; the claws horn.

Compared with examples from New South Wales (*australis*), my birds have a slightly shorter wing length and the ground-colour of the under parts a rather less rufescent buff.

Coturnix chinensis colletti (Mathews)

Excalfactoria chinensis colletti Mathews, *Aust. Avian Rec.* 1 (3): 73 (28 June 1912). Glencoe, Northern Territory.

Oenpelli: 1 ♂, 2 ♀, 5, 16, 16 October.

The king quail was common on the Oenpelli plains where the grass had not been completely burned away, and was often flushed almost from under the wheels of our lorries.

A male, with inactive gonads, had the irides deep crimson; the bill plumbeous, with the culmen and tip of the mandible blackish; the feet and toes bright yellow; the claws horn.

It should be noted that, when both *Synoicus* and *Excalfactoria* are combined with the genus *Coturnix*, as in this report, the name for the southern race of the king quail must be derived from *Excalfactoria chinensis victoriae* Mathews, 1912, since *Excalfatoria* [*sic*] *australis* Gould, 1865, becomes preoccupied by *Perdix australis* Latham, 1801.

The Territorian king quail is separable from *Coturnix chinensis victoriae* by its lesser size.

TURNICIDAE

Turnix maculosa pseutes Mathews

Turnix maculosa pseutes Mathews, *Novit. Zool.* 18 (3): 180 (31 Jan. 1912). Parry's Creek, near Wyndham, Kimberley Division, Western Australia.

Oenpelli: 3 ♂, 1 ♀, 16, 16, 16, 22 October.

On the plains near Oenpelli, where by October most of the grass had been converted to charred stubbles, occasional patches, too moist to have been consumed by the flames, remained like islets to afford shelter to this hemipode, the king quail, and the landrail. Because of the limited area of the coverts and their wide separation from each other, the populations of herbicolous birds inhabiting them were often of extraordinary density.

My specimens, all with inactive gonads, had the irides white; the bill horny brown, yellowish at the base of the mandible; the feet and toes yellowish; the claws horn.

Turnix castanota castanota (Gould)

Hemipodius castanotus Gould, *Proc. Zool. Soc. Lond.* 7: 145 (Mar. 1840). Northwest coast of Australia (= Port Essington, Northern Territory?).

Groote Eylandt (Umbakumba): 1 ♂, 24 June.

Although a quail was said by the aborigines to be common at Umbakumba, I never once saw it there, and my only specimen was shot by David H. Johnson as it walked along a pad at the edge of the stringybark forest.

This bird, with the gonads enlarged, had the irides yellow; the bill plumbeous; the feet and toes yellow; the claws horny yellow.

GRUIDAE

Grus rubicunda (Perry)

Ardea rubicunda Perry, *Arcana* 6: pl. [22] (1810). Botany Bay, New South Wales.

Oenpelli: 1 ♀, 15 October.

Brolgas were reported to me by members of our party from the shores of Thompson Bay on Groote Eylandt, and from the interior of the Arnhem Peninsula, but the species did not come to my personal notice until we arrived at Oenpelli. Here, at moist places on the short-grass plains, the bird occurred in bands composed of dozens of individuals. Perhaps owing to a certain degree of persecution from the buffalo shooters, these northern cranes seemed to me even wilder than members of the family usually are.

I have recorded that the irides of my specimen were deep yellow.

Mathews has named *'Mathewsia' rubicunda argentea* from Derby, W.A.; it is alleged to differ from the nominate race 'in its much paler coloration above and below'. Since the Oenpelli skin is darker, above and below, than either of two skins in Washington from New South Wales, further comment would be superfluous!

RALLIDAE

Rallus philippensis mellori (Mathews)

Eulabeornis philippensis mellori Mathews, *Novit. Zool.* 18 (3): 192 (31 Jan. 1912). Sandy Hook Island, Western Australia.

Oenpelli: ♂, 16, 16 October.

During our stay at Oenpelli, Mr L. H. Hillier reported to me the existence of a fairly large bird on the burned-over plains that 'looked as if it were just learning to fly'. A special trip in search of the mysterious fowl disclosed the presence of the landrail.

No better name could have been applied to the Oenpelli population, the representatives of which, in October, were confined to the small 'islands' of unburned grass that dotted vast expanses of blackened stubbles; these shelters they shared with the king quail and the hemipode. The two specimens taken were found fully half a mile from any noticeable water, and it is difficult to understand why they had not either moved to the vicinity of some permanent lagoon or altogether left the country.

With the exception of birds from Cape York (*yorki*), all Australian populations seem to belong to a single subspecies. When *Hypotaenidia* is treated as simply a subgenus of *Rallus*, von Pelzeln's name *australis* (New South Wales) becomes preoccupied by *Ralus australis* Sparrman, 1786 (New Zealand), and *mellori*, the name next available, must be used in its place.

Porphyrio porphyrio (Linnaeus)

[*Fulica*] *Porphyrio* Linnaeus, *Systema naturae* ed. 10, 1: 152 (1758). Asia, America.

One individual was observed flying over the extensive freshwater marsh at Yirrkalla, during the latter half of August.

Fulica atra australis Gould

Fulica australis Gould, *Proc. Zool. Soc. Lond.* 13: 2 (Apr. 1845). Western Australia.

Oenpelli: 2 ♀, 8 October, 6 November.

These two, collected on the large Oenpelli Lagoon, were the only examples of the coot seen in the Territory.

They had the irides red; the bill plumbeous; the frontal plate plumbeous white; the feet and toes plumbeous. The bird of 6 November had the ovaries slightly enlarged.

These specimens, theoretically *ingrami* of Mathews, differ in no way from those of New South Wales, and one may doubt the existence of more than one race in Australia.

OTIDIDAE

Choriotis australis (J. E. Gray)

Ot[*is*]. *australis* J. E. Gray, in Griffith's *Animal kingdom* 8 [Aves, 3]: 305 (1829). Australia.

Oenpelli: 1 ♂, 10 October.

The plain turkey was fairly common, singly or in pairs, from September to November, on the short-grass plains between Oenpelli and the East Alligator River, but the one specimen collected was too fat to be made into a study skin, and I finished by preserving only the bones of the extremities.

JACANIDAE

Irediparra gallinacea novaehollandiae (Salvadori)

Hydralector novae hollandiae Salvadori, *Ornitologia della Papuasia e delle Molucche* 3: 309 (1882). Australia.

Oenpelli: 1 ♂, 1 ♀, 4, 29 October.

The lotus-bird was found only at Oenpelli, but there it was common on all the permanent lagoons in which water lilies grew.

The specimen of 29 October had the gonads slightly enlarged. The one of 4 October is an immature moulting into the first adult plumage.

I cannot distinguish *'rothschildi'* (Kimberley Division) and *'melvillensis'* (Melville Island) from the populations of eastern Australia.

HAEMATOPODIDAE

Haematopus ostralegus longirostris Vieillot

Hæmatopus longirostris Vieillot, *Nouveau dictionnaire d'histoire naturelle,* nouv. éd., 15: 410 (Sept. 1817). Australasie = New South Wales.

Arnhem Peninsula (shore of Melville Bay): 1 ♀, 15 July.

The pied oyster-catcher was not observed in the Darwin area, but was noted at most other sea-side points visited by the expedition (including Croker Island on 23 April).

Near the centre of a sort of crater, extensive and bare of vegetation, behind the beach at Umbakumba, I found two eggs of this species during the month of June. The floor and walls of the crater were composed of coarse red sand (decomposed laterite), and the eggs, which matched well with their background, were deposited in an unlined depression in the gravel. If one peered cautiously over the rim, the spot was easily seen by the strikingly marked brooding bird, but as soon as one stood silhouetted against the sky, she (or he, for each of the pair showed interest in the nest), in a crouching posture, would quietly leave the eggs and only when as much as 100 feet away from them take to the air. While one parent was at the site, the other could always be found at a reef jutting from the beach outside the crater. Before the eggs had hatched, the curiosity of the aborigines (who would gladly have eaten them) led me to discontinue my visits.

The only specimen collected had the irides and eye ring red; the bill orange-red, becoming yellowish at the tip; the feet and toes crimson-rose; the claws black.

Mathews has called Territorian birds *H. o. picatus* King, claiming that they differ from *longirostris* 'in having white on the base of the primaries'. Since no white at all appears on the primaries of my example, it is evident that the character is inconstant and probably unworthy of nomenclatorial recognition.

Haematopus fuliginosus Gould

Hæmatopus fuliginosus Gould, *Birds of Australia,* pt. 18, [vol. 6]: pl. 8 (1845). Tasmania.

One example was seen on 15 or 16 May on a rocky islet in Port Bradshaw. A pied oyster-catcher was on the islet simultaneously, but the two birds were at no time seen together.

CHARADRIIDAE

Lobibyx miles (Boddaert)

Tringa miles Boddaert, *Table des planches enluminéez d'histoire naturelle*: 51 (1783). 'La Louisiane', *ex* d'Aubenton, pl. 835, error; type locality corrected to Australia, by Hartert, *Novit. Zool.* 12: 200 (1905), and to Timor Laut, by Mathews, ibid. 18: 215 (1912).

Groote Eylandt (Umbakumba): 1 ♂, 12 June.
Oenpelli: 1 ♀, 23 September.

The masked plover was not noted at all in the Darwin area, but was numerous at Croker Island on 23 April, and observed almost daily at Umbakumba, Yirrkalla, and Oenpelli.

It is a species of the short grass and the more open wattle scrub, and during our stay in the Territory was seen only in loose flocks, sometimes composed of dozens of individuals. These bands were never far from fresh water and, during the hotter hours of the day, spent much of the time resting quietly at the edge of some lagoon. At the first sign of a human trespasser, however, there arose a chorus of strident calls that effectively alarmed every other bird in the neighbourhood, and then the plover would fly in twos and threes to the farthest shore of the pond. The species is active at night, and the unmistakable rasping cries might often be heard from the sky, especially when the moon was bright.

My specimens had the irides, palpebral lappets, and wing spurs primrose yellow; the bill the same, tipped with rose; the feet and toes crimson-rose; the claws horny brown.

It may be mentioned that the facial appendages, usually referred to as 'wattles', are in fact outgrowths of the fleshy eye ring characteristic of so many members of the Charadriidae. In Mathews' *Birds of Australia* 3: pl. opposite p. 52, 1913, and again in Cayley's *What bird is that?* ed. 12: pl. 31, 1947, these lappets are incorrectly depicted as very tumid and as if scallopped along the upper margin. In fact, although soft and flexible in the living bird, they are no thicker than ordinary skin and are nowhere scallopped-edged.

In regard to correction of Boddaert's type locality, Mathews has pointed out (*Birds of Australia* 3: 54, 1913) that 'there can be little doubt that no bird came from north Australia to Europe before 1760, while there was communication between the Spice Islands and Europe at that early date.

Charadrius alexandrinus ruficapillus Temminck

Charadrius ruficapillus Temminck, in Temminck and Laugier, *Nouveau recueil de planches coloriés d'oiseaux* 5 [livr. 8]: pl. 47, fig. 2 (Mar. 1821). Océanie = New South Wales.

Darwin (Lee Point): 1 ♂, 1 ♀, 18, 18 April.

Groote Eylandt (Umbakumba): 1 ♂, 3 ♀, 15, 15, 15, 25 June.

The red-capped dotterel, restricted to sandy sea beaches, was locally common near Darwin and plentiful at Umbakumba. Its apparent absence from the shores of the Arnhem Peninsula is difficult to explain, since conditions there seemed eminently suitable to its requirements.

Nuptial dress is worn by all my specimens except the female of 18 April, and of these, all but a female of 15 June had the gonads slightly enlarged.

Charadrius r. tormenti, described by Mathews from the Kimberley Division, but said to occur also in the Northern Territory, is averred to differ from southern populations 'in its much paler coloration, which is especially noticeable on the red cap'. When northern and southern birds, in similar plumage, are compared *sex for sex*, no difference whatsoever can be seen.

Charadrius melanops Vieillot

Charadrius melanops Vieillot, *Nouveau dictionnaire d'histoire naturelle,* nouv. éd., 27: 139 (Dec. 1818). Terres-Australes.

Arnhem Peninsula (between Yirrkalla and Melville Bay): 1 ♂, 1 ♀, 31, 31 July.

Oenpelli: 1 ♂, 30 September.

This little plover was seen only at the shingly or muddy margins of fresh-water ponds and lagoons. It occurred at only one locality on the Yirrkalla Peninsula, but proved to be widely distributed in the vicinity of Oenpelli.

My female of 31 July is sub-adult; it has the head markings not fully developed, and the black of the breast band mixed with brown. The adult male of the same date had the irides brown; the eye ring red; the bill coral red, tipped black; the feet and toes flesh; the claws black. None of my specimens is in breeding condition.

I find no subspecific differences between birds from northern and southern Australia.

Charadrius mongolus stegmanni Stresemann

Charadrius mongolus stegmanni Stresemann, *Orn. Mber.* 48 (2): 55 (22 Apr. 1940). New name for *Charadrius mongolus litoralis* Stegmann, *Orn. Mber.* 45 (1): 25 (4 Jan. 1947) (Bering Island, Commander Islands), not *Charadrius littoralis* Bechstein, 1809.

Darwin (Nightcliff): 1 ♂, 3 ♀, 7, 7, 7, 17 April.

This dotterel was common in March and April on rocky foreshores of the Darwin area, associated with other migratory plovers and sandpipers.

Three of my four skins are in almost complete nuptial plumage and are inseparable from a long series of topotypical specimens of *stegmanni*. It may be found that *Charadrius m. mongolus*, with which Australian birds have hitherto been placed, does not occur at all in this part of the world.

Charadrius leschenaultii leschenaultii Lesson

Charadrius leschenaultii Lesson, *Dict. sci. nat.*, éd. Levrault, 42: 36 (1826). Pondicherry, Coromandel coast of India.

Darwin (Nightcliff): 2 ♀, 7, 7 April; (Lee point): 1 ♂, 18 April.
Oenpelli: 1 ♂, 10 October.

The large sand-dotterel was a fairly common bird of passage at Darwin from 19 March to 19 April, and was seen at Croker Island on 23 April. It haunted both rocky and sandy shores, usually in company with other waders, and especially the very similar Mongolian sand-dotterel. The only record made away from the sea was at Oenpelli, where, on 10 October, a single individual appeared amongst the migrating sandpipers on the mud of a drying lagoon.

The two specimens of 7 April are begining to acquire the nuptial plumage, which is complete in the one of 18 April.

This species is one of the ten or more Palaearctic shore birds given wholly unnecessary new names by Mathews, without the *least* indication of how the Australian-taken, non-breeding type specimens were supposed to differ from all others of their kind. It may be noted, without further comment, that, in the present case, the Australian example was made the type of a full species (*Pagoa zanda*) and, as would be obvious to any ornithologist, was *not*, as claimed (*Birds of Australia* 3: 101, 1913) 'in full breeding-plumage'.

Charadrius asiaticus veredus Gould

Charadrius veredus Gould, *Proc. Zool. Soc. Lond.* 16: 38 (14 Nov. 1848). Northern Australia = Port Essington, Northern Territory, *ex* Gould, *Handbook to the birds of Australia* 2: 230 (1865).

Oenpelli: 1 ♂, 1 ♀, 28 September, 2 November.

The Oriental dotterel first appeared, in small numbers, on the crushed-laterite aerodrome at Oenpelli on 28 September, and during the subsequent weeks of our stay there became fairly common in the neighbourhood of the camp, feeding amongst the tents in company with little whimbrels and the magpie-larks.

I never saw this species really at the edge of any body of water, but always on the dry, baked ground away from it, where even the vegetation was sparse. In winter dress, on the reddish soil, it was extraordinarly well concealed, and advantage of this fact was taken by the birds, which, at one's near approach, would stand stock-still until almost trodden upon, rather than seek safety in flight.

Erythrogonys cinctus Gould

Erythrogonys cinctus Gould, *Synopsis of the birds of Australia*, pt. 4: pl. [18] and text (Apr. 1838). New South Wales.

Oenpelli: 1 ♂, 1 ♀, 28 September, 5 October.

The red-kneed dotterel was found only at Oenpelli, where, in small numbers, it occurred on some of the softest and slimiest banks imaginable, at the margins of certain drying lagoons and the larger buffalo wallows. It was never seen on the muddy flats of the East Alligator River.

My specimens seem to be wholly adult, but have the front and crown concolourous with the mantle, with a blackish suffusion only on the ear coverts and across the nape. The black hood is doubtless worn only by the oldest birds.

The male had the irides dark brown; the bill deep rose, with the culmen and dertrum black; the tibiae and uppermost third of the tarsi deep rose; the rest of the tarsi, and the toes, pinkish brown; the claws horny brown.

The putative characters of *Erythrogonys cinctus mixtus* Mathews, named from north-western Australia, seem to represent nothing more important than individual variation. I have been unable to perceive any real difference between topotypes of *mixtus* and the birds of New South Wales.

Arenaria interpres morinella (Linnaeus)

[*Tringa*] *Morinella* Linnaeus, *Systema naturae*, ed. 12, 1: 249 (1766). *Ad maris littora* Americæ *septentrionalis*; & Europæ; type locality restricted to coast of Georgia, *ex* Catesby, *Natural history of Carolina* 1: 72, pl. 72 (1731).

Darwin (West Point): 1 ♂, 31 March.

Turnstones were noted in small numbers on many of the rocky headlands in the Darwin area.

My study of extensive series of ruddy turnstones has led to complete agreement with Mathews's treatment of the species (*Birds of Australia* 3: 10, 1913); the race *oahuensis*, not accepted by most recent authors, is, in my opinion, perfectly distinct from *interpres* and *morinella*.

This unique specimen, a male in full nuptial plumage, is apparently *not*, however, *oahuensis*, the only form recognized by Mathews from Australia, but matches the *most* highly coloured and extensively chestnut examples of *morinella*; it cannot be picked out from a series of comparable American birds, but is strikingly different from any one of my series of the two other races.

The problem of subspecific identification of such a single specimen has been recently discussed by Rand (*Auk* 65: 416-32, 1948), who would perhaps recommend that my bird be considered an individual variant of one of the Old World forms. In this case, one cannot certainly affirm that *morinella* does not winter fairly commonly in Australia; the *published* ranges of the several forms are usually thoroughly unreliable by the failure of authors to recognize *oahuensis* and fit it into their schemes. Amongst the handful who accept *oahuensis*, we have only the authority of Mathews for the belief that this race alone reaches Australia, and against this authority stands the fact that the alleged portrait of the type specimen of *A. i. nova* (*Aust. Avian Rec.* 3: 69, 1917, new name for bird 'figured and described in my *Birds of Australia*, vol. III, pl. 125, p. 6') seems to represent *morinella*, and not at all the form I and my colleagues in Washington unanimously agree is *oahuensis*. When, in the *Systema avium australasianarum*: 152, 1927, we find Mathews synonymizing *nova* with *oahuensis*, does it mean that the portrait is wrong? that Mathews was misapplying the name *oahuensis*? or that the synonymization is due only to a preconceived idea that *oahuensis* alone could visit Australia? The problem must wait until some Australian ornithologist has visited the northern coast in March and April and collected further material in nuptial dress for comparison with properly identified specimens.

SCOLOPACIDAE

Numenius minutus Gould

Numenius minutus Gould, *Proc. Zool. Soc. Lond.* 8: 175 (July 1841). Maitland on the Upper Hunter, New South Wales.

Oenpelli: 3 ♀, 23 September, 5, 17 October.

The little whimbrel first came to my attention at Oenpelli on 23 September, when a flock of about 100 was seen. During the ensuing days and weeks, the species became continually more abundant, so that by the middle of October it was possible to observe many tens, if not hundreds, of thousands of individuals during a morning's drive across the plain; the swarms of birds rising, one after the other, in wide arcs ahead of the car resembled elongated whips of dark smoke. It seems probable that most of the world population of *Numenius minutus* was then concentrated in that favoured spot.

With increasing numbers it became progressively more fearless; whereas to shoot one specimen from the flock of 23 September required hours of stalking, during the last weeks of our stay dozens of little whimbrels fed daily among our tents, associated with the Oriental dotterels and the magpie-larks and yielding nothing to these last in tameness.

I found these birds only on the drier portions of the plain, where it was clothed with short grass or burned stubble. Here they acted like so many golden plover, the feeding flocks spread thin over a great area and almost invisible against the ground.

I have noted that the example of 23 September had the irides dark brown; the bill black, with the base of the mandible fleshy; the feet and toes plumbeous; the claws black.

Numenius phaeopus variegatus (Scopoli)

Tantalus (*variegatus*) Scopoli, *Deliciae florae et faunae insubricae*, pt. 2: 92 (1786). No locality given = Luzón, Philippine Islands, *ex* Sonnerat.

Arnhem Peninsula (shore of Melville Bay): 1 ♂, 31 July.

This curlew was common on mudflats and foreshores in the Darwin area during my first stay there (19 March to 19 April), and was later seen at Croker Island (23 April). The single specimen taken was one of three that were first noted on 28 July at a point on Melville Bay where no shore birds had been found during earlier visits; since it seemed to be a bird of the year and was very lean, I believe that this small flock represented the vanguard of the south-bound whimbrel migration.

Numenius arquata (Linnaeus)

[*Scolopax*] *arquata* Linnaeus, *Systema naturae*, ed. 10, 1: 145 (1758). Europe.

This curlew was seen on a number of occasions during the last week of March and the first week of April on the rocky headlands at Nightcliff, consorting with *N. madagascariensis* and other shore birds. Although I was not successful in taking a specimen, I have no hesitation in affirming that this easily identified bird should be added to the Australian list: it is as large and as long-billed as *N. madagascariensis*, but differs at a glance by its white rump patch.

The race *orientalis* C. L. Brehm regularly migrates to the coasts of Java, and thus might be expected to reach northern Australia.

Numenius madagascariensis (Linnaeus)

[*Scopolax*] *madagascariensis* Linnaeus, *Systema naturae*, ed. 12, 1: 242 (1766). Madagascar, error; type locality corrected to Makassar, Celebes, by Neumann, *Anz. Orn. Ges. Bayern* 2: 150 (1932).

Groote Eylandt (Umbakumba): 1 ♀, 21 June.

During the last week of March and the first week of April, the eastern curlew was fairly common on the rocky headlands at Nightcliff, sometimes associated with the closely related *Numenius arquata* (a bird not yet collected in Australia). Three examples occurred along the sandy shores at Umbakumba in May and June; these were clearly individuals that had for some reason failed to migrate northward with others of their kind. The species was excessively shy, and my one specimen was obtained only by moonlight stalking.

It had the irides dark brown; the maxilla dark horny brown, becoming black toward the tip; the mandible violet-rose, becoming black toward the tip; the feet and toes plumbeous-grey; the claws black.

Limosa limosa (Linnaeus)

[*Scolopax*] *limosa* Linnaeus, *Systema naturae*, ed. 10, 1: 147 (1758). Europe.

One individual was observed on October 21, amongst other shore birds at a drying lagoon near Oenpelli.

Limosa lapponica novaezealandiae G. R. Gray

Limosa lapponica, var. *Novæ Zealandiæ* G. R. Gray in Richardson and J. E. Gray, *Zoology of the voyage of H.M.S. Erebus and Terror*, pt. 9 (Birds): 13 (1845). Waikouaiti, South Island, New Zealand.

Darwin (Nightcliff): 1 ♀, 7 April.

The bar-tailed godwit occurred in small numbers amongst the migrating shore birds at Darwin between 24 March and 7 April. My specimen was one of five or six individuals associated with whimbrels and stints resting on a rocky foreshore at high tide.

This bird, in non-breeding plumage, had the irides dark brown; the bill black, becoming violet-rose on the basal half; the feet and toes slate; the toes black.

Portenko (*Auk* 53: 194-7, 1936) has divided the Siberian bar-tailed godwits into two races, but has given no clues to how they may be separated in winter dress; his study is otherwise unsatisfactory in many particulars (see Mayr, ibid.: 367-8). Since *Limosa baueri* Naumann, the name most applicable to Australian birds on purely geographical grounds, is, *fide* Hartert and Steinbacher (*Vögel der paläarktischen Fauna, Ergänzungsband*, hft. 6-7: 481, 1938), a *nomen nudum*, Gray's designation, given a New Zealand specimen, may be tentatively employed.

Tringa stagnatilis (Bechstein)

T[*otanus*]. *stagnatilis* Bechstein, *Ornithologisches Taschenbuch von und für Deutschland*, pt. 2: 292, pl. 29 (1803). Germany.

Oenpelli: 1 ♂, 21 October.

Half a dozen marsh sandpipers were seen at West Point, near Darwin, at the close of March, and about the same number appeared at a drying lagoon near

Oenpelli at the middle of October. I found them very shy and considered it good luck to obtain even one specimen.

Tringa nebularia (Gunnerus)

Scolopax nebularia Gunnerus in Leem, *Beskrivelse over Finmarkens Lapper*: 251 (1767). District of Trondheim, Norway.

Cobourg Peninsula (Black Rock Point): 1 ♀, 1 October.

The greenshank was noted at a temporary pond near Nightcliff on 6 April and on several succeeding days, and at Oenpelli a few individuals were seen about the drying lagoons between 22 September and mid-October, but these were too wild to be collected. My one specimen was shot by D. H. Johnson.

Tringa glareola Linnaeus

[*Tringa*] *glareola* Linnaeus, *Systema naturae,* ed. 10, 1: 149 (1758). Europe; type locality restricted to Sweden, *apud* Hartert.

Oenpelli: 1 ♀, 17 October.

On 17 October, at the drying lagoon near Oenpelli, where the majority of my shore bird records for that area were made, I heard the unmistakable alarm notes of the wood sandpiper and quickly picked out the bird from a flock of some hundreds of sharp-tails. It proved to be excessively shy and was collected only after hours of stalking from one end of the pond to the other.

Actitis hypoleucos (Linnaeus)

[*Tringa*] *hypoleucos* Linnaeus, *Systema naturae,* ed. 10, 1: 149 (1758). Europe; type locality restricted to Sweden, *apud* Hartert.

Oenpelli: 1 ♂, 5 October.

The common sandpiper was by no means so frequently seen as it would have been in countries farther north, but was nevertheless generally distributed in season along the Territorian foreshores (especially where these were composed of rock), at pools behind the beaches, and, on occasion, among the mangroves. About the freshwater lagoons at Oenpelli, on the other hand, it occurred in but small numbers. The earliest migrant from Asia appeared at Yirrkalla on 21 August.

Heteroscelus incanus brevipes (Vieillot)

Totanus brevipes Vieillot, *Nouveau dictionnaire d'histoire naturelle,* nouv. éd., 6: 410 (Dec. 1816). Pays inconnu; type specimen from Timor, *fide* Pucheran in *Rev. Mag. Zool.*: 370 (1851).

Darwin (Nightcliff): 1 ♂, 3 ♀, 24 March, 3, 3, 10 April.

The grey-tailed tattler was one of the commonest of the migrant waders in the Darwin area, occurring on every reef and rocky headland, usually in association with other species. It was the only one of the family that I found tame and easy to collect.

Of my four specimens, only the bird of 24 March is not in more or less complete nuptial plumage.

Capella hardwickii (J. E. Gray)

Scolopax hardwickii J. E. Gray, *Zool. Misc.* no. 1: 16 (1831). Tasmania.

Darwin (Nightcliff): 1 ♀, 24 March.

Snipe were found in small numbers in March and April at swales in the open forest near Nightcliff, but the height of the spear grass at that season made collection all but impossible. My one specimen I owe to D. H. Johnson.

It is interesting that the bird is not *G. megala*, the species previously recorded from the Territory, but *hardwickii*, as was shown by its having only sixteen rectrices. The past tense is used advisedly, since the skin, while drying, had a misadventure with a dog, which resulted in entire loss of its tail. The number is nevertheless correct, for I had, in accordance with my usual practice, counted the feathers at the time of preparation and written the figure upon the label. It is not yet certain whether *hardwickii* is merely a straggler so far to the west, or has simply been overlooked by earlier collectors.

Ereunetes ruficollis ruficollis (Pallas)

Trynga ruficollis Pallas, *Reise durch verschiedene Provinzen des russischen Reichs* 3: 220, 700 (1776). Circa Lacus salsos Dauuriae campestris = Kulussutai, southern Transbaikalia, *ex* p. 220.

Darwin (Nightcliff): 1 ♀, 3 April.
Cobourg Peninsula (Black Rock Point): 1 ♀, 30 September.
Oenpelli: 1 ♂, 1 ♀, 28 September, 9 October.

This stint was fairly common in the Darwin area in March and April, amongst the other waders. On the return migration it appeared in relatively small numbers about the lagoons at Oenpelli, but this may have resulted from its preference for the margins of the sea.

The bird of 3 April was moulting into the nuptial dress. All my specimens had the bill black, the feet and toes black or brownish black.

Ereunetes acuminatus (Horsfield)

Totanus acuminatus Horsfield, *Trans. Linn. Soc. Lond.* 13 (1): 192 (May 1821). Java.

Oenpelli: 2 ♂, 3 ♀, 28 September, 10, 13, 17, 21 October.

The sharp-tailed stint was only fairly numerous on the reefs and beaches of the Darwin area in March and April, but proved to be the commonest of the shore birds on the more suitable terrain about Oenpelli from September to November. At the latter locality it occurred in flocks of hundreds on the mud of drying lagoons and at marshy places on the short-grass plains. It was surprising to find my specimens extraordinarily fat, although they must but a short time before have completed the long flight from northern Asia.

An adult had the tarsi olive brown, but a bird of the year had these parts yellowish olive.

Ereunetes ferrugineus (Pontoppidan)

[*Tringa*] *ferrugineus* Pontoppidan, *Danske Atlas* 1: 624 (1763). No locality given = Iceland and Kristiansö (an islet in the Baltic Sea north of Bornholm), *ex* Brünnich, *Ornithologia Borealis*: 53 (1764).

Oenpelli: 2 ♂, 2 ♀, 17, 18, 18, 21 October.

Some seven or eight curlew sandpipers were found at a drying lagoon near Oenpelli, where they associated with other migratory waders.

These specimens had the irides dark brown; the bill black; the naked portion of the tibiae brown; the tarsi and toes deep brown; the claws black.

For the use of Pontoppidan's specific name, see Stresemann, *Orn. Mber.* 49: 21, 1941.

RECURVIROSTRIDAE

Himantopus himantopus leucocephalus Gould

Himantopus leucocephalus Gould, *Proc. Zool. Soc. Lond.* 5: 26 (21 Nov. 1837). Australia.

Arnhem Peninsula (Yirrkalla): 1 ♂, 29 August.
Oenpelli: 1 ♂, 10 October.

On 29 August, a party of four stilts appeared on the Yirrkalla beach at low tide, and one of these was collected. The species was otherwise seen only about Oenpelli, where, however, it occurred in flocks of hundreds, on the muddy shores of the East Alligator when the tide was out, at other times in marshy areas on the short-grass plains.

The specimen of 29 August, an adult with inactive gonads, had the irides red; the bill black; the feet and toes rose; the claws black.

For a discussion of Mathews's putative races of this species in Australia, see Mayr (*Amer. Mus. Novit.*, No. 1007: 14-15, 1938).

BURHINIDAE

Burhinus magnirostris rufescens Mathews

Burhinus magnirostris rufescens Mathews, *Novit. Zool.* 18 (3): 225 (31 Jan. 1912). Parry's Creek, Kimberley Division, Western Australia.

Skirmish Point: 1 ♂, 6 May.
Groote Eylandt (Umbakumba): 1 ♀, 24 June.
Oenpelli: 1 ♀, 29 September.

The lesser stone-curlew was not seen at Darwin, but proved to be common at Umbakumba, Yirrkalla, and Oenpelli. It is a bird of the wattle scrub behind the sea beaches and of parklike forest, where the herbage is thin.

This species is seldom observed during the heat of the day, when it rests quietly, often in aggregations of a dozen or more individuals, under clumps of bushes or in wooded hollows of the dunes. At twilight, however, it moves toward fresh water, and then and all through the night, its true numbers are indicated by the ubiquity of its sweet, mournful notes. These melancholy sounds periodically swell into a chorus, as all the birds in the vicinity seem to try to outdo each other.

The specimen of 29 September, with enlarged ovaries and well-marked brood patches, was collected while setting on two recently laid eggs. The 'nest' was an unlined shallow depression in crumbling laterite, scarcely shaded by a thin, stunted eucalypt tree growing beside it. The mother attempted to escape detection by extending her neck flat along the ground, then, still in this position, began to creep away.

The eggs, rich Vinaceous-Buff, heavily blotched with Natal Brown (colours as used by Ridgway, *Color standards and color nomenclature*, 1912, pl. XL), seem to represent what North (*Nests and eggs of birds of Australia and Tasmania* 4 (3): 249, 1913) described as 'a rare variety . . . of a faint reddish-buff, thickly freckled, spotted and blotched with various shades of umber-brown'. Their principal interest lies, however, in the fact that, so closely did their colours correspond with those of the sand and gravel upon which they were deposited, that the eggs were all but invisible even as one stood looking down at them. A vial of sand and gravel from the nesting site now lies before me; the laterite granules are for the most part reddish buff, but a small percentage of them are deep red-brown or blackish brown. One might surmise that the 'rare variety' described by North is, in fact, the usual colour-type from areas of reddish soils. Collectors of this stone-curlew's eggs should in future take a sample of the soil, in order to discover whether there is a definite correlationship between the colours of earth and shell.

Burhinus magnirostris rufescens seems to be separable from the nominate race (New South Wales) by the characters given at the first description.

Esacus magnirostris magnirostris (Vieillot)

Œdicnemus magnirostris 'Geoffrey-S. Hilaire', Vieillot, *Nouveau dictionnaire d'histoire naturelle*, nouv. éd., 23: 231, pl. G39, fig. 1 (Sept. 1818). No locality given = Nouvelle-Hollande, *ex* 28: pl. G39, fig. 1 (1898).

Cape Cockburn: 1 ♀, 25 April.
Groote Eylandt (shore of Thompson Bay): 1 ♂, 1 ♀, 16, 23 June.

The beach stone-curlew occurs in pairs, probably at all the rocky foreshores of the Territory, as well as on the off-lying islands. In addition to those listed above, I have records for this species from Nightcliff, the shores of Melville Bay, and an islet in Port Bradshaw.

Although this is really a bird of the reefs, it may often be seen on flats of mud and sand in their vicinity at periods of ebb tide. It is excessively wild and difficult to approach. When alerted, its appearance can only be described as grotesque; it reminded me of a large duck, clumsily and woodenly mounted by an inept taxidermist. The only note I heard from it was a soft *peet, peet,* so comically disproportionate to the size of its author that, on one occasion, I was convinced that I was in immediate proximity to a brood of newly hatched dotterels.

The male of 23 June had the gonads enlarged. My specimens had the irides and eye ring yellow; the bill black, olive-plumbeous at the base of the maxilla; the feet and toes olive-plumbeous; the claws black.

Mayr (*Amer. Mus. Novit.*, No. 1007: 15, 1938) has shown that this stone-curlew, considered as a species, cannot be racially divided. My use of a trinomial name here reflects the fact that I consider *Esacus recurvirostris* (Cuvier), a form of the rivers of tropical continental Asia, conspecific with the bird of the Australasian beaches (see also Mayr, *Amer. Mus. Novit.*, No. 1417: 30, 1949).

GLAREOLIDAE

Stiltia isabella (Vieillot)

Glareola isabella Vieillot, *Analyse d'une nouvelle ornithologie élémentaire* 69 (1816). Australasie.

Groote Eylandt (Umbakumba): 1 ♂, 1 ♀, 24, 30 May.
Oenpelli: 1 ♂, 1 ♀, 21 September, 21 October.

At Umbakumba the swallow-plover occurred in very small numbers along a broad beach composed of coarse red sand, and one or two were noted in August on the gravelly shore of a freshwater pond near Melville Bay. But at Oenpelli, where it was known to the white Australians by the curious name of 'Arnhem Land grouse', the pratincole proved, throughout our stay, to be one of the commonest birds.

In the last locality this species spent the heat of the day standing solitary, at more or less even intervals of about 50 feet across the short-grass plains; an individual might occasionally run or fly in pursuit of a passing insect, but after such exercise it would resume its statuesque position. The only tendency toward flocking appeared late in the afternoons, when some hundreds of wheeling pratincoles often appeared over the aerodrome to feed on swarming ants and termites.

The specimen of 24 May was a juvenile on the wing, so young as to suggest its having been bred in the neighbourhood; from the adults it differs strikingly by the soft rufescent-buff coloration of its upper parts. The female of 30 May, with its white throat bounded by a broken line of streaks, is presumably sub-adult.

LARIDAE

Larus novaehollandiae forsteri (Mathews)

Bruchigavia novæ-hollandiæ forsteri Mathews, *Birds of Australia* 2 (4): 457 (1 Nov. 1912). New Caledonia.

Groote Eylandt (Umbakumba): 1 ♂, 22 May.
Arnhem Peninsula (shore of Melville Bay): 1 ♂, 19 July.

The silver gull was decidedly uncommon but apparently evenly distributed along the Territorian coast from Darwin to the Gulf of Carpentaria.

The bird of 22 May had the testes enlarged. The one of 19 July is in natal down, and was confiscated from aboriginal children, who had dislocated the wings and otherwise abused it.

The adult had the irides white; the eye ring coral red; the bill blood red, darker at the tip; the feet and toes coral red; the claws black.

My specimen differs in no consequential manner from topotypes of *forsteri*.

Chlidonias hybrida fluviatilis (Gould)

Hydrochelidon fluviatilis Gould, *Proc. Zool. Soc. Lond.* 10: 140 (Feb. 1843). Rivers and lakes of the interior of New South Wales.

Oenpelli: 2 ♂, 1 ♀, 10, 18, 18 October.

The marsh tern was found only at Oenpelli, where it occurred in small numbers at even the less extensive lagoons.

Although no specimen showed any gonadal activity, the two of 18 October are in full nuptial dress. These differ from the third, an adult male in eclipse plumage, also in having the bill coral red, rather than red-tinged black.

'*Hydrochelidon leucopareia rogersi*', named by Mathews from north-western Australia, was alleged to be 'lighter on the abdomen, when in full breeding plumage'. It is hardly necessary to say that the distinction is inappreciable to the less acute eyes of other ornithologists.

Gelochelidon nilotica macrotarsa (Gould)

Sterna macrotarsa Gould, *Synopsis of the birds of Australia* 2: pl. [18], fig. 2, and text (Apr. 1837). Tasmania.

Groote Eylandt (Umbakumba): 1 ♀, 10 June.
Oenpelli: 1 ♀, 17 October.

The gull-billed tern was noted once in the harbour of Darwin (19 April), occurred in a flock of some dozens on off-shore sand bars on the inner bay at Umbakumba, and appeared in very small numbers on the drying lagoons of Oenpelli in October.

My specimen of 10 June is in full nuptial plumage, but showed no gonadal activity; the one of 17 October is an adult in worn non-breeding dress.

These birds had the irides brown and the bill, feet, toes, and claws black.

Hydroprogne caspia (Pallas)

Sterna caspia Pallas, *Novi Comment. Acad. Sci. St-Pétersb.* 14 (1): 582, pl. 22, fig. 2 (1770). Caspian Sea.

Darwin (Nightcliff): 1 ♀, 17 April.

The Caspian tern occurred in very small numbers in the neighbourhood of Darwin, was seen once along the shores of Melville Bay, and was represented by a single individual at Oenpelli (21 October).

My specimen, found fishing over the surf near the mouth of Rapid Creek, is a female, presumably sub-adult, with white-streaked crown. It had the irides dark brown; the bill coral red, with blackish tip; the feet, toes, and claws black.

It is improbable that races of this species can be distinguished anywhere in the world (see Witherby, *Handbook of British birds* 5: 18, 1941).

Sterna bergii cristata Stephens

Sterna cristata Stephens in Shaw, *General zoology* 13 (1): 146 (Feb. 1826). China.

Cape Cockburn: 1 ♀, 24 April.

This is by far the commonest tern of the Territorian reefs and foreshores; it was seen at all suitable points visited, except on the inner bay at Umbakumba.

The specimen taken is a female in fresh nuptial dress; it had the ovaries enlarged.

My example differs neither in dimensions nor in colour of mantle from seasonally comparable topotypes of *cristata*. I find the pallid mantle attributed by some authors to Australian populations only in birds wearing more or less worn plumage.

Sterna bengalensis Lesson

Sterna bengalensis Lesson, *Traité d'ornithologie,* livr. 8: 62 (June 1831). Côtes de l'Inde = Bengal.

This species was seen in some numbers with roseate and black-naped terns in Boucaut Bay on 29 April, and a small flock was often noted near the shores of the inner bay at Umbakumba in May and June.

Sterna dougallii gracilis Gould

Sterna gracilis Gould, *Proc. Zool. Soc. Lond.* 13: 76 (Oct. 1845). Houtman's Abrolhos, off Western Australia.

Boucaut Bay: 1 ♂, 29 April.

On a reef exposed at low tide in Boucaut Bay, at no great distance from the one on which the barge *Phoenix* lay grounded, a flock of many hundred terns was found resting on 29 April. They were mainly of three species: the roseate, the black-naped, and the lesser crested. The roseate tern was not observed elswhere in the Territory.

My specimen, with gonads slightly enlarged, had the irides brown; the bill wholly black; the feet and toes coral red; the claws black. A roseate tinge that suffused the under parts in life is now (October 1949) all but vanished.

Sterna sumatrana sumatrana Raffles

Sterna sumatrana Raffles, *Trans. Linn. Soc. Lond.* 13 (2): 329 (not earlier than Nov. 1822). Sumatra.

Boucaut Bay: 1 ♀, 29 April.

This tern was seen only in Boucaut Bay, where it occurred with the roseate tern, but in smaller numbers.

My specimen, with gonads inactive, had the irides brown; the bill, feet, toes, and claws black.

Sterna albifrons Pallas

Sterna (albifrons) Pallas, *in* Vroeg, *Beredeneerde catalogus, Adumbratiunculae*: 5 (1764). No locality given = the Netherlands, *ex* 't over Maassche Land, p. 27.

The little tern occurred along sandy shores in the vicinity of Darwin (Nightcliff, Lee Point) in March and April, and on the inner bay at Umbakumba in May and June.

COLUMBIDAE

Ptilinopus regina ewingii Gould

Ptilinopus ewingii Gould, *Birds of Australia,* pt. 7 [vol. 5]: pl. [8] (1 June 1842). Cobourg Peninsula, Northern Territory.

Darwin (Nightcliff): 1 ♂, 19 March.
Arnhem Peninsula (Yirrkalla): 1 ♀, 27 August.

The beautiful rose-crowned pigeon was noted at Umbakumba and at Yirrkalla only in small numbers, but was locally very common at Nightcliff in March and April. It haunted fruiting trees and vines at the edge of monsoonal forest; near Darwin it was one of many species feeding on the berries of a *Celtis,* and at Yirrkalla was seen gorging on ripening figs.

The specimen of 19 March had the gonads enlarged, while that of 27 August had them slightly enlarged.

The female differs from the male by having the violet abdominal patch smaller and the under parts less extensively suffused with a much less vivid orange.

Ducula spilorrhoa melvillensis (Mathews)

Myristicivora bicolor melvillensis Mathews, *Aust. Avian Rec.* 1 (2): 27 (2 Apr. 1912). Melville Island, Northern Territory.

Oenpelli: 2 ♀, 25, 25 September.
Cobourg Peninsula (shore of Popham Bay): 1 ♂, 5 October.

The first nutmeg pigeon was seen at Yirrkalla early in the morning of 20 August, perched in a sterile wild nutmeg tree; thereafter, the species was not recorded until we travelled up the East Alligator on 20 September, when many individuals were heard and observed in the mangroves on the banks. Near our camp at Oenpelli, this pigeon proved to be a regular visitor to certain large fruiting fig trees, and by the end of our stay a number of pairs were breeding in their neighbourhood. The two birds of 25 September had the gonads slightly enlarged.

My specimens had the irides dark brown; the orbital skin plumbeous; the bill olive-yellow; the feet and toes plumbeous; the claws black. It may be worthy of mention that, in the dried skins, the bill has become slaty, with the dertrum horn-colour.

I follow Mayr (*List of New Guinea Birds* 44, 1941) in recognizing *D. s. melvillensis* as distinct from the nominate race, but my material does not permit me to state in what characters the two differ.

Geopelia humeralis apsleyi Mathews

Geopelia humeralis apsleyi Mathews, *Aust. Avian Rec.* 1 (2): 27 (2 Apr. 1912). Melville Island, Northern Territory.

Darwin (Nightcliff): 2 ♂, 10, 15 April.
Croker Island (Mission Station): 1 ♀, 23 April.
Groote Eylandt (Umbakumba): 1 ♀, 22 May.
Arnhem Peninsula (Yirrkalla): 2 ♂, 1, 1 August.
Oenpelli: 1 ♀, 21 September.

The bar-shouldered dove was exceedingly common at all localities visited by the expedition; it lived mainly in, or at the edge of, the monsoonal forest, but at Umbakumba occurred also in the wattle scrub, and at Oenpelli, in the open eucalypt and paperbark groves.

Of my series, only a single male of 1 August had the gonads inactive; all females were breeding, and those of 23 April and 22 May carried oviduct eggs. The long duration of the reproductive season sufficiently accounts for the species' abundance.

In July and August, at Yirrkalla, non-breeding individuals gathered into flocks of many hundreds to feed on grass seeds about our camp, especially where there had been a recent burn; birds at Nightcliff, in April, were feeding upon the fruits of a *Celtis*. Its versatile appetite would be a factor contributing to this dove's conspicuous success.

Mathews has correctly stated the difference between *G. h. inexpectata* of north-western Australia (*Novit. Zool.* 18: 186, 1912) and *humeralis* of southern Queensland, and between *G. h. apsleyi* and *inexpectata*, but, in his usual fashion, has failed to inform us how *apsleyi* and the very similar *humeralis* are to be distinguished. Since my material of *humeralis* is old and possibly discoloured, I hesitate to pronounce an opinion on apparent differences in hue, but find nevertheless that *apsleyi* may be separated from eastern birds by its generally smaller dimensions. I am decidedly not in agreement with those recent authors who have 'lumped' all Australian populations under the subspecific name *humeralis*.

Geopelia striata placida Gould

Geopelia placida Gould, *Proc. Zool. Soc. Lond.* 12: 55 (Sept. 1844). Port Essington, Northern Territory.

Darwin (Delissaville): 1 ♂, 27 March.
Groote Eylandt (Umbakumba): 1 ♂, 26 May.
Arnhem Peninsula (Yirrkalla): 1 ♂, 4 August.
Oenpelli, 1 ♀, 12 October.

The peaceful dove was common everywhere, although never seen in such numbers as *G. h. apsleyi*; it occurred in the open eucalypt forest, along roadsides, and in the neighbourhood of human habitation.

The breeding season of this species is as long as, or longer than, that of the bar-shoulder; only my specimen of 12 October had the gonads inactive.

A certain amount of individual variation appears in the depth of coloration, above and below, of any Australian population, and it is doubtful whether any character but size may be used to separate northern from southern birds.

Geopelia cuneata (Latham)

C[olumba]. cuneata Latham, *Supplementum indicis ornithologici*: 61 (1801). New South Wales.

Oenpelli: 1 ♀, 24 September.

The diamond dove was found only at Oenpelli, where, at one spot on the plain, a flock of 100 or more, associated with *Geopelia s. placida,* haunted a short-grass area where a quantity of branches and cut-off underbrush had been formed into piles for future burning.

My specimen had the irides yellow; the orbital skin red; the feet and toes flesh; the claws horn.

Whatever slight differences appear between this specimen and examples from New South Wales may be accounted for by the probability that the northern bird is sub-adult.

Chalcophaps indica longirostris Gould

Chalcophaps longirostris Gould, *Introduction to the birds of Australia,* 8vo, ed.: 79 (1 Aug. 1848). Port Essington, Northern Territory.

Groote Eylandt (lower slopes of Central Hill): 1 ♀, 1 July.

In addition to a pair found in a watered gully at the base of Central Hill, a very few green-winged pigeons were noted in dense stands of monsoonal forest at Yirrkalla and Nightcliff.

This was one of the shyest of its family; I usually saw it only as a green flash darting low across a jungle trail, immediately to be lost to sight in the heavy undergrowth.

My specimen, although sexed by me as a female with inactive ovaries, seems, by plumage, to be a male. It has the outermost primaries partly ensheathed.

Chalcophaps i. longirostris seems to be a perfectly distinct race, easily separable from *C. i. chrysochlora* of New South Wales by the bluish lilac (not deep vinaceous) coloration of the head, nape, upper back, and breast.

Phaps chalcoptera consobrina Mathews

Phaps chalcoptera consobrina Mathews, *Novit. Zool.* 18 (3): 188 (31 Jan. 1912). Parry's Creek, Kimberley Division, Western Australia.

Groote Eylandt (Umbakumba): 1 ♂, 1 ♀, 27 May, 1 June.

The bronzewing occurred in small numbers in the open eucalypt forest and in the wattle scrub at Umbakumba, but was seen at no other of the expedition's camps.

I found this pigeon usually in pairs, feeding on the ground; it would at first freeze to avoid detection, then, with a loud whirring of wings, fly into a tree, whence it would eye the intruder for a moment before disappearing into the bush. The plump body, streperous flight, and banded flabelliform tail gave it a startling resemblance to a gallinaceous bird. The call, often heard from the trees around a waterhole near my tent, both late in the evening and early in the morning, was a low, but far-carrying, *coo.*

Territorian specimens are distinguishable from those of New South Wales (*P. c. chalcoptera*) by having a slightly shorter wing and the under parts less strongly suffused with vinaceous-rose.

Petrophassa rufipennis Collett

Petrophassa rufipennis Collett, *Proc. Zool. Soc. Lond.* 1898: 354, pl. 28 (Oct. 1898). South Alligator River, Northern Territory.

Oenpelli (Tableland): 2 ♀, 26, 26 September.

The rock-pigeon was seen only in the vicinity of pools left in the bed of a wet-season stream atop the Arnhem Land Plateau near Oenpelli. There it occurred concentrated in some numbers, although not actually in flocks.

This is a terrestrial species which, sunning itself upon the broad rock faces, was almost invisible; when startled, it rose with a curiously lame flight and loud whirring of wings, to travel low over the ground for but a short distance, then to alight and either lie still with outspread pinions or run for cover amongst the tumbled stones and clumps of spinifex, amongst which it soon concealed itself. On occasion I watched individuals moving up rather steep rock walls, assisting their progress by a fluttering of expanded wings. The plump body appears, in the field, almost black; the contrasting colour of the primaries (cinnamon-rufous, *not* chestnut) is lost as soon as the wings are folded.

Each of these specimens has the feathers generally worn, but those of the rather long tail are specially so; it is probable that fraying takes place very rapidly in so rupicolous a bird.

They had the irides dark brown; the orbital skin dark plumbeous; the bill black; the feet and toes blackish brown; the claws black.

The plate of this pigeon in Mathews's *Birds of Australia* 1: opposite page 159, 1911, represents it poorly: the coloration is shown too rufescent-brown, especially on the tail; the conspicuous faintly green-glossed ashy markings are missing from the foreparts; the light metallic greenish or violet wash over secondaries and rectrices is not indicated. It must be admitted, however, that the nuances of hue in this peculiar species do not lend themselves well to the artist's brush.

Ocyphaps lophotes whitlocki Mathews

Ocyphaps lophotes whitlocki Mathews, *Novit. Zool.* 18 (3): 191 (31 Jan. 1912). East Murchison Division, Western Australia.

Finke River (Langra Well): 1 ♀, 27 November.

My only specimen of the crested pigeon was collected by D. H. Johnson at a point twelve miles south of Horseshoe Bend.

The white tips on the second outermost tail feathers of this bird have a width of 9 mm., which places it with the race *whitlocki* (see Mayr, *Emu* 51: 142, 1951).

Lophophaps plumifera leucogaster Gould

Lophophaps leucogaster Gould, *Birds of Australia Supplement*, pt. 4: pl. [9] (1867). Machrihanish Station, South Australia.

Finke River (Horseshoe Bend): 1 ♂, 27 November.

This pigeon was not seen at all by me; the specimen was collected by H. L. Hillier for D. H. Johnson.

Geophaps smithii (Jardine and Selby)

Columba smithii Jardine and Selby, *Illust. Orn.*, ser. 1, 2 (7): pl. 104 and text (Dec. 1830). New Holland, *ex* text to pl. 103; type locality restricted to northwest Australia, by Mathews, *Novit. Zool.* 18: 190 (31 Jan. 1912); corrected to Northern Territory, by Mathews, *Aust. Avian Rec.* 1: 29 (2 Apr. 1912).

Oenpelli: 1 ♀, 26 September.

The partridge-pigeon was observed only at Oenpelli, where it was locally not uncommon in open groves of paperbarks, from which tall grass had recently been burned away.

In all its behaviour, this species greatly resembles a partridge. The coveys, of a dozen or more individuals, stayed on the ground until one was virtually in their midst; then, when a bird was about to be trodden upon, it would fly, with an extraordinarily loud beating of wings, to a branch of a near-by tree. These dull-coloured pigeons are almost invisible on the sandy soil, and I was constantly astonished, when I thought the last had risen, to flush another and yet another.

My specimen had the irides white; the orbital skin rose-vermilion; the bill black; the feet and toes plumbeous slate; the claws black.

Mathews has named *G. s. blaauwi* from the Kimberley Division; it is said to differ from the Territorian form by having the orbital region yellow. I cannot find that any collector has noted this in the field, and suspect that the describer was misled by the fact that the red orbital skin rapidly becomes yellow in the dried skin. In the circumstances, I treat the species binomially.

PSITTACIDAE

Trichoglossus haematodus rubritorquis Vigors and Horsfield

[*Trichoglossus*] *rubritorquis* Vigors and Horsfield, *Trans. Linn. Soc. Lond.* 15 (1): 291 (1827). Australia; type locality restricted to north-west Australia, by Mathews, *Novit. Zool.* 18 (3): 259 (1912); further restricted to Derby, Western Australia, by Mathews, *Birds of Australia* 6 (1): 27 (1916).

Darwin (Nightcliff): 1 ♀, 2 April.
Croker Island: 1 ♀, 23 April.
Groote Eylandt (Umbakumba): 2 ♀, 26, 27 May.
Elcho Island: 1 unsexed, 14 September.
Oenpelli: 2 ♀, 4, 4 October.

The red-collared lorikeet was common to abundant at all the expedition's stations except Yirrkalla, and even there it appeared in small numbers during the period of anthesis of a scarlet-flowered mangrove. So long as nectariferous flowers were present, this species was indifferent to the type of forest growth in which it found itself.

My specimens had the irides ruby red; the bill orange-red; the feet and toes plumbeous slate; the claws dark horny brown. Gonadal activity was shown only by the female on 23 April.

This is merely a strongly marked subspecies of the widely distributed *Trichoglossus haematodus* (Linnaeus). '*Trichoglossus rubritorquis melvillensis*' is a variety that existed only in the author's imagination in 1912, and not even there a few years later (see Mathews, *Birds of Australia* 6 (1): 30, 1916).

Trichoglossus versicolor Vigors

[*Trichoglossus*] *versicolor* Vigors, in Lear, *Illustrations of the family of Psittacidae or parrots,* no. 7: pl. 36 (1 Sept. 1831). No locality given; type locality fixed as Cape York, Queensland, by Mathews, *Novit. Zool.* 18: 259 (1912).

Darwin (Nightcliff): 2 ♂, 1 ♀, 6, 6, 6 April.
Elcho Island, 1 ♀, 13 September.
Oenpelli: 1 ♂, 2 ♀, 22, 22, 22 October.

The varied lorikeet was first noted in the vicinity of Darwin at the end of March, flying high and swift in small parties; on 6 April it suddenly became common, and many could then have been shot from the trees. At Yirrkalla a very few were seen, again in small flocks high in the air. At the beginning of September, Johnson found it on Elcho Island, where gums were flowering, in countless myriads. At Oenpelli, it was again rare; the family party collected were the only ones observed.

The movements of this lorikeet seem to be governed wholly by the antheses of paperbarks and the various species of eucalypt, for it is one of the greediest of the nectarivorous birds. A tree occupied by a flock of them presents an animated spectacle, as the individuals cling in all possible positions to the flowering sprays, plentifully bespattering their plumage with honey at the same time. Their calls are piercing shrieks, which can be heard clearly at a great distance.

A male of 6 April had the gonads slightly enlarged. A juvenile female, accompanied by a pair of adults, was taken on 22 October; the top of its head is green, except for a dull red frontal band.

An adult female had the irides brown; the orbital skin white; the bill dull orange; the cere plumbeous blue; the feet and toes plumbeous blue.

So much individual variation appears in this species, that Mathews was well advised to synonymize his own subspecies, described from the Northern Territory and the Kimberley Division.

The genus *Psitteuteles* cannot be separated from *Trichoglossus* by any character whatsoever of more than specific value.

Calyptorhynchus magnificus macrorhynchus Gould

Calyptorhynchus macrorhynchus Gould, *Proc. Zool. Soc. Lond.* 10: 138 (Feb. 1843). Port Essington, Northern Territory.

Oenpelli: 1 ♂, 2 ♀, 10, 19, 31 October.

Black cockatoos were first noted in the neighbourhood of Darwin on 6 April, and pairs and small parties were frequently seen there until I left on the 19th. The species was not observed again until October, during which month the birds appeared in numbers at Oenpelli; indeed, one flock, which regularly haunted the course of an almost dried-up stream, was composed of some hundreds of individuals.

This cockatoo was usually found in the open eucalypt forest, but once at Darwin occurred inside, although near the edge of, monsoonal bush. It was most difficult to collect, since at least one of each aggregation seemed to stand guard and warn the others of human approach; however, when the female of 10 October was wounded, a male left the flock and flew back and forth low above her, offering an easy target, had I had the heart to shoot him.

The male of 31 October, although full-sized, seems to be sub-adult; new glossy feathers are everywhere appearing among the older unglossed ones.

The material available to me does not really confirm the validity of *macrohynchus,* a race I accept only tentatively. In using the specific name *magnificus,* I have followed Peters (*Check-list of birds of the world,* 3: 172, 1937), who, in turn, has been governed by Mathews (*Birds of Australia,* 12 (10): 429, 1927).

Kakatoe galerita fitzroyi (Mathews)

Cacatoes galerita fitzroyi Mathews, *Novit. Zool.* 18 (3): 264 (31 Jan. 1912). Fitzroy River, Kimberley Division, Western Australia.

Darwin (Nightcliff): 1 ♂, 13 April.
Arnhem Peninsula (Yirrkalla): 1 ♀, 11 August.
Oenpelli: 1 ♂, 23 October.

The large cockatoo was fairly common at Oenpelli, but occurred only in small numbers at other localities. I found the species in monsoonal forest, in paperbark swamps, and, less often, in the open eucalypt forest.

A pair were breeding near Yirrkalla in July in a hollow, perhaps 40 feet above ground, of a gigantic paperbark, one of an ancient grove. It was interesting to observe how these conspicuous birds, usually so vehemently noisy, came and went silently about the tree in question, only scolding the intruder when at a safe distance from the nest.

My specimens agree perfectly with a topotype of *K. g. fitzroyi*. From southern examples (*K. g. galerita*) they are easily separable by lesser size and by the complete absence of yellow coloration from the ear coverts.

Kakatoe sanguinea sanguinea (Gould)

Cacatua sanguinea Gould, *Proc. Zool. Soc. Lond.* 10: 138 (Feb. 1843). North coast of Australia = Port Essington, Northern Territory.

Groote Eylandt (Umbakumba): 1 ♂, 8 June.
Oenpelli: 1 ♀, 21 September.

The corella was found in but small numbers at Umbakumba, but at Oenpelli from September to November occurred in flocks of many thousands. Indeed, the species roosted in a paperbark grove at the edge of the Oenpelli Lagoon in such numbers as to whiten the trees, and one of the avine spectacles at this camp was the sight and sound of the great bands flying in formation at morning and evening to and from their feeding grounds on the marshy plains.

Each of my specimens, presumably adult, has the naked orbital region as in the lower figure in Mathews's plate (*Birds of Australia*, 6: pl. opposite p. 198, 1917).

Kakatoe roseicapilla (Vieillot)

Cacatua roseicapilla Vieillot, *Nouveau dictionnaire d'histoire naturelle*, nouv. éd., 17: 12 (Dec. 1817). Dans les Indes = New South Wales.

Although apparently wholly absent from the Darwin area in March and April, the galah was found to be common there at the middle of September; this is evidently another bird of the interior that reaches the coasts during the dry season. It was otherwise seen in the Territory only at Tennant Creek, where, in mid-November, it occurred in uncounted thousands.

Aprosmictus erythropterus coccineopterus (Gould)

Ptistes coccineopterus Gould, *Handbook to the birds of Australia* 2: 39 (1865). Port Essington, Northern Territory.

Darwin (Nightcliff): 2 ♀, 21, 21 March; (West Point): 1 ♂, 31 March.
Groote Eylandt (Umbakumba): 4 ♂, 29 May, 1, 13, 13 June.
Cobourg Peninsula (Cape Don): 1 ♀ ?, 22 September.
Oenpelli: 1 ♂, 2 October.

The red-winged parrot was common in the open eucalypt forest and at cultivated clearings about Darwin, and at Umbakumba, where it was found also in the wattle scrub; on the Arnhem Peninsula, however, where a very few occurred between Yirrkalla and Melville Bay, and at Oenpelli, where some small flocks were noted only near the foot of the Arnhem Land Escarpment, I considered it rare and local.

This species has a slow, heavily flapping, yet graceful, flight, reminiscent of that of a gull of equal size, and can be identified by this alone as far as it can be seen.

A female of 21 March had the irides brown; the bill orange, with the mandible paler than the maxilla; the feet and toes dark grey. Males had the coloration of the soft parts the same, except that the irides were garnet in adults, in younger birds orange.

Examples from Derby, W.A. (*'parryensis'* of Mathews) are indistinguishable from Territorian specimens.

Platycercus venustus (Kuhl)

Psittacus venustus Kuhl, *Nova Acta Phys. Acad. Leop. Carol.* 10: 52 (1820). Nova Hollandia = Arnhem Bay, Northern Territory.

Darwin (Nightcliff): 1 ♀, 9 April.
Arnhem Peninsula (between Yirrkalla and Melville Bay): 1 ♂, 1 ♀, 15, 15 July; (Yirrkalla): 1 ♂, 2 ♀, 23 July, 31, 31 August.
Cobourg Peninsula (Black Rock Point): 1 ♂, 30 September.

A small band of rosellas was regularly seen near Nightcliff, another was noted on Milingimbi Island, and the species was rather common near Yirrkalla; at all these places it inhabited the open eucalypt forest, feeding upon the seeds of grasses on the ground and flying, when molested, into trees, where it would sit quietly until the disturbance had passed.

My four adult females differ in a number of characters from the three adult males: they have the basic colour of the crown and mantle less blackish, more brownish; a small dark-red spot at each side of the front, the two sometimes connected by a very narrow, obsolescent reddish frontal band; a narrow, indistinct nuchal collar of yellow and/or orange-red, clearly separating the crown from the mantle; the concealed dark bases and the terminal fringes of the yellow feathers of the under parts more grey, less blackish; an amorphous red spot at the centre of the lower abdomen; the under tail coverts a less vivid red.

In addition to sexual dichromatism, there is in either sex a fair amount of merely individual variation. Thus, a male from the Cobourg Peninsula and a female from Yirrkalla have blue invading the white of the cheeks (as in the putative race *hilli* described from north-western Australia); a male from Yirrkalla has the ventral feathers suffused with greenish blue; one male and three females have some of the feathers of the upper breast suffused with red; one male has several greyish-buff feathers on the crown; one female has many of the sincipital feathers tipped with buffy orange and/or buffy green.

Further to complicate the picture, there are colour changes induced by wear; blues in the plumage become more greenish or greyish, and the yellows fade to buff or buffy white.

Since even topotypical material shows so much variability, it may be doubted whether any of the subspecies set up by Mathews is valid.

Mayr and Serventy (*Emu* 44: 36, 1944) consider *Platycercus venustus* a semi-species with *P. adscitus*. This concept cannot be indicated by conventional nomenclature.

CUCULIDAE

Cuculus saturatus Blyth

C[*uculus*]. *saturatus* 'Hodgson' Blyth, *J. Asiat. Soc. Beng.* 12 (2): 942 (Nov. 1843). No locality given = Nepal.

One was seen near Darwin (West Point) on 30 March, and two or three others were noted near Nightcliff during the first week of April.

Cuculus pallidus (Latham)

C[olumba]. pallida Latham, *Supplementum indicis ornithologici*: 60 (1801). New Holland = New South Wales.

Oenpelli: 1 ♂, 23 October.

The pallid cuckoo was first observed (and collected) on the day following the night in which the koel was first heard; a second specimen was noted one day later, when the koel was first collected. Whence these two species of cuckoo had so suddenly appeared must remain a mystery.

My example, with inactive gonads, is in a very browned adult plumage, but has acquired fresh grey secondaries in its right wing. It had the irides brown; the eye ring light greenish yellow; the maxilla black; the mandible horny yellow, tipped black; the interior of the mouth rich orange; the feet and toes horny brown; the claws black.

Northern birds seem to differ in no important manner from those of Tasmania and New South Wales.

Cacomantis variolosus dumetorum (Gould)

Cuculus dumetorum Gould, *Proc. Zool. Soc. Lond.* 13: 19 (Apr. 1845). Port Essington, Northern Territory.

Darwin (Nightcliff): 1 ♀, 13 April.

Groote Eylandt (Umbakumba): 1 ♂, 1 ♀, 3, 13 June.

Arnhem Peninsula (Yirrkalla): 5 ♀, 14, 19, 23, 26 July, 9 August; (between Yirrkalla and Melville Bay): 2 ♂, 1 ♀, 29, 29 July, 17 August.

Oenpelli: 1 ♀, 9 October.

During my first stay at Darwin (19 March-18 April), but a single brush cuckoo was seen (13 April), and on Groote Eylandt only two individuals were noted (3 and 13 June); on the Arnhem Peninsula, however, it was a common species, recorded almost every day between 7 July and 31 August. At Oenpelli it again proved to be rare, and the bird collected was the only one observed.

Not one of my series of twelve (3 ♂, 9 ♀) showed any gonadal activity, so it may be supposed that breeding takes place during the rainy summer months.

The moults and plumages of this species, which probably requires two years to achieve the colour characters of the old adult, are worthy of study, but far more material than lies before me must be used to elucidate the problem. As is often the case in Australia, the picture is clouded by the fact that representatives of so many distinct age groups may be collected at the same time and place, a result of the indefinitude of breeding seasons in much of the continent.

The distinctions between *Cacomantis v. dumetorum* and *C. v. variolosus* are given by Hartert (*Novit. Zool.* 32: 169, 1925). Mayr (*Bull. Amer. Mus. Nat. Hist.* 83: 149-50, 1944) has shown that the name *tymbonomus*, sometimes used in place of *dumetorum*, must be restricted to the population of Timor.

Cacomantis pyrrophanus prionurus (Lichtenstein)

C[uculus]. prionurus 'Ill.' Lichtenstein, *Verz. Doubl. Zool. Mus. Berl.*: 9 (1823). New South Wales.

Groote Eylandt (Umbakumba): 1 ♀, 24 May.

The only example of this cuckoo seen was collected in a small paperbark swamp. The species has apparently not previously been reported from the Northern Territory, to which it is doubtless merely a straggler on migration.

My specimen, in adult plumage except for some juvenal secondaries, had the irides brownish grey; the eye ring yellow; the bill black, yellowish at the base of the mandible; the interior of the mouth orange; the feet and toes horny yellow; the claws black.

Chrysococcyx basalis (Horsfield)

Cuculus basalis Horsfield, *Trans. Linn. Soc. Lond.* 13 (1): 179 (May 1821). Java.

Groote Eylandt (Umbakumba): 1 ♂, 6 July.

Oenpelli: 1 ♂, 22 October.

The two listed above are the only examples of this bronze cuckoo seen during the expedition's stay in the Territory. The bird of 6 July was found in wattle scrub; the other was in a stand of tall sedges at a wet area on the great Oenpelli plain, acting so much like a reed warbler that I was surprised to learn that I had shot a cuckoo.

The strange site in which the latter bird was taken, and the date (*cf.* the first appearances of *Cuculus pallidus* and *Eudynamys scolopacea*) suggest that it may have been a newly arrived migrant. Through some mishap it had suffered the loss of all but three of the rectrices, and the replacement feathers are appearing, but the specimen is not truly in moult.

Most recent authors have considered this species monotypic, but it should be noted that Whittell and Serventy (*Systematic list of the birds of Western Australia* 57, 1948), without explanation, call all Australian populations, '*Chalcites basalis modestus* (Diggles)'.

Chrysococcyx malayanus minutillus Gould

Chrysococcyx minutillus Gould, *Proc. Zool. Soc. Lond.* 27: 128 (May 1859). Port Essington, Northern Territory.

Darwin (Ngungiol): 1 ♂, 27 March; (West Point): 1 ♂, 31 March; (Nightcliff): 1 ♂, 13 April.

Milingimbi Island: 1 ♂, 10 May.

Groote Eylandt (Umbakumba): 1 ♀, 6 June.

Arnhem Peninsula (Yirrkalla): 2 ♀, 18, 22 July.

The little bronze cuckoo was frequently seen moving quietly and alone through the forest, at all our stations except Oenpelli. It occurred in the wattle scrub, the stringy-bark forest, and the monsoonal bush.

Although a soft, piping song was heard from certain individuals at Darwin (March) and Yirrkalla (July, August), no specimen proved to be in breeding condition. The one of 6 June is an immature, with the under parts wholly unbarred.

I have noted that the males had a fleshy coral-red eye ring.

Much remains to be learned about this bird, as I have indicated in an earlier paper (*Emu* 49: 167, Jan. 1950).

Eudynamys scolopacea cyanocephala (Latham)

C[*uculus*]. *cyanocephalus* Latham, *Supplementum indicis ornithologici*: 30 (1801). New Holland = New South Wales.

Oenpelli: 1 ♂, 24 October.

In the night of 22-3 October, the unmistakable cries of a koel were first heard; on the 24th, at a distance of some miles from camp, a second individual was seen and collected. From that time until we left Oenpelli, the calls were heard nightly, less often in daylight, and at least two other males were observed, although no female ever came to my attention.

My specimen is an adult male, with the gonads inactive. It had the irides blood red; the bill pale horny green; the feet and toes plumbeous; the claws slate.

Mayr (*Amer. Mus. Novit.* No. 939: 2-3, 1937) states that birds from northern Australia (*'subcyanocephala'*) are smaller than those from the south, but even the limited material before me indicates that there is so much individual variation in size as to make this asseveration doubtful. The length of the flattened wing of the Oenpelli skin measures 204 mm.; two adult males from a point on the railway 100 miles south of Darwin have this measurement 207 and 219 mm.; a topotype of *'subcyanocephala'* has it 214 mm., *fide* Mathews (*Birds of Australia* 7: 371, 1918). On the other hand, an adult male from Grafton, N.S.W., has a wing length of 218 mm., and is therefore smaller than one of the Territorian specimens and very little larger than the topotype of *'subcyanocephala'*. In the circumstances, I must agree with Whittell and Serventy (*Systematic list of the birds of Western Australia*: 58, 1948) that *'subcyanocephala'* should be synonymized with *cyanocephala*.

Centropus phasianinus macrourus Gould

Centropus macrourus Gould, *Birds of Australia*, pt. 29 [vol. 4]: pl. 92 (1 Dec. 1847). Port Essington, Northern Territory.

Darwin (Nightcliff): 1 ♀, 11 April.
Groote Eylandt (Umbakumba): 1 ♂, 1 ♀, 14, 23 June.
Arnhem Peninsula (Yirrkalla): 1 ♂, 25 July.
Oenpelli: 1 unsexed, 27 September.

The swamp-pheasant was common at all our localities, occurring most frequently in swampy areas overgrown with tall grass and pandanus, but at Oenpelli seen also clambering about the branches of the larger fig trees.

None of my specimens is in breeding condition, which probably comes on in most individuals during the rainy season; this surmise is supported by the fact that my bird of 11 April is a juvenile. Likewise, none really wears the so-called 'black plumage', although moult into this dress is farthest advanced in certain Territorian skins taken in September and October, near the end of the dry weather. I strongly suspect that the pheasant-coucal is not truly dimorphic, as was supposed by Mathews, but that the 'black plumage' is the wet-season breeding dress, the 'brown plumage' the dry-season eclipse (*cf.* the well-known plumage cycle of the Asiatic *Centropus bengalensis*). Notes on skins in the Australian Museum, kindly sent me by Mr Keith A. Hindwood of Sydney, and field observations made by him, while not conclusive, yet lend authority to this view.

Territorian birds seem to be separable from topotypical *phasianinus* (New South Wales) by greater size. Since Gould's folio *Birds of Australia* is not readily available to me in Washington, I am unable personally to decide whether the northern race should be called *melanurus* (Derby, W.A.), the name used by Mathews (*Systema avium Australasianarum* 1: 421, 1927), or *macrourus*, the one used by Peters (*Check-list of birds of the world* 4: 69, 1940). In the circumstances I favour *macrourus*, which, having been based upon the population of Port Essington, is more nearly certainly applicable to my specimens.

TYTONIDAE

Tyto sp.

The screech of a barn owl, indistinguishable from the sound I have elsewhere heard uttered by forms of *Tyto alba*, was frequently audible among the paperbarks below our camp at Umbakumba, but the author of the cry was never seen. An unidentified barn owl was frightened by D. H. Johnson late in October from a crevice in a cliff on one of the tors near Oenpelli.

STRIGIDAE

Ninox novaeseelandiae macgillivrayi Mathews

Ninox boobook macgillivrayi Mathews, *Aust. Avian Rec.* 1 (8): 194 (20 Mar. 1913). Patison Creek, Cape York, Queensland.

Darwin (Delissaville): 1 ♂, 27 March.

Groote Eylandt (Umbakumba): 1 ♂, 2 ♀, 2, 6, 15 June.

Arnhem Peninsula (between Yirrkalla and Melville Bay): 1 ♀, 27 August.

The boobook owl was heard nightly at Umbakumba, but was rather uncommon at other localities. My specimens were taken in monsoonal bush, in wattle scrub, paperbark swamps, and the open eucalypt forest.

Allowing for the individual variation usual in any owl population, this series is fairly uniform, and each bird can be matched by skins in the Mathews Collection from other Territorian stations. The female from the Arnhem Peninsula is the most aberrant in its somewhat larger dimensions, its darker and less rufescent plumage, and in having had the irides brown instead of golden. Since it carried an oviduct egg, it is clear that the brown eye colour cannot be explained by the specimen's youth.

On the basis of comparative material available, these skins agree best with *macgillivrayi,* and I have here followed Mayr in the assumption that that race is separable from *ocellata* (Raffles Bay), although I am doubtful whether this will actually prove to be the case. For a full discussion of the difficulties attendant upon taxonomic study of the Territorian boobook owl, see Mayr, *Emu* 43: 12-17, 1943.

Ninox connivens occidentalis Ramsay

Ninox connivens occidentalis Ramsay, *Proc. Linn. Soc. N.S.W.*, ser. 2, 1 (4): 1086 (22 Feb. 1887). Derby, Kimberley Division, Western Australia.

Oenpelli: 4 ♂, 1 ♀, 24, 27 September, 2, 6, 19 October.

A pair of these owls inhabited a hollow spout in a large dead tree at the edge of monsoonal forest near our camp at Yirrkalla and staged a curious performance each evening at twilight, when, bobbing and bowing side by side on a branch, they spent several minutes 'talking', one answering the other in a voice distinctly more highly pitched. Since the species was unknown to me, these strigine neighbours were christened 'dog-barking owls' by members of our party, quite unaware that almost every observer beginning with Gilbert has heard in their notes the cries of dogs. Later, the species was found to be fairly common in the vicinity of Oenpelli, where it was closely associated with certain very large examples of a *Ficus,* the dense foliage of which gave it ample concealment.

Four of my birds had the stomach empty, but the fifth had been feeding upon insects. All had the irides golden yellow; the cere and basal half of the culmen olive-yellow, the rest of the bill black; the toes yellow; the claws black.

These specimens are worn and faded, but most of them show new feathers of a deeper hue, appearing among the old. Mathews's *N. c. suboccidentalis* from Port Keats, N.T., is said to differ from *occidentalis* by its darker coloration; it is probable that the two series (or individuals) seen were not in really comparable plumage.

PODARGIDAE

Podargus strigoides phalaenoides Gould

Podargus phalænoïdes Gould, *Proc. Zool. Soc. Lond.* 7: 142 (Mar. 1840). North-west coast of Australia = Port Essington, Northern Territory.

Darwin (Nightcliff): 2 ♂, 22, 22 March; (West Point): 1 unsexed, 29 March.
Arnhem Peninsula (Yirrkalla): 1 ♂, 26 July.
Oenpelli: 2 ♂, 30 September, 9 October.

The tawny frogmouth, referred to by Territorians as 'little owl', was a common species, but was naturally more often heard than seen. All found by me in daylight were resting in the open eucalypt forest, usually high in the trees. The two examples taken near Nightcliff had been feeding on a highway and were flushed by the approach of a motor-car; each took refuge on a telephone wire. Their stomachs contained, in one case, centipedes, in the other, beetles and a grasshopper.

A bird of 22 March had the outermost primary still partly ensheathed. The one of 9 October had the gonads enlarged. All of my specimens had the irides golden yellow; the inside of the mouth chartreuse green; the feet and toes light brown.

For taxonomic discussion of Territorian frogmouths, see Deignan, *Emu* 51 (1): 71-3 (31 July 1951).

Podargus strigoides lilae Deignan

Podargus strigoides lilae Deignan, *Emu* 51 (1): 72 (31 July 1951). Ambukwamba = Umbakumba, Groote Eylandt, Northern Territory.

Groote Eylandt (Umbakumba): 4 ♂, 3 ♀, 17, 22 May, 1, 6, 6, 18, 19 June.

This small, silvery insular population of the frogmouth seemed to have identical habits with its representative of the mainland, resting in the open forest and entering clearings or the wattle scrub after nightfall. The two specimens of 6 June were found sleeping on a huge branch that had been carried to the ground by a high wind.

AEGOTHELIDAE

Aegotheles cristatus leucogaster Gould

Ægotheles leucogaster Gould, *Proc. Zool. Soc. Lond.* 12: 106 (Oct. 1844). Port Essington, Northern Territory.

Arnhem Peninsula (Purlokoimi Creek, south of Melville Bay): 1 ♀, 20 August; (Birluoja Creek, *ca.* 9 miles SW. of Yirrkalla): 1 ♀, 21 August.

The owlet-nightjar was never observed by me in the Territory; these two examples were collected by D. H. Johnson, who frightened them from holes in stringy-bark gums while hammering for possums. It is probably common enough throughout the areas visited.

Both specimens, with inactive gonads, are in the grey-brown phase of plumage.

CAPRIMULGIDAE

Caprimulgus macrurus yorki Mathews

Caprimulgus macrurus yorki Mathews, *Novit. Zool.* 18 (3): 291 (31 Jan. 1912). Cape York, Queensland.

Darwin (Nightcliff): 1 ♂, 1 ♀, 20 March, 14 April.
Arnhem Peninsula (Yirrkalla): 2 ♂, 16, 21 August.

This nightjar was recorded only from the two localities listed above, where it occurred in small numbers in, or at the edge of, stands of monsoonal forest. Its monotonous call, *tock-tock-tock*, often repeated, was heard nightly in March and April at the Nightcliff tourist hostel, and individuals were sometimes seen at dusk in its clearing.

The birds of 16 and 21 August had the gonads greatly enlarged.

My specimens are inseparable from topotypical *yorki*.

Eurostopodus guttatus gilberti Deignan

Eurostopodus guttatus gilberti Deignan, *Emu* 50 (1): 21 (31 July 1950). Ambukwamba = Umbakumba, Groote Eylandt, Northern Territory.

Groote Eylandt (Umbakumba): 2 ♂, 2 ♀, 4, 6, 18, 24 June.
Arnhem Peninsula (Yirrkalla): 1 ♂, 5 August.

The spotted nightjar was rather common at Umbakumba, spending the sunny hours among the spinifex and wattles on the white-sand plain, and at dusk appearing also along the beaches and over the dunes. Its extraordinary call, well rendered by Keartland as *caw, caw, caw, gobble, gobble, gobble,* was occasionally heard at night.

At the time of day my birds were necessarily collected, their stomachs were usually empty, but the specimen of 5 August had been feeding on Orthoptera.

A skin in the American Museum of Natural History shows that this recently described subspecies inhabits the northern coasts as far west as Darwin.

I am not at all satisfied with the specific nomenclature adopted above, but until the type of *guttatus* in London shall have been critically compared with certainly identified juveniles of the two Australian nightjars, it seems to me best to follow the usage of Peters (*Check-list of birds of the world* 4: 90, 1940). I must also wonder whether *Eurostopodus* should not be considered a mere subgenus of *Caprimulgus*, in view of the fact that the strong rictal bristles, characteristic of species of the latter group, are absent in their immature stages.

APODIDAE

Apus pacificus (Latham)

H[irundo]. pacifica Latham, *Supplementum indicis ornithologici*: lviii (1801). New Holland = New South Wales.

Small flocks of this species were seen flying high over Nightcliff during the latter half of March.

ALCEDINIDAE

Alcedo azurea ruficollaris (Bankier)

Alcyone ruficollaris Bankier, *Ann. Mag. Nat. Hist.* 6: 304 (Jan. 1841). Port Essington, Northern Territory.

Darwin (Nightcliff): 1 unsexed, 9 April.
Groote Eylandt (Umbakumba): 1 ♀, 10 June.
Arnhem Peninsula (Yirrkalla): 1 ♀, 14 August.
Oenpelli: 1 ♂, 26 September.

All azure kingfishers at our coastal camps were observed in the mangrove belt, but the Oenpelli bird was found at a shaded pool in the bed of a dried-up stream, where small fishes were concentrated. The species seems to occur much more rarely on fresh water in the north than in the south.

The specimens of 9 April and 14 August are younger than the others, differing by having the crown obsolescently barred transversely with black; the wings and mantle black, but lightly washed with blue; the under parts of a lighter colour, more ferruginous, less chestnut; the blue along the flanks much less extensive.

Of the two adults, the male is separable from the female mainly by his longer and more robust bill.

Alcyone azurea alisteri, named by Mathews from north-western Australia, seems to be based upon the male of *Alcedo a. ruficollaris* compared with females of the same race.

Alcyone ruficollaris Bankier antedates *Alcyone pulchra* Gould, 1846, by five years; Mathews erred in making Bankier's name a synonym of *Halcyon sancta* Vigors and Horsfield (see Hartert, *Novit. Zool.* 36: 99, 1913).

Alcedo pusilla (Temminck)

Ceyx pusilla Temminck, in Temminck and Laugier, *Nouveau recueil de planches coloriées d'oiseaux,* livr. 100, 4: pl. 595, fig. 3, and text (1836). Baie de Lobo = Triton Bay, south-western New Guinea.

A solitary bird was seen on 12 April at an artificial clearing in the mangrove near Nightcliff.

Dacelo leachii cervina Gould

Dacelo cervina Gould, *Birds of Australia and adjacent islands* 2: pl. [2] and text (Feb. 1838). Port Essington, Northern Territory.

Darwin (Nightcliff): 1 ♀, 12 April.
Groote Eylandt (Umbakumba): 2 ♂, 2 ♀, 5, 13, 18, 23 June.
Arnhem Peninsula (between Yirrkalla and Melville Bay): 1 ♀, 3 August.
Oenpelli: 1 ♂, 24 September.
Cobourg Peninsula (Black Rock Point): 1 ♂, 25 September.

The kookaburra was a common bird of the open eucalypt forest and inhabited clearings at all our camps, but especially so at Umbakumba, where its morning and evening choruses (the former beginning before sunrise, the latter continuing into the darkness) were extraordinarily clamourous.

The female of 23 June had the gonads slightly enlarged. I noted that the irides were buffy white in either sex.

A fair amount of individual and seasonal variation may be seen in my series, but all agree well enough with the bird of the Cobourg Peninsula, which is topotypical. The colour of the upper parts is darkest in the most newly moulted specimens; younger examples seem to have the under parts a deeper buff, more strongly barred.

The differences between the sexes are, to my eyes, those seen by Bowdler Sharpe, as quoted by Mathews (*Birds of Australia* 7: 137-8, 1918).; those pointed out so confidently by Mathews (ibid.) are, I fear, either imaginary in origin or a result of careful selection of specimens to prove a point.

Halcyon pyrrhopygia Gould

Halcyon pyrrhopygia Gould, *Birds of Australia,* pt. 1 [vol. 2]: pl. 22 (1840). New South Wales.

Groote Eylandt (Umbakumba): 1 ♂, 1 ♀, 21 May, 27 June.
Arnhem Peninsula (near shore of Melville Bay): 1 ♂, 31 July.
Oenpelli: 1 ♀, 5 October.

This kingfisher is another bird of the arid interior that, during the Territorian dry season, moves in small numbers towards the coasts. The specimens from Umbakumba were taken within sound of the sea.

The plumage of *Halcyon pyrrhopygia*, a species of unshaded plains, rapidly bleaches in the intense sunlight. Mathews (*Birds of Australia*, 7 (2): 178, 1918) has stated: 'This dull-coloured species shows little variation save in depth of the dullness [*sic*!], and, as this varies seasonally and long series had never been brought together, no value was given to the differences observed until I compiled my *Reference List* in 1912. . .'. My own experience has been that, the longer the series viewed, the more obviously individual is the variation shown. It may probably safely be said that Mathews's predecessors were perfectly correct in their opinion, and that his compilation of the *Reference List* of 1912 did not alter the facts one iota!

Halcyon macleayii distinguenda Mathews

Halcyon macleayii distinguendus Mathews, *Novit. Zool.* 18 (3): 288 (31 Jan. 1912). South Alligator River, Northern Territory.

Darwin (Delissaville): 1 ♀, 28 March.
Groote Eylandt (Umbakumba): 1 ♂, 1 ♀, 28 May, 6 June.
Arnhem Peninsula (Yirrkalla): 1 ♂, 3 ♀, 11, 17 July, 14, 15 August.
Oenpelli: 2 ♀, 28 September, 9 October.

The beautiful forest kingfisher was common in all parts of the Territory visited, occurring in mangroves, paperbark swamps, and in open eucalypt forest.

At Oenpelli the species was breeding in September and October, the eggs being laid in holes excavated in the nests of tree ants; the parent birds drew attention to these holes by their own actions, flying about and calling in alarm, and later the cries of the nestlings were constantly heard from them. A pair were one day collected for me by aborigines at a point some eight miles distant from our camp, the female dead, but the male uninjured; to save the helpless young, I released the living bird, which flew straight as an arrow in the direction whence he had been carried and, I doubt not, soon arrived back at the nest site.

Material before me, never examined by Mathews, fully supports his views (*Birds of Australia* 7 (2): 167-70, 1918) on the validity of the Australian races *macleayii* (southern Queensland and New South Wales), *barnardi* (northern Queensland), and *distinguenda* (Northern Territory). *Halcyon m. barnardi* differs from the nominate form by having the mantle a purer, less greenish, blue. *Halcyon m. distinguenda* has the tone of the mantle as in *barnardi*, but is separable from each of the eastern races by having the bill and wing shorter.

Mayr (*List of New Guinea birds*: 90, 1941) and Peters (*Check-list of birds of the world* 5: 202, 1945), the most recent revisers, give the type locality for *macleayii* as Port Essington, by restriction (von Berlepsch, *Senckenbergiana* 34: 75, 1911). If these authors are followed, the Territorian race must be called *macleayii*, with *distinguenda* as its synonym, but my own view is that Mathews was perfectly justified in correcting von Berlepsch's ill-considered and improbable locality to Moreton Bay, for the reasons he has given (*Birds of Australia* 7 (2): 169, 171, 1918). Restriction of type locality possesses no right of priority unless it has been done on thoroughly reasonable grounds.

Halcyon sancta sancta Vigors and Horsfield

[*Halcyon*] *Sanctus* Vigors and Horsfield, *Trans. Linn. Soc. Lond.* 15 (1): 206, 1827. New Holland = New South Wales.

Darwin (Nightcliff): 1 ♀, 21 March.
Groote Eylandt (Umbakumba): 1 ♀, 26 May.

Arnhem Peninsula (Yirrkalla): 3 ♂, 13 July, 28, 31 August.
Oenpelli: 2 ♂, 3, 14 October.

The sacred kingfisher was common at Darwin and Yirrkalla in or near mangrove swamps. Only one example was seen on Groote Eylandt (among paperbarks), and only one at Oenpelli (at a pool on the Tableland); these were possibly dry-season visitors to the Territory.

In comparing Territorian birds with those of New South Wales, I have found great individual variation, but no characters that might be considered to have racial value.

Halcyon chloris sordida Gould

Halcyon sordidus Gould, *Proc. Zool. Soc. Lond.* 10: 72 (Dec. 1842). North coast of Australia; type locality restricted to Cape York, Queensland, by Mathews, *Novit. Zool.* 18: 289 (1912).

Darwin (Nightcliff): 1 ♂, 1 ♀, 20 March, 11 September.

The mangrove kingfisher was locally common near Darwin at the edge of stands of mangroves along the foreshores; it was otherwise noted in the Territory only at a dense growth of mangroves between Yirrkalla and Melville Bay (22 August).

The specimen of 20 March is a well-grown immature; its stomach contained small hermit crabs. The bird of 11 September, a fully adult male, had the gonads slightly enlarged.

MEROPIDAE

Merops ornatus Latham

M[erops]. ornatus Latham, *Supplementum indicis ornithologici*: 35 (1801). New South Wales.

Darwin (Nightcliff): 1 ♀, 5 April.
Groote Eylandt (Umbakumba): 2 ♂, 27 May.
Arnhem Peninsula (Yirrkalla): 1 ♂, 1 ♀, 12, 12 July.
Oenpelli: 1 ♀, 9 October.
Finke River (Horseshoe Bend): 1 ♀, 24 November.

The bee-eater was common at all our camps, along roadsides, on the beaches, and at clearings in the open eucalypt forest.

The specimen of 5 April is an immature in postjuvenal moult and was therefore hatched during the preceding summer months. Since this would be the season when torrential downpours would prevent a subterranean nester from rearing its young in coastal areas, it may be assumed that the bee-eater is another of the several species that move from the more arid interior to the sea during the dry weather and withdraw at the breaking of the rains. At no time during our stay, however, was evidence of migration noted.

The rainbow bird, which spends so great a proportion of its time in intense sunlight, is unusually liable to decolorization, and *M. o. shortridgei,* described by Mathews from the Strelly River, Western Australia, is doubtless based upon one or more faded specimens. Of the series before me, only an example of 27 May is in reasonably good plumage, while the worn bird of 24 November might be mistaken by a careless worker for a member of a pale desert race.

CORACIIDAE

Eurystomus orientalis pacificus (Latham)

C[oracias]. pacificus Latham, *Supplementum indicis ornithologici*: 27 (1801). Port Jackson, New South Wales.

Darwin (Delissaville): 1 ♂, 28 March.

Oenpelli: 1 ♂, 22 September.

The roller was by no means common at the expedition's several camps; indeed, a few in the Darwin district and three birds at Oenpelli were the only ones observed.

The Oenpelli specimen was one of a pair that appeared regularly late in the afternoon to perch high in a dead paperbark at the edge of the Lagoon, whence they made flights after insects. The remaining individual was not molested and was soon joined by another; until the end of our stay in November they utilized the same tree for their evening meal. They were noisy birds and often swooped and soared about in a manner that suggested play rather than feeding activity.

The example of 22 September had the gonads unenlarged, but it was probably killed just at the beginning of the season for sexual activity, since McLennan found a nest with young along the King River in December. The Darwin specimen is immature, with the bill largely black.

Eurystomus o. bravi, named by Mathews from north-western Australia, is inseparable from *pacificus* (see Ripley in *Proc. Biol. Soc. Wash.* 55: 176, 1942).

PITTIDAE

Pitta iris Gould

Pitta iris Gould, *Birds of Australia,* pt. 6 [vol. 4]: pl. 3 (1 Mar. 1842). Port Essington, Northern Territory.

Darwin (Nightcliff): 1 ♀, 2 April.

The only pitta seen was flushed from the edge of a trail through dense monsoonal forest; it alit for a moment upon a low branch, thus enabling me to collect it. Upon the expedition's return to Nightcliff early in September, I heard frequently from the scrub a loud, clear double whistle, quite like the song of the Asiatic *Pitta brachyura moluccensis,* which I attributed to the rainbow pitta, although I was never able to see the singer.

My specimen is, I believe, sub-adult; moult appears in the primaries and a patch of feathers at the right side of the breast.

ALAUDIDAE

Mirafra javanica söderbergi Mathews

Mirafra javanica söderbergi Mathews, *Aust. Avian Rec.* 4 (6): 137 (1 Aug. 1921). New name for *Mirafra javanica nigrescens* Mathews, *Novit. Zool.* 18 (3): 426 (31 Jan. 1912). Eureka, Northern Territory, not *Mirafra nigrescens* Reichenow, 1900.

Oenpelli: 2 ♀, 23 September, 9 October.

This species was seen only in the near vicinity of the Oenpelli mission station at a small expanse of laterite gravel and short grass; it was by no means common even there. As in the case of the Territorian pipit, it may be surmised that the annual burning of the grasslands has reduced the numbers of the bush-lark.

HIRUNDINIDAE

Hirundo ariel (Gould)

Collocalia ariel Gould, *Birds of Australia,* pt. 9 [vol. 2]: pl. [15] (1 Dec. 1842). New South Wales.

Darwin (Nightcliff): 1 ♂, 1 ♀, 3, 3 April.

The only two fairy martins observed in the Territory appeared from over the sea at Nightcliff late in the afternoon of 3 April, struggling to reach the land against a violent blow that accompanied an approaching tropical storm. There is no reason to suppose, however, that they were migrating from over the water; they were very immature (skulls still unossified), and it is probable that, flying along the shore, they had been swept away by the gale.

The much paler head and back, supposed by Mathews to distinguish *Petrochelidon ariel conigravi* (north-western Australia) from birds of New South Wales, are simply two of the characters of immaturity.

Hylochelidon nigricans nigricans (Vieillot)

Hirundo nigricans Vieillot, *Nouveau dictionnaire d'histoire naturelle,* nouv. éd., 14: 523 (Sept. 1817). Nouvelle-Hollande = Hobart, Tasmania.

Darwin (Nightcliff): 1 ♀, 3 April.
Groote Eylandt (lower slopes of Central Hill): 1 unsexed, 2 July.
Arnhem Peninsula (Flying Boat Base at Melville Bay): 1 ♀, 15 July.
Oenpelli: 1 ♀, 13 October.

The tree-martin was common or periodically even abundant at every locality visited by the expedition. I found it along the beaches, over short-grass plains, and in the open eucalypt forest.

No specimen was in breeding condition, but the two taken in July are sub-adult; they differ from older birds in having the front almost concolorous with the crown and in having the secondaries tipped with white or buff.

My topotype of Mathews's *rogersi,* from Darwin, differs in no character whatsoever from a southern series of the species. The trinomial name is here used, because a valid race of this swallow inhabits Timor.

CAMPEPHAGIDAE

Lalage sueurii tricolor (Swainson)

Ceblepyris tricolor Swainson, *Zool.* 1 (4): 467 (Jan. 1825). Australia = New South Wales.

Darwin (Nightcliff): 1 ♀, 11 April.
Skirmish Point: 2 ♂, 1 ♀, 4, 4, 4 May.
Groote Eylandt (Umbakumba): 3 ♂, 5 ♀, 22, 26 May, 14, 21, 23, 23 June, 6, 6 July.
Arnhem Peninsula (Yirrkalla): 1 ♀, 18 July.
Oenpelli: 1 ♂, 1 ♀, 21 September, 14 October.

The white-winged triller occurred in flocks wherever the expedition touched; specimens were taken in stands of speargrass, in wattle scrub, mangroves, and open eucalypt forest.

The species was first seen in small numbers at Darwin in the last week of March, but became commoner during April; thereafter, it proved to be numerous everywhere until the beginning of November. My observations are in exact agreement with those of Gilbert, who at Port Essington did not see a single individual from the early part of November to the month of March.

Since nine out of my fifteen examples were so young as to have the skull incompletely ossified, I am inclined to believe that the populations that spend the dry season along the Territorian coasts have come from no great distance. This triller is to a considerable degree a terrestrial feeder, and the overwhelming growth of tall speargrass during the wet weather, which renders the regions of heaviest rainfall very unsuitable for the bird's requirements, is no doubt responsible for its moving back into a somewhat more arid zone for the duration of the breeding season.

Eight of my nine immatures (2 ♂, 6 ♀), taken between 4 May and 18 July inclusive, are in what Mayr (*Emu* 40: 113-15, 1940) calls 'first-year plumage'. The exceptional specimen, a male with unossified skull collected on 4 May, while showing no evidence of moult, has already acquired such characters of the adult as black rectrices (most of them fading to brown at the extreme tip) and *some* black remiges (among both the primaries and the secondaries, and not symmetrically!).

At any season may be seen a very few individuals in what appears to be the nuptial plumage of the male. Such a bird, collected on 14 October (non-moulting), proved to have a worn nuptial-*type* dress, in which the mantle is slightly sullied by the presence of scattered brown or partly brown feathers, and the median upper wing coverts are longitudinally centred with black. It may be surmised that the exceptional immature specimen mentioned in the preceding paragraph would have become one of these adults that have lost the true eclipse plumage.

Lalage leucomela rufiventris (G. R. Gray)

Campephaga rufiventris G. R. Gray, *Genera of birds* 1: 283 (1846). Raffles Bay, Northern Territory.

Darwin (Nightcliff): 1 ♂, 1 ♀, 4 April.
Groote Eylandt (Umbakumba): 1 ♂, 4 ♀, 22, 30, 31 May, 18, 18 June.
Arnhem Peninsula (Yirrkalla): 1 ♀, 10 July.

The varied triller was not seen at Oenpelli, but was common at all other localities visited. Since I took specimens in monsoonal forest, in open eucalypt forest, in wattle scrub, and amongst mangroves, the species cannot be considered 'restricted to tropical forest (scrub)', as averred by Mayr (*Emu* 40: 116, 1940).

This triller was usually found feeding on fruits; about Nightcliff it was most numerous where a species of *Celtis* was prolifically bearing, and at Yirrkalla, frequented fruiting fig trees.

A female in advanced postnatal moult was collected at Umbakumba on 18 June; its under parts show short blackish streaks, while the feathers of crown and mantle are narrowly tipped with pale brownish to give a barred appearance. An otherwise apparently adult female of 4 April retains some light-tipped feathers on the occiput.

Mayr (loc. cit., p. 116) states that the species 'is absent both from the Gulf of Carpentaria and from north-western Australia', and (p. 117) that Mathews's record of this race from north-western Australia 'is a slip'. In fact, Groote Eylandt, where the bird is common, lies in the Gulf of Carpentaria, and, in the National Museum at Melbourne, I have examined a pair collected by G. F. Hill at Parry Harbour, 12 and 15 April 1910 (*ex* Coll. H. L. White). The extensions of range do not, however, alter Mayr's broad conclusions, since both Groote Eylandt and Parry Harbour have an annual rainfall of at least forty inches.

Coracina tenuirostris obscura Mathews

Coracina tenuirostris obscura Mathews, *Novit. Zool.* 18 (3): 328 (31 Jan. 1912). Cairns, Queensland.

Darwin (Nightcliff): 1 ♂, 12 April.
Cobourg Peninsula (Cape Don): 1 ♂, 23 September.

The cicada-bird seemed to be very rare in the parts of the Territory visited by us; in addition to those listed above, my only record is for a male at Yirrkalla on 14 July. The species was never seen outside of paperbark swamps.

The specimen collected by David H. Johnson at Cape Don is sub-adult.

Coracina t. obscura is separable from the nominate race by its lesser size, but is apparently identical with *C. t. melvillensis* Mathews, named from Melville Island, Northern Territory.

Coracina novae hollandiae didima Mathews

Coracina novæ-hollandiæ didimus [*sic*] Mathews, *Aust. Avian Rec.* 1 (2): 42 (2 Apr. 1912). Melville Island, Northern Territory.

Darwin (Nightcliff): 1 ♀, 9 April.
Groote Eylandt (Umbakumba): 1 ♂, 5 ♀, 29, 29 May, 1, 1, 2, 14 June.
Arnhem Peninsula (Yirrkalla): 1 ♂, 1 ♀, 22 July, 8 August.
Cobourg Peninsula (Cape Don): 1 ♂, 22 September.
Oenpelli: 1 ♂, 13 October.

The large cuckoo-shrike was very common in open eucalypt forest at all localities visited by the expedition; it seems to be non-migratory in the coastal Northern Territory. The stomach of one of my specimens contained seeds and a large mantis, and the species was regularly represented at the feasts of insects produced by man-made bush fires.

My series is composed of 3 adult males, 3 adult females, 1 immature male, and 4 immature females. Within each age group I find no character by which the sexes may be certainly distinguished, although the length of wing averages shorter in females than in males. Only one subspecies can be recognized amongst the 11 birds, so it may be supposed that members of the migratory southern race do not reach Arnhem Land.

Material before me from various parts of Australia casts grave doubts on the separability of '*melanops*' from the Tasmanian nominate race. According to the authorities, the latter is differentiable *only* by a slightly smaller bill; two small-billed birds from 19 miles SW. of Nowra, N.S.W. (5 July and 9 August) might be taken for winter visitors from Tasmania, but two equally small-billed examples from 50 miles S. of Busselton, W.A. (19 June) can only represent the resident form of southern continental Australia ('*melanops*', *fide* Whittell and Serventy, *Systematic list of the birds of Western Australia*: 69, 1948). I suspect that series of breeding continental birds will show that bill size is too individually variable to serve as a subspecific character and that '*melanops*' must fall into the synonymy of *novæ hollandiæ*. Small-billed continental migrants might account for the report of members of a Tasmanian race at Port Moresby (see Mayr, *List of New Guinea birds*: 103, 1941).

Outside of *subpallida* (not examined by me, but recognized by Whittell and Serventy, loc. cit.), I can see but two continental subspecies, distinguishable only by a slight, but obvious, difference in wing length. Six adult males of '*melanops*' have this measurement from 202 to 210 mm., 3 adult females have it from 196 to 202 mm.; on the other hand, 3 adult males of *didima* have the wing length from

187 to 197 mm., while 3 adult females have it from 185 to 190 mm. Specimens of the shorter-winged form lie before me from 32 miles E. of Derby, W.A., and from the Atherton Tableland in Queensland, as well as from the Territory.

The characters erroneously believed by Mathews to be general in northern birds result from his type specimen having been immature. Those attributed by Hartert to *kuehni* (a synonym of *didima*) are equally worthless, since the colour tones of upper and under parts are variable, as is the extent of black on the throat and upper breast.

Coracina papuensis hypoleuca (Gould)

Graucalus hypoleucus Gould, *Proc. Zool. Soc. Lond.* 16: 38 (14 Nov. 1848). Port Essington, Northern Territory.

Darwin (Nightcliff): 1 ♀, 6 April.
Groote Eylandt (Umbakumba): 2 ♂, 1 ♀, 30 May, 9, 16 June.
Arnhem Peninsula (Yirrkalla): 1 ♀, 10 July.
Oenpelli: 1 ♀, 1 October.

This cuckoo-shrike was even commoner than its larger congener at all our localities; it was found both in open eucalypt forest and in stands of *Casuarina* along the beaches.

Females are distinguishable from males, even in the field, by having the lores and front a dull blackish slate, rather than an intense black; female skins before me, believed to be adult, have also, in unworn plumage, conspicuous white edgings to the secondaries, which are absent from these feathers in the males.

Names bestowed by Mathews upon birds from north-western Australia and Melville Island are certainly synonyms of *hypoleuca*.

Pteropodocys maxima (Rüppell)

Ceblepyris ('*typus*') *maxima* Rüppell, *Mus. Senckenb. Abh.* 3 (1): 28, pl. 3 (1839). Neuholland = New South Wales.

Three ground cuckoo-shrikes were seen at the boundary of the aerodrome near Alice Springs on 18 March.

MOTACILLIDAE

Anthus novaeseelandiae rogersi Mathews

Anthus australis rogersi Mathews, *Aust. Avian Rec.* 1 (8): 193 (20 Mar. 1913). Melville Island, Northern Territory.

Darwin (Nightcliff): 1 ♀, 11 April.
Groote Eylandt (Umbakumba): 1 ♀, 24 June.

This pipit occurred in small numbers in the vicinity of Darwin, on expanses of laterite that had been cleared and smoothed by the military during the war years. At all other places visited along the northern coasts, it was either apparently absent or a bird of surpassing rarity. Only single examples were noted at Yirrkalla and at Oenpelli, and since each of these places seemed to offer most suitable habitats for ground-dwelling birds, it may be supposed that the wanton burning-over of the countryside, year after year, has adversely affected the status of this and other grassland species.

The specimen of 11 April, in fine fresh plumage, had the ovaries slightly enlarged. It is probable that the breeding season along the Territorian shores falls in the dry season, since the heavy growth of spear grass would seem to preclude successful nesting during the rains.

Anthus novaeseelandiae subaustralis Mathews

Anthus australis subaustralis Mathews, *Novit. Zool.* 18 (3): 425 (31 Jan. 1912). Lake Way, Western Australia.

Finke River (Horseshoe Bend): 1 ♀, 26 November.

This specimen was collected for me by D. H. Johnson.

Although Whittell and Serventy (*A systematic list of the birds of Western Australia*: 99, 1948) combine all Australian populations of the ground-lark under the name *australis*, it is certain that some, at least, of the twelve subspecies accepted by Mathews (*Birds of Australia* 12 (4): 140, 1925) are perfectly valid. Without examination of all the forms there recognized, it is not possible to say what names may be mere synonyms, but material before me indicates that *'australis'*, *'bistriatus'*, *'adelaidensis'*, *'subaustralis'*, and *'rogersi'* are all distinct from each other.

TIMALIIDAE

Pomatostomus temporalis rubeculus (Gould)

Pomatorhinus rubeculus Gould, *Proc. Zool. Soc. Lond.* 7: 144 (Mar. 1840). North-west coast of Australia = Port Essington, Northern Territory.

Darwin (Nightcliff): 1 ♂, 22 March.
Oenpelli: 2 ♀, 26 September, 15 October.

The grey-crowned babbler was found near Darwin in monsoonal bush, as well as in the open eucalypt forest, but at Oenpelli only in the latter type of growth. At both localities, clusters of its large, globular nests were conspicuous in low trees, although no breeding seemed to be taking place at the times of my visits.

The bird of 22 March is in unworn plumage, but those of September and October have the feathers badly frayed.

The nesting habits of the Australian scimitar babblers are so different from those of their Asiatic cousins, that I prefer to keep the genus *Pomatostomus* Cabanis distinct from *Pomatorhinus* Horsfield.

Pomatostomus temporalis browni Deignan

Pomatostomus temporalis browni Deignan, *Emu* 50 (1): 20 (31 July 1950). In the Cape Arnhem Peninsula between Yirrkalla and Melville Bay, Northern Territory.

Arnhem Peninsula (between Yirrkalla and Melville Bay): 1 ♂, 1 ♀, 1 unsexed, 25, 25 July, 17 August.

Flocks of this form of the grey-crowned babbler were frequently seen in the open eucalypt forests of the Arnhem Peninsula.

I have noted that the male of 17 August, with gonads slightly enlarged, had the irides straw yellow; unfortunately, no record was made of the eye colour of other specimens.

Pomatostomus temporalis mountfordae Deignan

Pomatostomus temporalis mountfordae Deignan, *Emu* 50 (1): 19 (31 July 1950). Ambukwamba = Umbakumba, Groote Eylandt, Northern Territory.

Groote Eylandt (Umbakumba): 3 ♂, 1 ♀, 10, 23, 23, 23 June.

In the thin wattle scrub that clothed an extensive plain of white sand near our camp at Umbakumba, the Groote Eylandt race of the grey-crowned babbler was fairly common.

The specimen of 10 June is a well-grown juvenile; it differs from its parents by a shorter bill and by having the plumage, above and below, more richly coloured.

Of the three taken on 23 June, two, a male and a female, had the irides dark brown, but the third had them greyish white.

SYLVIIDAE

Malurus lamberti dulcis Mathews

Malurus dulcis Mathews, *Bull. Brit. Orn. Cl.* 21: 100 (30 May 1908). Ten miles E. of Alligator River = in the granite ranges ten miles east of South Alligator River, Northern Territory, *fide* Hartert, *Novit. Zool.* 12: 223 (1905).

Oenpelli: 3 ♂, 1 ♀, 22, 26 September, 27, 27 October.

This beautiful wren was fairly common in the vicinity of Oenpelli, where, however, it was strictly confined to the Tableland and its isolated outliers.

I found the species always, even when the gonads showed activity, in small, loose bands of about a half-dozen individuals. They were in constant motion, dashing from one clump of spinifex to another or fluttering up the sheer faces of cliffs like so many brilliant butterflies. They are inquisitive little birds and easily observed.

All my specimens are adult, and the three males had the testes enlarged.

The males had the irides dark brown; the bill black; the feet and toes fleshy horn. The female differed in having the bill pinkish brown.

Mack (*Mem. Nat. Mus. Vict.*, no. 8, 1934), in reviewing this genus, has divided the chestnut-shouldered wrens among five species, but since his distributional map (p. 122) shows no two overlapping forms, I prefer to treat all as races of *lamberti*.

Malurus leuconotus Gould

Malurus leuconotus Gould, *Proc. Zool. Soc. Lond.* 1865: 198 (May 1865). Interior of Australia; precise locality unknown = South Australia.

Finke River (Horseshoe Bend): 1 ♂, 27 November.

One fine adult male was collected by H. L. Hillier for D. H. Johnson.

Malurus melanocephalus cruentatus Gould

Malurus cruentatus Gould, *Proc. Zool. Soc. Lond.* 7: 143 (Mar. 1840). North-west coast of Australia = Port Essington, Northern Territory.

Cape Cockburn: 1 ♂, 25 April.

Groote Eylandt (Umbakumba): 3 ♂, 1 ♀, 21, 21, 28, 31 May.

Arnhem Peninsula (Yirrkalla): 1 ♂, 10 July; (between Yirrkalla and Melville Bay): 1 ♂, 25 July.

Oenpelli: 1 ♂, 9 October.

The red-backed wren was common in family parties at all points visited by the expedition, occurring in tall grass, in the shrubs and sedges at the edge of marshland, in open eucalypt forest, and in stands of pandanus. It is a fearless species, and if its blood-red back is usually turned away from an observer, this is perhaps due rather to the bird's inquisitiveness than to any attempt to conceal the brilliant colour.

Specimens of 25 April and 21 May had the gonads greatly enlarged; the one of 25 July had them still enlarged, but had begun the postnuptial moult, and this moult is even further advanced in the earlier-taken example of 10 July. The male of 31 May had the skull incompletely ossified; it wears a worn female-type plumage.

It should be noted that the harsh foliage amongst which this wren lives tends to fray and alter the feathers, so that the colour of the upper parts may become more brown than red. Specimens before me from Derby, which Ramsay named *boweri*, seem nowise to differ from Territorian ones.

It may be mentioned here that, in addition to *M*[*uscicapa*]. *melanocephala* Latham, 1801 (the earliest name given a member of this species), there is [*Muscicapa*] *Melanocephala* Forster (*Zoologica Indica*: 41, 1781), a *nomen nudum* for some bird of Ceylon. So far as I have been able to learn, Forster's name has never been allocated.

Amytornis woodwardi Hartert

Amytornis woodwardi Hartert, *Bull. Brit. Orn. Cl.* 16: 30 (29 Nov. 1905). Ten miles east of South Alligator River, about 85 miles from the coast, Northern Territory.

Oenpelli: 1 ♂, 3 October.

This fascinating bird occurred on the Escarpment and Plateau to the east of Oenpelli and on at least two of the isolated tors in the vicinity. Owing to its secretive manners, it would be difficult to assess its numbers.

The first one seen by me, on the crag named 'Argaluk', seemed to be a black-backed, white-bellied bird, and was mistaken for a willie wagtail as it vanished over the brink of a precipice. Later, on the Plateau, one stood before me for a second at the top of a boulder before disappearing into a mass of tumbled rocks. My third observation was made as I ascended the Escarpment: an excited *pit-pit-pit* led to my discovery of a grass-wren's head peering at me over a large rock, and a quick, scarcely aimed, shot brought down my one specimen. In future I was to see the species several times again, running and dancing, with tail held at a right angle to the body line, among rocks and dense clumps of spinifex, and in the brief moments during which it was in view, I was once more impressed by its apparent resemblance to the black-and-white fantail; so rapidly does it cross an open space that the striations of its plumage can scarcely be noted.

My specimen, an adult with inactive gonads, had the irides dark brown; the bill dark slate; the feet, toes, and claws slate.

I have read, over and over, Mathews's 'reasons' for setting up *Magnamytis alligator, nomen nov.* (*Birds of Australia* 10 (4): 212, 1923), based upon a topotypical female paratype of Hartert's *Amytornis woodwardi*, and am as unable to fathom his intention as before!

Aphelocephala nigricincta (North)

Xerophila nigricincta North, *Ibis*, ser. 7, 1: 340 (July 1895). Missionary Plains, Northern Territory.

A pair was seen near the aerodrome at Alice Springs on 18 March.

Epthianura crocea tunneyi Mathews

Epthianura crocea tunneyi Mathews, *Novit. Zool.* 18 (3): 342 (31 Jan. 1912). Arnhem Land = South Alligator River, Northern Territory.

Oenpelli: 1 ♂, 1 ♀, 23 September, 18 October.

The yellow chat occurred only on the marshy plains about Oenpelli, where it became ever less numerous as the hot sun and the burning of the grass reduced the extent of its preferred habitat.

This pretty bird was seen always at moist depressions, surrounded by, or overgrown with, a rank stand of tall grass and a gnarled leafless shrub. Here it acted much like the fantail-warbler, which inhabited the same places, sometimes showing itself briefly at the top of a bush, but usually creeping about invisibly in the tangled vegetation. When really flushed, it tended to fly to a completely new swale, perhaps a hundred yards away, so that collecting a specimen proved to be far from easy.

The male of 23 September had the gonads enlarged. I have noted that this one had the irides greyish white, while the female had them pale grey.

When compared with *Epthianura c. crocea* (Norman River, Qld.), the Territorian race is separable by its much more golden head, nape, and under parts, but no difference appears in the colour of the lores, as was claimed by Mathews at the original description. I have seen no material of *boweri*, described from the Fitzroy River by the same author, so cannot discuss the validity of that putative form.

Smicrornis brevirostris flavescens Gould

Smicrornis flavescens Gould, *Proc. Zool. Soc. Lond.* 10: 134 (Feb. 1843). Port Essington, Northern Territory.

Arnhem Peninsula (between Yirrkalla and Melville Bay): 1 ♂, 1 ♀, 15, 15 July.
Oenpelli (Tableland): 1 ♂, 1 October.

The weebill occurred locally and in small numbers on the Arnhem Peninsula, but was fairly common on the Plateau near Oenpelli.

I saw this species only in the stringy-bark forest, where pairs or small parties swept from treetop to treetop, so swiftly as to make observation difficult. Its food is gathered mainly from the leaves, and the search is accompanied by an endless twittering. The weebill's song is of a volume and richness that would do credit to a bird ten times its size.

The specimens of 15 July are adults in very worn plumage and with the gonads inactive. The one of 1 October, although fully fledged, had the skull still incompletely ossified. All had the irides white.

Megalurus timoriensis alisteri Mathews

Megalurus alisteri alisteri Mathews, *Novit. Zool.* 18 (3): 345 (31 Jan. 1912). Napier Broome Bay, Kimberley Division, Western Australia.

Oenpelli: 1 ♂, 2 ♀, 5, 6, 27 October.

The tawny grassbird was found only near Oenpelli, at a place on the plain where grasses and sedges grew thick and tall. In such a habitat the bird was almost impossible to see, but could be brought into view for a fleeting moment by squeaking, a sound to which the commoner fantail-warbler responded in much greater numbers.

Since specimens from north-eastern New South Wales seem to be inseparable from Territorian ones, it may be doubted whether this species shows subspecific variation within Australia. Ashby has named *mayi* (Pine Creek, N.T.) as distinct from *melvillensis* (Melville Island), but has not stated how his skins compared with topotypical *alisteri*. It should be noted that herbicolous warblers of this type are taxonomically useful only when they wear reasonably fresh plumage.

Cisticola exilis lineocapilla Gould

Cysticola lineocapilla Gould, *Proc. Zool. Soc. Lond.* 15: 1 (29 Mar. 1847). Port Essington, Northern Territory.

South Goulburn Island: 1 ♂, 26 April.

Oenpelli: 2 ♂, 1 ♀, 22, 22 September, 6 October.

Near the shore of South Goulburn Island, on 26 April, at an extensive clearing among the mangroves where grew a bushy samphire and coarse grasses, I found a family party of the fantail-warbler (two adults with three or four juveniles). Early in September, a few were noted at Nightcliff. During our stay at Oenpelli from late in September to the first week of November, the species was common at marshy areas where the introduced Kupang grass has become established.

The April example is very young, with skull incompletely ossified; it wears the juvenal plumage as described by Lynes (*Ibis*, ser. 12, 6: 190, 1930). The male of 22 September, although likewise with incompletely ossified skull, has a dress quite like that of the remaining specimens, which are 'winter' adults with the gonads wholly inactive.

Gerygone chloronota chloronota Gould

Gerygone chloronotus Gould, *Proc. Zool. Soc. Lond.* 10: 133 (Feb. 1843). Port Essington, Northern Territory.

Darwin (Nightcliff): 1 ♂, 1 ♀, 21, 25 March.

Groote Eylandt (Umbakumba): 2 ♂, 2 ♀, 24, 25, 31 May, 21 June.

Arnhem Peninsula (Yirrkalla): 2 ♀, 1 unsexed, 10, 13, 27 July.

My nine specimens were taken in monsoonal and open eucalypt forest and in paperbark swamps; I have no record of this warbler's frequenting the mangrove belt, although my predecessors in the Territory found it most numerous in that type of growth. In any case, the species seems scarcely to occur beyond sight of the sea, unless along the tidal rivers.

The examples of 25 March and 25 May had the gonads enlarged, and those of 24 May and 13 July were so young as to have the skull still incompletely ossified. I have noted that adults had the irides garnet or red.

A trinomial name is used, because races of the green-backed warbler are found on islands to the north of Australia; I am not prepared to accept the validity of forms proposed by Mathews from Melville Island and Western Australia.

Gerygone magnirostris magnirostris Gould

Gerygone magnirostris Gould, *Proc. Zool. Soc. Lond.* 10: 133 (Feb. 1943). Port Essington, Northern Territory = Greenhill Island, Sir George Hope Islands, Van Diemen Gulf, *ex* Gould, *Handbook to the birds of Australia* 1: 270 (1865).

Darwin (Nightcliff): 1 ♂, 21 March; (Ngungiol): 1 ♂, 27 March.

The large-billed warbler was seen nowhere except in the neighbourhood of Darwin, where, however, it was fairly common. One of my specimens was taken in monsoonal bush, the other in open eucalypt forest at no great distance from a mangrove swamp. Each of them had the gonads enlarged.

Gerygone levigaster levigaster Gould

Gerygone levigaster Gould, *Proc. Zool. Soc. Lond.* 10: 133 (Feb. 1843). Port Essington, Northern Territory = Cobourg Peninsula, *ex* Gould, *Handbook to the birds of Australia* 1: 270 (1865).

Darwin (Nightcliff): 1 ♂, 11 April.

This warbler was found only near Darwin, where I noted it in small numbers in certain stands of mangrove. In life, it is distinctly smaller and, apparently, greyer than the large-billed warbler.

To judge from my few specimens, the name 'buff-breasted warbler' would be more appropriately applied to *Gerygone magnirostris*.

Although, at the first description, Gould gave Port Essington as the type locality, in *Handbook to the birds of Australia* (loc. cit. *supra*) we read: 'Gilbert killed several specimens of this little bird on the Cobourg Peninsula, and on the islands in Van Diemen's Gulf, and sometimes observed a solitary individual among the mangroves near the settlement of Port Essington'. From this we may judge that not a single specimen from Port Essington formed part of Gould's original series, and that, accordingly, 'Port Essington' or 'near Port Essington' cannot be considered the correct type locality.

MUSCICAPIDAE

Rhipidura rufifrons dryas Gould

Rhipidura dryas Gould, *Proc. Zool. Soc. Lond.* 10: 132 (Feb. 1843). Port Essington, Northern Territory.

Darwin (Nightcliff): 1 ♂, 1 unsexed, 10, 10 April.
Groote Eylandt (Umbakumba): 1 ♂, 23 May.
Arnhem Peninsula (Yirrkalla): 1 ♂, 14 July.

The rufous fantail was strictly confined to stands of monsoonal forest and was, accordingly, local and rather uncommon at the expedition's camps. It is a very tame and inquisitive species, spending much of its time near the ground and coming near to investigate one's intrusion into its haunts.

The two birds of 10 April are in worn plumage; one has lost the entire tail and is just acquiring new rectrices. The example of 23 May is far advanced in the moult, with some of the tail feathers still ensheathed. The one of 14 July wears clean, fresh plumage.

Rhipidura rufiventris isura Gould

Rhipidura isura Gould, *Proc. Zool. Soc. Lond.* 8: 174 (July 1841). North-west coast of Australia = Port Essington, Northern Territory.

Darwin (Nightcliff): 1 ♂, 22 March.
Groote Eylandt (Umbakumba): 3 ♂, 1 ♀, 18, 20 May, 5, 13 June.
Arnhem Peninsula (Yirrkalla): 1 ♀, 14 July.
Oenpelli: 2 ♂, 12, 24 October.

The northern fantail proved to be one of the commonest species at all points visited by the expedition. It was found in mangrove swamps, in wattle scrub, in open eucalypt and monsoonal forests.

Breeding specimens were taken at Oenpelli in October; at that season the birds wore very frayed and brownish plumage. It may be surmised that Mathews's *melvillensis* from Melville Island ('brown not blue-grey above') was based upon similarly worn examples.

Rhipidura leucophrys picata Gould

Rhipidura picata Gould, *Introduction to the birds of Australia*, 8vo, ed.: 40 (1 Aug. 1848). Port Essington, Northern Territory.

Skirmish Point: 1 ♀, 6 May.

Arnhem Peninsula (Yirrkalla): 1 ♀, 24 July.
Oenpelli: 1 ♀, 5 October.

The willy wagtail is rare and local along the Territorian coasts and excessively shy, owing to persecution by the aborigines. It inhabited wattle scrub in the country of the Liverpool River tribe, but elsewhere was a bird of the open eucalypt forest.

Each of my specimens possesses at least a few white-tipped feathers on the throat; this is probably a mark of immaturity.

During the first week of November a pair of willy wagtails were found nesting in a low tree on the bank of the East Alligator; the nest was as conspicuous an object as could be imagined, having been constructed wholly of tufts of white fibres drawn from ripened bolls of the cotton plant, which has become sparingly naturalized on the Oenpelli plain.

Seisura inquieta nana Gould

Seisura nana Gould, *Ann. Mag. Nat. Hist.*, ser. 4, 6: 224 (Sept. 1870). Northern Australia.

Groote Eylandt (Umbakumba): 1 ♂, 1 ♀, 1 unsexed, 22, 28 May, 6 July.
Oenpelli: 1 ♂, 25 September.

The scissors-grinder was rare and local at localities visited by the expedition; a few pairs at Umbakumba living at the edge of a swamp overgrown with young paperbarks, and a pair at the margin of the large Oenpelli Lagoon, were the only ones seen.

The specimen of 25 September had the testes greatly enlarged. The unsexed bird of 22 May had the skull incompletely ossified, but outwardly might be taken for an adult female.

An example from Derby (type locality of *S. i. rogersi* Mathews) is nowise smaller than Territorian skins of the same sex.

Monarcha alecto nitida (Gould)

Piezorhynchus nitidus Gould, *Proc. Zool. Soc. Lond.* 8: 171 (July 1841). North-west coast of Australia = Port Essington, Northern Territory.

Groote Eylandt (Umbakumba): 2 ♂, 2 ♀, 23, 25, 26 May, 18 June.
Arnhem Peninsula (Yirrkalla): 1 ♀, 23 July.

The shining flycatcher was found in small numbers only at the two localities named above. It occurred both in mangrove swamps and in monsoonal forest, but in the latter type of growth was never far from the edge of water. Probably because the old male is more timid, I rarely saw a bird in the glossy blue-black plumage. The male of 26 May, with skull incompletely ossified, resembles the female, except for having the white under parts anteriorly sullied with fuscous.

My specimens had the irides dark brown; the bill plumbeous; the interior of the mouth bright orange; the feet, toes, and claws plumbeous.

Myiagra rubecula concinna Gould

Myiagra concinna Gould, *Birds of Australia*, pt. 31 [vol. 2]: pl. 90 (1 June 1848). Port Essington, Northern Territory.

Darwin (Nightcliff): 1 ♂, 1 ♀, 21, 25 March; (West Point): 1 unsexed, 30 March.
Cuthbert Point: 1 ♀, 28 April.
Groote Eylandt (Umbakumba): 1 ♂, 2 unsexed, 18, 19, 21 May.

Arnhem Peninsula (Yirrkalla): 1 ♂, 14 July.
Cobourg Peninsula (Cape Don): 1 ♂, 22 September.
Oenpelli: 1 ♀, 12 October.

The leaden flycatcher was one of the commonest of its family at all our camps; it was collected in mangrove and paperbark swamps, in open eucalypt and monsoonal forests.

No specimen is in breeding condition. The three birds taken in March are in postnuptial moult, with some of the rectrices still partly ensheathed.

Myiagra ruficollis mimikae Ogilvie-Grant

Myiagra mimikæ Ogilvie-Grant, *Bull. Brit. Orn. Cl.* 29: 26 (21 Nov. 1911). Mouth of the Mimika River, New Guinea.

Groote Eylandt (Umbakumba): 1 ♀, 25 May.

A sub-adult female, with skull incompletely ossified, collected at a small mangrove swamp on Groote Eylandt, was the only example of the species observed in the Territory.

Myiagra 'latirostris' cooperi Mathews (Melville Island) is inseparable from *Myiagra 'latirostris' kempi* Mathews (Cape York), which, in turn, is a synonym of *mimikae.*

Microeca leucophaea pallida De Vis

Micræca pallida De Vis, *Proc. Roy. Soc. Qld.* 1 (3): 159 (1884). Kimberley = Karumba, Queensland.

Arnhem Peninsula (Gove): 2 ♂, 2, 2 August.

The brown flycatcher was found only on the Arnhem Peninsula, and there was restricted to the immediate vicinity of the aerodrome at Gove, where it occurred in but small numbers.

My specimens are exceedingly worn, but agree well enough with birds from Normanton, which are virtually topotypical of *pallida.*

Microeca flavigaster flavigaster Gould

Micræca flavigaster Gould, *Proc. Zool. Soc. Lond.* 10: 132 (Feb. 1843). Port Essington, Northern Territory.

Darwin (Nightcliff): 1 unsexed, 20 March.
Groote Eylandt (Umbakumba): 2 ♂, 1 ♀, 1 unsexed, 23, 31, 31 May, 9 June.
Arnhem Peninsula (Yirrkalla): 1 ♀, 14 July.
Oenpelli: 1 ♀, 24 September.

While at Darwin I saw this flycatcher only among mangroves, it was elsewhere mainly a bird of the open eucalypt forest.

No specimen was in breeding condition, but the unsexed example of 23 May is sub-adult, with the skull not yet completely ossified and a few juvenal feathers still showing on the crown.

As the dry season advances, the yellow of the under parts becomes paler, the olive-green of the upper parts darker and browner, and the light edgings of the remiges tend to disappear. Thus, my skins of 20 March and 24 September might seem, to an uncritical eye, to belong to different races. Mathews's *melvillensis* was probably founded upon material in abraded plumage.

Pachycephala rufiventris falcata Gould

Pachycephala falcata Gould, *Proc. Zool. Soc. Lond.* 10: 134 (Feb. 1843). Port Essington, Northern Territory.

Groote Eylandt (Umbakumba): 2 ♂, 3 ♀, 2 unsexed, 18, 18, 19, 19, 19 May, 4, 8 June.
Arnhem Peninsula (Yirrkalla): 1 ♀, 27 July.
Oenpelli: 2 ♂, 4 ♀, 24 September, 12, 15, 17, 23, 27 October.

The rufous whistler was not seen at all in the vicinity of Darwin, but was common at Umbakumba, Yirrkalla, and Oenpelli. It inhabited mainly the open eucalypt forest, but on Groote Eylandt occurred also in the wattle scrub.

A male of 19 May had the gonads greatly enlarged, while a female of 19 May and a male of 23 October had them but slightly enlarged. Immatures with the skull incompletely ossified were taken on 19 May, 4 and 8 June; other immatures, with the skull wholly ossified, were collected on 24 September, 15 and 17 October.

I noted that adults of either sex had the irides red; the bill black; the feet and toes slate; the claws black. Juveniles differed in having the irides brown, and the bill horny brown, more yellowish on the basal half of the mandible. These last are easily separable from the adult females, not only by the colours of the soft parts, but also by having the secondaries edged with rufescent and the dark shaft streaks on the breast broader and more numerous.

Specimens from Groote Eylandt (nearest the type locality of '*pallida*') are the most richly coloured of the series; all are freshly moulted birds. With each succeeding month, my examples become progressively paler beneath, more frayed and brownish above. When seasonal change is taken into account, it appears that only two races occur in Australia: a larger one (*rufiventris*), ranging from southern Queensland to southern Western Australia; a smaller one (*falcata*), ranging from northern Queensland to the Kimberley Division.

Pachycephala simplex simplex Gould

Pachycephala simplex Gould, *Proc. Zool. Soc. Lond.* 10: 135 (Feb. 1843). Port Essington, Northern Territory.

Darwin (Nightcliff): 2 ♂, 19 March, 14 April; (East Point): 1 ♀, 26 March.
Groote Eylandt (Umbakumba): 1 ♂, 1 ♀, 6, 21 June.
Arnhem Peninsula (between Yirrkalla and Melville Bay): 1 ♂, 15 July.

This whistler was very common in the Darwin area, somewhat less numerous and more local at the other localities listed above. It inhabited monsoonal forest, paperbark and mangrove swamps.

All of my male specimens had the gonads large or even greatly enlarged. The bird of 26 March is sub-adult; it differs from older examples only by having the inner secondaries edged with bright rufescent.

I use the trinomial name, not because Mathews's *riordani* from Melville Island is thought to be valid, but because races of this species occur in Queensland and New Guinea.

Pachycephala parvula (Gould)

Colluricincla parvula Gould, *Proc. Zool. Soc. Lond.* 13: 62 (Sept. 1845). Port Essington, Northern Territory.

Skirmish Point: 1 unsexed, 4 May.
Arnhem Peninsula (Yirrkalla): 2 ♂, 4 ♀, 13, 19, 21, 23, 24 July, 7 August.
Cobourg Peninsula (Black Rock Point): 1 ♀, 25 September.

The little shrike-thrush was fairly common at Yirrkalla in stands of monsoonal forest and once was found there amongst mangroves. It is a retiring bird of the densest undergrowth and tangled vineries, scarcely visible in the heavy umbrage, and I seldom shot, or even saw, a specimen, until my attention had been drawn to it by the loud, melodious song. Although shy, it responded readily, like many other species of similar habits, to squeaking sounds.

The example of 24 July, taken in a mangrove swamp, had the gonads greatly enlarged; that of 7 August had them merely enlarged. Those of 4 May, 21 July, and 25 September, with rufous-washed wings, were obviously immature, although one of them (21 July) had the skull already completely ossified.

The putative characters relied upon by Mathews for dividing this bird into races are explainable by individual variation and seasonal change.

I find no criteria whatsoever, whether of morphology or of life habits, by which the 'genera' *Myiolestes* and *Colluricincla* can be separated from *Pachycephala.*

Pachycephala phaea pallidirostris (Bowdler Sharpe)

Collyriocincla pallidirostris Bowdler Sharpe, *Catalogue of the birds in the British Museum* 3: 293 (1877). Port Essington, Northern Territory.

Cape Cockburn: 2 ♀, 25, 25 April.

Groote Eylandt (Umbakumba): 5 ♂, 6 ♀, 19, 21, 23, 24, 27, 28, 28, 31 May, 2, 3, 11 June; (near foot of Central Hill): 1 ♀, 30 June.

Arnhem Peninsula (Yirrkalla): 3 ♂, 11, 27 July, 27 August.

Oenpelli: 1 ♀, 28 September.

The larger shrike-thrush was almost abundant at Umbakumba, where it occurred in monsoonal bush, in wattle scrub, in paperbark swamps, and in the stringy-bark forest. At our mainland localities it was less numerous, and only a single example was found at Oenpelli.

A male with the gonads greatly enlarged was taken on 11 June, another with them only slightly enlarged on 27 August. Juveniles were collected on 25 April, 19 and 28 May, 27 July, and 28 September. These last were similar in plumage to the old female, but easily separable from her by their narrowed rectrices and the rufescent edgings to the feathers of both wings and tail.

Adult males had the irides maroon; the bill and interior of the mouth black; the feet, toes, and claws plumbeous. Adult females and immatures of either sex differed in having the bill and interior of the mouth plumbeous.

I am not prepared to recognize any Australian races of the quondam *'Colluricincla brunnea'*. Specimens in the Mathews Collection from Melville Island are much browner above than mainland ones, but there is in this bird a strong tendency toward 'foxing', and it seems to me improbable that Zeitz's *'melvillensis'* should be distinct from the form of Port Essington.

When *'Colluricincla'* is held to be a synonym of *Pachycephala,* the new combination *'Pachycephala brunnea'* is preoccupied by *Pachycephala brunnea* Wallace, 1865. The name next available for the Territorian bird is *pallidirostris.*

Mayr and Serventy (*Emu* 44: 36, 1944) consider *'Colluricincla brunnea'* semispecific with *'C. harmonica'* of the *R.A.O.U. Checklist* (ed. 2, rev.), but I prefer to treat them as ordinary subspecies. I have adopted the specific name *phaea* from Serventy and Whittell (*A handbook of the birds of Western Australia*: 53, 1948).

Pachycephala woodwardi (Hartert)

Colluricincla woodwardi Hartert, *Novit. Zool.* 12 (1): 228 (11 Feb. 1905). Ten miles east of South Alligator River, Northern Territory.

A most melodious song sometimes heard on the slopes of the Escarpment near Oenpelli was attributed by me to this species, but I never succeeded in sighting the singer. A single example seen in the same neighbourhood on 3 October was so wild that there was no possibility of collecting it, as it darted from boulder to boulder.

ARTAMIDAE

Artamus leucorynchus leucopygialis Gould

Artamus leucopygialis Gould, *Birds of Australia,* pt. 6 [vol. 2]: pl. 33 (1 Mar. 1842). Namoi River, New South Wales.

Darwin (Nightcliff): 1 ♂, 2 April.
Groote Eylandt (Umbakumba): 2 ♂, 1 ♀, 9, 9, 9 June.
Arnhem Peninsula (Yirrkalla): 1 ♂, 9 July.
Oenpelli: 1 ♂, 23 October.

This wood-swallow was very common in flocks at all our coastal localities, but was seen only once or twice and in small numbers at Oenpelli. When not circling about high in the sky, it is a bird of the sea beaches and eucalypt forest clearings.

The specimen of 23 October had the gonads slightly enlarged.

Mathews has named *A. l. harterti* from north-western Australia, affirming that it is much paler than *leucopygialis* above. I have examined the original series in New York and found that their pallor is due simply to the fact that all were collected in February and are in exceedingly worn plumage.

Artamus cinereus venustus Bowdler Sharpe

Artamus venustus Bowdler Sharpe in Rowley's *Ornithological miscellany* 3: 198 (May 1878). North-western Australia = Victoria River, Northern Territory.

Arnhem Peninsula (Yirrkalla): 2 ♂, 1 ♀, 24, 24 July, 5 August.

This seems to be another bird of the more arid interior that moves outward to the coasts during the height of the dry season. I noted it only in the open stringy-bark forest near Yirrkalla, where it occurred in small flocks that flew before me from tree to tree, accompanied by flycatchers and honey eaters.

My specimens agree well with two taken 100 miles south of Darwin, which are thus nearly topotypical of *venustus.* The species is badly in need of a general revision, to determine the limits of individual variation within the several populations.

Artamus cinereus melanops Gould

Artamus melanops Gould, *Proc. Zool. Soc. Lond.* 1865: 198 (May 1865). St A'Becket's Pool, South Australia.

Finke River (Langra Well, 12 miles south of Horseshoe Bend): 1 ♂, 27 November.

One perfectly typical specimen of this form was collected for me by David H. Johnson.

Artamus minor Vieillot

Artamus minor Vieillot, *Nouveau dictionnaire d'histoire naturelle,* nouv. éd., 17: 298 (Dec. 1817). Terres Australes = New South Wales.

Darwin (Nightcliff): 1 ♂, 1 ♀, 12, 13 April.
Groote Eylandt (Umbakumba): 1 ♂, 27 June; (lower slopes of Central Hill): 1 ♂, 29 June.
Oenpelli: 2 ♂, 1 ♀, 9, 20, 20 October.

A family party of the little wood-swallow occurred at a clearing in the eucalypt forest near Nightcliff, and a small flock were seen on a sandy wattle-covered plain near Umbakumba, but the species was not otherwise noted out of sight of craggy hillsides. It proved to be common at Central Hill on Groote Eylandt, and abundant around the tors in the neighbourhood of Oenpelli.

This little bird was breeding near Oenpelli in October, but its nests were built in narrow clefts of such high and precipitous cliffs, that I was never able to reach them. A male of 9 October had the gonads enlarged. The two specimens of 20 October are nestlings, taken for me by an aborigine. The one of 13 April is sub-adult.

The plumage of *Artamus minor*, throughout its range, becomes individually variably paler with wear, and *Artamus minor derbyi* Mathews (north-western Australia), said to have 'a paler head and mantle', was named through the author's ignorance of this fact.

ORIOLIDAE

Oriolus flavo-cinctus flavo-cinctus (Vigors)

[*Mimetes*] *Flavo-cinctus* Vigors in King, *Survey intertropical coasts Australia* 2: 419 (15 Apr. 1826). Australia = Northern Territory.

Darwin (Nightcliff): 2 ♂, 25, 25 March.
Groote Eylandt (Umbakumba): 1 ♂, 2 ♀, 18, 24 May, 8 June.
Arnhem Peninsula (Yirrkalla): 2 ♂, 2 ♀, 10 July, 8, 8, 9 August.
Oenpelli: 1 ♀, 4 October.

This oriole was common at all our localities, occurring in both monsoonal and open eucalypt forest, mangroves, and paperbark swamps, depending on where ripening wild fruits were obtainable.

But five (2 ♂, 3 ♀) of my ten specimens are adults with reddish bills, and only a male of 25 March had the gonads enlarged. The other example of the same date is in the ultimate stage of the postnatal moult.

Oriolus sagittatus affinis Gould

Oriolus affinis Gould, *Introduction to the birds of Australia*, 8vo. ed.: 57 (1 Aug. 1848). Port Essington, Northern Territory.

Skirmish Point: 1 ♀, 4 May.
Groote Eylandt (Umbakumba): 1 ♂, 2 ♀, 3, 12, 14 June.
Arnhem Peninsula (Yirrkalla): 1 ♀, 20 July.
Oenpelli: 1 ♂, 27 October.

The olive-backed oriole was found in the same plant associations as its congener, sometimes even in the same tree, but was by no means common anywhere.

I am puzzled by the plumages exhibited in this series. All had the skull completely ossified, but all wear a dress that combines in an extraordinary manner characters of the adult with those of the immature. Thus, while the irides were red, the bills were blackish, merely tinged reddish on the mandible; all the feathers of the wings are broadly margined with buffy rufous, and the rectrices are narrowly tipped with buffy white; the colour of the head and mantle varies individually between olivaceous grey and greyish olive. I should take this immature type of plumage to be characteristic of the subspecies, if Mathews (*Birds of Australia* 12: pl. opposite p. 266, 1926) had not figured a bird from Melville Island that looks quite like adult specimens of *O. s. sagittatus* from New South Wales.

Sphecotheres flaviventris ashbyi Mathews

Sphecotheres flaviventris ashbyi Mathews, *Novit. Zool.* 18 (3): 436 (31 Jan. 1912). Alligator River, Northern Territory.

Darwin (Nightcliff): 1 ♀, 2 April.
Arnhem Peninsula (Yirrkalla): 1 ♀, 8 August.
Oenpelli: 2♂, 22 September, 15 October.

The figbird was very common in flocks at Nightcliff, feeding on the fruit of a species of *Celtis* at the edge of monsoonal forest; seen once at Umbakumba (one adult male in monsoonal forest); seen once at Yirrkalla, in a fruiting fig; common in flocks at Oenpelli, again in fruiting figs.

The specimens of 22 September and 15 October had the gonads respectively greatly enlarged and enlarged. The one of 2 April is moulting from the juvenal into the adult plumage.

The ineffable author of *ashbyi* has separated it from *flaviventris* (Cape York) as being 'much larger . . . and darker green above and below', and has neglected to indicate from which of the several Alligator Rivers his type has come.

Comparison of three adult Territorian males with three from Cape York shows that the wing length in the former series measures 143-149 mm., in the latter 142-146 mm. I recognize *ashbyi,* however, because my material, although indistinguishable above, yet has the under parts a decidedly richer yellow (not 'green') than *flaviventris.*

Mathews (*Birds of Australia* 9: 158, 1921), having set up a so-called family Sphecotheridae, immediately proves to his own satisfaction that *Sphecotheres* is a member of the Campephagidae! Nothing in the behaviour, voice, coloration, or plumage texture of the figbirds permits me to believe that they are anything but slightly aberrant orioles.

GRALLINIDAE

Grallina cyanoleuca (Latham)

C[*orvus*]. *cyanoleucus* Latham, *Supplementum indicis ornithologici*: 25 (1801). New South Wales.

Cape Cockburn: 1 ♀, 25 April.
Groote Eylandt (Umbakumba): 1 ♂, 1 ♀, 26 May, 11 June.
Arnhem Peninsula (Yirrkalla): 1 ♀, 4 August.
Oenpelli: 1 ♂, 1 ♀, 1, 31 October.

The magpie-lark does not reach the northern coasts until the rainy season is definitely ended; thus, up to 18 April not a single example was observed in the Darwin area, although the species was common enough there in September. The first seen by me was at Cape Cockburn on 25 April, and from that date onward the peewee was noted almost daily at every locality visited. It is a dooryard bird as in the South, but visits also the rocky foreshores.

A numerous band inhabited our camp at Oenpelli, walking boldly amongst the tents in company with *Numenius minutus* and *Charadrius asiaticus veredus.* So fearless does this species become, that some individuals would actually enter the restricted space between my tent and its fly to forage for insects and spiders, scrambling the whole length of the ridgepole between the sloping canvas walls.

No specimen showed any gonadal activity. A male of 11 June is sub-adult, possessing still some unglossed brown feathers on crown and mantle and a few white flecks on the throat.

Mathews has named *G. c. neglecta* from the Kimberley Division; it is alleged to differ from topotypical *cyanoleuca* by smaller size. Comparison of north-western and Territorian skins with those from New South Wales shows no size differences whatsoever, *if* the specimens compared are of the same age and sex.

CRACTICIDAE

Cracticus torquatus argenteus Gould

Cracticus argenteus Gould, *Proc. Zool. Soc. Lond.* 8: 126 (July 1841). North-west coast of Australia = Port Essington, Northern Territory.

Darwin (Nightcliff): 2 ♂, 9, 11 April.

The grey butcherbird was seen once or twice on the Arnhem Peninsula, but was otherwise noted only in the neighbourhood of Darwin. It is a shy and rather uncommon species of the open eucalypt forest.

I have found no reference in literature to the fact that *argenteus* has the lores black, rather than white or whitish, as in the southern forms.

Mathews's *Cracticus torquatus colletti* from the Mary River, N.T., is inseparable from my specimens and may safely be consigned to the limbo of imaginary races.

Cracticus cassicus picatus Gould

Cracticus picatus Gould, *Introduction to the birds of Australia,* 8vo. ed.: 35 (1 Aug. 1848). Port Essington, Northern Territory.

Darwin (Delissaville): 1 ♂, 28 March; (Nightcliff): 1 ♂, 6 April.
Groote Eylandt (Umbakumba): 1 ♂, 3 June.
Arnhem Peninsula (Gove): 2 ♂, 4, 5 August.
Oenpelli (Tableland): 1 ♂, 30 September.

The black-throated butcherbird was fairly common about Darwin and around our camp on Groote Eylandt, less numerous at other places visited by the expedition. It was never seen outside of the open eucalypt forest.

At Umbakumba this butcherbird and the black-backed magpie occurred together and gave us a combined chorus mornings and evenings. At that place it was impossible to distinguish the notes of one from those of the other species.

The specimen of 4 August had the gonads slightly enlarged; since its plumage is very worn, one may suppose that its breeding season was just ending. Brownish juveniles were collected on 5 August and 30 September.

Adults had the irides brown; the bill plumbeous, tipped black; the feet and toes slate.

The Australian forms usually combined in the 'species' *nigrogularis* seem to me conspecific with *C. cassicus* of New Guinea.

Cracticus tibicen eylandtensis (H. L. White)

Gymnorhina tibicen eylandtensis H. L. White, *Emu* 21 (3): 163 (1 Jan. 1922). Groote Eylandt, Northern Territory.

Groote Eylandt (Umbakumba): 1 ♂, 1 ♀, 28 May, 3 June.

The magpie occurred only in the immediate vicinity of our camp at Umbakumba, where it was almost restricted to the tall eucalypt forest. In its timidity and penchant toward the tops of the highest trees, it differed strikingly from its relatives of the southern states. Its song also proved to be quite distinct from theirs, but could not be separated from that of the cohabitant population of the butcherbird.

The bird of 28 May is a fine adult male, with inactive gonads (irides: brownish orange). The fact that the feathering of its tibiae is wholly black (*not* 'mottled'), and that its wing measures 241 mm. (*not* 224 mm.), seems to show that White's male cotype was by no means, as he supposed, fully adult. My specimen of 3 June is a sub-adult female, with the skull still incompletely ossified; its plumage differs in no wise from that described by White for the adult female.

The author of *eylandtensis* compared his material with, presumably, the race of New South Wales, rather than the geographically nearer *terraereginae* (Bartle Frere, Qld.). Two adult male topotypes of *terraereginae* in New York have the wing length 249 and 250 mm.; two from Normanton, Qld. which must probably be given the same name, have this measurement 240 and 265 mm. That I recognize *eylandtensis* as distinct is due wholly to the fact that my adult male from Groote Eylandt has a decidedly longer, and relatively slenderer, bill, than even the birds from Normanton.

Cracticus quoyi spaldingi Ramsay

Cracticus spaldingi 'Masters' Ramsay, *Proc. Linn. Soc. N.S.W.* 2 (2): 211 (Jan. 1878). Port Darwin, Northern Territory.

Darwin (Nightcliff): 2 ♂, 11, 11 April.

The black butcherbird was noted only in the neighbourhood of Darwin, where it was local, but not uncommon, in the mangrove belt. At least twice I saw small parties perched at the top of a gigantic tree jutting above monsoonal forest, but these were probably resting on the way from one mangrove swamp to another.

The flocks, or family groups, are exceedingly difficult to observe in the dense leafage of the mangroves or among the tangled roots, where they forage for crabs, but once seen, whether in the trees or in flight across a tidal creek, they resemble nothing so much as small crows. Several writers have accredited this species with the musical notes of its relatives, but I heard only a loud *ah-oo-ah, ah-oo-ah,* with the second syllable stressed, unlike any other bird cry known to me.

No brown example was ever found. One of my specimens is a juvenile and is moulting into the glossy plumage of the adult from an unglossed blackish dress such as might be worn by an immature drongo.

DICRURIDAE

Dicrurus hottentottus bracteatus Gould

Dicrurus bracteatus Gould, *Proc. Zool. Soc. Lond.* 10: 132 (Feb. 1843). Eastern and northern coasts of Australia = Queensland.

Darwin (Nightcliff): 2 ♀, 20 March, 15 April.
Groote Eylandt (Umbakumba): 1 ♂, 1 ♀, 3, 5 June.
Arnhem Peninsula (Yirrkalla): 1 ♀, 13 July.
Elcho Island: 1 ♂, 13 September.
Oenpelli: 1 ♀, 3 October.

The drongo was a common species at all localities visited in the Territory, where it seems to be sedentary. It was found both in monsoonal and open eucalypt forests, and is one of the birds that regularly harry the insects driven from cover by man-made bush fires.

None of my specimens shows any gonadal activity, but breeding must have taken place during our stay in the North, since the example of 3 October is undergoing postjuvenal moult.

The taxonomy of this form, with discussion of the synonymous *Dicrurus bracteatus baileyi* Mathews (South Alligator River), is exhaustively treated in Vaurie's Revision of the bird family Dicruridae in *Bull. Amer. Mus. Nat. Hist.* 93: 288-90, 1949.

CORVIDAE

Corvus orru cecilae Mathews

Corvus coronoides cecilae Mathews, *Novit. Zool.* 18 (3): 442 (31 Jan. 1912). Napier Broome Bay, Kimberley Division, Western Australia.

Groote Eylandt (Umbakumba): 3 ♂, 1 ♀, 2, 11, 15, 16 June; (near foot of Central Hill): 1 ♂, 2 July.

Arnhem Peninsula (shore of Rocky Bay): 1 ♀, 17 July.

The intermediate crow was very common at all the coastal localities visited by the expedition, where it was, apparently, the only species of crow present. Certain individuals at Oenpelli seemed smaller and *may* have represented *Corvus bennetti*, but my lack of success in shooting them makes it impossible to be sure.

This crow was found in every sort of terrain, from rocky foreshores to monsoonal forest, but was specially numerous about aboriginal encampments, where it was quite fearless *until* someone approached with a gun, the significance of which it seemed thoroughly to comprehend.

Birds collected on 2 and 15 June had the gonads enlarged, while one taken on 2 July had them greatly enlarged. My specimens, all adult, had the irides white and the bill, feet, toes, and claws black.

Corvus bennetti North

Corvus bennetti North, *Vict. Nat.* 17 (9): 170 (10 Jan. 1901). Moolah, New South Wales.

Alice Springs (Wigley Waterhole): 1 ♂, 21 November.

My only specimen of the little crow was shot by H. L. Hillier and prepared by David H. Johnson. It is a worn bird, with some of the remiges and rectrices in moult.

Stresemann (*Journ. f. Orn.* 91: 125, 1943) has rashly named *'Corvus difficilis'* on one moulting skin from the neighbourhood of Cloncurry, Queensland. The *main* characters by which the new form is separated from *bennetti* are its more rounded and longer wing (345 mm., against 288-320 mm.). The example listed above, with wing length of 333 mm., might easily be *'difficilis'* by all characters that are ascertainable in its condition of moult, except that the wing is quite as pointed as in *bennetti*. I consider *'Corvus difficilis'* a highly dubious species.

PTILONORHYNCHIDAE

Chlamydera nuchalis melvillensis Mathews

Chlamydera nuchalis melvillensis Mathews, *Aust. Avian Rec.* 1 (2): 52 (2 Apr. 1912). Melville Island, Northern Territory.

Darwin (West Point): 1 ♂, 31 March.
Groote Eylandt (Umbakumba): 2 ♂, 2 ♀, 30 May, 2, 3, 9 June.
Arnhem Peninsula (Yirrkalla): 3 ♂, 13, 20 July, 8 August.
Oenpelli: 1 ♂, 1 October.

The bowerbird was common at all the expedition's camps, in small stands of monsoonal forest, in wattle scrub, and in dense, brushy thickets; in the neighbourhood of Darwin it came fearlessly into gardens and plantations of fruit trees.

The species evidently breeds during the rainy season; immatures with the skull incompletely ossified were found from 31 March to 20 July. A young male of 8 August is acquiring the rose-violet nuchal crest of the adult for the first time.

An apparently old male of 1 October is moulting the rectrices; its gonads were slightly enlarged.

Bowers, some old and broken down by weather, others in good repair, were frequently seen; birds were always in their vicinity, but no use of them was noted. They are difficult to overlook because of the great masses of white or whitish objects, pebbles, bleached shells, and small bones, heaped at each end of the tunnel, but indeed, so fearless of man is this large species that bowers are sometimes found almost under the windows of occupied houses.

Comparison of 15 Territorian specimens with five topotypes of *oweni*, named from north-western Australia by Mathews, *Novit. Zool.* 18: 440, 1812, shows that *melvillensis* is a valid race, characterized by lesser size and greyer upper and under parts, almost without fulvous suffusion.

SITTIDAE

Neositta chrysoptera leucoptera (Gould)

Sittella leucoptera Gould, *Proc. Zool. Soc. Lond.* 7: 144 (Mar. 1840). North-west coast of Australia = Port Essington, Northern Territory.

Arnhem Peninsula (Yirrkalla): 2 ♂, 18, 22 July; (between Yirrkalla and Melville Bay): 1 ♀, 29 July.

The sittella was observed only on the Arnhem Peninsula, where it occurred locally amongst the stringy-bark trees and was once found in stunted paperbarks. Since this bird, in pairs or small parties, usually kept to the tops of the highest trees and, when frightened, flew for long distances through the forest, its collection was by no means easy.

The male of 18 July, with the gonads slightly enlarged, had the irides and eye ring light yellow; the bill light yellow, tipped black; the feet and toes chrome yellow; the claws horn.

The Australian nuthatch, in plumage pattern, colours of soft parts, voice, and life habits, seems to be closely related to *'Poecilositta' azurea* of Malaysia, but since it is not yet known whether the latter species breeds in holes or on the outside of a tree, it would be unwise to combine them generically at this time.

MELIPHAGIDAE

Melithreptus albogularis Gould

Melithreptus albogularis Gould, *Proc. Zool. Soc. Lond.* 15: 220 (29 Mar. 1848). Northern and Eastern Australia = Port Essington, Northern Territory.

Darwin (Nightcliff): 1 ♂, 20 March; (West Point): 1 ♂, 30 March.

Groote Eylandt (Umbakumba): 1 ♂, 19 May; (near foot of Central Hill): 1 ♂, 30 June.

Arnhem Peninsula (Yirrkalla): 2 ♂, 1 ♀, 2 unsexed, 10, 18 July, 13, 27, 30 August; (between Yirrkalla and Melville Bay): 1 ♂, 15 July.

Oenpelli: 1 ♂, 24 October.

This little honeyeater was one of the commonest birds at each of the expedition's camps. I found it in monsoonal bush, open eucalypt forest, mangrove and paperbark swamps, and in cultivation.

All adults taken between 30 March and 13 August had the gonads active. Five specimens collected between 10 July and 30 August are juveniles; the one of 10 July is scarcely beyond the nestling stage. All of these have the pileum brown, with the nuchal band clearly indicated only at the sides, and the upper parts strongly suffused with rufescent.

Old birds had the irides garnet; the orbital skin slaty or plumbeous blue; the bill black; the feet and toes livid flesh. Immatures differed in having the irides bright brown and the bill yellowish or fleshy at the base.

Examples shot from May to August show the effects of wear in the darker and duller green of the upper parts. It is possible that '*Melithreptus lunatus gradus*', named by Mathews from Melville Island, was based upon just such a worn specimen.

Entomyzon cyanotis albipennis (Gould)

Pl. 1

Entomyza albipennis Gould, *Proc. Zool. Soc. Lond.* 8: 169 (1841). Port Essington, Northern Territory.

Darwin (Nightcliff): 1 ♂, 23 March.
Cape Cockburn: 1 ♂, 25 April.
Arnhem Peninsula (Yirrkalla): 1 ♂, 1 ♀, 28 July, 10 August.
Oenpelli: 1 ♂, 23 September.

The blue-faced honeyeater was not uncommon, but very locally distributed, in parts of the Territory visited by the expedition, excepting Groote Eylandt, whence it has not yet been reported. It was found in mangrove swamps, open eucalypt forest, and, near Darwin, in an abandoned coconut plantation.

The bird of 25 April had the gonads enlarged; that of 23 September had them slightly so. The male of 10 August had the skull incompletely ossified, but is wearing a plumage quite like that of the adult, except for a single brown feather on the occiput.

Ramsayornis fasciatus fasciatus (Gould)

Glyciphila fasciata Gould, *Proc. Zool. Soc. Lond.* 10: 137 (Feb. 1843). Port Essington, Northern Territory.

Darwin (Nightcliff): 2 ♂, 5 ♀, 2 unsexed, 12, 13, 13, 13, 14, 14, 14, 15, 15 April.
Cape Cockburn: 1 ♂, 24 April.
Arnhem Peninsula (Yirrkalla): 2 ♂, 11, 16 August.
Oenpelli: 1 unsexed, 26 September.

The white-breasted honeyeater was found only at blooming paperbarks, and its numbers at a given place varied directly with the abundance of these flowers. Thus, near Nightcliff it was locally abundant in April, and equally so at Oenpelli in September and October, but decidedly uncommon at Yirrkalla in August, where only one small grove had its anthesis.

This species acted quite like other small honeyeaters in the flowering trees, fluttering before the sprays or hanging in grotesque positions from them. The only note I heard it utter was a soft *mew*.

My specimens had the irides tan; the bill fleshy brown; the feet and toes fleshy orange.

At West Point, in the Darwin area, a nest was found on 31 March; it was a pendant structure of paperbark, fastened to a paperbark sapling growing in a shallow forest pool, and hung about two feet above the water. The breeding season at Darwin was now almost finished, and only three of my birds (12, 15, and 24 April) showed even slight enlargement of the gonads. A skin of 13 April, with incompletely ossified skull, wears the juvenal dress, in which the breast is broadly streaked instead of barred, and the primaries are edged with yellowish buff; a few barred feathers of the adult plumage are appearing among the juvenal striated ones of the forecrown. A specimen of 26 September, likewise with unossified skull, is in advanced postjuvenal moult; it has the crown transversely barred as in the adult and is acquiring the fasciated breast feathers of the old bird.

The two adults from Yirrkalla are in fine fresh plumage; they differ from more worn examples by having the upper wing coverts and primaries narrowly edged with light brown, instead of greyish brown or grey.

Mathews's *R. f. apsleyi* from Melville Island was set up as a race having the primaries edged with yellowish buff, not grey, but, as I have indicated above, the yellowish buff is characteristic of immaturity in mainland birds, and probably also in those of Melville Island, while the greyish edgings are associated with a worn plumage.

Conopophila albogularis albogularis (Gould)

Entomophila ? albogularis Gould, *Proc. Zool. Soc. Lond.* 10: 137 (Feb. 1843). Port Essington, Northern Territory.

Darwin (Nightcliff): 1 unsexed, 22 March; (East Point): 1 ♀, 26 March; (West Point): 1 ♂, 30 March.

Groote Eylandt (Umbakumba): 1 ♂, 1 ♀, 27, 27 May.

Oenpelli: 1 ♀, 24 September.

This honeyeater was abundant at Darwin, rare and local on Groote Eylandt, fairly common at Oenpelli. I found it not only in monsoonal forest, but also in high grass, whether this was the rainy-season growth of spear grass or the strong stand at the edges of bodies of open water.

The bird of 30 March had the gonads enlarged. The two of 27 May are juveniles: one of them is in postjuvenal moult and is acquiring the rusty pectoral band; the other has the anterior under parts infuscated, with the throat hardly paler than the breast.

See remarks, under the following form, on the two sympatric species of *Conopophila*.

Conopophila rufogularis rufogularis (Gould)

Entomophila ? rufogularis Gould, *Proc. Zool. Soc. Lond.* 10: 137 (Feb. 1843). North coast of Australia = Derby, Kimberley Division, Western Australia.

Arnhem Peninsula (Yirrkalla): 1 ♂, 18 July.

An immature male, with skull incompletely ossified, was shot from a tall eucalypt near Yirrkalla; it appeared to be an example of *C. a. albogularis* and was so listed in my field catalogue, but study in the museum has convinced me that it is, in fact, a specimen of *C. r. rufogularis*.

It seems certain that only fully adult individuals of these two closely related species can be differentiated in the field. Since I have found in the literature no clues to separating the immatures even in the hand, the results of my examination may be of interest.

Conopophila a. albogularis seems to be a slightly smaller species than *rufogularis*: the wing length of eight specimens of the former (adults and full-grown juveniles) ranges from 62 to 64 mm.; that of two specimens of the later (adult and full-grown juvenile), from 67 to 69 mm. The only colour distinction I find between immatures of the two is that, whereas *albogularis* has the whitish throat patch extending to the upper breast and bordered by a fuscous hue, *rufogularis* has it restricted to the area later to be covered with rufous feathers and bordered by the palest grey.

Myzomela obscura obscura Gould

Myzomela obscura Gould, *Proc. Zool. Soc. Lond.* 10: 136 (Feb. 1843). Port Essington, Northern Territory.

Darwin (Nightcliff): 3 ♂, 19 March, 14, 15 April.
Arnhem Peninsula (Yirrkalla): 1 ♂, 1 ♀, 2 unsexed, 11, 13 July, 12, 13 August.
Oenpelli: 1 ♂, 10 October.

The dusky honeyeater was locally very common at Nightcliff, uncommon at Yirrkalla, very rare at Oenpelli, and not observed at all on Groote Eylandt. It occurred indifferently in monsoonal forest, in mangroves, and in paperbark swamps, requiring only that there be fresh water and blooming trees in the neighbourhood.

A male of 19 March had the gonads enlarged, while others of 14 and 15 April and of 10 October had them slightly enlarged. The specimen of 12 August is a juvenile with the skull incompletely ossified. It is evident that the breeding season is of long duration in the Territory.

Myzomela erythrocephala erythrocephala Gould

Myzomela erythrocephala Gould, *Proc. Zool. Soc. Lond.* 7: 144 (Mar. 1840). North-west coast of Australia = Port Essington, Northern Territory.

Darwin (Nightcliff): 2 ♂, 2 ♀, 19, 19, 21 March, 6 April; (Delissaville): 1 ♂, 28 March.
Groote Eylandt (Umbakumba): 1 ♂, 26 May.
Arnhem Peninsula (Yirrkalla): 1 ♂, 13 July; (between Yirrkalla and Melville Bay): 1 ♀, 22 August.
Cobourg Peninsula (Cape Don): 1 ♂, 19 September.

This honeyeater is almost confined to the mangrove belt, but at Nightcliff entered also into immediately adjacent stands of monsoonal forest. Although abundant in the Darwin area, it was rare and local at Umbakumba and Yirrkalla and, of course, not seen at all at Oenpelli.

Males of 28 March and 13 July had the gonads enlarged, and a female of 22 August had them slightly enlarged. Two males of 19 March and 6 April are in postjuvenal moult, with a number of worn brown feathers still present on the throat and pileum.

Meliphaga albilineata (H. L. White)

Ptilotis albilineata H. L. White, *Emu* 16 (3): 165 (15 Jan. 1917). King River, Northern Territory.

Oenpelli (Tableland): 1 ♂, 1 ♀, 1, 3 October.

This honeyeater of restricted range occurred in limited numbers on the Arnhem Land Plateau and its Escarpment, and also on some of the larger hills that rise like islands from the plain as one approaches the Tableland.

The bird is conspicuous only in its song, which alone brought it to my attention. I never heard the 'loud, clear whistle' reported by McLennan, the discoverer, but instead, a most melodious series of disconnected phrases uttered in a rich, flutelike voice, unlike that of any other of the family known to me; in American ornithological terms, it might have been described as the song of a vireo in the voice of an oriole. These harmonious sounds were infrequent and decidedly ventriloquial, and caused me to spend many a hot and weary hour picking my way over house-sized boulders and crumbling rock faces toward some isolated fig from which the bird was apparently singing, only at the last minute to hear the song again from a distant tree on the opposite wall of a canyon. I did not, as did McLennan, find the bird confined exclusively to deep and narrow ravines; it was likely to appear wherever a fruiting tree might attract it. The two specimens taken were shot from the tops of such trees, after long waits beneath for a fleeting glimpse of the small, obscurely coloured body amongst the heavy foliage.

My birds were adults with inactive gonads. They had the irides grey; the bill black; the rictus and interior of the mouth bright yellow; the feet, toes, and claws plumbeous slaty.

This species has been unfortunate in its portrayers. Cayley's picture (*What bird is that?*, ed. 2, pl. 14, fig. 6, 1931) shows some wholly imaginary head markings; Grönvold's, in Mathews (*Birds of Australia* 11: plate opposite p. 451, 1924), indicates that the white streak at the side of the head is interrupted, a result of the preparator's having not sufficiently extended the skin of the neck. The portrait in *Emu* 16: pl. 45, 1917, is, however, very satisfactory.

Lichmera indistincta rufescens (Mathews)

Stigmatops indistincta rufescens Mathews, *Novit. Zool.* 18 (3): 402 (31 Jan. 1912). Crawford Springs Station, Northern Territory.

Darwin (Nightcliff): 3 ♂, 1 ♀, 1 unsexed, 20, 21, 22, 23 March, 13 April; (Delissaville): 1 ♂, 27 March; (West Point): 1 ♂, 30 March.

Groote Eylandt (Umbakumba): 2 ♀, 21 May, 4 June.

Arnhem Peninsula (between Yirrkalla and Melville Bay): 1 ♂, 15 July; (Yirrkalla): 1 ♀, 23 July.

Cobourg Peninsula (Cape Don): 1 unsexed, 19 September.

Oenpelli: 2 ♂, 6, 18 October.

This sweet-voiced honeyeater was the most abundant land bird in the Darwin area and at Umbakumba, and, although less numerous at our other localities, was nevertheless common at all of them. It occurs indifferently in all types of vegetation, even when nectariferous flowers are not present, and is one of the few species that enters into the tall stands of spear grass, where it creeps about invisibly, indicating its presence only by a churring note similar to that of a small reed warbler. Its extraordinary adaptability is sufficient to explain its numerosity.

All adults collected in March and April had the gonads active, as did also one of 15 July. Juveniles, on the wing, were taken on 22 and 23 March; they differ from the adults by having the base of the bill yellow, the under parts strongly washed with bright yellow, the rump and upper tail coverts rufescent and the mantle overlaid with the same colour.

The plumage of the brown honeyeater becomes exceedingly worn and decolorized, and the great majority of specimens in the Mathews Collection, including the types, are quite unsuitable for taxonomic study. Territorian birds are somewhat smaller than southern ones and may be separated, but it is not easy to decide which of the many Mathewsian names must be used for them. The oldest name, *rufescens*, was given to a unique skin that, from long submersion in alcohol, has lost the lipochrome from its feathers and is therefore worthless. The indiscretion shown in naming such a specimen is compounded by the fact that the author himself was not certain of the whereabouts of the type locality: at the first description we find 'Northern Territory (Crawford Springs)', but in *Birds of Australia* 11 (7): 421, 1924, we read 'Crawford Springs Station, Northern Territory (?)'. Obviously, until the location of Crawford Springs is learned, we cannot decide to what population the name *rufescens* is applicable, and I use it here with misgivings.

Mr W. Rainbow, Librarian at the Australian Museum, has courteously investigated for me this and certain other obscure type localities. Of Crawford Springs, he writes: 'This place cannot be located, although there is a Crawford Waterhole (which could also be a spring) situated on a branch creek of the Victoria River, about ten miles north of the Victoria River Downs homestead. There is also a Crawford Creek which runs into the Hodgson River, a tributary of the Roper River.'

Meliphaga unicolor unicolor (Gould)

Ptilotis unicolor Gould, *Proc. Zool. Soc. Lond.* 10: 136 (Feb. 1843). Port Essington, Northern Territory.

Darwin (Nightcliff): 2 ♀, 22, 25 March.
Skirmish Point: 1 ♂, 6 May.
Groote Eylandt (Umbakumba): 1 ♀, 26 May.
Arnhem Peninsula (Yirrkalla): 1 ♂, 1 ♀, 13 July, 13 August.
Oenpelli: 1 unsexed, 1 October.

The white-gaped honeyeater was common wherever the expedition touched along the northern coasts; it occurred in small bands in monsoonal and open eucalypt forest, in mangroves and wattle scrub, at old clearings of abandoned cultivation, and in vineries at the edge of the bush.

The gonads were slightly enlarged in the specimen of 25 March, enlarged in the one of 6 May, and greatly enlarged in those of 26 May, 13 July, and 13 August.

My examples had the irides grey-brown; the bill black; the naked skin at the rictus ivory; the feet and toes brownish or slaty.

In its close flocking, its unmusical chattering, its liking for fruits, and its modest coloration, this bird bears an extraordinary resemblance to some of the smaller Asiatic bulbuls, although, of course, there can be no question of its true familial relationship.

Myzantha flavigula alligator Mathews

Myzantha flavigula alligator Mathews, *Novit. Zool.* 18 (3): 418 (31 Jan. 1912). Alligator River, Northern Territory.

Darwin (Delissaville): 1 ♂, 28 March.
Arnhem Peninsula (Mount Dundas [northern slope]): 1 ♂, 1 ♀, 21, 21 July; (between Yirrkalla and Melville Bay): 1 unsexed, 17 August.

The yellow-throated miner was of very local distribution, but in those favoured spots at which a flock had established residence, might always be seen within a restricted area. A bird of the open eucalypt forest, it was noisy and conspicuous wherever it occurred.

My specimens had the irides grey-blue; the facial skin, bill, feet, and toes yellow; the claws yellowish horn.

Examination of the Mathews Collection in New York has shown that this species breaks into a number of recognizable races, although probably not so many as Mathews himself supposed. *Myzantha flavigula lutea* (Derby, W.A.) is a valid form, characterized by pallid coloration. *Myzantha flavigula alligator*, of which *melvillensis* seems to be a synonym, is very near *flavigula* of New South Wales, but distinguishable by the greater extent of yellow on the forecrown and its more sullied under parts.

Philemon citreogularis sordidus (Gould)

Tropidorhynchus sordidus Gould, *Introduction to the birds of Australia,* 8vo. ed.: 64 (1 Aug. 1848). Cobourg Peninsula, Northern Territory.

Darwin (Nightcliff): 2 ♀, 23 March, 2 April.
Groote Eylandt (Umbakumba): 1 ♂, 20 May.
Arnhem Peninsula (near shore of Melville Bay): 1 ♀, 31 July; (Yirrkalla): 1 ♀, 10 August.
Oenpelli: 2 ♀, 22 September, 14 October.

The little friarbird, while not rare, was yet not nearly so common as its congeners, with which it associated in small numbers at flowering trees in mangrove swamps, in open eucalypt forest, and at the edge of monsoonal forest.

The specimen of 23 March had the ovaries enlarged. The one of 22 September had the skull incompletely ossified and wears on the upper breast the scattered yellow feathers of immaturity.

P. c. sordidus is separable from *citreogularis* (New South Wales) by its usually shorter wing, but usually more robust bill.

P. c. occidentalis, described by Ramsay from Derby, W.A., seems to be a valid race; two topotypes before me differ from Territorian specimens by their longer wing and decidedly longer and heavier bill.

Philemon moluccensis gordoni Mathews

Pl. 1

Philemon buceroides gordoni Mathews, *Aust. Avian Rec.* 1 (4): 102 (18 Sept. 1912). Gordon Point, Melville Island, Northern Territory.

Darwin (Nightcliff): 1 ♀, 19 March.
Arnhem Peninsula (Yirrkalla): 12 ♂, 1 ♀, 7, 7, 10, 10, 10, 11, 13, 13, 14, 14, 15, 16, 18 August.
Oenpelli: 1 ♂, 10 October.

This friarbird, far from being restricted to Melville Island, as has been supposed, is widely distributed over the Territorian mainland, at least in the maritime or submaritime districts. Excepting only the specimen from Oenpelli, which was in a fruiting fig tree, all of my series were taken in the mangrove swamps.

At Yirrkalla it occurred in numbers with the silver-crowned friarbird at a stand of blooming *Lumnitzera.* Like its congener, it was hard to find among the mangrove leaves, but could be located by a diagnostic call: *watch out, watch out* (with the first syllable stressed). Since every specimen shot as a result of these notes proved to be a male, it would seem that they are uttered by only one of the sexes.

One bird of 10 August is an immature, with incompletely ossified skull; it differs from adults mainly by having some of the feathers of the mantle narrowly fringed with white, and the primaries and outer secondaries edged with olive-yellow. Those of 15, 16, and 18 August had the gonads slightly enlarged, while the one of 10 October had them greatly enlarged.

My specimens had the irides garnet; the orbital skin and lores plumbeous slate; the naked skin at the sides of the neck slate; the bill black; the feet and toes slate; the claws dark horny brown.

This form was separated by Mathews from the race of Cape York on, among others, the character of 'its much smaller size generally'. There is, however, in the species a considerable range of individual variation in bodily dimensions, and some Territorian skins have a greater wing length than some from Cape York. Mathews's ignorance of the fact led him to have *gordoni* depicted (*Birds of*

Australia 12 (3): plate opposite p. 103, 1925) as only two-thirds the size of *yorki*, and in this he was followed by Cayley (*What bird is that?*, ed. 2: pl. 12, 1931), who, incidentally, is fantastically in error in all his portraits of the northern friarbirds.

Philemon moluccensis gordoni, while perfectly distinct from *yorki* in colour characters and development of the casque, is only slightly differentiated from populations of some of the islands to the north of Australia, with which its describer failed to compare it. I agree with Mayr (*Bull. Amer. Mus. Nat. Hist.* 83: 167, 1944) that the helmeted friarbirds may properly be combined in the species *moluccensis*.

Philemon argenticeps argenticeps (Gould)

Pl. 1

Tropidorhynchus argenticeps Gould, *Proc. Zool. Soc. Lond.* 7: 144 (Mar. 1840). North-west coast of Australia = Port Essington, Northern Territory.

Groote Eylandt (Umbakumba): 1 ♂, 1 ♀, 25 May, 7 June.

Arnhem Peninsula (near shore of Rocky Bay): 1 ♂, 24 July; (Yirrkalla): 8 ♂, 1 ♀, 9, 10, 11, 12, 12, 13, 17, 18, 18 August.

Elcho Island: 1 ♀, 14 September.

Oenpelli: 2 ♂, 26, 26 September.

The silver-crowned friarbird was common everywhere, including Darwin, where, by oversight, no specimen was taken. It was most numerous in the mangrove belt, but also entered the open eucalypt forest.

This is a very noisy species, but among its many notes I learned to recognize as diagnostic only a harsh, reiterated *more tobacco, uh, more tobacco, uh.* At Yirrkalla, where I made a special effort to collect it in series, I found the bird difficult to see in the dense mangrove foliage, but by means of this call succeeded in shooting no less than nine, of which eight proved to be males (the female was silent and obtained by a lucky accident); it appears then that the *tobacco*-call is given only by the male.

Although I have seen only Territorian examples of the species, I treat it trinomially, since it is very probable that races named by Mathews from north-western Australia and Cape York will be found valid.

DICAEIDAE

Dicaeum hirundinaceum hirundinaceum (Shaw)

Motacilla hirundinacea Shaw, *Naturalist's miscellany* 4: pl. 114 and text (Sept. 1792). New Holland = New South Wales.

Darwin (Nightcliff): 1 ♂, 1 ♀, 6, 17 April.

Groote Eylandt (Umbakumba): 1 ♂, 20 May.

Arnhem Peninsula (Yirrkalla): 1 ♂, 14 July.

Oenpelli: 1 ♂, 1 ♀, 12, 14 October.

Since we visited no places where species of *Loranthus* were not common, this flower-pecker was, of course, numerous at all our camps. I failed to meet with it in monsoonal forest, but noted it in all other types of vegetation, including the mangrove swamps.

Specimens of 6 April and 14 July had the gonads enlarged. The bird of 17 April is a juvenile; it had the bill orange, with culmen and tip horny brown.

Territorian birds seem to be quite inseparable from those of New South Wales. My use of a trinomial is in recognition of the existence of a valid race in the Aru Islands.

PLATE 1

Entomyzon cyanotis albipennis
Philemon moluccensis gordoni
Philemon argenticeps argenticeps

Pardalotus melanocephalus uropygialis Gould

Pardalotus uropygialis Gould, *Proc. Zool. Soc. Lond.* 7: 143 (Mar. 1840). Northwest coast of Australia = Port Essington, Northern Territory.

Darwin (Nightcliff): 1 ♂, 2 unsexed, 25 March, 10, 15 April.
Groote Eylandt (Umbakumba): 2 ♂, 18 May, 5 June.
Arnhem Peninsula (Yirrkalla): 2 ♂, 11, 27 July.

The pardalote was fairly common at the three localities listed above; it occurred indifferently at the edge of monsoonal bush, in open eucalypt forest, in wattle scrub, and at cultivated clearings, but nowhere did I find it out of sight and sound of the sea.

None of my specimens is in breeding condition, but several pairs were nesting in June in tunnels excavated by the birds themselves in the high, wattle-covered dunes that backed the beach at Umbakumba.

Although Whittell and Serventy (*Systematic list of the birds of Western Australia*: 89, 1948) accept no races in this species, my examination of material in the National Museum, Melbourne, showed that not less than three are recognizable. There are *P. m. melanocephalus,* with rufescent rump (Tweed River to Cardwell); *P. m. uropygialis,* with golden rump (Borroloola to Darwin); and *P. m. inexpectatus,* differing from *uropygialis* by its distinctly paler grey mantle and slightly paler golden rump (Napier Broome Bay to Derby).

ZOSTEROPIDAE

Zosterops lutea lutea Gould

Zosterops luteus Gould, *Birds of Australia,* pt. 11 [vol. 4]: pl. 83 (1 June 1843). Greenhill Island, Van Diemen Gulf, Northern Territory.

Darwin (Nightcliff): 1 ♀, 19 March.

The silvereye was noted only in the vicinity of Darwin, where it was common in the mangrove belt. Expecting to find it again at other localities, I foolishly neglected to take more than one specimen, which is a young female, almost fully fledged but with unossified skull.

An interesting discussion of the relationships of this species is given by Mayr, *Bull. Amer. Mus. Nat. Hist.* 83: 168-9, 1944.

ESTRILDIDAE

Poephila phaeton phaeton (Hombron and Jacquinot)

Fringilla phaeton Hombron and Jacquinot, *Ann. Sci. Nat.,* ser. 2, 16: 314 (Nov. 1841). Raffles Bay, Northern Territory.

Arnhem Peninsula (Yirrkalla): 2 ♂, 1 ♀, 9, 25 July, 9 August.
Oenpelli: 2 ♂, 21 September, 1 October.

The crimson finch was locally common in small flocks at both Yirrkalla and Oenpelli. This species seemed to be more dependent upon the immediate proximity of water than others of its family, and at Yirrkalla was scarcely seen except at the grass- and pandanus-lined edge of an extensive marsh.

No sexual activity was shown by any of my specimens. The bird of 25 July is a juvenile male; it differs sensibly from the adult by having the bill black; the head and body, above and below, brown (paler on the abdomen); the upper tail coverts and the rectrices dull red and the wings suffused with this colour. A male of 1 October, in advanced postjuvenal moult, is quite like the adult, except that only a few blackish feathers have yet appeared among the brown of the crown.

Old birds had the irides tan; the bill orange-red; the feet and toes brownish yellow.

Poephila p. iredalei, named by Mathews from Queensland, seems to be a valid race. *Poephila p. fitzroyi,* of the same author, from the Kimberley Division, although stated to have a brown, not black, head, is both portrayed and described in *Birds of Australia* 12: 258 and plate opposite p. 257, 1926, as having the crown 'greyish-black'. Since the young male, as appears above, loses last of all the brown crown of the juvenal plumage, it is probable that Mathews's type specimen was not fully adult.

Poephila guttata (Vieillot)

Fringilla guttata Vieillot, *Nouveau dictionnaire d'histoire naturelle,* nouv. éd., 12: 233 (June 1817). Iles Moluques = Timor.

This finch was noted in the Territory only at the aerodrome near Alice Springs, where it was numerous on 18 March.

Delacour (*Zoologica* 28: 80, 1943) has shown that the Australian zebra finch is conspecific with the one of Timor and Flores.

Poephila bichenovii annulosa (Gould)

Amadina annulosa Gould, *Proc. Zool. Soc. Lond.* 7: 143 (Mar. 1840). North-west coast of Australia = Cobourg Peninsula, Northern Territory.

Darwin: 1 ♂, 10 April.
Groote Eylandt (Umbakumba): 3 ♂, 2 ♀, 23, 23, 23, 23 May, 22 June.
Arnhem Peninsula (between Yirrkalla and Melville Bay): 1 ♂, 3 August.
Oenpelli: 1 ♀, 12 October.

This was by far the commonest finch met with in the Territory. At Darwin I saw it on a samphire flat, but it was more usually seen in the open eucalypt forest where, along the banks of some seasonal stream, grew coarse grasses and sedges and clumps of pandanus.

A male of 10 April had the gonads slightly enlarged, but actual breeding was taking place at Umbakumba as late as June, when nests were sometimes found concealed amongst the dead leaves of the pandanus plants. My specimens of 23 June were shot from a party of two adults with perhaps half a dozen juveniles not long on the wing.

Poephila acuticauda hecki Heinroth

Poëphila hecki Heinroth, *Orn. Mber.* 8 (2): 22 (Feb. 1900). Australia; type locality restricted to Port Darwin, Northern Territory, by Mathews, *Novit. Zool.* 18: 432 (1912).

Arnhem Peninsula (between Yirrkalla and Melville Bay): 3 ♂, 1 unsexed, 28, 28 July, 3, 3 August.
Oenpelli: 1 ♂, 21 September.

The long-tailed finch was locally common in small flocks on the Arnhem Peninsula and at Oenpelli, in grassy areas of the open eucalypt forest. This and the masked finch were found in much the same places, although the two species were not noted actually associating.

The two birds of 28 July had the gonads enlarged. The unsexed specimen of 3 August is a juvenile; it differs from the adult in just the ways the immature of the masked finch differs from its parent, except that in this case almost no blackish colour at all appears on the lower flanks.

Adult examples had the irides red; the bill orange; the feet and toes deep rose; the claws horn. At the time of writing (August 1949), their bills are still more orange than yellow, but the feet of all but one have changed to yellow, as have the feet of my series of *P. p. personata.*

Poephila personata personata Gould

Poëphila personata Gould, *Birds of Australia,* pt. 6 [vol. 3]: pl. 91 (1 Mar. 1842). Port Essington, Northern Territory.

Arnhem Peninsula (between Yirrkalla and Melville Bay): 2 ♂, 18 July, 22 August. Oenpelli: 1 ♂, 1 ♀, 29 September, 3 November.

The masked finch occurred very locally and in small numbers on the Arnhem Peninsula, but proved to be slightly less uncommon at Oenpelli. It was seen only at grassy areas in the open eucalypt forest.

The male of 22 August had the gonads slightly enlarged, but a juvenile male on the wing had already been collected as early as 18 July at the same place. The latter specimen differs from the adult mainly by having the dark portions of the front, lores, throat, and lower flanks merely blackish, rather than black; the upper parts duller; the under parts buffy brown, almost without vinaceous suffusion; the bill largely dark horny brown.

Adults had the irides blood red; the bill yellow; the feet and toes deep rose; the claws horn.

Lonchura castaneothorax castaneothorax (Gould)

Amadina castaneothorax Gould, *Synopsis of the birds of Australia* 2: pl. [2], fig. 2, and text (Apr. 1837). Australia.

Darwin (West Point): 1 ♀, 31 March.

The chestnut-breasted finch was recorded only in the vicinity of Darwin, during March and April; it occurred then in pairs or flocks at the edge of extensive stands of spear grass.

On 21 March, a nest was found near Nightcliff, in a clearing where the spear grass had been broken down by storms to make a dense tangle of culms and blades. Set about two feet above the ground, it was a bulky globular structure of grass, with a projecting entrance tunnel.

My specimen from West Point, with the ovaries enlarged, had the irides brown; the bill, feet, and toes plumbeous; the claws horn.

Mathews has named *asimilis* (Eureka, N.T.) on the basis of differences shown by Territorian birds in fresh feather, compared with eastern ones in worn plumage. *Worn* Territorian specimens are identical with worn *L. c. castaneothorax*; whether *assimilis* may yet be recognized must be decided by comparison of newly moulted examples of the two populations.

(*Manuscript received, February 1953*)

6

MAMMALS OF THE ARNHEM LAND EXPEDITION

by David H. Johnson*

The specimens of mammals collected in the Northern Territory of Australia by the American-Australian Scientific Expedition to Arnhem Land constitute the primary basis for this report. The expedition, under the general direction of Charles P. Mountford, was in the field throughout the dry season of 1948, from March until November, and made investigations in various branches of anthropology and natural history. The main base camps were on Groote Eylandt, at Yirrkalla near Cape Arnhem, and at Oenpelli near the East Alligator River. Collections of mammals were made at those places by the writer, who was attached to the expedition as mammalogist, and additional specimens were obtained on various side trips and at stops *en route*. Altogether they provide a good representation of the Recent mammals of coastal Arnhem Land.

After the primary field work of the expedition was finished, opportunity was taken on the return trip to stop for a few days each at Alice Springs and Horse Shoe Bend Station in central Australia, where further specimens were collected. Also available for study in the United States National Museum was an important and mostly unreported collection made in 1920 by Charles M. Hoy at Brocks Creek and Douglas River, about 100 miles south of Darwin. Inclusion of this extra but pertinent material has extended the scope of the present report to include the mammals of the entire Northern Territory of Australia, as far as they are now represented in the U.S. National Museum at Washington.

Two preliminary papers describing new species and subspecies collected by the expedition have been published by the author in *Proc. Biol. Soc. Wash.* 67: 77-80 (1954) and 72: 183-8 (1959). Now that the Arnhem Land Expedition specimens have been studied as a unit, a representative set will be sent to the Australian Museum in Sydney, where they will be available to mammalogists in Australia.

The amassing of a collection of scientific specimens of mammals is a co-operative affair, especially in an area such as Arnhem Land, where distances are great, where mammals are not abundant in species or in numbers of individuals, and where standard collecting techniques often fail. Many persons helped in bringing together the present series, and I am grateful to all of them. Every member of the expedition staff assisted at one time or another, either by saving specimens or by reporting observations that led to successful collecting. William E. Harney and H. L. Hillier were indispensable as guides, interpreters, and

* Smithsonian Institution, Washington, D.C.

co-collectors, the former on the Cape Arnhem Peninsula and at Oenpelli, the latter at Oenpelli and in central Australia. The success of field work at places away from the main expedition and its facilities was enhanced by generous hospitality offered by Mr K. Davidson, chief lighthouse keeper at Cape Don, Mr and Mrs Harold Shepherdson at the Elcho Island mission station, and Mrs Golder at the Horse Shoe Bend cattle station. Finally I owe a great debt of gratitude to many aborigines, some never learned by name, who for little or no reward collected specimens, showed me where to look for different species, guided me on walkabout trips, and tried earnestly to impart some of their profound knowledge of animal life.

ITINERARY IN 1948

Field work began in late March while the expedition had temporary headquarters at Nightcliff, a few miles north-east of Darwin. Late wet season rains and the imminence of departure limited activities at this place to occasional brief excursions, and only a few specimens were collected.

Nearly five weeks, from mid-April to late May, were spent on board the motorized barge *Phoenix* travelling from Darwin to Groote Eylandt, with many stops *en route*. Brief collecting was possible at some of the places visited by the *Phoenix*, and one or more specimens of mammals were obtained at Croker Island on 23 April, Skirmish Point on 8 May, Milingimbi on 10 May, Elcho Island on 17 May, and Cape Wilberforce on 20 May.

The expedition's first base camp, Umbakumba on Port Langdon, Groote Eylandt, served as base of operations from 24 May to 8 July. On the sandy dunes and flats of this north-eastern corner of Groote Eylandt mammals were fairly numerous and a new species of the hopping mouse *Notomys* was discovered. In late June an excursion was made to the rocky country about Central Hill, where an entirely different series of habitats was available and several different species were added to the collections.

On 8 July the move from Groote Eylandt to the second base camp, Yirrkalla on the Cape Arnhem Peninsula, was made by RAAF flying boat. Field work was carried out at Yirrkalla and at other places in that vicinity until 12 September. Several walkabout trips were made out from Yirrkalla, with the result that most of the Cape Arnhem Peninsula was covered and specimens were collected at localities that are well known to the local aborigines but are not permanently inhabited and are not shown on maps. In the accounts that follow, all localities in this area are designated by the local name along with 'Cape Arnhem Peninsula'. They include: *Melville Bay,* west of Yirrkalla, specimens coming from the vicinity of Drimmie Head; *Barnambarnga* or 'Dugong Bay', on the coast between Yirrkalla and Melville Bay, opposite Bremer Island; *Gove Airfield,* about three miles south-west of Yirrkalla, an important and populated site during the Second World War, abandoned but still usable in 1948; *Rocky Bay,* on the coast two miles south of Yirrkalla; *Kaulpulmi, Danganboy, Kanbulabulaui, Duripulmi,* and *Larporkuru,* sites successively farther south from Rocky Bay along the trail leading from

Yirrkalla to Port Bradshaw, which passes behind Cape Arnhem and meets the northern end of a long white-sand beach at Larporkuru; *Birluoja Creek, Nakala Creek, Kwaiturumuru Creek,* and *Purlonkoimi Creek,* all small streams that flow toward the southern end of Melville Bay in the area south-west of Gove Airfield and successively farther from that field in the order named. Birluoja Creek is an upper tributary of Nakala Creek.

On 12 September, I separated from the main party and left Yirrkalla on the mission boat *Larapan,* arriving at Elcho Island the next day. The expedition supply boat *Victory,* bound from Yirrkalla to the new base camp at Oenpelli, picked me up at Elcho Island on 16 September and put me ashore at Cape Don on 18 September.

The following three weeks were spent on the Cobourg Peninsula with headquarters at the Cape Don lighthouse. Local trips were made by boat to Popham Bay on the north shore of the peninsula and to Black Rock Point on the south shore. From the latter locality a trip was made overland on foot to Knocker Bay on Port Essington.

On 8 October, I left Cape Don and walked overland to join the main party at Oenpelli. This twelve-day trip provided an intimate view of the entire length of the Cobourg Peninsula, with its remarkable fauna of large introduced mammals, and of the plains country about the mouths of Murgenella Creek and Cooper Creek. Work was continued at Oenpelli until the expedition departed on 9 November. The well-developed rock habitat present on the hills around Oenpelli produced the richest collecting ground encountered on the entire expedition.

With the completion of the work at Oenpelli, the expedition group returned to Darwin and thence south to Adelaide. At Tennant Creek on 16 November, I separated again from the main group, and with H. L. Hillier proceeded to Alice Springs. There we spent five days, making collections on the plains south of Alice Springs and at Wigley Waterhole in the Macdonnell Range north of the town. On 22 November we went south on the weekly train to Rumbalara and thence to Horse Shoe Bend cattle station on the Finke River. A week of collecting along the dry bed of the Finke River and in the adjacent red sand-ridge country completed the field work, and we proceeded on to Adelaide.

HISTORY OF MAMMALOGY IN THE NORTHERN TERRITORY

Study of the expedition collections has made it necessary to review the rather scanty mammalogical literature pertaining to the Northern Territory. The published articles that have been found to contain information directly concerning mammals in this area are listed in the bibliography that concludes this report. The highlights of the collecting activities that took place prior to the Arnhem Land Expedition, as revealed in the pages of the cited papers, are as follows.

The first mammals to be collected for scientific purposes are probably the dugongs obtained around 1827 by Dr Robert Tytler in Raffles Bay and deposited in the Indian Museum at Calcutta (see Blyth 1859: 14).

The real pioneer of Arnhem Land mammalogy was John Gilbert, who in 1840 and 1841 collected in the vicinity of Port Essington, which then was an important settlement, specimens that were used by Gould in writing his monumental three-volume *Mammals of Australia* (1845-60). As a result of Gilbert's collecting, Port Essington became the type locality of numerous species of mammals and is well known in mammalogical literature. Four years later Gilbert was again headed for Port Essington with an overland expedition led by Ludwig Leichhardt, when he was killed by natives in northern Queensland. Leichhardt and his party eventually reached Port Essington by way of the Roper River and the Alligator River country, but his journal contributes little concerning mammals other than descriptions of fruit bat colonies and feral water buffaloes, which even then were abundant about the mouth of Cooper Creek.

The 'North Australian Expedition' contributed a few specimens of mammals to the British Museum, most of them being collected by Dr J. R. Elsey around 1856 near the mouth of the Victoria River.

A large and important collection was made by the Norwegian Knut Dahl in 1894 and 1895 south of Port Darwin, from the area of the Daly and Victoria Rivers east to the headwaters of the South Alligator River. The specimens went to the area and west as far as Brocks Creek in 1902 and 1903. Thomas (1904a) published by Robert Collett (1897). Field notes were published separately by Dahl (1897), and many years later he published an English language version of a book describing his experiences under the title *In Savage Australia* (1926). Some of Dahl's specimens were later acquired by the British Museum and were mentioned in various papers by Oldfield Thomas.

J. T. Tunney collected for Walter Rothschild in the South Alligator River area and west as far as Brocks Creek in 1902 and 1903. Thomas (1904a) published an account of this collection, which appears thereafter to have been widely scattered. Besides the British Museum specimens, some have been reported in the Rothschild collection at Tring, some in the Western Australian Museum at Perth, and a few specimens (included in the present report) are in the U.S. National Museum in Washington.

In 1920, Charles M. Hoy, collecting for William L. Abbott on behalf of the United States National Museum, made a collection of about 110 specimens at Douglas River and Brocks Creek. These were labelled simply '100 miles south of Port Darwin' and were reported under that locality in the few papers that have mentioned them, including the description of *Melomys cervinipes albiventer* by Kellogg (1945) and incidental notes on other specimens by Sanborn (1931) and Tate (1951). With the aid of Hoy's notebooks it has been possible to reconstruct enough of his itinerary to determine which specimens came from Brocks Creek and which from Douglas River, and those localities are used in the present accounts.

A major collecting contribution was made in 1924 and 1925 by G. H. Wilkins, who obtained specimens at Groote Eylandt, Roper River, Milingimbi, Goulburn Islands, and King River, as well as at other localities, for the British Museum.

These specimens were never fully reported, although three papers were published on special groups of mammals by Oldfield Thomas in 1926. In 1929 Wilkins published a popular book, *Undiscovered Australia,* in which his travels are described.

I have not attempted to review the early collecting in the southern part of the Northern Territory, which in most of the literature has been designated as 'Central Australia', but the outstanding scientific expedition to that area was the Horn Scientific Expedition to Central Australia, the mammals of which were reported in a series of papers by Waite (1896, 1898) and Spencer (1895-7). The Horn Expedition worked mainly at localities along the Finke River and about the James and Macdonnell Ranges west of Alice Springs. They also obtained a number of specimens from Charlotte Waters. More recently collections have been made by H. H. Finlayson of the South Australian Museum at many localities, mostly south of the latitude of Alice Springs. Excellent reports have been made by Finlayson on some of these collections.

The vast middle section of the Northern Territory, between Arnhem Land and the Alice Springs area, is virtually uncollected, except for specimens obtained by W. Stalker at Alexandria and elsewhere on the Barkly Tableland for the British Museum and reported by Thomas (1906e).

In spite of the various collections mentioned, including that made by the Arnhem Land Expedition, it can be said that the mammalogy of the Northern Territory of Australia is yet in a pioneer stage. The distances between collecting localities are still so vast, and the collections that have been made are so scattered in different museums around the world, that the patterns of distribution and geographic variation can be seen only dimly for most species and not at all for some. A number of species are known only from single specimens or from single series, and there are areas of thousands of square miles where no mammalogist has ever worked.

INTRODUCED MAMMALS IN THE NORTHERN TERRITORY

In addition to the native and some non-native species of mammals for which formal accounts are presented in the following pages, a number of conspicuous and biologically important species were observed but were not collected as specimens. Most of these are domestic strains of mammals that escaped captivity or were intentionally released. With varying degrees of success they have established self-perpetuating colonies. The formal zoological literature of Australia tends to ignore these newcomers, with the result that it is not possible to learn from regular sources, for example, how far south water buffaloes have spread, or how far north rabbits have gone.

Directly or indirectly as a result of the white human invasion of the Northern Territory within historic times, the following kinds of exotic mammals became established as members of the fauna of the area:

European rabbit, *Oryctolagus cuniculus*
House rat or roof rat, *Rattus rattus*

House mouse, *Mus musculus*
Horse, *Equus caballus*
Ass, *Equus asinus*
Pig, *Sus scrofa*
Dromedary camel, *Camelus dromedarius*
Indian sambar deer, *Rusa unicolor*
Water buffalo, *Bubalus bubalis*
Banteng, *Bos banteng*

The dog, *Canis familiaris,* as represented by the dingo, is also essentially an exotic species, apparently having arrived in Australia with human assistance somewhat in advance of the others.

Information on rabbits, house rats, house mice, and dogs will be found in the main body of this report. Foxes, house cats, camels, horses, and asses are important in the southern part of the Northern Territory; some data on their status and numbers are given by Finlayson (1943a).

The following five species of ungulates occurred in 1948 on the Cobourg Peninsula, where, under the protection of that area as a faunal reserve and the total absence of a resident human population, they appeared to be flourishing:

Horses of a small strain known as 'Timor ponies' were abundant. They are supposed to be descendants of stock imported in the active days of the Port Essington settlement, more than a century ago. Bands of these nimble little horses roamed through the thickets and meadow-like glades bordering the stream courses on the western side of the Port. Each band that I encountered was led by a dominant stallion who enacted a miniature version of the characteristic equine routine, snorting and stamping at the human intruder in menacing fashion, then nipping his brood of mares into a galloping dash away through the forest.

Pigs were seen at the south end of Port Essington and at the mouth of James Creek on 11 October, wallowing in the few remaining water-holes at the end of the dry season. The origin, duration, and geographic extent of this population are not known to me. Dahl (1926) mentioned that there were populations of feral pigs south of Darwin in 1893, supposed to have descended from domestic stock brought in by Chinese settlers.

Sambar deer were seen twice, and tracks many other times on the western and southern shores of Port Essington. At Knocker Bay a female and a half-grown fawn were startled from a stunted thicket of mangroves and ran out into the shallow water of the bay for a distance before turning back to shore and disappearing into other thickets.

Water buffaloes were brought from Timor to Melville Island, Raffles Bay, and Port Essington in the period from 1825 to 1838, according to the *Australian Encyclopaedia* (1925 ed., 1: 212-3). They were 'too poor to kill on arrival and too wild to retain in captivity when once ashore', and escaped animals soon built up large populations and spread widely. In 1948 signs of buffaloes could be seen at nearly every place a stop was made along the coast from Yirrkalla to Cape Don.

The small stream valleys apparently support correspondingly small populations. At Knocker Bay Creek and other places about Port Essington during the late dry season, only solitary old bulls were seen in the isolated meadows. On the open black-soil plains of Murgenella Creek and Cooper Creek, and in the low forested plain between, buffaloes were present in bands of up to twenty individuals. A harvest of buffalo hides was being reaped as a commercial venture by the mission at Oenpelli; a description of this activity has been published by Frederick D. McCarthy (1954), a member of the Arnhem Land Expedition party. The extensive marshy plains about the south shore of Van Diemen Gulf, from Cooper Creek west to the Adelaide River, have proved to furnish ideal water buffalo habitat. In his book *Ourselves Writ Strange*, Alan Marshall (1948) has given a vivid account of the buffalo herds there as viewed from the air in the wet season.

Banteng were seen about the upper end of Port Essington, on the opposite side of the peninsula near James Creek, and at various places for about twenty miles farther east, nearly to the base of the Cobourg Peninsula. Solitary dark grey bulls with conspicuous whitish rump patches and white stockings were seen on three different occasions, and two bands of ten or twelve reddish-tan cows and calves were also encountered.

ACCOUNTS OF SPECIES AND SUBSPECIES

The following accounts cover only the kinds of mammals that were collected or sufficiently well observed for certain identification by the Arnhem Land Expedition, together with a few kinds not found by the expedition but represented by specimens available for scientific study in the United States National Museum. The Museum's register numbers are added after the collection date of each series.* The accounts are intended as contributions to knowledge about the various kinds of mammals, not as complete compilations of scientific knowledge about them. For information on species and subspecies not listed here, the reader is referred to papers cited in the terminal bibliography. An especially useful combination is Iredale and Troughton's (1934) checklist and Troughton's (1941) *Furred Animals of Australia.*

MONOTREMATA

TACHYGLOSSIDAE

Tachyglossus aculeatus acanthion (Collett)

Pl. 1

Echidna acanthion Collett, *Forh. Vidensk-Selsk. Krist.* 13: 1 (1884 or 1885). (See also: Collett, *Proc. Zool. Soc. Lond.* 1885 (1): 150, June 1885.) Type locality: Gracemere, near Rockhampton, Queensland (restricted by Tate, *Bull. Amer. Mus. Nat. Hist.* 98: 577 [1952]). Collett based *Echidna acanthion* on nine specimens in the Museum of the University of Christiania (now Oslo) collected by Lumholz and designated respectively as no. 1 (a young male caught at Gracemere near Rockhampton in July 1881) and nos. 2 to 9 (caught at Coomoo-

* Ridgway's *Color standards and color nomenclature* (1912) has been adopted for the description of the specimens, the names of the colours being indicated by a capital initial.

boolaroo, 80 miles west of Rockhampton in February and March 1884). The statement by Tate (*loc. cit.*) that 'Collett described the type of *acanthion* from Gracemere, near Rockhampton, and listed eight others from Coomooboolaroo . . .' may be taken as a selection of specimen no. 1 of Collett as lectotype and restriction of the type locality to Gracemere.

Melville Bay, Cape Arnhem Peninsula, 1. Female in alcohol (skull cleaned); 2 Sept. 1948 (283963).

West side of Lake Hubert, Groote Eylandt, 2. Male and female, skins with skulls; 28 June 1948 (283961-62).

Other records for Northern Territory: Daly River (Collett 1897: 335); Nellie Creek (Thomas 1904a: 229); Mary River (Collett 1897: 335).

Judging from the statements of settlers and aborigines, and the few published records, spiny anteaters are uniformly but sparsely distributed over all of Arnhem Land and perhaps the remainder of the Northern Territory as well. No other mammalian species except the dingo seems to be so little influenced by the extreme climatic differences between the arid interior and the relatively humid northern coast. Locally, the animals show a decided preference for the vicinity of bluffs or rock outcroppings; in the absence of these, hard-packed soil provides adequate burrowing sites.

At several places in the vicinity of Cape Arnhem and near Delissaville opposite Darwin, aborigines pointed out crevices and small caves in the bases of ironstone bluffs, which they said were occupied by 'porcupines'. It seems unlikely that, in spite of their phenomenal digging powers, these animals could actually excavate tunnels in this extremely hard rock, though they might succeed in enlarging already existing crevices. The aborigines considered it useless to try to dig them out of such places.

At a point about two miles west of the shore of Lake Hubert, Groote Eylandt, natives dug an adult pair, male and female, out of a burrow for me at about noon on 28 June. This was in an area of rather stunted woodland, where the soil was hard-packed, reddish in colour, and heavily intermixed with pebbles of ironstone. The soil in surrounding areas was appreciably more sandy. Numerous excavations were apparent, most of them being deep pits and parts of old caved-in tunnels. The natives showed interest in only one of these, an open tunnel that entered the ground on a slight slant and seemed to have been recently used. In cross-section it was about eight inches wide and six inches high. One of the men poked a long stick into the tunnel to locate its direction, then dug into the ground with a sharpened stick at a point about five feet from the entrance. He finally broke through into the tunnel at a place where its roof was nine inches below the surface of the ground. At that point the tunnel swerved to the right, and the probing and digging process was repeated. The second vertical hole broke through at a depth of fourteen inches, at the end of the tunnel and directly over the spiny anteaters. There was no appreciable terminal enlargement nor any lateral branching of the tunnel. The female was nearest the hole and was taken out first. This operation seemed to take all the strength of two men, as the animal wedged itself into the cavity and apparently braced its quills against the hard earthen walls. Some of the quills were broken but none was pulled out of the skin in this rough handling. The larger male proved even more difficult to extract, and the men succeeded in doing so only after disabling the animal by spearing it repeatedly with a sharp wire heated over a fire.

The female of this pair had a rudimentary pouch with glandular nipple areas and surprisingly large milk glands, but there were no young in the pouch and no

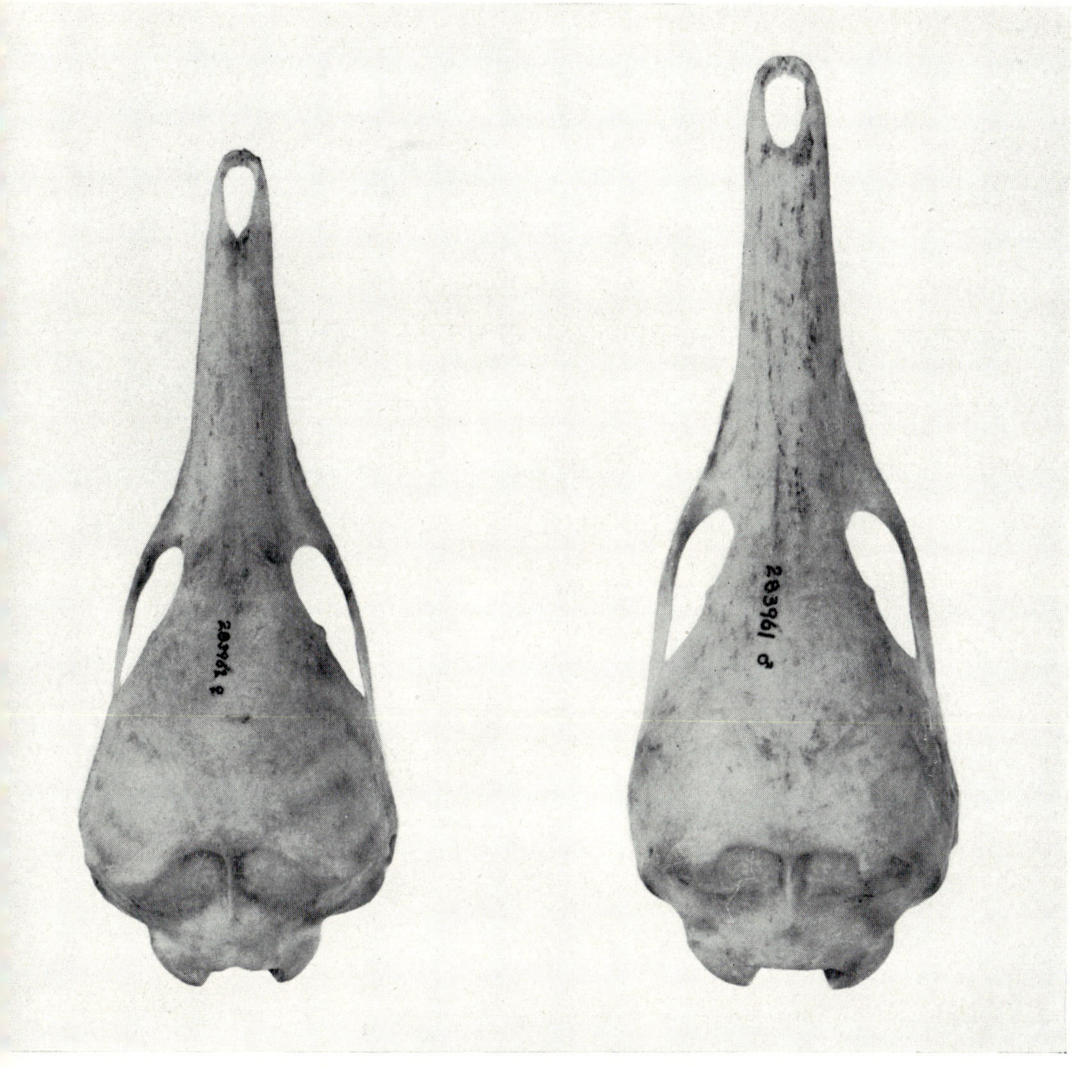

Plate 1. *Tachyglossus aculeatus acanthion*

Skulls from Groote Eylandt, x 1

Above, ♂, no. 283961; below, ♀, no. 283962

developing eggs in the reproductive tract. The flesh had a characteristic odour, resembling that of crushed ants. The natives ate the skinned bodies eagerly, and even salvaged strips of fat and mammary tissue that were trimmed from the skins. The thick layer of dermal muscles in which the bases of the quills were imbedded showed remarkable reflex activity after the skin was completely removed from the body; when the skin was touched contractions of the muscles caused slow writhing movements of entire sections.

An immature female spiny anteater was brought to me from rocky country near Melville Bay on 2 September. It had been found and speared by a native woman during the middle of the day.

In 1894 and 1895 Dahl encountered these animals frequently in the area about the Daly River and the Mary River. He reached the interesting conclusion that the sharp restriction of the animals to the broken rocky regions was the result of persecution by the aborigines, whom Dahl considered to be their only enemies. One individual that he kept alive in a bag for more than two weeks without food remained fat and apparently in good health (Dahl 1897: 199-200).

Systematic notes. Until more is known about the extent of individual variation, allocation of specimens to subspecies must remain largely provisional. Size, colour, relative hairiness, and various cranial characters, all of which have been used by different authors to distinguish races, are subject to an undetermined degree of variation within local populations. The even-aged pair obtained on Groote Eylandt indicates a pronounced disparity in size between males and females (see measurements, Table 1), but this factor has been given scant attention by previous workers. The extent to which colour and degree of hairiness are influenced by seasonal climatic changes and the differences between individuals of different age-groups are yet to be worked out in detail. In this connection, Thomas (1885) analysed the characters of the 31 skins and 20 skulls available to him, but these specimens represented a wide variety of localities and many of them lacked essential data.

Among the various subspecific names of *Tachyglossus aculeatus* available, Collett's *acanthion* appears to be most nearly applicable to our specimens from north-eastern Arnhem Land. The narrow and smoothly sloping braincase characterizing that race is shown by the three specimens from Arnhem Land, which represent both sexes and young as well as old individuals, in marked contrast to 12 other skulls in the United States National Museum from more eastern and southern localities. Whether this same character is shown by specimens from western Arnhem Land is not known. It seems probable that the spiny anteaters inhabiting the southern part of the Northern Territory may belong to a different race, but I have seen no specimens from the latter region.

The two specimens from Groote Eylandt are alike in external appearance. The heavy spines on the back and sides have no intermixture of hairs; the latter are restricted to isolated tufts of short bristles subtending the weak and flattened spines on the legs and underparts. In colour, the dorsal spines are black-tipped for about half in inch on the average, the result being an almost equal mixture of black and straw colour. The short hairs and spines of the face, legs, and underparts are dark brown. The exposed skin on the beak and underparts is darkly pigmented. The third claw on the hind foot is equal in size to the fourth. Each hind foot of the male bears a vestigial spur three mm. long and one mm. in diameter at the base.

The young female from the mainland near Melville Bay differs in having the dorsal spines more straw-coloured; only the extreme tips are faintly pigmented. The hairs and spines on the underparts are likewise pale-coloured, and the skin appears unpigmented. Because such differences are to be noted among specimens from the same population in other places, these are not thought to indicate a racial difference between mainland and Groote Eylandt animals. The cranial features, except for those associated with age and sex, are remarkably similar in all three specimens.

TABLE 1

EXTERNAL AND CRANIAL MEASUREMENTS OF *Tachyglossus*
NORTHERN TERRITORY, AUSTRALIA

	Total length	Hind foot with claw	Greatest length of skull	Greatest breadth of skull	Least interorbital breadth	Depth of cranium*
Tachyglossus aculeatus acanthion						
Lake Hubert, Groote Eylandt						
♂ 283961	555	77	124·4	46·5	15·0	28·8
♀ 283962	404	62	111·9	44·8	14·9	27·2
Melville Bay, Cape Arnhem Pen.						
♀ 283963	390	55	99·8	40·4	13·6	25·5

* Measured from palate to crown in median plane

MARSUPIALIA

DASYURIDAE

Antechinus bellus (Thomas)

Pl. 2

Phascogale bella Thomas, *Novit. Zool.* 11: 229 (March 1904). Type locality: South Alligator River, Northern Territory.

Duripulmi, Cape Arnhem Peninsula, 1. Male skin with skull; 11 Aug. 1948 (283968).

Purlonkoimi Creek, Cape Arnhem Peninsula, 1. Male skin with skull; 20 Aug. 1948 (283969).

Other records for Northern Territory: South Alligator River (Thomas 1904a: 229; Tate 1947a: 133); 20 miles west of South Alligator River (Tate 1947a: 133).

The present records of this large marsupial mouse from near Cape Arnhem extend the known range of the species about 300 miles eastward from the type locality at South Alligator River.

The specimen from Duripulmi was caught during the night in a rat trap that had been set beside a granite boulder on a grass-covered knoll. No similar habitat was seen elsewhere in the general vicinity. The grassy area was surrounded by other plant associations, including stringy-bark forest, monsoonal forest, paper-bark forest, and samphire flat. The specimen from Purlonkoimi Creek was caught alive by aborigines during midday in a hilly area characterized by stringy-bark forest and laterite ledges. I was unable to obtain a description of the exact type of habitat in which it was found.

Systematic notes. The two skins from near Cape Arnhem seem to be somewhat darker than the type specimen, which Thomas (1904a: 229) described as 'pale pinkish fawn, nearest to, but darker than, Ridgway's "ecru-drab".' Colour description of the specimen from Duripulmi: Upper parts between Drab and Light Drab, slightly darker on midback than on sides; facial region slightly more brownish,

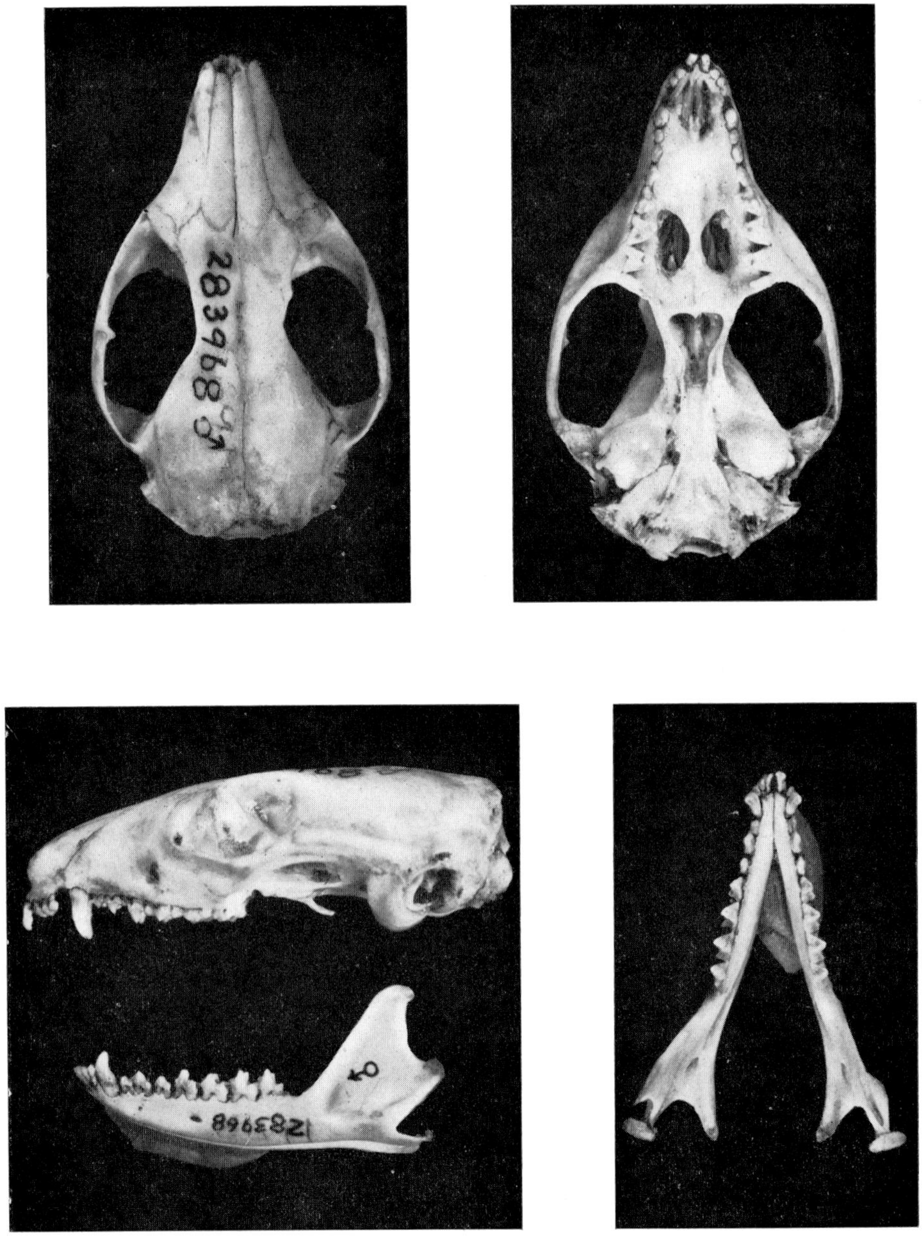

PLATE 2. *Antechinus bellus*

Skull from Duripulmi, Cape Arnhem Peninsula; no. 283968, ♂, x 2

darkest in median line and with faintly paler supra-orbital streak; postauricular areas not differentiated by colour; underparts soiled whitish, with grey bases of hairs showing through; hairs stained yellowish about glandular area in middle of throat; line of demarcation between dorsal and ventral colour areas indistinct; feet whitish; tail bicoloured on basal two-thirds (dorsal dark area a narrow brownish strip, darker at edges than in midline), tip uniformly dull brown and slightly penicillate. The upper parts of the specimen from Purlonkoimi Creek are darker, near Drab.

The skulls, compared with those of other species of *Antechinus*, appear massive and broad. The zygomatic breadth in these specimens is actually and relatively greater than in the type specimen. There are noticeable postorbital swellings on the frontals (well developed on the right side but absent on the left in no. 283968) and thin spinous postorbital processes on the zygomata. There is a single pair of large palatal vacuities opposite the first three molars; the pterygoids have thin, spinous, reflected hamular processes. The last permanent premolar is conspicuously larger than the anterior two; the first incisors are strongly developed; the smaller second, third, and fourth incisors are of uniform size. The alisphenoid bullae are prominent.

The darker colour and broader skulls of the specimens from eastern Arnhem Land may indicate geographic variation in this species, but it is evident with only two specimens at hand that there is also a great deal of individual variation.

The question raised by Tate (1947a: 133) as to whether the pads on the feet in this species are striated or granular may be answered by these specimens. Both the fore and hind feet bear striated pads similar in general form and arrangement to those of *Antechinus flavipes*.

Antechinus bilarni Johnson
Pl. 3, 16

Antechinus bilarni Johnson, *Proc. Biol. Soc. Wash.* 67: 77 (22 March 1954). Type locality: Oenpelli, East Alligator River, Northern Territory.

Oenpelli, 1. Female skin with skull; 29 Oct. 1948 (283970). The type specimen, previously recorded (Johnson 1954: 77).

Other records for Northern Territory: ? Port Essington (Thomas 1888: 292); ? Daly River (Collett 1897: 334).

It is thought that this species is restricted to rocky areas, but the actual evidence available concerning its habitat preference is meagre. The type specimen was trapped in a well-shaded and relatively cool situation at the base of a cliff on the north-eastern side of Unbalanja Hill at Oenpelli. This is a small outlier of the rocky tableland that makes up most of central Arnhem Land. The extreme flattening of the skull is a feature of many rock-ledge inhabiting species of mammals, seeming to be an adaptation for squeezing into narrow fissures.

The type specimen has been described and discussed in detail in the original description of the species (Johnson 1954). Outstanding features are the pale dorsal coloration (grizzled Hair Brown), conspicuous cinnamon-coloured pale patch behind each ear, large ears projecting well beyond the fur, long and slightly penicillate tail, and flattened skull. Measurements are given in Table 2.

The nearest relative seems to be *Antechinus melanurus* of New Guinea. None of the previously described Australian species of *Antechinus* bears much resemblance to *A. bilarni* in external features. The flattened skull suggests the condition found in *Planigale ingrami*, but this resemblance is thought to be the result of parallel adaptations for fissure-living rather than to indicate close genetic relationship.

In size, *A. bilarni* resembles *A. flavipes,* which is a common species in wooded parts of eastern and southern Australia. Individual specimens of some kind of *Antechinus* collected at Port Essington and Daly River were referred by Thomas (1888: 292) and Collett (1897: 334), respectively, to the race now known as *Antechinus flavipes leucogaster*. However, it has been shown by Shortridge (1910: 838-9) that *leucogaster* has a restricted range in the south-western part of Western Australia. The chance that the Port Essington and Daly River specimens really belong to that race is very remote, and it seems far more likely that they represent *A. bilarni*. I have not seen either specimen.

The species was named in honour of William E. Harney, who acted as guide and interpreter through most of the work of the Arnhem Land Expedition. To the aborigines he is known far and wide as 'Bilarni'.

Antechinus maculatus sinualis (Thomas)

Phascogale minutissima sinualis Thomas, *Ann. Mag. Nat. Hist.*, ser. 9, 17: 634 (June 1926). Type locality: vicinity of Emerald River, Groote Eylandt, Northern Territory.

Port Langdon, Groote Eylandt, 4. Three male skins with skulls, one female in alcohol; 9-16 June 1948 (283964-67).

Other record for Northern Territory: Groote Eylandt = vicinity of Emerald River, Groote Eylandt (Thomas 1926b: 634; Tate 1947a: 131).

One of the four specimens of this tiny dasyurid was trapped at the expedition camp site at Umbakumba, near the crest of the sandy ridge paralleling the shore, where wattle scrub is the dominant vegetation. The remaining three were caught by natives somewhere in the same general vicinity, but possibly in a different environmental association.

This subspecies was known previously only from the type specimen, an immature male collected by G. H. Wilkins near Emerald River, on the western coast of Groote Eylandt, 19 January 1925. The only near relative, *Antechinus maculatus maculatus,* inhabits the border area of Queensland and New South Wales, more than a thousand miles from Groote Eylandt.

Systematic notes. The coloration is uniform in the three skins from Port Langdon: Upper parts Buffy Brown somewhat darkened by black hair tips; underparts almost white (slightly lighter than Pale Olive-Buff); feet whitish; outer sides of forearms, legs, and proximal halves of tarsal areas with dark dorsal colour; tail dark above, very slightly paler below. These specimens differ from Gould's illustration of '*Antechinus minutissimus*' (Gould 1852: pl. 45) only in having a slightly darker dorsal colour and more white on the feet.

The three skulls are in good condition and have complete dentition. There is no evidence of flattening of the braincase such as is found in *Planigale.* The frontals have no postorbital processes or temporal ridges, but in the two older skulls there are weakly developed occipital and lambdoidal crests. The alisphenoid elements of the bullae are only moderately inflated. The palatal vacuities are comparatively small, being exceeded in area by the anterior palatine foramina. The permanent Pm^4 is a robust tooth, larger than either of the two anterior premolars; the permanent Pm_4 is minute and single-rooted.

Some remarks on tooth succession are made possible by the fact that one of the three skulls (no. 283964) retains part of the milk dentition. In this specimen the last upper deciduous 'premolar' is in place on the left side and has been displaced by the tip of the irrupting permanent premolar on the right side. It has a worn, multicusped crown, measuring 0·4 x 0·25 mm. in diameter, and in general form

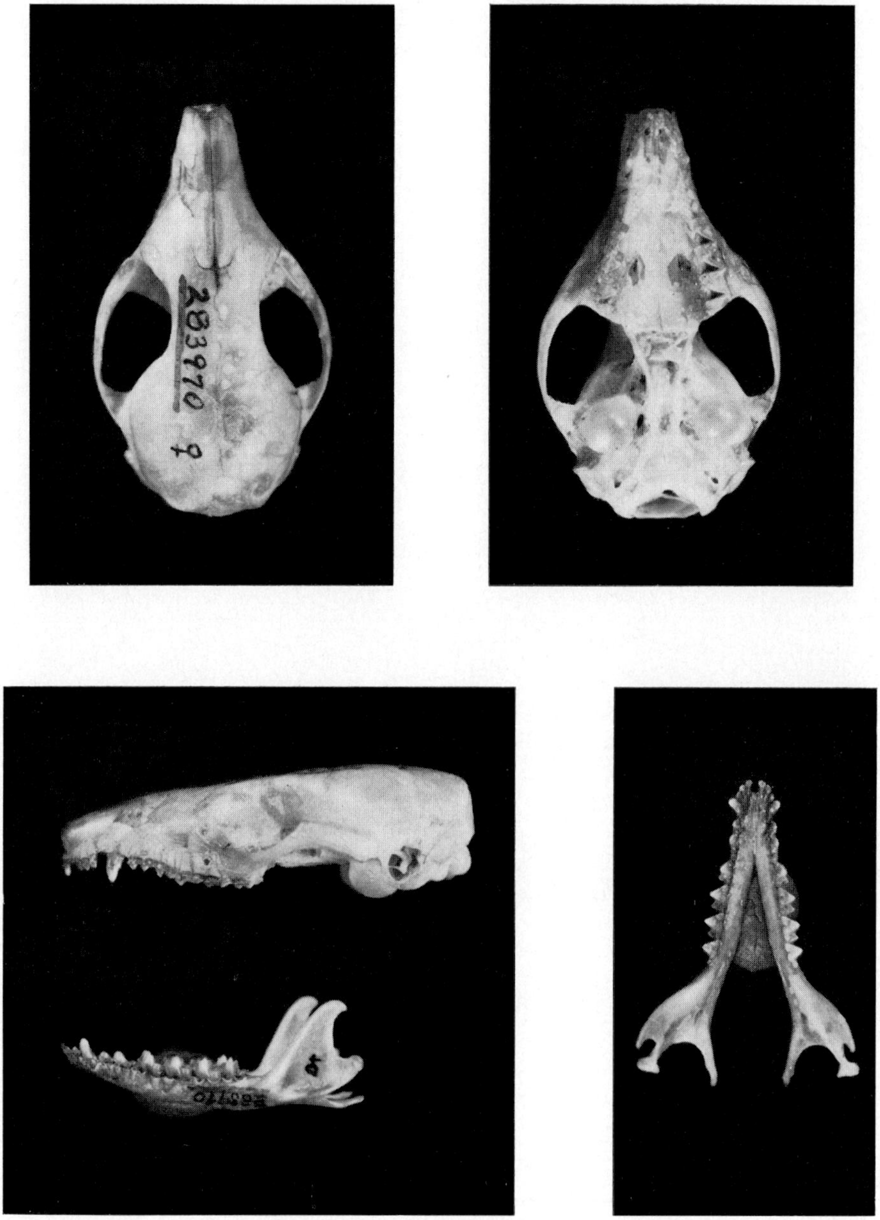

PLATE 3. *Antechinus bilarni*
Skull of type specimen, no. 283970, ♀, x 2

resembles the large molar teeth posterior to it. In each mandible, the Pm_4 persists as a tiny pointed spicule placed on the outer side of the alveolar line and crowded between the first molar and the median premolar, the crowns of which almost touch over it. The other two skulls have no trace of the milk teeth, and the crowns of the median premolars and the first molars, both above and below, are farther apart, as though displaced by the growth of the bulkier permanent last premolars.

TABLE 2

EXTERNAL AND CRANIAL MEASUREMENTS OF *Antechinus* AND *Sminthopsis* NORTHERN TERRITORY, AUSTRALIA

	Head and body	Tail	Hind foot with claw	Ear	Condylobasal length	Zygomatic breadth	Depth of braincase	Diameter of alisphenoid bulla	Tooth-row M1-M3
Antechinus bellus									
Duripulmi									
Cape Arnhem Pen.									
♂ 283968	148	122	26	22	33·3	20·6	8·5	4·2	6·0
Purlonkoimi Creek									
Cape Arnhem Pen.									
♂ 283969	135	111	23	22	31·3	19·3	7·9	4·2	6·0
Antechinus bilarni									
Oenpelli									
♀ 283970	106	109	19	20	27·2	16·1	7·0	3·8	5·7
Antechinus maculatus sinualis									
Port Langdon									
Groote Eylandt									
♂ 283964	67	64	12	11	18·9	10·6	4·9	2·4	3·9
283965	70	65	13	12	19·8	11·4	5·3	2·3	3·9
283966	78	67	13	12	20·3	10·6	5·0	2·4	4·1
♀ 283967	68	67	11	8	—	—	—	—	—
Sminthopsis larapinta nitela									
Oenpelli									
♂ 283971	94	87	20	17	25·8	14·9	7·5	3·5	5·0

In the specimen preserved in alcohol (no. 283967), the number and position of the pads on the feet are as described by Thomas (1888: 292) for the Queensland race, the minute accessory seventh pad on the outer edge of the hind foot being clearly shown. The pads, however, in this and the other three specimens, are smooth rather than striated. The only definite statement concerning the species in this regard is Finlayson's (1934: 227) comment that specimens from Serpentine Creek, near Rockhampton, had 'seven sole pads with distinctly transverse striate tubercles'. The nature of these pads seems to be more variable than the older literature indicates. A peculiar feature of the feet of this same individual from Groote Eylandt is that the claws and one or more terminal phalanges are missing from each toe of both the fore feet and the hind feet. The pouch has an opening approximately four millimetres in diameter, opening toward the rear, and with an almost circular arrangement of five pairs of undeveloped nipples.

Sminthopsis larapinta nitela Collett

Sminthopsis nitela Collett, *Proc. Zool. Soc. Lond.* 1897: 334 (August 1897). Type locality: Daly River, Northern Territory.

Oenpelli, 1. Male in alcohol (skull cleaned); 30 Sept. 1948 (283971).

Other record for Northern Territory: Daly River (Collett 1897: 335; Thomas 1906e: 543; Tate 1947a: 124).

The only specimen the Expedition obtained was brought into camp by aborigines at Oenpelli and given to William E. Harney, who preserved it in alcohol. This specimen agrees closely with Collett's original description of *nitela* from Daly River, except that in the Oenpelli specimen the tail is shorter and the small pad at the base of the hallux is a little larger. The tail is rather thick at the base, but it is not conspicuously incrassated. Colours (assumed to be somewhat faded from long immersion in alcohol): back grizzled Olive-Brown; sides heavily washed with Cinnamon-Buff; belly whitish; all body pelage grey at the base; head with conspicuous dark median dorsal stripe (narrow on nose and between eyes, broad on crown) and an inconspicuous dark stripe beneath each eye; sides of face between Cinnamon-Buff and Clay Color; lips, chin, and feet whitish. For measurements, see Table 2.

As compared with the illustration of typical *larapinta* published by Spencer (1896a: pl. 2) the specimen from Oenpelli shows conspicuous differences, including more pointed ears, a more slender and less hairy tail, and a pronounced median facial stripe. The facial stripe is said by Thomas (1906e: 542) to be present in *larapinta* but to have been overlooked by Spencer. I have followed Tate (1947a: 124) in listing *nitela* as a subspecies of *Sminthopsis larapinta*.

Satanellus hallucatus hallucatus (Gould)
Pl. 16

Dasyurus hallucatus Gould, *Proc. Zool. Soc. Lond.* 10: 41 (November 1842). Type locality: Port Essington, Northern Territory.

Brocks Creek, 1. Skull, probably male, picked up 17 Nov. 1920 (237749).

Douglas River, 15. Two male and four female skins with skulls, one adult female and eight young in alcohol; 12 Oct.-1 Nov. 1920 (237743-48, 237892-900).

Oenpelli, 4. Three female skins with skulls and one female skull only; 23 Oct.-4 Nov. 1948 (283995-98).

Black Rock Point, Cobourg Peninsula, 1. Female skin with skull; 26 Sept. 1948 (283994).

Yirrkalla, 8. Two male and three female skins with skulls, three lots of pouch-young in alcohol; 28 Aug.-1 Sept. 1948 (283986-93).

Other records for Northern Territory: Daly River (Collett 1897: 333); Port Darwin (Thomas 1888: 271); Adelaide River (Collett 1897: 333); Mount Shoebridge (Collett 1897: 333); Union Town (Collett 1897: 333); Nellie Creek (Thomas 1904a: 228); Mary River (Collett 1897: 333); South Alligator River (Thomas 1904a: 228, 1926c: 543); Port Essington (Gould 1842d: 41; Gray 1843: 98; Thomas 1888: 270, 1909a: 150, 1926c: 543; Tate, 1947a: 143); King River (Thomas 1926c: 543); Alexandria (Thomas 1906e: 540, 1926c: 543).

These small spotted native cats showed a definite preference for a rocky habitat but were not necessarily restricted to it. At Yirrkalla they were found in the broken terrain of the laterite bluff area between the mission station and Gove Airfield. On 28 August three adult females were caught there in cage type live traps, and on 1 September two adult males were caught in the same area. Trapping elsewhere in the vicinity of Yirrkalla failed to produce specimens of this species.

At Oenpelli they were trapped about the cliffs and talus of Unbalanja Hill, and the animals themselves were sometimes seen by the anthropologists who camped there to study aboriginal cave paintings. At Douglas River in 1920, Hoy found them inhabiting hollow trees and limestone caves.

The species seems to have a well-defined annual breeding cycle, the young being produced during the dry season. In three females of the subspecies *nesaeus* from Groote Eylandt, collected on 7, 24 and 30 June respectively, the pouch area was inactive, its borders only faintly defined, the nipples small, and glandular swelling and discoloration lacking. Three females of the subspecies *hallucatus* collected on 28 August at Yirrkalla all had very small young attached to the nipples, which were enlarged and enclosed within a marsupial area bordered laterally by well-defined folds of skin; the fur and skin in this area were deeply reddish in colour. Other adult females from Douglas River (12 to 17 October) and from Oenpelli (23 October to 4 November) all showed evidence of having nursed recently but did not have young attached to the nipples.

A family of eight young (seven males and one female) collected with their mother from a hollow tree at Douglas River on 15 October 1920 averaged near 85 mm. in length of head and body. Two larger males collected in the same vicinity on 1 November measured 183 and 184 mm. in head-and-body length and had reached the stage of development where the third upper molar teeth were about to irrupt.

With the series of specimens now available for study in the United States National Museum it is possible to evaluate some of the characters used and the conclusions reached by Thomas (1926c), who last revised the species. That author recognized four geographical subspecies, which are, from west to east: *exilis*, a small pallid race from the north-eastern part of Western Australia, type locality Parrys Creek near Wyndham; typical *hallucatus* from the northern coastal part of Arnhem Land, type locality Port Essington; *nesaeus*, a small and short-tailed race restricted to Groote Eylandt; and *predator*, a large dark-coloured subspecies from northern Queensland, type locality Utinga on Cape York Peninsula.

In general the present study confirms the above arrangement, but the various races are not so sharply defined as Thomas seemed to think when he based them mainly on differences in colour and size. Colour is actually subject to seasonal variation that exceeds in each individual the whole range of geographic variation of the species. Specimens taken in the dry season (May to September) are in old pelage with most of the buffy bands of the guard hairs worn away. As a result, the dark underfur is exposed and these specimens appear almost black with the white spots standing out sharply in contrast. Toward the end of the dry season there is a moult in which this worn pelage is replaced by a fresh coat of long buff-banded hair characteristic of specimens taken in October and November. Paradoxically, the specimens in worn pelage have a dark, sleek appearance, whereas those in the longer new pelage appear shaggy and drab-coloured.

There is a pronounced difference in size between the sexes, males being consistently larger than females (see measurements, Table 3).

Taking these factors into account, the various populations of *Satanellus hallucatus* in Arnhem Land seem to be remarkably constant, and except for the Groote Eylandt population of *nesaeus* they are all here considered to represent the subspecies *hallucatus*. It is quite possible that the western form *exilis* also occurs in the Northern Territory in the Victoria River region, as the type locality is not far across the border in Western Australia. When series of specimens are available from central Arnhem Land and from the areas bordering the deserts on the south, it will be interesting to know how they compare with this northern coastal series.

TABLE 3

EXTERNAL AND CRANIAL MEASUREMENTS OF *Satanellus* NORTHERN TERRITORY, AUSTRALIA

	Head and body	Tail	Hind foot, with claw	Ear from notch	Condylobasal length	Zygomatic breadth	Interorbital breadth	Tooth-row M1-M3
Satanellus hallucatus hallucatus								
Brocks Creek								
♂ 237749	—	—	—	—	69·1	—	14·6	12·8
Douglas River								
♀ 237743	246	285	51	—	60·9	34·2	13·2	12·8
237744	260	279	49	—	61·9	35·3	13·2	13·1
237745	256	276	50	—	59·4	34·4	12·2	13·0
237748	280	272	51	—	61·7	36·6	13·7	12·7
Oenpelli								
♀ 283995	247	266	52	40	59·6	36·2	13·2	12·7
283996	272	293	52	39	62·7	36·5	13·4	13·0
283997	—	—	—	—	60·5	34·9	13·6	13·3
283998	278	255	49	41	62·0	36·5	14·4	12·7
Black Rock Pt, Cobourg Pen.								
♀ 283994	268	307	52	40	63·6	37·6	14·2	13·5
Yirrkalla								
♂ 283992	299	272	59	44	71·6	44·2	15·4	13·7
283993	293	282	58	39	66·6	41·3	14·9	12·9
283986	254	275	49	43	64·7	38·5	15·1	13·4
♀ 283988	271	252	48	42	63·3	37·0	14·4	12·7
283990	269	245	47	39	63·9	37·2	13·7	13·5
Satanellus hallucatus nesaeus								
Groote Eylandt								
♂ 283972	297	271	51	37	—	39·6	13·9	12·7
283973	285	240	49	37	64·0	38·4	14·7	13·3
283974	277	245	50	37	65·5	39·2	13·8	12·7
283975	264	246	46	36	62·6	36·6	13·0	13·2
283976	299	282	54	39	—	—	—	—
283978	265	239	48	36	64·3	36·8	13·6	12·7
283979	289	281	52	38	68·6	41·5	15·2	13·0
283980	262	265	53	35	63·3	38·2	14·7	12·8
283981	294	269	52	37	67·6	39·9	15·0	13·0
283984	273	228	45	35	63·1	38·0	13·8	12·4
283985	269	245	50	—	61·7	38·6	14·3	12·1
♀ 283977	254	215	43	37	59·6	34·9	11·6	13·0
283982	252	240	45	34	60·7	35·0	13·6	13·3
283983	242	226	43	35	57·1	33·2	12·4	12·2

Satanellus hallucatus nesaeus (Thomas)

Dasyurus hallucatus nesaeus Thomas, *Ann. Mag. Nat. Hist.*, ser. 9, 18: 544 (November 1926). Type locality: Groote Eylandt, Northern Territory.

Port Langdon, Groote Eylandt, 11. Eight male and two female skins with skulls (one with body skeleton), one female in alcohol; 26 May-24 June 1948 (283972-82).

Central Hill, Groote Eylandt, 3. One male and one female, skins with skulls; one male skull only; 30 June-2 July 1948 (283983-85).

Other record for Northern Territory: Groote Eylandt (Thomas 1926c: 544; Tate 1947a: 143).

Most of the specimens were caught in meat-baited cage traps set near logs and the bases of trees in eucalypt forest. There are no rocky habitats in the immediate vicinity of Umbakumba on Port Langdon where most of the trapping was done, but the species appeared to be common nevertheless. At Central Hill the habitat was of the more characteristic rock-ledge type.

The series collected on the Arnhem Land Expedition shows a body size somewhat greater than that described by Thomas (1926c: 544) for the three original specimens, which were collected by G. H. Wilkins in March 1925. As the measurements (Table 3) show, however, Groote Eylandt specimens are consistently smaller than those from the mainland. The tail is relatively shorter, equalling on the average about 91 per cent of the head-and-body length, as compared with nearer 100 per cent in the mainland populations. The coloration is essentially as described by Thomas.

NOTORYCTIDAE

Notoryctes typhlops (Stirling)

Psammoryctes typhlops Stirling, *Trans. Roy. Soc. S. Aust.* 12: 158 (December 1889). (Name based on description in the same journal, 11: 21-24 [April 1889].) Type locality. Idracowra Cattle Station, Finke River, Northern Territory.

Idracowra Cattle Station, 1. Skeleton, prepared from a specimen in alcohol received in 1899 from Professor D'Arcy W. Thompson, University College, Dundee, Scotland (49545). First recorded from this locality by Stirling (1889a: 21) and subsequently by many authors. Variant spellings are Idracowie and Indracowrie.

Other records for Northern Territory: Undoolya Station, near Alice Springs (report cited by Stirling 1891a: 156); Tempe Downs (report cited by Stirling 1891a: 156); Liddle's Hills, Basedow Range area (Finlayson 1947: 185); Crown Point, Finke River (Spencer 1896a: 43; Troughton 1941: 57); Charlotte Waters (Spencer 1896c: 52).

The possibility of collecting marsupial moles was one of the chief reasons for visiting Horse Shoe Bend Station on the Finke River after the principal work of the Arnhem Land Expedition was completed in 1948. Unfortunately this period, 22 to 29 November, came in the midst of a prolonged dry spell. No rain had fallen for several months and the sand ridges inhabited by the moles were excessively dry. Both white and aboriginal inhabitants of the area declared that it was quite futile to attempt to find the animals except immediately after the infrequent rains, when tracks and disturbances caused by burrowing might be seen on the surface of the damp sand near the ridge tops.

PERAMELIDAE

Isoodon macrourus macrourus (Gould)

Pl. 4

Perameles macroura Gould, *Proc. Zool. Soc. Lond.* 10: 41 (November 1842). Type locality: Port Essington, Northern Territory.

Nightcliff, 7 miles north-east of Darwin, 3. One female skin with skull, one skull only (sex unknown), and set of four pouch-young in alcohol; 5-13 April 1948 (283999-4001).

Brocks Creek, 1. Male skin with skull; 13 Nov. 1920 (237739).

Douglas River, 7. Five male and one female skins with skulls (one with body skeleton), and one male skull only; 12-31 Oct. 1920 (237732-38).

Oenpelli, 3. Male skin with skull, male and female complete skeletons; 1-3 Nov. 1948 (284017-19).

Black Rock Point, Cobourg Peninsula, 3. Two male skins with skulls, one female complete skeleton; 26-8 Sept. 1948 (284013-15).

James Creek, Cobourg Peninsula, 1. Male skull only; 12 Oct. 1948 (284016).

Elcho Island, 2. Male and female skins with skulls; 14-15 Sept. 1948 (284011-12).

Rocky Bay, Cape Arnhem Peninsula, 1. Female skin with skull; 26 Aug. 1948 (284010).

Port Langdon, Groote Eylandt, 8. Two male and three female skins with skulls, one male skull only, one male and set of two pouch young in alcohol; 5-16 June 1948 (284002-09).

Other records for Northern Territory: Daly River (Collett 1897: 333); Mount Shoebridge (Collett 1897: 333); Katherine River (Collett 1897: 333); The Granites (Tate 1948b: 339); South Alligator River (Thomas 1904a: 227); Port Essington (Gould 1842d: 41; Thomas 1888: 235; Tate 1948b: 339). A supposed record from Victoria River (Thomas 1888: 235) has been shown to refer to *Isoodon obesulus* (Thomas 1904a: 228).

Throughout the area investigated by the Arnhem Land Expedition this seems to be the only species of bandicoot, and it is evidently distributed uniformly over the mainland and most if not all of the major islands. *Isoodon obesulus* a somewhat smaller species that has been recorded from various localities south and south-west of Arnhem Land and as far north as the South Alligator River is not represented in the material reported here.

The practice of burning off the ground vegetation that is indulged in every year by aborigines throughout Arnhem Land destroys the habitat of most kinds of ground-dwelling small mammals. The bandicoots seem better able than other kinds to survive this burning, perhaps because they customarily take shelter in hollow logs that are only slightly scorched by the brief fires. There is no doubt, however, that the bandicoot population is greatly reduced as a result of the burning. In the small patches of forest understorey that escape the fires, nests and runways of bandicoots can nearly always be found.

A typical nest was found in the stringy-bark forest south of Melville Bay on 18 August. It was on the ground in an area where subsurface moisture had produced a slightly better cover of grass and small shrubs than was present in the surrounding forest. The nest was a flattened mass of grass and twigs about 18 inches in diameter and 6 inches deep lying in a shallow depression which left about half the nest above ground level. A well marked runway extended into the surrounding growth of grass from either end of the nest, which was completely roofed over and seemed to have a single shallow central chamber. One of the aborigines demonstrated his method of catching bandicoots by cautiously approaching the nest until he was within about ten feet of it, then hurling a 'tomahawk' into the nest with such force that any animal there would surely have been disabled. The nest happened to be empty, but I was assured that this method of hunting was often successful, as was an alternative method of leaping suddenly on the nest and pinning its occupants down with the feet and hands.

In April two bandicoots were found dead on the road near Nightcliff, north-east of Darwin, having been killed by passing cars as they ran across between banks of high speargrass. At Oenpelli bandicoots were trapped in a large patch of coarse grass growing on dry soil near the upper end of the billabong. Otherwise all observations of bandicoots were made in forested areas.

They were active mainly at night, nearly all trapped animals being caught in the period between dusk and dawn. At James Creek on the Cobourg Peninsula on 12 October, a male bandicoot made a thorough and persistent investigation of my camp site about an hour after dark, showing no fear of the dying campfire and numerous articles of equipment. In the vicinity of Umbakumba on Groote Eylandt an occasional bandicoot was seen and heard bounding off through the sparse shrubby undergrowth in the daytime. These were thought to be animals that were disturbed from diurnal resting places.

There does not seem to be a sharply defined breeding season. Pouch-young were found in females collected at Nightcliff on 13 April and on Groote Eylandt on 12 June. All other adult females collected during the 1948 dry season had nursed young recently, including one from Elcho Island on 15 September which was so young that on the basis of size and tooth development it would have been judged immature. Partly grown animals were caught throughout the season from early June to November.

The skull of a large male (no. 284011) from Elcho Island is remarkably

Table 4

External and Cranial Measurements of *Isoodon*

Northern Territory, Australia

	Head and body	Tail	Hind foot, with claw	Ear from notch	Condylobasal length	Zygomatic breadth	Length of nasals	Tooth-row M1-M3
Isoodon macrourus macrourus								
Nightcliff								
♀ 284000	366	160	72	35	—	—	31·0	12·4
Brocks Creek								
♂ 237739	360	162	75	—	73·3	33·8	32·4	12·6
Douglas River								
♂ 237732	384	170	74	—	74·0	34·5	30·8	13·7
237734	300	159	69	—	—	—	—	12·0
237736	362	179	77	—	73·8	—	—	12·8
237737	355	168	71	—	73·9	35·2	31·0	12·3
237738	365	158	76	—	73·5	—	—	12·1
♀ 237735	278	155	65	—	—	—	25·9	11·8
Oenpelli								
♂ 284017	261	128	65	31	64·5	30·0	25·5	12·0
284019	—	—	69	33	65·7	30·9	26·1	11·5
Black Rock Pt, Cobourg Pen.								
♂ 284013	346	153	70	35	—	32·3	31·4	12·7
284014	436	184	86	40	78·2	38·0	32·5	13·4
♀ 284015	347	138	73	—	68·9	30·9	30·8	12·8
James Creek, Cobourg Pen.								
♂ 284016	345	—	75	35	71·2	34·2	31·0	12·8
Elcho Island								
♂ 284011	424	191	84	37	79·1	36·9	33·8	13·3
Rocky Bay, Cape Arnhem Pen.								
♀ 284010	340	159	71	36	70·3	31·8	29·4	13·0
Port Langdon, Groote Eylandt								
♂ 284003	382	188	78	36	74·7	34·0	31·1	12·8
284006	344	163	71	35	68·9	31·6	29·3	12·1
284009	329	163	71	34	67·1	30·6	29·2	11·5

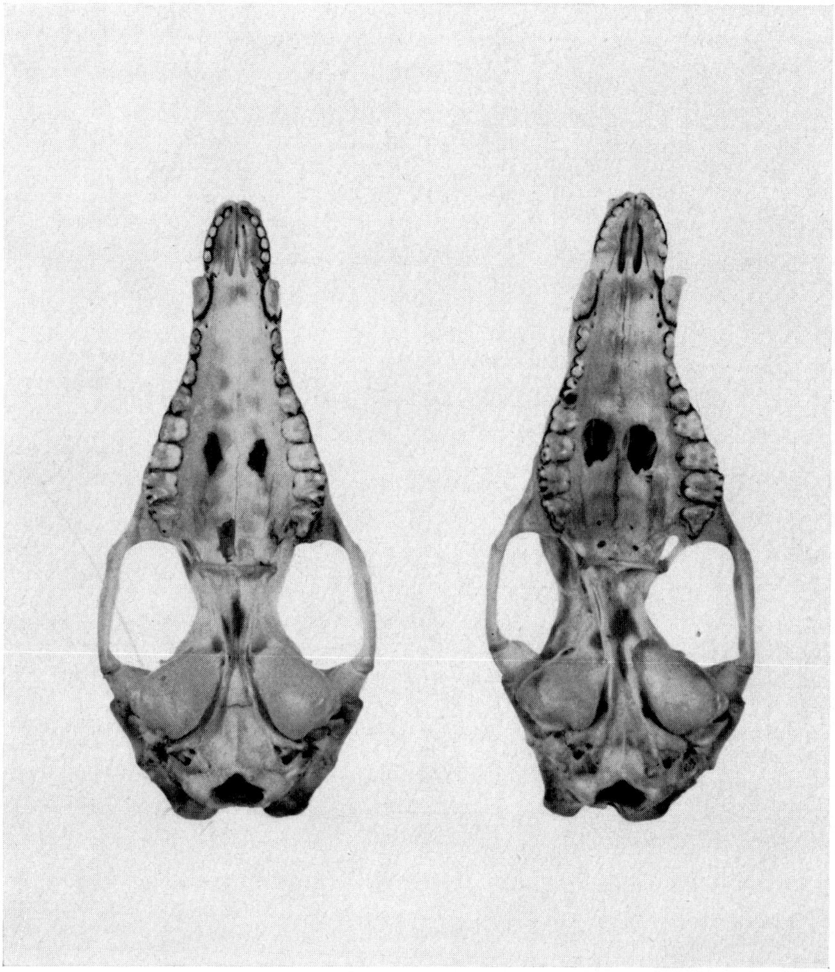

PLATE 4. *Isoodon macrourus macrourus*—ventral view of skulls

Left, normal skull from Black Rock Point, no. 284014, ♂, x 1
Right, asymmetrical skull from Elcho Island, no. 284011, ♂, x 1

asymmetrical. The orbito-temporal fossa on the left side is larger and is placed farther posteriorly than that on the right, and the main axis of the skull is deflected to give a slight concavity on the left side. There is no evidence of injury to explain the asymmetry (Pl. 4).

There is considerable variation in size among the specimens at hand (see measurements, Table 4), but apparently it is all attributable to age, sex, or purely individual variation. I find no consistent pattern of geographic variation. The Groote Eylandt series is indistinguishable from mainland series.

PHALANGERIDAE

Trichosurus vulpecula arnhemensis Collett

Trichosurus vulpecula var. *arnhemensis* Collett, *Proc. Zool. Soc. Lond.* 1897: 328 (August 1897). Type locality (restricted by Iredale and Troughton 1934: 31): Daly River, Northern Territory.

Port Darwin, 5. Two male and one female skins with skulls, one male in alcohol, 1 Oct. 1920 (237698-700, 237890); skull only picked up on beach, 24 Nov. 1920 (237836). Other specimens recorded by Thomas (1888: 189).

Bathurst Island, 2. Immature male and female skins with skulls; Dec. 1920 (237751-52).

Cape Don, Cobourg Peninsula, 1. Male skin with skull; 20 Sept. 1948 (284031).

Croker Island, 2. Male skins with skulls; 23 Apr. 1948 (284020-21).

Elcho Island, 1. Male skin with skull; 14 Sept. 1948 (284030).

Yirrkalla, 2. Male skin with mandibles and male pouch-young in alcohol; 24 July and 4 August 1948 (284022-23).

Rocky Bay, Cape Arnhem Peninsula, 1. Male skin with skull; 25 August 1948 (284029).

Purlonkoimi Creek, Cape Arnhem Peninsula, 2. Male skins with skulls; 20 Aug. 1948 (284027-28).

Kwaiturumuru Creek, Cape Arnhem Peninsula, 3. Male skins with skulls; 18-19 Aug. 1948 (284024-26).

Other records for Northern Territory: Daly River (Collett 1897: 328); Hall's Creek (Thomas 1904a: 227); Eureka (Thomas 1904a: 227); Katherine River (Collett 1897: 328); South Alligator River (Thomas 1904a: 227); Port Essington (Thomas 1888: 189); Alexandria (Thomas 1906e: 540).

These generally arboreal animals occur regularly, but in my experience not abundantly, throughout coastal Arnhem Land, including the larger nearby islands except Groote Eylandt. Two collecting methods that proved successful were cutting into likely-looking hollow trees in the daytime and eye-shine hunting with a strong light at night. Both methods required many miles of walking for each specimen collected.

In the area south of Melville Bay, where 'walkabout' collecting was done from 17 to 21 August, male possums, but no females, were found in the hollow thin-walled boles of a kind of smooth-barked gum tree. Specimens of the large arboreal rat, *Mesembriomys gouldii,* were found in identical situations but not in the same individual trees.

On 1 October 1920, Hoy collected specimens near Darwin 'from trees along sea cliff'.

Well advanced single pouch-young were obtained at Yirrkalla on 4 August 1948 (male, 150 mm. in head-and-body length) and near Darwin on 1 October 1920 (female, 185 mm.). Immature animals, all males with the last upper molars still growing, were taken at Elcho Island on 14 September, at Cape Don on 20 September, and at Bathurst Island sometime in December 1920.

I find no geographic variation in colour or cranial characters within the Arnhem Land area. Adult males have more reddish-coloured underfur than the younger individuals, in which the underfur is drab grey. A very old male from Rocky Bay, near Cape Arnhem, is rusty-orange on the neck and foreback and has a conspicuous reddish-brown throat patch.

TABLE 5

EXTERNAL AND CRANIAL MEASUREMENTS OF *Trichosurus* NORTHERN TERRITORY, AUSTRALIA

Trichosurus vulpecula arnhemensis	Head and body	Tail	Hind foot, with claw	Ear	Condylobasal length	Zygomatic breadth	Length of nasals	Tooth-row M1-M3
Port Darwin								
♂ 237699	358	302	65	—	72·2	49·1	30·1	14·7
237700	364	276	66	—	76.7	—	31·5	13·5
♀ 237698	264	266	61	—	72·3	47·2	27·2	14·4
Croker Island								
♂ 284020	354	256	63	47	69·0	44·9	27·7	13·6
284021	338	244	64	45	66·3	43·4	27·0	14·2
Yirrkalla								
♂ 284022	385	260	65	52	—	—	—	—
Rocky Bay, Cape Arnhem Pen.								
♂ 284029	363	243	60	48	76·3	48·0	32·2	13·7
Purlonkoimi Creek Cape Arnhem Pen.								
♂ 284027	334	248	60	52	67·3	45·2	30·0	13·2
284028	338	246	60	50	70·6	42·1	30·5	13·9
Kwaiturumuru Creek Cape Arnhem Pen.								
♂ 284024	400	267	62	52	76·4	49·0	33·2	14·2
284025	376	266	58	48	71·0	48·4	30·4	14·6
284026	345	264	58	49	68·3	45·5	28·3	14·0

Habitat conditions for *Trichosurus* are evidently more favourable in the forested northern part of the Northern Territory than in the arid southern part where there are few large trees except in broken lines along stream beds. The status of possums in the latter area is vague. No recent opinion seems to have been expressed as to the subspecies represented by specimens from localities south of Alexandria, and no specimens are now available to me from that area. Spencer (1896a: 16-17) recorded 'var. *typicus*' as 'occurring everywhere amongst the eucalypts which border the river-beds' in central Australia, but he did not mention specimens or specific localities. Jones (1924: 199) described but did not name a 'small form of light blue-grey colour' inhabiting central Australia. Finlayson (1947: 186) provided a definite locality, Wollara in the Basedow Range area, but did not identify the subspecies.

Petaurus breviceps ariel (Gould)

Belidea ariel Gould, *Proc. Zool. Soc. Lond.* 10: 11 (June 1842). Type locality: Port Essington, Northern Teritory.

Cape Don, Cobourg Peninsula, 5. One male and four female (two young) skins with skulls; 20 Sept. 1948 (284052-56).

Black Rock Point, Cobourg Peninsula, 1. Female skin with skull; 25 Sept. 1948 (284057).

Skirmish Point, near mouth of Liverpool River, 1. Young female skin with skull; 8 May 1948 (284032).

Elcho Island, 7. Two male and two female skins with skulls, 17 May 1948; one male skin with skull and two young male skulls only, 15 Sept. 1948 (284033-39).

Yirrkalla, 1. Female skin with skull; 23 July 1948 (284051).

Port Langdon, Groote Eylandt, 10. Four male and three female skins with skulls; one male and one female skins only, one skull only of unknown sex; 29 May-21 June 1948 (284040-49).

Central Hill, Groote Eylandt, 1. Female in alcohol; 1 July 1948 (284050).

Other records for Northern Territory: Mount Shoebridge (Collett 1897: 332); Mary River (Collett 1897: 332); Katherine River (Tate 1945b: 9); South Alligator River (Thomas 1904a: 227; Tate 1945b: 9); Melville Island (Finlayson 1934: 223); Port Essington (Gould 1842c: 11; Gray 1843: 84; Thomas 1888: 157; Tate 1945b: 9); Groote Eylandt (Tindale 1925: 80; Finlayson 1934: 223).

This is one of the best-known small mammals in Arnhem Land, more because of its attractiveness in captivity than because of abundance in the wild. At the permanent settlements along the coast some of the aborigines carry on a small trade of selling souvenirs to the infrequent visitors, usually under the supervision of the missionaries or other station managers. 'Flying squirrels' are popular items in the trade, and the present series from Cape Don and Elcho Island were obtained in this way. The others were either shot by eye-shine hunting at night or were taken from hollow limbs of trees by native helpers.

When Dahl collected mammals south of Darwin in 1894 and 1895 he found this species to be uncommon and difficult to obtain (Dahl 1897: 205-6). Collecting in the same area in 1920, Hoy failed to obtain any specimens of it.

In the vicinity of Port Langdon on Groote Eylandt they were comparatively abundant, individuals being collected at night both in eucalypt forest on the flats and in wattle scrub on the sand dunes. It proved most successful to hunt alone on windless nights, when a soft grunting call-note made by the animals could be heard at considerable distance.

At Cape Don, Mr K. Davidson, chief lighthouse keeper, told of an episode in which a captive flying squirrel escaped and climbed the outside wall of the lighthouse as far as the observation railing near the top. When pursued there it leaped off into space and sailed away over the edge of the cliff, finally disappearing into the forest near the water's edge far below.

There is no evidence of a restricted breeding season. Two pouch-young 40 mm. in head-and-body length were being carried by a female collected at Central Hill on Groote Eylandt on 1 July. These were in a well-defined pouch, which in the alcohol preserved specimen is about 40 mm. in diameter and has an opening barely 20 mm. long. This is in contrast to the statement by Dahl (1897: 206) that '*Petaurus* having no pouch, the young simply adhere to the teats of the mother'.

Partly grown young were collected at Skirmish Point on 8 May (only the first upper molar tooth functional), at Cape Don on 20 September (two molars functional), and at Elcho Island on 15 September (three molars functional).

Other than a slight tendency for the specimens from Elcho Island to appear larger and darker in colour, there is no appreciable geographic variation. The differences in the Elcho Island series are slight and could possibly result from chance selection. However, they do suggest the interesting possibility of a saturate race on the chain of Wessel Islands, which extend north-eastward from Elcho Island and are mammalogically unknown. I can find no indication of difference between Groote Eylandt and other populations.

TABLE 6

EXTERNAL AND CRANIAL MEASUREMENTS OF *Petaurus*
NORTHERN TERRITORY, AUSTRALIA

	Head and body	Tail	Hind foot, with claw	Ear from notch	Condylobasal length	Zygomatic breadth	Length of nasals	Tooth-row M1-M3
Petaurus breviceps ariel								
Cape Don, Cobourg Pen.								
♂ 284055	152	176	26	25	33·8	25·8	12·0	5·6
♀ 284054	140	157	25	24	31·5	23·7	11·0	5·4
284056	149	179	27	26	33·5	24·6	12·3	5·6
Black Rock Pt, Cobourg Pen.								
♀ 284057	138	169	23	26	32·3	24·0	11·5	5·4
Elcho Island								
♂ 284033	173	228	32	26	38·2	27·6	14·3	5·7
284035	180	215	31	26	37·6	27·1	13·5	—
♀ 284034	162	221	29	27	36·7	24·9	13·5	5·8
284036	156	211	32	28	35·5	25·3	13·3	5·5
Yirrkalla								
♀ 284051	152	177	26	25	32·7	24·6	12·1	5·5
Port Langdon, Groote Eylandt								
♂ 284043	156	190	28	26	—	—	—	—
284045	158	197	27	28	34·8	25·4	12·1	5·3
284048	163	181	26	28	35·0	25·4	11·5	5·2
284040	170	179	28	26	35·0	25·6	12·7	5·1
♀ 284044	156	164	26	27	33·3	24·1	11·7	5·5
284046	153	180	27	26	33·1	24·2	11·7	5·2
284047	155	190	27	26	33·6	23·7	11·6	5·4
Central Hill, Groote Eylandt								
♀ 284050	170	202	30	25	—	—	—	—

Petropseudes dahlii (Collett)

Pl. 16

Pseudochirus dahlii Collett, *Zool. Anz.* 490: 464 (December 1895). Type locality: Mary River, Northern Territory.

Oenpelli, 2. Female skins with skulls (one with body skeleton); 30 Oct. and 3 Nov. 1948 (284058-59).

Other records for Northern Territory: Union Town (Collett 1897: 329); Nellie Creek (Thomas 1904a: 227; Tate 1945a: 22); Mary River (Collett 1895: 464, 1897: 329); South Alligator River (Thomas 1904a: 227).

The specimens from Oenpelli provide an easternmost record of occurrence for this seldom-collected phalanger. As Dahl (1897: 203) has pointed out, it probably occurs generally throughout the rocky central plateau of Arnhem Land. There is no indication that it occurs anywhere outside that area.

Its presence at Oenpelli was first discovered on 30 October 1948, by Captain Walter Murphy, who had volunteered to assist the expedition by searching for aboriginal cave drawings. He came upon two of the animals during the daytime well back in a sheltered rock crevice about two miles east of the mission. One of them escaped by retreating out of sight into the crevice. The other, which was shot, proved to be an adult female. Shortly after dark on 3 November, I shot an immature female in the same general area while eye-shine hunting for rock wallabies.

Neither of the females had young in the pouch or showed signs of having carried young recently.

Measurements of the two specimens (adult female 284058 and young female 284059), respectively, are as follows: Head and body 328, 316; tail 225, 224; hind foot with claw 54, 55; ear from notch 28, 26; condylobasal length 65·3, 56·9; zygomatic breadth 36·6, 36·5; length of nasals 28·3, 21·7; tooth-row M^1-M^3 13·7, 13·1.

MACROPODIDAE

Petrogale venustula Thomas

Pl. 16

Petrogale venustula Thomas, *Ann. Mag. Nat. Hist.*, ser. 9, 17: 628 (June 1926). Type locality: King River, Northern Territory (near lat. 11° 56′ S., long. 133° 25′ E.; not the King River on the Upper Daly as Thomas supposed; see discussion below).

Oenpelli, 9. Three male and three female adult and sub-adult skins with skulls (one with body skeleton), one male adult skull only, one pouch-young skin with skull, and one pouch-young in alcohol; 22 Oct.-3 Nov. 1948 (284060-68).

Other record for Northern Territory: King River (Thomas 1926b: 629).

Rock wallabies were conspicuous on the rocky hills about Oenpelli, especially in the mornings and late afternoons. They were seen creeping through narrow fissures, bounding through darkened galleries beneath fallen rock slabs or along the tops of exposed ridges, and scurrying through sparse ground vegetation between isolated rock outcrops. At least some nocturnal activity is indicated by the fact that one was shot well after dark in the course of eye-shine hunting.

Two pouch-young were preserved. One taken on 23 October, 200 mm. in head-and-body length, is well haired and exhibits the complete colour pattern of the adult; another taken 28 October measures 90 mm. and is practically hairless and without pigment except on the muzzle.

These animals are so closely restricted to the rock habitat that their presence or absence in a given area can be predicted with a high degree of accuracy. About Oenpelli, wherever rock layers were sufficiently exposed to have formed fissures, ledges, or talus heaps, rock wallabies or recent signs of their presence could be found. When Cooper Creek was followed upstream from the black soil plain into the hills, rock wallabies were seen at the first rock bluff along the way.

In contrast to the favourable habitat provided by the quartzitic sandstone exposures in the Oenpelli area, the bluffs of red laterite that occur frequently at Yirrkalla and elsewhere near the coast do not support rock wallabies.

At Central Hill on Groote Eylandt, old signs (presumably of *Petrogale longmani*) were abundant, but, because the population was then low or for some other reason, none of the animals could be found in two days of hunting there in late

June of 1948. No rock wallabies were found in the Cape Arnhem area, a fact that caused no surprise because a suitable habitat was lacking. Wilkins (1929: 260) reported seeing rock wallabies in 1925 on Mount Alexander, between Port Bradshaw and Caledon Bay.

Except for slightly greater average size (see measurements, Table 7), the specimens from Oenpelli agree in detail with the description of typical *venustula*. The pure white shoulder patches are especially notable. Lacking specimens of the various neighbouring kinds for comparison I am unable to determine whether or not *venustula* intergrades with any of them.

The several kinds of rock wallabies recorded from the northern third of the Northern Territory present a confusing pattern of distribution. This is partly cleared up when it is realized that, in describing *Petrogale venustula*, Thomas (1926b: 629) misplaced the King River of Wilkins' expedition, erroneously identifying it with the King River tributary to the upper Daly River at about lat. 14° 40′ S., long. 132° 30′ E. From the published narrative of his Australian collecting, it is clear that in early November of 1924 when the type specimen was taken, Wilkins was camped at about lat. 11° 56′ S., long. 133° 25′ E., on the King River that reaches the north coast of Arnhem Land near the Goulburn Islands. The site is described as being in rocky hills some fifteen miles inland (Wilkins 1929: 200). The remarkable resemblance of the present series from Oenpelli to typical *venustula* is thus explained, the two localities being less than forty miles apart and on the fringe of the same rock escarpment.

From specimens collected by Wilkins, Thomas also named two other species: *wilkinsi*, a larger and more yellowish wallaby from Roper River, and *longmani*, a greyer form with less contrasting pattern from Groote Eylandt. These three seem to share many characters in common and to differ from the short-eared rock wallabies related to *Petrogale brachyotis*, which have been recorded from South Alligator River and other localities in the south-western part of Arnhem Land.

Petrogale lateralis Gould

Petrogale lateralis Gould, *A Monograph of the Macropodidae*, pt. 2, pl. 24 (1842). Type locality: Swan River, Western Australia (restricted through designation of lectotype by Thomas 1922c: 128; Tate 1948a: 272 erroneously gives the type locality as Liverpool Plains, New South Wales).

Wigley Water Hole, 4 miles north of Alice Springs, 2. Female skin with skull, and pouch-young in alcohol; 21 Nov. 1948 (284069-70).

Other records for Northern Territory: Barrow Creek (Tate 1948a: 273); Simpson's Gap (Finlayson 1930: 179); Alice Springs (Spencer 1896a: 15; Finlayson 1930: 179); Paisley's Bluff, 37 miles west of Alice Springs (Spencer 1896a: 15; Finlayson 1930: 179); Koporilya Spring, 10 miles west-south-west of Hermannsburg (Finlayson 1930: 179); Glen Edith, west of James Range (Finlayson 1947: 188); King's Creek, George Gill Range (Finlayson 1947: 188).

The adult female was shot on a narrow ledge of the vertical wall of one of the small gorges that converge at Wigley Water Hole. Its single pouch-young measures 20 mm. in head-and-body length. Measurements of the adult are given in Table 7.

Finlayson (1930: 179) has commented on the probability that rock wallabies from the desert ranges of central Australia are at least subspecifically distinct from typical *Petrogale lateralis* of the south-western part of Western Australia, and in addition he has described what appears to be geographic variation among populations within the central area. Lack of comparative specimens prevents any further investigation of these interesting problems at present.

After comparing the present specimen with examples of typical *Petrogale penicillata* from coastal New South Wales, I am reluctant to follow Tate (1948a: 273-4) in his treatment of *lateralis* as a subspecies of *penicillata*. The specimen from the Northern Territory is much smaller and differently coloured, and it has noticeably more inflated bullae and smaller third upper incisors. There still remains the question, however, of whether or not the population it represents has been correctly identified as *Petrogale lateralis*.

Peradorcas concinna canescens Thomas

Peradorcas concinna canescens Thomas, *Ann. Mag. Nat. Hist.*, ser. 8, 4: 198 (September 1909). Type locality: Nellie Creek, Northern Territory.

Brocks Creek, 8. Four male (two young) and two female skins with skulls, two pouch-young in alcohol; 15-23 Nov. 1920 (237690-93; 237712-13, 237865-66). Previously recorded by Tate (1948a: 284, '100 miles south of Port Darwin').

Other records for Northern Territory: Daly River (Collett 1897: 325); Nellie Creek (Thomas 1904a: 226, 1909c: 198); Mary River (Collett 1897: 325; Thomas 1909c: 198); King River (Thomas 1926b: 630); Milingimbi, Crocodile Islands (Thomas 1926b: 630).

The Brocks Creek series was collected by Charles M. Hoy 'from among rocks on a hill'. He caught most of the specimens in raisin-baited traps. Each of the two adult females carried a single young in the pouch, one being a male 45 mm. long (head-and-body length) on 22 November, the other a female 110 mm. long on 23 November. The latter animal has well-developed eyelashes and mystacial,

TABLE 7

EXTERNAL AND CRANIAL MEASUREMENTS OF *Petrogale* AND *Peradorcas* NORTHERN TERRITORY, AUSTRALIA

	Head and body	Tail	Hind foot, with claw	Ear from notch	Condylobasal length	Zygomatic breadth	Length of nasals	Tooth-row M1–M3
Petrogale venustula								
Oenpelli								
♂ 284060	381	348	110	43	74·4	42·6	32·3	5·6*
284064	411	372	124	48	83·5	46·1	37·8	7·3
284065	442	436	120	45	90·7	48·9	42·2	6·8
284068	436	—	120	47	82·6	46·4	34·0	6·9
♀ 284061	407	380	110	48	82·6	46·2	37·4	6·4
284063	414	403	108	46	—	—	37·7	6·9
284066	384	365	110	45	82·8	46·5	39·5	6·7
Petrogale lateralis								
Wigley Water Hole								
♀ 284069	471	450	131	62	85·6	46·8	41·5	6·9
Peradorcas concinna canescens								
Brocks Creek								
♂ 237690	325	345	110	—	—	—	—	14·2
237691	337	344	109	—	70·6	43·5	33·5	—
237712	204	231	84	—	46·4	32·4	17·2	—
237713	205	242	79	—	46·9	32·7	16·9	14·0
♀ 237692	308	314	103	—	64·8	30·0	29·6	—
237693	315	318	101	—	68·2	41·0	29·1	—

* Milk tooth

genal, pre- and postorbital, and carpal vibrissae; at first glance its body appears hairless, but under magnification very fine scattered hairs can be seen.

A detailed study of the skulls and teeth of this series has been published by Tate (1948a: 282-284). External and cranial measurements are given in Table 7.

Wallabia agilis agilis (Gould)

Pl. 5

Halmaturus agilis Gould, *Proc. Zool. Soc. Lond.* 9: 81 (March 1842). Type locality: Port Essington, Northern Territory.

Nightcliff, 6 miles north of Darwin, 4. One male and two female skins with skulls, one skull only of unknown sex; 4-15 Apr. 1948 (284071-74).

Douglas River, 13. Two male and four female skins with skulls, one male and one female skulls only, five pouch-young in alcohol; 13-24 October 1920 (237608-10, 237612, 237618-19, 237623-24, 237856-60). Previously recorded by Tate (1948a: 301, '100 miles south of Port Darwin').

Cape Don, Cobourg Peninsula, 3. Male and female skins with skulls, pouch-young in alcohol; 19-21 Sept. 1948 (284106-08).

Popham Bay, Cobourg Peninsula, 1. Male skin with skull; 5 Oct. 1948 (284111).

Black Rock Point, Cobourg Peninsula, 2. Female skin with skull, pouch-young in alcohol; 30 Sept. 1948 (284109-10).

Milingimbi, 12. Female skull only, 11 May 1948 (284075); five skulls with mandibles, two skulls without mandibles, one pair of mandibles, and three single mandibles, all from miscellaneous skeletal material picked up in August 1948 (284076-86).

Elcho Island, 1. Male skin with skull; 13 Sept. 1948 (284105).

Cape Wilberforce, 1. Skull of unknown sex picked up 20 May 1948 (284087).

Yirrkalla, 7. Two male and three female skins with skulls, one male skin only, two pouch-young in alcohol; 13 July-4 Aug. 1948 (284089-95).

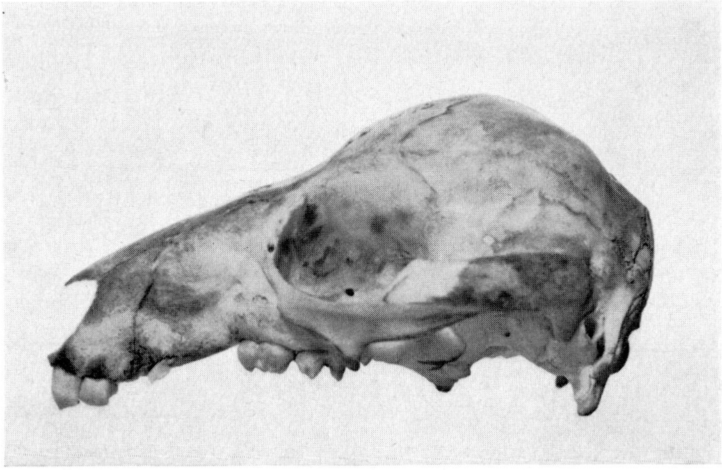

PLATE 5. *Wallabia agilis agilis*

Skull of young animal, showing deciduous 'prelacteal incisor', no. 284095, ♂, x 1

Rocky Bay, Cape Arnhem Peninsula, 2. Female skull only and pouch-young in alcohol; 2 Sept. 1948 (284103-04).

Duripulmi, Cape Arnhem Peninsula, 2. Female skin and skull, pouch-young in alcohol; 11 Aug. 1948 (284099-100).

Kanbulabulaui, Cape Arnhem Peninsula, 2. Female skin with skull, pouch-young in alcohol; 11 Aug. 1948 (284096-97).

Larporkuru, Cape Arnhem Peninsula, 1. Male skin with skull; 8 Aug. 1948 (284098).

Kwaiturumuru Creek, Cape Arnhem Peninsula, 2. Female skin with skull, pouch-young in alcohol; 19 Aug. 1948 (284101-02).

West side of Lake Hubert, Groote Eylandt, 1. Male skin with skull; 28 June 1948 (284088).

Other records for Northern Territory: Victoria River (Collett 1897: 324); Daly River (Collett 1897: 324); Port Darwin (Collett 1897: 324); Katherine River (Collett 1897: 324); South Alligator River (Thomas 1904a: 225); Port Essington (Gould 1842a: 81; Thomas 1888: 44; Tate 1948a: 301); Raffles Bay, Cobourg Peninsula (Dumont d'Urville 1844: 44-5); Groote Eylandt (Tindale 1925: 80; Finlayson 1931a: 71).

This wallaby is the wild mammal most frequently seen by the human population in the coastal area of Arnhem Land. The species is encountered with such regularity as to suggest that it is continuously distributed all across the northern part of the Northern Territory from Western Australia to Queensland. This is certainly true as far as the low-lying coastal areas are concerned, but the absence of specimens from some well-worked inland localities such as Brocks Creek and Nellie Creek suggests that the species is rare or absent in the drier and more rugged terrain of the interior.

Recorded observations of habitat and numbers generally fall into one or the other of two categories: (a) mobs of from dozens to hundreds of individuals on plains near the large rivers, or (b) isolated individuals or small groups in the vicinity of the minor streams and billabongs and about patches of monsoonal and mangrove forest. My own observations are all in the latter category, but in his field notes Charles M. Hoy summarized his experiences with the species along the Douglas River in October 1920 as follows: 'During the day they frequent the scrub that lines the river banks, sleeping or noisily fighting or playing about. At night they spread over the adjoining plains to feed. They keep together in immense mobs, over two hundred being seen together at one time.'

The grass-and-pandanus border that characteristically separates streams and billabongs from the dry stringy-bark forest provides the favoured habitat in the rougher terrain of the coastal strip east of Darwin. Although they make short excursions into the dry forest areas they never seem to go far from water. Small patches of monsoonal forest such as occur near Nightcliff and Yirrkalla are frequented by the wallabies but are not essential for their existence. Their tracks are also to be seen regularly on sandy beaches bordering such areas, and when the tide is low they venture well out into the muddy mangrove swamps.

Aborigines told me that these animals can swim well, but I did not see any actually in deep water. Wallabies surprised at the edge of a billabong sometimes splashed through shallow water in escaping. I was told that a swimming wallaby had once been overtaken by a boat in the salt water strait between Elcho and Howard Islands.

During the wet season, any terrestrial animal that occupies this wallaby's habitat must of necessity be quite tolerant of moisture. Near Nightcliff in early April they foraged during and between intense rain showers in a meadow of rank grass where a human intruder was soon thoroughly soaked. This unusual tolerance of moisture and the ability to swim probably provide the explanation of the presence of the species on most if not all of the coastal islands, including Elcho Island and Groote Eylandt which have not been reached by the larger kangaroos.

Feeding activity was greatest in early morning and late afternoon, and it presumably continued after dark. Eye-shine hunting at night was moderately successful as a means of collecting specimens, but this species produces a relatively dim eye reflection, and far more wallabies were heard than were seen in the course of night hunting.

The species is prolific. Nearly every adult female, and some that were sub-adult, had a young wallaby in the pouch. One instance of twins was noted. The condition of the nipples and occasional instances in which adult females carrying small young in the pouch were accompanied by half-grown young suggested that a series of offspring may be produced by each female during the dry season. Present observations and specimens cover only the period from April to October.

The course of development of the young wallabies, from small pouch-young to the virtually adult stage where the deciduous third and fourth premolars are replaced by the permanent premolar, is outlined in the following list of specimens collected, arranged according to increasing head-and-body length:

284093, 24 mm. male, 31 July, Cape Arnhem Peninsula. Attached to nipple, eyes and ears closed, hairless, unpigmented, fore limbs much larger than hind limbs.

237858, 33 mm. male, 17 October, Douglas River. Attached to nipple, eyes closed, ear pinnae folded down but free at tips, hairless, unpigmented, fore limbs slightly larger than hind limbs.

284097, 38 mm. male, 7 August, Cape Arnhem Peninsula. Attached to nipple, eyes closed, ear pinnae folded down but wholly free, hairless, eyes (visible through skin of closed lids) and tip of claws pigmented, hind limbs longer but slenderer than fore limbs.

284100, 50 mm. female, 11 August, Cape Arnhem Peninsula. Attached to nipple, eyes closed, ear pinnae free and erect, ear orifice not open, hairless, pigment and limb proportions as in the preceding specimen.

237859-60, 145 mm. twins, female and male, 24 October, Douglas River. Attached to separate nipples, eyes closed, ears free and with visible orifice, hairless but with definite papillae indicating position of mystacial vibrissae, pigment as in preceding two specimens, hind limbs in these and all the following specimens exceeding fore limbs.

284102, 125 mm. female, 19 August, Cape Arnhem Peninsula. Free from nipple (as are all the following specimens), eyes closed but with slit-like orifice well defined, ears (as in all following specimens, erect and with open orifice, eye-lashes and vibrissae (mystacial, genal, pre- and postorbital, and carpal) present in this and all following specimens, fine hairs visible under magnification on head and ears and on anal protuberance; skin pigmented on muzzle, ears, fore and hind legs beyond middle of radial and tibial sections, and end of tail.

284110, 140 mm. female, 30 September, Cobourg Peninsula. As in the preceding specimen but with hairs visible at bases of hind claws.

237857, 145 mm. female, 17 October, Douglas River. As in the preceding specimen but fine hairs present on fore and hind legs and about shoulders; lower incisors not visible in this or the preceding specimens.

237856, 185 mm. male, 17 October, Douglas River. As in the preceding specimen but body hair slightly more extensive and pigment extending higher on legs; tips of lower incisors visible.

284107, 190 mm. male, 21 September, Cobourg Peninsula. Eyes in this and all following specimens open; skin pigmented on face and top of head, on forelegs to elbows and on hind legs nearly to knees, and on all of tail; tips of lower incisors visible.

284104, 255 mm. female, 2 September, Cape Arnhem Peninsula. Hair present in some degree on all parts of body, colour pattern well defined; lower incisors conspicuous and tips of median upper incisors barely visible.

284095, 349 mm. male, 4 August, Cape Arnhem Peninsula. At in-and-out-of-pouch stage, fully haired and externally generally similar to adult; molariform deciduous fourth premolars functional, first molars forming and at alveolar level in cleaned skull, a pair of small bilobed teeth resembling the 'prelacteal incisors' described by Tate (1947b) present behind the emerging third upper incisors (Pl. 5).

237612, 386 mm. female, 14 October, Douglas River. Externally as in preceding specimen; first molar teeth barely functional, no 'prelacteal incisors'.

284108, 472 mm. male, 21 September, Cobourg Peninsula. Third upper incisors and first molars fully functional, second molars visible on cleaned skull just below alveolar level.

237619, 483 m. male, 13 October, Cobourg Peninsula. Similar to the preceding specimen.

284091, 552 mm. female, 22 July, Cape Arnhem Peninsula. Second molars barely functional, third molars forming well below alveolar level.

284073, 575 mm. female, 13 April, Nightcliff. Second molars fully functional; third molars visible below alveolar level.

284074, 593 mm. male, 15 April, Nightcliff. Second molars barely functional, less developed than in the preceding specimen in spite of greater size (explainable on the basis of sexual dimorphism).

284075, 605 mm. female, 11 May, Milingimbi. Second molars fully functional and third molars emerged to slightly beyond alveolar level.

237609, 650 mm. male, 13 October, Douglas River. Similar to the preceding specimen but third molars slightly more advanced.

Measurements of specimens in which the permanent premolar is present, considered here to be adult, are given in Table 8.

There seems to be no significant variation in colour or cranial characters correlated with geographic distribution in this area. The single specimen from Groote Eylandt is the darkest in the collection, but it can almost be matched by specimens from the Cape Arnhem Peninsula, and each large series shows considerable individual variation in colour.

Osphranter antilopinus Gould

Pl. 6

Osphranter antilopinus Gould, *Proc. Zool. Soc. Lond.* 9: 80 (March 1842): Type locality: Port Essington, Northern Territory.

Douglas River, 5. Three male and two female skins with skulls (one with body skeleton); 15 Oct.-3 Nov. 1920 (237605-07; 237611, 237613). This and the following series were reported previously by Tate (1948a: 329) from 'about 100 miles south of Port Darwin'.

Brocks Creek, 4. Two male and one female skins with skulls, one pouch-young in alcohol; 13-19 Nov. 1920 (237614-15, 237617, 237861).

Yirrkalla, 3. Male skin with skull, male skull only without mandibles, and part skeleton of unknown sex; 13-30 Aug. 1948 (284113, 284119, 284121).

Gove Airfield, Cape Arnhem Peninsula, 3. Male skin with skull and legbones, male skin only, and female skin with skull; 25 July-9 Sept. 1948 (284112, 284122, 284124).

Kaulpulmi, Cape Arnhem Peninsula, 6. One male skull only, two female skins only, one female skin with skull, two pouch-young in alcohol; 16-29 Aug. 1948 284114-118, 284120).

Cape Arnhem, Cape Arnhem Peninsula, 1. Female skin with skull; 2 Sept. 1948 (284123).

Other records for Northern Territory: Victoria River (Collett 1897: 323); Hermit Hill (Collett 1897: 323); Margaret River (Thomas 1904a: 224); Glencoe (Collett

TABLE 8

EXTERNAL AND CRANIAL MEASUREMENTS OF *Wallabia*
NORTHERN TERRITORY, AUSTRALIA

	Head and body	Tail	Hind foot, with claw	Ear from notch	Condylobasal length	Zygomatic breadth	Length of nasals	Tooth-row M1–M3
Wallabia agilis agilis								
Nightcliff								
♀ 284072	636	604	217	73	122·7	67·1	55·0	24·7
Douglas River								
♂ 237623	—	—	—	—	143·1	74·4	64·4	26·9
♀ 237608	606	658	233	—	122·0	70·6	54·1	26·0
237610	639	719	233	—	124·0	70·7	55·7	24·6
237618	637	657	243	—	122·3	69·8	54·8	26·7
237624	—	—	—	—	124·6	73·3	53·7	25·3
Cobourg Peninsula								
♂ 284111	702	799	252	85	133·6	71·5	57·0	26·8
♀ 284106	625	658	235	79	123·0	66·0	55·8	25·1
284109	693	637	235	81	131·8	73·6	58·5	23·8
Milingimbi								
sex ? 284076	—	—	—	—	143·2	75·0	62·9	23·7
284078	—	—	—	—	132·4	69·5	59·4	26·1
284079	—	—	—	—	135·1	73·6	58·2	24·2
284080	—	—	—	—	135·4	71·0	65·8	26·8
284081	—	—	—	—	124·5	67·8	53·1	26·5
284982	—	—	—	—	135·4	70·3	—	27·1
Elcho Island								
♂ 284105	520	580	218	72	102·8	61·1	44·7	23·5
Cape Wilberforce								
sex ? 284087	—	—	—	—	135·2	66·5	—	25·3
Cape Arnhem Peninsula								
♂ 284089	710	760	278	80	—	—	—	—
284090	635	728	254	87	125·5	68·0	58·9	27·6
284098	689	733	255	80	128·7	69·4	57·9	25·5
♀ 284092	619	657	222	—	113·0	62·0	47·6	25·0
284094	639	673	227	83	125·5	68·5	61·3	25·7
284096	635	665	229	76	117·9	66·2	56·2	26·3
284099	652	659	240	77	125·4	65·7	54·1	24·8
284101	616	693	243	80	121·7	65·9	49·9	25·3
284103	—	—	—	—	126·4	67·3	—	24·7
W. side Lake Hubert, Groote Eylandt								
♂ 284088	798	640	240	70	127·4	69·9	58·1	25·1

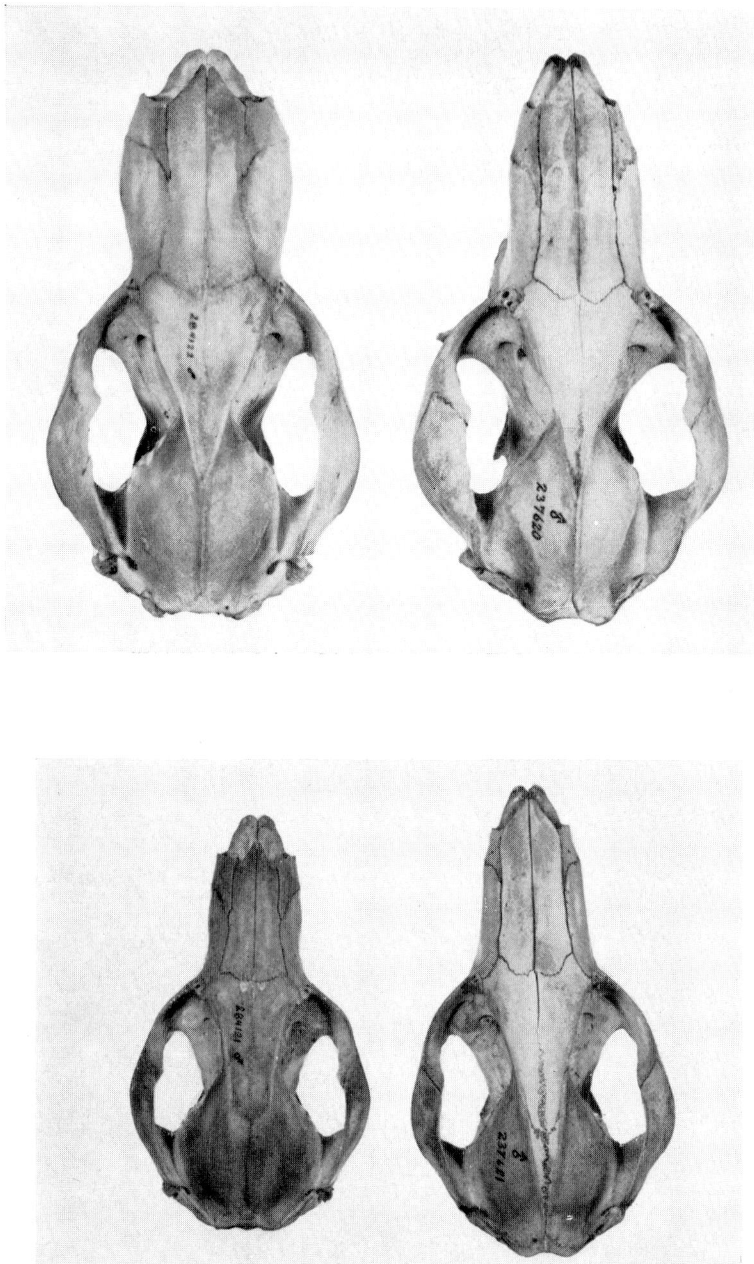

PLATE 6. *Osphranter antilopinus* (left) and *O. robustus alligatoris* (right)
Comparison of skulls to show difference in shape of rostrum
Above, old males, nos. 284122 and 237620; below, young males, nos. 284121 and 237681; x 0·5

1897: 323); Burrundie (Thomas 1904a: 224); Eureka (Thomas 1904a: 224); Edith River (Thomas 1904a: 224); South Alligator River (Thomas 1904a: 224); Port Essington (Gould 1842a: 80; Thomas 1888: 22).

These large kangaroos were seen regularly near Yirrkalla and at other places on the Cape Arnhem Peninsula. Proximity of rocky ravines and bluffs seemed to be a requirement for their local occurrence in this general area. On several occasions individuals or pairs were disturbed from midday resting places in the sparse shade of shrubby thickets in the eucalypt forest bordering Gove Airfield, on level ground. If frightened or pursued they inevitably ran off toward the broken country at the near by heads of ravines. Others, encountered in the evening near streams below the ironstone bluffs, where they had apparently come to drink, retreated upslope to the protection of the rough terrain. The large males in particular were fascinating to watch as they moved with a sort of ponderous agility across the talus and up over ledges where a man could not follow.

According to Dahl (1897: 213) this species was numerous both in hilly country and on the forested plains in the vicinity of the Victoria and Daly Rivers in the early 1890s. In 1920 Hoy collected specimens at Douglas River and at Brocks Creek. At the former locality they occurred on the plain in habitats described as 'light scrub', 'grass plain', and 'river bank', no mention being made of rocky terrain. At Brocks Creek they were collected along with *Osphranter robustus alligatoris* 'among rocks on a hill'. Pouch-young that were preserved have the following head-and-body measurements: Cape Arnhem Peninsula, 14 August, 85 mm. female; same area, 27 August, 220 mm. male; Brocks Creek, 17 November, 200 mm. male.

The problem of whether or not there is geographic variation in *Osphranter antilopinus* from eastern to western Arnhem Land will bear further investigation. The specimens from Cape Arnhem Peninsula are deeper red in colour than those from Douglas River and Brocks Creek. Lack of topotypes from the geographically intermediate Port Essington area prevents further study of this difference at present.

Iredale and Troughton (1934:51) and Tate (1948a: 329) have treated *Osphranter antilopinus* as being conspecific with other wallaroos of the Northern Territory, but in doing so they have minimized its distinctive morphological features and appear to have overlooked the fact that in some places it occurs along with *Osphranter robustus alligatoris* without any evidence of intergradation. The treatment of these as distinct species by earlier authors (Thomas 1904a; Schwarz 1910; Le Souef 1928) better fits the evidence derived from study of the present material.

Characters that distinguish *Osphranter antilopinus* from *O. robustus* are: deeper red colour of males at all ages (from Fawn Color to Verona Brown of Ridgway, 1912); higher, broader, and relatively shorter skull; shorter and broader nasal bones; greater bulge at anterior frontal region, followed posteriorly by broader and deeper intertemporal depression in roof of skull; more abrupt convergence of temporal crests near fronto-parietal suture; and greater lateral swelling of rostrum. The last character has been discounted by some workers as being too variable to be reliable, but in the present series it provides a positive means of distinguishing the species (Pl. 6). In every specimen of *Osphranter antilopinus* (and in none of *O. robustus alligatoris*) there is a distinct lateral bulge on the rostrum at the anterior borders of the premaxillaries, set off by a constriction of the rostrum behind the bulge in the region of the openings of the infraorbital canals. In all specimens of *O. antilopinus* in which the third molar is fully grown the breadth of the rostrum is more than 25 per cent of the condylobasal length.

In young specimens the rostrum is relatively narrower, but the bulge set off by a constriction is present. The rostrum of *O. robustus alligatoris*, in contrast, tapers continuously throughout its length. Measurements of both species are given in Table 9.

It is evidently this species rather than *Megaleia rufa* that is called 'Red Kangaroo' by Thomson (1949: 65).

Osphranter robustus alligatoris (Thomas)

Pl. 6

Macropus robustus alligatoris Thomas, *Novit. Zool.* 11: 224 (March 1904). Type locality: South Alligator River, Northern Territory.

Brocks Creek, 6. Two male and one female skins with skulls (one with body skeleton), two male and one female skulls only; 14-22 Nov. 1920 (237616, 237620-22, 237654-55).

Other records for Northern Territory: South Alligator River (Collett 1897: 325; Thomas 1904a: 224).

The specimens were collected by Hoy on a rocky hill where *O. antilopinus* was also found. In his notes the collector states that the habits were similar to those of wallaroos in northern Queensland.

In taxonomic literature this race of wallaroo has been confused on the one hand with *Osphranter antilopinus* (see discussion above) and on the other with *O. bernardus*. The present material confirms its existence as a geographical subspecies of *O. robustus* as it was originally described by Thomas. *O. bernardus*, the small black wallaroo of central Arnhem Land, is not represented in this series, judging from the measurements of that apparently distinct species published by Rothschild (1903) and Thomas (1904a: 225).

The three skins show clearly the controversial character of a whorl in the middle of the foreback, the hair from there to the crown of the head being directed forward. This feature is not present in any of the specimens discussed in the foregoing account (*O. antilopinus*) or in that which follows (*O. r. erubescens*). The cranial characters are essentially as in *O. r. robustus* and *O. r. erubescens*.

Osphranter robustus erubescens (Sclater)

Macropus erubescens Sclater, *Proc. Zool. Soc. Lond.* 1870: 126 (2 June 1870). Type locality: vicinity of Lake Hope, 200 miles inland from Port Augusta, South Australia.

Wigley Water Hole, 4 miles north of Alice Springs, 7. One male and three female skins with skulls, one male complete skeleton, one male skull only, one pouch-young in alcohol; 18-19 Nov. 1948 (284125-31).

Various authors (including Spencer 1896a: 14, and Finlayson 1943a: 58) have written of this kangaroo as being widely distributed in the rocky ranges of the southern part of the Northern Territory, but precise locality records seem to be lacking in the literature.

The specimens from Wigley Water Hole were collected in the middle of a long dry period, when water was scarce and dust storms were frequent. The kangaroos stood singly or gathered in small groups about the waterhole and in shaded spots in the rocky gorges leading back into the hills from it. In general they seemed spiritless and apathetic in behaviour in contrast to the more active red wallaroos of the Arnhem Land coastal areas.

A female joey in the pouch of an old female killed on 18 November measures

70 mm. in head-and-body length. Two young animals collected on the following day each had the first molar tooth fully functional but the second molars still unirrupted. Measurements of adults are given in Table 9.

TABLE 9
EXTERNAL AND CRANIAL MEASUREMENTS OF *Osphranter*
NORTHERN TERRITORY, AUSTRALIA

	Head and body	Tail	Hind foot, with claw	Ear from notch	Condylobasal length	Zygomatic breadth	Length of rostrum	Tooth-row M1–M3
Osphranter antilopinus								
Douglas River								
♂ 237605	1007	977	364	—	169·6	97·6	44·5	32·0
237606	873	863	342	—	150·1	93·3	41·6	32·3
237613	1005	956	350	—	171·1	101·4	46·0	32·6
♀ 237611	798	744	288	—	149·3	88·7	38·8	29·8
Brocks Creek								
♂ 237614	898	840	342	—	148·8	92·2	39·7	33·4
237615	1053	1072	383	—	184·4	102·6	48·6	30·0
♀ 237617	830	777	302	—	—	88·8	41·7	31·3
Cape Arnhem Peninsula								
♂ 284119	—	—	—	—	174·5	97·5	46·8	32·0
284120	—	—	—	—	190·6	102·5	55·5	33·0
284122	1077	943	358	119	174·3	97·2	54·8	34·2
♀ 284114	741	696	301	113	—	—	—	—
284117	782	700	288	100	134·2	84·2	35·0	26·8
284123	—	—	275	110	131·9	78·6	29·7	26·5
284124	854	725	303	110	100·9	92·5	44·0	29·1
Osphranter robustus alligatoris								
Brocks Creek								
♂ 237620	1003	987	323	—	172·5	99·0	41·1	33·4
237655	—	—	—	—	147·7	83·5	33·5	33·1
♀ 237616	698	667	246	—	133·7	77·2	32·0	30·7
237622	—	—	—	—	134·5	79·9	28·7	30·8
Osphranter robustus erubescens								
Wigley Water Hole								
♂ 284128	987	855	315	125	167·8	97·4	40·7	31·8
♀ 284125	852	731	270	116	146·4	88·2	32·3	31·8
284130	826	674	277	121	142·0	81·8	32·0	32·4

Megaleia rufa rufa (Desmarest)

Kangurus rufus Desmarest, *Tabl. Encycl. Meth. Mamm.* 2: 541 (6 July 1822). Type locality: Blue Mountains, New South Wales.

Wauchope, 1. Female skin with skull, 17 Nov. 1948 (284132).

58 miles north of Alice Springs, 1. Female skin with skull; 17 Nov. 1948 (284133).

Horse Shoe Bend, Finke River, 1. Male skin with skull; 26 Nov. 1948 (284134).

Other records for Northern Territory: Alexandria (Thomas 1906e: 540; Schwarz 1910: 89); Burt Plains (Spencer 1896a: 13); Horn Valley, Macdonnell Range (Spencer 1896a: 12); Wollara, Basedow Range area (Finlayson 1947: 190); Mount Olga (Spencer 1896a: 12); Ayers Rock (Spencer 1896a: 13).

In November 1948 red kangaroos were common and conspicuous on the sandy plains beside the highway between Tennant Creek and Alice Springs and along the railway south of Alice Springs. Mechanized traffic on these thoroughfares seemed to disturb the animals very little, whereas some miles away from the railway line they proved to be quite wary when any attempt was made during daylight to approach them in a car or on foot. This was true at a seldom used emergency aeroplane landing field between Rumbalara and Horse Shoe Bend, which was a favourite feeding ground.

Many kangaroos are killed by cars on the paved highway north of Alice Springs, especially at night when the animals often stand in the road, apparently unaware of the danger represented by approaching headlights. I was told that the crashes resulting from the custom of fast night driving and the obstinacy or bewilderment of large kangaroos on the road sometimes cause serious injury to cars and passengers as well as to the kangaroos.

Near Tea Tree Well there were heaps of skinned carcasses of red kangaroos that had been killed by professional hide-hunters. These hunters used a system of shooting at night with the aid of a spotlight mounted on a small truck.

CHIROPTERA

PTEROPODIDAE

Two kinds of large fruit bats, *Pteropus alecto gouldii* and *Pteropus scapulatus,* occur together throughout coastal Arnhem Land (Pl. 7). In size, general appearance, and habits they are enough alike that a distinction between the species is seldom made in the field. Even experienced collectors such as Tunney and Hoy seem not to have been aware that they were observing and collecting two species. The published accounts of the natural history of fruit bats in Arnhem Land are, therefore, nearly all composite and it is only a matter of guesswork to try to determine now which species was represented in a given observation.

The following account of *'Pteropus funereus'* quoted from Gould's *Mammals of Australia* (Gould 1849: text of vol. 3, pl. 30) describes the impression made by the fruit bats on the explorers of a century ago and is equally applicable to modern conditions. It is fairly certain that both species were involved in these as in more recent observations.

> Mr Gilbert's notes inform me that 'it is extremely abundant in all parts of the Cobourg Peninsula; during the day it may be seen in great numbers suspended from the upper branches of the mangroves overhanging the creeks: while living it emits a very strong and disagreeable odour which is perceptible at a considerable distance; it becomes very active at night, and while flying about in search of food utters a loud, trembling, but shrill whistle.' Frequent mention is made of this species in Dr Leichhardt's Journal of his Expedition from Moreton Bay to Port Essington, and despite of its disagreeable odour, it often formed for himself and party an excellent and welcome article of food. Like the other species, it feeds upon fruit and the honey of the various flowers; in one instance Dr Leichhardt found them feeding upon the blossoms of the tea-tree, and remarked that they were more than usually fat and delicate, while those that had been revelling among the blossoms of the gum-trees were not so fat, and had a strong, unpleasant odour. So numerous did they become towards the latter part of the journey, that 'twelve were brought in for luncheon, thirty more were procured during the afternoon, and at least fifty were left wounded and hanging to the trees; upon another occasion they were seen clustering in such numbers, that the branches of the low trees drooped with their weight so near the ground, that they could easily be killed with cudgels. In the neigh-

bourhood of the River Roper, myriads were suspended in thick clusters on the highest trees, in the most shady and rather moist parts of the valley; they started as the travellers passed, and the flapping of their large membranous wings produced a sound like that of a hail-storm.' Dr Leichhardt went the next day with two of his party to the spot where they had seen the greatest number, and while he was examining the neighbouring trees his companions shot sixty-seven, of which fifty-five were brought to the camp, and served for dinner, breakfast, and luncheon, each of the party receiving eight: the animal here lived upon a small, blue, oval stone-fruit, of an acid taste, with a bitter kernel, growing on a tree of moderate size.

As seems to be generally true with the large kinds of fruit bats, far inland records are lacking. For some reason and in spite of their ability to fly long distances, they seldom venture more than a few miles from the coastal swamps and estuaries or the banks of the larger rivers.

An attempt was made in 1948 to distinguish habitat differences between the two species, with only partial success. Such observations as were made led to the general conclusion that when actively foraging *Pteropus alecto* was more likely to be seen singly or in small groups about the canopies of dense monsoonal forest or near the large trees along stream courses, whereas the more sociable *Pteropus scapulatus* formed large flocks in the dried eucalypt forest. In the Daly and Victoria River areas Dahl (1897: 191) found *Pteropus scapulatus* in small numbers mixed with the more abundant populations of the other species and was unable to define a habitat difference.

Reports of large colonies of fruit bats roosting during the day in mangrove or other trees, as described by Dahl and by Gould (loc. cit.), were heard from time to time. One such site in a mangrove swamp near Black Rock Point on the Cobourg Peninsula was investigated on 29 September, but that roost was found to be abandoned.

Pteropus alecto gouldii Peters
Pl. 7

Pteropus gouldii Peters, *Mber. K. Preuss. Akad. Wiss.* 1867: 703 (after 28 November 1867). Type locality: Rockhampton, Queensland.

Nightcliff, 6 miles north of Darwin, 3. Adult male and female and young female, skins with skulls; 20 March 1948 (284135-37).

Darwin, 1. Female in alcohol; 1 Oct. 1920 (237986). Other specimens recorded by Collett (1897: 319).

Douglas River, 15. Two male skins with skulls, two male and one female skulls only, six males and four females in alcohol (skin of one mounted for exhibition); 12 Oct.-2 Nov. 1920 (237816-18, 237820-21, 237988-97).

Oenpelli, 3. One male and two female skins with skulls; 23-30 Oct. 1948 (284144-46).

Cape Don, Cobourg Peninsula, 2. Female skins with skulls; 21 Sept. 1948 (284142-43).

Yirrkalla, 4. Two female skins with skulls, one female skull only, one female complete skeleton; 11 July-26 Aug. 1948 (284138-41).

Other records for Northern Territory: Victoria River (Dahl 1897: 191); Daly River (Collett 1897: 319); Quail Island (Dobson 1878: 61; Andersen 1912: 373); 'Alligator R.' [probably South Alligator River] (Andersen 1912: 373); Port Essington (Gray 1843: 37, 1870: 104; Gould 1849: text of pl. 30; Andersen 1912: 373).

At Yirrkalla and at Cape Don single individuals of *Pteropus alecto* were found hanging quietly at midday in small patches of dense monsoonal forest.

An adult female shot at Nightcliff on 20 March 1948 had a 120 mm. young female clinging to it. The mouth of the young bat was clamped tightly over one of the nipples. This accords with the observation by Dahl (1897: 191) that at Victoria River in March and April females had young clinging to them. No other attached young were found during the remainder of the 1948 season. An excessively fat male, judged from greatly worn teeth to be very old, was also collected at Nightcliff on 20 March.

Pteropus scapulatus Peters
Pl. 7

Pteropus scapulatus Peters, *Mber. K. Preuss. Akad. Wiss.* 1862: 574 (after 14 Aug. 1862). Type locality: Cape York, Queensland.

Darwin, 1. Male in alcohol; 1 Oct. 1920 (237987).

Douglas River, 1. Male skull only; 24 Oct. 1920 (237819).

Oenpelli, 7. Six male skins with skulls, one male in alcohol; 24 Oct.-6 Nov. 1948 (284150-56).

Elcho Island, 3. Male skins with skulls; 14 Sept. 1948 (284147-49).

Other records for Northern Territory: Daly River (Dahl 1897: 191; Collett 1897: 318); South Alligator River (Thomas 1904a: 222; Andersen 1912: 407).

The frequency with which this species is confused with *Pteropus alecto* is discussed above. Characters which distinguish *P. scapulatus* are: smaller size (head-

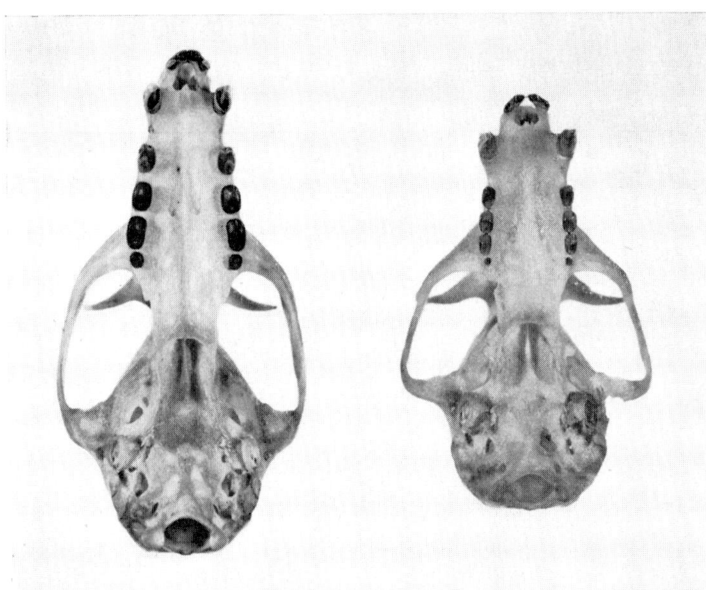

PLATE 7. *Pteropus alecto gouldii* (left) and *P. scapulatus* (right)

Comparison of skulls of two species in Arnhem Land, ventral view
P. alecto gouldii, no. 284136, ♀, from near Darwin, x 1
P. scapulatus, no. 284150, ♂, from Oenpelli, x 1

and-body length near 220 rather than 265 mm.; forearm near 140 rather than 170 mm.); coloration predominantly brownish rather than blackish; and teeth smaller, particularly the cheek teeth.

At Elcho Island on 14 September 1948, a flock estimated to contain more than a hundred individuals was observed at about 9 p.m. by bright moonlight feeding on flowers of eucalypt trees near the mission landing field. The flock was fairly well scattered out, not more than four or five individuals being in a single tree at one time. Most of the trees (not specifically identified) in this section of the forest were blossoming, and on the preceding day it was noted that flocks of small parrots had gathered to tear at the same flowers.

At intervals the bats uttered single sharp call notes, suggesting the 'yip' of a small dog, this sound coming monotonously from a large area of the forest and

TABLE 10

EXTERNAL AND CRANIAL MEASUREMENTS OF ADULT *Pteropus*
NORTHERN TERRITORY, AUSTRALIA

	Head and body	Hind foot, with claw	Ear from notch	Forearm	Condylobasal length	Zygomatic breadth	Breadth at intertemporal constriction	Upper tooth-row C–M2
Pteropus alecto gouldii								
Nightcliff								
♂ 284135	267	53	35	173	69·7	40·8	7·1	26·0
♀ 284136	251	53	37	177	67·6	36·2	8·0	24·6
Douglas River								
♂ 237816	271	63	—	191	68·8	36·6	8·0	25·8
237817	268	64	—	174	68·6	38·4	6·9	25·6
237818	—	—	—	—	67·4	39·1	8·6	25·7
237821	—	—	—	—	69·2	40·4	7·4	26·8
237989	278	66	—	185	—	—	—	—
237990	280	68	—	182	—	—	—	—
237993	256	62	—	—	—	—	—	—
237995	261	57	—	176	—	—	—	—
237997	252	62	—	176	—	—	—	—
♀ 237992	244	56	—	168	—	—	—	—
237994	261	58	—	175	—	—	—	—
237996	258	59	—	—	—	—	—	—
Oenpelli								
♂ 284145	268	54	36	173	67·2	39·8	8·4	24·4
Yirrkalla								
♀ 284139	262	57	35	171	70·2	36·4	6·9	25·7
284140	252	58	35	178	70·2	34·7	7·6	25·4
284141	—	—	—	167	—	35·6	9·2	24·8
Pteropus scapulatus								
Douglas River								
♂ 237819	—	—	—	—	53·9	34·3	9·1	18·4
Oenpelli								
♂ 284150	225	42	39	139	55·2	35·6	10·0	18·2
284155	213	46	36	142	52·9	34·4	9·7	18·5
284156	214	42	33	138	53·3	34·6	8·9	17·5
Elcho Island								
♂ 284147	221	46	34	141	53·8	34·0	8·0	18·1
284148	217	41	34	135	54·1	32·9	9·8	17·2
284149	212	44	34	—	54·1	—	—	—

helping to define the extent of the flock. To my ear it sounded higher pitched, shorter in duration, and less harsh than calls attributed to *Pteropus alecto.* Three bats shot out of this flock proved to be adult males.

At Oenpelli near the end of October these bats flew regularly at dusk over the lagoon, dipping low as if to drink and usually leaving a long ripple on the surface of the water. It was noted that after making this pass over the lagoon some of the bats flew to a row of trees about a quarter of a mile away and settled in the branches. When shot, they proved to have the fur of the chest and belly soaked with water. It seemed that, rather than trying to drink on the wing as was first supposed, they were saturating the thick ventral fur and then licking the water from it at leisure while hanging in the trees.

EMBALLONURIDAE

Taphozous georgianus Thomas

Taphozous australis georgianus Thomas, *J. Bombay Nat. Hist. Soc.* 24: 62 (30 September 1915). Type locality: King George Sound, Western Australia.

Oenpelli, 6. One male and three female skins with skulls, one male and one female in alcohol; 26 Sept.-2 Nov. 1948 (284157-62).

Other record for Northern Territory: Mary River (Dahl 1897: 192; Collett 1897: 320; Thomas 1915c: 62).

The local distribution of *Taphozous* in Arnhem Land apparently coincides with that of the rocky escarpment that forms the central plateau. Our party encountered this habitat only at Oenpelli and at Central Hill on Groote Eylandt, and at each of those places bats of this genus were found. They were not encountered at the coastal collecting sites, even where laterite cliffs were present.

In the caves and fissures of the rocky hills near Oenpelli they were fairly common in October and early November. During the day they clung to vertical rock faces, usually in dimly lighted situations adjacent to dark and narrow crevices, into which they retreated when disturbed. From two to about ten bats were seen in each favourable roosting site. There was no great concentration of numbers and no obvious accumulation of guano. They remained in these places at least until dark, and nothing was learned of their foraging activities.

In the narrow intertemporal constriction and the large basial pits the specimens from Oenpelli agree well with the description of *georgianus* and not with that of *australis,* as these were differentiated by Thomas (1915c: 62-3). I have no material at present to determine whether these two names apply to different species as arranged by Troughton (1925) or merely to geographic races of one species. No specimens as large as those from Queensland described as *Taphozous troughtoni* by Tate 1952: 607) are present in this series.

An incipient gular pouch is barely suggested by a slight glandular area on the throat in males and is quite lacking in females, exactly as described on the basis of other specimens by Troughton (1925: 338, pl. 47). For measurements see Table 11.

On the escarpment forming the northern face of Central Hill on Groote Eylandt, a bat that was certainly of the genus *Taphozous* and possibly of this same species was accidentally frightened from a dark recess under overhanging rocks at about 4 p.m. on 29 June. It was pursued and flushed repeatedly from similar situations along the same ledge. It finally escaped and was last seen circling high in the air in bright sunlight, apparently hawking for insects along the face of the cliff in company with several wood swallows (*Artamus*).

Saccolaimus flaviventris (Peters)

Taphozous flaviventris Peters, *Proc. Zool. Soc. Lond.* 1866: 430 (April 1867). Type locality: Australia.

Danganboy, Cape Arnhem Peninsula, 2. Male and female skins with skulls; 7 Aug. 1948 (284164-65).

Port Langdon, Groote Eylandt, 1. Male skin with skull; 3 June 1948 (284163).

Other record for Northern Territory: Port Darwin (Troughton 1925: 320).

In its local distribution this tree-roosting species has a range complementing that of the cave and rock-fissure roosting *Taphozous georgianus*. Dead hollow limbs of fairly tall trees are favoured daytime roosts for *Saccolaimus*. A male and a female were flushed during midday from such a limb at Danganboy, a few miles south of Yirrkalla on 7 August when sticks were thrown against the branch, which was about thirty feet above the ground and extended obliquely upward from the main trunk. A similar situation was described by aborigines who brought in a single male at Port Langdon on Groote Eylandt; they said that more than two or three were never found in one tree. This roosting site seemed to be identical with that of a nocturnal bird, the owlet nightjar (*Aegotheles cristatus*).

MEGADERMATIDAE

Macroderma gigas (Dobson)

Megaderma gigas Dobson, *Proc. Zool. Soc. Lond.* 1880: 461, pl. 46 (October 1880). Type locality: Mount Margaret, Wilson's River, south-western Queensland.

Records for Northern Territory: Oenpelli (Marshall 1948: 255-7); Alice Springs (Spencer 1896a: 11).

Although no specimens were collected, there was evidence of the presence of this large carnivorous bat.

On 9 April at Darwin, where a large colony of bent-winged bats (*Miniopterus*) roosted in the cave-like refrigerating chambers of an abandoned meat works, numerous severed and desiccated wings of *Miniopterus* were found scattered on the concrete floor near the sheltered doorway through which most of the bats entered and left the rooms. It is thought that this must be a regular feeding site for one or more individuals of *Macroderma*, which is known to prey on smaller bats and to discard the wings of its victims in this manner.

At Oenpelli in late October a false vampire bat visited a site on the veranda of sleeping quarters at the mission on at least two occasions. Such visits were frequent occurrences according to the mission staff, the bats bringing live prey which they killed and ate at this place. Discarded items here included the wings of birds and the hind feet of large frogs. Alan Marshall, who visited Oenpelli in March 1946, kindly sent me a photograph of a false vampire bat captured there. This photograph has been published in his book *Ourselves Writ Strange* with a description of the feeding habits of the bat, apparently at the same site that was being used two years later (Marshall 1948: 255-7, illus. opp. p. 289).

HIPPOSIDERIDAE

Hipposideros bicolor gilberti Johnson

Pl. 8

Hipposideros bicolor gilberti Johnson, *Proc. Biol. Soc. Wash.* 72: 183 (4 November 1959). Type locality: Oenpelli, East Alligator River, Northern Territory.

Douglas River, 3. Male, female, and nursing young in alcohol (skull of male cleaned); 30 Oct. 1920 (237956, 237961, 237964).

Oenpelli, 4. Three male skins with skulls, one female in alcohol; 23 Oct.-1 Nov. 1948 (284167-70).

The specimens from Douglas River were collected by Hoy 'from a limestone cave'. For many years they have been misidentified as *Hipposideros stenotis* Thomas, a larger species with peculiar club-shaped median processes on the nose leaves. Details of the characters that ally them, along with the Oenpelli specimens, with the widely ranging *Hipposideros bicolor* are given in the original description (Johnson 1959: 183).

TABLE 11

EXTERNAL AND CRANIAL MEASUREMENTS OF *Taphozous, Saccolaimus, Hipposideros* AND *Rhinonicteris* NORTHERN TERRITORY, AUSTRALIA

	Head and body	Tail	Hind foot, with claw	Ear from notch	Forearm	Condylobasal length	Zygomatic breadth	Interorbital constriction
Taphozous georgianus								
Oenpelli								
♂ 284161	75	18	11	21	64	19·4	12·1	4·1
♀ 284158	83	30	15	22	70	—	—	—
284159	78	22	13	21	58·7	18·3	12·3	4·1
284160	73	23	12	20	—	18·3	11·7	4·0
Saccolaimus flaviventris								
Cape Arnhem Peninsula								
♂ 284164	96	29	18	24	75	24·3	16·7	5·2
♀ 284165	104	25	17	24	77	24·8	16·9	5·2
Groote Eylandt								
♂ 284163	96	21	13	25	77	24·9	17·3	5·1
Hipposideros bicolor gilberti								
Douglas River								
♂ 237961	38	27	8	17	37	13·8	7·9	2·2
♀ 237956	39	26	8	17	37	—	—	—
Oenpelli								
♂ 284167	46	25	8	20	39	14·0	8·3	2·2
284168	44	27	8	19	38	13·7	7·9	2·3
284170 (type)	45	24	8	21	37	13·7	8·0	2·1
♀ 284169	46	29	9	19	39	14·2	8·3	2·2
Rhinonicteris aurantia								
Oenpelli								
♂ 284166	52	30	9	12	47	—	—	—

At Oenpelli these bats were found hanging during the day in dark cave-like recesses among the tumbled rock slabs on the sides of Unbalanja Hill.

A hairless young male with a head-and-body length of 28 mm. was clinging to the breast of a female collected at Douglas River on 30 October 1920. The adult female taken at Oenpelli on 30 October 1948 had in its uterus a single foetus 16·5 mm. in crown-rump length.

This distinct subspecies has only one Australian ally, *H. b. albanensis* of the Cape York Peninsula, a slightly larger and much darker-coloured form. The Arnhem Land race was named for John Gilbert in recognition of his pioneer collecting at Port Essington. Measurements are given in Table 11.

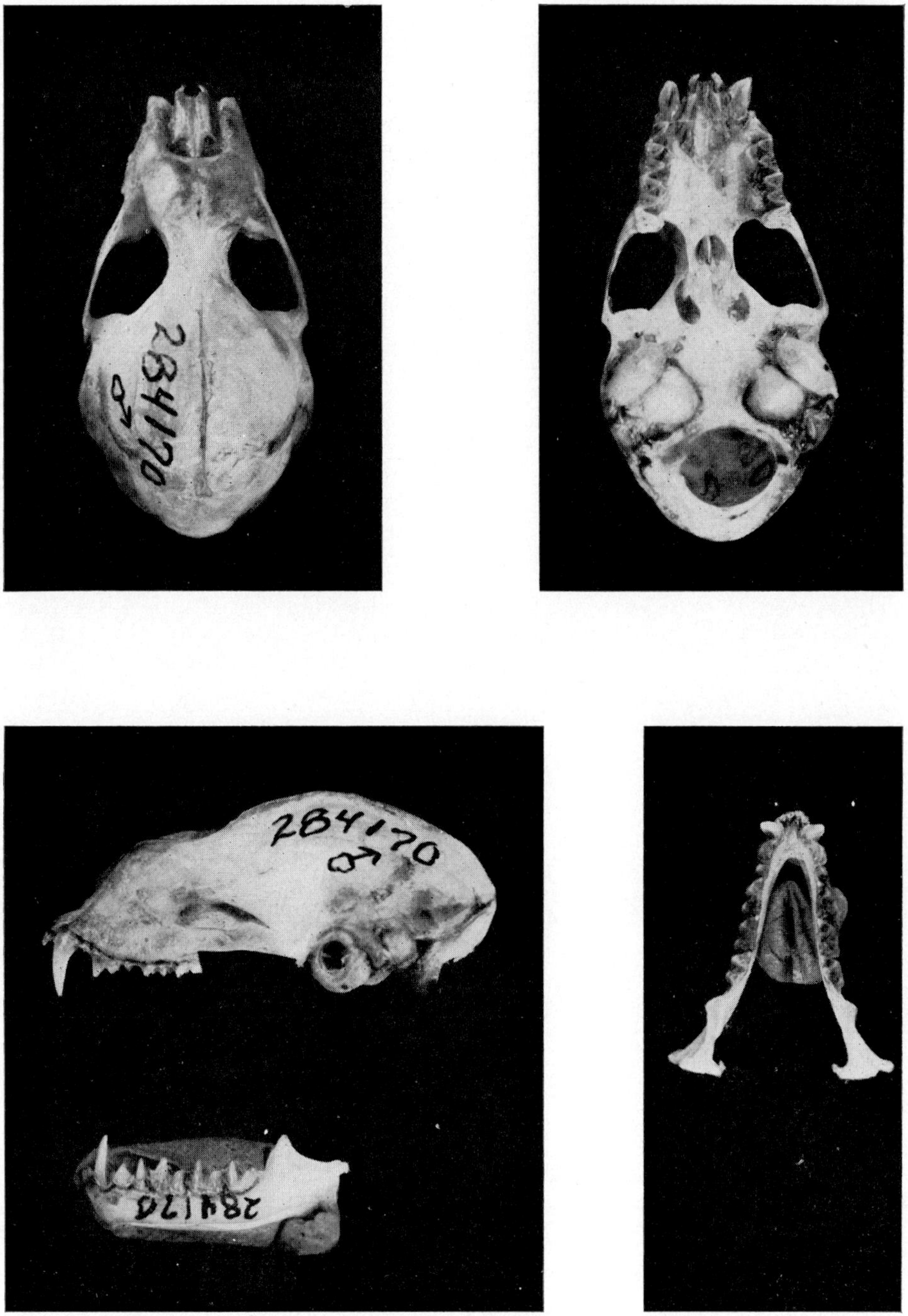

PLATE 8. *Hipposideros bicolor gilberti*
Skull of type specimen, no. 284170, ♂, x 4

Rhinonicteris aurantia (Gray)

Rhinolophus aurantius Gray in Eyre, *Journals of expeditions of discovery into central Australia . . . in the years 1840-41,* 1 (appendix): 405, pl. 1 (1845). Type locality: Port Essington, Northern Territory.

Oenpelli, 1. Male in alcohol; 28 October 1948 (284166).

Other records for Northern Territory: Port Essington (Gray 1845: 405; Gould 1851, text of vol. 3, pl. 35; Gould 1863: introd., p. xxxviii; Peters 1871: 314); exact locality uncertain, North Australian Expedition, J. R. Elsey (Dobson 1878: 127).

This is one of the most beautiful and at the same time one of the least known of the Australian bats. The specimen obtained by the Arnhem Land Expedition was captured by H. L. Hillier during the evening in his quarters at the Oenpelli Mission Station.

Nothing definite seems to have been published on the habits of the species in the Northern Territory since the brief comment by Gould (1851, text of pl. 35, vol. 3) that it is abundant on the Cobourg Peninsula and 'retires during the day-time to the hollow spouts and holes of the various species of *Eucalypti*'. According to Gray (1845: 405) the specimen on which he founded the species 'was found flying near the hospital at Port Essington, by Dr Sibbald, R.N.'

In the specimen from Oenpelli, which is preserved in alcohol, the nose leaf and associated structures agree well with those described and illustrated by Dobson (1878: 126, pl. 8, fig. 2). The illustration given by Jones (1925b: 466, fig. 308) of a bat referred to this species from an unspecified locality shows larger ears, a nose leaf with a hairless and more deeply incised dorsal margin, and other minor differences. The latter specimen is also described as having pale eyelashes and uniformly coloured body hairs, whereas in the specimen from Oenpelli the eyelashes are darker than the surrounding fur and the hairs on the upper parts are banded with darker colour near their tips.

VESPERTILIONIDAE

Eptesicus pumilus caurinus Thomas

Eptesicus pumilus caurinus Thomas, *Ann. Mag. Nat. Hist.*, ser. 8, 13: 439 (April 1914). Type locality: Drysdale, Kimberley district, Western Australia.

Brocks Creek, 1. Male in alcohol; 13 Nov. 1920 (237968).

Oenpelli, 4. Male skin with skull, two adult females and one young male in alcohol; 23 Oct.-5 Nov. 1948 (284183-86).

Bremer Island, off Cape Arnhem Peninsula, 1. Female in alcohol; August 1948 (284182).

Larporkuru, Cape Arnhem Peninsula, 3. Male and female skins with skulls, female in alcohol; 9 August 1948 (284179-81).

Port Langdon, Groote Eylandt, 1. Male skin with skull; 15 June 1948 (284178).

Although this species is generally said (as by Troughton 1941: 348) to be distributed over all of Australia I find no previous specific records of its occurrence in the Northern Territory.

The specimen collected by Hoy at Brocks Creek was flying over a waterhole at dusk. The three from Larporkuru on the Cape Arnhem Peninsula were found at midday crowded together in a small hole in the ceiling of a cave. The one from Port Langdon on Groote Eylandt was flying about at night in Mr Fred Gray's house at Umbakumba.

A female collected on Bremer Island sometime around mid-August had a foetus

12 mm. long in the uterus. A female collected at Oenpelli on 5 November had a young male (head-and-body length 26 mm.) clinging to it.

For external and cranial measurements see Table 12.

Nycticeius balstoni balstoni (Thomas)

Scoteinus balstoni Thomas, *Abstr. Proc. Zool. Soc. Lond.* no. 31 (8 May 1906). (Description amplified in *Proc. Zool. Soc. Lond.* 1906: 472 (10 October 1906).) Type locality: North Pool, Laverton, Western Australia.

Horse Shoe Bend, Finke River, 7. One male and four female skins with skulls, one male and one female in alcohol; 23-4 Nov. 1948 (284171-77).

These bats came at dusk to drink by skimming just above the surface of the water in a large open tank at the Horse Shoe Bend cattle station. Most of the specimens were caught by stretching a fine wire about an inch above the water surface. Bats that struck the wire and fell into the water were caught by hand as they swam to the edge of the tank.

Nycticeius balstoni caprenus (Troughton)

Scoteinus balstoni caprenus Troughton, *Aust. Zool.* 8: 279 (12 March 1937). Type locality: Roebuck Bay, Kimberley region, Western Australia.

Darwin, 1. Female in alcohol (skull removed and cleaned); 3 Oct. 1920 (237973).

The specimen has been labelled *Scoteinus greyi* and was previously reported under that name by Sanborn (1931: 26) from the non-existent locality 'Lincoln, North Territory'. This error resulted from faulty labelling, the correct locality being Darwin, where Hoy collected this and a few other specimens of mammals in early October 1920 before beginning his field work farther south at Douglas River and Brocks Creek.

In size and other characters this specimen agrees well with the description of *caprenus*. Measurements are given in Table 12. It is larger than *Nycticeius greyi*, judging from measurements of that species given by Dobson (1878: 264).

Chalinolobus gouldii venatoris Thomas

Chalinolobus gouldi venatoris Thomas, *Ann. Mag. Nat. Mist.*, ser. 8, 2: 372 (October 1908). Type locality: Alexandria, Northern Territory.

Douglas River, 2. Females in alcohol (one with skull removed and cleaned); 25-31 Oct. 1920 (237954, 237967).

Wigley Waterhole, 4 miles north of Alice Springs, 1. Female skin with skull; 20 Nov. 1948 (284187).

Horse Shoe Bend, Finke River, 1. Female skin with skull; 24 Nov. 1948 (284188).

Other records for Northern Territory: Alexandria (Thomas 1906e: 537, 1908: 372); Bluff Hole, Alexandria (Thomas 1906e: 537); Hermannsburg (specimens seen in Australian Museum, Sydney); Charlotte Waters (specimen seen in Victorian National Museum, Melbourne).

Little was learned of the habits of this bat except that like many others it becomes active at dusk and visits waterholes to drink at that time. Hoy recorded one of the specimens from Douglas River as 'flying through light scrub at dusk'. The female collected at Wigley Water Hole was one of several bats that circled in the evening over the water, which was then limited to three small pools in a narrow cliff-walled gorge and was the only open water within several miles. The female from Horse Shoe Bend, which on 24 November had two small foetuses in the uterus, visited the same open water tank used for drinking by *Nycticeius balstoni*.

Chalinolobus picatus nigrogriseus (Gould)

Scotophilus nigrogriseus Gould, *Mammals of Australia*, pt. 8, vol. 3, pl. 44 (1856). Type locality: vicinity of Moreton Bay, south-eastern Queensland.

Brocks Creek, 2. Males in alcohol (one with skull removed and cleaned); 18-27 Nov. 1920 (237953, 237955).

Yirrkalla, 8. Six male skins with skulls and two males in alcohol; 16 July-16 Aug. 1948 (284190-97).

Port Langdon, Groote Eylandt, 1. Male skin with skull; 6 June 1948 (284189).

Other record for Northern Territory: Port Essington (Dobson 1878: 252; Thomas 1909a: 150).

Like *Chalinolobus gouldii* this species was generally seen in flight at dusk in the vicinity of open fresh water, but its interest in this environment appeared to involve foraging for flying insects as well as drinking.

At Yirrkalla and at other places on the Cape Arnhem Peninsula, *Chalinolobus picatus* was a familiar sight in the evenings, darting about the tops of trees that line the creek banks and the borders of marshy areas. They were not seen in the dry eucalypt forest. The only specimen collected on Groote Eylandt was flying at dusk over wattle scrub on the stationary sand dunes near Umbakumba.

All of the eleven specimens now at hand from the Northern Territory are males. I do not know whether this is a result of coincidence or reflects some kind of sexual segregation in the species.

In *Chalinolobus picatus* there is a trend from east to west across northern Australia toward increasing length and contrast of the white hair tips on the dorsal surface, which gives the animals a characteristic frosted appearance. This culminates in the small, pale, and heavily frosted subspecies *rogersi* in the Kimberley district of Western Australia. The specimens from Arnhem Land are somewhat intermediate in this character between *rogersi* and typical *nigrogriseus* of eastern Queensland but in general are closer to the latter. The present material confirms the opinion of Thomas (1909a: 150) that the most abrupt transition from dark to pale individuals takes place near the boundary between the Northern Territory and Western Australia. It might be expected that the small blackish race *picatus* occurs in the southern part of the Northern Territory, but as yet there are no records to confirm this surmise.

The small upper premolar in *Chalinolobus* is crowded into the angle between the canine and the first large premolar on the lingual side of the tooth-row. In *C. picatus* this tooth is so reduced as to be almost invisible and in some individuals is lacking. In old specimens the tooth may lose its bony alveolus and appear to be anchored only in the soft tissue of the palate. Consequently it is liable to be lost when skulls are thoroughly cleaned. Of nine skulls here studied the small premolar is present on both sides in three, on one side only in four, and entirely absent in two.

Measurements are given in Table 12.

Miniopterus schreibersii orianae Thomas

Miniopterus orianae Thomas, *Ann. Mag. Nat. Hist.*, ser. 9, 10: 616 (December 1922). Type locality: Casuarina Bay, Port Darwin, Northern Territory.

Darwin, 41. Six male and eighteen female skins with skulls, one male complete skeleton, six males and ten females in alcohol; 9 April 1948 (284198-238).

Other record for Northern Territory: Casuarina Bay, Port Darwin (Thomas 1922e: 616).

TABLE 12

EXTERNAL AND CRANIAL MEASUREMENTS OF *Eptesicus, Nycticeius* AND *Chalinolobus*
NORTHERN TERRITORY, AUSTRALIA

	Head and body	Tail	Hind foot, with claw	Ear from notch	Forearm	Condylobasal length	Zygomatic breadth	Intertemporal constriction
Eptesicus pumilus caurinus								
Brocks Creek								
♂ 237968	45	35	8	—	32	—	—	—
Oenpelli								
♂ 284183	46	32	7	11	32	11·3	7·9	3·3
♀ 284184	42	30	7	11	32	—	—	—
284185	45	32	7	12	34	—	—	—
Bremer Island								
♀ 284182	44	34	8	12	34	—	—	—
Cape Arnhem Peninsula								
♂ 284179	45	35	8	12	33	12·1	8·4	3·5
284181	48	36	7	11	33	—	—	—
♀ 284180	54	37	7	12	33	12·3	8·7	3·6
Groote Eylandt								
♂ 284178	46	31	7	11	30	11·5	7·9	3·8
Nycticeius balstoni balstoni								
Horse Shoe Bend								
♂ 284172	61	36	9	12	36	—	—	—
284174	59	36	10	15	35	14·3	10·4	3·4
♀ 284171	67	34	10	12	37	14·8	10·5	3·7
284173	61	38	9	15	37	—	—	—
284175	68	38	9	14	37	14·3	10·9	3·6
284176	65	39	11	13	37	14·9	10·9	3·5
284177	66	35	11	14	38	14·9	10·7	3·6
Nycticeius balstoni caprenus								
Darwin								
♀ 237973	—	—	—	—	34	13·0	10·1	3·7
Chalinolobus gouldii venatoris								
Douglas River								
♀237954	63	47	10	—	43	14·2	—	4·5
237967	63	55	10	—	43	—	—	—
Wigley Water Hole								
♀ 284187	63	47	10	13	42	14·6	10·7	4·8
Horse Shoe Bend								
♀ 284188	67	44	10	15	43	15·3	11·3	4·7
Chalinolobus picatus nigrogriseus								
Brocks Creek								
♂ 237953	—	—	—	—	35	12·0	8·6	3·8
237955	—	—	—	—	33	—	—	—
Yirrkalla								
♂ 284190	55	34	8	10	35	12·3	8·8	3·9
284191	52	43	9	12	35	12·6	—	3·9
284192	52	42	8	12	35	—	9·2	3·8
284193	48	40	8	11	—	—	—	—
284194	49	39	10	11	—	—	—	—
284195	50	39	9	11	36	12·2	8·6	3·8
284196	54	35	7	12	33	12·6	8·7	3·8
284197	52	34	8	12	35	12·4	8·5	3·7
Port Langdon, Groote Eylandt								
♂ 284189	—	41	8	9	—	—	—	3·6

The entire series of 41 specimens was collected at one time with hand nets in the concrete-walled storage chambers on one of the upper floors of a long-abandoned meat works at Darwin. It was estimated that at least a thousand bats were present in these completely dark and cave-like rooms in the daytime, but no accurate count was made and the entire extensive building was not explored. Only the one species was seen, but about an outside doorway scattered wings of *Miniopterus* gave evidence that the carnivorous false vampire bat was also present and was

TABLE 13
EXTERNAL AND CRANIAL MEASUREMENTS OF *Miniopterus* AND *Nyctophilus*
NORTHERN TERRITORY, AUSTRALIA

	Head and body	Tail	Hind foot, with claw	Ear from notch	Forearm	Condylobasal length	Zygomatic breadth	Intertemporal breadth
Miniopterus schreibersii orianae								
Darwin								
♂ 284199	59	53	10	12	44	14·7	8·8	4·0
284200	63	56	11	13	46	14·8	8·7	3·9
284203	60	54	10	12	45	15·2	8·8	3·9
284207	64	57	11	12	46	15·1	9·1	3·9
284212	63	52	11	12	45	14·8	8·8	4·0
284235	60	55	10	12	44	14·6	8·7	3·9
284238	63	51	10	12	45	14·6	8·6	3·9
♀ 284198	60	50	11	12	44	14·6	8·5	3·9
284201	62	54	10	12	45	14·8	8·8	3·8
284202	63	57	11	12	44	14·8	8·8	4·0
284204	59	55	11	12	45	14·6	8·6	3·7
284205	61	56	11	12	45	14·8	8·5	3·8
284206	64	52	10	12	45	14·9	8·5	3·8
284208	63	54	11	13	46	15·0	8·8	3·8
284209	63	52	11	12	45	15·0	8·7	3·8
284210	62	52	10	12	44	14·7	8·6	4·0
284211	64	54	10	12	45	14·6	8·6	3·8
284213	62	53	10	12	46	14·9	8·7	3·9
284214	61	57	10	11	45	14·9	8·5	3·8
284231	63	51	10	12	44	14·6	8·7	3·9
284232	59	55	11	12	44	14·8	8·7	4·0
284233	58	56	10	12	44	14·8	8·7	4·0
284234	58	56	11	12	44	14·7	8·5	3·7
284236	63	51	10	12	44	14·6	8·4	3·8
284237	60	55	10	12	45	14·6	8·5	3·9
Nyctophilus bifax daedalus								
Brocks Creek								
♂ 237958	—	—	—	—	42	15·3	10·6	3·5
♀ 237957	61	45	11	—	43	—	10·6	3·6
Nyctophilus arnhemensis								
Brocks Creek								
♂ 237960	—	—	—	—	—	14·0	—	—
Cape Arnhem Peninsula								
♂ 284240	51	43	9	21	37	14·0	10·0	3·5
284242	54	36	9	20	36	13·6	—	3·3
♀ 284241	57	43	9	21	37	—	—	—
Groote Eylandt								
♂ 284239	51	44	10	21	37	14·2	10·5	3·8

preying on this species (see account of *Macroderma gigas* above). The bats clung in irregular clusters to the concrete walls and ceilings. The floors beneath were covered with a mixture of dust and guano up to an inch in depth.

The three specimens on which the description of this race was based were collected near Darwin in July and were characterized by a strikingly pale colour for a member of this group, described by Thomas as 'nearly uniform pale brown (near, but not quite as dark as "sayal brown")'. This description exactly matches seven of the twenty-four skins in the present series, which was collected on 9 April. Of the remaining seventeen specimens, seven are uniformly coloured near Hair Brown, a colour which approaches that normally encountered in Oriental and Australasian races of *Miniopterus schreibersii* but which still distinguishes *orianae* as one of the palest of the eastern races. Among the skins now available in the United States National Museum, these specimens find their nearest colour counterparts in French specimens of *M. schreibersii schreibersii*, which are slightly paler and glossier. In the series from Darwin there remain ten specimens which combine the two colour types, being Hair Brown on the body and Sayal Brown on the head and neck in varying degrees. The colours contrast sharply and appear to represent a pattern of moulting, the darker fur being replaced by the brighter reddish fur, the process beginning on the head. This seems to indicate a seasonal change of pelage (in April) at the end of the wet season, the dull greyish coat being moulted and replaced by a brighter brown coat. There is no correlation between these colour types and sex or age in the specimens examined. No young bats are included, the entire series being remarkably uniform in age and dimensions (see measurements in Table 13).

Nyctophilus bifax daedalus Thomas

Nyctophilus daedalus Thomas, *Ann. Mag. Nat. Hist.*, ser. 8, 15: 498 (May 1915). Type locality: Daly River, Northern Territory.

Brocks Creek, 2. Male and female in alcohol (skulls removed and cleaned); 19 and 23 Nov. 1920 (237957-58).

Other records for Northern Territory: Daly River (Dahl 1897: 193; Collett 1897: 319; Thomas 1915b: 498; Tate 1941b: 593, 595); Melville Island (Thomas 1915b: 498); Port Essington (Dobson 1878: 174; Thomas 1915b: 498).

Of the two specimens collected by Hoy in 1920, the male was flying over a waterhole at dusk and the female was found clinging to the bark of a tree in daytime. In 1894, Dahl found them abundant in the evening around the Uniya Mission Station on the Daly River, noting that they never rose to any great height and that their flight appeared feeble and fluttering (Dahl 1897: 193, under '*Nyctophilus timoriensis*').

Measurements are given in Table 13.

Nyctophilus arnhemensis Johnson
Pl. 9

Nyctophilus arnhemensis Johnson, *Proc. Biol. Soc. Wash.* 72: 184 (4 November 1959). Type locality: Rocky Bay, Cape Arnhem Peninsula, Northern Territory.

Brocks Creek, 1. Male in alcohol (skull removed and cleaned); 19 Nov. 1920 (237960).

Rocky Bay, Cape Arnhem Peninsula, 3. Two male and one female skins with skulls; 12-24 Aug. 1948 (284240-42).

Port Langdon, Groote Eylandt, 1. Male skin with skull; 18 June 1948 (284239).

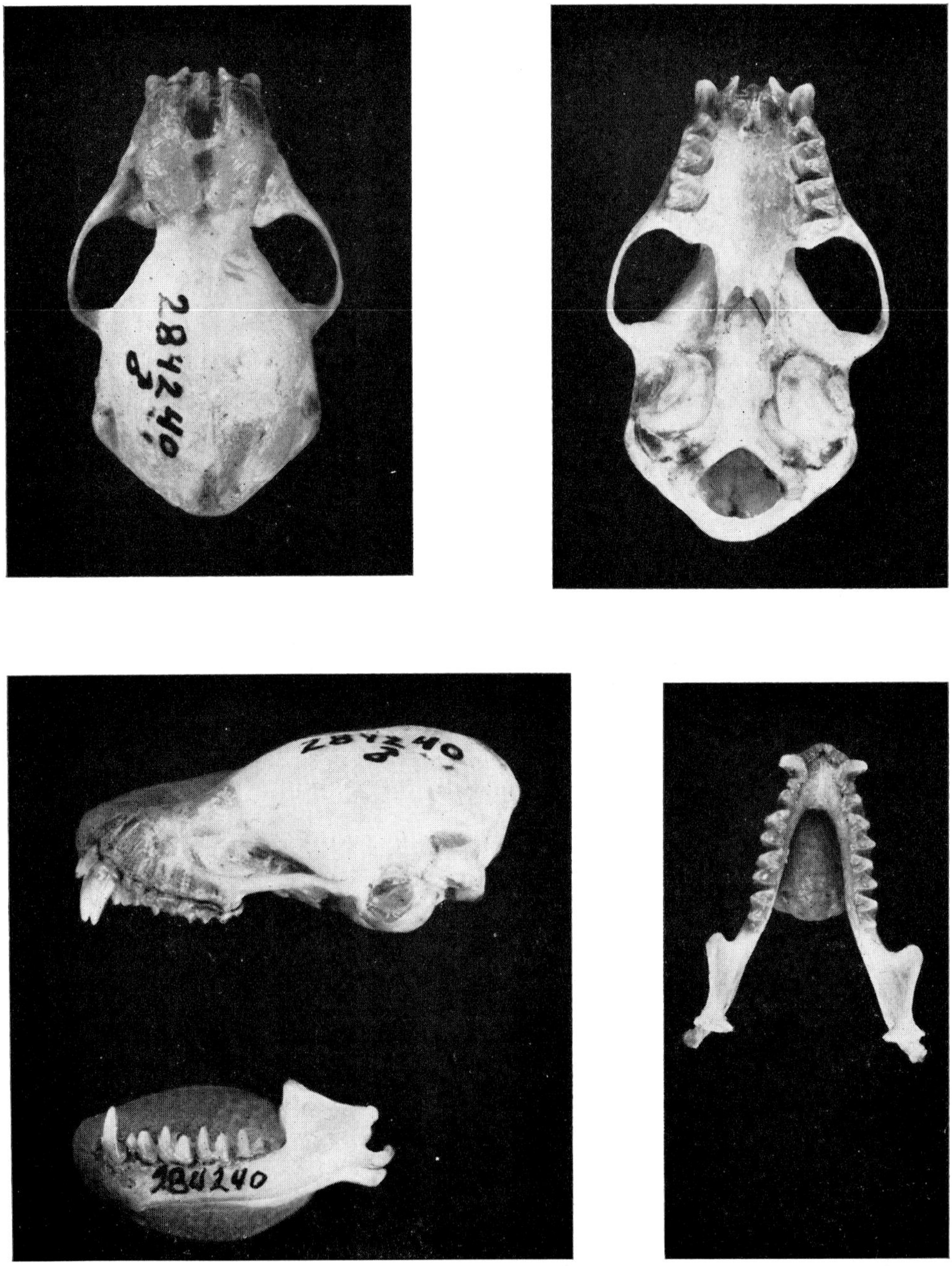

PLATE 9. *Nyctophilus arnhemensis*
Skull of type specimen, no. 284240, ♂, x 4

In foraging this bat flutters close about the foliage in the more densely forested areas. It is likely to be encountered at night in the tunnel-like passages where trails or roadways penetrate the isolated patches of monsoonal forest. Compared with most other bats, its flight is slow and weak. Daytime refuge sites are provided by hanging flaps of the soft, flexible bark of large paperbark trees (*Melaleuca leucadendra*) in low-lying areas. On two occasions at Rocky Bay these bats were found beneath 'dog-eared' sections of the bark about eight feet from the ground. The specimen from Brocks Creek, according to Hoy's notebook, was flying over a waterhole at dusk, and the specimen from Groote Eylandt was caught at night in Mr Fred Gray's house at Umbakumba on Port Langdon.

Characters that distinguish this species from *N. geoffroyi* and *N. walkeri* are detailed in the original description (Johnson 1959: 184). Its nearest relative is probably *Nyctophilus microtis* of New Guinea, but no direct comparison of specimens has been made to confirm this guess. Measurements are given in Table 13.

MOLOSSIDAE

Tadarida loriae cobourgiana Johnson

Pl. 10

Tadarida loriae cobourgiana Johnson, *Proc. Biol. Soc. Wash.* 72: 185 (4 November 1959). Type locality: Black Rock Point, Cobourg Peninsula, Northern Territory.

Black Rock Point, Cobourg Peninsula, 1 (the type specimens). Female skin with skull; 25 Sept. 1948 (284243).

The single specimen was shot from a group of about ten bats, seemingly all of the same kind, that flew about one end of a meadow-like opening in the forest about a mile inland from Black Rock Point. This was shortly after sundown, while the sky was still bright. The bats flew very rapidly and the group dispersed soon after it was discovered. It was thought that the bats must have emerged in a group from one of several standing dead trees in the vicinity. As is discussed in detail in the original description (Johnson 1959: 185) this bat shows closer affinity to *Tadarida loriae* of Papua and the Cape York Peninsula than it does to *Tadarida norfolkensis* of south-eastern Australia.

External and cranial measurements of the type specimen: length of head and body 53 mm., tail 36, hind foot with claw 8·5, ear from notch 14, forearm 34, length of skull from anterior projection of premaxillaries to summit of occipital protuberance 14·7, inter-temporal breadth 3·8, upper tooth-row C-M^3 6·0.

LAGOMORPHA

LEPORIDAE

Oryctolagus cuniculus (Linnaeus)

Lepus cuniculus Linnaeus, *Systema Naturae*, ed. 10, 1: 58 (1758). Type locality, Germany.

Alice Springs, 1. Male skin with skull; 20 Nov. 1948 (284245).

7 miles south-west of Alice Springs, 1. Male skin with skull; 17 Nov. 1948 (284244).

Rabbits were seen commonly about Alice Springs and Horse Shoe Bend cattle station in November of 1948.

In the vinicity of Alice Springs they seemed equally at home on the flat plains country and in the rocky hills and ravines of the Macdonnell Range. Near the aeroplane landing field several miles south of the town the rabbits had conspicuous

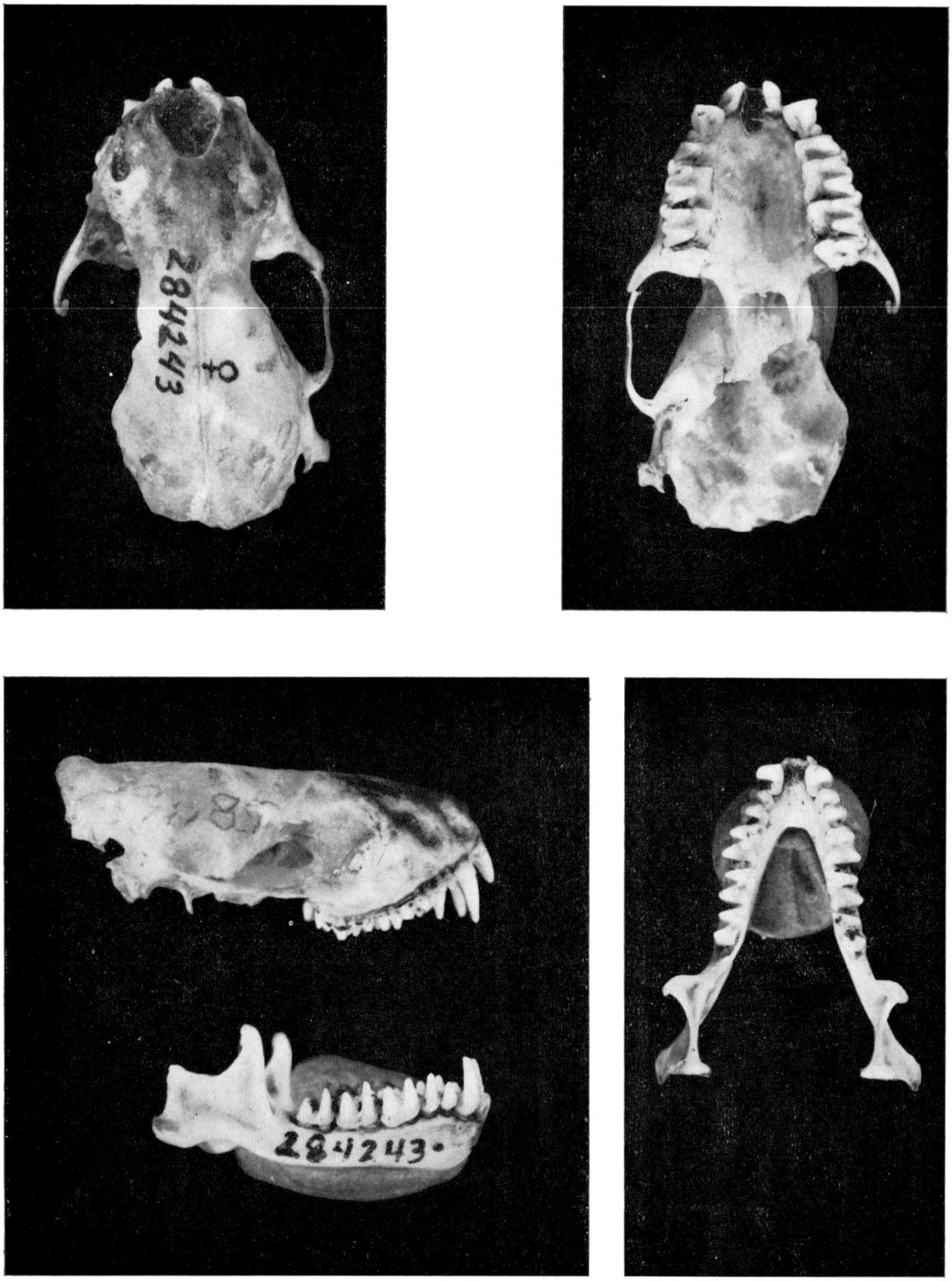

PLATE 10. *Tadarida loriae cobourgiana*
Skull of type specimen, no. 284243, ♀, x 4

warrens scattered at intervals over the open plains, which were then very dry and supported only sparse vegetation in the form of clumps of yellowed grass. No rabbits were visible about the warrens at midday, but after sundown and before dark they emerged from the warrens in surprising numbers. The maximum count of rabbits in view at one time about a warren was 30.

In the rocky country north of Alice Springs, near Wigley Waterhole, rabbits had well-beaten paths through dry grass and resting forms in sheltered places behind boulders, but no conspicuous warrens. There was no indication that water was available for drinking in the vicinity.

Of approximately 500 rabbits seen about Alice Springs all had normal coloration except one that was yellowish in colour and another that had a conspicuous piebald mixture of white and the normal agouti pattern.

Measurements of the two adult males: length of head and body 419 and 412, length of tail 69 and 69, hind foot with claw 106 and 102, ear from notch 78 and 81, condylobasal length of skull 70·4 and 72·5, zygomatic breadth 38·7 and 40·2, inter-orbital breadth 17·7 and 21·2, alveolar length of upper tooth-row 16·1 and 15·5.

Successful as they obviously are in the arid southern part of the Northern Territory, rabbits have failed to build up populations in Arnhem Land. No rabbits were seen or heard of there in the course of our work. A factor that might prevent their colonization of that area is the prolonged wet season which would offer special hazards to a burrowing animal.

RODENTIA

MURIDAE

Conilurus penicillatus penicillatus (Gould)

Mus penicillatus Gould, *Proc. Zool. Soc. Lond.* 10: 12 (June 1842). Type locality: Port Essington, Northern Territory.

South Alligator River, 3. Two males and one of undetermined sex, skins with skulls; May-June 1903 (141485-87).

Other records for Northern Territory: Hermit Hill (Collett 1897: 321); Daly River (Collett 1897: 321); Adelaide River (Collett 1897: 321); Mary River (Collett 1897: 321); Port Essington (Gould 1842c: 12; Gray 1843: 109); Groote Eylandt (Tate 1951: 271).

These specimens, which were received in exchange from the British Museum in 1905, are part of the series of forty collected by J. T. Tunney in the vicinity of the South Alligator River and reported on by Thomas (1904a: 223). They were later studied by Tate in the preparation of his monograph on the rodents of Australia and New Guinea (Tate 1951: 271).

All three specimens have black tails, and one of the three has a white spot on the forehead. Measurements are given in Table 14. Two still carry the original field labels, which bear data confirming the impression received from other sources that the locality 'South Alligator River' as cited by Thomas covers a broad area and a great variety of habitats. No. 1425 (USNM no. 141486) is an adult male that was caught in a hollow tree during the day on 6 May, fifteen miles from the head of the South Alligator River. No. 1371 (USNM no. 141485), a young male, was collected on June 30 in a hollow tree near the mountains ten miles east of the South Alligator River.

Zyzomys Thomas

Zyzomys Thomas, *Ann. Mag. Nat. Hist.*, ser. 8, 3: 372 (April 1909). Type: *Mus argurus* Thomas.

Laomys Thomas, *Ann. Mag. Nat. Hist.*, ser. 8, 3: 373 (April 1909). Type: *Laomys woodwardi* Thomas.

The three species included in the genus *Zyzomys* are all rather small, harsh-furred rats having a tendency toward fatty thickening of the skin of the tail. They are restricted to northern and central Australia, and locally they are to be found only in the vicinity of cliffs or rocky hills. All three species occur in the Northern Territory, *argurus* and *woodwardi* together in the stony interior of Arnhem Land and *pedunculatus* alone in the arid ranges of the southern parts of the territory. One of the northern species, *argurus*, is found on Groote Eylandt, and Tate (1951: 266) has reported it from localities in northern Queensland.

Incrassation of the tail has been used by some authors as a character of importance in diagnosing species and genera, and some of the confusion in the nomenclature of the present group may be traced to this source. The fat-tailed condition is characteristic of all the species of *Zyzomys* under some conditions, and apparently is not found to any pronounced degree in other Australian rodents. It is not shown, however, in all specimens of *Zyzomys*. The factors that determine the expression of this character are not well known, but it is probably significant that in the specimens of *Zyzomys argurus indutus* from Groote Eylandt the degree of incrassation of the tail is directly correlated with the age of the individual rat as indicated by wear of the cheek teeth. The oldest individuals have the fattest tails and the youngest show only a slight thickening of the skin.

The generic name *Laomys*, originally based on the species *woodwardi* and also including *pedunculatus*, has little or no standing. The characters that distinguish these species from *argurus* seem to be of specific rather than generic value. The same conclusion has already been reached by Ellerman (1947: 260; 1949: 25, 35) and by Tate (1951: 265), although the latter author prefers to retain *Laomys* as a subgenus. The species of *Zyzomys* are known from only scanty material. The total of twenty-four specimens obtained by the Arnhem Land Expedition exceeds the combined totals of all the previously collected specimens that have been recorded in the literature.

Key to the Species of *Zyzomys*

1 Larger, length of head and body more than 140 mm. condylobasal length of skull more than 33 mm. Range: northern parts of Western Australia and Northern Territory . . . . *Zyzomys woodwardi*

Smaller, length of head and body less than 140 mm., condylobasal length of skull less than 33 mm. 2

2 Teeth small, upper tooth-row 4·7-5·3 mm. Range: northern parts of Western Australia, Northern Territory, and Queensland . . . . . . *Zyzomys argurus*

Teeth larger, upper tooth-row 6·0-6·6 mm. Range: southern part of Northern Territory . . . . *Zyzomys pedunculatus*

Zyzomys argurus indutus (Thomas)

Pl. 11, 16

Mesembriomys argurus indutus Thomas, *Ann. Mag. Nat. Hist.*, ser. 8, 3: 151 (February 1909). Type locality: Parry's Creek, near Wyndham, Kimberley district, Western Australia.

Oenpelli, 12. Five male and three female skins with skulls, male and female complete skeletons, male and female in alcohol; 28-9 Oct. 1948 (284329-40).
Central Hill, Groote Eylandt, 9. Four male and two female skins with skulls, two male skulls only, one male in alcohol; 30 June-2 July 1948 (284320-28).

These small rock rats were trapped on Unbalanja Hill near the Oenpelli Mission Station. Here as elsewhere the species seemed to be restricted to a rock-ledge habitat. Unbalanja Hill is one of several outliers of the stony Arnhem Land escarpment in the vicinity of Oenpelli. It is completely surrounded by sandy plain and black-mud plain associations and is about a half-mile distant from similar small rocky hills. In 1948 it appeared to have a large population of these rats, whereas traps set in similar habitats on two of the neighbouring rocky hills caught none.

On Groote Eylandt the rats were trapped on the lower slopes of the abrupt north face of Central Hill. Individuals were caught in a variety of situations, including: on rock ledges, under the edges of large isolated boulders, in talus areas, about spinifex clumps, and under pandanus. The common factor in all these situations was the rocky substrate, which appears to be necessary for the presence of this species. All the rats were caught at night, even though the traps were left set through the day.

The skin is very thin and easily torn, especially on the rump and about the back of the head. The tails break off easily.

These specimens agree well with the description of *indutus*. There is no significant difference between the series from Oenpelli and that from Groote Eylandt. The mammary formula is 0-2: 4. For measurements see Table 14.

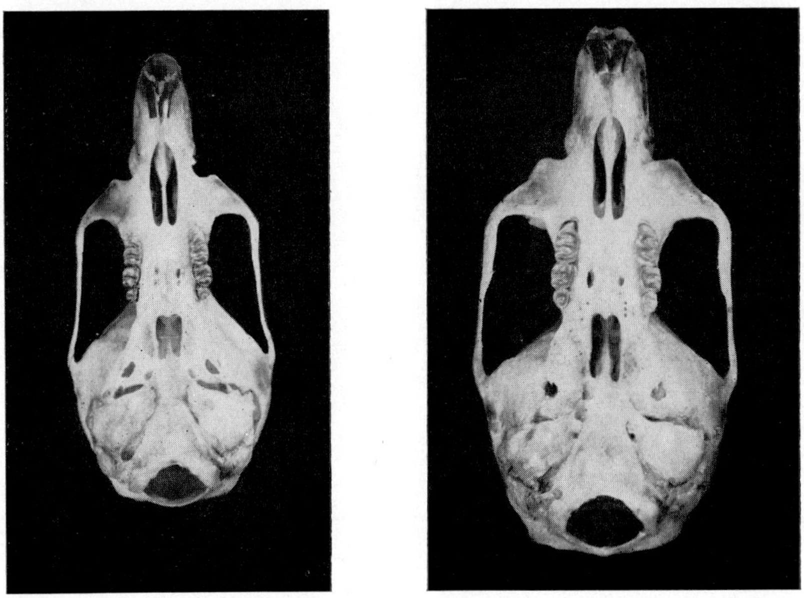

PLATE 11. Comparison of skulls of two species of *Zyzomys* in Arnhem Land

Left, *Z. argurus indutus*, no. 284335, ♂, from Oenpelli, x 2
Right, *Z. woodwardi*, no. 284343, ♂, from Oenpelli, x 2

Zyzomys woodwardi (Thomas)
Pl. 11

Laomys woodwardi Thomas, *Ann. Mag. Nat. Hist.*, ser. 8, 3: 373 (April 1909). Type locality: Parry's Creek, near Wyndham, Kimberley district, Western Australia.

Oenpelli, 3. Two male skins with skulls, one male in alcohol; 29 Oct.-3 Nov. 1948 (284341-43).

Heretofore the species has been known only from the two specimens collected in 1908 by J. P. Rogers at Parry's Creek, Western Australia. The present record thus extends the known range of *Zyzomys woodwardi* some 300 miles to the eastward and into the Northern Territory.

The specimens were trapped on Unbalanja Hill in a habitat that appeared identical with that of the more abundant *Zyzomys argurus*. Only adult males were caught, and none of them had a complete tail. The skin, as in *Zyzomys argurus*, was very thin and easily torn by traps and by handling in the process of preparing specimens. For measurements see Table 14.

Mesembriomys gouldii gouldii (Gray)

Mus hirsutus Gould, *Proc. Zool. Soc. Lond.* 10: 12 (June 1842). Type locality: Port Essington, Northern Territory. Preoccupied by *Mus hirsutus* Elliott in *Madras J.* 10: 213 (1839).

Hapalotis gouldii Gray, *List of the specimens of Mammalia in the collection of the British Museum*, p. 116 (1843). New name for *Mus hirsutus* Gould.

South Alligator River, 2. Female skins with skulls; 10 May and 29 Aug. 1903 (141488-89).

Gove Airfield, Cape Arnhem Peninsula, 2. Female and male skins with skulls; 9 July and 10 Sept. 1948 (284344 and 284349).

Nakala Creek, Cape Arnhem Peninsula, 1. Male skin only; 17 Aug. 1948 (284345).

Birluoja Creek, Cape Arnhem Peninsula, 3. Young male in alcohol, young female and adult male skins with skulls; 21 Aug. 1948 (284346-48).

Other records for Northern Territory: Hermit Hill (Collett 1897: 322; Tate 1951: 268); Daly River (Collett 1897: 322); Mount Shoebridge (Collett 1897: 322); Katherine River (Collett 1897: 322); South Alligator River (Thomas 1904a: 223; Tate 1951: 268); East Alligator River (Tate 1951: 268); Port Essington (Gould 1842c: 12; Gray 1843: 116; Tate 1951: 268).

There seems to be no adequate vernacular name in general use for this distinctive species. The terms 'rabbit-rat' (from the large ears) and 'jerboa-rat' (from the long and white-tipped tail) are particularly inappropriate and misleading when applied to a strictly arboreal animal with habits and habitat anything but rabbit- or jerboa-like. The word 'mantbul', used by the aborigines of the Cape Arnhem Peninsula, is suggested as a fitting replacement for these meaningless book names.

In addition to the specimens collected by the Arnhem Land Expedition on the Cape Arnhem Peninsula, the United States National Museum obtained, through exchange with the British Museum in 1905, two specimens collected by J. T. Tunney in 1903 in the vicinity of the South Alligator River. One of these is no. 1468 of the series recorded as *Conilurus hirsutus* by Thomas (1904a: 222). It is a young female with the following information in the collector's handwriting on the original label: 'South Alligator River, 15 miles from head. Caught in hollow tree with mother during day.' The other specimen, an adult female, has the data: 'In hollow tree at the head of Tidal waters on Alligator River'.

At Gove Airfield on the Cape Arnhem Peninsula, the rats had established themselves about the temporary buildings that were erected in the stringy-bark forest near the landing field during World War II. By 1948 these structures had been abandoned by their human occupants and were gradually falling apart. The rats climbed at night about the walls and exposed rafters. Service men stationed there during the war were said to have been startled by the sight of enormous rats in the buildings.

In the course of a 'walkabout' trip in the area south of Melville Bay mantbuls were encountered at Nakala Creek and one of its tributaries, Birluoja Creek. On the afternoon of 17 August an adult male was captured in a hollow tree beside Nakala Creek about a mile above the head of tidewater. On 21 August at Birluoja Creek another male was captured in the same kind of tree. In another tree a few yards away there was a nest containing two dead and partly decomposed young. The trees frequented by these rats have a smooth grey outer bark and a deceptively sturdy appearance. Actually the main trunk and the larger branches are thin shells of bark and living wood surrounding a capacious branching cavity that provides shelter for the rats and for other mammals, including *Trichosurus vulpecula*, as well. The trees are easily felled by chopping with a light axe through the inch-thick walls. The rats stayed in the hollow limbs after the trees crashed to the ground, and were captured by hand. The abandoned nest consisted of a plug of leaves, bark, and other trash choking an upper section of the main trunk cavity about twelve feet from the ground. The dead young ones were lying in a depression on top of this accumulation.

The specimens from the South Alligator River have a slightly more yellowish suffusion on the body both above and below than do those from the Cape Arnhem Peninsula. The latter specimens have a pectoral area of grey-based fur, which may indicate gradation toward the wholly grey-bellied *M. g. rattoides* of northern Queensland. The pattern of pale markings on the hind feet, shown with some variation in all the specimens, consists of whitish hairs at the bases of the claws, an irregular cross-band over the distal ends of the second, third, and fourth metatarsals, and a streak (in some specimens discontinuous) along the second metatarsal, this streak expanding into a broad white patch on the upper part of the foot. The subspecies *M. g. rattoides* and *M. g. melvillensis* are described as having wholly dark feet.

There is an unusual degree of individual variation in the detailed cranial characters, especially in the length of the anterior palatine foramina and in the size of the bullae. A larger series would be required to determine the limits of this variation. It appears that females are larger than males. One of the specimens from Gove Airfield has a fourth upper molar tooth, an unusual occurrence in any placental mammal. This specimen and its significance have been discussed in a separate paper (Johnson 1952).

Rattus rattus (Linnaeus)

Mus rattus Linnaeus, *Systema Naturae*, ed. 10, 1: 61 (1758). Type locality: Uppsala, Sweden.

A slate-coloured rat, thought to be a representative of the 'black' colour phase of the introduced European roof rat, was seen in a warehouse at Bagot, on the outskirts of Darwin on 13 November 1948. From local reports, the species is common about the town.

A white-bellied rat, apparently of this species, was recorded from the Daly River by Collett (1897: 323).

TABLE 14

EXTERNAL AND CRANIAL MEASUREMENTS OF *Conilurus, Zyzomys,* AND *Mesembriomys* NORTHERN TERRITORY, AUSTRALIA

	Head and body	Tail	Hind foot with claw	Ear from notch	Condylobasal length	Zygomatic breadth	Interorbital breadth	Upper tooth-row M^{1-3}
Conilurus penicillatus penicillatus								
South Alligator River								
♂ 141485	170	180	44*	—	—	—	5·9	8·4
141486	190	—	47*	—	—	24·0	6·2	8·4
Zyzomys argurus indutus								
Oenpelli								
♂ 284330	113	116	24	18	—	13·4	4·5	5·1
284333	111	120	23	18	27·7	14·5	4·5	5·1
284334	105	121	25	18	27·1	13·7	4·2	5·0
284335	110	—	24	18	28·6	13·2	4·0	4·8
284336	110	—	24	18	27·1	14·0	4·4	5·1
284337	107	116	25	—	27·9	13·9	4·5	5·3
284339	110	120	24	18	—	—	—	—
♀ 284329	113	122	25	18	27·6	13·6	4·3	4·9
284331	104	116	24	—	—	—	4·6	5·1
284332	122	136	25	18	28·5	13·6	4·0	5·4
284338	118	115	24	—	27·6	14·2	4·8	4·8
284340	108	—	24	17	—	—	—	—
Groote Eylandt								
♂ 284321	118	—	25	20	—	—	—	5·0
284322	122	106	26	19	—	14·6	4·8	5·4
284323	108	—	25	21	28·1	13·9	4·6	4·9
284324	104	116	26	18	—	—	4·8	4·9
284325	—	118	26	19	—	15·1	4·6	5·2
284326	119	118	24	20	—	—	—	—
284327	104	111	25	16	26·7	14·0	4·5	5·0
♀ 284320	—	121	25	—	28·9	14·6	4·8	4·9
284328	117	119	26	19	30·1	14·7	4·5	4·9
Zyzomys woodwardi								
Oenpelli								
♂ 284341	138	—	31	21	—	—	—	—
284342	156	—	32	21	36·8	18·5	5·3	6·3
284343	153	—	31	22	33·7	18·0	5·5	6·0
Mesembriomys gouldii gouldii								
South Alligator River								
♀ 141489	310	310	70	—	61·1	31·7	11·0	11·8
Cape Arnhem Peninsula								
♂ 284345	299	343	78	44	—	—	—	—
284348	313	343	74	43	62·5	31·5	11·0	11·4

* On dry skin

Rattus tunneyi tunneyi (Thomas)

Mus tunneyi Thomas, *Novit. Zool.* 11: 223 (March 1904). Type locality: Mary River, Northern Territory.

Douglas River, 3. Two female skins with skulls, one male in alcohol; 12-17 Oct. 1920 (237780-81, 237925).

Brocks Creek, 6. Two male and three female skins with skulls, one female skeleton only; 13 Nov. 1920 (237788, 237790-94).

Oenpelli, 1. Part of body (in alcohol) and fragments of skull (cleaned) of male; 3 Nov. 1948 (284251).

Other records for Northern Territory: Mary River (Thomas 1904a: 223; Brazenor 1936a: 6; Ellerman 1949: 47, 67; Tate 1951: 345). The locality '100 miles south of Port Darwin' of Tate (1951: 345) is equivalent to Douglas River and Brocks Creek combined.

Hoy's field notebook gives the following information concerning the habitat of the rats from Douglas River: 'From bank of creek', 'From dry creek bed', and 'From hole under tree'. The specimens from Brocks Creek were trapped on the bank of a dry creek, along with *Melomys cervinipes albiventer*.

At Oenpelli on 3 November 1948, a whistling eagle (*Haliastur sphenurus*) flying in the open forest at the base of Unbalanja Hill was seen to have the hindquarters of a rat dangling from its claws. The bird was shot, and fragments of the skull of the rat, including all of the teeth except the left upper incisor, were found in its crop. These meagre remains are of an adult male rat having the large cheekteeth, long incisive foramina, short tail, spiny pelage, white hind feet, and creamy underparts of *Rattus tunneyi*. It is provisionally assigned to the typical race, but its rather large size (see measurements, Table 15) suggest possible relationship with the insular *R. t. melvilleus.*

The series of eight specimens collected by Hoy at Douglas River and Brocks Creek provide near topotypes of *tunneyi*. Tate (1951) has recently recorded and discussed these specimens. Colour of an adult male (no. 237788): Upper parts Ochraceous-Tawny mixed with black; sides between Pinkish Buff and Cinnamon Buff, somewhat dulled by underlying grey basal colour; underparts pale Cream Buff; upper surfaces of feet white; hairs on upper side of tail darker than those below, but in general effect tail appears almost unicolour. Fur of body both above and below thickly set with flattened whitish spines, which do not extend beyond fur and consequently do not give skins an outward appearance of spininess; upper parts with numerous long hairs extending beyond fur, these black on back, black with pale tips on sides and rump.

There are two anterior and three posterior pairs of mammae (clearly to be seen in no. 237780). This character, and others including the greater size, longer tail, and paler colour, readily distinguish *tunneyi* from *colletti* of the same general area. Whatever the relationships of these two rats to forms occurring outside the Northern Territory may eventually prove to be, they are readily distinguishable from one another in Arnhem Land, and Ellerman (1949: 47-8, 67) apparently erred in listing both as subspecies of *Rattus sordidus*. Both *tunneyi* and *colletti* are, however, members of the large-toothed, short-headed '*sordidus* group' of *Rattus*, which seems to include two or more species in Australia.

Rattus gestri colletti (Thomas)

Mus colletti Thomas, *Novit. Zool.* 11: 599 (September 1904). Type locality: South Alligator River, Northern Territory.

South Alligator River, 1. Female skin with skull; about 1903 (141483).

This rather small, dusky-coloured, short-tailed rat seems to be known only from the original series of twenty-five specimens collected by Tunney and studied by Thomas. The specimen now in the United States National Museum is one of that series, having been received in exchange from the British Museum in 1905. The field label is missing.

The specimen agrees well with Thomas's original diagnosis of *colletti* except that the colour of the backs of the hands and feet is dark brown like the upper parts of the body rather than 'dull whitish or brownish white'. The mammae are in three anterior and three posterior pairs. The fur is soft and without stiff spines.

In listing *colletti* as a subspecies of *gestri*, I have followed Tate (1951). It is clearly a species distinct from *Rattus tunneyi*, which occurs in the same general area in Arnhem Land.

TABLE 15

EXTERNAL AND CRANIAL MEASUREMENTS OF *Rattus*
NORTHERN TERRITORY, AUSTRALIA

	Head and body	Tail	Hind foot with claw	Ear from notch	Condylobasal length	Zygomatic breadth	Interorbital breadth	Upper tooth-row M1–3
Rattus tunneyi tunneyi								
Douglas River								
♀ 237780	149	125	30	—	33·9	18·9	5·0	7·6
237781	136	124	31	—	33·9	—	5·0	7·5
Brocks Creek								
♂ 237788	151	147	32	—	35·1	19·6	5·1	7·9
237791	146	143	33	—	36·7	—	5·2	8·1
♀ 237790	141	127	30	—	34·5	19·3	4·8	7·5
237792	142	129	31	—	34·7	19·6	4·7	7·9
237793	142	140	33	—	35·7	18·8	4·7	7·9
237794	141	128	31	—	34·7	19·2	5·0	7·9
Oenpelli								
♂ 184251	—	150	35	—	—	—	—	7·9
Rattus gestri colletti								
South Alligator River								
♀ 141483	—	—	—	—	—	16·5	4·8	6·8

Melomys cervinipes albiventer Kellogg

Pl. 12

Melomys cervinipes albiventer Kellogg, *Proc. Biol. Soc. Wash.* 58: 69 (7 May 1945). Type locality: Douglas River, about 100 miles south of Port Darwin, Northern Territory.

Douglas River, 7. One male and two female skins with skulls, one female complete skeleton, two female skulls only, set of two foetuses in alcohol; 20 Oct.-9 Nov. 1920 (237782-87, 237932).

Brocks Creek, 1. Male skin with skull; 13 Nov. 1920 (237789).

Popham Bay, Cobourg Peninsula, 4. Male and female skins with skulls, male complete skeleton, female skull only; 6 Oct. 1948 (284425-28).

Black Rock Point, Cobourg Peninsula, 2. Male and female skins with skulls; 30 Sept. 1948 (284423-24).

Elcho Island, 1. Female skin with skull; 15 Sept. 1948 (284422).

Yirrkalla, Cape Arnhem Peninsula, 17. Six male and four female skins with skulls, three female skins only, two female skulls only, one male complete skeleton, one female in alcohol; 9 July-26 Aug. 1948 (284399-411, 284418-21).

Rocky Bay, Cape Arnhem Peninsula, 1. Male skin with skull; 24 Aug. 1948 (284417).

Nakala Creek, Cape Arnhem Peninsula, 2. Male and female skins with skulls; 18 Aug. 1948 (284412-13).

Kwaiturumuru Creek, Cape Arnhem Peninsula, 3. Two male and one female skins with skulls; 20 Aug. 1948 (284414-16).

It is remarkable that the presence of *Melomys* on the mainland of the Northern Territory was not recognized until Kellogg described the present form in 1945, because in fact it is the commonest and most generally distributed rodent along the coast of Arnhem Land. This is largely because its preferred habitat, the grass-pandanus association bordering freshwater streams and ponds, recurs at close intervals throughout that area.

Nests are arboreal, and as far as my observations go, they are always in pandanus. Foraging activities are terrestrial, at least in large part. Many specimens were trapped in grassy places on damp ground at the edges of marshes or lagoons, and some under tangled shrubby vegetation at the borders of monsoonal forest. They were not found either on wholly dry ground or in marshy areas where water stood on the surface. During the long dry season, as the ponds and streams gradually dry up, the terrestrial foraging activities must be shifted locally to follow the area of damp ground. During the wet season large areas inhabited by the mice are under water and their activities must be correspondingly more arboreal.

The thick mass of dead leaves that hangs on the stem beneath the green top of a growing pandanus provides a copious shelter for these mice. The leaves are arranged on the stem in spiral fashion, the flat bases of the individual leaves overlap one another, and blades of the leaves are bent downward and shredded to form a dense outer thatch. A flat spiral passageway, well protected from weather and most enemies, is thus left next to the stem, and in this the mice build their nests and apparently spend most of the daylight hours. The dry leaves are highly combustible, and the aboriginal method of capturing *Melomys* is to set fire to the lower leaves, then stand by to catch the mice as they leap from their furiously burning home.

Three nests were obtained from pandanus trees at Rocky Bay on 24 August. Each is a flattened structure from six to eight inches in diameter and from one and one-half to two inches deep. The depth of the nest is determined by the depth of the opening between successive spirals of the leaves, as in each instance the nest completely filled a section of this area next to the stem. The nests are composed entirely of shredded pandanus leaves, the long fibres arranged concentrically about a central nest cavity. There is no clearly defined inner lining of the cavity.

At Yirrkalla, *Melomys* sometimes entered the outbuildings of the mission station. On 27 August, two were found drowned in a half-filled bucket of water that had been left on the floor in a storage shed.

Specimens in adult pelage show a considerable range in size, and in the large series from Cape Arnhem Peninsula collected in July and August there are so clearly a larger and a smaller size group as to suggest that two species might be represented. The difference, however, proves to be entirely a matter of age, all the large individuals being older as indicated by well worn molar teeth. There is no significant variation in colour between eastern and western Arnhem Land. None of the specimens, even among those from the Cobourg Peninsula, has the dark ventral colour described for *M. c. melicus* of Melville Island.

Melomys cervinipes mixtus Troughton

Melomys mixtus Troughton, *Rec. Aust. Mus.* 19: 257 (19 September 1935). Type locality: Groote Eylandt, Northern Territory.

Port Langdon, Groote Eylandt, 28. Thirteen male and seven female skins with

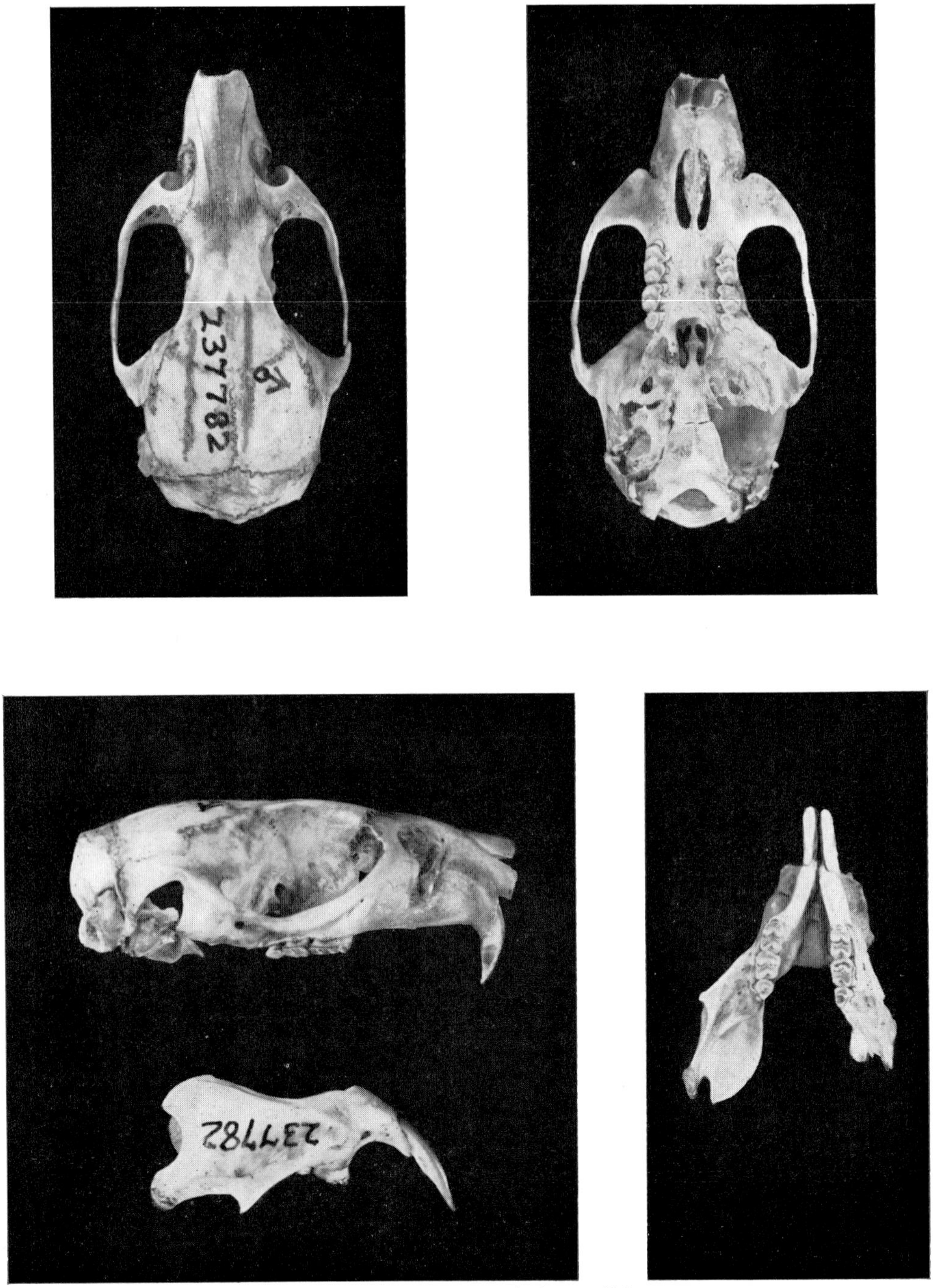

PLATE 12. *Melomys cervinipes albiventer*
Skull of type specimen, no. 237782, ♂, x 2

skulls, male skull only, male and female complete skeletons, female adult in alcohol, two male and two female young in alcohol; 5 June-7 July 1948 (284367, 284369, 284372, 284374-98).

Nearly the entire series came from the borders of the marshy area south-east of Mr Fred Gray's house at Umbakumba.

Aside from smaller size (see measurements, Table 16) there are no constant characters to distinguish this Groote Eylandt race from *Melomys cervinipes albiventer* of the adjacent mainland, and I am convinced that both belong to the same species. In his comprehensive review of the genus, Tate (1951) has assigned the Groote Eylandt subspecies to a distinct species, *Melomys lutillus.* This presents a difficult taxonomic situation which must be left to the future for final solution. For the present both races are treated as subspecies of *cervinipes,* that being the oldest name involved in the immediate problem and the one generally in use for other races of *Melomys* in the Northern Territory.

Mus musculus, subspecies undetermined

Horse Shoe Bend, Finke River, 5. Male skins with skulls; 24-8 Nov. 1948 (284315-19).

House mice were caught in traps set about the bases of shrubs in the alluvial flats of grey sandy soil in the bend of the Finke River and among clumps of porcupine grass on the red sand ridges north-east of the Horse Shoe Bend Station.

In colour the five specimens are strikingly pale as compared with house mice from coastal localities in Queensland, New South Wales, South Australia, and southern Western Australia. On the basis of this material it seems that the house mice of arid central Australia belong to a recognizably paler race than those of the more humid coastal areas. The way in which this difference has come about, whether by evolution of a comparatively homogenous population under extreme environmental conditions or by differences in survival rates among previously evolved races separately introduced, is a matter of speculation and will probably always remain so. Until there are enough specimens available to demonstrate the boundaries of the supposed races, and perhaps until current evolutionary changes have reached a state of relative equilibrium, there is considerable hazard and no great advantage to be gained in attempting to apply subspecific names. The coastal specimens in general resemble the subspecies *Mus musculus domesticus* of north-western Europe, whence most of the introductions into Australia were undoubtedly made. These have dark-coloured underparts, the bases of the hairs plumbeous and the tips faintly or strongly yellowish. Of the Finke River specimens, one has this type of coloration and the remaining four have the underparts creamy white with reduced dark basal colour, thus resembling the Mediterranean subspecies *M. m. brevirostris.* There is no particular resemblance to the typical race *M. m. musculus* of Sweden and eastern Germany, although that name in a restricted sense is often mistakenly applied to house mice in various parts of the world. Measurements are given in Table 17.

Leggadina delicatula delicatula (Gould)

Mus delicatulus Gould, *Proc. Zool. Soc. Lond.* 10: 13 (June 1842). Type locality: Port Essington, Northern Territory (lectotype specimen designated by Thomas 1921a: 432).

Mus albirostris Gray, *List of the specimens of Mammalia in the collection of the British Museum*: 112 (13 May 1843). Name quoted from Gould manuscript as synonym of *Mus delicatulus.*

TABLE 16

EXTERNAL AND CRANIAL MEASUREMENTS OF *Melomys*
NORTHERN TERRITORY, AUSTRALIA

	Head and body	Tail	Hind foot with claw	Ear from notch	Condylobasal length	Zygomatic breadth	Interorbital breadth	Upper tooth-row M^{1-3}
Melomys cervinipes albiventer								
Douglas River								
♂ 237782 (type)	136	165	31	—	30·4	16·9	4·8	6·0
♀ 237783	116	133	28	—	29·3	15·4	4·7	5·9
237784	109	139	28	—	—	—	4·7	5·8
237785	—	—	—	—	29·6	16·2	4·7	6·2
237786	121	—	25	—	—	15·2	4·6	6·1
237787	125	133	28	—	29·2	15·3	4·8	6·2
Brocks Creek								
♂ 237789	139	161	30	—	31·4	—	4·9	6·3
Cobourg Peninsula								
♂ 284423	151	149	32	20	31·3	16·8	5·2	6·2
284425	118	125	29	16	28·8	15·3	4·6	6·2
284427	127	129	29	17	29·0	15·4	4·9	6·0
♀ 284424	143	156	31	19	30·5	16·2	4·8	6·1
284426	115	118	26	16	27·6	14·4	4·8	5·9
284428	116	122	27	—	27·7	14·8	5·0	6·3
Elcho Island								
♀ 284422	106	115	27	16	26·4	14·3	4·7	6·1
Cape Arnhem Peninsula								
♂ 284399	132	145	28	16	29·0	15·7	4·7	5·6
284401	119	130	28	18	27·6	15·4	5·0	6·1
284404	114	116	27	16	27·4	14·4	4·7	5·7
284411	121	139	27	17	31·3	15·8	4·9	5·6
284412	116	133	28	16	28·8	15·0	4·8	5·7
284414	127	138	29	16	28·6	14·8	5·0	5·4
284415	116	129	29	16	27·9	14·8	4·7	5·8
284417	105	116	21	16	25·6	14·3	4·6	5·5
284418	139	—	28	16	29·4	15·9	4·8	5·7
284419	135	146	28	16	30·1	16·0	5·3	6·0
284421	130	148	29	17	29·4	15·9	5·1	5·7
♀ 284400	120	135	26	16	28·0	14·8	4·6	5·7
284402	139	152	27	18	29·9	16·1	4·6	5·9
284405	120	134	26	16	—	15·4	5·3	5·6
284407	135	134	26	17	—	—	—	—
284408	123	123	26	15	27·5	14·9	4·7	5·7
284409	126	138	26	17	—	—	—	—
284413	113	134	28	16	26·8	14·5	4·5	5·6
284416	100	127	27	16	25·4	14·2	4·4	6·0
Melomys cervinipes mixtus								
Groote Eylandt								
♂ 284375	118	131	28	16	28·5	15·0	4·7	5·9
284376	117	139	27	16	28·1	15·3	5·1	6·0
284381	99	109	25	15	24·6	13·7	4·3	5·7
284382	101	110	26	15	25·1	13·8	4·5	5·7
284385	100	118	25	15	25·4	13·6	4·2	5·5
284387	121	130	28	16	27·0	15·1	4·6	5·8
284388	105	117	26	17	26·1	14·0	4·2	5·5
284391	99	112	25	15	24·6	14·2	4·6	5·5
284394	113	117	26	17	26·7	14·5	4·5	6·0
284395	—	—	27	15	25·8	14·7	4·6	5·5
♀ 284374	105	114	25	15	24·6	13·9	4·4	5·4
284383	105	120	26	16	25·6	14·8	4·7	5·3
284384	121	125	26	17	28·0	15·0	4·8	6·0
284389	105	—	25	—	25·0	14·3	4·5	5·2
284390	123	126	27	16	27·3	15·4	4·8	5·8
284393	112	118	24	15	25·6	13·9	4·4	5·7
284398	101	115	25	16	25·1	14·2	4·4	5·6

Oenpelli, 1. Male in alcohol; 1 Nov. 1948 (284314).

Yirrkalla, 41. Eleven male and twelve female skins with skulls, one male skull only, two male complete skeletons, three male and three female adults in alcohol, nine young in alcohol; 7 July-30 Aug. 1948 (284264-95, 284305-13).

Rocky Bay, Cape Arnhem Peninsula, 9. One male and three female skins with skulls, two female skulls only, three sets of nestling young in alcohol; 24 Aug. 1948 (284296-304).

Other record for Northern Territory: Port Essington (Gray 1843: 112; Thomas 1921a: 432; Tate 1951: 252).

This tiny mouse inhabits unforested areas where the soil is sandy and well drained and where there is sparse herbaceous vegetation. Such situations are comparatively rare in Arnhem Land and along the coast are confined principally to sand dune areas which occur in scattered patches alternating with rocky headlands, swamps, and marshy plains.

Two sites favourable for *Leggadina* occur near Yirrkalla, one being the low grass-covered ridge of old dunes just west of the mission station, on which the Arnhem Land Expedition camped in 1948, and the other being the recent dunes behind Rocky Bay. At the Yirrkalla site the soil was hard-packed and only a few burrows were found. Concentrated trapping there produced a small but steady yield of specimens over a four-week period.

At Rocky Bay, about two miles south of Yirrkalla, a series of parallel curving ridges of loose sand filled the low-lying area between the shore and a paperbark-pandanus swamp. On the slopes of these ridges and in the depressions between them burrows of *Leggadina* were numerous. Each burrow had a low mound of sand at the entrance and extended into the ground at a shallow angle. Several were dug out on 24 August. None extended more than six feet from the entrance and none reached a depth of more than about ten inches. Only the freshest-looking burrows were excavated and in most of them one or more mice were found. Three burrows had nests with young mice, numbering two, three, and four young per nest, respectively.

The only previous account of the habitat of *Leggadina delicatula* gives a similar description. Gould (1855: text of vol. 3, pl. 23) quoted John Gilbert's journal written at Port Essington in 1840 or 1841, as follows:

> I only met with this species on one occasion, on the Native Companion plains near Point Smith, at the entrance of the harbour, when I found four in a hole which ran along a few inches below the surface for about five feet in a zigzag manner, and terminated in a circular space, wherein was a nest of fine dried grass, in which I captured them.

Leggadina delicatula mimula (Thomas)

Pseudomys (*Leggadina*) *delicatulus mimulus* Thomas, *Ann. Mag. Nat. Hist.*, ser. 9, 17: 634 (June 1926). Type locality: Groote Eylandt, Northern Territory.

Port Langdon, Groote Eylandt, 12. Two male and three female skins with skulls, one male and five females in alcohol, set of three nestling young in alcohol; 7-27 June 1948 (284252-63).

The generally sandy terrain of the part of Groote Eylandt about Port Langdon would seem to be ideal for *Leggadina*, but in 1948 the population was small compared with that concentrated in the isolated patches of suitable habitat on the Cape Arnhem Peninsula. Trapping was wholly unsuccessful as far as the catching of this mouse was concerned, and all of the specimens obtained were dug out of burrows by native women and children.

The status of *mimula* as a recognizable subspecies is confirmed by the present material, which shows on the average a greater size and darker colour for specimens of that race (see measurements, Table 17). The difference is not great, however, and selected specimens from Groote Eylandt would fit well into the series from Yirrkalla.

TABLE 17

EXTERNAL AND CRANIAL MEASUREMENTS OF *Mus* AND *Leggadina*
NORTHERN TERRITORY, AUSTRALIA

	Head and body	Tail	Hind foot with claw	Ear from notch	Condylobasal length	Zygomatic breadth	Interorbital breadth	Upper tooth-row M1–3
Mus musculus								
Horse Shoe Bend								
♂ 284315	83	73	18	12	20·6	10·9	3·6	3·4
284316	76	79	18	13	20·0	10·7	3·6	3·4
284317	74	72	18	10	19·3	9·9	3·4	3·4
284318	80	70	19	12	19·7	10·4	3·5	3·5
284319	72	70	17	11	18·5	9·9	3·6	3·3
Leggadina delicatula delicatula								
Cape Arnhem Peninsula								
♂ 284265	67	69	18	11	18·2	10·5	3·4	3·6
284268	64	63	19	11	17·0	9·7	3·2	3·4
284269	63	63	18	10	16·9	9·5	3·1	3·5
284270	65	65	18	11	—	—	—	—
284271	67	—	19	11	18·2	10·3	3·5	3·4
284272	61	63	19	11	17·8	9·8	3·2	3·6
284284	61	68	17	11	17·1	9·7	3·2	3·4
284285	60	58	18	10	16·6	9·4	3·4	3·5
284295	64	67	17	11	—	—	—	—
♀ 284266	68	73	18	11	19·2	10·3	3·4	3·4
284267	57	61	17	10	16·4	9·7	3·1	3·3
284275	64	61	18	10	17·4	10·1	3·4	3·6
284276	64	65	18	10	17·4	10·0	3·3	3·4
284279	65	71	18	11	17·9	10·3	3·1	3·5
284281	59	63	18	11	16·9	10·4	3·4	3·4
284282	70	78	19	11	18·7	10·3	3·3	3·6
284283	71	68	17	11	18·5	10·4	3·4	3·5
284297	68	61	18	11	17·7	10·5	3·4	3·4
284300	70	65	17	11	18·4	9·6	3·1	3·4
284302	68	65	17	11	18·4	10·0	3·2	3·4
284306	70	—	19	11	—	—	—	—
284309	68	73	18	11	18·3	10·7	3·5	3·5
Leggadina delicatula mimula								
Groote Eylandt								
♂ 284253	62	77	19	11	17·7	9·9	3·5	3·4
284257	66	64	19	11	18·0	10·2	3·4	3·5
♀ 284252	71	74	20	11	19·0	10·5	3·4	3·7
284254	71	75	19	12	18·8	10·2	3·2	3·6
284255	75	72	18	12	18·0	9·9	3·3	3·6
284298	64	71	18	11	17·8	9·8	3·5	3·5
284308	61	—	17	11	—	—	—	—
284310	67	71	18	11	18·1	10·4	3·3	3·5
284311	68	80	18	11	18·9	10·4	3·2	3·6
284312	60	63	18	11	17·1	9·8	3·2	3·3
284313	66	69	19	11	17·9	10·3	3·2	3·5

Notomys alexis alexis Thomas

Notomys alexis Thomas, *Ann. Mag. Nat. Hist.*, ser. 9, 9: 316 (April 1922). Type locality: 35 miles south-west of Alroy, Alexandria region, Northern Territory.

Horse Shoe Bend, Finke River, 4. One male and two female skins with skulls, set of five foetuses in alcohol; 27-9 November 1948 (284363-66).

Other records for Northern Territory: Vicinity of Alexandria Downs Station (Troughton 1941: 318); 35 miles south-west of Alroy (Thomas 1906e: 539, 1921b: 539, 1922a: 316; Brazenor 1934: 80; Finlayson 1940: 131; Tate 1951: 261); Tennant Creek (Brazenor 1934: 81; Tate 1951: 261); Tea Tree Well (Brazenor 1934: 81; Tate 1951: 261); Reedy Creek, Gill Range (Spencer 1896c: 75; Brazenor 1934: 81; Tate 1951: 261); Alice Springs (Brazenor 1934: 81); Idracowra, Finke River (Finlayson 1940: 128); Mount Burrell, 50 miles north-west of Idracowra (Finlayson 1940: 128); Ayers Rock (Finlayson 1940: 127); between Wollara and Basedow Range, lat. 24° 55′ S., long. 132° 25′ E. (Finlayson 1940: 127); Charlotte Waters (Spencer 1896a: 10; Finlayson 1940: 128). A specimen from Henbury, Finke River, has been referred by Tate (1951: 262) to *N. a. everardensis*, which according to Finlayson (1940: 128) does not occur north of the Musgrave Range, about on the southern border of the Northern Territory.

At Horse Shoe Bend, hopping mice were trapped near the crests of semi-permanent sand ridges within a mile north-east of the station buildings. A few characteristic tracks and burrow mounds were seen on the packed red sandy soil. A burrow system that was excavated had a pattern generally similar to that described in the account of *Notomys carpentarius* of Groote Eylandt.

A young adult female caught on 29 November was in advanced pregnancy, with five foetuses averaging near 21 mm. in crown-rump length.

Notomys carpentarius Johnson
Pl. 13

Notomys carpentarius Johnson, *Proc. Biol. Soc. Wash.* 72: 186 (4 Nov. 1959). Type locality: Umbakumba, Port Langdon, Groote Eylandt, Northern Territory.

Port Langdon, Groote Eylandt, 13. Four male and seven female skins with skulls, male and female in alcohol; 6-29 June 1948 (284350-62).

The north-eastern corner of Groote Eylandt, including most of the area surrounding Port Langdon, is primarily sandy. Large active wind-blown dunes of buff-coloured sand extend along the eastern coast, between Port Langdon and the open sea, and partly stabilized dunes with wattle-scrub vegetation and a greyish-white superficial layer of sand line the Little Lagoon at Umbakumba. Behind the dunes and extending to the rocky escarpment of Central Hill are sandy flats more or less overgrown with mixed forest. This predominance of a sandy substrate helps to explain the quite unexpected occurrence of hopping mice on Groote Eylandt, far out of the regular range of the genus in arid central Australia.

The site of the first camp of the Arnhem Land Expedition at Umbakumba proved to be in the midst of a flourishing colony of *Notomys carpentarius*. Burrows were numerous all along the sandy ridge, and tracks of the mice could be seen in the mornings about our tents. Conventional trapping methods were utterly unsuccessful with this species. The mice showed no interest in any of various baits tried, and their only reaction to traps was to kick sand on them. All of the specimens were obtained by digging out burrow systems, at which activity the aboriginal women and children were proficient.

Each of several burrow systems excavated in searching for the mice and for study

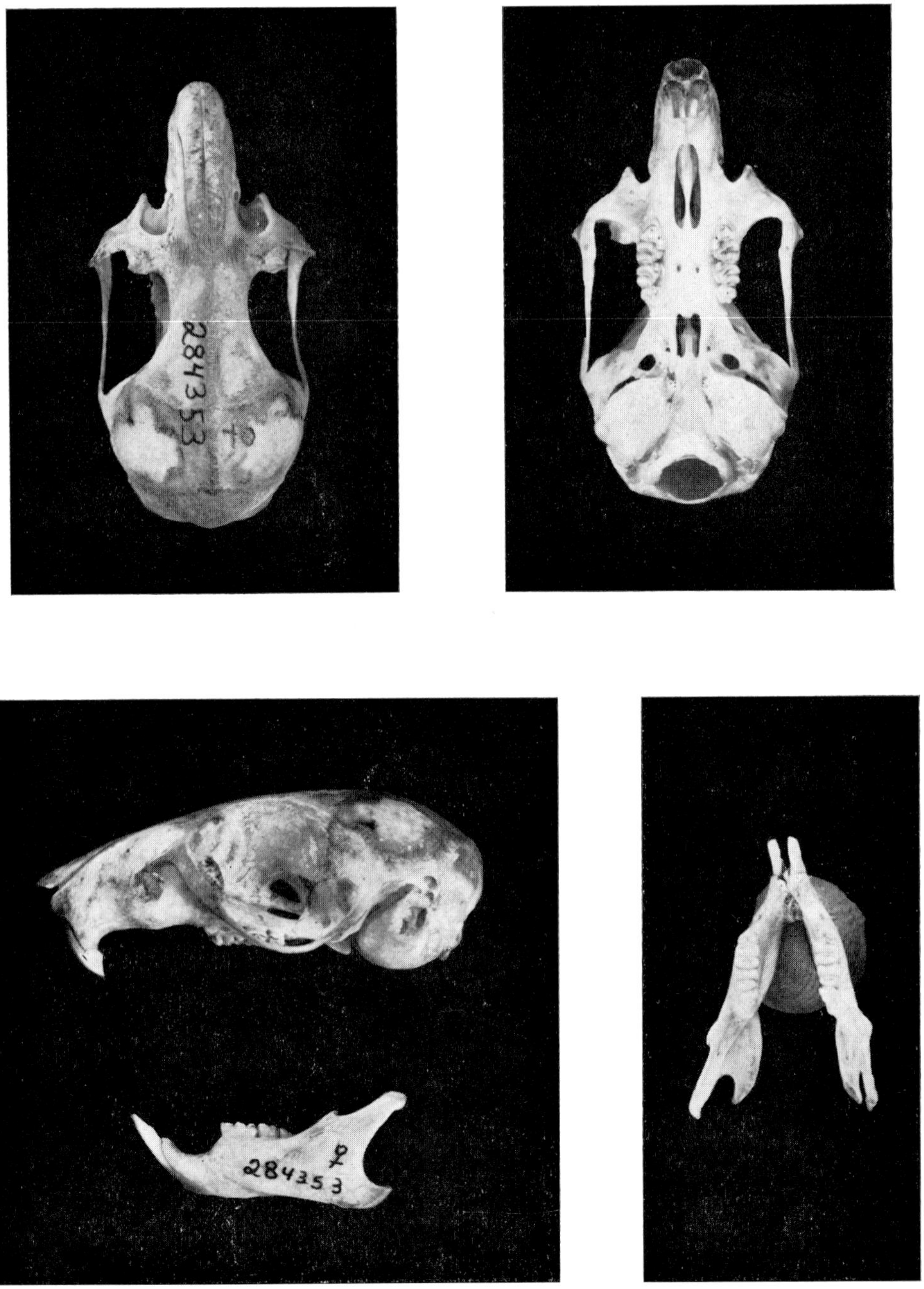

PLATE 13. *Notomys carpentarius*
Skull of type specimen, no. 284353, ♀, x 2

proved to be same variation of a common plan. The mounds of reddish sand brought up from below the surface by the mice showed conspicuously on the surface layer, which was grey in colour from weathering and intermixture of humus. These mounds lay on the surface at the ends of long and gently sloping tunnels that were themselves plugged for several feet with sand from deeper parts of the burrows. Apparently having once served the purpose of original entrance sites, these tunnels were filled with excess dirt and left sealed off. From the deeper parts of each burrow system one or more shafts rose vertically and appeared to end at a point about an inch below the surface of the ground. These vertical shafts were apparently the regular entrances and exists for the mice, being kept plugged at the surface when not in use. They were comparatively small in diameter and their walls had a smooth blackish-grey surface more like the grey upper layer of soil than the reddish deeper layers through which they principally passed. Occasionally one of the vertical shafts would be found with the entrance open, but any effort to follow its course by excavating from the top resulted in filling the shaft with sand. It is difficult to imagine how the mice manage to enter the narrow vertical shaft at the surface and then close it behind them without having the plug of surface sand fall down the shaft. That some of the grey surface sand does fall into the burrow system is shown by the fact that pockets of it can sometimes be found pushed back into unused sections of the tunnel.

A burrow system was excavated in its entirety at Umbakumba on 3 June. The site was on gently sloping terrain just south of the summit of the large sand ridge. The mound of freshly excavated subsurface sand by means of which the burrow system was first detected lay beneath the edge of a low wattle bush and partly covered some of the prostrate branches. The mound was about six inches deep at the centre, and spread out in an irregular shape over about five and one-half square feet of ground surface. The main tunnel, which was nine feet long and for the first five feet was packed with sand, was irregularly elliptical in cross-section, being from two to two and one-half inches wide and from two and one-half to three inches high. It reached a depth of three and one-half feet, and toward its inner end was entered by three vertical shafts spaced about eighteen inches

TABLE 18

EXTERNAL AND CRANIAL MEASUREMENTS OF *Notomys*
NORTHERN TERRITORY, AUSTRALIA

	Head and body	Tail	Hind foot with claw	Ear from notch	Condylobasal length	Anterior zygomatic breadth	Interorbital breadth	Upper tooth-row M1–3
Notomys alexis alexis								
Horse Shoe Bend								
♂ 284363	90	115	31	21	23·2	14·5	5·2	4·6
♀ 284364	97	135	33	21	24·4	14·3	4·9	4·8
284365	101	127	32	23	25·8	14·6	5·3	5·0
Notomys carpentarius								
Port Langdon, Groote Eylandt								
♂ 284350	91	159	38	18	25·4	13·9	5·7	5·4
284360	95	—	38	21	26·6	14·4	5·2	5·4
♀ 284351	98	—	38	21	26·6	13·9	5·3	5·6
284352	102	158	40	20	26·9	15·1	5·2	5·6
284353 (type)	112	173	39	20	27·6	16·1	5·5	5·9

apart. The vertical shafts were circular in cross-section and nowhere more than one and one-fourth inches in diameter; they were lined with a smooth, blackish, almost polished surface. The terminal vertical shaft and part of the adjacent horizontal tunnel were tightly packed with grey surface sand. No nesting or storage chambers were found in this way or any of the other burrows that were excavated.

Immature mice were taken along with adults in mid-June, and two females collected on 8 June had four and five uterine foetuses, respectively.

Notomys alexis of the Barkly Tableland area appears to be the nearest relative of *N. carpentarius,* both geographically and in structure. There are notable differences from that species however, as outlined in the description of the Groote Eylandt species (Johnson 1959: 186).

Hydromys chrysogaster beccarii Peters

Hydromys beccarii Peters, *Ann. Mus. Stor. Nat. Genova* 6: 303 (1875). Type locality: Weri, Kei Islands.

Hydromys caurinus Thomas, *Ann. Mag. Nat. Hist.*, ser. 8, 4: 197 (September 1909). Type locality: Parry's Creek, near Wyndham, Kimberley district, Western Australia.

Hydromys melicertes Thomas, *Ann. Mag. Nat. Hist.*, ser. 9, 8: 430 (October 1921). Type locality: Biro, Apsley Strait, Melville Island, Northern Territory.

Hydromys grootensis Troughton, *Rec. Aust. Mus.* 19: 252 (19 September 1935). Type locality: Groote Eylandt, Northern Territory.

Brocks Creek, 1. Female skin with skull; 13 Nov. 1920 (237717).

Oenpelli, 3. Male and female skins with skulls, male in alcohol; 5-31 Oct. 1948 (284248-50)

Yirrkalla, 1. Female skin with skull; 23 July 1948 (284247).

Port Langdon, Groote Eylandt, 1. Male skin with skull; 9 June 1948 (284246).

Other records for Northern Territory: Daly River (Collett 1897: 323; this reference was misquoted by Thomas 1909c: 197 as 'Mary River'); Biro, Apsley Strait, Melville Island (Thomas 1921a: 430); Groote Eylandt (Troughton 1935: 253).

In his comprehensive review of *Hydromys,* Tate (1951: 226-36) has included the names *caurinus, melicertes,* and *grootensis,* among others, as synonyms of *Hydromys chrysogaster beccarii.* This race he regards as having a range that includes the Kei and Aru Islands, the entire south coast of New Guinea, the northern part of the Cape York Peninsula, the north coast of the Northern Territory including Groote Eylandt and Melville Island, and the Kimberley district of Western Australia. This arrangement is adopted for the present, even though it involves a radical revision of the concepts of variation in the water rats as expressed in the earlier literature. Present material is not adequate to evaluate Tate's work on this group, but the subject should be reopened when larger series of specimens have been gathered from the north coast of Australia.

Hoy noted that his specimen from Brocks Creek was trapped at the edge of a waterhole. Of the three specimens from Oenpelli one was accidentally caught and drowned several feet beneath the surface of the lagoon in a gill net set by the ichthyologist, Dr Robert R. Miller, one was caught in a rat trap set at the edge of the lagoon, and the third was killed somewhere in the vicinity by aborigines. The specimens from Yirrkalla and Groote Eylandt were trapped on grassy banks of freshwater streams in marsh areas. Fragments of freshwater crustaceans called 'yabbies' left by a water rat along a stream at Port Langdon on Groote Eylandt have been identified by Dr Fenner A. Chace, jun., of the United States National Museum as *Cherax quadricarinatus* (von Martens).

The five skins agree generally in having dark-coloured underparts and in having the white colour of the tail restricted to the terminal inch or two. The specimen from Yirrkalla has a somewhat paler belly than the others and the specimen from Groote Eylandt has a more yellowish dorsum. The latter character seems to confirm the validity of Troughton's *grootensis* as an amber-toned race, but the other characters mentioned by Troughton (broad feet, strong claws, long nasals, long palatal foramina) are not shown in this specimen. Measurements are given in Table 19.

TABLE 19

EXTERNAL AND CRANIAL MEASUREMENTS OF *Hydromys* NORTHERN TERRITORY, AUSTRALIA

Hydromys chrysogaster beccarii	Head and body	Tail	Hind foot with claw	Ear from notch	Greatest length of skull	Zygomatic breadth	Length of nasals	Upper toothrow M1–2
Brocks Creek								
♀ 237717	289	315	70	—	56·8	—	19·0	8·8
Oenpelli								
♂ 284249	286	—	66	21	—	—	—	8·4
♀ 284250	277	271	61	20	57·4	28·7	19·0	8·5
Yirrkalla								
♀ 284247	274	251	60	—	53·0	26·0	17·9	9·6
Groote Eylandt								
♂ 284246	271	203	58	20	52·1	26·5	15·9	7·4

CETACEA

DELPHINIDAE

Sousa, species undetermined

Yirrkalla. Four mandibles picked up on the beach in August 1948 (284431-34).

Among bones gathered for me by native children along the beach and about camp sites at Yirrkalla are fragments of four mandibles of narrow-snouted dolphins of the genus *Sousa*. Identification of these with one or another of the named species of this genus (including *plumbeus* Cuvier 1829, *lentiginosus* Gray 1866, *sinensis* Flower 1870, *borneensis* Lydekker 1901, and perhaps others) would be little more than guesswork under the present circumstances.

Distinctive characters shown by these mandibles and seemingly applicable to the various species in the genus as well are: ramus as viewed from above uniformly curved, its lateral margin concave except at the extreme tip; lower margin as viewed from the side convex throughout the distal half, the curvature increasing at the anterior end without indication of an angular symphyseal process; symphysis elongate, its rugose area extending from one-fourth to one-third the length of the mandible; teeth about 32 on each side, fairly large, closely set, with smooth and conical crowns; alveoli separated by narrow bony septa about one millimeter thick.

There are 32 alveoli in the only specimen (no. 284434) in which the series is complete. A few teeth remain in the anterior parts of this specimen and no. 284431, which is probably the opposite mandible of the same individual. Each of these teeth has the greatest diameter of its crown transverse to the line of the tooth-row and is anchored firmly in its alveolus by an anteroposterior enlargement of its root; to accommodate this expansion the base of each bony interalveolar septum is pierced by a circular fenestration through which the roots of adjacent teeth come nearly or quite into contact.

The third mandible (no. 284432) apparently represents a younger individual. Its ramus is more slender and the alveoli lack the basal expansion and fenestration of the intervening septa.

The fourth mandible (no. 284433) is more massive and apparently represents a larger individual than the others. It is so different that at first it was thought to belong to another kind of dolphin. I am indebted to Dr Francis Fraser of the British Museum for finally determining it as another mandible of *Sousa*. Measurements of all four specimens are given in Table 20.

TABLE 20

MEASUREMENTS OF MANDIBLES OF *Sousa* FROM YIRRKALLA, CAPE ARNHEM PENINSULA

Specimen	No. 284434 left side		No. 284431 right side		No. 284432 right side	No. 284433 right side
Greatest length of ramus	404		—		—	—
Length of rugose area of symphysis	122		—		—	—
Depth of ramus (10 mm. behind posterior end of symphysis)	24·4		24·5		18·4	24·4
Thickness of ramus (at same place)	17·6		17·3		13·8	17·3
Alveolar length of tooth-row	245		—		—	—
Average distance between centres of teeth (near middle of tooth-row)	8·5		8·6		8·4	9·4
Diameters of crowns of individual teeth (*a*, transverse; *b*, longitudinal)	*a*	*b*	*a*	*b*		
Tooth no. 8 (approximate)	6·3	5·8	6·0	—		
9	6·4	6·0	—	—		
10	6·5	6·0	—	—		
11	6·6	6·2	—	—		
12	—	6·2	6·8	6·2		
13	6·9	6·2	6·9	6·1		
14	7·1	6·1	7·0	6·2		
15	7·1	6·0	7·2	6·2		
16	7·1	6·2	7·3	6·2		

Orcaella brevirostris (Gray)

Pl. 14, 15

Orca brevirostris Gray, *Catalogue of Seals and Whales in the British Museum*: 285 (March 1866). Type locality: Harbour of Vizagapatam, east coast of Indian Peninsula.

Phocaena (*Orca*) *brevirostris* Owen, *Trans. Zool. Soc. Lond.* 6: 24 (15 August 1866). Based on same specimen as *Orca brevirostris* Gray.

Melville Bay, Cape Arnhem Peninsula, 2. Skulls of unknown sex, one with partial body skeleton, picked up at native camp site on 16 July 1948 (284429-30).

The short-beaked porpoises of this species have long been known to occur along coasts, in estuaries, and well up the larger rivers of south-eastern Asia, westward

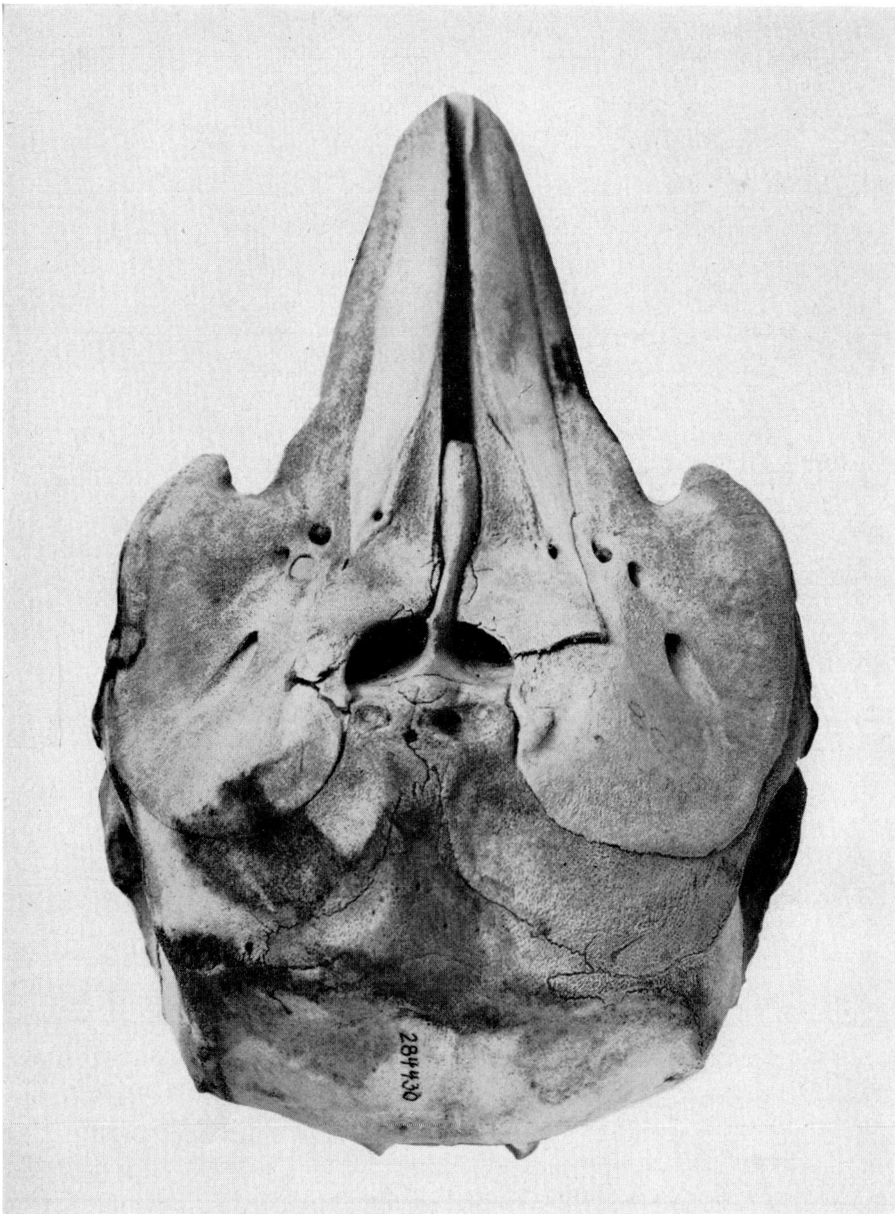

PLATE 14. *Orcaella brevirostris*
Skull from Melville Bay, no. 284430, dorsal view, x 0·5

S

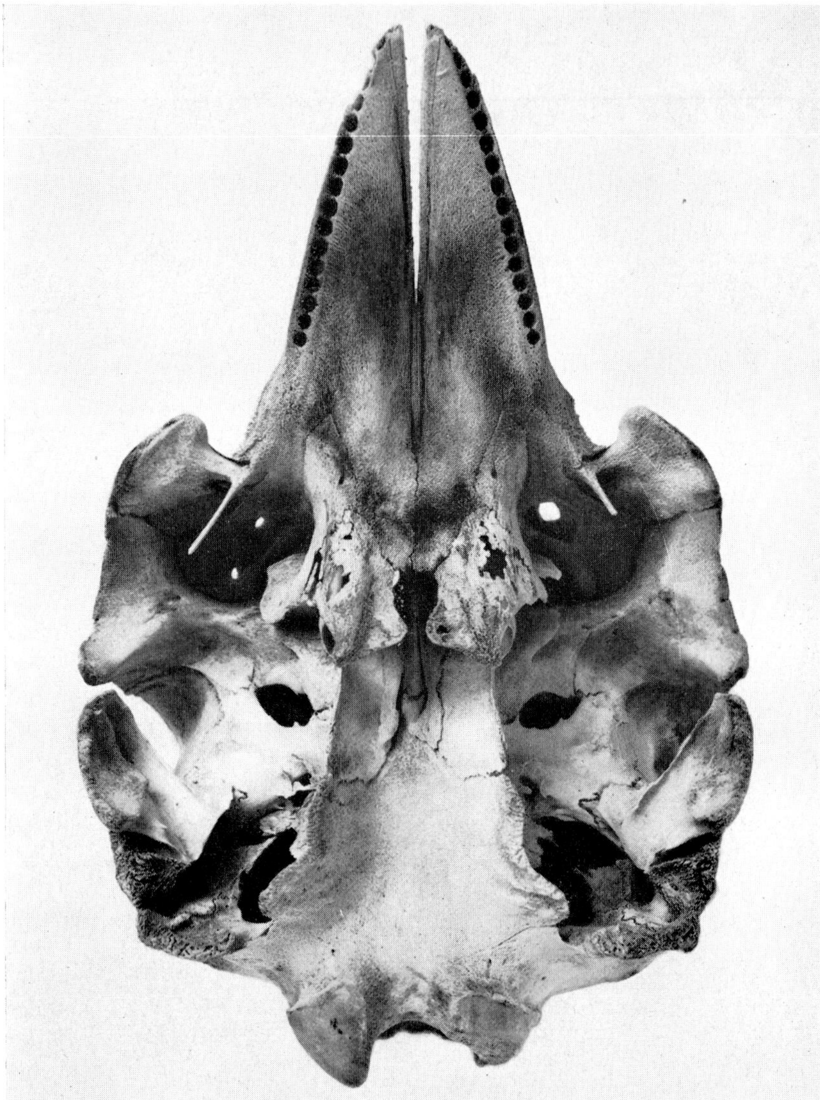

PLATE 15. *Orcaella brevirostris*
Skull from Melville Bay, no. 284430, ventral view, x 0·5

as far as the western shore of the Bay of Bengal. No records east of Borneo have been recorded, however, and the present specimens thus extend the range eastward some 1600 miles and add a new genus and species to the list of Australian mammals.

This is an inshore species, more likely to be found in sheltered bays and brackish estuaries than in the open ocean. It has no aversion to fresh water, being found as far as 900 miles up the Irrawaddy River in Burma.

A group of four or five slow-swimming blackish porpoises sighted about reefs in Boucaut Bay near the mouth of the Liverpool River in early May of 1948 were probably of this species. Aborigines at Yirrkalla were familiar with the species, distinguishing it from at least two other, longer-beaked, porpoises. The two skulls from Melville Bay were remains of porpoises that had been eaten by natives, and had very likely been killed in the protected waters of that bay.

The skulls are characterized by the broad and short braincase and the short, triangular rostrum with 16 or 17 teeth on each side. Measurements of the two skulls in millimeters are as follows: greatest length 296, 292; greatest breadth . . ., 190; anterior border of nares to tip of rostrum, 171, 169.

CARNIVORA

CANIDAE

Canis familiaris dingo Meyer

Canis antarticus Kerr, *Animal Kingdom* 1: 136 (1792). Based on the 'Dog of New South Wales', *The Voyage of Governor Phillip*. . .: 274, pl. xlv. Type locality, therefore: vicinity of Sydney, New South Wales. (Name suppressed by Opinion 451 of International Commission on Zoological Nomenclature, 8 March 1957.)

Canis dingo F. A. A. Meyer, *Systematisch-summarische Uebersicht der neuesten zoologischen Entdeckungen in Neuholland und Afrika*: 33 (1793). Based on the 'Dingo or Dog of New South Wales' of White, and the 'Dog of New South Wales' of Phillip. Type locality, therefore: vicinity of Sydney, New South Wales.

Oenpelli, 1. Male skin with skull; 25 Oct. 1948 (284438).

Yirrkalla, 1. Skull only picked up on beach in July 1948 (284439).

Rocky Bay, Cape Arnhem Peninsula, 1. Skull only picked up 17 July 1948 (284436).

Kwaiturumuru Creek, Cape Arnhem Peninsula, 1. Female skin with skull; 20 Aug. 1948 (284437).

Port Langdon, Groote Eylandt, 1. Female skin with skull and body skeleton; 12 June 1948 (284435).

No sharp distinction can be made between the camp dogs of the aborigines and the wild dogs of the Arnhem Land area. The camp dogs are more varied in colour, with black, brown, white, or piebald types predominating, whereas most of the wild dogs are of the reddish brown 'dingo' colour. The camp population regularly receives fresh blood from the wild stock through adoption of pups from litters discovered by women on food gathering trips and by men on hunting expeditions. A tendency to favour the odd-coloured individuals in selecting pups from wild litters may explain the greater variety of colours among camp dogs. In this area there was very little evidence of intermixture with domestic breeds of European origin. The aborigines have a generally tolerant, even friendly, attitude toward the dogs in general, whether wild or tame, and about the camps there are obvious demonstrations of reciprocal affection between individual dogs and their owners. My trapping of wild dingoes for scientific specimens seemed to be viewed with mild disfavour by the aborigines, who apparently do not consider these animals to be in any sense a menace to themselves or their economy. There are many stories of the cleverness shown by the natives in protecting camp dogs from extermination in the clean-up raids periodically carried out by white officials.

The two skins of adult females from Groote Eylandt and the Cape Arnhem Peninsula are strikingly similar in details of colour pattern. Each has the muzzle dusky adjacent to the rhinarium, the chin almost black laterally with a white median spot which tapers posteriorly to a narrow streak, an irregular midventral white streak from throat to belly, white feet, and a few white hairs at the tip of the tail. Of seven other skins in the United States National Museum collection, only one from the Atherton Tableland of Queensland (no. 238468) has a similar pattern; the remaining skins, from New South Wales or lacking locality data, have the muzzle paler rather than darker than the general body colour.

Measurements of the Arnhem Land specimens are given in Table 21. The female from Groote Eylandt, which was very thin and had recently nursed young, weighed 21 pounds.

TABLE 21

EXTERNAL AND CRANIAL MEASUREMENTS OF *Canis*
NORTHERN TERRITORY, AUSTRALIA

	Head and body	Tail	Hind foot with claw	Ear from notch	Condylobasal length	Zygomatic breadth	Mastoid breadth	Length of upper carnassial tooth P4
Canis familiaris dingo								
Oenpelli								
♂ 284438 (young)	—	—	—	—	139	75·6	57·4	—
Cape Arnhem Peninsula								
♀ 284437	812	303	176	93	167	90·8	61·8	17·6
sex ? 284436	—	—	—	—	184	104·0	67·2	19·6
284439	—	—	—	—	161	—	58·5	18·0
Groote Eylandt								
♀ 284435	800	337	181	98	168	91·3	63·9	19·1

SIRENIA

DUGONGIDAE

Dugong dugon (Müller)

Trichecus dugon P. L. S. Müller in Linné's *Vollständingen Natursyst.*, Suppl.: 21 (1776). Type locality: Indian Ocean (Cape of Good Hope to the Philippine Islands).

Cape Don, Cobourg Peninsula, 2. Skulls, found 7 Oct. 1948 (284458-59).

Popham Bay, Cobourg Peninsula, 2. Skull fragments and mandibles, found 6 Oct. 1948 (284456-57).

Black Rock Point, Cobourg Peninsula, 13. Five skulls and eight sets of mandibles, found 29 Sept.-1 Oct. 1948 (284443-55).

Barnambarnga, Cape Arnhem Peninsula, 1. Part of skeleton, found 15 July 1948 (284440).

Melville Bay, Cape Arnhem Peninsula, 1. Skull and part of skeleton, found 15 July 1948 (284441).

PLATE 16. Mammals of the rocky hills near Oenpelli

Petropseudes dahlii *Petrogale venustula*

Satanellus hallucatus hallucatus *Zyzomys argurus indutus* *Antechinus bilarni*

Port Bradshaw, Cape Arnhem Peninsula, 1. Skull, found 25 July 1948 (284442).
Other records for Northern Territory: Raffles Bay (Blyth 1859: 14); north and west coasts of Groote Eylandt (Tindale 1925: 78).

Dugongs are common all along the coast of the Northern Territory. Nearly all the information concerning them has been obtained from the coastal aborigines, to whom the dugong is an important source of food, either by direct questioning or by observation of hunting methods and the refuse at camp sites. Tindale (1925: 78-9) has described the method of harpooning dugongs from dugout canoes on Groote Eylandt. From what I was told by natives on the Cape Arnhem and Cobourg Peninsulas essentially the same methods are in use there.

Locally the dugongs are most commonly found in comparatively sheltered bays and channels rather than on the exposed shores where wave action is greater. Tindale found this to be true on Groote Eylandt, where the dugongs were most common on the sheltered north and west coasts. Sites that were reported to me to be especially good for dugong-hunting were Barnambarnga and Melville Bay on the Cape Arnhem Peninsula, Cadell Strait between Elcho Island and the mainland, and Popham Bay and Black Rock Point on opposite sides of the Cobourg Peninsula. At all these places comparatively shallow areas of clear salt water were present. At Black Rock Point I was told that at periods of low tide the dugongs stayed in deeper water beyond a reef on which the waves then broke; at high tide they crossed the reef and foraged in the protected area nearer shore in water ten or twelve feet deep. This foraging was said to be done where there were considerable beds of green underwater vegetation, but I have no evidence that they actually fed on this material.

Unlike the manatees of the Atlantic Basin, dugongs do not inhabit murky or brackish water, and they do not ascend rivers. Surprisingly little is actually known

TABLE 22
CRANIAL MEASUREMENTS OF *Dugong dugon*
NORTHERN TERRITORY, AUSTRALIA

	Greatest zygomatic breadth	Mastoid breadth	Length of zygomatic arch	Length of premaxillary symphysis	Breadth of rostrum	Length of narial aperture	Breadth of narial aperture	Greatest length of mandible	Least depth of mandible	Length of mandibular symphysis
Cobourg Peninsula										
284443	212	159	187	190	80	98	67	—	—	—
284444	—	—	—	—	—	—	—	252	75	121
284445	—	—	—	—	—	—	—	—	71	117
284446	—	—	—	—	—	—	—	247	79	119
284447	—	—	—	—	—	—	—	—	77	118
284449	—	—	184	—	70	—	62	—	—	—
284450	209	156	189	—	—	—	69	—	—	—
284451	183	—	—	162	65	102	63	—	—	—
284452	—	—	—	—	—	—	—	240	70	109
284454	—	—	—	—	—	—	—	—	66	116
284455	199	—	—	182	83	115	65	—	—	—
284456	—	—	—	—	—	—	—	240	74	115
284458	—	158	181	174	74	101	68	258	71	122
284459	192	—	180	170	76	115	63	—	—	—
Cape Arnhem Peninsula										
284441	—	—	—	191	74	110	69	261	74	121

of their life history, because all of their normal activities are carried out under water. Dugongs may have been the first mammals to be collected as scientific specimens in what is now the Northern Territory of Australia. Edward Blyth (1859: 14), commenting on specimens in the Indian Museum at Calcutta, quoted the following extract concerning Dr Robert Tytler from Corbyn's *Indian Review* of 1838:

> During his various expeditions, Dr Tytler made some valuable collections of natural curiosities, of which he largely and liberally contributed to the Asiastic Society of Calcutta. In 1827, he read a paper on the Duyong, or Dayoung. The bones of four different individuals were picked up by Dr Tytler at Raffles Bay, on the north coast of New Holland. In one instance they were sufficiently numerous to form nearly an entire skeleton of the animal. This creature is not uncommon in the eastern archipelago, but its existence on the coast of New Holland was made known for the first time by Dr Tytler.

For lack of comparative material, I have not attempted to settle the question of whether or not the Australian dugong is taxonomically distinct from that of the Indian Ocean. Most authorities who have had an opportunity to compare specimens from the two areas think not (i.e. Gray 1866a: 362-3; Pocock 1940). Some measurements of the various skulls found in 1948 are given in Table 22.

REFERENCES

Andersen, K. (1912). *Catalogue of the Chiroptera in the collection of the British Museum.* Second ed., Vol. I: *Megachiroptera.*

Blyth, E. (1859). On the great rorqual of the Indian Ocean, with notices of other cetals, and of the Syrenia or marine pachyderms. *J. Asiat. Soc. Beng.*, 28: 1-18.

Brazenor, C. W. (1934). A revision of the Australian jerboa mice. *Mem. Nat. Mus. Vict.* 8: 74-89, pls. 5-7.

——— (1936a). Two new rats from Central Australia. Ibid., 9: 5-8, pl. 1.

——— (1936b). An extended description of *Pseudomys novaehollandiae*, with remarks on its affinity to *P. hermannsburgensis.* Ibid., 9: 9-13, pl. 2.

Cabrera, A. (1919). *General mammalium. Monotremata, Marsupialia.*

Collett, R. (1885). On *Echidna acanthion* from Northern Queensland. *Proc. Zool. Soc. Lond.*, 1885: 148-61, pl. 10.

——— (1895). On a new Pseudochirus from N.W. Australia. *Zool. Anz.*, 18: 464-8.

——— (1897). On a collection of mammals from North and North-west Australia. *Proc. Zool. Soc. Lond.*, 1897: 317-36, pl. 23.

Dahl, K. (1897). Biological notes on North-Australian Mammalia. *Zoologist*, ser. 4, 1: 189-216.

——— (1926). *In savage Australia. An account of a hunting and collecting expedition to Arnhem Land and Dampier Land.*

Dobson, G. E. (1875). On the genus *Scotophilus*, with description of a new genus and species allied thereto. *Proc. Zool. Soc. Lond.*, 1875: 368-73.

——— (1878). *Catalogue of the Chiroptera in the collection of British Museum.*

Dumont d'Urville, J. (1844). *Voyage au Pôle Sud et dans l'Océanie sur les corvettes l'Astrolabe et la Zélée . . . pendant . . . 1837-1840*, Vol. 6.

Ellerman, J. R. (1941). *The families and genera of living rodents.* Vol. 2: *Family Muridae.*

——— (1947). Notes on some Asiatic rodents in the British Museum. *Proc. Zool. Soc. Lond.*, 117: 259-71.

——— (1949). *The families and genera of living rodents.* Vol. 3 (1).

Finlayson, H. H. (1930). Notes on some South and Central Australian mammals. Pt. 1: *Petaurus breviceps, Thalacomys lagotis, Petrogale lateralis. Trans. Roy. Soc. S. Aust.,* 54: 177-80.

——— (1931a). On mammals from the Dawson Valley, Queensland. Pt. 1. Ibid., 55: 67-89, pls. 1-3.

——— (1931b). Notes on some South and Central Australian mammals. Pt. 2. Ibid., 55: 161-2.

——— (1932). Preliminary descriptions of two new mammals from South Australia. Ibid., 56: 168-71.

——— (1933). On mammals from the Lake Eyre Basin. Pt. 1: The Dasyuridae. Ibid., 57: 195-202.

——— (1934). On mammals from the Dawson and Fitzroy Valleys; central coastal Queensland. Pt. 2. Ibid., 58: 218-31.

——— (1935). On mammals from the Lake Eyre Basin. Pt. 2: The Peramelidae. Ibid., 59: 227-36.

——— (1936). On mammals from the Lake Eyre Basin. Pt. 3: The diprotodont marsupials and Ornithodelphia. Ibid., 60: 157-61.

——— (1939). On the mammals from the Lake Eyre Basin. Pt. 4: The Monodelphia. Ibid., 63: 88-118, pls. 4-5.

——— (1940). On Central Australian mammals. Pt. 1: The Muridae. Ibid., 64: 125-36, pls. 14-15.

——— (1941). On Central Australian mammals. Pt. 2: The Muridae. Ibid., 65: 215-32, pls. 10-13.

——— (1942). A new Melomys from Queensland with notice of two other Queensland rats. Ibid., 66: 243-7, pls. 8-9.

——— (1943a). *The Red Centre. Man and beast in the heart of Australia.*

——— (1943b). A new species of Lagorchestes (Marsupialia). *Trans. Roy. Soc. S. Aust.,* 67: 319-21, pls. 33-4.

——— (1947). On the weights of some Australian mammals. Ibid., 71: 182-94.

——— (1958). On Central Australian mammals (with notice of related species from adjacent tracts). Pt. 3: The Potoroinae. *Rec. S. Aust. Mus.,* 13: 235-302, pls. 27-31.

Gadow, H. (1892). On the systematic position of *Notoryctes typhlops. Proc. Zool. Soc. Lond.,* 1892: 361-70.

Gilbert, L. (1948). Zoological notes on the Northern Territory (June 1944 to September 1945). I. Savannah woodland. *Vict. Nat.,* 65: 96-102.

Glauert, L. (1934). The distribution of the marsupials in Western Australia. *J. Roy. Soc. W. Aust.,* 19: 17-32.

Gould, J. (1841a). Observations on *Dasyurus maugei* and *D. viverrinus* of Geoffroy, and description of a new species. *Proc. Zool. Soc. Lond.,* 8: 151.

——— (1841b). Exhibition of skulls of a new species of kangaroo. Ibid., 9: 77.

——— (1842a). Descriptions of four new species of kangaroos. Ibid., 9: 80-3.

——— (1842b). On a new species of *Petrogale—P. inornata.* Ibid., 10: 5-6.

——— (1842c). On some new species of Australian mammals. Ibid., 10: 10-14.

——— (1842d). Characters of a new species of *Perameles,* and a new species of Dasyurus. Ibid., 10: 41-2.

——— (1842e). On two new species of kangaroo (*Petrogale concinna* and *Halmaturus binoë*). Ibid., 10: 57-8.

——— (1845). *The mammals of Australia.* Pt. 1, including: vol. 1, pls. 4, 5, 6, 7, 10, 23, 24, 29, 30, 31, 32, 39, 47; vol. 3, pls. 8, 9. (Collation of the parts of Gould's *Mammals of Australia* follows F. H. Waterhouse, 'The dates of publication of some of the zoological works of the late John Gould, F.R.S.', pp. 1-4; 1885.)

——— (1849). Ibid. Pt. 2, including: vol. 1, pls. 3, 8, 15, 25, 27, 28, 46; vol. 2, pls. 54, 55, 56, 58; vol. 3, pls. 6, 21, 28, 30.

——— (1851). Ibid. Pt. 3, including: vol. 1, pls. 44, 48, 49, 50, 51, 52, 53, 54; vol. 2, pls. 66, 68, 70; vol. 3, pls. 5, 11, 29, 35.

——— (1852). Ibid. Pt. 4, including: vol. 1, pls. 2, 41, 42, 43, 45; vol. 2, pls. 12, 13, 18, 19, 20, 21, 61, 62; vol. 3, pls. 14, 43.

——— (1853). Ibid. Pt. 5, including: vol. 1, pl. 22; vol. 2, pls. 6, 7, 39, 40, 60; vol. 3, pls. 1, 2, 7, 10, 24, 25, 26, 27, 36 (this, from the text, was issued after Mar. 9, 1858, when Tomes revision in PZS pt. 26, p. 29 referred to here, was issued), 37.

——— (1854). Ibid. Pt. 6, including: vol. 1, pls. 13, 14, 34, 36, 37, 38, 40; vol. 2, pls. 14, 15, 63, 67, 69; vol. 3, pls. 13, 32, 34.

——— (1855). Ibid. Pt. 7, including: vol. 1, pls. 1, 11, 55, 56; vol. 2, pls. 37, 38, 43, 44, 64, 65; vol. 3, pls. 19, 23, 40, 41, 47.

——— 1856). Ibid. Pt. 8, including: vol. 1, pls. 12, 16, 17, 18, 19, 33; vol. 2, pls. 16, 17, 28, 48, 49, 50; vol. 3, pls. 22, 33, 44.

——— (1857). Ibid. Pt. 9, including: vol. 2, pls. 1, 2, 22, 23, 24, 25, 26, 27, 31, 32, 41, 42; vol. 3, pls. 4, 18, 46.

——— (1858a). Ibid. Pt. 10, including: vol. 1, pl. 20; vol. 2, pls. 5, 8, 9, 10, 11, 51; vol. 3, pls. 3, 15, 16, 17, 20, 31, 45, 48.

——— (1858b). On four new species of Mus and one of Hapalotis from Australia. *Proc. Zool. Soc. Lond.*, 1857: 241-3.

——— (1859). *The mammals of Australia.* Pt. 11, including: vol. 1, pls. 9, 57, 58; vol. 2, pls. 29, 30, 47, 52, 53, 57; vol. 3, pls. 12, 38, 39, 51, 52.

——— (1860). Ibid. Pt. 12, including: vol. 1, pls. 21, 26, 35; vol. 2, pls. 3, 4, 33, 34, 35, 36, 45, 46, 59; vol. 3, pls. 42, 49, 50.

——— (1863). Ibid. Introduction, vol. 1: i-xliv.

Gray, J. E. (1842). Description of two new species of Mammalia discovered in Australia by Captain George Grey, Governor of South Australia. *Ann. Mag. Nat. Hist.*, 9: 39-42.

——— (1843). *List of the specimens of Mammalia in the collection of the British Museum.*

——— (1845). Descriptions of some new Australian animals. *In* Eyre, J. E., *Journals of expeditions of discovery into central Australia . . . in the years 1840-41.* Vol. 1 (appendix): 405-11, pls. 1-2.

——— (1847a). *List of the osteological specimens in the collection of the British Museum.*

——— (1847b). Characters of six new genera of bats not hitherto distinguished. *Proc. Zool. Soc. Lond.*, 15: 14-16.

——— (1862). Notice of a new species of dolphin (*Delphinus catalania*), discovered in north Australia by Mr John Macgillivray. Ibid., 1862: 143-5.

——— (1866a). *Catalogue of seals and whales in the British Museum.* Second ed.

——— (1866b). A revision of the genera of pteropine bats (*Pteropidae*), with the descriptions of some apparently undescribed species. *Proc. Zool. Soc. Lond.*, 1866: 62-7.

——— (1866c). Notes on the skulls of dolphins, or bottle-nose whales, in the British Museum. Ibid., 1866: 211-16.

——— (1870). *Catalogue of monkeys, lemurs, and fruit-eating bats in the collection of the British Museum.*

Iredale, T., and Troughton, E. le G. (1934). A check-list of the mammals recorded from Australia. *Mem. Aust. Mus.*, 6; xi + 122.

Johnson, D. H. (1952). The occurrence and significance of extra molar teeth in rodents. *J. Mammal.*, 33: 70-2.

——— (1954). A new marsupial of the genus Antechinus from northern Australia. *Proc. Biol. Soc. Wash.*, 67: 77-80.

——— (1959). Four new mammals from the Northern Territory of Australia. *Proc. Biol. Soc. Wash.*, 72: 183-8.

Jones, F. Wood (1923). *The mammals of South Australia.* Pt. 1: The monotremes and the carnivorous marsupials.

——— (1924). *The mammals of South Australia.* Pt. 2: The bandicoots and the herbivorous marsupials.

——— (1925a). A revision of the South Australia jerboa mice, with the description of a new species. *Rec. S. Aust. Mus.*, 3: 1-7.

——— (1925b). *The mammals of South Australia.* Pt. 3: The Monodelphia.

——— (1949). A study of a generalized marsupial *Dasycercus cristicauda* (Krefft). *Trans. Zool. Soc. Lond.*, 26: 409-501, pls. 1-2.

Kellogg, R. (1945). A new Australian naked-tailed rat (Melomys). *Proc. Biol. Soc. Wash.*, 58: 69-71.

Krefft, G. (1867). On the classification of the small *Dasyuridae* of Australia, with descriptions of two new genera and one new species. *Proc. Zool. Soc. Lond.*, 1866: 431-5, pl. 36.

Le Souef, A. S. (1928). The *Macropus robustus* group of kangaroos. *Aust. Zool.*, 5: 247-56, pl. 30.

Lydekker, R. (1896). *A hand-book to the Marsupialia and Monotremata.*

Marshall, A. (1948). *Ourselves writ strange.*

Matschie, P. (1915a). Einige Beitrage zur Kenntniss der Gattung *Pseudochirus* Ogilb. *S.B. Ges. Naturf. Fr. Berl.*, 1915 (4): 83-95.

——— (1915b). Der Dingo-Hund des Macdonnell-Gebirges. Ibid., 1915 (4): 101-7.

——— (1916). Die Verbreitung der Beuteltiere auf Neuguinea mit einigen Bermerkungen über ihre Einteilung in Untergattungen. *Mitt. Zool. Mus. Berl.*, 8: 259-308, map.

McCarthy, F. D. (1954). Buffalo hunting in Arnhem Land. *Aust. Mus. Mag.*, 11: 252-6, illus.

Ogilby, J. D. (1891). *Advance copy of portion of hand-list of Australian mammals.*

——— (1892). Catalogue of Australian mammals with introductory notes on general mammalogy. *Aust. Mus. Cat.*, no. 16.

Ogilby, [W.] (1838). On a new species of marsupial animal found by Major Mitchell on the banks of the River Murray in New South Wales. *Proc. Zool. Soc. Lond.*, 6: 25-7.

Ogilby, W. (1841). Observations upon a small Australian rodent exhibited by Mr Gould and supposed to be identical with the *Dipus Mitchellii. Proc. Zool. Soc. Lond.*, 8: 151-2.

Owen, R. (1839). Outlines of a classification of the Marsupialia. *Proc. Zool. Soc. Lond.*, 7: 5-19.

——— (1866). On some Indian *Cetacea* collected by Walter Elliot, Esq. *Trans. Zool. Soc. Lond.*, 6: 17-48, pls. 3-14.

Peters, W. (1867). On *Taphozous flaviventris* Gould, a new species of bat from Australia. *Proc. Zool. Soc. Lond.*, 1866: 430.

——— (1871). Uber die Gattungen und Arten der Hufeisennasen, *Rhinolophi. Mber. K. Preuss. Akad. Wiss.*, 1871: 301-32.

Pocock, R. I. (1926). The external characters of *Thylacinus, Sarcophilus,* and some related marsupials. *Proc. Zool. Soc. Lond.,* 1926: 1037-84.

——— (1940). Some notes on the dugong. *Ann. Mag. Nat. Hist.,* ser. 11, 5: 329-45.

Price, A. Grenfell (1930). *The history and problems of the Northern Territory, Australia.*

Ridgway, R. (1912). *Color standards and color nomenclature.*

Rothschild, W. (1903). Preliminary diagnosis of a new genus and species of kangaroo. *Novit. Zool.,* 10: 414.

——— (1904). Note on *Dendrodorcopsis woodwardi.* Ibid., 10: 543.

——— (1905a). Note on *Macropus rufus* Desm., with description of a new subspecies. Ibid., 12: 508.

——— (1905b). Notes on two kangaroos from the 'Northern Territory of South Australia', with description of a new species. Ibid., 12: 509-10.

Sanborn, C. C. (1931). Bats from Polynesia, Melanesia, and Malaysia. *Zool. Ser. Field Mus. Nat. Hist.,* 18 (2): 7-29.

Schwarz, E. (1910). Die grossen Kanguruhs und ihre geographischen Formen. *Novit. Zool.,* 17: 86-109, pls. 1-4.

Shortridge, G. C. (1910). Account of the geographical distribution of the marsupials and monotremes of South-West Australia, having special reference to the specimens collected during the Balston Expedition of 1904-1907. *Proc. Zool. Soc. Lond.,* 1909: 803-48.

Spencer, Baldwin (1895). Preliminary notice of two new species of marsupials from Central Australia. *Proc. Roy. Soc. Vict.,* new ser., 7: 222-4.

——— (1896a). Mammalia. *Rept. Horn Scientific Expedition to Central Australia,* pt. 2 (Zoology): 1-52, pls. 1-5.

——— (1896b). Preliminary description of certain new marsupials from Central Australia, together with remarks upon the occurrence and identity of *Phascologale cristicauda. Proc. Roy. Soc. Vict.,* new ser., 8: 5-13.

——— (1896c). Through Larapinta Land: a narrative of the Horn Expedition to Central Australia. *Rept. Horn Scientific Expedition to Central Australia,* pt. 1: 1-136, pls. 1-11.

——— (1896d). Summary of the zoological, botanical and geological results of the Expedition. Ibid., pt. 1: 137-99. September, 1896. (Mammals, pp. 143-6.)

——— (1896e). Additions to the fauna. Ibid., pt. 1 (suppl): 210-11.

——— (1897). Description of two new species of Marsupials from Central Australia. *Proc. Roy. Soc. Vict.,* new ser., 9: 5-11, pl. 2.

Stirling, E. C. (1888). A new Australian mammal. *Nature,* 38: 588-9.

——— (1889a). Preliminary notes on a new Australian mammal. *Trans. Roy. Soc. S. Aust.,* 11: 21-4.

——— (1889b). No title. ('Dr Stirling announced that he would shortly submit a full diagnosis of the newly-discovered mammal, referred to in last year's volume of the Transactions, to which he gives the name of *Psammorcyctes typhlops,* gen. et sp. nov.') Ibid., 12: 158.

——— (1891a). Description of a new genus and species of Marsupialia, *'Notoryctes typhlops'.* Ibid., 14: 154-87, pls. 2-9.

——— (1891b). On the newly-discovered Australian mammal *Notorcytes typhlops. Proc. Zool. Soc. Lond.,* 1891: 327-9.

——— (1892). Exhibition of, and remarks upon an original drawing of the new Australian mammal (*Notoryctes typhlops*). Ibid., 1891: 466.

Tate, G. H. H. (1941a). A review of the genus *Hipposideros* with special reference to Indo-Australian species. *Bull. Amer. Mus. Nat. Hist.,* 78: 353-93.

——— (1941b). Notes on vespertilionid bats of the sub-families Miniopterinae, Murininae, Kerivoulinae, and Nyctophilinae. Ibid., 78: 567–97.

——— (1942). Review of the vespertilionine bats, with special attention to genera and species of the Archbold collections. Ibid., 80: 221–97.

——— (1945a). The marsupial genus *Pseudocheirus* and its subgenera. *Amer. Mus. Novit.*, no. 1287.

——— (1945b). Notes on the squirrel-like and mouse-like possums (Marsupialia). Ibid., no. 1305.

——— (1947a). On the anatomy and classification of the Dasyuridae (Marsupialia). *Bull. Amer. Mus. Nat. Hist.*, 88: 97–156.

——— (1947b). An example of 'prelacteal incisors' in advanced pouch young of *Macropus. J. Mammal.*, 28: 399–400.

——— (1948a). Studies on the anatomy and phylogeny of the Macropodidae (Marsupialia). *Bull. Amer. Mus. Nat. Hist.*, 91: 233–352.

——— (1948b). Studies in the Peramelidae (Marsupialia). Ibid., 92: 313–46.

——— (1951). The rodents of Australia and New Guinea. Ibid., 97: 183–430.

——— (1952). Mammals of the Cape York Peninsula, with notes on the occurrence of rain forest in Queensland. Ibid., 98: 563–616.

Thomas, O. (1885). On the characters of the different races of *Echidna. Proc. Zool. Soc. Lond.*, 1885: 329–39.

——— (1887). On the specimens of *Phascologale* in the Museo Civico, Genoa, with notes on the allied species of the genus. *Ann. Mus. Stor. Nat. Genova,* ser. 2, 4: 502-11.

——— (1888). *Catalogue of the Marsupialia and Monotremata in the collection of the British Museum (Natural History).*

——— (1892). Description of a third species of the genus *Nyctophilus. Ann. Mag. Nat. Hist.*, ser. 6, 9: 405–6.

——— (1897). The mammals collected in British New Guinea by Dr Lamberto Loria. *Ann. Mus. Stor. Nat. Genova,* ser. 2, 18: 606–22.

——— (1898). Descriptions of three new mammals from the East Indian Archipelago and Australia. *Novit. Zool.*, 5: 1–4.

——— (1904a). On a collection of mammals made by Mr J. T. Tunney in Arnhem Land, Northern Territory of South Australia. Ibid., 11: 222–9.

——— (1904b). New species of *Pteropus, Mus,* and *Pogonomys* from the Australian region. Ibid., 11: 597–600.

——— (1906a). On the generic arrangement of the Australian rats hitherto referred to *Conilurus,* with remarks on the structure and evolution of their molar cusps. *Ann. Mag. Nat. Hist.*, ser. 7, 17: 81–5.

——— (1906b). New mammals from the Australian region. Ibid., ser. 7, 17: 324–32.

——— (1906c). [Description of *Mus forresti* and *Phascogale ingrami.*] *Abstr. Proc. Zool. Soc. Lond.*, no. 32: 6.

——— (1906d). On mammals collected in south-west Australia for Mr W. E. Balston. *Proc. Zool. Soc. Lond.*, 1906: 468–78.

——— (1906e). On mammals from northern Australia, presented to the National Museum by Sir Wm. Ingram, Bt., and the Hon. John Forrest. Ibid., 1906: 536–43.

——— (1908). New bats and rodents in the British Museum collection. *Ann. Mag. Nat. Hist.*, ser. 8, 2: 370–5.

——— (1909a). Some mammals from N. E. Kimberley, northern Australia. Ibid., ser. 8, 3: 149–52.

——— (1909b). On the N. Australian rats referred to the genus *Mesembriomys*. Ibid., ser. 8, 3: 372–4.

——— (1909c). Two new mammals from N. Australia. Ibid., ser. 8, 4: 197–8.

——— (1910). The generic arrangement of the Australian murines hitherto referred to '*Mus*'. Ibid., ser. 8, 6: 603-7.

——— (1913a). Two new Australian mammals. Ibid., ser. 8, 11: 79-80.

——— (1913b). On new mammals obtained by the Utakwa expedition to Dutch New Guinea. Ibid., ser. 8, 12: 205–12.

——— (1913c). Some new species of *Uromys*. Ibid., ser. 8, 12: 212-17.

——— (1915a). Two new species of *Leuconoe*. Ibid., ser. 8, 15: 170–2.

——— (1915b). Notes on the genus *Nyctophilus*. Ibid., ser. 8, 15: 493–9.

——— (1915c). Notes on *Taphozous* and *Saccolaimus*. *J. Bombay Nat. Hist. Soc.*, 24: 57–63.

——— (1921a). Notes on Australasian rats, with a selection of lectotypes of Australasian *Muridae*. *Ann. Mag. Nat. Hist.*, ser. 9, 8: 425–33.

——— (1921b). Notes on the species of *Notomys*, the Australian jerboa-rats. Ibid., ser. 9, 8: 536–41.

——— (1922a). Two new jerboa-rats. Ibid., ser. 9, 9: 315–17.

——— (1922b). The bandicoots of Nuyts Archipelago, S. Australia, and Cape York, N. Queensland. Ibid., ser. 9, 9: 677–9.

——— (1922c). A selection of lectotypes of the typical Australian marsupials, in the British Museum collection. Ibid., ser. 9, 9: 127–8.

——— (1922d). A new species of *Mastacomys* from a cave in South Australia. Ibid., ser. 9, 10: 550–1.

——— (1922e). A new bat of the genus *Miniopterus* from N. Australia. Ibid., ser. 9, 10: 616–17.

——— (1923). On some Queensland *Phalangeridae*. Ibid., ser. 9, 11: 246–50.

——— (1926a). Two new rock-wallabies (*Petrogale*) discovered by Capt. G. W. Wilkins in Northern Australia. Ibid., ser. 9: 17: 184–7.

——— (1926b). On various mammals obtained during Capt. Wilkins's expedition in Australia. Ibid., ser. 9, 17: 625–35.

——— (1926c). The local races of *Dasyurus hallucatus*. Ibid., ser. 9, 18: 543-4.

Thomas, O., and Dollman, G. (1909). Mammals from Inkerman, North Queensland, presented to the National Museum by Sir William Ingram, Bt., and the Hon. John Forrest. *Proc. Zool. Soc. Lond.*, 1908: 788–94.

Thomson, D. F. (1949). Arnhem Land: explorations among an unknown people. Part III. On foot across Arnhem Land. *Geogr. J.*, 114: 53–67.

Tindale, N. B. (1925). Natives of Groote Eylandt and of the west coast of the Gulf of Carpentaria. *Rec. S. Aust. Mus.*, 3: 61–102.

Troughton, E. Le G. (1925). A revision of the genera *Taphozous* and *Saccolaimus* (Chiroptera) in Australia and New Guinea, including a new species, and a note on two Malayan forms. *Rec. Aust. Mus.*, 14: 313–41.

——— (1928). A new genus, species, and subspecies of marsupial mice (Family Dasyuridae). Ibid., 16: 281–8.

——— (1931). Three new bats of the genera *Pteropus, Nyctimene,* and *Chaerephon* from Melanesia. *Proc. Linn. Soc. N.S.W.*, 56: 204-9.

——— (1932). On five new rats of the genus *Pseudomys*. *Rec. Aust. Mus.*, 18: 287–94.

(1935). Five new rats of the genera *Hydromys* and *Melomys* from northern Australia. Ibid., 19: 251–8.

——— (1937). Six new bats (Microchiroptera) from the Australasian region. *Aust. Zool.*, 8: 274–81.

——— (1941). *Furred animals of Australia.*

Waite, E. R. (1896). Muridae. *Rep. Horn Scientific Expedition to Central Australia,* pt. 2, Zoology: 114–28, pls. 5–6.

——— (1898). Observations on Muridae from Central Australia. *Proc. Roy. Soc. Vict.,* new ser., 10: 114–28.

Wilkins, G. H. (1928). *Undiscovered Australia, being an account of an expedition to tropical Australia to collect specimens of the rarer native fauna for the British Museum, 1923–1925.*

(Manuscript received, December 1959)

INDEX

Brown, Prior, Anderson Pty. Ltd., 430 Little Bourke Street, Melbourne, C.1